Heat and Mass Transfer

This book series publishes monographs and professional books in all fields of heat and mass transfer, presenting the interrelationships between scientific foundations, experimental techniques, model-based analysis of results and their transfer to technological applications. The authors are all leading experts in their fields.

Heat and Mass Transfer addresses professionals and researchers, students and teachers alike. It aims to provide both basic knowledge and practical solutions, while also fostering discussion and drawing attention to the synergies that are essential to start new research projects.

José R. Simões-Moreira · Elí W. Zavaleta-Aguilar

Heat Transfer

An Engineering Course

Springer

José R. Simões-Moreira
Mechanical Engineering Department
Escola Politécnica of University of São Paulo
São Paulo, Brazil

Elí W. Zavaleta-Aguilar
Engineering Department
Institute of Sciences and Engineering
of São Paulo State University (UNESP)
São Paulo, Brazil

ISSN 1860-4846 ISSN 1860-4854 (electronic)
Heat and Mass Transfer
ISBN 978-3-031-80317-8 ISBN 978-3-031-80318-5 (eBook)
https://doi.org/10.1007/978-3-031-80318-5

The original submitted manuscript has been translated into English. The translation was done using artificial intelligence. A subsequent revision was performed by the author(s) to further refine the work and to ensure that the translation is appropriate concerning content and scientific correctness. It may, however, read stylistically different from a conventional translation.

Translation from the Portuguese language edition: "Fundamentos de Transferência de Calor para Engenharia", by José R. Simões-Moreira and Elí Wilfredo Zavaleta Aguilar, © The authors 2023. Published by LTC. All Rights Reserved.

This Springer imprint is published by the registered company Springer Nature Switzerland AG
The registered company address is: Gewerbestrasse 11, 6330 Cham, Switzerland

Then King Nebuchadnezzar leaped to his feet
in amazement and asked his advisers,
"Weren't there three men that we tied up and
threw into the fire?"
They replied, "Certainly, Your Majesty."
He said, "Look! I see four men walking
around in the fire, unbound and unharmed,
and the fourth looks like a son of the gods."
Daniel 3:24–25

To our beloved families

Preface

The subject of Heat and Mass Transfer typically follows Engineering Thermodynamics or Classical Thermodynamics in engineering curricula. The latter discipline introduces concepts such as substance properties, energy, entropy, mass and energy balances, their application in thermal machines and processes, and energy transformation equipment. Thermodynamics is based on the assumption of thermal, mechanical, and chemical equilibrium states. However, in non-equilibrium thermal systems, heat transfer occurs from regions of higher temperature to lower temperature regions, and mass or chemical species transfer from higher to lower concentration regions. Practical design of thermal equipment, such as size, shape, and operation, falls outside the scope of Thermodynamics. For example, this discipline allows for evaluating a condenser's thermal capacity in terms of operating temperature, working substance, and condensation load. However, it does not address questions such as how to size the equipment, its construction details, and the number and the tubing diameter required to achieve the specified thermal load. Should fins be installed in a tube or should another technique to enhance heat transfer be employed? Another practical and familiar example that distinguishes the two subjects is the phenomenon of body cooling. Suppose a body is at a given temperature and is placed in a cold chamber, like a refrigerator cabinet. Thermodynamics allows for determining the total heat exchanged in the cooling process, from its initial temperature to the final temperature, when it equalizes with the cabinet, nevertheless, it does not provide any information on how long it will take to reach this new thermal equilibrium state. Nor does it have the theoretical tools to indicate which techniques can be employed to increase or decrease the cooling time. Therefore, there is a clear distinction between the scopes of these two subjects.

In this book, we have included all relevant and fundamental topics clearly and concisely for engineering courses, since the subject is part of the curriculum in almost all engineering majors. The addressed topics aim at an introductory course, but the book also has sufficient depth for a more advanced course. Heat Transfer topics are typically divided into three fundamental modes: conduction, convection, and thermal radiation. On the mass transfer side, diffusion and convective mass transfer occur. Following many engineering curricula, less space is devoted to mass transfer, and thus, the topic has been condensed into the final chapter, allowing a continuous discussion and analysis of heat transfer modes in the first 11 chapters, with the presentation of mass transfer in a separate chapter.

Chapter 1 provides a broad introduction to heat and mass transfer phenomena. One-dimensional steady-state heat conduction or diffusion is discussed in Chap. 2. The subject is expanded to two- and three-dimensional geometric cases in Chap. 3, including an introduction to numerical solution methods. Chapter 4 addresses transient problems for elementary situations, such as lumped systems, and for more complex geometries using both analytical solutions and numerical techniques. Forced heat convection on external and internal surfaces is covered in Chaps. 5 and 6, respectively. The case of natural or free convection is addressed in Chap. 7. Chapter 8 is dedicated to heat transfer with phase change, condensation, and vaporization. One of the major applications of heat transfer is in the thermal design of heat exchangers, which is discussed in detail in Chap. 9. Thermal radiation phenomena are covered in Chaps. 10 and 11, with the first focusing on fundamentals and the second on engineering applications, including the theory of shape factors. Finally, the last chapter covers mass diffusion and convection phenomena.

The authors recognize that every text is subject to corrections and adjustments and would like to express thanks in advance for any suggestions and comments to improve the book, which can be sent to the authors. As a final note, it should be emphasized that this book results from the author's teaching experience on the subject. The first author has over 30 years of experience in teaching heat and mass transfer. Both professors took many solved and proposed problems from exams. The authors are indebted to the following professors, who contributed by reviewing original chapters and providing problems and examples for the book: Caridad Noda N. Pérez (UFG), Cláudio R. Freitas Pacheco (Poli—USP), Edson Cordeiro do Valle (UFRGS), Elaine M. Cardoso (UNESP), Gherhardt Ribatski (EESC-USP), Jorge R. H. Guerrero (UFPE), Lilian Cardoso de Melo (IPEN-SP), Marcelo da Silva Rocha (IPEN—SP), and Marcos de Mattos Pimenta (Poli—USP).

José R. Simões-Moreira
Full Professor
SISEA—Renewable and Alternative Energy Systems Lab.
Mechanical Engineering Department
Escola Politécnica of University of São Paulo
São Paulo, Brazil

Elí W. Zavaleta-Aguilar
Assistant Professor
SES—Sustainable Energy Systems Lab
Engineering Department
São Paulo State University (UNESP)
São Paulo, Brazil

Contents

About the Authors

José R. Simões-Moreira earned his Ph.D. in Mechanical Engineering from Rensselaer Polytechnic Institute (RPI) in New York in 1994, and his undergraduate and Master's degrees in Mechanical Engineering from the Escola Politécnica of the University of São Paulo (Poli-USP), Brazil, where he is presently a Full Professor. He teaches various undergraduate and graduate courses in the fields of Thermal Engineering and Energy. He coordinates a specialization course in Renewable Energy, Distributed Generation, and Energy Efficiency, he is the head of Laboratory of Renewable and Alternative Energy Systems (SISEA), a research group at the institution. He has also led several basic and applied research projects in Thermal Engineering and Energy funded by companies and research agencies. He also works as a consultant on engineering projects.

He is the author of *Fundamentals and Applications of Psychometrics* and the editor of *Renewable Energy, Distributed Generation, and Energy Efficiency* in Portuguese in addition to more than a hundred technical-scientific articles and chapters published in international books. He frequently gives lectures, grants interviews, and delivers presentations on topics within his expertise. His research areas include concentrated solar energy, supersonic gas separation, ammonia-water heat absorption cycles, and black carbon and hydrogen production by plasma.

Elí W. Zavaleta-Aguilar earned his Ph.D. in Energy from the Institute of Energy and Environment at the University of São Paulo (IEE-USP), Brazil, his master's degree in Mechanical Engineering from Poli-USP, and his undergraduate degree in Mechanical Engineering from the National University of Trujillo (UNT), Peru. He is an Assistant Professor at the Institute of Science and Engineering—Itapeva Campus, at São Paulo State University (UNESP), where he teaches courses in the fields of Thermal Systems, Fluids, and Mass Transfer, in Brazil. He coordinates research and University Extension projects in the field of Thermal Engineering and Energy. He is a researcher recognized by the National Council of Science, Technology, and Technological Innovation (CONCYTEC) of Peru.

He is the author of several scientific articles book, and book chapters. His research areas include concentrated solar energy, ammonia-water absorption refrigeration cycles, and applications of photovoltaic solar energy.

Symbols

A	Area (m^2)
Bi	Biot number (–)
Bo	Boiling number (–)
Bi'	Numerical Biot number (–)
c	Speed-of-light (m/s)
C	Specific heat (J/kgK); thermal capacity (W/K); thermal capacities ratio (–); concentration (mol/m^3)
C_f	Friction coefficient (–)
Co	Convection number (–)
COP	Coefficient of performance (–)
D	Diameter (m); mass diffusivity coefficient (m^2/s)
E	Energy (J); emissive power (W/m^2)
$\dot{E}$	Energy time rate (W)
E_b	Black body emissive power (W/m^2)
$\dot{E}_{st}$	Store internal energy rate (W)
f	Friction factor (–)
F	Shape factor (–), force (N)
Fo	Fourier number (–)
Fo'	Numerical Fourier number (–)
Fr	Froude number (–)
g	Gravity acceleration (m/s^2)
G	Irradiation (W/m^2)
Gr	Grash of number (–)
Gr^*	Modified Grash of number (–)
h	Coefficient of convective heat transfer (W/m^2K or $W/m^2{}^\circ C$); Planck constant (Js), specific enthalpy (J/kg)
I	Electrical current (A); radiation intensity (W/m^2sr)
j	Mass flux (kg/m^2s); Stanton factor (–)
J	Radiosity (W/m^2), Bessel function (–); molar diffusive flux (mol/m^2s)
k	Thermal conductivity (W/mK)
k_m	Coefficient of convective mass transfer (m/s)

L	Length (m)
$LMTD$	Logarithmic mean temperature difference (K, °C)
m	Mass (kg)
$\dot{m}$	Mass flow (kg/s)
M	Molecular mass (kg/kmol)
N	Tube number (–); molar flow (mol/m^{2}s)
n	Mass flow (kg/m^{2}s), number of mols (mol);
N_L	Number of tubes in longitudinal direction (–); N_T number of tubes in transversal direction (–)
NTU	Number of transfer units (–)
Nu	Nusselt number (–)
p	Pressure (Pa)
P	Perimeter (m)
Pr	Prandtl number (–)
q	Heat flux (W/m^2)
q'''	Thermal energy (heat) generated by volume unit (W/m^3)
Q	Heat transfer (J)
$\dot{Q}$	heat transfer rate (W)
R	Thermal resistance (K/W); surface resistance (1/m^2); contact resistance (1/m^2); universal ideal gas constant (kJ/kmol.K)
R_e	Electrical resistance (Ω)
R_d	Incrustation resistance (m^2 K/W)
$R_{t,c}$	Contact thermal resistance (m^2 K/W)
r	Sphere or cylinder radius (m)
r, ϕ, z	Cylindrical coordinates
r, θ, ϕ	Spherical coordinates
Ra	Rayleigh number (–)
Ra^*	Modified Rayleigh number (–)
Re	Reynolds number (–)
Sc	Schmidt number (–)
Sh	Sherwood number (–)
St	Stanton number (–)
t	Time (s)
T	Temperature (K, °C)
u	Specific internal energy (J/kg)
u, v, w	Velocity components in cartesian coordinate (m/s)
U	Global heat transfer coefficient (W/m^{2}K or W/m^2°C); internal energy (J)
$\forall$	Volume (m^3)
w	Mass fraction (–)
$\dot{W}$	Work power (W)
We	Weber number (–)
x	Liquid molar fraction (–)
x, y, z	Cartesian coordinate
y	Gas molar fraction (–)

Greek Letters

α Thermal diffusivity (m^2/s); absorptivity (–); thermodynamic factor (–)
β Volumetric expansion coefficient (1/K)
Γ_x Mass flow rate by length unit (kg/ms)
δ Boundary layer thickness (m)
Δ Difference (–)
ε Fin effectiveness (–); heat exchanger effectiveness (–); emissivity (–); roughness (m)
η Efficiency (–); similarity variable (–)
θ Temperature difference (K, °C), dimensionless temperature (–)
λ Wave length (μm), sum constant (–)
μ Dynamic viscosity (kg/ms)
ρ Density (kg/m^3); reflectivity (–); electrical resistivity (Ωm); mass concentration (kg/m^3)
σ Stefan-Boltzmann constant (W/m^2K^4), surface tension (N/m)
τ Shear stress (Pa); transmissivity (–); time constant (s)
ν Kinematic viscosity (m^2/s)
ϕ Association factor (–)
ω Solid angle (sr)

Subscript

A Solute
crit Critical
e Exit
env Environment
f Film
G Generated
H Hydrodynamics
i in
max Maximum
o Out
sat Saturation
t Thermal
w Wall
λ Spectral
∞ Far field

Chapter 1
Introduction: Modes of Heat and Mass Transfer

Abstract The chapter begins by showing the fundamental differences between the disciplines of Heat Transfer and Thermodynamics with the aid of ordinary examples, which will help the students to perceive those differences. The chapter briefly presents the three modes of heat transfer, which are conduction, convection, and thermal radiation, in addition to the diffusive and convective modes of mass transfer. The physical principles governing heat conduction, Fourier's law of conduction, and the definition of thermal conductivity, k, are presented. Newton's law of convection is introduced, when the heat transfer coefficient, h, is defined. The third form of heat transfer by thermal radiation is presented next, with the definition of the Stefan-Boltzmann law. Finally, diffusive mass transfer, given by Fick's law, and convective mass transfer are presented. The diffusion coefficients, D_{AB}, and convection of mass, k_m, are defined. Several solved exercises are presented, and at the end, proposed problems are presented.

Keywords Conduction · Convection · Thermal radiation · Mass diffusion · Mass convection

1.1 Heat Transfer x Thermodynamics

Normally, the course *Heat and mass transfer* follows the course *Thermodynamics* or *Engineering Thermodynamics* in the curriculum of engineering and technology. Naturally, the following doubt arises among students: what is the difference between these two courses as both deal with *heat*? Or, is there any difference between them? To properly answer those questions, consider the following examples.

(a) **Thermal equilibrium—bottle in the refrigerator**

Consider a bottle, initially at room temperature, that is, $T_b = T_{sur}$. Next, that bottle is placed inside a refrigerator, which has a temperature T_R, as illustrated in Fig. 1.1.

J. R. Simões-Moreira and E. W. Zavaleta-Aguilar, *Heat Transfer*, Heat and Mass Transfer, https://doi.org/10.1007/978-3-031-80318-5_1

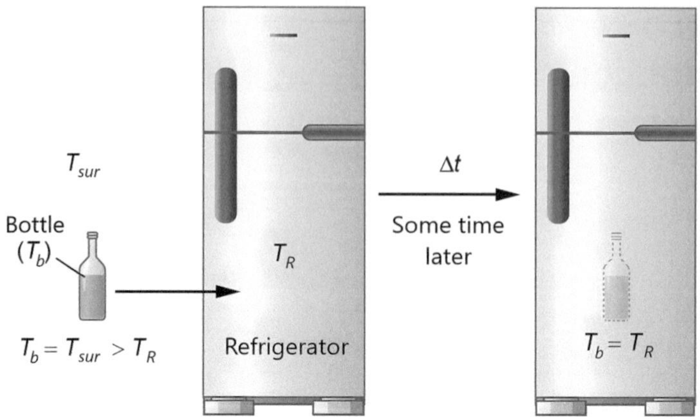

Fig. 1.1 Bottle placed inside the refrigerator

The aforementioned cooling problem allows the following analyses, each within the scope of each discipline:

Thermodynamics: The First Law of Thermodynamics states that $Q_T = mC\Delta T$, where Q_T is the total heat transferred from the bottle to cool it from the initial temperature, T_{sur}, to the refrigerator temperature, T_R. As seen, the calculation is based on its mass, m, initial and final temperature difference, ΔT, and average specific heat, C, of the bottle.

Heat transfer: Answers other questions of interest that *Thermodynamics* does not address, such as: how much time, Δt, will it take for the bottle to reach thermal equilibrium with its new environment (refrigerator cabinet), that is, for $T_b = T_R$? Another question: What parameters interfere with the increase or decrease in the cooling time?

Hence, *Thermodynamics* does not allow to establish the time interval for the new state of thermal equilibrium to be reached, although it informs the amount of heat Q_T is necessary to remove from the bottle so that a new thermal equilibrium is established. On the other hand, the *Heat transfer* allows one to estimate that equilibrium time interval, Δt, as well as to define what controlling parameters interfere with the elapsed time for the equilibrium and, consequently, it allows one to increase or decrease it as necessary by adjusting the controlling parameters.

(b) **Another example—operation of a refrigeration cycle**

Consider a standard refrigeration cycle like the one schematized in Fig. 1.2. The analysis of the cycle within the scope of each discipline is summarized in the following points:

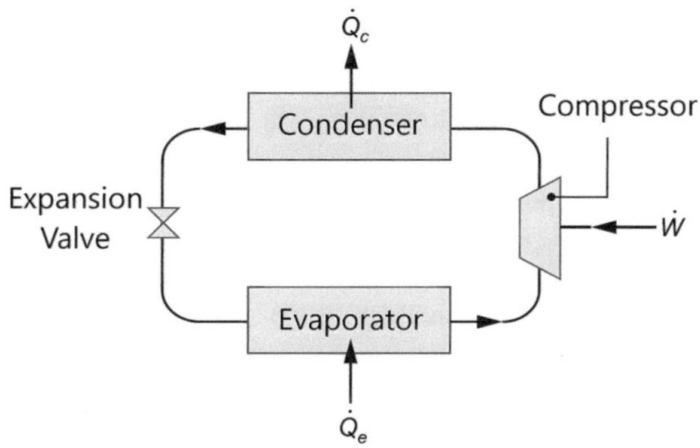

Fig. 1.2 Scheme of a vapor compression refrigeration cycle

Thermodynamics: First Law of Thermodynamics: $\dot{W} = \dot{Q}_e - \dot{Q}_c$ allows one to know or establish the relationship between the driving compression power, $\dot{W}$, and the heat transfer rates in the evaporator and condenser, $\dot{Q}_e$ and $\dot{Q}_c$, respectively. However, it does not allow dimensioning of the cycle components to establish, for example, the size and diameter of the condenser and evaporator coils. Thus, it only deals with the values of energy rates involved, as well as with the energy behavior of the system, such as the coefficient of performance, *COP*, of the cycle, given by

$$COP = \frac{\dot{Q}_e}{\dot{W}} \tag{1.1}$$

Heat transfer: It allows the sizing of the thermal equipment of the cycle, answering the following questions:

- What is the size of the evaporator/condenser?
- What is the tubing diameter and length?
- How to achieve a higher/lower rate of heat exchange?
- What properties of the coil materials are relevant to increase (or decrease) the heat transfer process?

Broadly speaking, the key problem of heat transfer is the knowledge and determination of the heat transfer rate. Generally, whenever there are gradients or finite differences in temperature or phase change (condensation or boiling), there will also be heat transfer. Heat transfer can occur within a body or system or at the interface of that body with a medium.

Knowledge of heat transfer mechanisms allows:

- to increase heat transfer rate: design of condensers, evaporators, boilers, heat exchangers, and other thermal energy transfer equipment;
- to decrease heat transfer rate: to avoid or reduce losses during the "transport" of thermal energy, for example, in chilled water tubes of refrigeration circuits and steam transport pipes;
- to control temperature: internal combustion engines, turbine blades, and heaters, to name a few examples.

1.2 Physical Mechanisms of Heat Transfer

Heat transfer occurs through three physical mechanisms: conduction, convection, and thermal radiation. Each of these mechanisms is presented below.

(a) **Heat conduction**

- *Gases, liquids*—heat transfer occurs from the high-temperature region to the low-temperature region by molecular collision from the more energetic particles to the less energetic ones.
- *Solids*—energy is transferred by vibration of the molecular network (less effective) as well as by the movement of free electrons (more effective), in the case of good electrical conductors. Generally, good electrical conductors are also good heat conductors, and vice versa. Often, electrical insulators are also thermal insulators.

Figure 1.3 illustrates the cross-section of a wall whose faces are kept at different temperatures T_1 and T_2, which results in a heat transfer through its thickness L. It can be said that the rate of heat transfer, $\dot{Q}_x$, in the direction x is directly proportional to the perpendicular area where the heat transfer occurs and to the temperature gradient in that direction at a given position x, that is

$$\dot{Q}_x \propto A \frac{dT}{dx} \tag{1.2}$$

where A is the area perpendicular to the heat transfer rate and T is the temperature at position x. The proportionality constant in Eq. (1.2) is the *conductivity* or *thermal conductivity* of the material, k. Thus, this equation becomes:

$$\dot{Q}_x = -kA \frac{dT}{dx} \tag{1.3}$$

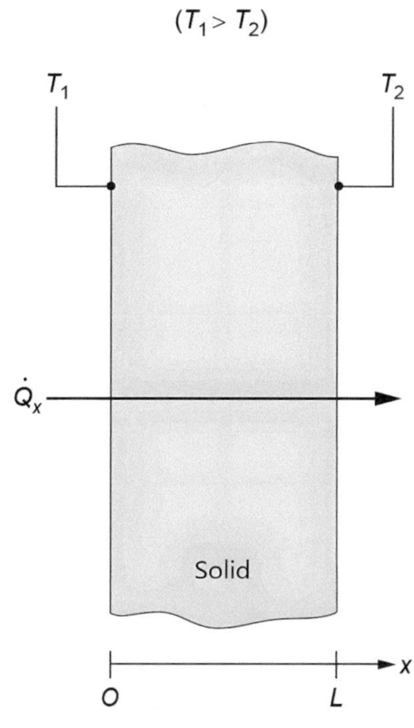

Fig. 1.3 Heat transfer by conduction within a solid wall

The units of the quantities in Eq. (1.3) in the international system (SI) are:

$$\begin{aligned}
[\dot{Q}_x] &= \text{W} \\
[A] &= \text{m}^2 \\
[T] &= \text{K or } ^\circ\text{C} \\
[x] &= \text{m}
\end{aligned}$$

The unit of thermal conductivity k is $[k] = $ W/mK or W/m°C. Thermal conductivity is a transport property of the material. As a rule, thermal conductivity increases in the following sequence according to the material: gases and vapors, which have the lowest values (order of 0.1 W/m°C), thermal insulating materials, liquids, metallic solids, metallic alloys, pure metals, and some types of crystals, including graphene (order of 1000 W/m°C). Therefore, the range of variation of thermal conductivity for materials is about 10,000 times or more. Tabulated values for various materials are in Appendix A.

Equation (1.3) is known as *Fourier's Law* of heat conduction, which was established by J. Fourier, announcing it for the first time in 1822. Note that the heat transfer rate ($\dot{Q}_x$), the perpendicular area (A), the temperature gradient (dT/dx), and k can vary along the x direction.

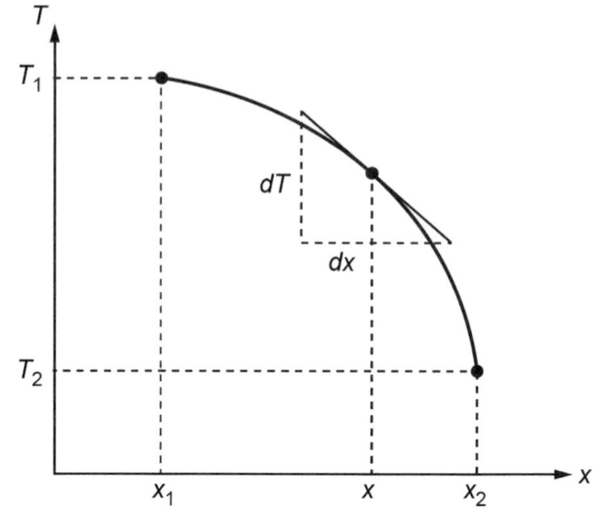

Fig. 1.4 Direction of heat transfer according to temperature distribution

Need for the negative signal in Fourier's Law

Figure 1.3 shows that there is heat transfer $\dot{Q}_x$ in the positive x direction because $T_1 > T_2$. Thus, the temperature is decreasing in the x direction within the material, as indicated in Fig. 1.4 and, therefore, $dT/dx < 0$. In addition, thermal conductivity and area are always positive magnitudes. Therefore, it is concluded that it is necessary to insert the negative signal (–) in the heat conduction expression (Fourier's Law) to maintain the convention that the positive conduction heat rate is, $\dot{Q}_x > 0$, in the increasing x direction. The analysis of the reciprocal of temperature distribuition, that is $T_1 < T_2$, will also demand a negative signal in Fourier's Law.

Under steady-state conditions, without thermal energy generation (heat) inside a solid, constant properties, and unidirectional heat transfer, the temperature gradient is given by Eq. (1.4).

$$\frac{dT}{dx} = \frac{T_2 - T_1}{L} \tag{1.4}$$

where L is the material thickness. The multidimensional case of heat conduction will be studied in Chap. 3.

Example 1.1 (*Heat conduction*)

The iron base illustrated in the figure below is 0.5 cm thick, 150 cm^2 in area, and its thermal conductivity is 15 W/mK (stainless steel). Assuming that all the electrical power is transformed into thermal power and transferred to the iron base, evaluate the temperature of the iron base's inner surface (T_i). The iron base outer temperature is kept at $T_o = 200\,°\text{C}$. Redo the calculation considering that the base material is an aluminum alloy with a thermal conductivity of 168 W/mK.

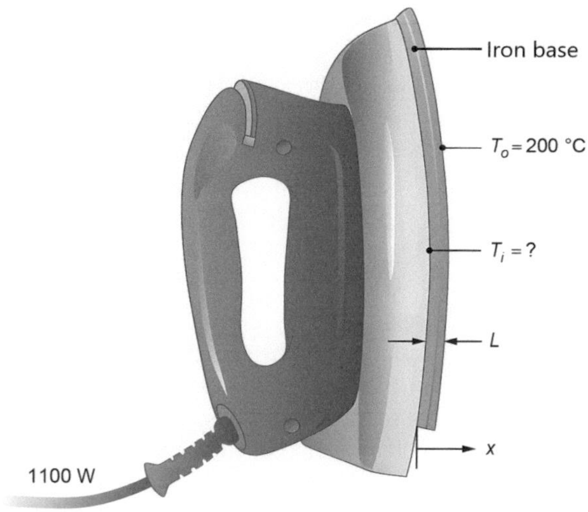

Solution

Problem data:

- Heat transfer area: $A = 0.015\,\text{m}^2$.
- Base thickness: $L = 0.005\,\text{m}$.
- Conductivity of materials (properties from Table A.2):

  stainless steel $\rightarrow k_{ss} = 15\,\text{W/mK}$;
  aluminum $\rightarrow k_{al} = 168\,\text{W/mK}$;

- Temperature on the outer face: $T_o = 200\,°\text{C}$.

  Assumptions
- Steady-state;
- Unidimensional conduction.

  From Fourier's Law: $\dot{Q} = -kA\frac{dT}{dx} = -kA\frac{(T_o - T_i)}{L-0}$.
  Therefore, $T_i = T_o + \frac{\dot{Q}L}{kA}$.
  For stainless steel: $T_i = 200 + \frac{1100\times0.005}{15\times0.015} = 224.4\,°\text{C}$.
  For aluminum: $T_i = 200 + \frac{1100\times0.005}{168\times0.015} = 202.2\,°\text{C}$.

Note that, because the conductivity of aluminum is higher than that of stainless steel, the temperature differences between T_i and T_o are smaller and, therefore, the temperature distribution within the iron base is less accentuated.

(b) **Heat Convection**

Heat convection is based on *Newton's cooling law* (1701), which mathematically is expressed according to Eq. (1.5).

$$\dot{Q} = hA(T_w - T_\infty) \tag{1.5}$$

being h the *convection heat transfer coefficient*, sometimes also called *film coefficient*, A the heat exchange surface area [m^2], T_w the surface or wall temperature [°C or K], and T_∞ the far away fluid temperature [°C or K].

The process of heat transfer by convection occurs between a surface and a fluid that surrounds it, as indicated in Fig. 1.5. In this, the surface temperature T_w transfers heat by convection to the fluid that has a lower temperature, T_∞, far away. In the figure, the fluid is in motion (forced convection) with speed u_∞ parallel to the plate, but convection also occurs when the fluid is at rest (natural convection). Boundary layers are also illustrated.

The central problem of heat transfer by convection is the determination of the convection heat transfer coefficient, h, which depends on several factors, among them: fluid-surface contact geometry (surface area, roughness, and geometry), transport properties of the fluid, temperature levels, speeds, to name some of the intervening factors. Table 1.1 shows some illustrative typical values of that coefficient according to the physical state (liquid, vapor, and phase change) and for natural or forced convection.

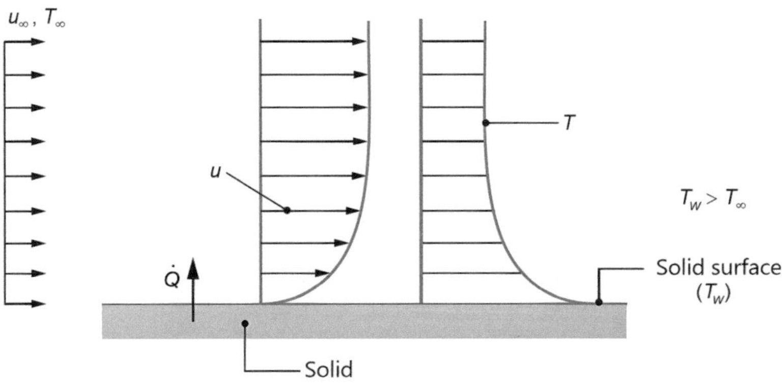

Fig. 1.5 Heat transfer by forced convection from a surface to a moving fluid. Notice that boundary layers are established

Table 1.1 Range of values of the heat transfer coefficient, h

Form	Physical medium	h (W/m^2°C)
Natural convection	Air and gases	1–50
	Liquids	50–2000
Forced convection	Air and gases	20–500
	Liquids	50–10,000
Boiling water		3000–100,000
Condensing steam		5000–100,000

Example 1.2 (*Heat convection*)

In the previous example, consider that the air temperature surrounding the iron is 15 °C. Find the value of the heat transfer coefficient by convection.

Solution

Ideally, all the thermal energy generated in the electrical resistance is transferred as heat to the base of the iron and from there to the surrounding air. Newton's cooling law states that

$\dot{Q} = hA(T_o - T_\infty)$, therefore $h = \frac{\dot{Q}}{A(T_o - T_\infty)}$

$$h = \frac{1100}{0.015 \times (200 - 15)}$$

$$h = 396.4\,\mathrm{W/m^2K}.$$

Example 1.3 (*Heat convection*)

The external surface wall of a house is at 14 °C and is subjected to cold air at 7 °C at such a speed that the convection heat transfer coefficient is 27 W/m^2 K. Evaluate the heat rate loss from the house wall to the external environment (considering the heat losses in doors and windows as the same as those of the wall).

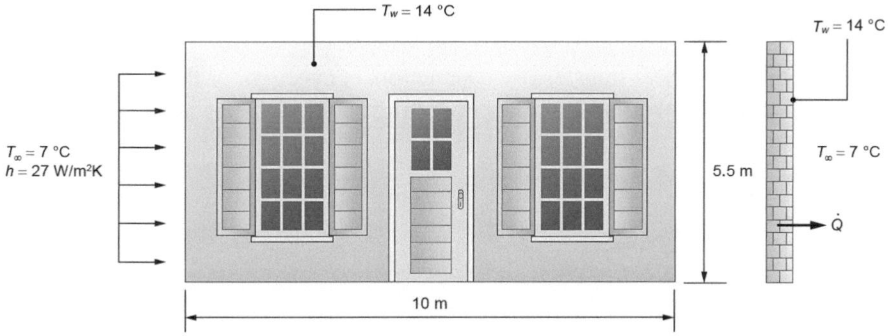

Solution

- Wall area: $A = 10 \times 5.5 = 55\,\mathrm{m}^2$.
- Exchange heat rate: $\dot{Q} = hA(T_w - T_\infty) = 27 \times 55 \times (14 - 7) = 10{,}395\,\mathrm{W}$.

(c) Thermal radiation

Thermal radiation is the third form of heat transfer and is governed by the *Stefan-Boltzmann Law* which, for a black body, is indicated in Eq. (1.6). Stefan obtained it empirically (1879) and Boltzmann theoretically (1884).

$$\dot{Q} = \sigma A T^4 \tag{1.6}$$

where σ is the Stefan-Boltzmann constant (5.669×10^{-8} W/m^{2}K^4), A is the surface area of the black body [m^2], and T is the absolute temperature of the surface of the black body [K].

A black body is also called a *perfect irradiator of thermal radiation*, which irradiates the maximum possible thermal radiation at a given temperature. It also absorbs all incident thermal radiation. In real bodies (gray), where the average emissivity of its surface is ε, the heat rate emitted from its surface is:

$$\dot{Q} = \varepsilon \sigma A T^4 \tag{1.7}$$

The physical mechanism of heat transfer by thermal radiation is not yet fully understood. It is known that this form of thermal energy transport can occur in the form of electromagnetic waves or photons, depending on the physical model adopted—the so-called *wave-particle duality*. However, regardless of the physical model adopted, the transfer of heat by thermal radiation does not need any material medium to propagate. Because of this form of heat transfer, life exists on Earth, due to the energy in the form of heat and light coming from solar radiation that reaches our planet.

For any mode of heat transfer (by conduction, convection, or thermal radiation), the *heat flux* is defined as the ratio between the rate of heat transferred and the perpendicular cross-sectional area, Eq. (1.8).

$$q_x = \frac{\dot{Q}_x}{A} \tag{1.8}$$

Example 1.4 (*Combined heat transfer by convection and thermal radiation*)

In Example 1.3, calculate now the heat exchange by thermal radiation with the surroundings around the iron, to which a temperature of 15 °C is attributed. The emissivity of the iron base is 0.55. Find the portions of heat exchanged by pure convection and by thermal radiation.

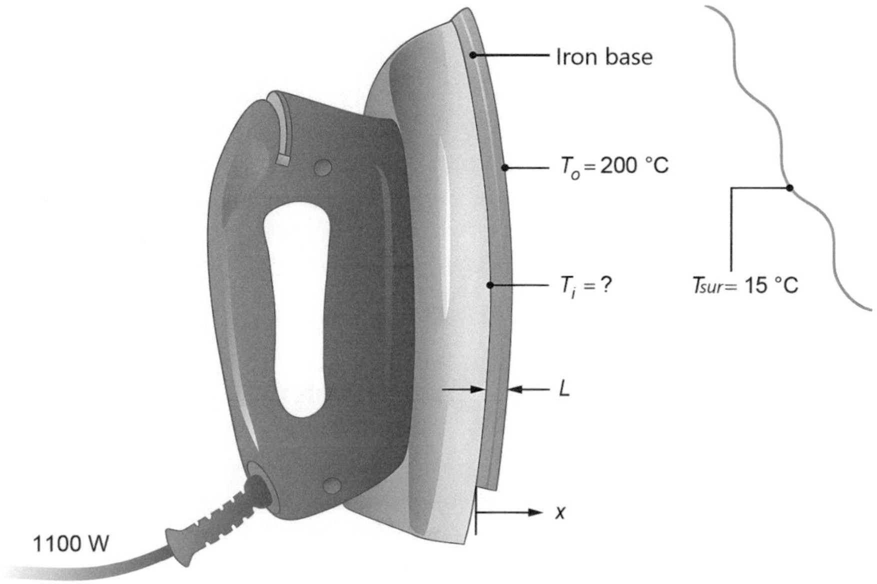

Solution

The net heat transfer (emitted minus received) by thermal radiation from the base of the iron to the surroundings is:

$\dot{Q}_r = \varepsilon A\sigma\left(T_o^4 - T_{sur}^4\right)$, or.

$\dot{Q}_r = h_r A(T_o - T_{sur})$ where $h_r = \varepsilon\sigma\left(T_o^2 + T_{sur}^2\right)(T_o + T_{sur})$ is called the *coefficient of heat transfer by thermal radiation.*

Thus, the total rate of heat exchanged by the iron base will be given by two terms, one by convection to the air and the other by radiation to the surroundings:

$\dot{Q} = \dot{Q}_{conv} + \dot{Q}_r = hA(T_o - T_\infty) + h_r A(T_o - T_{sur})$, if $T_\infty = T_{sur}$, it follows that

$\dot{Q} = h_{comb}A(T_o - T_\infty)$, where $h_{comb} = h + h_r$ is the *combined coefficient for convection and thermal radiation.* As all the heat is transferred to the environment, in Example 1.2 the coefficient obtained was called the heat transfer coefficient by convection, however, in fact, that value obtained of 396.4 W/m^{2}K is the combined convection and thermal radiation coefficient, since a global energy balance was performed in the iron base. Thus, the heat transfer coefficient only by radiation is $h_r = 0.55 \times 5.67 \times 10^{-8} \times \left(473.2^2 + 288.2^2\right) \times (473.2 + 288.2) = 7.3$ W/m^{2}K.

Finally, the heat coefficient by convection will be the difference between them:

$$h = 396.4 - 7.3 = 389.1 \ \text{W/m}^2\text{K}$$

The purely convective portion of heat exchange will therefore be:

$$\dot{Q}_{conv} = 389.1 \times 0.015 \times (200 - 15) = 1079.7\,\text{W}$$

The thermal radiation portion will be:

$$\dot{Q}_{rad} = 7.3 \times 0.015 \times (200 - 15) = 20.3\,\text{W}.$$

Observations:

- In many practical applications, the portion corresponding to heat exchange by radiation is disregarded for the cases where there are small temperature differences involved. However, this portion becomes increasingly relevant as the temperature differences increase substantially.

1.3 Physical Mechanisms of Mass Transfer

The *potential* that controls and gives course to heat transfer is the temperature difference within a medium or between different mediums. Similarly, mass transfer will also occur whenever there are concentration differences of a species (A) within a medium or on the surface of a solid or liquid and any fluid medium. The analogy with heat transfer is practically perfect. There are two forms of mass transfer, as indicated below.

(a) **Mass diffusion (molecular)**

The mass transfer, in this case, represented by the diffusive mass flux, j_A [kg/m^{2}s], is defined as the mass of species A that crosses a surface per unit of time and occurs from the region of high concentration, ρ_{A1} [kg/m^3], to the region of low concentration, ρ_{A2}, of this species within a solid or quiescent fluid over a length z, as illustrated in Fig. 1.6.

The diffusive mass flux, j_A, in the direction z is mathematically defined by *Fick's Law* (Adolf Eugen Fick 1855) as equal to the binary diffusion coefficient (or diffusivity) of A diffusing into B, D_{AB} [m^2/s], times the concentration gradient of A, according to Eq. (1.9).

$$j_A = -D_{AB}\frac{d\rho_A}{dz} \tag{1.9}$$

The mass diffusivity D_{AB} is a transport property of the material that strongly depends on temperature and, the phase of the material, pressure, and concentration, among other properties, as it will be seen in Chap. 12. Thus, as it occurs with Fourier's Law, it is necessary to introduce the negative signal (–) to match the mass flux in the direction of the lower concentration. Fick's Law, as well as any forms of mass transfer, can be expressed in terms of mass (kg) or molar (mol).

Under steady-state conditions, without chemical reaction, rectangular coordinates, and unidirectional mass transfer, the concentration gradient is given by Eq. (1.10).

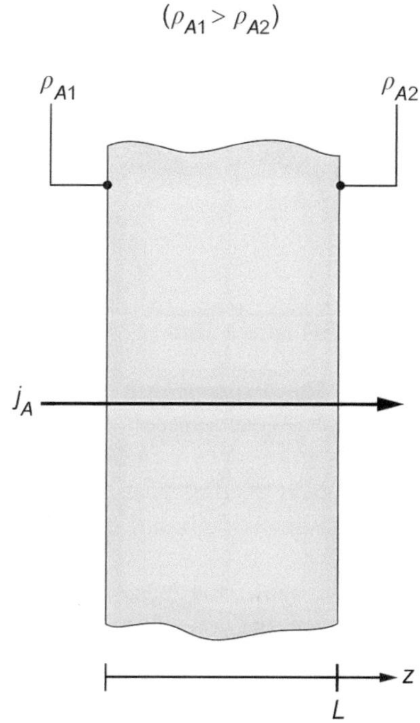

Fig. 1.6 Mass transfer by diffusion

$$\frac{d\rho_A}{dz} = \frac{\rho_{A1} - \rho_{A2}}{L} \tag{1.10}$$

Example 1.5 (*Diffusion of hydrogen gas*)

Gaseous hydrogen is stored in a pressurized steel cylinder with a wall thickness of 12.7 mm. Evaluate the diffusive molar flow of hydrogen through the wall, which means loss of hydrogen. Assume that the molar concentration of hydrogen at the internal solid interface is 1.5 kmol/m^3 and that it is negligible at the external wall; in addition, consider that the diffusion coefficient of hydrogen in steel is 2.6×10^{-11} m^2/s.

Solution

Assumptions:

- Steady-state, one-dimensional.
- Uniform molar concentration gradient.
- No chemical reactions of hydrogen.
- The wall thickness is small compared to the tank diameter.

Under these assumptions, the concentration gradient is constant, whose value is $\frac{dC_A}{dx} = \frac{C_{A2}-C_{A1}}{L-0}$, in which C_{A1} and C_{A2} are the molar concentrations of hydrogen on the inner and outer wall of the tank, respectively, and L is the thickness of the tank, thus: $\frac{dC_A}{dx} = -\frac{C_{A2}-C_{A1}}{L}$.

$$C_{A1} = 1.5\,\text{kmol/m}^3, \quad C_{A2} = 0 \text{ and } D_{AB} = 2.6 \times 10^{-11}\text{m}^2/\text{s}$$

Resulting in a molar diffusive flux of:

$$J_A = -2.6 \times 10^{-11}\left(\frac{0 - 1.5}{0.0127}\right)$$

$$J_A = 3.07 \times 10^{-9} \text{ kmol/m}^2\text{s}$$

(b) **Mass convection**

Unlike what happens in molecular mass transfer or diffusion where the movement of species A occurs randomly and in a generally quiescent medium, mass transfer by convection occurs in a convective motion medium (macroscopic movement). The process of mass transfer by convection occurs between a solid surface and a surrounding fluid (liquid or gas), or between a liquid and a gas, as indicated in Fig. 1.7.

In the figure, the mass concentration of A in the fluid next to the wall (ρ_{Aw}) is greater than the concentration of A far from the surface ($\rho_{A\infty}$) and therefore, there will be a mass flux of A (n_A [kg/m^{2}s]) transferred from the wall to the fluid, defined in a way analogous to Newton's cooling law, according to Eq. (1.11).

$$n_A = k_m(\rho_{Aw} - \rho_{A\infty}). \tag{1.11}$$

where k_m [m/s] is the convective mass transfer coefficient.

Example 1.6 (*Convective mass transfer of water*)

Estimate the water loss in liters and in cm of height per day from a small pond of 2 m by 5 m in which air with a water vapor concentration of 0.013 kg/m^3 circulates above

Fig. 1.7 Direction of mass transfer by convection

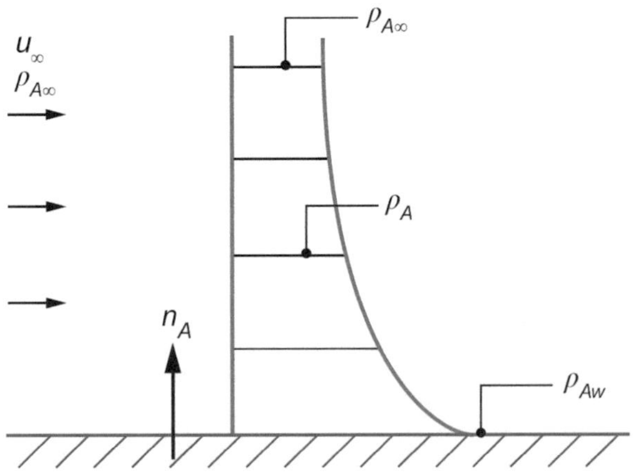

it. The convective mass transfer coefficient between the air and the water surface of the pond is 0.01 m/s and the mass concentration of water vapor at the air-water interface is 0.02 kg/m^3.

Solution

Disregarding the effects of radiation.

$$n_A = k_m(\rho_{Aw} - \rho_{A\infty}) = 0.01 \times (0.02 - 0.013) = 7 \times 10^{-5}\,\text{kg/m}^2\text{s}.$$

The rate of mass loss of evaporated liquid is:

$$\dot{m}_A = An_A = 2 \times 5 \times 7 \times 10^{-5} = 7 \times 10^{-4}\ \text{kg/s}$$

In a day there are 86,400 s, so the rate of liquid evaporated is $7 \times 10^{-4} \times 86,400 = 60.48\,\text{kg/day}$. Assuming the water density of 1000 kg/m^3; therefore, 60.68 L evaporates per day.

As the pond has a surface area of 10 m^2, the height H of water evaporated per day will be 10H = 60.68/1000; therefore, the height is H = 0.61 cm.

1.4 Proposed Problems

1.1 A room is kept at 20 °C with air conditioning on a summer day. One of the room's walls, 2.7 m × 5 m, has an external temperature of 30 °C, while the internal face is 20 °C, due to the circulation of cold air. Considering that the wall is 20 cm thick and made of a material with an average thermal conductivity of 0.5 W/m °C, determine the rate of heat loss through the wall in watts, as well as the heat flux in W/m^2.

1.2 The internal wall temperature of an oven is 800 °C. The oven is made up of a common brick wall. Determine the thickness that the brick wall should have so that the heat loss is less than 1000 W/m^2 and the external surface temperature does not exceed 50 °C. What could be done to reduce heat loss?

1.3 The temperature of a metal block varies according to the equation: $T = -50{,}000x^2 - x + 50$, where T is in °C and x in m. The thermal conductivity of the metal is 57 W/m^{2}K and its thickness is $L = 10$ mm. Assuming one-dimensional heat conduction through the thickness $x = L$, evaluate the temperature and heat flux at $x = 0$, $L/2$, and L.

1.4 Redo problem 1.3, knowing that the thermal conductivity of the metal varies with temperature according to: $k = 60 - 0.05T$, where k is in W/mK and T in °C. Evaluate the heat flux at $x = 0$, $L/2$ e L, considering the variation of k with temperature.

1.5 A thermodynamic analysis of a refrigeration cycle indicated that the refrigeration load of an evaporation coil is 2 kW when the fluid vaporizes at 10 °C and maintains this value on the coil wall. What should be the size of the coil tube, if its external diameter is 1 cm and the air is at an average temperature of 25 °C? The heat convection coefficient is 100 W/m^2°C.

1.6 A refrigerator has a *COP* of 2, and its compressor has a power rating of 1/4 hp. Given that the condenser has an external surface area of 1.5 m^2, the ambient air temperature is 17 °C, and the convective heat transfer coefficient between the condenser and the ambient air is 15 W/m^{2}K, calculate the average external temperature of the condenser.

1.7 A 1 cm-thick metal slab with a thermal conductivity of 15 W/mK is subjected to a heat flux of 10,000 W/m^2 on one side (Face A). Temperatures of 60 °C and 15 °C were measured on the opposite side (Face B) and in the adjacent ambient air, respectively. Assuming steady-state conditions, determine the temperature on Face A and the convective heat transfer coefficient on Face B.

1.8 The surface temperature of the human skin is about 35 °C. Compare the cooling, that is, the rate of heat removal per unit of the skin surface, when it is in contact with air at 30 °C in natural convection ($h = 6$ W/m^2°C) and when one is in front of a working fan ($h = 15$ W/m^2°C).

1.9 A 100-watt incandescent light bulb contains a tungsten filament that operates at temperatures between 2500 °C and 3000 °C. Determine the convective heat loss from the glass bulb to the surrounding environment, assuming the bulb has a spherical shape with a 60 mm diameter and maintains a surface temperature of 100 °C. The convective heat transfer coefficient between the glass bulb and air is 15 W/m^{2}K, and the ambient air temperature is 15 °C.

1.10 A person is standing in front of a large furnace. Assuming that the furnace wall temperature is 50 °C and its emissivity is 0.6, calculate the heat flux in W/m^2 that the person receives from the wall.

1.11 A space probe with an approximately spherical shape and 1.5 m diameter has an emissivity of 0.82. Due to electronic processing, its external temperature is −160 °C. Calculate the heat transfer rate, given that space has an approximate temperature of 2.7 K and there is no other thermal interaction with other bodies.

1.12 Pieces of glass heated in furnaces can be molded into different shapes. In a production process involving a 12 cm diameter sphere, glass at 850 °C is exposed to ambient air at 20 °C, while the surrounding walls are also maintained at 20 °C. Evaluate the thermal losses through convection, radiation, and the total combined heat transfer, given that the convective heat transfer coefficient between the glass and ambient air is 55 W/m^{2}K and the glass emissivity is 0.5.

1.13 Water can migrate through the walls of a house if they have different concentrations. On a cold day, the water concentration on the external side of a bedroom plaster wall is 0.2 mol/m^3; on the internal side, next to the insulation, consider it negligible. Knowing that the wall is made of *drywall* with dimensions of 2.5 m high, 7 m wide, and 12.5 mm thick and that the mass diffusivity of water in the plaster wall is 0.2×10^{-5} cm^2/s, evaluate the rate of water diffusing in the wall.

1.14 After rainfall, a 2 m × 3 m area becomes covered with a 3 mm thick water film. The convective mass transfer coefficient of water to air is 0.7 cm/s, with water concentration in the air adjacent to the water layer being 0.031 kg/m^3 and in the ambient air far from the ground being 0.017 kg/m^3. Calculate the time required for all this water to be transferred to the ambient air, assuming uniform evaporation across the entire area. The water density is 996 kg/m^3.

1.15 A solid salt (NaCl) plate of 10 cm × 12 cm is located in seawater where the water speed is 1 m/s and, under these conditions, the mass transfer coefficient is 1.6×10^{-5} m/s. Knowing that the concentration of salt at the salt-water interface on the liquid side is 350 kg/m^3 and the concentration of salt in the seawater far away is 35 kg/m^3, how long will it take to dilute this layer of salt that has a thickness of 5 mm, knowing that only one side of it is exposed to water and that the density of solid salt is 2165 kg/m^3?

Chapter 2
One-Dimensional Conduction in Steady State. Extended Surfaces or Fins

Abstract The general equation of heat conduction is derived in Cartesian coordinates first and next it is presented in cylindrical and spherical coordinates. The general heat conduction equation is simplified for steady-state cases without internal thermal energy generation (heat). Next, the cases of one-dimensional heat conduction in a steady state with and without internal thermal energy generation (heat) are presented for flat walls and cylindrical tubes using the appropriate coordinate systems. The case of composite walls where conduction occurs in several contacting walls is analyzed through electrical analogy. The electrical analogy is extended to composite cylindrical tubes. Heat transfer in flat and cylindrical electrical resistances are analyzed, to obtain the internal temperature distribution. The overall heat transfer coefficient is defined. The critical radius of thermal insulation is discussed, with the indication of application in electrical wire and tubing isolation systems. Fins or extended surfaces make up the last topic discussed in the chapter, first by the deriving of the general fin equation, followed by various situations and types of fins. The efficiency and effectiveness of a single fin and of a finned surface are presented.

Keywords General equation of heat conduction · One-dimensional conduction · Overall (or global) heat transfer coefficient · Electrical resistance · Critical radius · Fin

Heat transfer by conduction occurs within solid bodies and in fluids under certain conditions and consists of transferring heat from higher to lower temperature regions. The phenomenon of heat conduction is governed by *Fourier's Law* of 1722. Fourier's Law in the version of one-dimensional conduction is given by Eq. (2.1). By simple inspection of this equation, it is inferred that the rate of heat transfer, $\dot{Q}$ [W], is directly proportional to the temperature gradient in the direction where conduction occurs, x [m], and to the perpendicular area, A [m^2], as illustrated in Fig. 2.1.

$$\dot{Q} = -kA\frac{dT}{dx} \tag{2.1}$$

J. R. Simões-Moreira and E. W. Zavaleta-Aguilar, *Heat Transfer*, Heat and Mass Transfer, https://doi.org/10.1007/978-3-031-80318-5_2

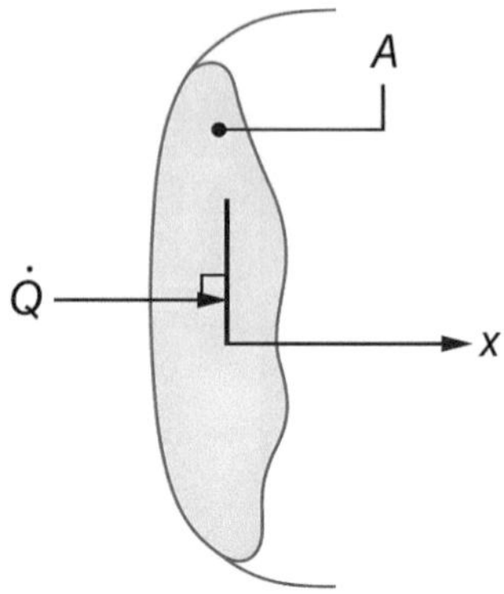

Fig. 2.1 Heat conduction rate $\dot{Q}$ occurs in the direction perpendicular to the area

where k is the *thermal conductivity* of the physical medium, which is a property of the medium discussed below.

Dimensional analysis of the previous equation reveals the units of k.

$$[k] = \frac{[\dot{Q}]}{[A]\left[\frac{dT}{dx}\right]} = \frac{\mathrm{W}}{\mathrm{m^\circ C}}, \text{ or } [k] = \frac{\mathrm{W}}{\mathrm{mK}}.$$

Note: the unit of thermal conductivity can be expressed as a function of temperature given in °C or K, since $1\Delta^\circ\mathrm{C} = 1\Delta K$.

Laboratory experiment for obtaining k

Suppose an experiment like the one outlined in Fig. 2.2a, in which an electric current (I) is supplied by a power supply (ΔV) to the electrical resistance (R_e), which is wound around the base of a rod. The rod is thermally insulated, so that the rate of thermal energy (heat) produced by the joule effect in the electrical resistance is transferred by one-dimensional conduction along the rod. By installing temperature sensors (e.g., thermocouples) along the rod length, the temperature profile can be obtained as a function of the position x from the base of the rod, as indicated in the graph in Fig. 2.2b.

As indicated in Fig. 2.2b, the temperature distribution is linear. Furthermore, it is known that the heat transfer rate is the dissipated electrical power itself, that is, $\dot{Q} = \Delta VI = R_e I^2$, where ΔV is the electric voltage difference [V], I is the electric current [A] and R_e, the ohmic resistance [Ω]. Given the cross-sectional area A of the rod, then, by Fourier's Law (Eq. 2.1), the thermal conductivity of the rod material is determined, k, in this case:

$$k = -\frac{\dot{Q}}{A\frac{\Delta T}{\Delta x}} \tag{2.2}$$

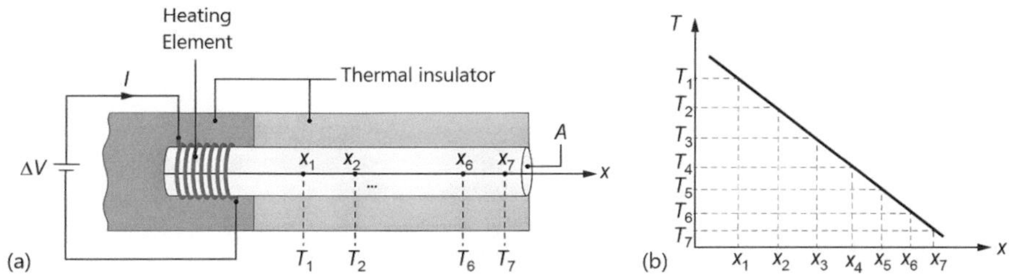

Fig. 2.2 **a** Experimental setup for obtaining thermal conductivity k; **b** temperature profile in the rod lengthwise

In the case of the cross-section A being constant, heat flux by conduction, q_x, can be defined as the ratio between the heat conduction rate and the area of the cross-section, i.e.

$$q_x = \frac{\dot{Q}}{A} \tag{2.3}$$

An important aspect of the phenomenon of heat conduction at the molecular level is that conduction is governed by more than one physical mechanism depending on the nature of the material and its physical state (gaseous, liquid or solid). The heat conduction mechanisms are presented below for the three physical states of matter.

Gases

The molecular collision of gaseous molecules allows the exchange of kinetic energy between the more energetic molecules and the less energetic ones. It is known that molecular kinetic energy is closely associated with the absolute temperature of the gas. The higher the temperature, the greater the molecular movement that will provide a higher rate of molecular collisions, and, therefore, thermal energy will flow more quickly within gases. It can be shown that thermal conductivity is related to temperature according to $k \propto \sqrt{T}$. For certain gases at moderate pressure, k depends solely on temperature T. Thus, tabulated data for a given temperature and pressure can be used for another moderate pressure as long as it is at the same temperature. This is not valid for values near the thermodynamic critical state.

Liquids

Qualitatively, the physical mechanism of heat conduction transport in liquids is the same as that of gases. Molecular vibration may also play some role. However, the situation is considerably more complex due to the lower mobility of the molecules.

Solids

Two forms govern the heat transfer by conduction in solids: vibration of the crystal lattice and transport by free electrons. The second mode is the most effective one and predominantly occurs in metallic materials. This explains why, in general, good

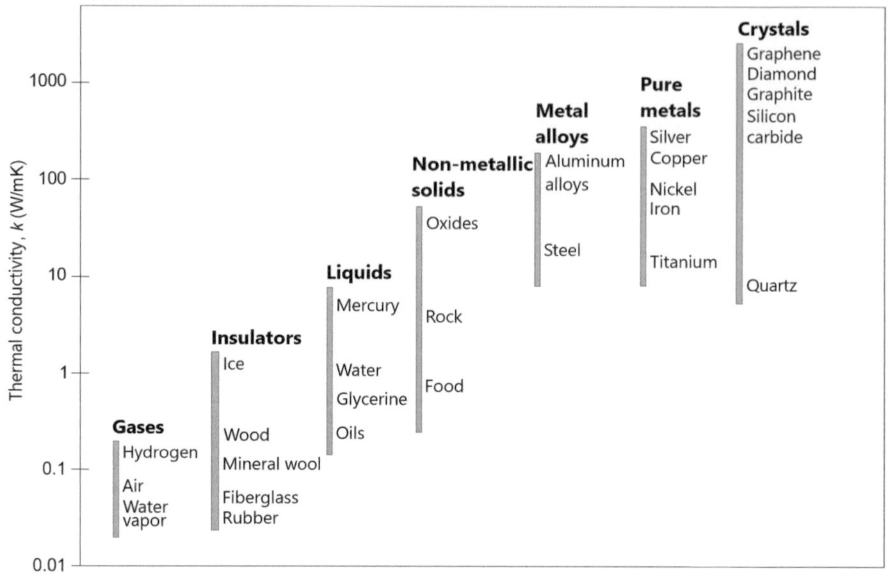

Fig. 2.3 Thermal conductivity in various materials

conductors of electricity are also good conductors of heat. The transfer of heat in thermal insulators occurs through the vibration of the crystal lattice, which is considerably less efficient.

The diagram in Fig. 2.3 qualitatively illustrates the order of magnitude of the thermal conductivity of various materials. It is noted that, in general, the thermal conductivity increases in the following ascending sequence: gases, liquids, and solids, with pure metals having high thermal conductivity and, finally, some crystals having very high thermal conductivity. Excretion for solid insulators that have lower conductivity than liquids because of the weak heat transport mechanism as mentioned.

2.1 General Equation of Heat Conduction in Cartesian Coordinates

The primary problem of heat conduction is to determine the rate of heat transferred within a given material in the x, y, and z directions, as well as the internal temperature distribution, that is, $T(x, y, z, t)$. To solve this problem, it is first necessary to establish the general heat conduction equation, which can be obtained from an energy balance in a differential control volume, as indicated in Fig. 2.4.

The energy balance (or First Law of Thermodynamics) for the differential control volume (CV) in Fig. 2.4 accounts for the rates of thermal energy involved, so that the following scheme can be established:

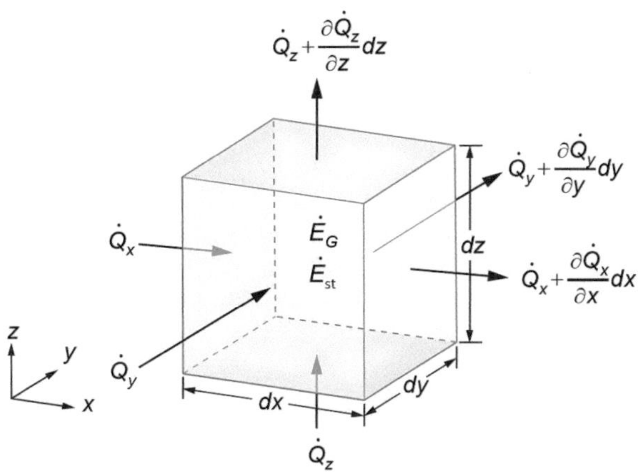

Fig. 2.4 Energy balance in the differential control volume

$$\underbrace{\left\{\begin{array}{c}\text{Rate of heat transfer}\\ \text{entering the CV}\end{array}\right\}}_{\text{I}} - \underbrace{\left\{\begin{array}{c}\text{Heat transfer rate}\\ \text{leaving the CV}\end{array}\right\}}_{\text{II}}$$

$$+ \underbrace{\left\{\begin{array}{c}\text{Thermal energy rate}\\ \text{(heat) generated in}\\ \text{the CV}\end{array}\right\}}_{\text{III}} = \underbrace{\left\{\begin{array}{c}\text{Rate of change}\\ \text{of internal energy}\\ \text{of the CV}\end{array}\right\}}_{\text{IV}}$$

Explicating the terms (I), (II), (III), and (IV), we have:

(I) *Heat transfer rate entering the CV in three directions from Fourier's Law—each elementary area (dxdy, dxdz, dydz) is perpendicular to the heat transfer rate direction:*

Direction x: $\dot{Q}_x = -k_x dydz \frac{\partial T}{\partial x}$.
Direction y: $\dot{Q}_y = -k_y dxdz \frac{\partial T}{\partial y}$.
Direction z: $\dot{Q}_z = -k_z dxdy \frac{\partial T}{\partial z}$.

(II) *Heat transfer rate leaving the CV in all three directions:*

For this, a Taylor series expansion of the heat transfer rate in the direction of interest should be performed, truncating at the first term of the expansion (the others tend to zero as they are of higher order):

Direction x: $\dot{Q}_x + \frac{\partial \dot{Q}_x}{\partial x} dx$
Direction y: $\dot{Q}_y + \frac{\partial \dot{Q}_y}{\partial y} dy$
Direction z: $\dot{Q}_z + \frac{\partial \dot{Q}_z}{\partial z} dz$

(III) *Thermal energy rate (heat) generated within the CV.*

The rate of thermal energy (heat) generation results from the transformation of one type of energy, for example, electrical energy into thermal energy by the Joule effect within the control volume, or by the curing of a resin or, still, by an exothermic or endothermic chemical reaction (in the endothermic case, the rate of thermal energy generation will be negative – minus signal) or nuclear reactions. It is assumed that the generation of thermal energy occurs uniformly throughout the differential control volume. The generated thermal energy will, subsequently, diffuse in the form of heat.

$$\dot{E}_G = q_G''' dxdydz$$

where q_G''' is the rate of thermal energy (heat) generated per unit volume [W/m^3].

(IV) *Rate of change of internal energy within the CV*

The energy stored within the CV is its own internal energy; thus, the rate of change of stored energy, $\dot{E}_{st}$, is just the time rate of change of internal energy U, which is:

$$\dot{E}_{st} = \frac{\partial U}{\partial t} = m\frac{\partial u}{\partial t} = \rho C dxdydz\frac{\partial T}{\partial t}$$

with C being the specific heat [J/kgK], m the mass of the elementary CV, ρ the density and u the specific internal energy. Strictly speaking, the specific heat C is the specific heat at constant volume which, in the case of solids and liquids, is very close to the specific heat at constant pressure and there is no need to make any distinction.

Combining the terms (I)–(II) + (III) = (IV) yields:

$$\dot{Q}_x + \dot{Q}_y + \dot{Q}_z - \left(\dot{Q}_x + \frac{\partial \dot{Q}_x}{\partial x}dx + \dot{Q}_y + \frac{\partial \dot{Q}_y}{\partial y}dy + \dot{Q}_z + \frac{\partial \dot{Q}_z}{\partial z}dz\right)$$
$$+q_G''' dxdydz = \rho C dxdydz\frac{\partial T}{\partial t}$$

The heat transfer rate terms cancel out to obtain the following simplified equation:

$$-\frac{\partial \dot{Q}_x}{\partial x}dx - \frac{\partial \dot{Q}_y}{\partial y}dy - \frac{\partial \dot{Q}_z}{\partial z}dz + q_G''' dxdydz = \rho C dxdydz\frac{\partial T}{\partial t}$$

and, substituting Fourier's Law (Eq. 2.1) for the heat transfer rate terms in each of the three directions, we get:

$$-\frac{\partial}{\partial x}\underbrace{\left(-k_x\frac{\partial T}{\partial x}dydz\right)}_{\dot{Q}_x}dx - \frac{\partial}{\partial y}\underbrace{\left(-k_y\frac{\partial T}{\partial y}dxdz\right)}_{\dot{Q}_y}dy - \frac{\partial}{\partial z}\underbrace{\left(-k_z\frac{\partial T}{\partial z}dxdy\right)}_{\tilde{Q}_z}dz$$
$$+q_G''' dxdydz = \rho C dxdydz\frac{\partial T}{\partial t}$$

Dividing both sides by the elementary control volume $dxdydz$, one obtains:

$$\frac{\partial}{\partial x}\left(k_x\frac{\partial T}{\partial x}\right)+\frac{\partial}{\partial y}\left(k_y\frac{\partial T}{\partial y}\right)+\frac{\partial}{\partial z}\left(k_z\frac{\partial T}{\partial z}\right)+q_G'''=\rho C\frac{\partial T}{\partial t} \tag{2.4}$$

The Eq. (2.4) is the *general heat conduction equation* in Cartesian coordinates. It is a second-order partial differential equation. There is no general analytical solution for all situations because the solution of this equation depends on the geometry of the material and the initial (temporal) and boundary (spatial) conditions. The study of this class of differential equations is called *boundary value problems*. Therefore, in the field of heat transfer, it is generally solved for various geometries and practical situations of interest. The solution of this differential equation is a solution of the type $T = T(x, y, z, t)$, that is, the spatial distribution of temperature within the medium and its progress over time. Some simplifications are adopted in order to solve many real practical situations, such as the cases analyzed below.

Cases of interest

(a) **Uniform thermal conductivity *k* (isotropic material) and constant (independent of *T*)**

The thermal conductivity k can be considered uniform in an isotropic material such as in ordinary metals. Such a hypothesis would not apply to the case of a composite material or natural fibers, where conductivity in one direction may be different from that in another direction—think about the case of a tree trunk where the fibers are aligned with the main axis. However, it can vary with temperature $k = k(T)$, but if this dependence is negligible, we have that:

$$k = k_x = k_y = k_z = \text{constant} \tag{2.5}$$

Therefore, substituting this information into Eq. (2.4), results in:

$$\frac{\partial^2 T}{\partial x^2}+\frac{\partial^2 T}{\partial y^2}+\frac{\partial^2 T}{\partial z^2}+\frac{q_G'''}{k}=\frac{1}{\alpha}\frac{\partial T}{\partial t} \tag{2.6}$$

or, in more synthetic mathematical notation, it is given by:

$$\nabla^2 T+\frac{q_G'''}{k}=\frac{1}{\alpha}\frac{\partial T}{\partial t} \tag{2.7}$$

in which $\nabla^2=\frac{\partial^2}{\partial x^2}+\frac{\partial^2}{\partial y^2}+\frac{\partial^2}{\partial z^2}$ is the mathematical operator *Laplacian* in the Cartesian coordinate system (more will be said about that mathematical operator ahead). By definition, α is the *thermal diffusivity*, whose unit in the SI is:

$$[\alpha] = \frac{[k]}{[\rho][c]} = \frac{\left[\frac{W}{\mathrm{m\cdot K}}\right]}{\left[\frac{\mathrm{kg}}{\mathrm{m}^3}\right]\left[\frac{\mathrm{J}}{\mathrm{kg\cdot K}}\right]} = \frac{\mathrm{m}^2}{\mathrm{s}}$$

The thermal diffusivity α indicates the ratio between the "capacity" to conduct heat and to store thermal energy. In practical terms, it indicates the "speed" of propagation of *heat* within the material. This tranport property is tabulated for several materials in the appendix section.

(b) **Absence of thermal energy (heat) generation and *k* uniform and constant**

The next case to be analyzed is no thermal energy (heat) generation with an isotropic and constant conductivity. These simplifications introduced in the general equation (Eq. 2.4) result in the *Fourier's equation*, expressed in Eq. (2.8).

$$\nabla^2 T = \frac{1}{\alpha}\frac{\partial T}{\partial t} \tag{2.8}$$

(c) **Steady-state and *k* uniform and constant**

In this case, the time rate term is null and, considering an isotropic and constant conductivity, results in *Poisson's equation*, expressed in Eq. (2.9).

$$\nabla^2 T + \frac{q_G'''}{k} = 0 \tag{2.9}$$

(d) **Steady state, *k* constant and uniform and no thermal energy (heat) generation**

Considering these conditions, one obtains *Laplace's equation*, expressed by Eq. (2.10).

$$\nabla^2 T = 0 \tag{2.10}$$

The way to write the heat conduction equation in the notation of Eq. (2.7) and its subsequent simplifications using the Laplacian operator is preferable, because, although the general heat conduction equation has been derived for the Cartesian coordinate system, the use of the symbolic formulation of the Laplacian operator is more general, being independent of the adopted coordinate system. Thus, if there is interest in using other coordinate systems, just replace the Laplacian operator with the coordinate system of interest, such as the cases of cylindrical coordinates, Eq. (2.11), or spherical, Eq. (2.12):

– Laplacian in cylindrical coordinates (r, φ, z):

$$\nabla^2 = \frac{1}{r}\frac{\partial}{\partial r}\left(r\frac{\partial}{\partial r}\right) + \frac{1}{r^2}\frac{\partial^2}{\partial \phi^2} + \frac{\partial^2}{\partial z^2} \tag{2.11}$$

– Laplacian in spherical coordinates (r; φ, θ):

$$\nabla^2 = \frac{1}{r^2}\frac{\partial}{\partial r}\left(r^2\frac{\partial}{\partial r}\right) + \frac{1}{r^2\sin\theta}\frac{\partial}{\partial\theta}\left(\sin\theta\frac{\partial}{\partial\theta}\right) + \frac{1}{r^2\sin^2\theta}\frac{\partial^2}{\partial\phi^2} \tag{2.12}$$

2.2 One-Dimensional Heat Conduction in Steady State Without Thermal Energy (Heat) Generation—Flat Plate or Wall Case

The simplest case of heat conduction is that of the wall or flat plate, in steady state, without thermal energy (heat) generation and the transport properties (thermal conductivity) are constant. This is the case illustrated in Fig. 2.5, where a wall or plate of thickness L has the left face kept at a temperature T_1 and the right face kept at temperature T_2. One could imagine that this is, for example, a wall that separates two environments of different temperatures. As it will be seen later, the temperature distribution $T(x)$ within the wall is linear, as sketched in this figure, with $T_1 > T_2$.

To solve this case, we will start from the general equation of heat conduction, that is, $\nabla^2 T + \frac{q'''_G}{k} = \alpha\frac{\partial T}{\partial t}$. Introducing the simplifications of the problem:

- There is no thermal energy (heat) generation: $q'''_G = 0$.
- Steady state: $\frac{\partial T}{\partial t} = 0$.
- One-dimensional (1D) in x direction: $\nabla^2 T = \frac{d^2T}{dx^2}$.

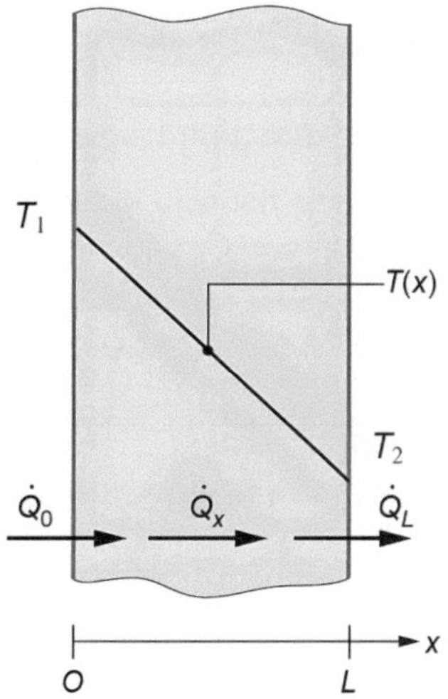

Fig. 2.5 Temperature distribution in a flat wall without thermal energy (heat) generation in steady state and constant thermal conductivity

Thus, with the introduction of these simplifications, the general equation becomes in the following elementary form.

$$\frac{d^2T}{dx^2} = 0 \tag{2.13}$$

To solve this second-order homogeneous differential equation with constant coefficients, consider the following change of variables: $\theta = \frac{dT}{dx}$

Therefore, substituting into the equation, one gets: $\frac{d\theta}{dx} = 0$.

Solving by indefinite integration using separation of variables, one obtains: $\int d\theta = \int 0dx + C_1$, that is, $\theta = C_1$.

However, as it was defined: $\theta = \frac{dT}{dx}$, therefore $\frac{dT}{dx} = C_1$.

Integrating the equation indefinitely once again: $\int dT = \int C_1 dx + C_2$, to obtain: $T = T(x) = C_1 x + C_2$, which is the equation of a straight line, that is, the temperature distribution is linear.

To obtain the integration constants C_1 and C_2, one must apply the boundary conditions (BC) which, in this example, are given by the surface temperatures of the two faces. In mathematical terms, this means that:

$$\begin{aligned} &\text{BC}_1: \text{ at } x = 0, T = T_1 \rightarrow T_1 = C_1(0) + C_2 \rightarrow C_2 = T_1 \\ &\text{BC}_2: \text{ at} x = L, T = T_2 \rightarrow T_2 = C_1(L) + T_1 \rightarrow C_1 = \frac{T_2 - T_1}{L} \end{aligned}$$

Thus, introducing the constants, the final distribution of temperatures inside the wall is given by:

$$T = T(x) = (T_2 - T_1)\frac{x}{L} + T_1 \tag{2.14}$$

Heat conduction transferred through a flat wall

Suppose a flat wall is subjected to different temperatures at its external surfaces as solved above. Consider that the temperature of the left face is greater than that of the right, that is, $T_1 > T_2$, as depicted in Fig. 2.5. To find the heat flux by conduction, q_x, in the wall, Fourier's Law (Eq. 2.1) must be used. Substituting the temperature distribution (Eq. 2.14) into this equation results:

$q_x = -k\frac{d}{dx}\left[(T_2 - T_1)\frac{x}{L} + T_1\right] = -k\frac{(T_2 - T_1)}{L}$, therefore

$$q_x = k\frac{(T_1 - T_2)}{L} \tag{2.15}$$

Note that the heat flux is constant for any x position inside the wall, as q_x is independent of x ($q = q_x = q_0 = q_L$). Furthermore, it is positive in the positive x direction, this means that the direction of heat conduction follows the positive x

direction. Finally, knowing the equation that governs the heat conduction through the wall, one can:

(a) Increase the heat flux q :

- with the use of good heat-conductive material, $k \uparrow$
- decreasing the wall thickness, i.e., $L \downarrow$

(b) or decrease the heat flux q:

- with the use of thermal insulating material, $k \downarrow$
- increasing the wall thickness, i.e., $L \uparrow$

The rate of heat transferred in the wall will be: $\dot{Q}_x = Aq_x$, where A is the perpendicular area (wall area) to the heat flux.

$$\dot{Q}_x = Ak\frac{(T_1 - T_2)}{L} \tag{2.16}$$

Example 2.1 (*Heat conduction in a flat wall*)

Steady-state heat conduction without thermal energy (heat) generation occurs in a plate of thickness L. Furthermore, assume that it occurs only in the x direction (one-dimensional). Determine the validity of the following cases:

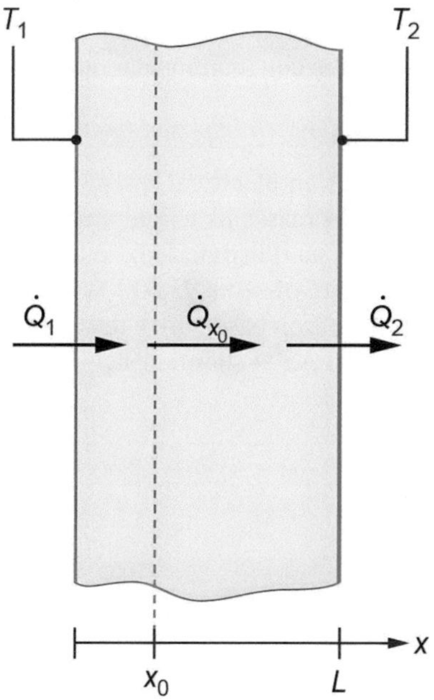

(a) $T_1 > T_2,\ \dot{Q}_1 > 0$.
(b) $T_1 < T_2,\ \dot{Q}_2 > 0$.

(c) $\dot{Q}_1 = \dot{Q}_2 = \dot{Q}_{x_0}$.
(d) The temperature distribution within the wall is linear

Solution

From Eq. (2.16): $\dot{Q}_x = q_x A = \frac{kA(T_1 - T_2)}{L}$.

(a) The heat conduction rate at face 1 is:

    $\dot{Q}_1 = \frac{kA(T_1 - T_2)}{L}$, as $T_1 > T_2$, therefore $\dot{Q}_1 > 0$, that is, the positive direction of heat conduction is to the right in agreement with the positive direction of the axis x. Thus, the statement is correct.
(b) $\dot{Q}_2 = \frac{kA(T_1 - T_2)}{L}$, if $T_1 < T_2$, therefore $\dot{Q}_2 < 0$.

    The statement is incorrect, as the heat conduction signal is negative; this means that its conduction direction is from right to left, i.e., on the opposition to the x positive direction.
(c) In this problem, the heat transfer $\dot{Q}_x = \frac{kA(T_1 - T_2)}{L}$ is independent of x, thus it is constant at any x position, so the statement is correct.
(d) The temperature distribution (Eq. 2.14) within the wall is: $T = T(x) = (T_2 - T_1)\frac{x}{L} + T_1$ and therefore that it is linear, thus the statement is correct.

Comment: Fourier's Law is a function of the temperature gradient, therefore, a positive heat conduction rate indicates that its direction is the same as x positive, and, when negative, it indicates that the conduction heat rate is in the negative x direction.

Example 2.2 (*Conduction with known imposed heat flux*)

An iron is basically made up of an electrical resistance (heating element), which, when connected to electricity, increases its temperature by the Joule effect transferring heat to the base of the iron, as illustrated in the diagram below. Assuming that the resistance transfers a heat flux of $q_0 = 35{,}000\ \text{W/m}^2$, calculate the temperature difference of the two faces of the iron base with a thickness of 0.5 cm, assuming that the material is (a) aluminum; (b) A304 stainless steel.

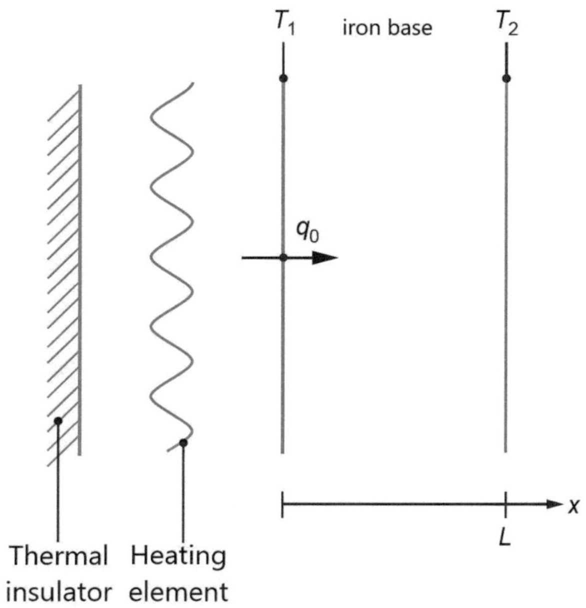

Solution

Assuming a perfect thermal insulation on the wall to the left of the electrical resistance and no heat loss on the edges of the iron, the conduction of heat transferred from the electrical resistance by the Joule effect completely reaches the base of the iron. Under these conditions, the heat flux at the base of the iron is defined as: $q_0 = k\frac{(T_1 - T_2)}{L}$. Therefore, the temperature difference $\Delta T = T_1 - T_2$ is evaluated by: $\Delta T = \frac{q_0 L}{k}$.

(a) Aluminum, $k = 237$ W/mK (Table A.2)

$$\Delta T = \frac{35,000 \times 0.005}{237} = 0.7\,^{\circ}\mathrm{C}$$

(b) Stainless steel A304, $k = 14.9$ W/mK (Table A.2)

$$\Delta T = \frac{35,000 \times 0.005}{14.7} = 11.7\,^{\circ}\mathrm{C}$$

It can be noted that stainless steel causes a greater temperature drop due to its lower thermal conductivity, which may be undesirable in some applications.

Example 2.3 (*Heat conduction in a solar collector glass cover*)

Solar energy can be used to heat water in a flat plate solar collector. The collector shown in the figure has a glass cover ($k = 1.4$ W/mK). Due to the temperature difference between the absorber plate and the heated air inside the collector, heat is lost to the environment. In steady state, this heat loss is 140 W/m^2. Evaluate the temperature difference between the top and bottom faces of the 4 mm thick glass cover.

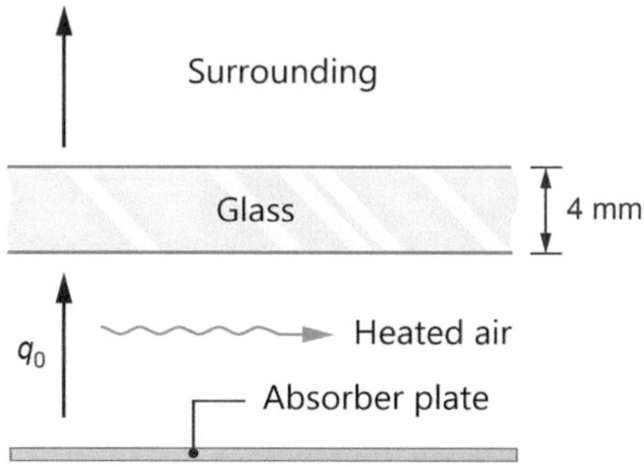

Solution

Heat flux by conduction can be evaluated from the relationship $q_0 = k\frac{(T_1 - T_2)}{L}$; thus, the temperature difference $\Delta T = T_1 - T_2$ is evaluated by: $\Delta T = \frac{q_0 L}{k}$

$$\Delta T = \frac{140 \times 0.004}{1.4} = 0.4\,^{\circ}\mathrm{C}$$

The temperature difference between the surfaces of the glass cover of solar collectors is generally small, so in some calculations, it is assumed that the glass cover temperature is uniform.

2.3 One-Dimensional Heat Conduction in Steady State Without Thermal Energy (Heat) Generation—Cylindrical Tube Case

In the case of heat conduction through the cylindrical wall of a tube, the cylindrical coordinate system must be used. So the solution of heat conduction in a cylindrical tube is equivalent to the previous case of a wall or flat plate with constant properties and without of thermal energy (heat) generation in steady state. Consider the cylindrical tube, as sketched in Fig. 2.6, with the internal temperatures T_i and external T_o established at the radial positions r_i and r_o, respectively. The tube length is L.

Starting from the general equation of heat conduction (Eq. 2.7), it is now convenient to use the Laplacian operator in cylindrical coordinates, Eq. (2.11), which results in:

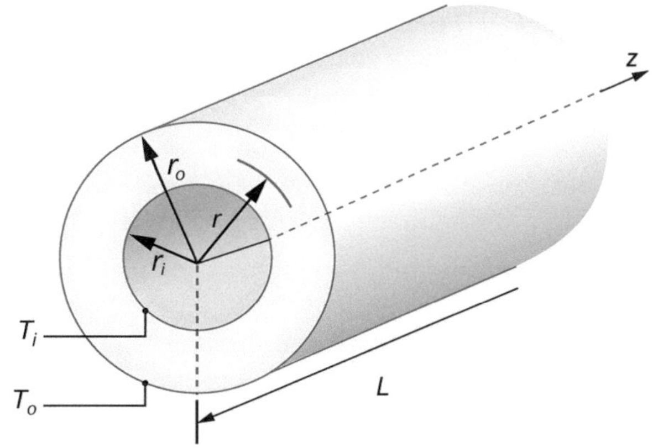

Fig. 2.6 Parameters for the analysis of one-dimensional heat conduction in a cylindrical tube

$$\frac{1}{r}\frac{\partial}{\partial r}\left(r\frac{\partial T}{\partial r}\right)+\frac{1}{r^2}\frac{\partial^2 T}{\partial \phi^2}+\frac{\partial^2 T}{\partial z^2}+\frac{q_G'''}{k}=\frac{1}{\alpha}\frac{\partial T}{\partial t} \tag{2.17}$$

Introducing the problem simplifications:

- There is no thermal energy generation (heat) :$q_G''' = 0$.
- Steady state: $\frac{\partial T}{\partial t} = 0$.
- One-dimensional (1D): temperature variation only with the radius (r).
  - valid for a long tube where the tube end conditions do not interfere (no edge effects), that is, T does not depend on z, that is, $\frac{\partial T}{\partial z} = 0$;
  - there is a radial symmetry, i.e., T does not depend on the angular position ϕ, that is, $\frac{\partial T}{\partial \phi} = 0$.

By applying all these simplifications, the result is Eq. (2.18):

$$\frac{d}{dr}\left(r\frac{dT}{dr}\right)=0 \tag{2.18}$$

To solve Eq. (2.18), it must be integrated twice:
1st integration: separate the variables and integrate once, to obtain:

$$\int d\left(r\frac{dT}{dr}\right)dr=\int 0dr+C_1 \rightarrow r\frac{dT}{dr}=C_1$$

2nd integration: after separating variables, comes:

$$\int dT=\int C_1\frac{dr}{r}+C_2 \rightarrow T=T(r)=C_1\ln r+C_2$$

Therefore, the temperature distribution in the case of the cylindrical tube is a logarithmic one in the direction of the radial coordinate and not linear, as in the case of the flat wall (Eq. 2.14). The determination of the constants C_1 and C_2 is done by applying the boundary conditions which, as indicated in Fig. 2.6, are:

BC_1: the internal surface is kept at a constant temperature, that is, $r = r_i, T = T_i$.

BC_2: the outer surface is also kept at another constant temperature, that is, $r = r_o$, $T = T_o$.

$$\mathrm{BC}_1: r = r_i, T = T_i \rightarrow T_i = C_1 \ln r_i + C_2$$
$$\mathrm{BC}_2: r = r_o, T = T_o \rightarrow T_o = C_1 \ln r_o + C_2$$

Subtracting the internal and external temperatures, one finds that $T_i - T_o = C_1 \ln\frac{r_i}{r_o}$, or $C_1 = \frac{T_i - T_o}{\ln\left(\frac{r_i}{r_o}\right)}$.

Substituting C_1 into BC_2, we get: $C_2 = T_o - \frac{(T_i - T_o)}{\ln\left(\frac{r_i}{r_o}\right)}\ln r_o$.

Finally, substituting C_1 and C_2 into the equation, yields:

$$T = T(r) = (T_i - T_o)\frac{\ln\left(\frac{r}{r_o}\right)}{\ln\left(\frac{r_i}{r_o}\right)} + T_o \tag{2.19}$$

Equation (2.19) represents the temperature distribution in the cylindrical wall of thickness $r_o - r_i$. Assuming $T_i > T_o$, its logarithmic behavior is indicated in Fig. 2.7.

The heat conduction rate in the radial direction is obtained by using Fourier's Law, $\dot{Q}_r = -kA_r\frac{dT}{dr}$, where the area A_r is the area perpendicular to the heat conduction flux, and it is not the tube cross-sectional area. It is the area of the cylindrical "shell" of radius r illustrated in Fig. 2.8.

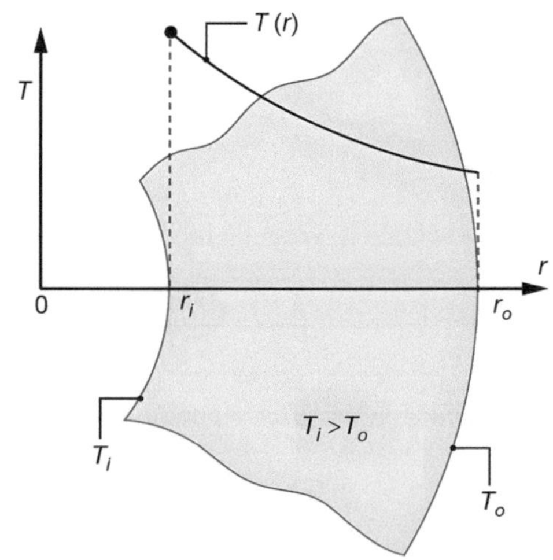

Fig. 2.7 Logarithmic temperature distribution in the cylindrical tube

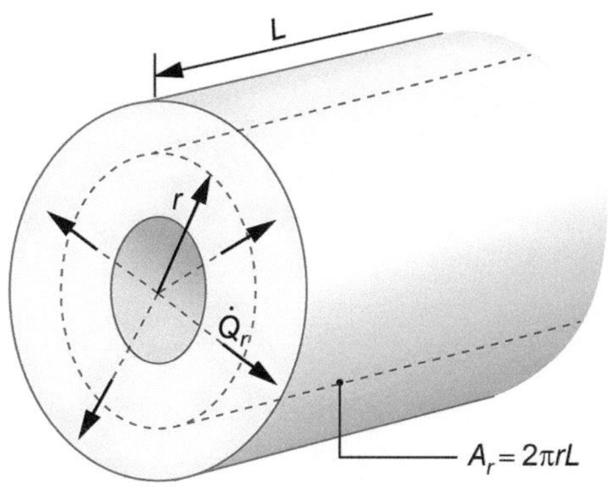

Fig. 2.8 Area of the cylindrical "shell"

With $A_r = 2\pi rL$, where L is the tube length.

Substituting the logarithmic temperature distribution into the Fourier equation, yields:

$$\dot{Q}_r = -k(2\pi rL)\frac{d}{dr}\left[(T_i - T_o)\frac{\ln\left(\frac{r}{r_o}\right)}{\ln\left(\frac{r_i}{r_o}\right)} + T\right] = -k(2\pi rL)\frac{(T_i - T_o)}{\ln\left(\frac{r_i}{r_o}\right)}\frac{\frac{1}{r_o}}{\frac{r}{r_o}}.$$

Finally:

$$\dot{Q}_r = 2\pi kL\frac{(T_i - T_o)}{\ln\left(\frac{r_o}{r_i}\right)} \tag{2.20}$$

It can be noted in Eq. (2.20) that the heat conduction rate does not depend on the radius (r), being constant across the cylindrical surfaces, that is, $\dot{Q} = \dot{Q}_i = \dot{Q}_r = \dot{Q}_o$. However, the heat conduction flux, q_r, does depend on the radial position, r, as indicated below:

$$q_r = \frac{\dot{Q}_r}{A_r} = \frac{2\pi kL\frac{(T_i - T_o)}{\ln\left(\frac{r_o}{r_i}\right)}}{2\pi rL}$$

or

$$q_r = \frac{k}{r}\frac{(T_i - T_o)}{\ln\left(\frac{r_o}{r_i}\right)} \tag{2.21}$$

Example 2.4 (*Heat conduction in a cylindrical tube*)

Steam flows inside a stainless steel tube ($k = 15$ W/mK) which has an internal radius of $r_1 = 8$ cm and an external radius of $r_2 = 9$ cm. By accident, 10 cm of thermal insulation is removed. The temperatures of the internal and external surfaces are $T_1 = 120$ °C and $T_2 = 102.5$ °C, respectively. Obtain the temperature distribution of the tube wall and determine the heat loss from the steam through the tube wall in this uncovered section of the pipe.

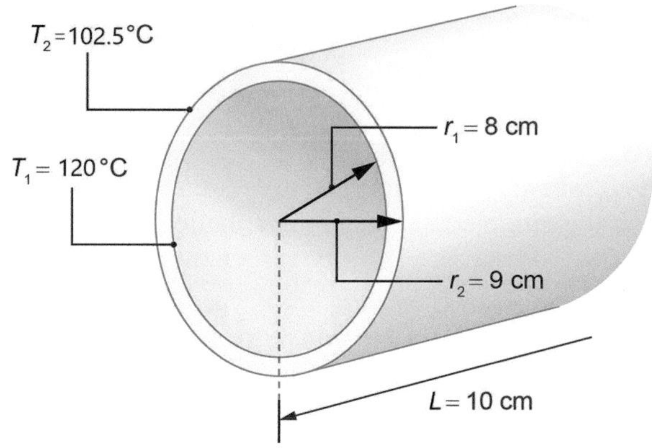

Solution

Considering steady state, one-dimensional heat conduction, constant transport properties, without thermal energy generation, the temperature distribution is the one indicated in Eq. (2.19). Substituting the given values of the problem, one obtains:

$$T(r) = (T_1 - T_2)\frac{\ln\left(\frac{r}{r_2}\right)}{\ln\left(\frac{r_1}{r_2}\right)} + T_2 = (120 - 102.5)\frac{\ln\left(\frac{r}{0.09}\right)}{\ln\left(\frac{0.08}{0.09}\right)} + 102.5, \text{ or}$$

$$T(r) = -148.6 \ln r - 255.3$$

With r in meters, it results in the sought distribution of temperatures (°C) in the wall thickness of the uncovered tube as a function of the radius.

The rate of heat transfer from the steam through the uncovered section of the metal cylinder wall is determined using Fourier's Law, indicated in Eq. (2.20):

$$\dot{Q} = 2\pi kL\frac{(T_1 - T_2)}{\ln\left(\frac{r_2}{r_1}\right)}$$

Substituting the numerical values, yields:

$$\dot{Q} = 2\pi \times 15 \times 0.1 \times \frac{(120 - 102.5)}{\ln\left(\frac{0.09}{0.08}\right)}$$

$$\dot{Q} = 1399\,\text{W}$$

This means that, in just 10 cm of uncovered pipe, the steam loses 1399 W of heat to the external environment.

2.4 Composite Flat Walls

In civil construction, furnace walls, refrigeration chambers, and insulation systems in general, the wall is made up of several layers of different materials, for example, the wall of a house. Figure 2.9 schematically shows a flat wall composed of three different materials.

To solve this type of composite wall configuration, first note that the transferred heat conduction rate $\dot{Q}$ is the same that crosses all walls (steady state, without thermal energy generation and one-dimensional). In this way, for each wall, the following equations can be written:

Wall 1: $\dot{Q} = k_1 A \frac{(T_1 - T_2)}{L_1} \rightarrow T_1 - T_2 = \frac{\dot{Q}L_1}{k_1 A}$

Wall 2: $\dot{Q} = k_2 A \frac{(T_2 - T_3)}{L_2} \rightarrow T_2 - T_3 = \frac{\dot{Q}L_2}{k_2 A}$

Wall 3: $\dot{Q} = k_3 A \frac{(T_3 - T_4)}{L_3} \rightarrow T_3 - T_4 = \frac{\dot{Q}L_3}{k_3 A}$

Adding up the temperature difference of the three walls, it results: $T_1 - T_4 = \dot{Q} \sum \frac{L_i}{k_i A}$, noticing that the temperatures of the intermediate faces have been canceled out. Finally, it may also be written as

$$\dot{Q} = \frac{T_1 - T_4}{R} \tag{2.22}$$

where $T_1 - T_4$ refers to the temperature difference of the *two outer faces of the composite wall. R* is known as the *thermal resistance* of the composite wall, given

Fig. 2.9 A composite flat wall schematics

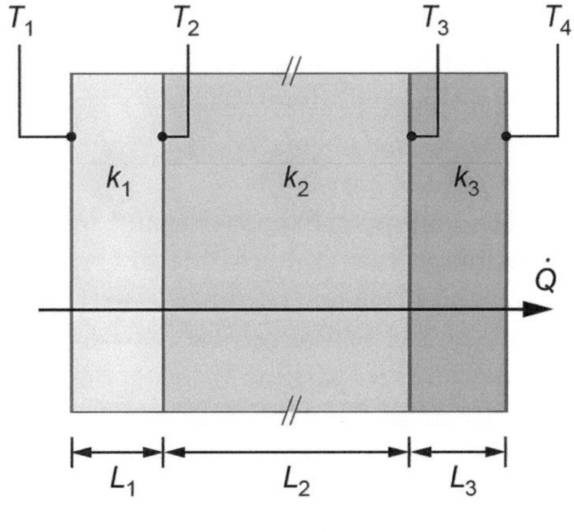

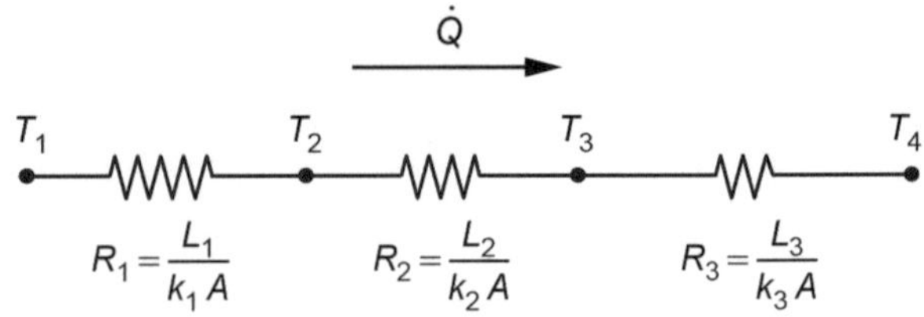

Fig. 2.10 Thermal circuit of 3 walls in series

by:

$$R = \sum \frac{L_i}{k_i A} \tag{2.23}$$

Electrical analogy

The technique allows a perfect *electrical analogy* to be established between ohmic electrical and thermal heat conduction phenomena, through the following correspondence:

$$i \to \dot{Q}, \quad U \to \Delta T, \text{ and } R_{\text{OHMIC}} \to R_{\text{THERMAL}}.$$

Thus, for example, for the composite wall shown in Fig. 2.9, its "thermal circuit" is indicated in Fig. 2.10.

By the electrical analogy, more complex wall configurations (in series and parallel) can be resolved, like the system indicated in Fig. 2.11. For this wall configuration, it is assumed that the interfaces between walls A, B, C, and D have a uniform temperature equal to T_2; similarly, the temperature of the face between walls B, C, D, and E has a uniform temperature equal to T_3.

The analogous electrical circuit is indicated in Fig. 2.12.

The heat conduction rate is given in Eq. (2.24):

$$\dot{Q} = \frac{\Delta T_{total}}{R_T} = \frac{T_1 - T_4}{R_T} \tag{2.24}$$

where the equivalent resistance R_T is given by:

$R_T = R_A + R_{//} + R_E$ and $\frac{1}{R_{//}} = \frac{1}{R_B} + \frac{1}{R_C} + \frac{1}{R_D}$, where $R_{//}$ is the thermal resistance of the parallel circuit.

The analogy technique is useful for solving heat transfer in composite walls. In the solution, it is still possible to use the concept of resistance to include the heat convection occurring on the external wall faces. Just as in heat conduction, thermal resistance can be defined by the heat convection phenomenon, which relates the difference in temperature between the environment and the surface with the rate of heat transferred on that surface according to Newton´s cooling law of convection (Eq. 1.5)

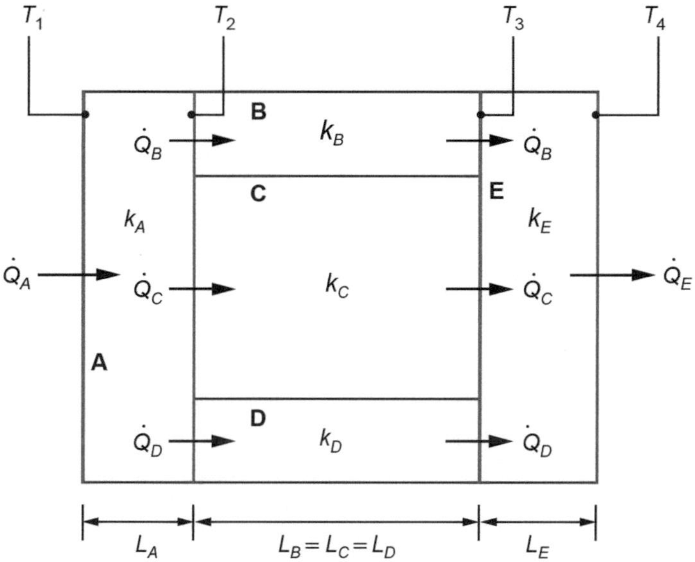

Fig. 2.11 Composite walls in series and parallel

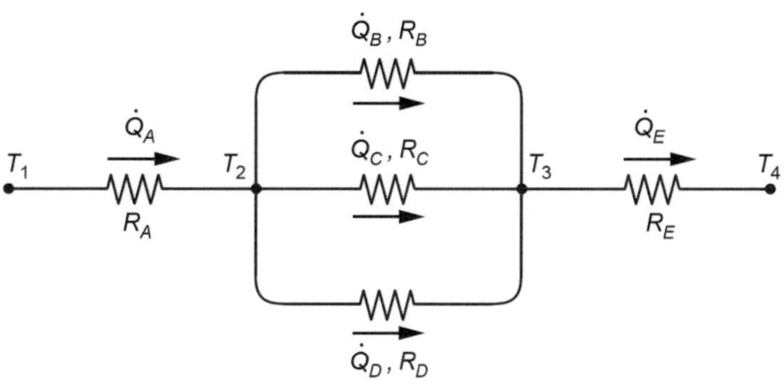

Fig. 2.12 Equivalent thermal analogic circuit of the composite wall arrangement indicated in Fig. 2.11

$$R_{conv} = \frac{1}{hA} = \frac{\Delta T}{\dot{Q}} \tag{2.25}$$

Example 2.5 (*Thermal insulation in walls*)

In cold regions, such as in Picotani, in the Peruvian Andes at 5000 m altitude, the environment temperature can reach –15 °C at night. It is proposed to install a composite plaster wall (k_p = 0.17 W/mK), an insulating material, and wood (k_w = 0.16 W/mK) as indicated in figure. It is intended to maintain the internal air temperature at 20 °C in a house that has an area of 5 × 7 m and a height of 2.5 m. In the house, there are four people producing a total thermal load of 400 W. Assuming a perfectly thermally insulated floor and that the ceiling is constructed of the same material as the composite walls, it is asked to evaluate: (a) the necessary thickness (L) if the insulating material is composed of polystyrene plates, (b) what thermal power should an extra device (e.g., electric heater) inside the house have to reduce this thickness by half and continue maintaining 20 °C in the interior of the house. Ignore the condensation/freezing of water vapor in the air on the walls.

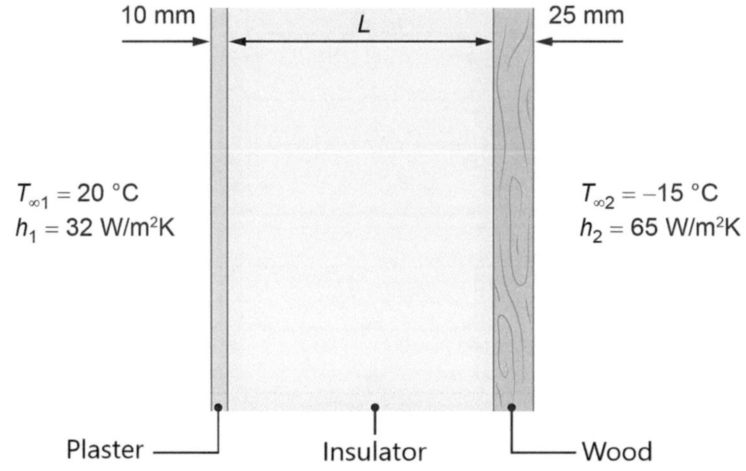

Solution

Assuming a steady state, one-dimensional conduction, without heat generation and constant transport properties, it is concluded that the heat transfer rate is constant through the wall, and, therefore, the concept of thermal resistance can be used. The problem can be represented in the analog thermal circuit indicated in the following scheme.

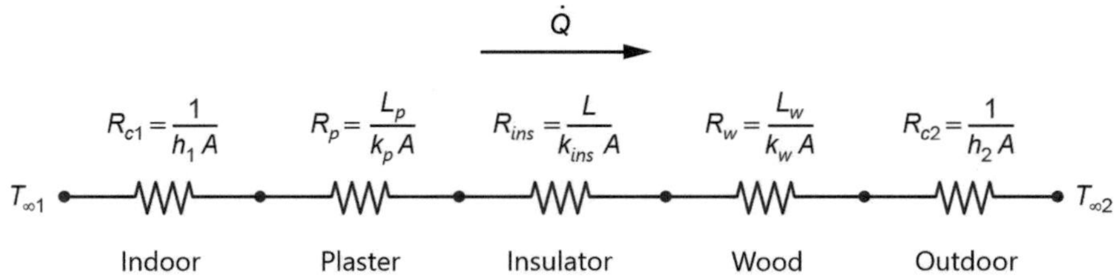

The side walls and the ceiling have a total area of:

$$A = 2 \times 5 \times 2.5 + 2 \times 7 \times 2.5 + 5 \times 7 = 95\,\text{m}^2$$

The heat transfer rate can be obtained by:

$$\dot{Q} = \frac{T_{\infty 1} - T_{\infty 2}}{R} = \frac{T_{\infty 1} - T_{\infty 2}}{\frac{1}{h_1 A} + \frac{L_g}{k_p A} + \frac{L}{k_{\text{ins}} A} + \frac{L_m}{k_w A} + \frac{1}{h_2 A}}$$

(a) Substituting values:

$$400 = \frac{20 - (-15)}{\frac{1}{32 \times 95} + \frac{0.010}{0.17 \times 95} + \frac{L}{0.027 \times 95} + \frac{0{,}025}{0.160.025 \times 95} + \frac{1}{65 \times 95}}$$

Thus, the thickness of the insulation wall should be: $L = 0.217\,\text{m}$ or $L = 21.7\,\text{cm}$.

(b) By reducing the thickness of the thermal insulation by half and maintaining the internal temperature at 20 °C, one would be transferring a heat rate from inside to outside the house of:

$$\dot{Q}_1 = \frac{20 - (-15)}{\frac{1}{32 \times 95} + \frac{0.010}{0.17 \times 95} + \frac{0.109}{0.027 \times 95} + \frac{0.025}{0.16 \times 95} + \frac{1}{65 \times 95}} = 773.5\,\text{W}$$

Thus, a device that produces a heat rate of 773.5–400 = 373.5 W is needed to maintain a 20 °C inside temperature in the house.

Note: if this additional heating equipment is not installed, the internal temperature of the house will be:

$$400 = \frac{T_i - (-15)}{\frac{1}{32 \times 95} + \frac{0.010}{0.17 \times 95} + \frac{0.109}{0.027 \times 95} + \frac{0.025}{0.16 \times 95} + \frac{1}{65 \times 95}} \Rightarrow$$

$$T_i = -1.9\ ^\circ\text{C}.$$

2.5 Contact Thermal Resistance

Whenever two solid surfaces are put into contact, the interfacial region between them may have a thermal resistance of contact, $R_{t,c}$, because there is no "perfect" contact between any two surfaces due to surface roughness, as illustrated in Fig. 2.13. As a consequence, a temperature difference at the interface is established. Therefore, heat transfer will occur by conduction at the contacting peaks of the roughness and by conduction through the trapped fluid (air) in cavities between the surfaces. Thermal radiation can also be relevant if high-temperature differences are present.

The heat transfer at the interface, between the contacting surfaces is $\dot{Q} = \frac{T_1 - T_2}{R} = \frac{T_1 - T_2}{\frac{R_{t,c}}{A}}$, thus the contact thermal resistance is given by Eq. (2.26):

$$R_{t,c} = \frac{T_1 - T_2}{q} \tag{2.26}$$

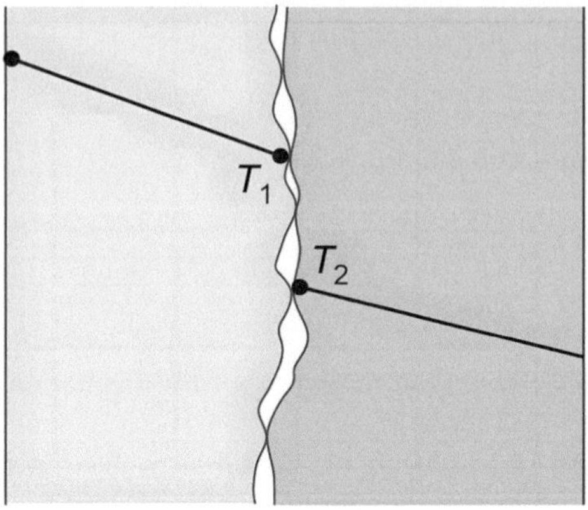

Fig. 2.13 Thermal resistance between solid surfaces

Table 2.1 Some values of thermal resistance at the interface between solids

Interface	$R_{t,c}$ [m^{2}K/W]
Brass/Brass with 15 μm tin solder insertion at 1–10 MPa (Yoyanovich and Tuarze 1969)	0.025×10^{-4}–0.14×10^{-4}
Copper/Aluminum with eupec grease insertion (Henton) (Sartre and Lallemand 2001)	1.47×10^{-4}
Copper/Aluminum with graphite sheet insertion (Furon® C695) (Sartre and Lallemand 2001)	6.25×10^{-4}
Copper/Aluminum with silicone sheet insertion with aluminum oxide (CHO-Therm® 1674) (Sartre and Lallemand 2001)	8.01×10^{-4}

The contact thermal resistance varies with the materials, surface finishing, and with the compression between surfaces. Some values are indicated in Table 2.1.

2.6 Thermal Resistance—Various Situations

Table 2.2 summarizes various geometries, the thermal circuit, the heat transfer rate, and the thermal resistance.

Table 2.2 Summary of thermal resistances for some situations of interest

Configuration	Thermal circuit	Heat transfer rate	Thermal resistances
Flat wall 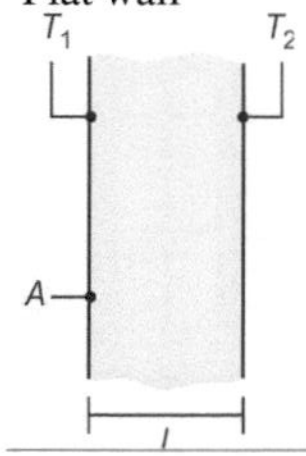	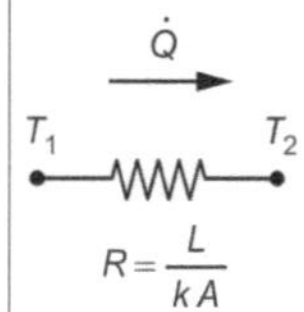	$\dot{Q} = \frac{T_1 - T_2}{R}$	$R = \frac{L}{kA}$
Flat wall with convection 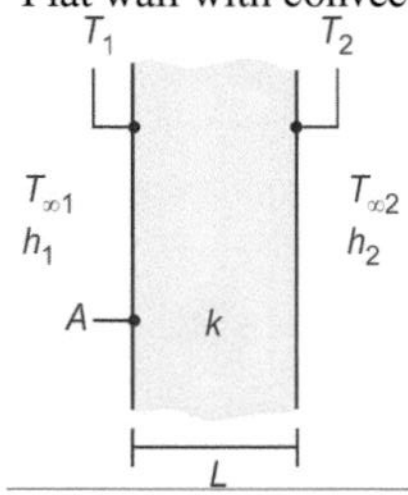	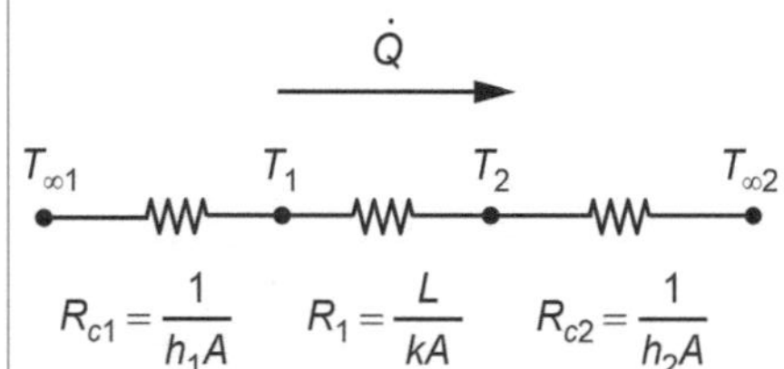	$\dot{Q} = \frac{T_{\infty 1} - T_{\infty 2}}{R}$	$R = R_{c1} + R_1 + R_{c2}$ $R = \frac{1}{h_1 A} + \frac{L}{kA} + \frac{1}{h_2 A}$

(continued)

Table 2.2 (continued)

Configuration	Thermal circuit	Heat transfer rate	Thermal resistances

Composite walls 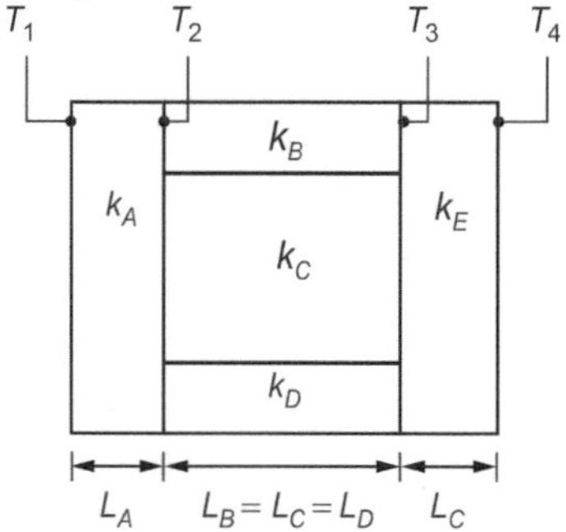	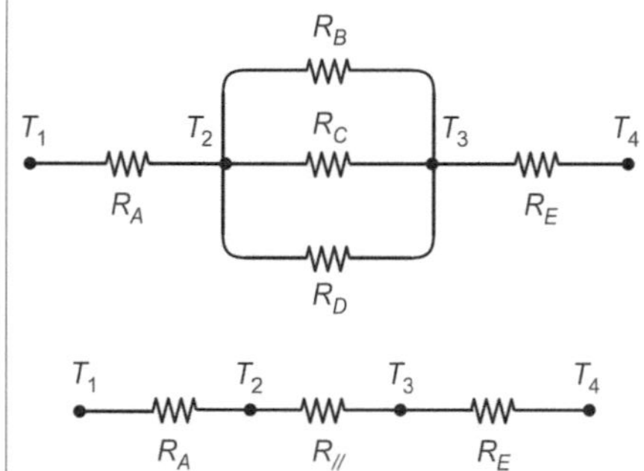	$\dot{Q} = \frac{T_1 - T_2}{R}$	$R_i = \frac{L_i}{k_i A_i}$
			 $R = R_A + R_{//} + R_E$
			 $\frac{1}{R_{//}} = \frac{1}{R_B} + \frac{1}{R_C} + \frac{1}{R_D}$

Cylindrical tube 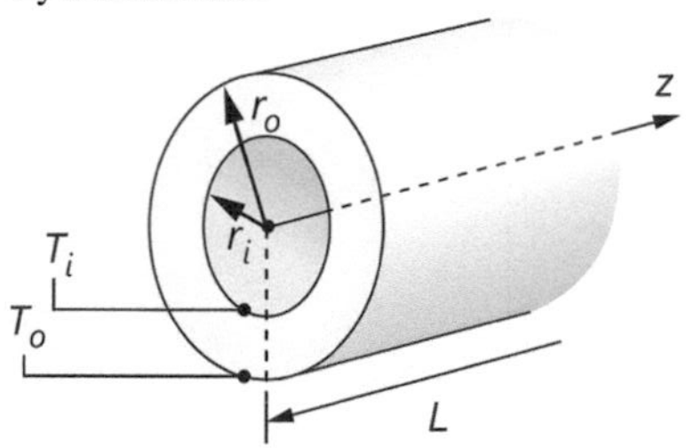	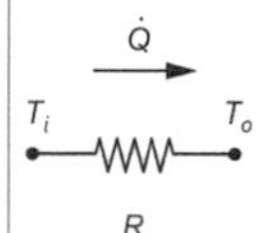	$\dot{Q} = \frac{T_i - T_o}{R}$	$R = \frac{\ln\left(\frac{r_o}{r_i}\right)}{2\pi k L}$

(continued)

Table 2.2 (continued)

Configuration	Thermal circuit	Heat transfer rate	Thermal resistances
Composite cylindrical tube 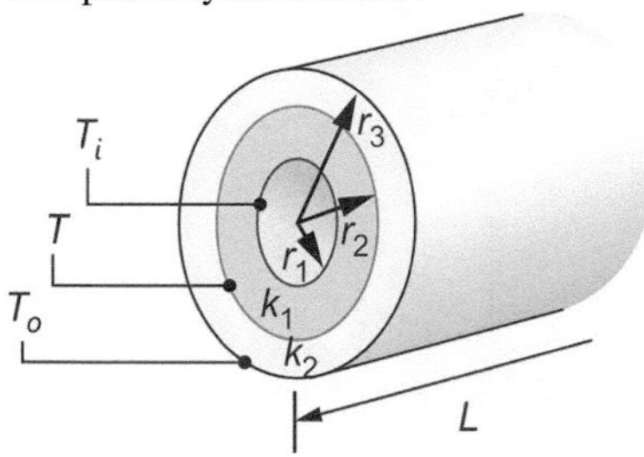	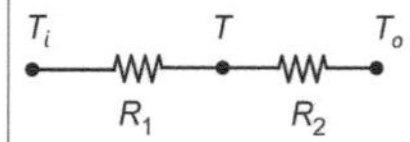	$\dot{Q} = \frac{T_i - T_o}{R}$	$R = \sum \frac{\ln\left(\frac{r_{i+1}}{r_i}\right)}{2\pi k_i L}$
External convection in cylindrical tube 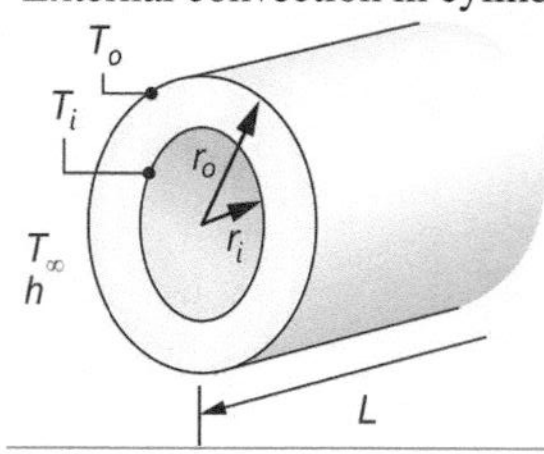	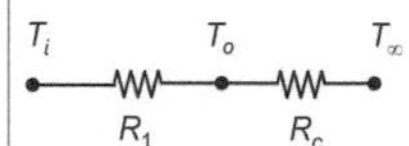	$\dot{Q} = \frac{T_i - T_\infty}{R}$	$R = R_1 + R_c$ $R = \frac{\ln\left(\frac{r_o}{r_i}\right)}{2\pi k L} + \frac{1}{hA_o}$

2.7 Overall Heat Transfer Coefficient *U*

The heat transfer rate between two environments or regions separated by a wall must consider not only the heat conduction through the wall, but also the heat convection that occurs on both sides of the wall, as seen in Example 2.5. For this, the *overall heat transfer coefficient*, U, is defined as:

$$\dot{Q} = UA\Delta T_{\text{total}} \tag{2.27}$$

Clearly, U is associated with the thermal resistance, as shown below.

Cases

1. *Flat wall with convection on both sides*

In this case the thermal resistance (Table 2.2) is given by $R = \frac{1}{h_1 A} + \frac{L}{kA} + \frac{1}{h_2 A}$. By comparing this expression with the definition of the global coefficient, Eq. (2.27), one obtains:

$$\dot{Q} = \frac{\Delta T}{R} = UA\Delta T$$

Thus, the relationship between the global coefficient and thermal resistance is:

$$UA = \frac{1}{R} \text{ or } U = \frac{1}{RA}$$

Therefore,

$$U = \frac{1}{\frac{1}{h_1} + \frac{L}{k} + \frac{1}{h_2}} \tag{2.28}$$

2. *Cylindrical tube with internal and external convection*

The procedure for relating the global coefficient to thermal resistance is the same. However, in the case of cylindrical tubes, there is a problem associated with the area of reference. In other words, it is necessary to specify whether U refers to the internal area of the tube, U_i, or the external area, U_o. However, the two values are interchangeable through the following expression: $U_o A_o \Delta T_{\text{total}} = U_i A_i \Delta T_{\text{total}}$, since the heat transfer rate must be the same, regardless of the adopted area of reference. Therefore, $U_o A_o = U_i A_i$. U_o referred to the external area is:

$$U_o = \frac{1}{\frac{A_o}{A_i h_i} + \frac{A_o \ln\left(\frac{r_o}{r_i}\right)}{2\pi kL} + \frac{1}{h_o}} \tag{2.29}$$

and U_i referred to the internal area:

$$U_i = \frac{1}{\frac{1}{h_i} + \frac{A_i \ln\left(\frac{r_o}{r_i}\right)}{2\pi kL} + \frac{A_i}{A_o h_o}} \tag{2.30}$$

2.8 Heat Conduction in a Flat Plate or Wall with Thermal Energy (Heat) Generation

The thermal energy (heat) generation is the result of the conversion of one form of energy into heat, as exemplified in the following cases:

1. *Conversion of electrical energy into heat (Joule effect)*

In this case, electrical power is converted into heat by Joule effect. The electrical power can be evaluated by Eq. (2.31).

$$P = R_e I^2 \tag{2.31}$$

where P is the electrical power [W], R_e is the ohmic resistance [Ω] and I is the electrical current [A]. Also, if V is the electrical potential difference [V], the electrical power is:

$$P = VI \text{ or } P = \frac{V^2}{R_e} \tag{2.32}$$

Usually, the volumetric density of thermal energy (heat) generation is defined as $q'''_G = \frac{P}{\forall_s}$ [W/m³], where $\forall_s$ is the volume of the solid in which heat is generated.

2. *Exothermic chemical reaction* ($q'''_G > 0$), for example, the heat released during the curing of resins and concrete. In the case of an *endothermic chemical reaction*, $q'''_G < 0$.

3. *Absorption of radiation, neutrons*, among others.

Note that the term "generation" should be understood as the transformation of one form of energy into thermal energy, commonly called heat in this text. As a first example, consider a flat wall (plate) with a uniform heat generation, like the flat electrical resistance indicated in Fig. 2.14. For the case where $W \gg L$, and $C \gg L$, where W is the width of the resistance, C is its length, and $2L$ is its thickness. Under these conditions, there will be approximately one-dimensional heat conduction in the direction of the material's thickness in order to dissipate to the surrounding environment the thermal energy in the form of internal heat generated.

The general equation of heat conduction (Eq. 2.7) is $\nabla^2 T + \frac{q'''_G}{k} = \frac{1}{\alpha}\frac{\partial T}{\partial t}$. Introducing the adopted hypotheses, one obtains:

- Steady state: $\frac{\partial T}{\partial t} = 0$.
- One-dimensional heat conduction in the direction x: $\nabla^2 T = \frac{d^2 T}{dx^2}$.

With these simplifications, the general equation results in $\frac{d^2 T}{dx^2} + \frac{q'''_G}{k} = 0$.

An important issue regarding the solution of heat transfer problems is establishing the proper boundary conditions that govern the problem. In the present case, the boundary conditions are:

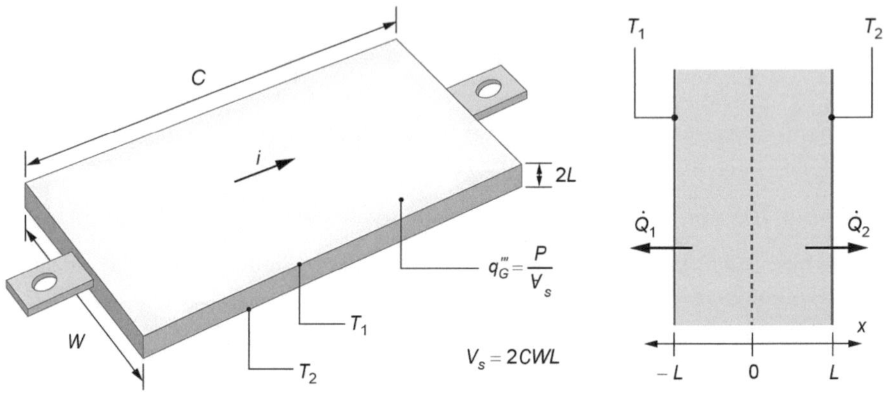

Fig. 2.14 Diagram of the main lengths of the thermal resistance and detail of the one-dimensional heat transfer

BC_1: at $x = -L$ (left resistance face), $T = T_1$.

BC_2: at $x = L$ (right resistance face), $T = T_2$.

Consider the following variable change (just for convenience): $\theta = \frac{dT}{dx}$, then $\frac{d\theta}{dx} = -\frac{q_G'''}{k}$.

Integrating the general equation by parts yields (after substitution of variable) $T(x) = -\frac{q_G'''}{2k}x^2 + C_1x + C_2$.

Comment: this is a parabolic temperature distribution. The heat generation due to the Joule effect is positive and as k is also a positive quantity, thus the term $-\frac{q_G'''}{2k}$ is a negative one, which means that the parabola is concave downwards. On the other hand, if q_G''' were negative, which can occur with curing processes of some resins (endothermic processes), the parabola concavity would be upwards.

Applying the two boundary conditions to the previous equation:

BC_1: $T_1 = -\frac{q_G'''L^2}{2k} - C_1L + C_2$.

BC_2: $T_2 = -\frac{q_G'''L^2}{2k} + C_1L + C_2$.

Adding up the above equations, one obtains:

$T_1 + T_2 = -\frac{q_G'''L^2}{k} + 2C_2$. Therefore, $C_2 = \frac{T_1+T_2}{2} + \frac{q_G'''L^2}{2k}$

Substituting into (BC_1) or (BC_2), to obtain $C_1 = \frac{T_2-T_1}{2L}$

Substituting the two constants, the final temperature distribution is given by:

$$T(x) = \frac{q_G'''}{2k}\left(L^2 - x^2\right) + (T_2 - T_1)\frac{x}{2L} + \frac{T_2 + T_1}{2} \tag{2.33}$$

Boundary conditions cases

Case A: suppose that the two faces are at the same temperature, T_w, as indicated in Fig. 2.15: $T_1 = T_2 = T_w$. From this, it follows that: $T(x) = \frac{q_G'''}{2k}\left(L^2 - x^2\right) + T_w$.

This situation constitutes a symmetrical temperature distribution. The maximum temperature, in this case, occurs in the central plane, where $x = 0$ (note the symmetry of the problem). If it were the case of an endothermic reaction, or $q_G''' < 0$, the

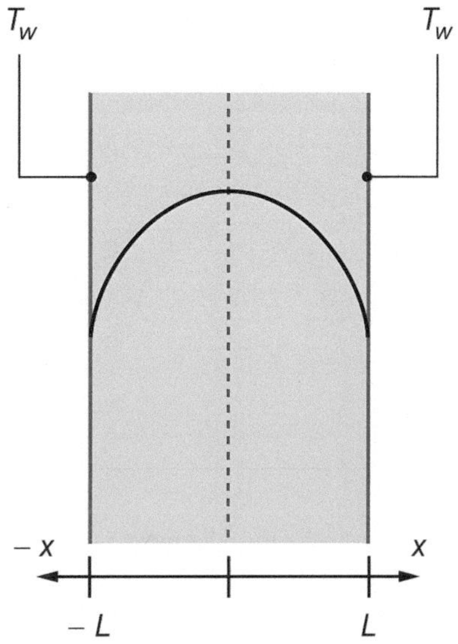

Fig. 2.15 Heat generation in steady state with equal face temperatures in a flat resistance

concavity would be upwards and, in the central plane, would have the minimum temperature.

$$T_{max} = T_0 = \frac{q_G''' L^2}{2k} + T_w \tag{2.34}$$

One could also arrive at this expression using: $\left.\frac{dT}{dx}\right|_{x=0} = 0$. The heat flux (Fourier's Law) depends on the position of the x-plane of interest, which results in Eq. (2.35):

$$q_x = -k\frac{dT}{dx} = -k\frac{d}{dx}\left[\frac{q_G'''(L^2 - x^2)}{2k} + T_w\right] = xq_G'''$$

$$q_x = xq_G''' \tag{2.35}$$

At the central plane ($x = 0$), the heat flux is null due to the symmetry of the problem and the boundary conditions. Thus, the central plane performs the case of an adiabatic wall, $q_0 = 0$, as schematically indicated in Fig. 2.16.

Case B: now suppose that the temperature of one of the faces is higher than the other. For example, $T_1 > T_2$, as illustrated in Fig. 2.17.

It is known that the heat flux is null at the position (x_{max}) where the temperature is a maximum:

$-k\left.\frac{dT}{dx}\right|_{x_{max}} = 0$, or $-k\frac{d}{dx}\left[\frac{q_G'''}{2k}\left(L^2 - x^2\right) + (T_2 - T_1)\frac{x}{2L} + \frac{T_1+T_2}{2}\right] = 0$,

which results in: $-\frac{q_G'''}{k}x_{max} + \frac{(T_2-T_1)}{2L} = 0$, whose solution is

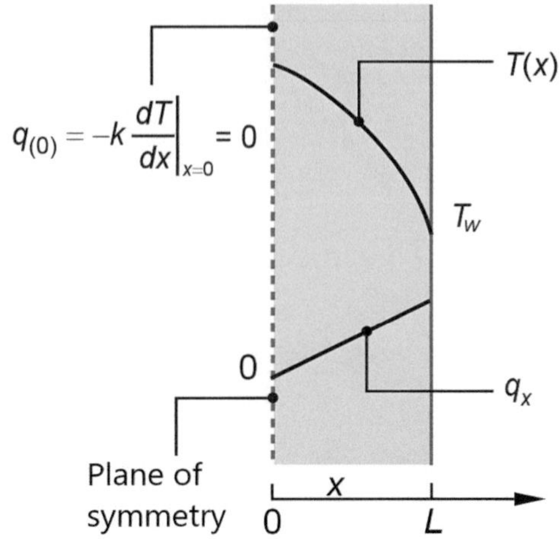

Fig. 2.16 Temperature and heat flux distribution ($T_1 = T_2 = T_w$) in a flat resistance

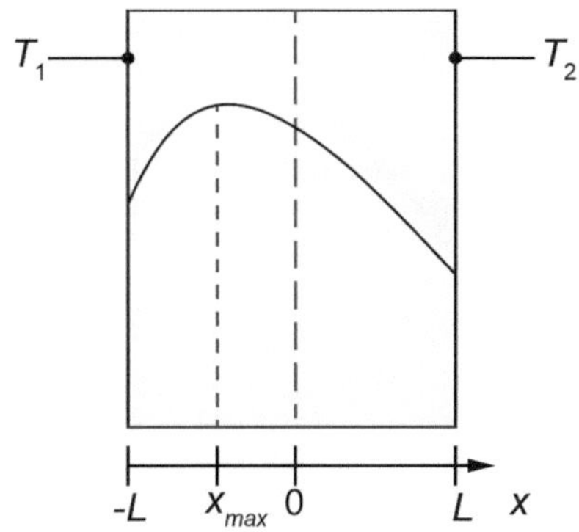

Fig. 2.17 Temperature distribution in which $T_1 > T_2$ in a flat resistance

$$x_{max} = \frac{(T_2 - T_1)k}{2Lq_G'''} \tag{2.36}$$

Substituting the value of x_{max} into the expression of the temperature distribution, one finds the value of the maximum temperature:

$$T_{max} = \frac{q_G'''L^2}{2k} + \frac{(T_2 - T_1)^2 k}{8L^2 q_G'''} + \frac{T_2 + T_1}{2} \tag{2.37}$$

Example 2.6 (*Wall with thermal energy (heat) generation*)

Consider a flat wall of nichrome (alloy with 80% Ni and 20% Cr, $k = 15$ W/mK) 8 cm thick that has a uniform thermal energy (heat) generation of 0.25 MW/m^3. One side of the wall (left) is thermally insulated, while the other is exposed to an environment of 20 °C and $h = 50$ W/m^{2}K. Determine the maximum temperature of the wall and the temperatures on both sides of it.

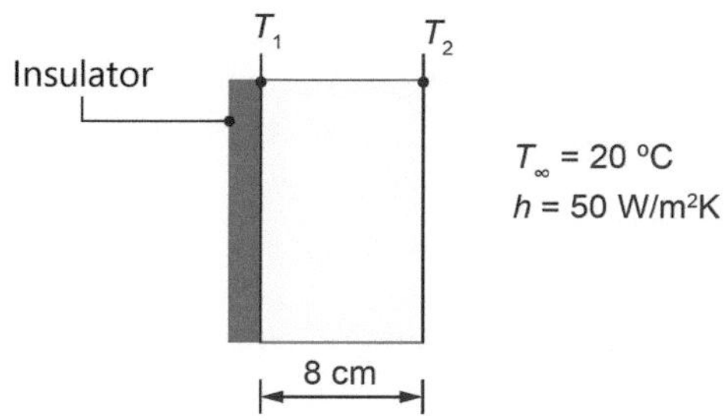

Solution

Assumptions:

- Steady state.
- Unidirectional conduction.
- Uniform thermal energy (heat) generation.

From the previous assumptions: $\frac{d^2T}{dx^2} + \frac{q_G'''}{k} = 0 \rightarrow \frac{dT}{dx} = -\frac{q_G'''}{k}x + C_1$.
BC_1: Internal wall (1) is adiabatic.

$$q_{x=0} = -k\left.\frac{dT}{dx}\right|_{x=0} = 0 \rightarrow -k\left(-\frac{q_G'''}{k}(0) + C_1\right) = 0 \rightarrow C_1 = 0$$

$$\frac{dT}{dx} = -\frac{q_G'''}{k}x \rightarrow T = -\frac{q_G'''}{2k}x^2 + C_2$$

BC_2: External wall (2) transfers heat by convection

$$q_L = -k\left.\frac{dT}{dx}\right|_{x=L} = h(T_2 - T_\infty)$$

$$-k\left(-\frac{q_G'''}{k}L\right) = h\left(-\frac{q_G'''}{2k}L^2 + C_2 - T_\infty\right) \rightarrow C_2 = \frac{q_G'''L}{h} + \frac{q_G'''L^2}{2k} + T_\infty$$

Therefore, the temperature distribution will be: $T(x) = \frac{q_G'''}{2k}\left(L^2 - x^2\right) + \frac{q_G'''L}{h} + T_\infty$
To find the position of the maximum temperature in the plate, the temperature distribution is derived in x:
$\frac{dT}{dx} = \frac{d}{dx}\left[\frac{q_G'''}{2k}\left(L^2 - x^2\right) + \frac{q_G'''L}{h} + T_\infty\right] = -\frac{2xq_G'''}{2k} = 0$, therefore, $x_{max} = 0$.
Nichrome has a thermal conductivity of $k = 15$ W/mK.

The maximum temperature on the wall will be:

$$T_{max} = T(0) = \frac{q'''_G L^2}{2k} + \frac{q'''_G L}{h} + T_\infty = \frac{250{,}000 \times 0.08^2}{2 \times 15} + \frac{250{,}000 \times 0.08}{50} + 20$$

$$T_{max} = T_1 = 473.3\,°\mathrm{C}$$

$$T_2 = T(L) = \frac{q'''_G L}{h} + T_\infty = \frac{250{,}000 \times 0.08}{50} + 20$$

$$T_2 = 420\,°\mathrm{C}$$

2.9 Heat Conduction in Cylinders with Thermal Energy (Heat) Generation

This section studies the case of thermal energy (heat) generation in solid cylinders. As an application example, the case of heat rate generated by the Joule effect by electrical current in resistances and electrical wires is presented, as illustrated in Fig. 2.18.

Starting from the general equation of heat conduction in cylindrical coordinates (Eq. 2.11), we have:

$$\frac{1}{r}\frac{\partial}{\partial r}\left(r\frac{\partial T}{\partial r}\right) + \frac{1}{r^2}\frac{\partial^2 T}{\partial \varphi^2} + \frac{\partial^2 T}{\partial z^2} + \frac{q'''_G}{k} = \frac{1}{\alpha}\frac{\partial T}{\partial t}$$

Assumptions:

- Radial symmetry: $\frac{\partial^2 T}{\partial \varphi^2} = 0$ (there is no heat conduction in the angular direction).
- Tube is very long: $\frac{\partial^2 T}{\partial z^2} = 0$ (negligible edge effects at the tube ends).
- Steady state: $\frac{\partial T}{\partial t} = 0$.

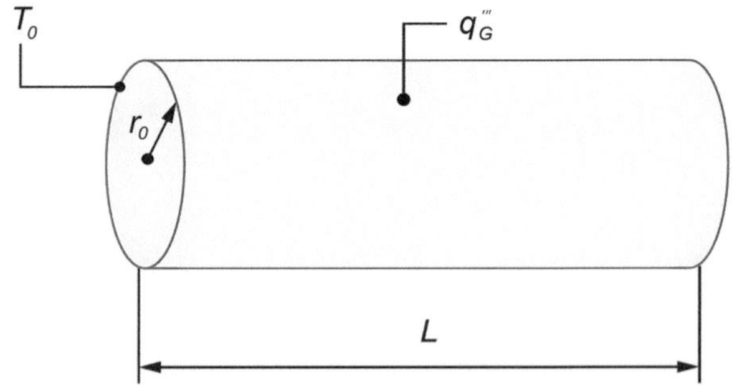

Fig. 2.18 Solid cylinder with thermal energy (heat) generation

Thus, the general equation is simplified to:

$$\frac{1}{r}\frac{d}{dr}\left(r\frac{dT}{dr}\right)+\frac{q_G'''}{k}=0$$

Integrating by parts assuming q_G''' constant:

$$r\frac{dT}{dr}=-\frac{q_G'''r^2}{2k}+C_1$$

Dividing by r and carrying out a second integration, yields:

$$T(r)=-\frac{q_G'''r^2}{4k}+C_1\ln r+C_2$$

Boundary conditions for obtaining constants C_1 and C_2:
BC_1: at $r=r_o$, $T=T_o$ the temperature of the external surface is known.
BC_2: at $r=0$, $\frac{dT}{dr}\big|_{r=0}=0$ radial symmetry at the center line.

The BC_2 implies that the heat flux is null at the center line and, as a result, it can also be stated that the maximum temperature occurs at the cylinder center. By carrying out the derivative an indetermination occurs for $r=0$, and therefore it is necessary to establish the limit of the equation when the radius approaches zero. Thus, from the second boundary condition, BC_2, it follows that:

$$\lim_{r\to 0}\left[-\frac{q_G'''r}{2k}+\frac{C_1}{r}\right]=0.$$

From which results $C_1=0$, so that the expression always remains null.
From the first boundary condition, BC_1, it follows that:

$$T_o=-\frac{q_G'''r_o^2}{4k}+C_2 \text{ or } C_2=T_o+\frac{q_G'''r_o^2}{4k}$$

.

Finally, the temperature distribution in the cylinder is:

$$T=\frac{q_G'''}{4k}\left(r_o^2-r^2\right)+T_o \tag{2.38}$$

which is a parabolic equation (second degree), as schematically indicated in Fig. 2.19. The maximum temperature at the center line is $T_{max}=\frac{q_G'''r_o^2}{4k}+T_o$

Think. Suppose that you, as an expert engineer, are called to prepare a technical opinion on a fire suspected of having originated in the overheating of the electrical system. How would you know whether or not an overheating occurred in light of the subject discussed in this topic? *Hint:* examine suspected wire cross-section.

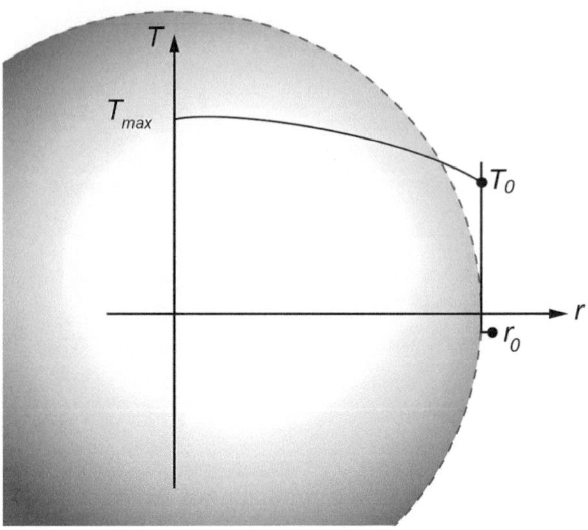

Fig. 2.19 Temperature distribution in the solid cylinder with internal thermal energy generation (heat)

Example 2.7 (*Heat conduction in electrical wires due to the Joule effect*)

It was applied a voltage of 1 V in a 3.2 mm diameter and 30 cm long stainless steel wire. The wire is kept in an environment at 25 °C and the heat transfer coefficient by convection is 100 W/m^{2}K. Calculate the temperature at the center line and at the wire surface. The resistivity ρ of the wire is 70 μΩcm and its thermal conductivity is 22.5 W/mK.

Solution

The rate of thermal energy (heat) generated per unit volume is the electrical power dissipated in the volume:

$$P = \frac{V^2}{R_e};\ R_e = \rho\frac{L}{A}$$

$$\rho = 70 \times 10^{-8}\ \Omega\text{m}$$

$$L = 0.3\text{m},\ A = \pi\frac{D^2}{4} = \pi\frac{\left(3.2 \times 10^{-3}\right)^2}{4} = 8.0425 \times 10^{-6}\ \text{m}^2$$

$$R_e = \rho\frac{L}{A} = \frac{70 \times 10^{-8} \times 0,3}{8.0425 \times 10^{-6}} = 2.6111 \times 10^{-2}\ \Omega$$

$$P = \frac{1^2}{2.6111 \times 10^{-2}} = 38.3\ \text{W}$$

$$q'''_G = \frac{P}{\forall} = \frac{P}{AL} = \frac{38.3}{8.0425 \times 10^{-6} \times 0.3} = 1.587 \times 10^7\ \frac{\text{W}}{\text{m}^3}$$

$$P = hA_L(T_w - T_\infty) \Rightarrow T_w = T_\infty + \frac{P}{hA_L}$$

being A_L the wire side area. Therefore, the surface wire temperature is

$$T_w = 25 + \frac{38.3}{100 \times \pi \times 3.2 \times 10^{-3} \times 0.3} = 152.0\ ^\circ\mathrm{C}$$

From Eq. (2.38) for $r = 0$, $T_0 = T_{max}$:

$$T_0 = T_w + \frac{q_G''' r_o^2}{4k} = 152.0 + \frac{1.587 \times 10^7 \times \left(1.6 \times 10^{-3}\right)^2}{4 \times 22.5}$$

$$\rightarrow T_0 = 152.45\ ^\circ\mathrm{C}$$

Comments: *The temperature within the wire is quite uniform due to the high conductibility of metals.*

Example 2.8 (*Insulated tubular resistance*)

Consider a 304 stainless steel cylindrical tube, length $L = 70$ cm, and radii $r_i = 10$ cm and $r_o = 15$ cm, coated with perfect thermal insulation on the external surface. Its internal surface is maintained at a constant temperature equal to $T_i = 100$ °C due to the removal of heat by the vaporization of water inside it. Also, consider that there is generation of thermal energy (heat) $q_G''' = 0.2$ MW/m^3 uniform in the tube. See the figure. It is asked:

(a) Calculate the radial temperature distribution.
(b) Determine the rate of heat transfer removed (from the inner wall).
(c) Determine the external surface temperature.

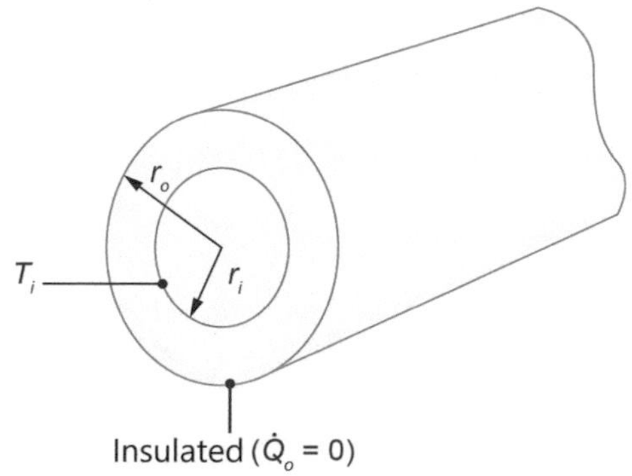

Solution

(a) Considering radial symmetry, long tube, without edge effects in the axial direction and steady state, the following heat conduction equation in cylindrical coordinates is obtained:

$$\frac{1}{r}\frac{d}{dr}\left(r\frac{dT}{dr}\right)+\frac{q_G'''}{k}=0.$$

As already seen (Sect. 2.9), the temperature distribution is:

$$T(r)=-\frac{q_G'''r^2}{4k}+C_1\ln r+C_2$$

C_1 and C_2 are found from the boundary conditions of the specific problem:
BC_1: $r=r_i$, $T=T_i$ constant internal temperature.
BC_2: $r=r_o$, $-kA_o\left.\frac{dT}{dr}\right|_{r_o}=0$ zero heat transfer rate on the external surface. Thus, one obtains C_1 and C_2, this is:

$$C_1=\frac{q_G'''r_o^2}{2k};\ C_2=T_i+\frac{q_G'''r_o^2}{4k}\left[\left(\frac{r_i}{r_o}\right)^2-2\ln(r_i)\right]$$

Therefore, the temperature distribution after substitution of C_1 and C_2 is:

$$T(r)=\frac{q_G'''r_o^2}{4k}\left[\frac{r_i^2-r^2}{r_o^2}+2\ln\left(\frac{r}{r_i}\right)\right]+T_i$$

Substituting values, for A304 steel $k=14.9$ W/mK (Table A.2),

$$T(r)=481.3-3355.7r^2+151.0\ln r$$

The temperature distribution in the radial direction in the tube wall is indicated in the following graph.

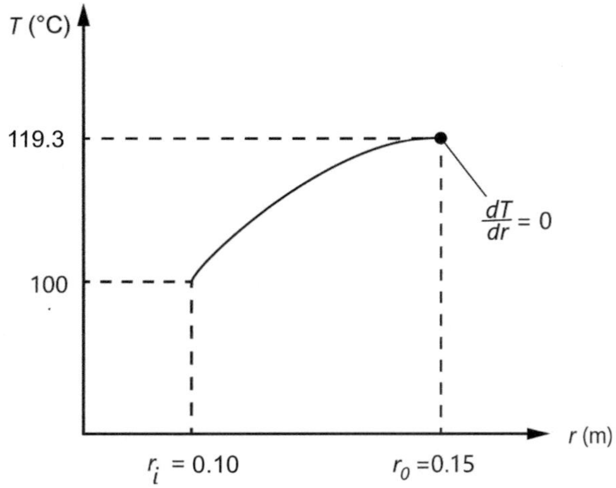

(b) The heat transfer rate removed from the inner wall is obtained by Fourier's Law:

$$\dot{Q}_i = -kA\left.\frac{dT}{dr}\right|_{r=r_i} = -kA\frac{d}{dr}\left\{\frac{q_G'''r_o^2}{4k}\left[\frac{r_i^2 - r^2}{r_o^2} + 2\ln\left(\frac{r}{r_i}\right)\right] + T_i\right\}$$

$$\dot{Q}_i = \pi L q_G'''\left(r_o^2 - r_i^2\right)$$

Substituting values,

$$\dot{Q}_i = 5497.8 \text{ W}.$$

(c) The temperature of the outer surface, which is also the maximum, is:

$$T_{max} = T_o = \frac{q_G'''r_o^2}{4k}\left[\frac{r_i^2 - r_o^2}{r_o^2} + 2\ln\left(\frac{r_o}{r_i}\right)\right] + T_i$$

Substituting values,

$$T_{max} = T_o = 119.3\,°\text{C}.$$

2.10 Critical Radius of Thermal Insulation

Pipes for transporting heated (or cooled) fluids should be thermally insulated (see Fig. 2.20) from the environment to avoid thermal losses, as this implies increased costs and inefficiencies. One might suppose that simply installing of layers of thermal

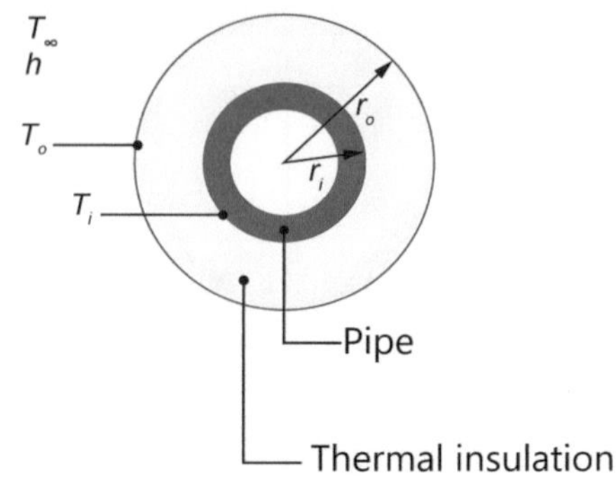

Fig. 2.20 Thermal insulation around pipes

insulation around the pipe would be sufficient. However, a detailed study will show the need to establish a technical criterion for carrying out this operation.

As seen, the heat transfer rate that the fluid transfers to the environment ($T_i > T_\infty$) occurs first by conduction through the pipe wall, and continues by conduction through the thermal insulation layer to transfer heat by convection from the external insulation surface to the environment. The following analysis occurs only on the thermal insulation and the external environment, so that the combined heat transfer rate is:

$$\dot{Q} = \frac{T_i - T_\infty}{\frac{\ln\left(\frac{r_o}{r_i}\right)}{2\pi kL} + \frac{1}{2\pi L r_o h}} \quad \text{or} \quad \dot{Q} = \frac{2\pi L(T_i - T_\infty)}{\frac{\ln\left(\frac{r_o}{r_i}\right)}{k} + \frac{1}{r_o h}}$$

Note that the outer radius of the insulation r_o that appears in both terms in the denominator indicates opposite contributions to the heat transfer rate. If the outer radius of the insulation r_o increases, the resistance to heat convection will decrease, while the thermal conduction resistance will increase. This is illustrated in the graph in Fig. 2.21 and gives rise to a maximization (or minimization) problem.

From calculus, it is known that the maximum heat transfer rate occurs when the derivative of this heat rate is zero, that is,

$$\frac{d\dot{Q}}{dr_0} = \frac{-2\pi L(T_i - T_\infty)}{\left[\frac{\ln\left(\frac{r_0}{r_i}\right)}{k} + \frac{1}{r_0 h}\right]^2}\left(\frac{1}{k\,r_0} - \frac{1}{h\,r_0^2}\right) = 0$$

Whose solution results in the following relationship between the quantities involved:

$$\frac{1}{k} = \frac{1}{hr_0}$$

that is, the critical outer radius, $r_0 = r_{crit}$, will be:

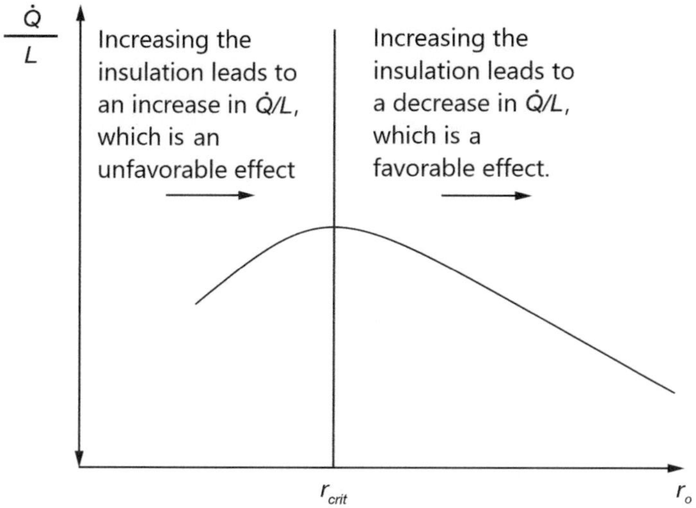

Fig. 2.21 Effect of the critical radius of insulation

$$r_{crit} = \frac{k}{h} \tag{2.39}$$

In Eq. (2.39), r_{crit} is the outer radius at which $\dot{Q}$ reaches a maximum value and it is called *the critical radius of thermal insulation.* If the outer insulation radius is *originally* less than $\frac{k}{h}$, the heat transfer will be increased by adding insulation layers up to the thickness given by the critical radius—as per the trend of the graph in Fig. 2.21. In this case, the opposite effect of reducing the heat transfer rate loss would occur. On the other hand, if the original outer radius of the insulation is greater than the critical radius, successive additions of insulating layers will indeed decrease heat losses.

To illustrate, consider a value of the heat transfer coefficient by convection of $h = 10\,\text{W/m}^2\,°\text{C}$ (natural convection). Table 2.3 indicates the critical insulation radii for some thermal insulators.

Table 2.3 Critical radii for some materials considering $h = 10\,\text{W/m}^2\,°\text{C}$

Material	k (W/mK)	r_{crit} (mm)
Teflon	0.350	35
Leather	0.159	15.9
Particleboard	0.087	8.7
Fiberglass blanket, paper coated	0.046	4.6
Extruded polystyrene	0.029	2.9
Aluminum foil and laminated glass paper (75 to 150 layers, in vacuum)	0.000017	0.0017
Plastic coating of electrical wire	0.18	18

Example 2.9 (*Critical radius of insulation*)

A 4 mm diameter electrical conductor wire is covered with a 2 mm thick plastic coating ($k = 0.18\,\text{W/mK}$). The wire is initially exposed to a quiescent air with a convective heat transfer coefficient $h = 10\ \text{W/m}^2\text{K}$. Later, a wind blows on the wire causing an increase in h to 100 $\text{W/m}^2\text{K}$. Analyze for the two situations whether or not the plastic coating improves the heat transfer to the environment.

Solution

As seen, the critical radius depends on the thermal conductivity of the insulation and the heat transfer coefficient by convection (Eq. 2.38). Thus, in this example, the same electrical wire has two critical radii associated with *h*,

$$r_{crit1} = \frac{0.18}{10} \rightarrow r_{crit1} = 18\,\text{mm}$$

$$r_{\text{crit2}} = \frac{0.18}{100} \rightarrow r_{\text{crit2}} = 1.8\,\text{mm}$$

As the external radius of the plastic insulation is 4 mm, therefore:

- If $h = 10\,\text{W/m}^2\text{K}$, the plastic coating aids in transfering heat to the environment, because $r_0 < r_{crit1}$.
- If $h = 100\,\text{W/m}^2\text{K}$,the plastic coating acts as a thermal insulator, because $r_0\ r_{crit2}$.

A scheme of this behavior is indicated in the following figure (the curves of heat transfer rate per unit length are without scale). Curve 1 is for $h = 10\ \text{W/m}^2\text{K}$; curve 2 is for $h = 100\,\text{W/m}^2\text{K}$.

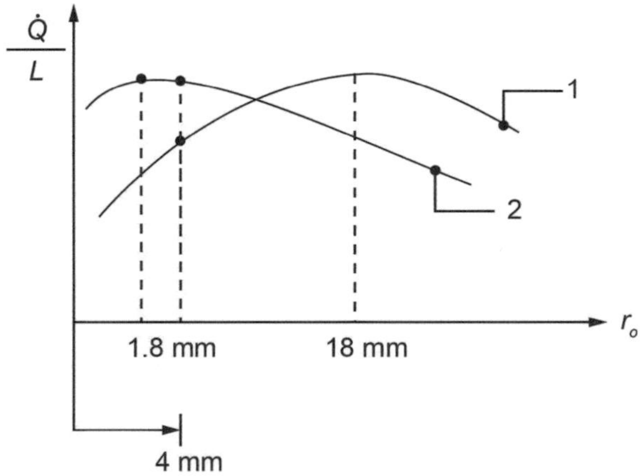

2.11 Fins or Extended Surfaces

Consider a surface from where heat transfer occurs to a surrounding fluid at temperature T_∞, as illustrated in Fig. 2.22. The surface is at temperature T_b.

From a simple analysis, it is known that heat transfer can be improved, for example, if the fluid relative velocity to the surface is increased, as happens when a fan is turned on in direction to the surface. With this, the value of the heat transfer coefficient h is increased, and consequently, the heat transfer rate. However, there is a price to pay, which is the fact that it will require the use of a piece of equipment to force the mechanical movement of the fluid, that is, fans and blowers (air) or pumps (liquids). On the other hand, the heat transfer rate can be increased passively by installing *fins* on the surface, as illustrated in Fig. 2.23. In this way, the original thermal exchange surface is increased, or extended, by increasing the exposed area or contact between the heated (or cooled) surface and the fluid. Therefore, fins are an extended heat exchange area.

Here are some examples of fin applications:

- cylinder jacket of air-cooled internal combustion engines, like the engine of the "old" beetle and motorcycle engines, lawnmowers;
- electric motor case;
- condensers and evaporators, like those in air conditioning units;
- heat sinks of electronic components and computer CPUs.

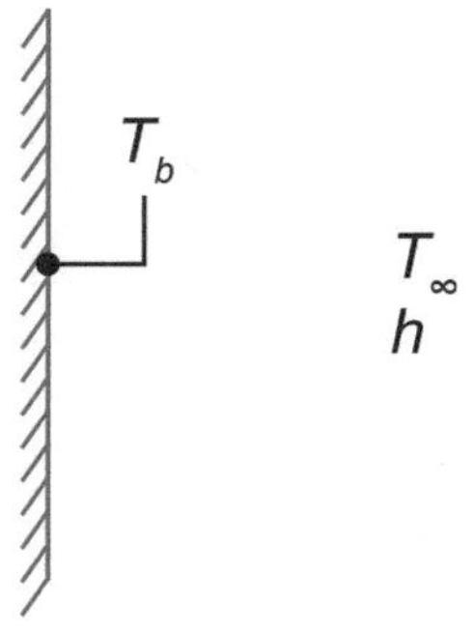

Fig. 2.22 Convective heat transfer from a surface at T_b to a fluid at T_∞

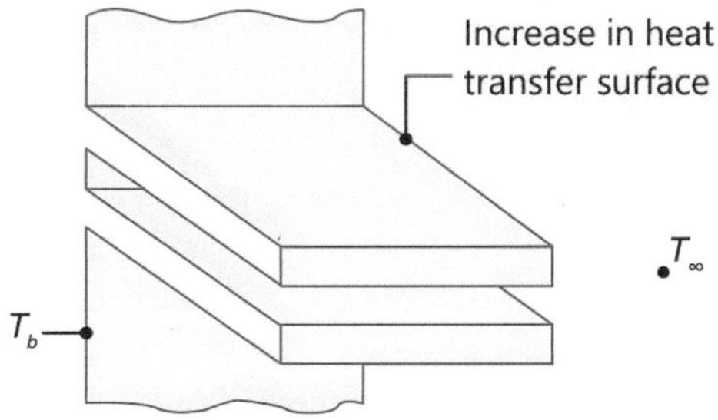

Fig. 2.23 A surface equipped with fins to increase heat exchange with a fluid

iStock.com/angusben iStock.com/KangeStudio

Fig. 2.24 Fins on the cylinder of an air-cooled motorcycle engine and on an electric motor case

Types of fins

Figure 2.24 shows the diagram of an air-cooled internal combustion motorcycle engine and a photograph of an electrical motor. Note that the fins are built as an integral part of the engine cylinder or the motor case. There are literally hundreds or thousands of fin shapes that are often associated with their construction process (extrusion, welding, among others). Figure 2.25 illustrates a series of constructive geometries of fins.

2.11.1 General Fin Equation

The general fin equation results from the balance of the heat transfer rate between the fin and the environment. For this, consider an elementary control volume, as indicated in Fig. 2.26, in which it is assumed that the cooling case with the fin transferring heat to the environment. Thus, the heat transfer rate by conduction that reaches the left face of the elementary CV will be equal to the sum of the rate of heat transferred on the right face of the CV added to the portion of heat lost by convection to the environment through the lateral surface of the fin, A_L, as well illustrated on the right side of Fig. 2.26.

The following main assumptions are valid:

- Steady state.
- Uniform temperature in the cross-section.
- Constant transport properties.
- The temperature of the lateral surface (superficial) for heat exchange by convection is the own temperature T of the cross-section.

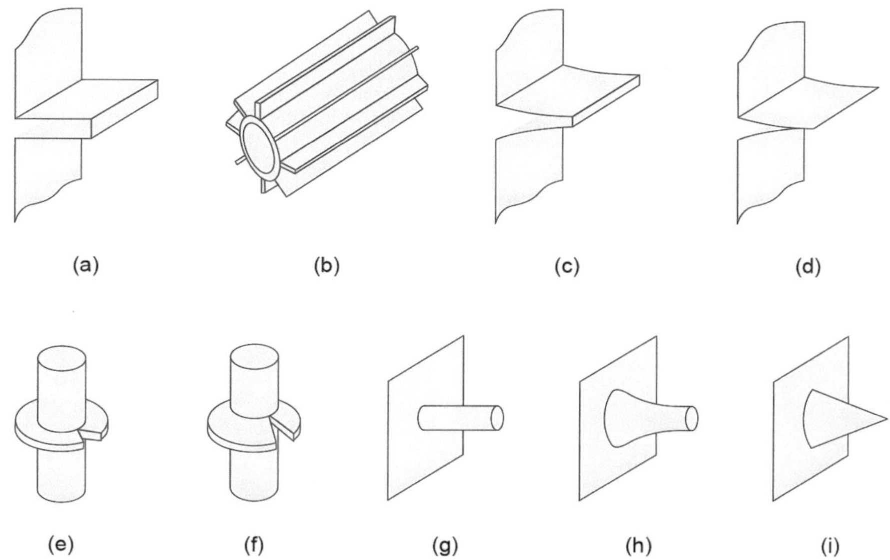

Fig. 2.25 Different types of finned surfaces: **a** longitudinal fin with rectangular profile; **b** cylindrical tube with rectangular profile fins; **c** longitudinal fin with parabolic profile; **d** longitudinal fin with parabolic profile ending in tip; **e** cylindrical tube equipped with radial fin; **f** cylindrical tube equipped with truncated conical profile radial fin; **g** cylindrical pin; **h** parabolic pin; **i** conical pin. Adapted from Kreith and Bohn (2001)

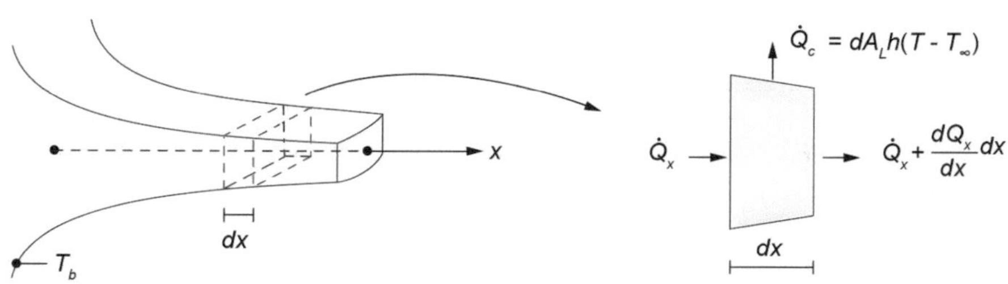

Fig. 2.26 Control volume for thermal analysis of the fin

Energy Balance

$$\underbrace{\left\{\begin{array}{c}\text{Heat transfer rate}\\ \text{entering the CV}\\ \text{by conduction}\end{array}\right\}}_{\text{I}} = \underbrace{\left\{\begin{array}{c}\text{Heat transfer rate}\\ \text{leaving the CV}\\ \text{by conduction}\end{array}\right\}}_{\text{II}} + \underbrace{\left\{\begin{array}{c}\text{Heat transfer rate}\\ \text{leaving the CV}\\ \text{by convection}\end{array}\right\}}_{\text{III}} \tag{2.40}$$

Explicit terms:

(I) $\dot{Q}_x = -kA_x\frac{dT}{dx}$
(II) $\dot{Q}_{x+dx} = \dot{Q}_x + \frac{d\dot{Q}_x}{dx}dx$ truncated Taylor series expansion.
(III) $\dot{Q}_c = hA_L(T - T_\infty)$

Substituting the previous equations into the global energy balance, yields:

$$\dot{Q}_x = \dot{Q}_x + \frac{d\dot{Q}_x}{dx}dx + hdA_L(T - T_\infty),$$

which simplifies to

$$\frac{d\dot{Q}_x}{dx}dx + hdA_L(T - T_\infty) = 0$$

Substituting Fourier's Law of conduction (Eq. 2.1) for $\dot{Q}_x$, yields: $\frac{d}{dx}(-kA_x\frac{dT}{dx})dx + hdA_L(T - T_\infty) = 0$, to finally obtain

$$\frac{d}{dx}\left(A_x\frac{dT}{dx}\right)dx - \frac{hdA_L}{k}(T - T_\infty) = 0 \tag{2.41}$$

Note that A_x is the cross-sectional area of the fin and A_L is its lateral area at the considered x position. Equation (2.41) constitutes the *general fin equation.*

Constant cross-section fin

The simplest fin geometry is the one with a constant cross-section like, for example, a prismatic fin with a rectangular or circular cross-section. When the cross-section is constant, the lateral area A_L is directly associated with the *wet perimeter* of the fin, P, through Eq. (2.42).

$$dA_L = Pdx \tag{2.42}$$

The wetted perimeter is the fin external perimeter in contact with the fluid. Thus, the general equation of the fin for prismatic fins with constant cross-section (A) is simplified in Eq. (2.43):

$$\frac{d^2T}{dx^2} - \frac{hP}{kA}(T - T_\infty) = 0 \tag{2.43}$$

Changing the variable: $\theta = T - T_\infty \quad \rightarrow \quad d\theta = dT$ and substituting into the previous equation, one obtains:

$$\frac{d^2\theta}{dx^2} - m^2\theta = 0 \tag{2.44}$$

where m is defined according to Eq. (2.45),

$$m^2 = \frac{hP}{kA} \tag{2.45}$$

The general solution of Eq. (2.44) is of the type:

$$\theta = C_1 e^{mx} + C_2 e^{-mx} \tag{2.46}$$

This general solution comes from the characteristic polynomial, which has two real and distinct roots (m and $-m$). See "Calculus Reminder"below.

Calculus Reminder

General solution of a homogeneous 2nd order differential equation with constant coefficients b and c:

$$\frac{d^2y}{dx^2} + b\frac{dy}{dx} + cy = 0$$

It is assumed that the solution is $y = e^{nx}$
Substituting this solution, yields

$$n^2 e^{nx} + bne^{nx} + ce^{nx} = 0$$

From which one obtains the characteristic polynomial, whose solution leads to three solutions depending on the roots n, as shown in the following three possible cases.

$$n^2 + bn + c = 0$$

Case 1: n_1 and n_2 are real and distinct

$$y = c_1 e^{n_1 x} + c_2 e^{n_2 x}$$

Case 2: n_1 and n_2 are equal reals

$$y = c_1 e^{n_1 x} + c_2 x e^{n_1 x}$$

Case 3: n_1 and n_2 are complex conjugates
$n_1 = p + qi$; $n_2 = p - qi$

$$y = e^{px}\left[c_1 \cos(qx) + c_2 \sin(qx)\right]$$

where $p = -\frac{b}{2}$ and $q = \frac{\sqrt{4c-b^2}}{2}$.

To solve Eq. (2.46) one must apply the boundary conditions and thus determine the constants C_1 and C_2, which are related to the heat transfer phenomena at the base $(x = 0)$ and at the fin tip $(x = L)$.

1st boundary condition

Initially, the imposed condition is that the fin base $(x = 0)$ is at temperature T_b, as illustrated in Fig. 2.26.

In this case, for $x = 0 \Rightarrow T(0) = T_b$, or $\theta(0) = \theta_b = T_b - T_\infty$. Therefore, from Eq. (2.46), comes $\theta_b = C_1 e^0 + C_2 e^{-0}$, or

$$C_1 + C_2 = \theta_b \tag{2.47}$$

2nd boundary condition

The 2nd BC is linked with the fin tip heat transfer condition. There are at least four practical cases (a), (b), (c), and (d) studied below:

(a) **Long fin**

In this case, it is assumed that the fin is very long, having transferred heat thoroughly lengthwise, so that at the tip the fin temperature reaches the fluid temperature, T_∞. From a mathematical point of view, a very long fin can be simplified as an "infinite" fin, that is:

$x \to \infty \quad \Rightarrow \quad T = T_\infty$ or $\theta = 0$.

Thus, from Eq. (2.46), one obtains:

$$\lim_{x\to\infty}\left[C_1 e^{mx} + C_2 e^{-mx}\right] = 0 \;\Rightarrow\; \lim_{x\to\infty}\left[C_1 e^{mx} + \frac{C_2}{e^{mx}}\right] = 0 \;\Rightarrow\; \lim_{x\to\infty}\left[C_1 e^{mx}\right] = 0 \;\Rightarrow\; C_1 = 0$$

The constant C_1 must be zero, otherwise the limit is unbound. Substituting in Eq. (2.47), it follows that $C_2 = \theta_b$.

So that the temperature distribution, in this case, is:

$$\theta = \theta_b e^{-mx} \quad \Rightarrow \quad \frac{\theta}{\theta_b} = e^{-\sqrt{\frac{hP}{kA}}x}$$

or, substituting the definition of θ comes:

$$\frac{\theta(x)}{\theta_b} = \frac{T(x) - T_\infty}{T_b - T_\infty} = e^{-\sqrt{\frac{hP}{kA}}x} \tag{2.48}$$

The graph of the temperature distribution of the long fin is schematized in Fig. 2.27.

Note in Fig. 2.27 that the rate of heat transferred by conduction through the fin base is $\dot{Q}_b$ and must necessarily be equal to the rate of total heat transferred along the fin by convection, $\dot{Q}_f$. This quantity can be calculated by two methods:

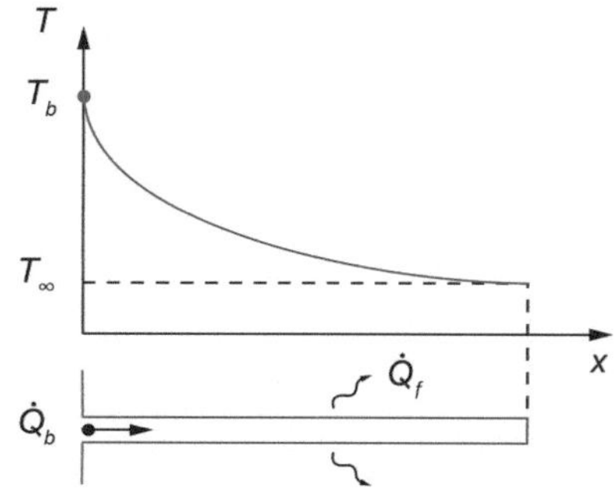

Fig. 2.27 Temperature distribution in a long fin

(I) $\dot{Q}_f = \dot{Q}_b$, that is, the total heat transfer rate is equal to the heat rate transferred by conduction at the base of the fin.

(II) $\dot{Q}_f = \int_0^\infty hP(T - T_\infty)dx$. Thus, the total heat transfer rate is the integral of the convective heat rate over the entire surface of the fin.

Using method (I), yields:

$$\dot{Q}_f = -kA_b \frac{dT}{dx}\bigg|_{x=0} = -kA_b \frac{d\theta}{dx}\bigg|_{x=0}$$

But, $A_b = A =$ constant for a prismatic fin,

$$\dot{Q}_f = -kA\frac{d}{dx}\left[\theta_b e^{-mx}\right] = -kA\theta_b(-m)e^{-mx}\big|_{x=0}$$

$$\dot{Q}_f = kA\theta_b\sqrt{\frac{hP}{kA}} = \theta_b\sqrt{hPkA}$$

which results in

$$\dot{Q}_f = \sqrt{hPkA}(T_b - T_\infty) \tag{2.49}$$

By method (II):

$\dot{Q}_f = \int_0^\infty hP\theta dx$; or $\dot{Q}_f = hP\int_0^\infty \theta_b e^{-mx}dx \Rightarrow$

$$\dot{Q}_f = hP\theta_b \lim_{\varsigma\to\infty}\int_0^\varsigma e^{-mx}dx = hP\theta_b \lim_{\varsigma\to\infty}\left(-\frac{e^{-mx}}{m}\right)_0^\varsigma$$

$$= -\frac{hP\theta_b}{m}\lim_{\varsigma\to\infty}\left(e^{-m\varsigma} - 1\right) = \frac{hP\theta_b}{m} = \sqrt{hPkA}\ \theta_b$$

or $\dot{Q}_f = \sqrt{hPkA}(T_b - T_\infty)$, which, evidently, reproduces the previous result indicated in Eq. (2.49).

(b) **Fin tip is adiabatic**

This case constitutes another boundary condition that indicates the case where the heat transfer at the fin tip is very small and negligible, so that the tip can be considered adiabatic in a practical sense. Thus, we have:
$\dot{Q}_L = -kA\frac{dT}{dx}\Big|_{x=L} = 0 \quad \Rightarrow \quad \frac{d\theta}{dx}\Big|_{x=L} = 0$ (adiabatic tip)or, substituting the general solution (Eq. 2.46), yields:

$$\frac{d}{dx}\left[C_1 e^{mx} + C_2 e^{-mx}\right]_{x=L} = 0$$

Resulting in:

$$C_2 = \theta_b \frac{e^{mL}}{\left(e^{mL} + e^{-mL}\right)}$$

The constant C_1 is obtained from Eq. (2.47):

$$C_1 = \theta_b \left[\frac{e^{-mL}}{e^{mL} + e^{-mL}}\right]$$

Thus, substituting the two constants into Eq. (2.46), we get:

$$\theta = \underbrace{\theta_b \frac{e^{-mL}}{e^{mL} + e^{-mL}}}_{C_1} e^{mx} + \underbrace{\theta_b \frac{e^{mL}}{e^{mL} + e^{-mL}}}_{C_2} e^{-mx}$$

which can be rearranged to obtain:

$$\frac{\theta}{\theta_b} = \frac{e^{-mL}}{e^{mL} + e^{-mL}} e^{mx} + \frac{e^{mL}}{e^{mL} + e^{-mL}} e^{-mx} = \frac{\frac{\left(e^{m(L-x)} + e^{-m(L-x)}\right)}{2}}{\frac{\left(e^{mL} + e^{-mL}\right)}{2}}$$

or, substituting the definitions of hyperbolic trigonometric functions (Table 2.4), one obtains:

$$\frac{\theta(x)}{\theta_b} = \frac{T(x) - T_\infty}{T_b - T_\infty} = \frac{\cosh[m(L-x)]}{\cosh(mL)} \tag{2.50}$$

Table 2.4 Basic hyperbolic functions

Function	Definition	Derivative
sinh(x)	$\frac{e^x - e^{-x}}{2}$	$\cosh(x)$
cosh(x)	$\frac{e^x + e^{-x}}{2}$	$\sinh(x)$
tanh(x)	$\frac{\sinh(x)}{\cosh(x)}$	$\text{sech}^2(x)$

Considering the method (I) for calculating the heat transfer rate transferred by the fin, we obtain:

$$\dot{Q}_f = -kA\frac{d\theta}{dx}\bigg|_{x=0} = -kA\frac{d}{dx}\left[\frac{\theta_b\cosh[(L-x)m]}{\cosh mL}\right]_{x=0} = \frac{-kA\theta_b\sinh(mL)}{\cosh(mL)}(-m)$$

$$\dot{Q}_f = \frac{-kA\theta_b\sinh(mL)}{\cosh(mL)}(-m) = kA\theta_b m\tanh(mL)$$

or, finally,

$$\dot{Q}_f = \sqrt{hPkA}(T_b - T_\infty)\tanh(mL) \tag{2.51}$$

(c) **Fin with specified temperature at the tip**

Some fins may have their end installed in a system that has the tip temperature ($x = L$) specified. Under these conditions, it is found that:

$$x = L \quad \Rightarrow \quad \theta_L = T_L - T_\infty = C_1e^{mL} + C_2e^{-mL}$$

but, on the other hand, from the first boundary condition, Eq. (2.47):

$$C_1 + C_2 = \theta_b.$$

Solving this system of two equations, the two constants C_1 and C_2 can be obtained. Substituting them into the equation of the temperature distribution, one obtains

$$\frac{\theta(x)}{\theta_b} = \frac{T(x) - T_\infty}{T_b - T_\infty} = \frac{\left(\frac{T_L - T_\infty}{T_b - T_\infty}\right)\sinh[mx] + \sinh[m(L-x)]}{\sinh(mL)} \tag{2.52}$$

And the rate of heat transferred by the fin, according to method (I) is

$$\dot{Q}_f = \sqrt{hPkA}(T_b - T_\infty)\frac{\cosh(mL) - \frac{T_L - T_\infty}{T_b - T_\infty}}{\sinh(mL)} \tag{2.53}$$

(d) **Fin with heat loss by convection at the tip**

This is the most realistic case, that is, the fin tip also transfers heat. In this situation, the heat conduction rate at the fin tip must be equal to the heat transfer rate by convection. Therefore,

at $x = L \Rightarrow -k\frac{dT}{dx}\Big|_{x=L} = h(T_L - T_\infty)$.

on the other hand:

$$C_1 + C_2 = \theta_b$$

Solving for these two conditions one obtains the temperature distribution:

$$\frac{\theta(x)}{\theta_b} = \frac{T(x) - T_\infty}{T_b - T_\infty} = \frac{\cosh[m(L-x)] + \left(\frac{h}{mk}\right)\sinh[m(L-x)]}{\cosh(mL) + \left(\frac{h}{mk}\right)\sinh(mL)} \tag{2.54}$$

And the heat transfer rate:

$$\dot{Q}_f = \sqrt{hPkA}(T_b - T_\infty)\frac{\sinh(mL) + \left(\frac{h}{mk}\right)\cosh(mL)}{\cosh(mL) + \left(\frac{h}{mk}\right)\sinh(mL)} \tag{2.55}$$

The temperature distribution and the heat transfer rate for the four cases analyzed are summarized in Table 2.5.

Table 2.5 Summary of the temperature distribution and the rate of heat transferred in fins with uniform cross-section (prismatic)

Temperature distribution	Heat transfer rate
(a) Very long fin $\frac{T(x)-T_\infty}{T_b-T_\infty} = e^{-\sqrt{\frac{hP}{kA}}x}$	$\dot{Q}_f = \sqrt{hPkA}(T_b - T_\infty)$
(b) Adiabatic fin tip $\frac{T(x)-T_\infty}{T_b-T_\infty} = \frac{\cosh[m(L-x)]}{\cosh(mL)}$	$\dot{Q}_f = \theta_b\sqrt{hPkA}\tanh(mL)$
(c) Specified fin tip temperature $\frac{T(x)-T_\infty}{T_b-T_\infty} = \frac{\left(\frac{T_L-T_\infty}{T_b-T_\infty}\right)\sinh[mx]+\sinh[m(L-x)]}{\sinh(mL)}$	$\dot{Q}_f = \sqrt{hPkA}(T_b - T_\infty)\frac{\cosh(mL)-\frac{T_L-T_\infty}{T_b-T_\infty}}{\sinh(mL)}$
(d) Convection at fin tip $\frac{T(x)-T_\infty}{T(x)-T_b} = \frac{\cosh[m(L-x)]+\left(h/mk\right)\sinh[m(L-x)]}{\cosh(mL)+\left(h/mk\right)\sinh(mL)}$	$\dot{Q}_f = \sqrt{hPkA}(T_b - T_\infty)\frac{\sinh(mL)+\left(h/mk\right)\cosh(mL)}{\cosh(mL)+\left(h/mk\right)\sinh(mL)}$

where $m = \sqrt{hP/kA}$

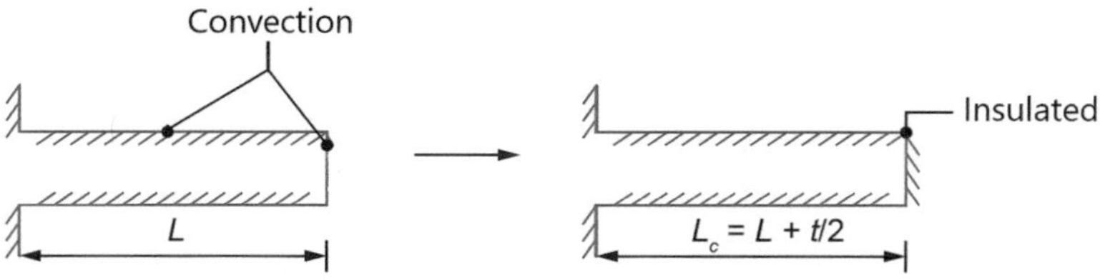

Fig. 2.28 Rectangular fin corrected length

(e) **Corrected Fin Length**

The most realistic case is that of heat convection at fin tip (case d—Table 2.5), but it involves more complex mathematical expressions. A practical way to evaluate the heat transfer from the fin is to adopt the solution of case (b) of adiabatic end by replacing the fin length L by the corrected fin length, L_c, defined as:

$$L_c = L + \frac{A}{P} \tag{2.56}$$

For a rectangular fin of width b and thickness t ($t \ll b$), it is given that:

$$\frac{A}{P} = \frac{bt}{(2b + 2t)} = \frac{t}{(2 + b)} \approx \frac{t}{2}$$

In the case of cylinders, replace t with the cylinder's radius, r. The error introduced by this approximation will be less than 8% as long as $\frac{ht}{k} < 0, 5$. This situation is illustrated in Fig. 2.28.

Example 2.10 (*Fins*)

Consider an electric soldering iron, as illustrated in the figure. For simplification, it was considered that the soldering tip is formed by a solid copper rod of length $L =$ 50 mm and diameter $D = 6$ mm, as shown in the illustration. A 40 W total heat rate is conducted to the rod base, while its right end is adiabatic with corrected length, L_c. Under these conditions, determine the temperature distribution along the copper tip. The convection heat transfer coefficient is 100 W/m^{2}K and the temperature environment, $T_\infty = 25$ °C.

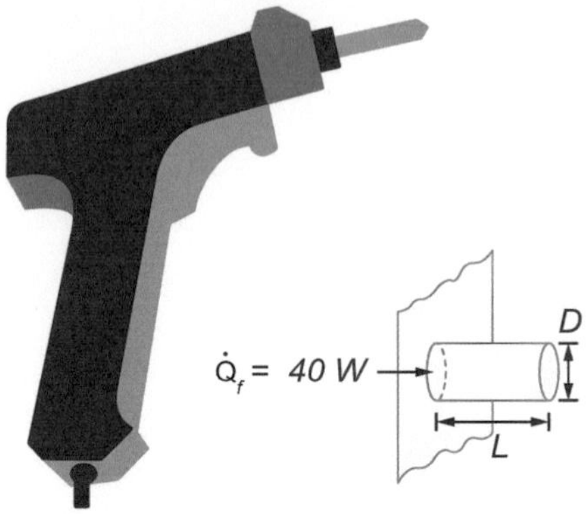

Solution

This is the case of the adiabatic tip shown in Table 2.5 with corrected length.

Copper: $k = 401$ W/mK (Table A.2)

$$P = \pi D \ \ ; \mathrm{A} = \frac{\pi D^2}{4} \Rightarrow \frac{P}{A} = \frac{4}{D}$$

$$m = \sqrt{\frac{hP}{kA}} = \sqrt{\frac{100}{401} \times \frac{4}{0.006}} = 12.89$$

$$L_c = L + \frac{r}{2} = 0.05 + 0.001 = 0.051\,\mathrm{m}$$

From Eq. (2.51), one obtains:

$$\theta_b = T_b - T_\infty = \frac{\dot{Q}_f}{\sqrt{hPkA} \times \tanh(mL_c)}$$
$$= \frac{40}{\sqrt{100 \times \frac{\pi^2 0.006^3}{4} \times 401} \times \tanh(12.89 \times 0.051)}$$
$$\theta_b = T_b - T_\infty = 471.3\,^\circ\mathrm{C}$$

The temperature distribution (in °C) is given by Eq. (2.50)

$$\frac{T(x)-T_\infty}{T_b-T_\infty} = \frac{\cosh[m(L_c-x)]}{\cosh(mL_c)} = \frac{\cosh[12.89\times(0.051-x)]}{\cosh(12.89\times0.051)} \Rightarrow$$
$$\frac{T(x)-25}{471.3} = 0,817\cosh[12.89 \times (0.051 - x)] \Rightarrow$$
$$T(x) = 25 + 385.1\cosh[12.89 \times (0.051 - x)]$$

The temperature distribution at the copper rod is as follows:

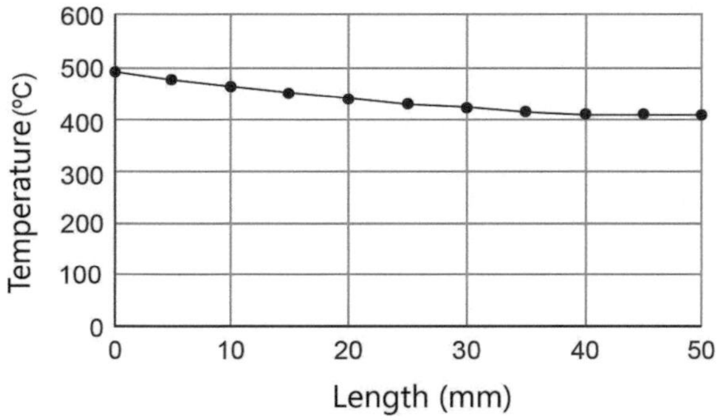

Example 2.11 (*Fins made of various materials*)

A 2 mm thick, 10 mm wide and 10 cm long flat rod is connected to two walls kept at 100 °C. The environment temperature is 20 °C, whose convection heat transfer coefficient is 20 W/m^{2}K. What is the heat lost rate to the environment from the rod assuming it is made of (a) aluminum, and (b) A304 stainless steel? For each material, (c) what will be the temperature in the fin middle point? (d) Make a schematic of the temperature distribution of the plate.

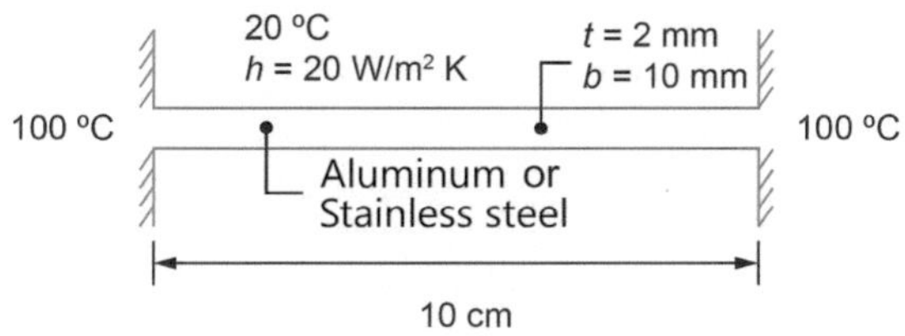

Solution

From figure, it can be noted that the fin is symmetrical from both the geometric point of view and boundary conditions in relation to the center line. Thus, at the center section, at $L = 5$ cm, there is no net heat transfer by conduction. Therefore, the central plane can be considered as adiabatic and the problem can be approached assuming an adiabatic fin tip, length $L = 5$ cm and base temperature at $T_b = 100$°C.

Calculating the fin geometric parameters, one obtains:

$$P = 2 \times (0.002 + 0.010) = 0.024\,\text{m}$$

$$A = 0.002 \times 0.010 = 2 \times 10^{-5}\,\text{m}^2$$

$$L = 0.05\,\text{m}.$$

(a) Aluminum: $k = 237$ W/mK

$$m = \sqrt{\frac{hP}{kA}} = \sqrt{\frac{20 \times 0.024}{14.9 \times 2 \times 10^{-5}}} = 40.134/\text{m}$$
$$\dot{Q}_f = \theta_b \sqrt{hPkA}\tanh(mL)$$
$$\dot{Q}_f = (100 - 20)\sqrt{20 \times 0.024 \times 14.9 \times 2 \times 10^{-5}}\tanh(40.134 \times 0.05)$$

$\dot{Q}_f = 1.77$ W. Therefore, the rod loses 3.54 W when considering both sides.

(b) Stainless steel (A304): $k = 14.9$ W/mK

$$m = \sqrt{\frac{hP}{kA}} = \sqrt{\frac{20 \times 0.024}{14.9 \times 2 \times 10^{-5}}} = 40.134/\text{m}$$
$$\dot{Q}_f = \theta_b \sqrt{hPkA}\tanh(mL)$$
$$\dot{Q}_f = (100 - 20)\sqrt{20 \times 0.024 \times 14.9 \times 2 \times 10^{-5}}\tanh(40.134 \times 0.05)$$

$\dot{Q}_f = 0.92$ W. Therefore, the rod transfers 1.84 W when considering both sides.

(c) $\frac{T_L - T_\infty}{T_b - T_\infty} = \frac{\cosh[m(L-L)]}{\cosh[mL]}$

Aluminum: $T_L = (100 - 20)\frac{\cosh(10.063 \times 0)}{\cosh(10.063 \times 0.05)} + 20 \Rightarrow T_L = 90.8\,°\text{C}$

Stainless steel: $T_L = (100 - 20)\frac{\cosh(40.134 \times 0)}{\cosh(40.134 \times 0.05)} + 20 \Rightarrow T_L = 41.1\,°\text{C}$

(d) $T(x) = (T_b - T_\infty)\frac{\cosh[m(L]}{\cosh[mL]} + T_\infty$

Aluminum: $T(x) = 70.84\cosh(0.5032 - 10.0630x) + 20$

Stainless steel: $T(x) = 21.13\cosh(2.0067 - 40.1340x) + 20$.

This distribution corresponds up to a distance of 5 cm; which is repeated on the other side of the fin, as indicated in the figure's graph.

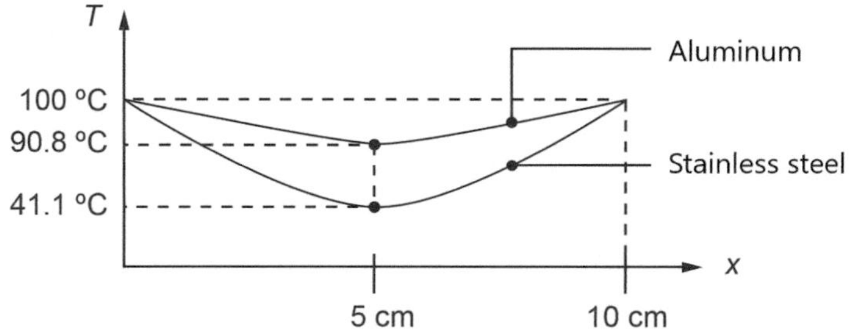

2.11.2 *Fin Efficiency*

The fin theory developed in the previous section applies to a detailed analysis of fin design. As seen, for prismatic fins, there are analytical solutions. Even irregular geometric sections, or those involving more complex boundary conditions, can be solved through numerical solution schemes of the general fin equation. However, there is a method of selecting types of fins called *the fin efficiency method*. This method establishes a technique of selecting, among various types and configurations of fins, the one(s) that satisfy certain operational conditions and efficiency.

The fin efficiency, η_f, is defined by the ratio between the actual heat transfer rate by a fin and the maximum possible heat rate exchange, which corresponds to the hypothetical case of the entire fin being at the base temperature:

$$\eta_f = \frac{\dot{Q}_f}{\dot{Q}_{f,\max}} \tag{2.57}$$

where $\dot{Q}_f$ is the actual heat rate transferred by the fin and $\dot{Q}_{f,\max}$ is the maximum hypothetical heat rate transferred that would be transferred if the entire fin were at the base temperature. In the ideal case, the maximum value of heat transfer rate can be easily obtained by means of:

$$Q_{f,\max} = hA_f\theta_b \tag{2.58}$$

where $\theta_b = T_b - T_\infty$ and A_f is the exposed area of the fin surface where convective heat transfer takes place.

A scheme showing the variation of the fin temperature in the real and ideal cases is shown in Fig. 2.29.

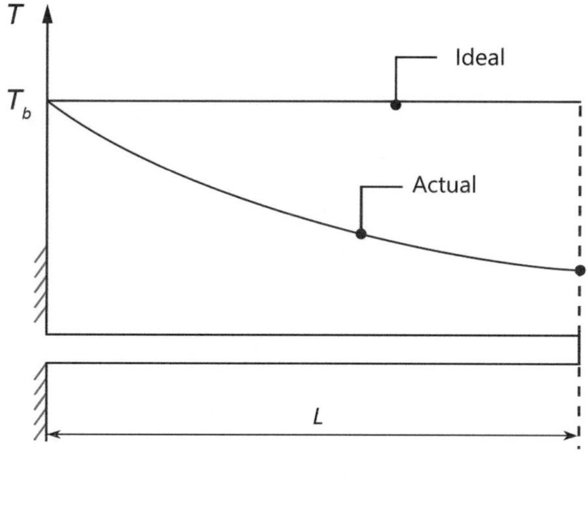

Fig. 2.29 Variation of the actual and ideal temperature of a fin

For a rectangular fin with adiabatic tip, the fin efficiency definition (Eq. 2.57) results from the ratio between the results of Eqs. (2.51) and (2.58):

$$\eta_f = \frac{\sqrt{hPkA}\theta_b \tanh(mL_c)}{hPL_c\theta_b}, \text{ or}$$
$$\eta_f = \frac{\tanh(mL_c)}{mL_c} \tag{2.59}$$

with $m = \sqrt{\frac{hP}{kA}}$, as previously defined (Eq. 2.45). On the other hand, the wetted perimeter of the fin (in contact with the fluid) is given by: $P = 2(b + t) \approx 2b$, in case of $t << b$ (thin fin), being the cross-area $A = bt$, from which one obtains:

$$mL_c = \sqrt{\frac{2h}{kt}} L_c \tag{2.60}$$

More complex cases of geometries have already been solved in literature, obtaining the fin efficiency as a function of several parameters. Figure 2.30 shows the case of the efficiency of four types of fins. So that, given a geometry and other fin constructive and operational parameters, through a graph like that of Fig. 2.30, the actual heat transfer rate can be obtained, from the very definition of efficiency, given by:

$$\dot{Q}_f = \eta_f h A_f \theta_b \tag{2.61}$$

Table 2.6 shows the theoretical expressions of efficiency for some types of fins. In this table, the modified Bessel functions appear, whose values are in Table C.1 in the appendix section.

Based on the efficiency of the rectangular fin profile, Eq. (2.59), some pertinent comments can be drawn:

1. Fins should be installed where the convective heat transfer coefficient h is low (usually in natural convection in gases, such as atmospheric air);
2. The fin should be made out of a material of high thermal conductivity, such as copper or aluminum. Aluminum is better due to its low cost and low density and, therefore, reduced final weight;
3. By inspecting the fin efficiency graphs of Fig. 2.30, the triangular fin ($y \sim x$) requires less material (volume) for the same heat dissipation load than the rectangular fin. However, the parabolic profile fin ($y \sim x^2$) is the one that has the best heat dissipation load per unit of volume ($\dot{Q}/\forall$), but it is only slightly superior to the triangular profile;
4. Annular fins are used in tubes;
5. In some cases, the contact resistance between the fin and the surface where it is placed should be considered. The contact resistance depends on the manufacturing process, as indicated in Fig. 2.31, where two manufacturing techniques

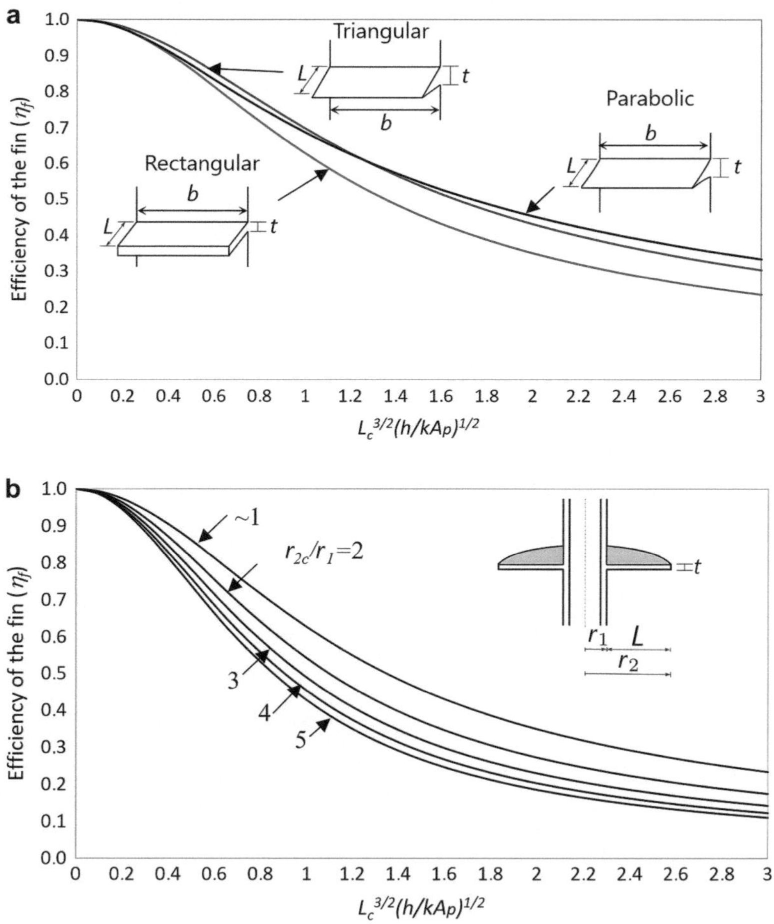

Fig. 2.30 Efficiency of **a** rectangular, triangular, and parabolic fin; **b** annular fin

are shown. In case (a), the fins are grooves machined in the tube itself. Case (b) shows the installation of fins on the tube by pressing. Sometimes, after installation, the tube undergoes a radial expansion by an internal hydraulic pressure. Welding techniques for fins fixing are also employed.

Example 2.12 (*Use of fins in tubes*)

Consider two types of circular fins. In case (a), the fins are made of low-chromium carbon steel, with $k = 37.7$ W/mK and, in case (b) they are made of an aluminum alloy ($k = 177$ W/mK). Their dimensions are indicated in the following sketch. The external temperature of the tube is 100 °C and the environment around the tube is at 15 °C. Assuming, for both cases, a heat transfer coefficient by convection between the fins and the environment of $h = 45$ W/m^{2}K, calculate the rate of heat transferred

Table 2.6 Theoretical expressions for some types of fins

Type	Parameters	Efficiency
Rectangular	$m = (2h/kt)^{\frac{1}{2}}, L_c = L + t/2,$ $A_t = L_c t, A_f = 2bL_c$	$\eta_f = \frac{\tanh(mL_c)}{mL_c}$
Triangular	$m = (2h/kt)^{\frac{1}{2}}, L_c = L,\ A_p =$ $\frac{L_c t}{2}, A_f = 2b\left[L^2 + (t/2)^2\right]^{\frac{1}{2}}$	$\eta_f = \frac{1}{mL}\frac{I_1(2mL)}{I_0(2mL)}$
Parabolic	$m = (2h/kt)^{\frac{1}{2}},\ L_c = L,$ $y = (t/2)(1 - x/L)^2$ $A_p = (t/3)L,$ $A_f = b\left[C_1 L + \frac{L^2}{t}\ln(t/L + C_1)\right],$ $C_1 = \left[1 + (t/L)^2\right]^{\frac{1}{2}}$	$\eta_f = \frac{2}{1+\sqrt{4m^2L^2+1}}$
Annular	$m = (2h/kt)^2, r_{2c} = r_2 + t/2,$ $A_p = L_c t, A_a = 2\pi\left[r_{2c}^2 - r_1^2\right]$	$\eta_f =$ $\frac{2r_1}{m(r_{2c}^2 - r_1^2)}\frac{K_1(mr_1)I_1(mr_{2c}) - I_1(mr_1)K_1(mr_{2c})}{I_0(mr_1)K_1(mr_{2c}) + K_0(mr_1)I_1(mr_{2c})}$

A_p: fin cross-sectional area; A_f: fin total exposed area; b: fin width; L_c: corrected length; t: thickness; I_0, I_1, K_0 and K_1: modified Bessel functions and indicated in Table C.1

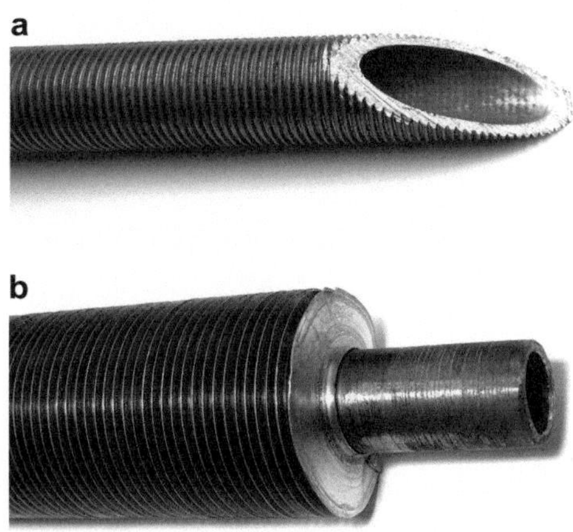

Fig. 2.31 Two methods of manufacturing and installing fins on tube surface **a** grooves; **b** direct tube installation

for each case. Neglect the contact resistance between the fin and the tube, as well as assume the same h between the fins and the environment.

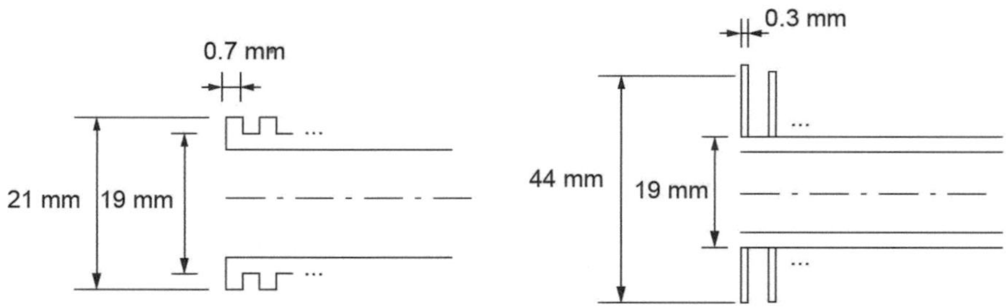

Solution

(a) Low-chromium carbon steel fin. The thermal conductivity of this material is 37.7 W/mK.

$$t = 0.0007\,\mathrm{m}$$
$$L = 0.001\,\mathrm{m};\ r_1 = 0.0095\,\mathrm{m};\ r_2 = 0.0105\,\mathrm{m}$$
$$L_c = L + \frac{t}{2} = 0.00135\,\mathrm{m}$$
$$A_P = L_c t = 0.00135 \times 0.0007 = 9.45 \times 10^{-7}\,\mathrm{m}^2$$
$$L_c^{\frac{3}{2}}\left(\frac{h}{kA_P}\right)^{\frac{1}{2}} = 0.00135^{\frac{3}{2}}\left[\frac{45}{\left(37.7 \times 9.45 \times 10^{-7}\right)}\right]^{\frac{1}{2}} = 0.06$$
$$r_{2c} = r_2 + t/2 = 0.0108\,\mathrm{m}$$
$$\frac{r_{2c}}{r_1} = \frac{0.0108}{0.0095} = 1.14$$

With these last two parameters, we obtain $\eta_f \approx 98\%$ (Fig. 2.30).
The exposed area of the fin (two sides) is: $A_f = 2\pi\left(r_{2c}^2 - r_1^2\right) = 1.657 \times 10^{-4}\ \mathrm{m}^2$.
Thus, the rate of heat exchanged by the fin will be:

$$\dot{Q}_f = \eta_f h A_a \theta_b = 0.98 \times 45 \times 1.657 \times 10^{-4} \times (100 - 15)$$
$$\dot{Q}_f = 0.62\ \mathrm{W}.$$

(b) Aluminum alloy fin

$$k = 177\,\mathrm{W/mK}$$
$$t = 0.0003\ \mathrm{m}$$
$$L = 0.0125\ \mathrm{m};\ r_1 = 0.0095\ \mathrm{m};\ r_2 = 0.022\ \mathrm{m}$$
$$L_c = 0.0127\ \mathrm{m}$$
$$A_P = 3.80 \times 10^{-6}\,\mathrm{m}^2$$

$$L_c^{\frac{3}{2}}\left(\frac{h}{kA_P}\right)^{\frac{1}{2}} = 0.37$$

$$r_{2c} = 0.0222\,\text{m}$$

$$\frac{r_{2c}}{r_1} = \frac{0.0222}{0.0095} = 2.33$$

$$\eta_f \approx 86\%$$

$$A_a = 2.528 \times 10^{-3}\ \text{m}^2$$

$$\dot{Q}_f = 8.32\ \text{W}.$$

In this example, although the efficiency of the aluminum fin is lower than that of the steel fin, however the heat dissipation rate of the aluminum fin is higher due to its larger area.

2.11.3 *Overall Efficiency of a Finned Surface*

Fins are usually installed in large quantities on surfaces, as indicated in Fig. 2.32. Therefore, it is relevant to consider the thermal performance of the surface taking into account all the installed fins. Therefore, efficiency should be evaluated for the entire surface considering the finned and non-finned parts of the surface as well.

The overall efficiency of a finned surface η_{gf} is defined as:

$$\eta_{gf} = \frac{\dot{Q}_T}{\dot{Q}_{\max}} \tag{2.62}$$

where $\dot{Q}_{\max}$ is the total heat rate exchanged by the finned surface if its surface temperature is T_b and $\dot{Q}_T$ is the actual total heat rate exchanged by the finned surface:

$$\dot{Q}_T = \dot{Q}_f + \dot{Q}_{nf} = N_f \eta_f h A_f \theta_b + h A_{nf} \theta_b \tag{2.63}$$

Fig. 2.32 Finned surface

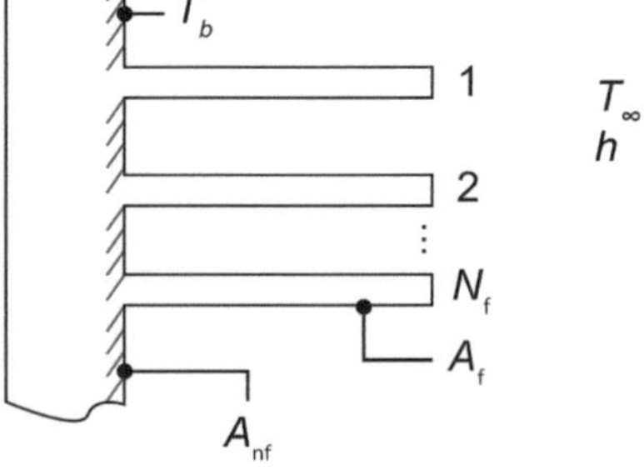

$\dot{Q}_f$ is the heat transfer rate by the N_f fins and $\dot{Q}_{nf}$ is the heat transfer rate from the base surface where there are no fins (A_{nf}). Thus, $A_{nf} = A_T - N_f A_f$, where A_T is the total heat exchange surface. Therefore,

$$\eta_{gf} = 1 - \frac{N_f A_f}{A_T}(1 - \eta_f) \tag{2.64}$$

2.11.4 *Fin Effectiveness*

The fin efficiency presented in Sect. 2.11.2 is a procedure for selecting types of fins from a table, graph, or equation that provides the efficiency of several types of fins. However, there is still a relevant question that needs to be addressed in the study of fins: will there be an actual increase in heat transfer as a result of fins installation? Of course, this piece of information is crucial for the designer to decide whether or not to install them. In order to make a technical decision, one must use the concept of *the fin effectiveness method*, ε. In this method, the heat transfer rate through the fin is compared with the heat transfer rate that would occur if the fin had not been installed (Fig. 2.33). Thus, the effectiveness ε is defined as the ratio between the heat transfer rate through the fin and the heat transfer rate through its base without a fin ($\dot{Q}_{bnf}$), that is:

$$\varepsilon = \frac{\dot{Q}_\mathrm{f}}{\dot{Q}_{bnf}} = \frac{\dot{Q}_\mathrm{f}}{hA_b\theta_b} \tag{2.65}$$

The heat transfer rate without the fins, $\dot{Q}_{bnf}$, is what would occur at the base of the fin, as illustrated in Fig. 2.33. As a general rule, the use of fins is justified when $\varepsilon > 2$.

For a rectangular fin with an adiabatic tip, the effectiveness is obtained from Eq. (2.51):

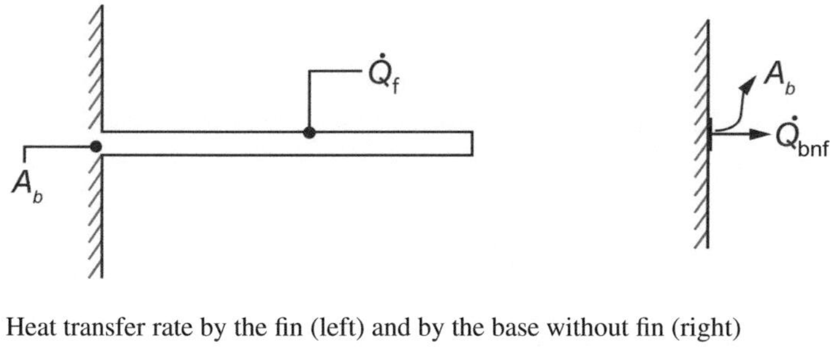

Fig. 2.33 Heat transfer rate by the fin (left) and by the base without fin (right)

$$\varepsilon = \frac{\sqrt{hPkA}\theta_b \tanh(mL_c)}{hA_b\theta_b}$$

In this case, $A = A_b$ and, therefore,

$$\varepsilon = \frac{\tanh(mL_c)}{\sqrt{hA/kP}} \tag{2.66}$$

The following two examples will illustrate the use of effectiveness.

Example 2.13 (*Effectiveness of fins*)

A stainless steel circular fin ($k = 19$ W/mK), $L = 5$ cm and $r = 1$ cm, is subjected to three cooling conditions, namely:

A: boiling water, $h = 5000$ W/m^{2}K.
B: forced air convection, $h = 100$ W/m^{2}K.
C: natural air convection, $h = 10$ W/m^{2}K.
Calculate the effectiveness of the fin.

Solution

Let's check the influence of h

For stainless steel: $k = 19$ W/mK.

$$L_c = L + \frac{r}{2}$$

$$mL_c = \sqrt{\frac{hP}{kA}}\left(L + \frac{r}{2}\right) = \sqrt{\frac{h2\pi r}{k\pi r^2}}\left(L + \frac{r}{2}\right) = \sqrt{\frac{2h}{kr}}\left(L + \frac{r}{2}\right)$$

$$mL_c = \sqrt{\frac{2h}{19(0.01)}}\left(0.05 + \frac{0,01}{2}\right) = 0.178\sqrt{h}$$

$$\sqrt{\frac{hA}{kP}} = \sqrt{\frac{h\pi r^2}{k2\pi r}} = \sqrt{\frac{hr}{2k}} = \sqrt{\frac{h0.01}{2 \times 19}} = 0.0162\sqrt{h}$$

Substituting these two results into the effectiveness expression, yields:

$$\varepsilon = \frac{\tanh(mL_c)}{\sqrt{hA/kP}} = \frac{\tanh\left(0.178\sqrt{h}\right)}{0.0162\sqrt{h}}$$

Now, analyzing the three cases (different values of h):

A: $h = 5000\,\text{W/m}^2\text{K}$ $\quad \varepsilon = \frac{\tanh\left(0.178\sqrt{5000}\right)}{0.0162\sqrt{5000}} = \frac{1}{1.145} = 0.873.$

B: $h = 100\,\text{W/m}^2\text{K}$ $\quad \varepsilon = \frac{\tanh\left(0.178\sqrt{100}\right)}{0.0162\sqrt{100}} = \frac{0.945}{0.162} = 5.833.$

C: $h = 10\,\text{W/m}^2\text{K}$ $\quad \varepsilon = \frac{\tanh\left(0.178\sqrt{10}\right)}{0.0162\sqrt{10}} = \frac{0.510}{0.051} = 10.0$

Comment: as seen, fin installation does not always improve the heat transfer. In case A, for example, the installation of fins degrades the heat transfer, as $\varepsilon < 1$. In cases B and C, the heat transfer is augmented. A basic criterion is that the ratio hA/Pk should be much less than 1 to justify the use of fins.

A: $\frac{hA}{kP} = 1.31$.

B: $\frac{hA}{kP} = 0.026$.

C: $\frac{hA}{kP} = 0.00262$.

The fin should be placed on the tube side of lower heat transfer coefficient, which is also the one with the highest thermal resistance.

Example 2.14 (*Fin effectiveness—continuation*)

Regarding the previous example, now consider that the fin is made of three different materials: copper ($k = 368$ W/mK), stainless steel ($k = 19$ W/mK) and aluminum ($k = 240$ W/mK). For each case, the convection heat transfer coefficient is $h = 100$ W/m^{2}K. Calculate the fin effectiveness for each material.

Solution

The materials evaluated are:

A: copper $k = 368$ W/mK.

B: stainless steel $k = 19$ W/mK.

C: aluminum $k = 240$ W/mK.

$m = \sqrt{\frac{2h}{kr}} = \sqrt{\frac{2\times100}{k\times0.01}} = \frac{141.4}{\sqrt{k}}$ and therefore, $mL_c = \frac{141.4}{\sqrt{k}}(0.05 + 0.01/2) = \frac{7.76}{\sqrt{k}}$.

In the denominator: $\sqrt{\frac{hA}{kP}} = \sqrt{\frac{hr}{2k}} = \sqrt{\frac{100\times0.01}{2k}} = \frac{1}{\sqrt{2k}}$.

Substituting both results, one obtains $\varepsilon = \sqrt{2k}\tanh\left(7.76/\sqrt{k}\right)$. Substituting the conductivity values to obtain.

A: copper $k = 368\,\mathrm{W/mK}\quad \varepsilon = 10.7$

B: stainless steel $k = 19\,W/\mathrm{mK}\quad \varepsilon = 5.8$

C: aluminum $k = 240\,\mathrm{W/mK}\quad \varepsilon = 10.1$

Comment: the fin manufacturing material also affects the effectiveness of a fin. One should seek to use a high thermal conductivity material (copper or aluminum). Generally, the material used is aluminum because it presents several advantages, such as:

- it is easy to work with and it can be extruded to make built in fins;
- it has a relatively low cost;
- it has a low density, which implies a lower final weight of the equipment;
- it has an excellent thermal conductivity.

In some situations, the fins may be part of the original equipment design and be cast together with the part, as it occurs with electric motor cases and cylinders of air-cooled internal combustion engines, for example. In this case, the fins are made of the same material as the motor block.

2.11.5 *Overall Effectiveness of a Finned Surface*

When one wants to assess whether or not a finned surface transfers enough heat to support its installation, the overall effectiveness should be used, defined as:

$$\varepsilon_{gf} = \frac{\dot{Q}_T}{\dot{Q}_{Tb}} \tag{2.67}$$

where $\dot{Q}_{Tb}$ is the heat exchange rate by the surface before the placement of the fins, $\dot{Q}_{Tb} = hA_{Tb}\theta_b$, where A_{Tb} is the total area of the surface before installing the fins. Substituting the Eqs. (2.62) in (2.67), one obtains an expression of the overall effectiveness as a function of the fin efficiency.

$$\varepsilon_{gf} = \frac{A_{nf} + N_f A_f \eta_f}{A_{Tb}} \tag{2.68}$$

Example 2.15 (*Effectiveness of a finned surface*)

To cool a computer processor (CPU) a heat sink is used, which consists of a 50 mm × 40 mm plate with 108 (12 × 9) cylindrical 2 mm diameter fins and 20 mm in length spaced 2 mm and symmetrically arranged. The fins are made out of aluminum, and the average heat transfer coefficient of the fins with the air at 20 °C is 50 W/m^{2}K. What is the effectiveness of the finned surface? Considering that the processor cannot exceed 50 °C, what is the heat rate that will dissipate at this temperature?

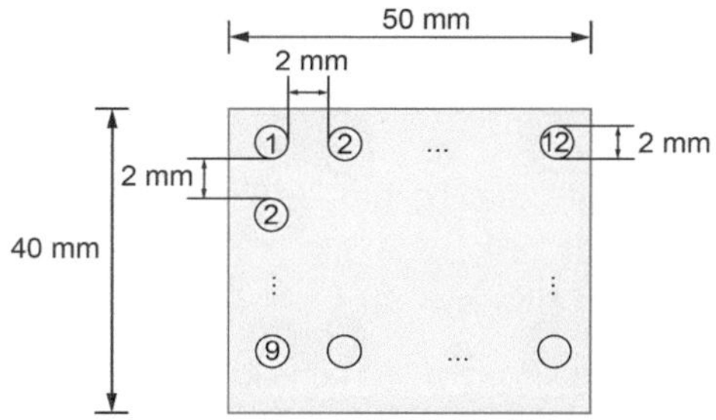

Solution

$k = 237\,\text{W/mK}$ (Table A.2), $L = 0.020\,\text{m}$, $d = 0.002\,\text{m}$,$h = 50\,\text{W/m}^2\text{K}$

$$L_c = L + \frac{r}{2} = 0.020 + \frac{0.001}{2} = 0.0205\,\text{m}$$

$$m = \sqrt{\frac{hP}{kA}} = \sqrt{\frac{h\pi d}{k\frac{\pi d^2}{4}}} = \sqrt{\frac{4h}{kd}} = \sqrt{\frac{4 \times 50}{237 \times 0.002}} = 20.54/\text{m}$$

$$\eta_f = \frac{\tanh(mL_c)}{mL_c} = \frac{\tanh(20.54 \times 0.0205)}{20.54 \times 0.0205} = 0.97$$

Base total area without any fin: $A_{Tb} = 0.05 \times 0.04 = 2 \times 10^{-3}\,\text{m}^2$
Surface area without fins: $A_{nf} = A_{Tb} - N_f A_f$

$$A_{nf} = 2 \times 10^{-3} - 108\pi \frac{0.002^2}{4} = 1.66 \times 10^{-3}\,\text{m}^2$$

Corrected surface area of a fin: $A_f = \pi dL_c$

$$A_f = \pi 0.002 \times 0.0205 = 1.287 \times 10^{-4}\text{m}^2$$

$$\varepsilon_{gf} = \frac{A_{nf} + N_f A_f \eta_f}{A_{Tb}} = \frac{1.66 \times 10^{-3} + 108 \times 1.287 \times 10^{-4} \times 0.97}{2 \times 10^{-3}} = 7.6$$

As the effectiveness is greater than 2, its use is justified the usage of fins.
The rate of heat dissipated by the finned surface will be: $\dot{Q}_T = \varepsilon_{gf} h A_{Tb}(T_b - T_\infty)$

$$\dot{Q}_T = 7.6 \times 50 \times 2 \times 10^{-3} \times (50 - 20) \quad \rightarrow \dot{Q}_T = 22.8\,\text{W}$$

2.12 Heat Pipes

Heat pipes are heat transfer devices for transferring heat very effectively at high rates from one heat source to another at a lower temperature. Its working principle is based on a fluid's evaporation–condensation process within a sealed tube as illustrated in Fig. 2.34a. As the figure shows, a heat source transfers heat to the device (heat in) to a liquid film next to the wall. As the liquid receives heat, it vaporizes and the vapor fills the tube core. In the other extremity, the tube transfers heat (heat out) to the low-temperature environment by condensing the vapor. The evaporating and condensing ends are separated by an adiabatic section. The condensed liquid returns to the other end in contact with the heat source through a wick next to the wall. The wick can be made from fiber, screen mesh, or sintering porous material.

The thermal modeling of a heat pipe can be carried out similarly to thermal resistance analysis. Figure 2.34b shows the thermal resistances.

$$R_{HP} = 2 \times R_W + R_{wi,e} + R_{wi,c} + R_e + R_c + R_{vapor} \tag{2.69}$$

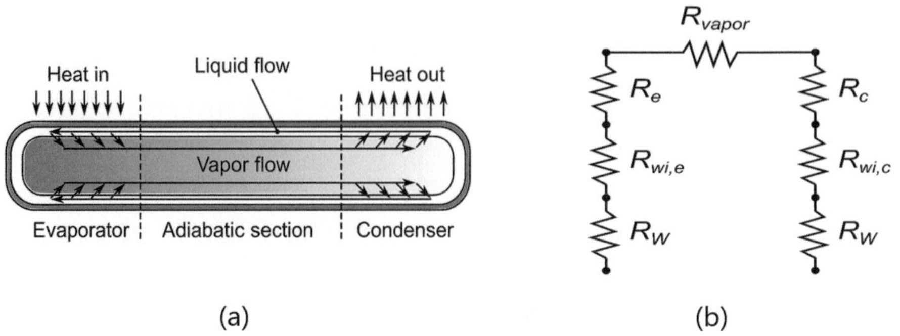

Fig. 2.34 Heat pipe **a** schematics; **b** thermal resistances

where

R_{HP} total or equivalent heat pipe thermal resistance;
$R_e; R_c$ thermal resistance for evaporation and condensation at liquid–vapor interface, respectively;
R_{vapor} vapor core resistance;
R_W pipe wall conduction resistance;
$R_{wi,e}$ wick thermal resistance for evaporation;
$R_{wi,c}$ wick thermal resistance for condensation;

The processes involving phase change (evaporation and condensation) usually have much lower thermal resistances as seen in Chap. 8. Also, assuming wick resistance $R_{wi} = R_{wi,e} \cong R_{wi,c}$, the total heat pipe thermal resistance is simplified to

$$R_{HP} \cong 2 \times (R_W + R_{wi}) \tag{2.70}$$

Equation (2.70) shows that the thermal resistance of a heat pipe is mostly dominated by the tube wall and wick structure radial resistances. The axial conduction of both the tube wall and wick structure was neglected in this simple analysis. The external thermal contact resistances between the heat source (evaporation) and the environment (condensation) were neglected, a simplification that should be applied carefully and not considered in Eq. (2.69). A more specific modeling should consider those effects as well. While R_W can be easily calculated, such is not the case of R_{wi}.

2.13 Proposed Problems

2.1 Determine the heat flux transferred through a flat wall of thickness $L = 50$ mm when the wall is made of: (a) concrete; (b) 1010 carbon steel; (c) pine wood. The temperatures of the external faces are 25 and 10 °C.

2.2 Part of a flat wall of unknown material is subjected to a laboratory test. It has a thickness $L = 150$ mm and a difference temperature $\Delta T = 10$ °C between the two external surfaces was measured. A heat flux $q = 10$ W/m^2 was also imposed. What is the thermal conductivity of the wall material? Consulting the tables in Appendix A, identify which material(s) the wall could be made of.

2.3 The temperature of the human body drops from the central level (*arterial blood level*), where the temperature is $T_a = 36.5$ °C, to the superficial level (relative to the skin) in a layer, called *subcutaneous*, which has a thickness δ of about 1 cm, whose thermal conductivity is close to 0.4 W/m°C. It is asked:

(a) Calculate the surface temperature of the skin (T_s) when the body is exposed to atmospheric air at 20 °C with a heat transfer coefficient of 25 W/m^2 °C.
(b) Calculate the surface temperature of the skin (T_s) when the body is exposed to water at 20 °C with a heat transfer coefficient of 500 W/m^2 °C.
(c) In which of the two cases would you "feel colder" and why?

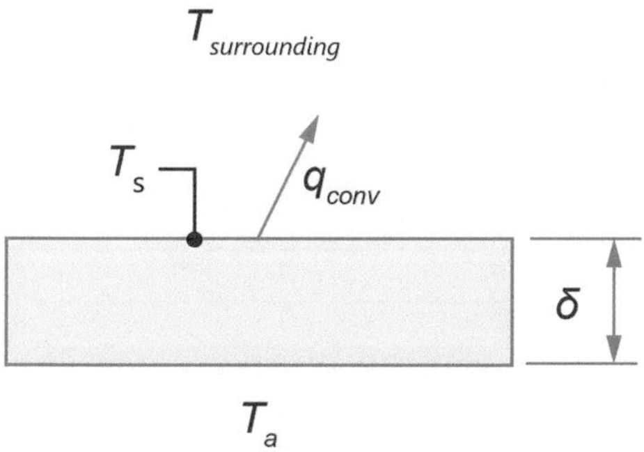

2.4 A 150 mm thick oven wall is made of refractory brick, whose thermal conductivity varies according to the temperature according to $k = 24.125 - 0.009375 \times T$, where T is in °C and k in W/m°C. Knowing that on the inner surface of the wall facing the oven the temperature is 1300 °C and on the outside it is 50 °C, determine the temperature distribution along the thickness of the wall and the heat flux lost.

2.5 The oven walls are made of a refractory brick layer with average thermal conductivity $k_r = 17.5$ W/m°C and thickness of 150 mm covered by common brick ($k_t = 0.72$ W/m°C) with a thickness of 500 mm. The internal temperature of the assembly is 1100 °C and the exterior surface, 50 °C. Calculate the heat loss rate per unit area (m^2) and the temperature on the surface (T_c) of perfect contact between the two layers of bricks.

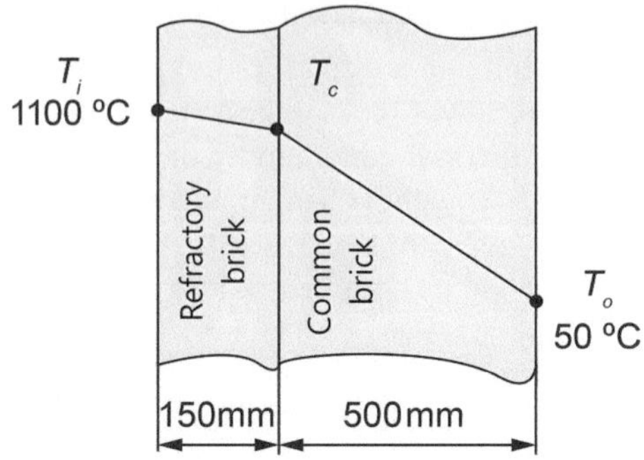

2.6 In order to decrease by 50% the brick layer thickness from Problem 2.5, it was decided to install a layer of diatomite between the two layers of bricks, as indicated in the figure, whose thermal conductivity is $k_d = 0.05$ W/m°C. What should be the thickness L_d of the diatomite layer so that the same heat loss flux is maintained and the new contact temperatures on the surfaces T_{c1} and T_{c2} indicated in the figure's scheme?

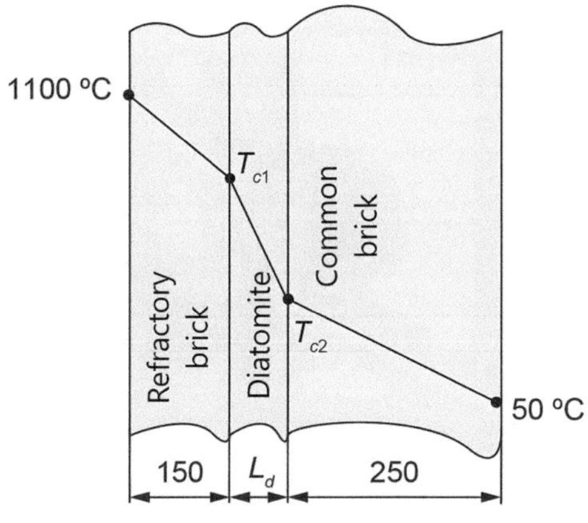

2.7 Write the general equation of heat conduction in Cartesian coordinates and simplify it for a steady-state, one-dimensional problem with constant and uniform properties and without heat generation. Justify in detail all the simplifications. Indicate what should be the temperature distribution $T(x)$ and the heat flux $q(x)$ in relation to the thermal conductivity k and the quantities indicated for each of the three cases shown in the following figures. A is the perpendicular area to the heat flux.

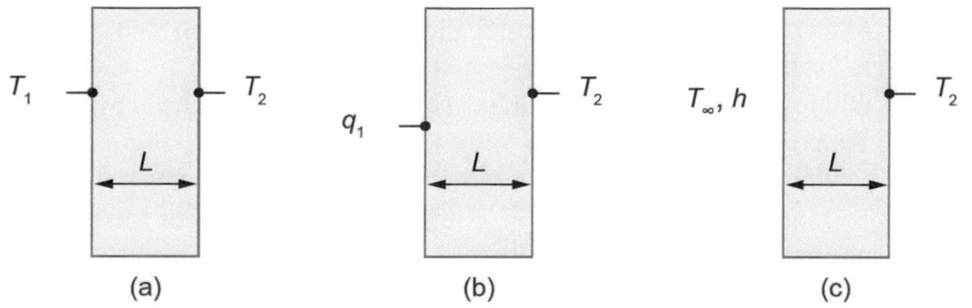

2.8 A wall insulation project from a cold climate country is schematized in the figure ahead. The wall will be exposed to an external environment ($T_o = -20$ °C and $h_o = 12.0$ W/m^2°C) and to an internal environment ($T_i = 24$ °C and $h_i = 8.0$ W/m^2°C). The data of the materials and thicknesses are indicated in the table. For these conditions, it is asked:

(a) Heat flux in the wall per unit area [W/m^2].
(b) Keeping the materials in the three layers and the thicknesses e_1 and e_3 constant, what design modifications could be made to layer 2 to reduce the flux calculated in item (a) by 30%?

Layer	Thickness [mm]	Thermal conductivity [W/m°C]
1	20	0.8
2	80	1.5
3	10	0.5

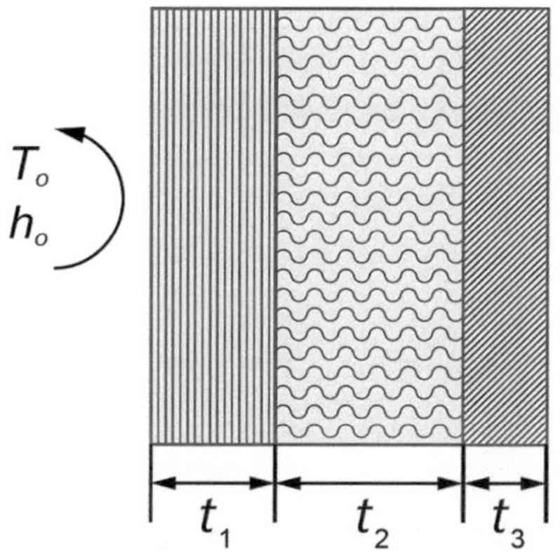

2.9 A composite flat wall has two layers with thermal conductivities k_A and k_B, which are separated by a thin electric heater. The contact resistances on the surfaces are negligible. The external wall faces are in contact with two environments of temperatures T_i and T_o, whose heat transfer coefficients are h_i and h_o, respectively, as shown in the figure. The heat flux rate produced in the heater is q_h. It is asked:

(a) Sketch the equivalent thermal circuit for the system and identify all resistances in terms of the relevant variables.
(b) Obtain an expression that can be used to determine the heater temperature T_h explicitly.
(c) Obtain an expression for the ratio between the heat fluxes to the external and internal fluids, that is, $\frac{q_A}{q_B}$. How can the variables of the problem be adjusted to minimize this ratio?

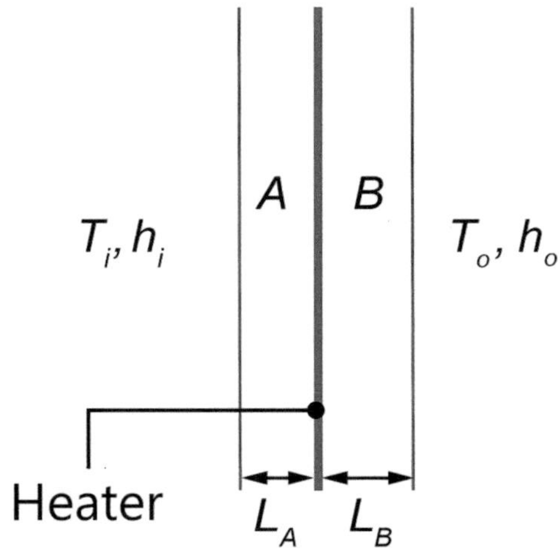

2.10 A large 316 stainless steel tank is used to store industrial water at $T_{\infty 1} = 20$ °C. On a sunny day, the outer surface receives solar irradiation G as indicated, which is fully absorbed by the wall. The external air temperature is $T_{\infty 2}$. Under these conditions, the temperature of the internal (T_1) and external (T_2) surfaces of the tank wall is requested (use at least one decimal place). The data are as follows: $t = 5$ mm; $G = 600$ W/m^2 and $T_{\infty 2} = 30$ °C.

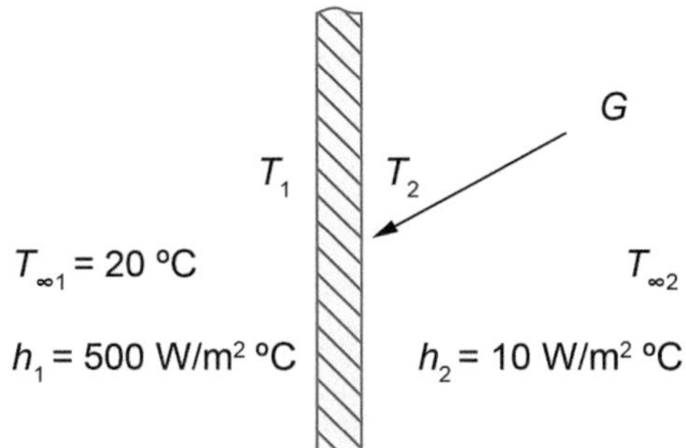

2.11 The coil of a steam reheater is made of a material with a thermal conductivity of $k = 14$ W/m°C and its internal and external diameters are, respectively, 32 mm and 42 mm. The temperatures of the external and internal surfaces are 580 °C and 400 °C, respectively. Determine the heat loss rate per meter of pipe.

2.12 A steel pipe with internal and external diameters of, respectively, 100 mm and 110 mm has a thermal conductivity $k_t = 50$ W/m°C and is coated with a 50 mm layer of thermal insulation, whose conductivity is $k_i = 0.07$ W/m°C. The pipe internal temperature is 250 °C and that of the outer surface of the insulation is 50 °C. Determine the heat loss rate in 1 m of pipe and the temperature of the pipe external surface (or internal of the thermal insulation).

2.13 A carbon steel pipe with a 12 mm internal diameter and a wall thickness of 2 mm is used to produce saturated steam. The water enters as saturated liquid at 100 °C and leaves the pipe as saturated steam at 100 °C, at a mass flux rate of 0.01 kg/s. The heating is done by the perpendicular circulation of combustion gases at 450 °C. The average heat transfer coefficient from the combustion products to the pipe is 40 W/m^2°C. Under these conditions, the required length L of the pipe is requested knowing that the internal temperature of the pipe is 100 °C.

2.14 Determine the overall U heat transfer coefficient for the pipe indicated in the scheme. Show that the heat loss rate from a metallic pipe (low thermal resistance) is dominated by the side that has the lowest coefficient h of heat transfer and that the temperature of the pipe wall is dominated (or imposed) by the side that has the highest heat transfer coefficient. Disregard the thermal conduction resistance of the pipe.

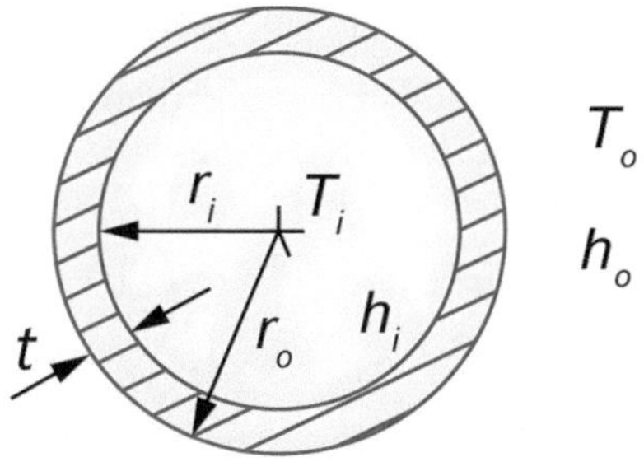

2.15 A flat wall with a thickness of 7.5 cm generates internal thermal energy (heat) at a rate of 10^5 W/m^3. One side of the wall is insulated, while the other side is exposed to an environment at 80 °C. The convection heat transfer coefficient of the exposed side to the air is 100 W/m^2 °C. If the thermal conductivity of the wall is 12 W/m°C, what is the maximum temperature of the wall?

2.16 Consider a pure copper electrical wire with a diameter of 4 mm that has an electrical resistance of 1.3×10^{-3} Ω/m and is traversed, in an overload situation, by an electric current of 100 A. This wire is coated with a plastic film that serves to electrically insulate it. The thermal conductivity of the plastic material is

$k_f = 0.20$ W/m°C and its thickness is 1 mm. The insulation is exposed to the ambient air, at a temperature $T_\infty = 20$ °C and the convection coefficient is $h = 9$ W/m^2°C.

(a) Write the heat conduction equation with the boundary conditions suitable for solving the problem.
(b) Determine the total dissipated power.
(c) Determine the equation that provides the temperature distribution in the wire, with the numerical values of the problem.
(d) Calculate the maximum temperature in the wire.
(e) Determine the wire external surface temperature.

2.17 A nuclear fuel element with thickness $2L$ is covered with a steel coating of thickness b on both sides. The thermal energy (heat) generated inside the nuclear fuel, q_G''', is removed by a fluid at temperature T_∞, adjacent to one of the surfaces, and the convection coefficient on this surface is h. The other surface is well thermally insulated. The fuel and the steel have thermal conductivities k_f and k_s, respectively. Sketch the temperature distribution $T(x)$ for the entire system and describe the main characteristics of the distribution.

For $k_f = 60$ W/mK, $L = 15$ mm, $b = 3$ mm, $k_s = 15$ W/mK, $h = 10{,}000$ W/m^{2}K and $T_\infty = 200$ °C, what are the highest and lowest temperatures in the fuel element if thermal energy is generated uniformly at a volumetric rate of $q_G''' = 2 \times 10^7$ W/m^3? What are the corresponding positions?

If the insulation is removed and equivalent convection conditions are maintained on each surface, what is the corresponding form of the temperature distribution in the fuel element? What are the highest and lowest temperature in the fuel? What are the corresponding positions?

2.18 In a cylindrical fuel rod of a nuclear reactor, heat at a given rate is internally generated according to the following expression:

$$q_G''' = q_o''' \times \left[1 - \left(\frac{r}{r_o}\right)^2\right]$$

where q_G''' is the rate of thermal energy (heat) generated per unit volume [W/m^3] at position r, q_o''' is the rate of thermal energy generated per unit volume on the centerline of the bar and r_o is the outer radius of the bar. Calculate the difference between the temperature of the centerline and the surface temperature for a bar with a diameter of 50 mm. The thermal conductivity is 25 W/m°C and the heat removal flux from the surface is 0.15 MW/m^2.

2.19 The protective shielding of a nuclear reactor can be idealized as a large flat wall of $L = 25$ cm in thickness, whose thermal conductivity is $k = 1$ W/m°C. The radiation from inside the reactor penetrates the shielding and produces an internal generation of heat q_G''' that decreases exponentially, from $q_i''' = 1000$ W/m^3 on the internal surface to a value of 100 W/m^3 at a distance of

12.5 cm from the internal surface. If the external surface is maintained at 40 °C by forced convection, determine:

(a) The coefficient λ of the internal heat generation expression, given below (see *Hint*).
(b) The equation of the temperature distribution inside the wall with the appropriate boundary conditions and numerical values.
(c) The temperature of the internal surface.

Hint: define the differential equation of heat conduction with internal heat generation that varies according to $q_G(x) = q_{Go}e^{-\lambda x}$. All the heat generated is transferred only through the outer wall.

2.20 The circular steel rod tip with a diameter of 40 mm and a length of 300 mm is connected to two isothermal walls maintained at 200 and 90 °C, as indicated in the figure. An air flux crosses the rod with a convection heat transfer coefficient $h = 20$ W/m^2°C, with the air temperature being 20 °C. Determine the rate of heat transferred to the air, the rate of heat transferred at the two isothermal bases, and the temperature distribution along the bar.

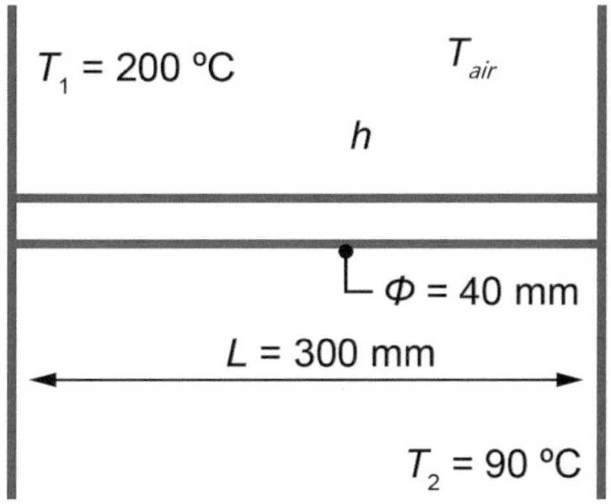

2.21 A circular fin made of an aluminum alloy is mounted on a heated tube with an external radius of 2.5 cm. The fin thickness is 0.4 mm and its external radius is 40 mm. The tube wall temperature is 150 °C, the surrounding air temperature is 20 °C and the convection heat transfer coefficient is 30 W/m^{2}K. It is asked to determine the rate of heat transferred by the fin.

(*Note*: use the transport properties of an alloy of aluminum from the table and assume that this value is constant.)

2.22 A carbon steel tube with an internal diameter of 10 mm and a wall thickness of 2 mm is used to produce steam water saturated, as illustrated. The water enters as liquid saturated state at 120 °C and leaves the tube as saturated steam at 120 °C. The heating is done by the circulation perpendicular to the tube of combustion gases at 470 °C. The average heat transfer coefficient from the combustion products to the tube is 40 W/m^2°C. Under these conditions, it is asked for the necessary tube length L and the temperature of the tube's external wall, T_w.

Now, consider the installation of aluminum circular fins spaced $\delta = 5$ mm, thickness t of 1 mm, and external diameter $D_2 = 27$ mm. Determine the new length of the tube. Assume that the temperature of the tube's external wall remains the same as in the case without fins, as well as the convection heat transfer coefficient h is also the same. Due to the high internal convection heat transfer coefficient and the conduction thermal resistance through the tube being negligible, it can be considered that the tube's external temperature is the actual boiling water temperature (120 °C). The specific enthalpy of water vaporization is h_{LV} 2202 kJ/kg and the mass flow rate is 0.01 kg/s.

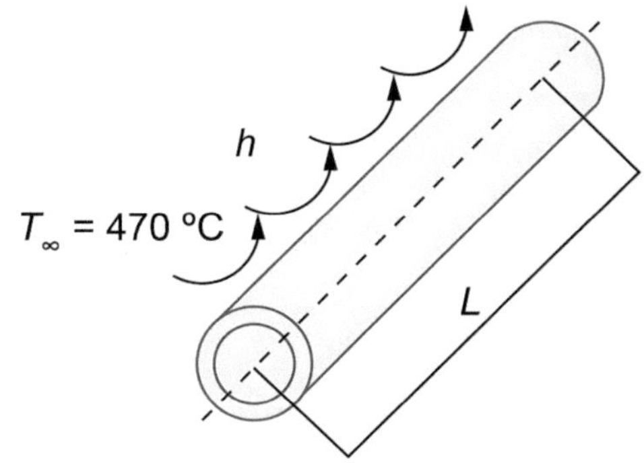

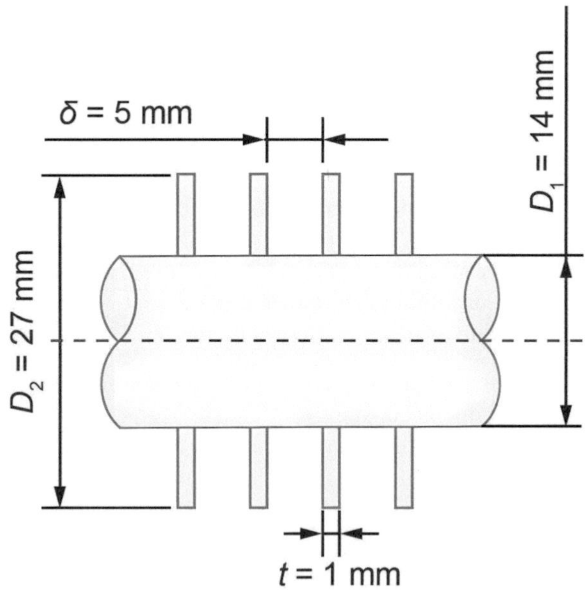

2.23 A transistor must dissipate 0.1 W without exceeding the maximum operating temperature of 150 °C. See other data in the figure. For the indicated conditions, check if the dissipation surface temperature of the transistor in contact with the air does not exceed the maximum specified value. Determine the surface temperature to check if it is within the operational conditions.

Alternatively, it is proposed to use a uniform and rectangular profile fin as indicated in the figure. For this new condition, calculate the base temperature of the fin. This temperature can be compared with the maximum specified. The contact resistance between the fin and the transistor is negligible, due to the assembly with thermal paste and the thickness of the fin base. Consider the heat exchange in the fins and in the region without fins. Check if this second design option meets the maximum temperature limit.

Data: $h = 9$ W/m^2°C; $k_{al} = 177$ W/m°C; $T_\infty = 25$ °C, measurements in mm.

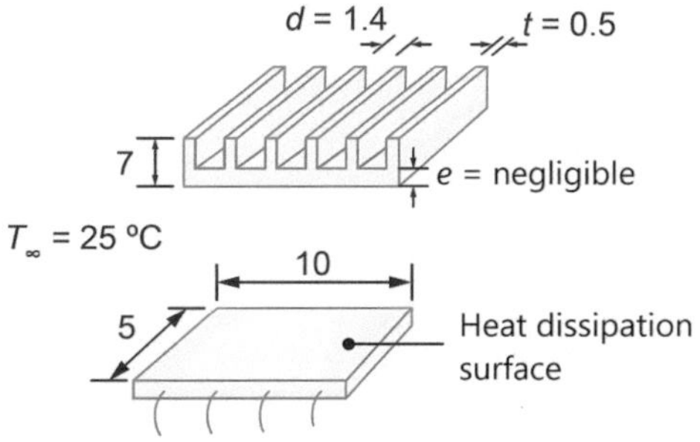

References

Kreith, F., and M.S. Bohn. 2001a. *Principles of heat transfer*, 6th ed. New York: Brooks/Cole.

Sartre, V., and M. Lallemand. 2001a. Enhancement of thermal contact conductance for electronic systems. *Applied Thermal Engineering* 21: 221–235.

Yoyanovich, M.M., and M. Tuarze. 1969a. Experimental evidence of thermal resistance at soldered joints. *Journal of Spacecraft and Rockets* 6 (7): 855–857.

Chapter 3
Multidimensional Steady-State Conduction

Abstract In this chapter, the case of multidimensional heat conduction is studied. The classic case of a rectangular plate with prescribed wall temperatures in a steady state without thermal energy (heat) generation is studied using the method of separation of variables in Cartesian coordinates x and y. The method is extended to analyze cylindrical geometries. The cases analyzed are illustrated through solved examples. The method of superposition of solutions is presented, which allows combining individual solutions of boundary conditions. In the second part of the chapter, the numerical method of finite differences is presented to solve the general equation of heat conduction. The continuous medium that makes up the solid or surface is discretized into nodal points from which an energy balance allows for establishing the temperature distribution within it. The boundary conditions are also discretized. The method is applied to cartesian and cylindrical coordinate systems with the presentation of solved examples. The chapter ends presenting of the conduction shape factors subject, which allow to solve the heat conduction in some specific geometries.

Keywords Analytical solution · Conduction shape factor · Method of separation of variables · Numerical method of finite differences

Chapter 2 was dedicated to a heat transfer analysis in one-dimensional steady-state conduction. However, many actual practical heat transfer problems are bi or three-dimensional, which requires the solution of the heat conduction equation beyond the one-dimensional case. However, analytical solutions to the heat conduction equation exist for a limited number of problems, whose boundary conditions and geometries are elementary. Cases of more complex geometries and boundary conditions, which are the most common, must be solved by numerical solution techniques, such as *the method of finite differences*, which is the subject of analysis in the second part of this chapter. Nevertheless, the student should know analytical solution methods for the most elementary cases. In this section, two heat conduction classic problems are analyzed by solving the governing partial differential equation using *the method of separation of variables*.

J. R. Simões-Moreira and E. W. Zavaleta-Aguilar, *Heat Transfer*, Heat and Mass Transfer, https://doi.org/10.1007/978-3-031-80318-5_3

3.1 Two-Dimensional Steady-State Conduction—An Analytical Solution by the Method of Separation of Variables

Consider the case of a $b \times L$ rectangular plate, whose sides are subjected to constant temperatures in a steady state, as illustrated in Fig. 3.1. Note that the plate sides are at the same temperature T_1, except one side, which is held at T_2. Thus, the solution of the general steady-state heat conduction equation without internal generation of energy (heat) will provide the spatial distribution of the plate temperature, that is, $T = T(x, y)$, from which the heat flux local vector can be obtained.

The general equation of heat conduction in cartesian coordinates with constant properties is given by Eq. (2.6):

$$\frac{\partial^2 T}{\partial x^2} + \frac{\partial^2 T}{\partial y^2} + \frac{\partial^2 T}{\partial z^2} + \frac{q_G'''}{k} = \frac{1}{\alpha}\frac{\partial T}{\partial t} \tag{2.6}$$

Introducing the following assumptions: steady-state, no internal generation of thermal energy (heat), and a two-dimensional problem, the general equation reduces to:

$$\frac{\partial^2 T}{\partial x^2} + \frac{\partial^2 T}{\partial y^2} = 0 \tag{3.1}$$

The boundary conditions of this problem are the fixed temperatures at plate sides expressed mathematically as:

BC_1: $T(0, y) = T_1$ temperature on the left face;

BC_2: $T(L, y) = T_1$ temperature on the right face;

BC_3: $T(x, 0) = T_1$ temperature on the lower face;

BC_4: $T(x, b) = T_2$ temperature on the upper face.

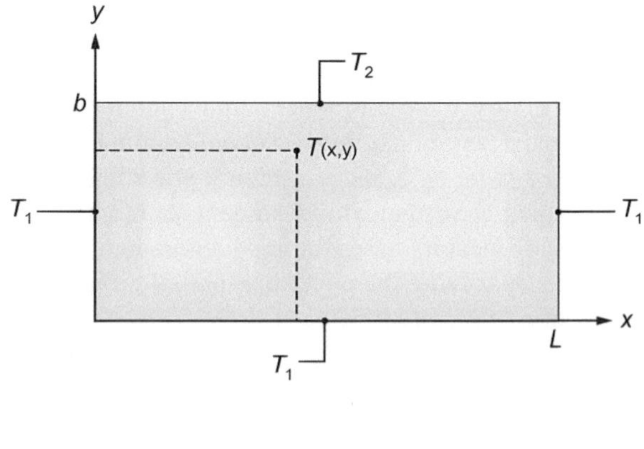

Fig. 3.1 Rectangular plate of dimensions $b \times L$ subjected to isothermal boundary conditions

It is convenient to perform a change of variables in order to obtain the dimensionless temperature, $\theta(x, y)$.

$$\theta = \frac{T - T_1}{T_2 - T_1} \tag{3.2}$$

In this way, the boundary conditions in the new variable $\theta = \theta(x, y)$ are:

BC_1: $\theta(0, y) = 0$;
BC_2: $\theta(L, y) = 0$;
BC_3: $\theta(x, 0) = 0$;
BC_4: $\theta(x, b) = 1$.

The elementary variation of temperature is: $\frac{dT}{T_2 - T_1} = d\theta$. Then, substituting in Eq. (3.1), we obtain:

$$\frac{\partial^2\theta}{\partial x^2} + \frac{\partial^2\theta}{\partial y^2} = 0 \tag{3.3}$$

This is the conduction equation in the new variable $\theta(x, y)$.

The classical solution of this type of partial differential equation is achieved by using the method of separation of variables. This technique assumes that the temperature distribution $\theta(x, y)$ is the product of two other functions $X(x)$ and $Y(y)$, which, in turn, are exclusive functions on the individual problem's original coordinates x and y, that is:

$$\theta(x, y) = X(x) \times Y(y) \tag{3.4}$$

Thus, the partial differentials in relationship to x of this new function are:

First derivative: $\frac{\partial\theta}{\partial x} = Y\frac{dX}{dx}$.

Second derivative: $\frac{\partial^2\theta}{\partial x^2} = Y\frac{d^2X}{dx^2}$.

Similarly with respect to y:

Second derivative: $\frac{\partial^2\theta}{\partial y^2} = X\frac{d^2Y}{dy^2}$.

Therefore, substituting these two partial derivatives into Eq. (3.3), we obtain:

$$Y\frac{d^2X}{dx^2} + X\frac{d^2Y}{dy^2} = 0$$

or, dividing by the product XY, yields:

$$\frac{1}{Y}\frac{d^2Y}{dy^2} = -\frac{1}{X}\frac{d^2X}{dx^2} \tag{3.5}$$

In Eq. (3.5) the left hand side is a function of coordinate y only, and the right hand side is a function of coordinate x. However, both sides are equal, no matter what the x and y values are. This implies that each side of the equation can neither be an isolated function of x nor y because otherwise, the equality would not hold. Therefore, the only possibility of keeping the equation valid is that the equality must be a constant magnitude independent of the variables x and y which, for mathematical convenience, is assigned the positive constant λ^2. Thus, we have:

$$\frac{1}{X}\frac{d^2X}{dx^2} = -\lambda^2 \tag{3.6}$$

and

$$\frac{1}{Y}\frac{d^2Y}{dy^2} = \lambda^2 \tag{3.7}$$

It is worth noting that the original partial differential equation, Eq. (3.5), gave rise to two other ordinary differential equations (Eqs. 3.6 and 3.7). Solutions of these two new equations are well known (see Calculus Reminder, in Sect. 2.11.1):

$$X(x) = C_1\cos(\lambda x) + C_2\sin(\lambda x) \text{ and}$$
$$Y(y) = C_3e^{-\lambda y} + C_4e^{\lambda y}$$

Substituting both solutions into Eq. (3.4), we have:

$$\theta(x, y) = [C_1\cos(\lambda x) + C_2\sin(\lambda x)]\left(C_3e^{-\lambda y} + C_4e^{\lambda y}\right)$$

Constants values C_1, C_2, C_3, and C_4 depend on the imposed boundary conditions which, in this case, are prescribed face temperatures. Thus,

From BC_1: $\theta(0, y) = 0$, or

$$\theta(0, y) = [C_1\cos(\lambda 0) + C_2\sin(\lambda 0)]\left(C_3e^{-\lambda y} + C_4e^{\lambda y}\right) = C_1\left(C_3e^{-\lambda y} + C_4e^{\lambda y}\right) = 0$$

From which it is concluded that the only possibility is $C_1 = 0$, otherwise the trivial solution would be obtained, therefore:

$$\theta(x, y) = C_2\sin(\lambda x)\left(C_3e^{-\lambda y} + C_4e^{\lambda y}\right)$$

From BC_3: $\theta(x, 0) = 0$

$$C_2\sin(\lambda x)\left(C_3e^{\lambda 0} + C_4e^{-\lambda 0}\right) = C_2\sin(\lambda x)(C_3 + C_4) = 0$$

from which it is obtained that: $C_3 + C_4 = 0 \rightarrow C_3 = -C_4$, because C_2 cannot be zero, therefore:

$$\theta(x, y) = C_2C_4\sin(\lambda x)\left(e^{-\lambda y} - e^{\lambda y}\right)$$

From BC$_2$: $\theta(L, y) = 0$, it follows that:

$$C_2C_4\sin(\lambda L)\left(e^{\lambda y} - e^{-\lambda y}\right) = 0$$

but, as the product of the two constants (C_2 and C_4) cannot be null, it implies that: $\sin(\lambda L) = 0$.

In this case, there is an unlimited number of particular solutions. The possible values of λ that satisfy this condition are: $\lambda L = n\pi$, that is, $\lambda_n = \frac{n\pi}{L}$, where $n = 1, 2, 3, \ldots$

The value of $n = 0$, therefore, $\lambda = 0$ results in a trivial solution and was not considered. The constants λ_n are the so-called *eigenvalues.*

Therefore, collecting the previous results, it is found that the temperature distribution so far is:

$$\theta_n(x, y) = \underbrace{2C_2C_4}_{C_n}\sin\left(\frac{n\pi x}{L}\right)\underbrace{\left(\frac{e^{\frac{n\pi y}{L}} - e^{-\frac{n\pi y}{L}}}{2}\right)}_{\sinh\left(\frac{n\pi y}{L}\right)}, \quad \text{that is,}$$

$$\theta_n(x, y) = C_n\sin\left(\frac{n\pi x}{L}\right)\sinh\left(\frac{n\pi y}{L}\right) \tag{3.8}$$

For each $n = 1, 2, 3, \ldots$ there is a corresponding particular θ_n solution. Also, the constant resulting from the product $2C_2C_4$ assumes n distinct values and, therefore, it was replaced by a new single constant C_n which also depends on the value of n. Thus, the general solution should be a linear combination of all possible nth particular solutions θ_n. That is,

$$\theta(x, y) = \sum_{n=1}^{\infty} C_n\sin\left(\frac{n\pi x}{L}\right)\sinh\left(\frac{n\pi y}{L}\right)$$

Finally, the constants C_n are obtained from the last boundary condition, BC$_4$: $\theta(x, b) = 1$, that is:

$$\sum_{n=1}^{\infty} C_n\sin\left(\frac{n\pi x}{L}\right)\sinh\left(\frac{n\pi b}{L}\right) = 1 \tag{3.9}$$

The last task is to find the C_n coefficients from the previous series to obtain the final distribution of temperatures. This task is achieved using the *orthogonal function theory*, reviewed in the following box.

Review of orthogonal functions

An infinite set of functions $g_1(x)$, $g_2(x)$, is said to be orthogonal in the domain $a \leq x \leq b$, if $\int_a^b g_m(x)g_n(x)dx = 0$, for $m \neq n$

Many functions exhibit the property of orthogonality, including the trigonometric functions sine and cosine, such as: $\sin\left(n\pi \frac{x}{L}\right)$ and $\cos\left(n\pi \frac{x}{L}\right)$ in the interval $0 \leq x \leq L$.

It is also verified that any function $f(x)$ can be expressed in an infinite series of orthogonal functions, that is:

$f(x) = \sum_{m=1}^{\infty} A_m g_m(x)$

To obtain the coefficients A_m proceed as follows:

(1) Multiply by $g_n(x)$ both sides of the equation:

$g_n(x)f(x) = g_n(x) \sum_{m=1}^{\infty} A_m g_m(x)$

(2) Integrate over the interval [a, b] of interest:

$\int_a^b g_n(x)f(x)dx = \int_a^b \left[g_n(x) \sum_{m=1}^{\infty} A_m g_m(x) \right] dx$

Using the orthogonality property in the interval, that is:

$\int_a^b g_m(x)g_n(x)dx = 0, \quad \text{if} \quad m \neq n$

The summation can be eliminated, as all other terms are null, except the mth term when $m = n$, so: $\int_a^b g_m(x)f(x)dx = \int_a^b A_m g_m^2(x)dx$.

Finally, the constants of the series A_m can be obtained:

$$A_m = \frac{\int_a^b g_m(x)f(x)dx}{\int_a^b g_m^2(x)dx}$$

By applying the orthogonality property to the following situation with $f(x) = 1$ and $g_n(x) = \underbrace{\sin\left(\frac{n\pi x}{L}\right)}_{\text{orthogonal function}}$; $n = 1, 2, 3, \ldots$

Therefore, expanding the function $f(x) = 1$, we obtain:

$$\sum_{n=1}^{\infty} A_n \sin\left(\frac{n\pi x}{L}\right) = 1$$

Thus, the coefficients of the series, A_n can be obtained as indicated in the review box above:

$$A_n = \frac{\int_0^L \sin\left(\frac{n\pi x}{L}\right)dx}{\int_0^L \sin^2\left(\frac{n\pi x}{L}\right)dx} = \frac{2}{\pi}\frac{(-1)^{n+1}+1}{n} \tag{3.10}$$

So,

$$\sum_{n=1}^{\infty} \frac{2}{\pi}\frac{(-1)^{n+1}+1}{n}\sin\left(\frac{n\pi x}{L}\right) = 1 \tag{3.11}$$

Comparing Eqs. (3.9) and (3.11), we obtain:

$$\sum_{n=1}^{\infty} C_n \sin\left(\frac{n\pi x}{L}\right)\sinh\left(\frac{n\pi b}{L}\right) = \sum_{n=1}^{\infty} \frac{2}{\pi}\frac{(-1)^{n+1}+1}{n}\sin\left(\frac{n\pi x}{L}\right)$$

So, by the rule of series equality, for two series to be equal each individual term must be equal, then we get:

$$C_n = \frac{2\left[(-1)^{n+1}+1\right]}{n\pi \sinh\left(\frac{n\pi b}{L}\right)}, n = 1, 2, 3, \ldots \tag{3.12}$$

So, the final solution is:

$$\theta(x, y) = \frac{2}{\pi}\sum_{n=1}^{\infty} \frac{(-1)^{n+1}+1}{n}\sin\left(\frac{n\pi x}{L}\right)\frac{\sinh\left(\frac{n\pi y}{L}\right)}{\sinh\left(\frac{n\pi b}{L}\right)} \tag{3.13}$$

The temperature distribution graph is shown in Fig. 3.2, for $b = L/2$.

To evaluate the heat flux in the x and y directions, q_x and q_y, it is necessary to keep in mind that the heat flux is a vector quantity, that is:

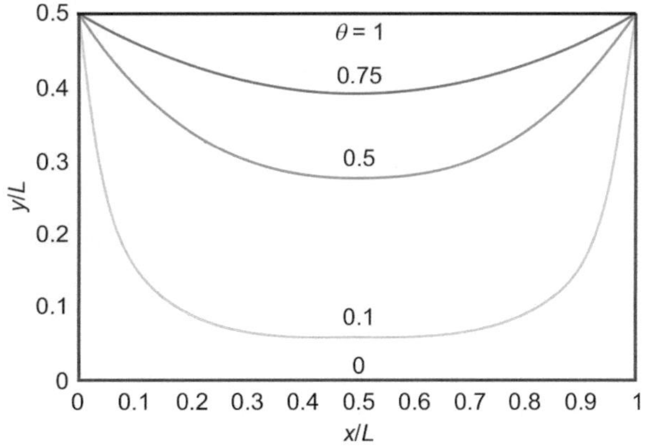

Fig. 3.2 Two-dimensional temperature distribution (isothermal lines) in a plate for $b = L/2$

$$\vec{q}_x = -k\frac{\partial T}{\partial x}\vec{i} \quad \text{and} \quad \vec{q}_y = -k\frac{\partial T}{\partial y}\vec{j}$$

where the total heat flux vector is $\vec{q} = \vec{q}_x + \vec{q}_y$, and the heat flux modulus is $q = \sqrt{q_x^2 + q_y^2}$ in W/m^2.

Therefore, it is possible to obtain the heat flux from the temperature distribution. $q_x = -k\frac{\partial T}{\partial x} = -k(T_2 - T_1)\frac{\partial \theta}{\partial x}$, substituting the solution $\theta(x, y)$ from Eq. (3.13), we obtain:

$$\frac{\partial \theta}{\partial x} = \frac{2}{\pi}\sum_{n=1}^{\infty}\frac{(-1)^{n+1}+1}{n}\left[\cos\left(\frac{n\pi x}{L}\right)\right]\left(\frac{n\pi}{L}\right)\frac{\sinh\left(\frac{n\pi y}{L}\right)}{\sinh\left(\frac{n\pi b}{L}\right)}, \quad \text{and}$$

$$\frac{\partial T}{\partial x} = (T_2 - T_1)\frac{2}{\pi}\sum_{n=1}^{\infty}\frac{(-1)^{n+1}+1}{n}\left[\cos\left(\frac{n\pi x}{L}\right)\right]\left(\frac{n\pi}{L}\right)\frac{\sinh\left(\frac{n\pi y}{L}\right)}{\sinh\left(\frac{n\pi b}{L}\right)},$$

Substituting in the x heat flux, yields,

$$q_x = -\frac{2k(T_2 - T_1)}{L}\sum_{n=1}^{\infty}\left[(-1)^{n+1}+1\right]\cos\left(\frac{n\pi x}{L}\right)\frac{\sinh\left(\frac{n\pi y}{L}\right)}{\sinh\left(\frac{n\pi b}{L}\right)} \tag{3.14}$$

Similarly, the heat flux in the y direction is obtained:

$$q_y = -\frac{2k(T_2 - T_1)}{L}\sum_{n=1}^{\infty}\left[(-1)^{n+1}+1\right]\sin\left(\frac{n\pi x}{L}\right)\frac{\cosh\left(\frac{n\pi y}{L}\right)}{\sinh\left(\frac{n\pi b}{L}\right)} \tag{3.15}$$

Example 3.1 (*Temperature and heat flux in a square plate*)

A 1 m side square plate is subjected to a 50 °C temperature on its three sides, and at 100 °C on the upper side as depicted in the diagram below. Find the temperatures and heat flux for the positions 1 through 6 indicated in the figure.

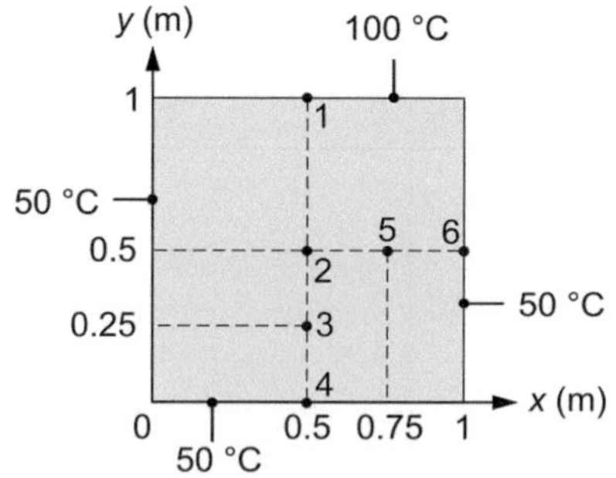

Solution

First, let us define the non-dimensional temperature $\theta = \frac{T - T_1}{T_2 - T_1} = \frac{T - 50}{50}$. Therefore, the problem in the new variable is exactly the one solved in this section. The temperature distribution is given by Eq. (3.13) and the heat fluxes are given by Eqs. (3.14) and (3.15). The solution for the given position is presented in the table below. Note that points 1 and 6 have prescribed temperatures. All *y* direction heat fluxes are negative, which is in accordance with the given boundary conditions (higher temperature on top). The *x*-heat flux is null at the symmetry line.

Point	θ	T (°C)	q_x (W/m^2)	q_y (W/m^2)	$\lvert q \rvert$ (W/m^2)
1	1	100	0	−49.1	49.1
2	0.25	62.5	0	−2736	2736
3	0.1	54.8	0	−1493	1493
4	0	50	0	−1134	1134
5	0.18	59.1	1761	−2095	2737
6	0	50	2737	0	2737

3.2 Analytical Solution in Cylindrical Coordinates

The general equation of heat conduction in cylindrical coordinates for constant transport properties is expressed by Eq. (2.17). In this section, the case of two-dimensional distribution is presented where the temperature variation occurs in the radial direction and lengthwise for a cylindrical tube in a steady state, without heat generation. In this solution, there is no temperature variation in the angular direction ϕ at the same radial position r. Thus, the temperature distribution to be determined is a function of r and z cylindrical coordinates, that is, $T = T(r, z)$. Starting from the general equation of heat conduction, with the simplifying assumptions, we obtain:

$$\frac{1}{r}\frac{\partial}{\partial r}\left(r\frac{\partial T}{\partial r}\right) + \frac{\partial^2 T}{\partial z^2} = 0 \tag{3.16}$$

Here, it is analyzed the case of a solid cylinder subjected to convection on the surface, as it occurs with circular fins as it was studied in Sect. 2.11.1, but now from the perspective of a two-dimensional problem. Obviously, the cylinder can be subjected to different heat transfer conditions on its surface, or different boundary conditions. In this section, the problem of a fin with conditions indicated in Fig. 3.3 is solved, namely convective heat loss on its surface and adiabatic (isolated) end.

The problem assumes that the base temperature (T_b) is known ($x = 0$) and isolated at the right end, $x = L$. The fin surface is subjected to heat convection to the surrounding environment with a known value h. Therefore, the boundary conditions are:

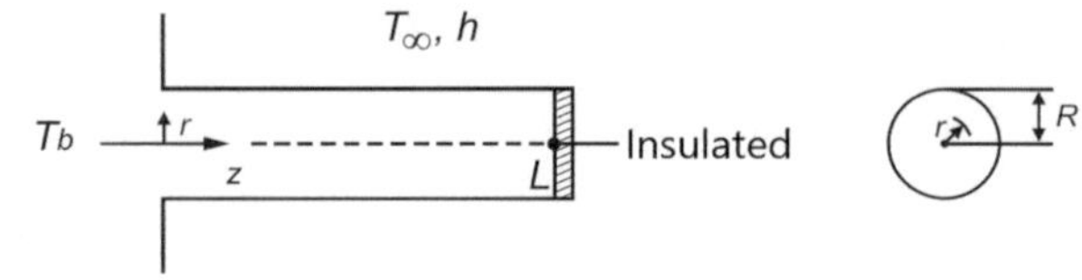

Fig. 3.3 Problem of a two-dimensional fin in cylindrical coordinates subjected to convective heat exchange

BC$_1$: $T(r, 0) = T_b$ temperature at the base of the fin,
BC$_2$: $\left.\frac{\partial T}{\partial z}\right|_{z=L} = 0$ end thermally isolated.
BC$_3$: $r = 0$, finite temperature at the central line (it eliminates the indeterminacy of the singularity at $r = 0$, see similar situation in BC$_2$ of Sect. 2.9).
BC$_4$: $-k\left.\frac{\partial T}{\partial r}\right|_{r=R} = h[T(R, z) - T_\infty]$ convection on the lateral surface of the fin.

To solve this class of problems, it is convenient to define the dimensionless temperature $\theta(r, z)$, in this case, it is:

$$\theta(r, z) = \frac{T(r, z) - T_\infty}{T_b - T_\infty}$$

Thus, Eq. (3.16) takes the following form in the new variable:

$$\frac{1}{r}\frac{\partial}{\partial r}\left(r\frac{\partial\theta}{\partial r}\right) + \frac{\partial^2\theta}{\partial z^2} = 0 \tag{3.17}$$

Using the method of separation of variables, two new functions are defined that depend on the coordinates independently, that is, $U = U(r)$ and $V = V(z)$, as seen in the previous section. Thus, the dimensionless temperature becomes the product of these two new functions.

$$\theta(r, z) = U(r)V(z) \tag{3.18}$$

Substituting into Eq. (3.16) and dividing by the product UV, we get:

$$\frac{1}{U}\frac{d}{dr}\left(r\frac{dU}{dr}\right) = -\frac{1}{V}\frac{d^2V}{dz^2} = -\lambda^2$$

From which we obtain the two ordinary differential equations, as indicated by the method:

$$r\frac{d^2U}{dr^2} + \frac{dU}{dr} + U\lambda^2 = 0 \quad \text{and}$$

$$\frac{d^2V}{dz^2} - V\lambda^2 = 0$$

whose general solutions are, respectively:

$$U = C_1 J_0(\lambda r) + C_2 Y_0(\lambda r), \text{ and}$$
$$V = C_3 e^{\lambda z} + C_4 e^{-\lambda z}$$

where J_0 is the Bessel function of the first kind of order 0 and Y_0 is the Bessel function of the second kind of order 0 (Appendix C contains the definitions of the Bessel functions). These two equations (U and V) are solved considering the boundary conditions already defined and modified in the dimensionless temperature.

BC_1: $\theta(r, 0) = 1$,
BC_2: $\left.\frac{\partial \theta}{\partial z}\right|_{z=L} = 0$,
BC_3: $r = 0$, finite temperature at the central line, and
BC_4: $-k\left.\frac{\partial \theta}{\partial r}\right|_{r=R} = h\theta(R, z)$.

For additional details of the solution, it is suggested to consult the work of Carslaw and Jaeger (1959), where the authors provide general solutions for several boundary conditions. Thus, the final solution is:

$$\theta(r, z) = \frac{2hL^2}{kR} \sum_{n=1}^{\infty} \frac{\cosh\left[\lambda_n\left(1 - \frac{z}{L}\right)\right] J_0\left(\frac{\lambda_n r}{L}\right)}{\cosh(\lambda_n) J_0\left(\frac{\lambda_n R}{L}\right)\left[\lambda_n^2 + \left(\frac{hL}{k}\right)^2\right]} \quad (3.19)$$

The eigenvalues λ_n must satisfy the following condition:

$$\lambda_n J_1\left(\lambda_n \frac{R}{L}\right) = \frac{hL}{k} J_0\left(\lambda_n \frac{R}{L}\right) \quad (3.20)$$

with J_1 being the Bessel function of the first kind of order 1. The eigenvalues λ_n depend on the process variables (R, L, h, k).

Once the temperature distribution is known, the heat flux can be obtained from Fourier's Law of conduction. In the radial direction, the heat flux r is:

$$q_r = -k\frac{\partial T}{\partial r}$$

Applying the partial derivative to Eq. (3.19) (see properties of derivation of the Bessel function in Appendix C), we get:

$$q_r = \frac{2hL(T_b - T_\infty)}{R} \sum_{n=1}^{\infty} \frac{\lambda_n \cosh\left[\lambda_n\left(1 - \frac{z}{L}\right)\right] J_1\left(\frac{\lambda_n r}{L}\right)}{\cosh(\lambda_n) J_0\left(\frac{\lambda_n R}{L}\right)\left[\lambda_n^2 + \left(\frac{hR}{k}\right)^2\right]} \quad (3.21)$$

Similarly, the heat flux in the axial direction z is evaluated by:

$$q_z = -k\frac{\partial T}{\partial z}, \quad \text{or}$$

$$q_z = \frac{2hL(T_b - T_\infty)}{R} \sum_{n=1}^{\infty} \frac{\lambda_n \sinh\left[\lambda_n\left(1 - \frac{z}{L}\right)\right] J_1\left(\frac{\lambda_n r}{L}\right)}{\cosh(\lambda_n) J_0\left(\frac{\lambda_n R}{L}\right)\left[\lambda_n^2 + \left(\frac{hR}{k}\right)^2\right]} \tag{3.22}$$

Example 3.2 (*Two-dimensional heat transfer in a cylindrical rod*)

A cylindrical A316 steel rod is cooled with water at 20 °C. The heat transfer coefficient by convection between the rod surface and the water is 350 W/m^2°C. One rod end is at a 90 °C uniform temperature, while the other is thermally insulated. Find and plot the distribution of dimensionless isotherms $\theta = 1; 0.7; 0.3; 0.1; 0.01; 0.001; 0.0001; 0.00001;$ and 0.000002.

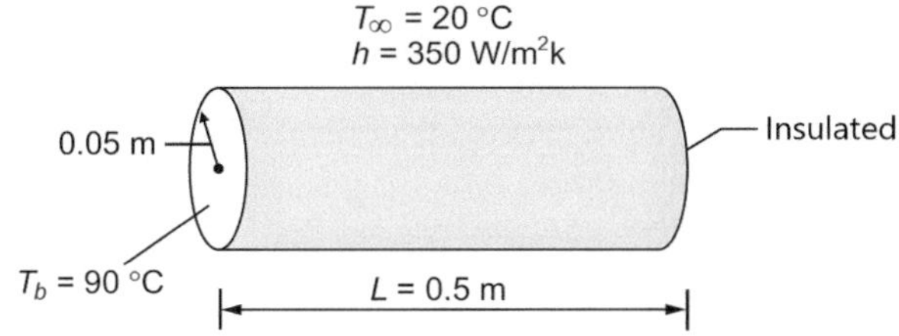

Solution

Initially, the eigenvalues must be evaluated. With the values from the example, Eq. (3.20) is expressed as: $\lambda_n J_1(0.1\lambda_n) = 13.06 J_0(0.1\lambda_n)$. To understand the behavior of this equation, see the following graph. In it, you can notice that the values λ where $z = 0$ correspond to the eigenvalues λ_n. This figure also indicates the eigenvalues of the exercise up to a limit of 700.

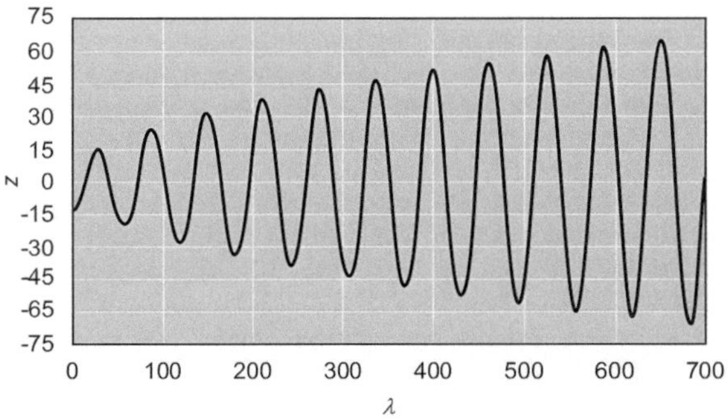

n	λ_n	n	λ_n
1	13.8773	13	385.0868
2	41.4843	14	416.4845
3	71.9736	15	447.8848
4	103.0036	16	479.2871
5	134.2105	17	510.6911
6	165.4957	18	542.0965
7	196.8223	19	573.503
8	228.1733	20	604.9105
9	259.54	21	636.3188
10	290.9173	22	667.7279
11	322.302	23	699.1375
12	353.6923		

Once known the eigenvalues, they are substituted into Eq. (3.19) - truncated in $n = 23$ for the sake of solving this problem, which results in:

$$\theta(r, z) = 261.19 \sum_{n=1}^{23} \frac{\cosh\left[\lambda_n\left(1 - \frac{z}{0.5}\right)\right] J_0\left(\frac{\lambda_n r}{0.5}\right)}{\cosh(\lambda_n) J_0(0.1\lambda_n)\left[\lambda_n^2 + 32.65\right]}$$

The example asks for the dimensionless isothermal lines, which can be found by substitution of appropriated values, for example $\theta = 0.7$ and so on, in the interval $0 \le r \le 0.05$ m and $0 \le z \le 0.5$ m. Graphical representation of several isothermal lines can be seen in following figure.

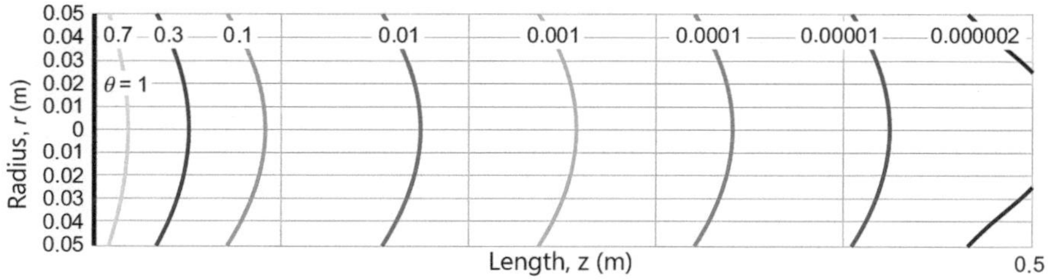

3.3 Any Boundary Condition and the Principle of Superposition of Solutions

Section 3.1 analyzed the case of a rectangular plate subjected to three boundary conditions held at the same temperature and one at a different temperature on its faces. Through an appropriate change of variables, the equation in the new dimensionless temperature variable θ made the three isothermal boundary conditions assumed zero (homogeneous), and the fourth boundary condition became $\theta = 1$. By expanding the fourth boundary condition, $f(x) = 1$ in a sine series, it was possible to find the coefficients of the infinite series expansion of the solution given by Eq. (3.14).

Using the theory of orthogonal functions, any function can be expanded into an infinite series of a given class of orthogonal functions in a given interval, such as the trigonometric functions sine and cosine. Thus, it is possible to solve the problem illustrated in Fig. 3.4, in which one of the faces has the dimensionless temperature described by a function $f(x) = \theta(x, b)$, which corresponds to the boundary condition BC_4, that is, any temperature distribution on the upper face with $y = b$.

The sine series expansion of the function $f(x)$ that describes the boundary condition is given by:

$$\sum_{n=1}^{\infty} A_n \sin\left(\frac{n\pi x}{L}\right) = f(x) \tag{3.23a}$$

where the expansion constants are given by:

$$A_n = \frac{2}{L}\int_0^L f(x)\sin\left(\frac{n\pi x}{L}\right)dx \tag{3.23b}$$

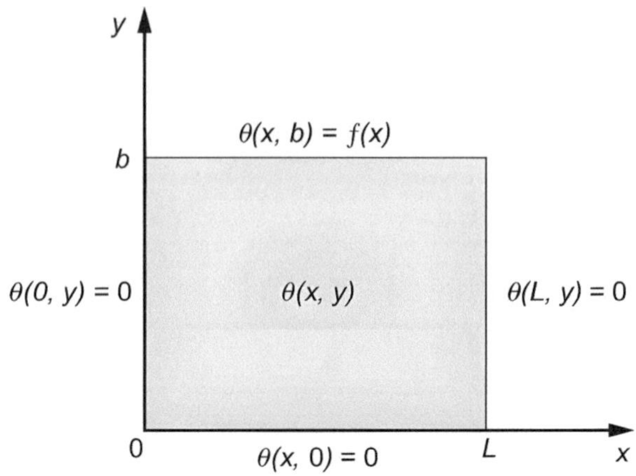

Fig. 3.4 Boundary condition for any distribution of temperature $f(x) = \theta(x, b)$

Comparing with Eq. (3.9), which comes from the fourth boundary condition, we have:

$$\sum_{n=1}^{\infty} C_n \sin\left(\frac{n\pi x}{L}\right)\sinh\left(\frac{n\pi b}{L}\right) = f(x) \tag{3.23c}$$

Comparing the previous equation with Eqs. (3.23a) and (3.23b), we get the final distribution of temperatures $\theta(x, y)$:

$$\theta(x, y) = \frac{2}{L}\sum_{n=1}^{\infty} \sin\left(\frac{n\pi x}{L}\right)\frac{\sinh\left(\frac{n\pi y}{L}\right)}{\sinh\left(\frac{n\pi b}{L}\right)}\int_0^L f(x)\sin\left(\frac{n\pi x}{L}\right)dx \tag{3.24}$$

The Laplace equation that governs two-dimensional steady-state conduction is linear. With this, it is possible to establish the solution of a case with more complex boundary conditions and combine the individual results. This is called the *principle of superposition.* To illustrate the method, consider the case of a rectangular plate $L \times b$ with the following boundary conditions prescribed on the four faces: $\phi(x, 0) = f_1(x)$; $\phi(0, y) = f_2(y)$; $\phi(x, b) = f_3(x)$; and $\phi(L, y) = f_4(y)$. Those boundary conditions are indicated in Fig. 3.5.

The final solution is given by the superposition of the four problems that, each individually, have three homogeneous boundary conditions (null). Note that each of the individual solutions of the cases illustrated in Fig. 3.5 is given by Eq. (3.24), taking care of swaping b and L whenever necessary. Therefore, each of the solutions ϕ_1, ϕ_2, ϕ_3, and ϕ_4 is determined. Finally, the solution of the original problem is given by the sum of the four solutions, that is:

$$\phi(x, y) = \phi_1(x, y) + \phi_2(x, y) + \phi_3(x, y) + \phi_4(x, y) \tag{3.25}$$

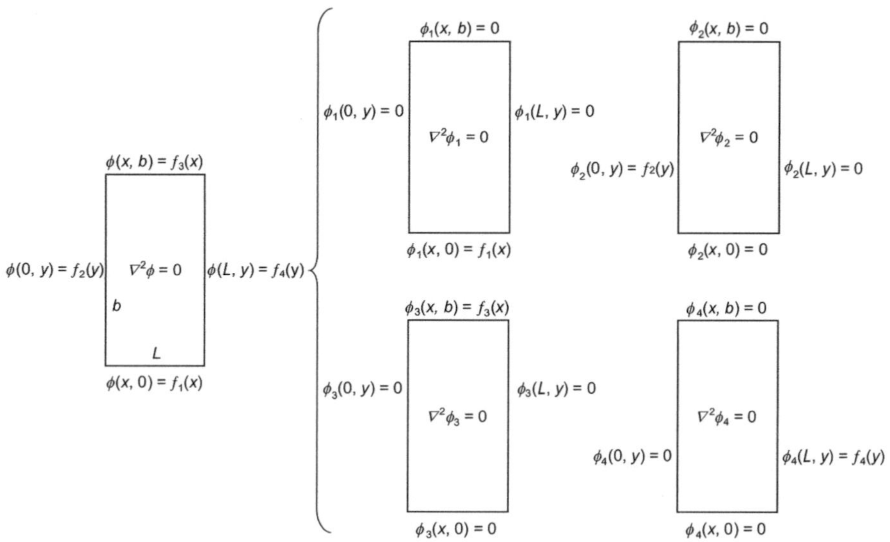

Fig. 3.5 Principle of superposition of solutions

Example 3.3 (*Solution by superposition*)

Consider a 1 m side square plate subject to the temperatures indicated in the figure. By using the principle of superposition of solutions, find the temperature at the plate center.

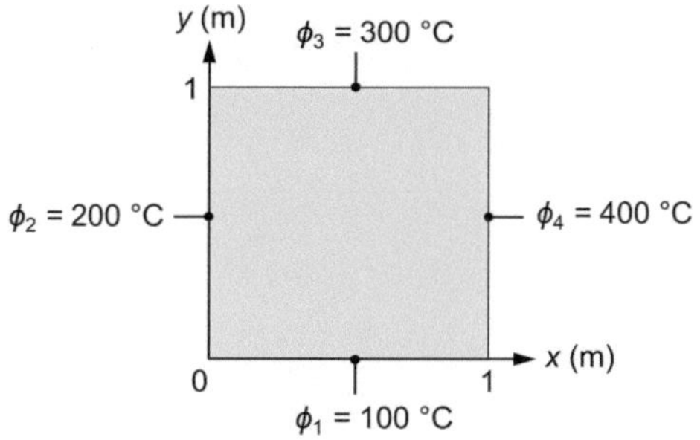

Solution

The original problem is split into four other subproblems as indicated below. Each one consists of the boundary condition of interest and three homogeneous boundary conditions as if it were an isolated problem.

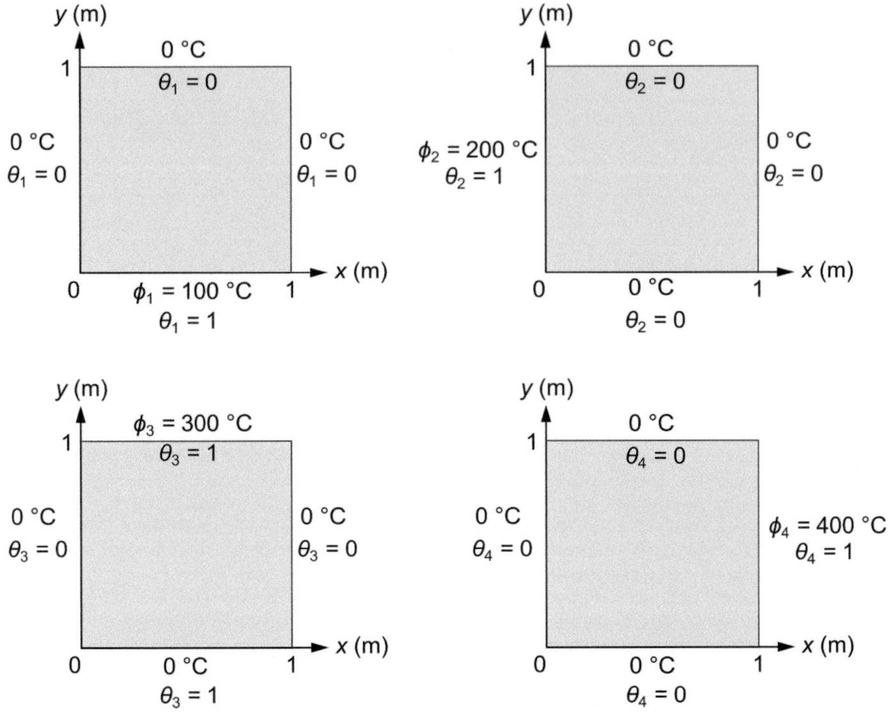

The following non-dimensional temperatures are defined:

$$\theta_1 = \frac{\phi_1}{100}, \theta_2 = \frac{\phi_2}{300}, \theta_3 = \frac{\phi_3}{300}, \theta_4 = \frac{\phi_4}{400}$$

For all cases $\theta_1 = \theta_2 = \theta_3 = \theta_4 = 0.25$ (see Example 3.1), therefore:

$$\phi_1 = 100 \times 0.25 = 25\,^\circ\mathrm{C}, \qquad \phi_2 = 200 \times 0.25 = 50\,^\circ\mathrm{C},$$

$$\phi_3 = 300 \times 0.25 = 75\,^\circ\mathrm{C}, \qquad \phi_4 = 400 \times 0.25 = 100\,^\circ\mathrm{C}$$

Finally, the temperature at the plate center is obtained

$$\phi = \phi_1 + \phi_2 + \phi_3 + \phi_4 = 250\,^\circ\mathrm{C}$$

3.4 Numerical Solution—Finite Differences

Analytical solutions of the heat conduction equation, analyzed in Sects. 3.1 to 3.3, in two-dimensional configurations can become quite complex, and, truly, in most cases of practical engineering problems, there are no analytical solutions. In this case, we resort to numerical solution methods. There is a certain variety of available methods, and here a relatively simple and intuitive method is examined which is known as *the finite differences method.*

The technique consists of dividing the original continuous solid into discrete points or *nodal points*, and applying an energy balance for each nodal point, as illustrated in Fig. 3.6. Thus, the original continuous medium (solid or surface) is transformed into a discrete system formed by a matrix of points with thermal properties that concentrate the "thermal information" of the original continuous medium at those nodal points. To illustrate the technique, consider the scheme of Fig. 3.6, where a generic nodal point (m, n) has as neighbors the nodal points $(m-1, n)$ to the left, $(m+1, n)$ to the right, $(m, n-1)$ below and $(m, n+1)$ above. The distance between the nodal points is Δx and Δy, in the two main directions.

The heat conduction equation in steady state, two-dimensional, without thermal energy (heat) generation, is given by Eq. (3.1). Thus, by discretizing the partial derivatives in each direction x and y, we obtain:

$\left.\frac{\partial T}{\partial x}\right|_{m-\frac{1}{2},n} \approx \frac{(T_{m,n}-T_{m-1,n})}{\Delta x}$ First derivative in the direction x—left face.

$\left.\frac{\partial T}{\partial x}\right|_{m+\frac{1}{2},n} \approx \frac{(T_{m+1,n}-T_{m,n})}{\Delta x}$ First derivative in the direction x—right face.

$\left.\frac{\partial^2 T}{\partial x^2}\right|_{m,n} \approx \frac{\left.\frac{\partial T}{\partial x}\right|_{m+\frac{1}{2},n} - \left.\frac{\partial T}{\partial x}\right|_{m-\frac{1}{2},n}}{\Delta x}$ Second derivative in the direction x—at the center.

Next, the first derivatives are replaced to obtain:

$$\left.\frac{\partial^2 T}{\partial x^2}\right|_{m,n} \approx \frac{T_{m-1,n} + T_{m+1,n} - 2T_{m,n}}{(\Delta x)^2} \tag{3.26a}$$

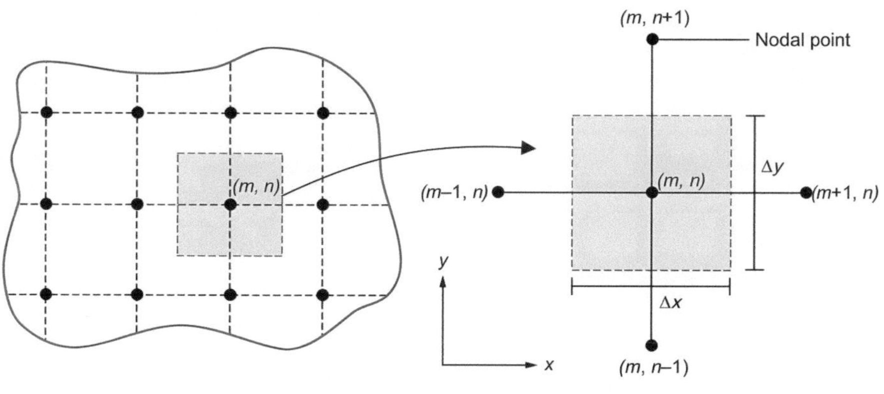

Fig. 3.6 Division of the continuous medium into a nodal network for a plane case

Similarly, in the y direction:

$$\left.\frac{\partial^2 T}{\partial y^2}\right|_{m,n} \approx \frac{T_{m,n-1} + T_{m,n+1} - 2T_{m,n}}{(\Delta y)^2} \tag{3.26b}$$

This technique is called "central difference". By substituting the discrete approximations of the nodal point (m, n) of the partial derivatives from the two previous equations into the original equation of differential heat conduction, Eq. (3.1), for the case where $\Delta x = \Delta y$, we get the following algebraic equation for the nodal point (m, n):

$$T_{m-1,n} + T_{m+1,n} + T_{m,n-1} + T_{m,n+1} - 4T_{m,n} = 0 \tag{3.27}$$

Equation (3.27) is the form of the heat equation in finite differences in cartesian coordinates and uniform mesh size ($\Delta x = \Delta y$). Note that the nodal temperature $T_{m,n}$ is the arithmetic mean of the four adjacent nodal points temperatures.

Analysis of boundary regions

The algebraic Eq. (3.27) is valid for an interior nodal point (m, n). The next step is to find the cases of the nodal points on the boundaries, that is, it is also necessary to discretize the boundary conditions. Let's start with the case of convection on the right face, as illustrated in Fig. 3.7. A nodal point (on the right) is located on the surface where convection occurs to the external environment kept at temperature T_∞.

When performing an energy balance for the nodal point in view, considering a unit of depth, the heat rates entering and leaving the nodal point (m, n) allow the following energy balance:

$$-k\Delta y\frac{\left(T_{m,n} - T_{m-1,n}\right)}{\Delta x} - k\frac{\Delta x}{2}\frac{\left(T_{m,n} - T_{m,n-1}\right)}{\Delta y} - k\frac{\Delta x}{2}\frac{\left(T_{m,n} - T_{m,n+1}\right)}{\Delta y}$$
$$= h\Delta y\left(T_{m,n} - T_\infty\right)$$

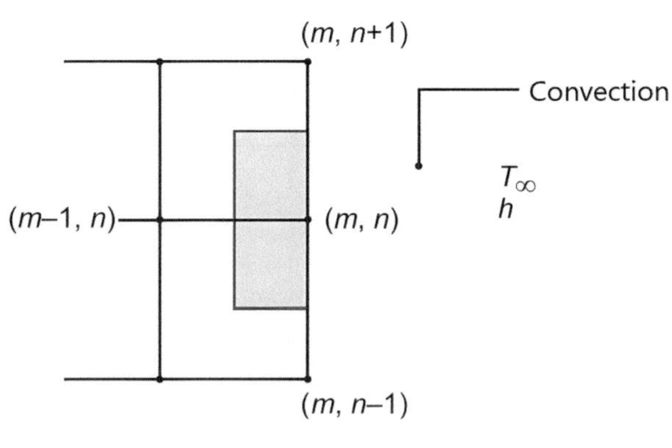

Fig. 3.7 Convective boundary condition

With the hypothesis that $\Delta x = \Delta y$. Thus, after rearranging, we obtain:

$$T_{m,n}\left(h\frac{\Delta x}{k} + 2\right) - \frac{1}{2}\left(2T_{m-1,n} + T_{m,n+1} + T_{m,n-1}\right) - h\frac{\Delta x}{k}T_\infty = 0 \tag{3.28}$$

For other boundary conditions, see Table 3.1.

Once all nodal points equations are collected, a system of N equations by N unknowns is obtained:

$$\begin{array}{llll} a_{11}T_1 & + a_{12}T_2 & + \cdots a_{1N}T_N & = c_1 \\ a_{21}T_1 & + a_{22}T_2 & + \cdots a_{2N}T_N & = c_2 \\ . & . & . & = . \\ . & . & . & = . \\ . & . & . & = . \\ a_{N1}T_1 & + a_{N1}T_2 & + \cdots a_{NN}T_N & = c_N \end{array}$$

or, in simplified matrix notation:

$$[A][T] = [C] \tag{3.29}$$

where

$$A = \begin{bmatrix} a_{11} & a_{12} & \ldots & a_{1N} \\ a_{21} & a_{22} & \ldots & a_{2N} \\ . & . & . & . \\ . & . & . & . \\ . & . & . & . \\ a_{N1} & a_{N2} & \ldots & a_{NN} \end{bmatrix}, \quad T = \begin{bmatrix} T_1 \\ T_2 \\ . \\ . \\ . \\ T_N \end{bmatrix}, \quad C = \begin{bmatrix} c_1 \\ c_2 \\ . \\ . \\ . \\ c_N \end{bmatrix}$$

If there is internal energy (heat) generation q_G''' it must be taken into account. An energy balance for an interior nodal point (m, n) will result in Eq. (3.30).

$$T_{m-1,n} + T_{m+1,n} + T_{m,n-1} + T_{m,n+1} - 4T_{m,n} + \frac{q_G'''(\Delta x)^2}{k} = 0 \tag{3.30}$$

The above equation is valid for an inner nodal point. For the remaining nodal points on the boundaries, such as those indicated in Table 3.1, an additional term for those listed nodal points equations must also be included.

There are several numerical techniques for solving a linear system of equations. There are many software, scientific calculators, and apps that solve a linear system of equations by using various techniques, including the Gaussian elimination method. In problems with many nodal points (tens of thousands), there are techniques that demand fewer RAM memory positions. It is suggested to refer to the literature on the subject for obtaining optimized solutions. The truncation error of this finite difference

Table 3.1 Finite differential equations for nodal points where $\Delta x = \Delta y$

Scheme	Equation
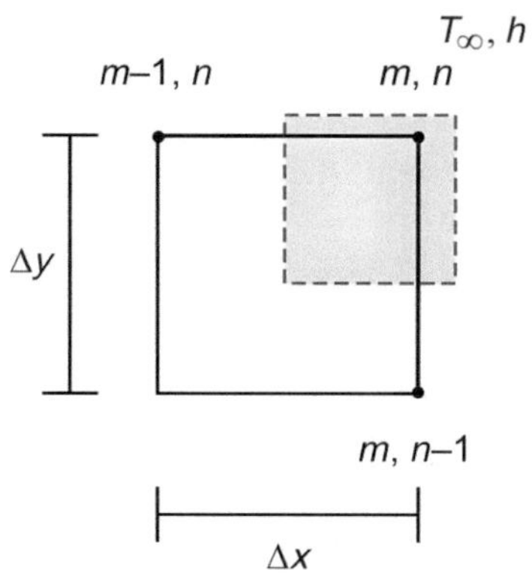	External corner with external convection $T_{m,n-1} + T_{m-1,n} - 2\left(\frac{h\Delta x}{k} + 1\right)T_{m,n} + \frac{2h\Delta x}{k}T_\infty = 0$
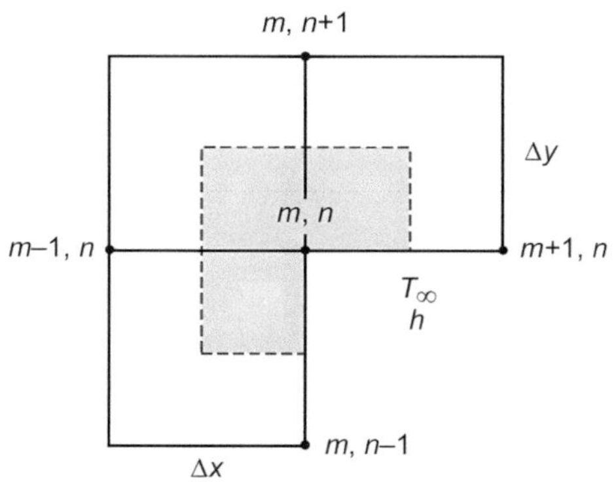	Internal corner with external convection $2(T_{m-1,n} + T_{m,n+1}) + (T_{m+1,n} + T_{m,n-1}) - 2\left(3 + \frac{h\Delta x}{k}\right)T_{m,n} + \frac{2h\Delta x}{k}T_\infty = 0$

(continued)

Table 3.1 (continued)

Scheme	Equation
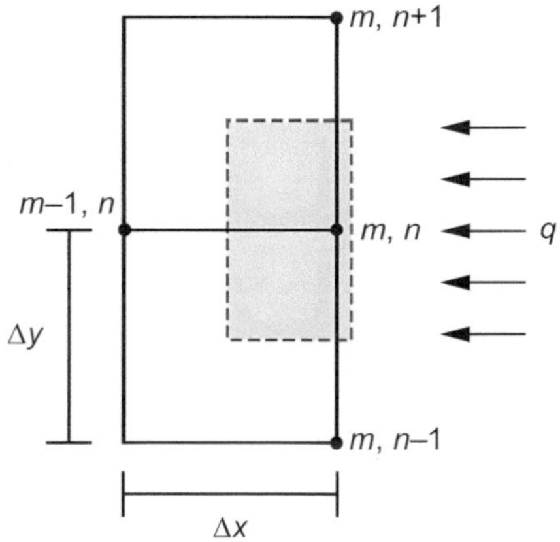	Flat part with constant heat flux $(2T_{m-1,n} + T_{m,n+1} + T_{m,n-1}) - 4T_{m,n} + \frac{2q\Delta x}{k} = 0$
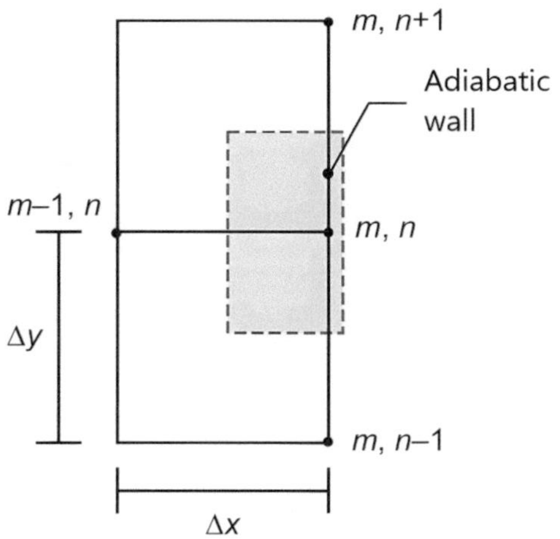	Adiabatic or symmetric flat part $(2T_{m-1,n} + T_{m,n+1} + T_{m,n-1}) - 4T_{m,n} = 0$

technique is in the order of $(\Delta x)^3$. Therefore, as the mesh size (Δx) decreases, the solution accuracy increases fast, but it demands more computational time. Thus, it is advised to carry out some mesh size tests and check not only the accuracy but also if the error increases, as a too small mesh size may increase the overall error because of computer round-off problems.

Example 3.4 (*Two-dimensional heat transfer in plates*)

A rectangular plate with thermal conductivity $k = 10$ W/m°C is subjected to the boundary conditions illustrated in the figure. Determine the temperature distribution at the indicated nodal points. The upper face exchanges heat by convection with a fluid at 20 °C and the heat transfer coefficient is 200 W/m^2°C, while the lateral and lower faces are at a constant temperature of 100 °C. Consider $\Delta x = \Delta y = 0.1$ m.

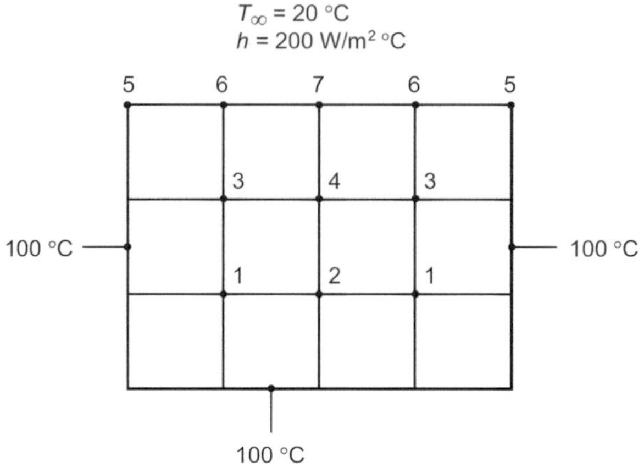

Solution

Note that because of the problem symmetry (boundary conditions and geometry), fewer nodal points are necessary.

For the internal nodal points (1–4) Equation (3.27) is valid:

$$T_{m-1,n} + T_{m+1,n} + T_{m,n-1} + T_{m,n+1} - 4T_{m,n} = 0$$

Therefore,

node 1: $-4T_1 + T_2 + T_3 + 2 \times 100 = 0.$
node 2: $2T_1 - 4T_2 + T_4 + 100 = 0.$
node 3: $T_1 - 4T_3 + T_4 + T_6 + 100 = 0.$
node 4: $T_2 + 2T_3 - 4T_4 + T_7 = 0.$
node 5: $T_5 = 100$ °C (known temperature).

At the nodal points with convection (6–7), the following equation applies:

$$T_{m,n}\left(h\frac{\Delta x}{k} + 2\right) - h\frac{\Delta x}{k}T_\infty - \frac{1}{2}\left[2T_{m,n-1} + T_{m-1,n} + T_{m+1,n}\right] = 0$$

Therefore,

$$\text{node 6}: T_6\left(\frac{200 \times 0.1}{10} + 2\right) - \frac{200 \times 0.1}{10}20 - \frac{1}{2}[2T_3 + 100 + T_7] = 0, \text{ or}$$

$$T_3 - 4T_6 + 0.5T_7 + 90 = 0$$

$$\text{node 7}: T_7\left(\frac{200 \times 0.1}{10} + 2\right) - \frac{200 \times 0.1}{10}20 - \frac{1}{2}[2T_3 + T_6 + T_6] = 0, \text{ or}$$

$$T_4 + T_6 - 4T_7 + 40 = 0$$

Combining all the nodal equations in a matrix form, we have the following system of linear equations:

$$\begin{pmatrix} -4 & 1 & 1 & 0 & 0 & 0 \\ 2 & -4 & 0 & 1 & 0 & 0 \\ 1 & 0 & -4 & 1 & 1 & 0 \\ 0 & 1 & 2 & -4 & 0 & 1 \\ 0 & 0 & 1 & 0 & -4 & 0.5 \\ 0 & 0 & 0 & 1 & 1 & -4 \end{pmatrix} \begin{pmatrix} T_1 \\ T_2 \\ T_3 \\ T_4 \\ T_6 \\ T_7 \end{pmatrix} = \begin{pmatrix} -200 \\ -100 \\ -100 \\ 0 \\ -90 \\ -40 \end{pmatrix}$$

The solution of the system by Gaussian elimination method results in:

$$\begin{aligned} T_1 &= 91.4\,^\circ\mathrm{C} \\ T_2 &= 88.3\,^\circ\mathrm{C} \\ T_3 &= 77.1\,^\circ\mathrm{C} \\ T_4 &= 70.5\,^\circ\mathrm{C} \\ T_5 &= 100\,^\circ\mathrm{C}\ \text{(prescribed)} \\ T_6 &= 46.7\,^\circ\mathrm{C} \\ T_7 &= 39.3\,^\circ\mathrm{C} \end{aligned}$$

Example 3.5 (*Two-dimensional heat transfer in a fin*)

A rectangular cross-section A304 stainless steel fin is subjected to an environment at 15 °C, as indicated in the figure. A fan promotes a 150 W/m^{2}K heat transfer coefficient between the fin and the environment. If the fin has a 100 °C temperature at the base, find the temperature distribution considering 20 nodal points, having the fin tip isolated. Also, the fin width is large enough to consider only the two-dimensional temperature distribution

.

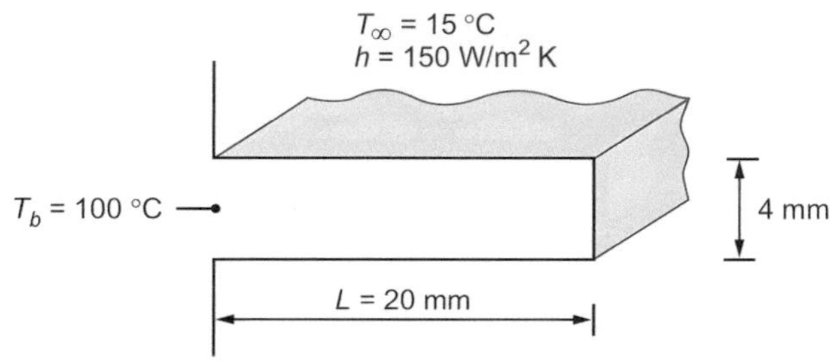

Solution

As the fin has nodal points that are subjected to the same boundary conditions on the upper and lower faces, it is possible to assign equal nodes by symmetry, as illustrated, with $\Delta x = 0.002$ m.

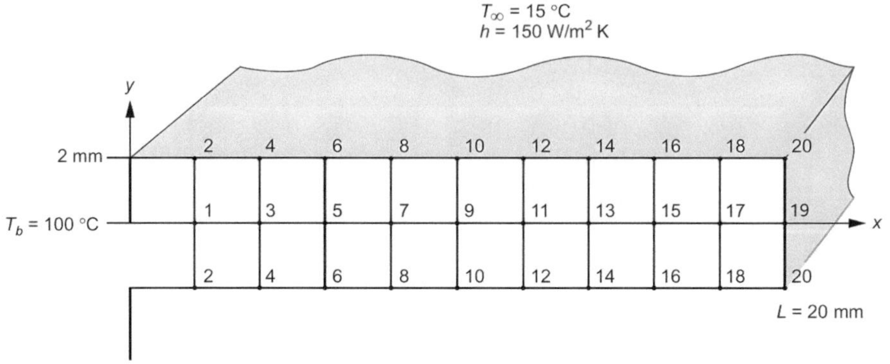

– For the internal nodes: 1, 3, 5, 7, 9, 11, 13, 15, 17, Eq. (3.27) is used, in which the equations are indicated in this order:

$$\begin{aligned}
100 + T_3 + T_2 + T_2 - 4T_1 &= 0 \\
T_1 + T_5 + T_4 + T_4 - 4T_3 &= 0 \\
T_3 + T_7 + T_6 + T_6 - 4T_5 &= 0 \\
T_5 + T_9 + T_8 + T_8 - 4T_7 &= 0 \\
T_7 + T_9 + T_8 + T_8 - 4T_7 &= 0 \\
T_9 + T_{13} + T_{12} + T_{12} - 4T_{11} &= 0 \\
T_{11} + T_{15} + T_{14} + T_{14} - 4T_{13} &= 0 \\
T_{13} + T_{17} + T_{16} + T_{16} - 4T_{15} &= 0 \\
T_{15} + T_{19} + T_{18} + T_{18} - 4T_{17} &= 0
\end{aligned}$$

– For the external nodes with convective losses: 2, 4, 6, 8, 10, 12, 14, 16, 18, Eq. (3.28) is used, according to the equations indicated below. The thermal conductivity of A304 is $k = 14.9$ W/mK (Table A.2).

$$T_2[150 \times 0.002/14.9 + 2] - 0.5(2T_1 + 100 + T_4) - [150 \times 0.002/14.9] \times 15 = 0$$

$$-T_1 + 2.020T_2 - 0.5T_4 = 50.302$$

$$T_4[150 \times 0,002/14.9 + 2] - 0.5(2T_3 + T_2 + T_6) - [150 \times 0.002/14.9] \times 15 = 0$$

$$-0.5T_2 - T_3 + 2.020T_4 - 0.5T_6 = 0.302$$

Similarly for the remaining nodal points:

$$-0.5T_4 - T_5 + 2.020T_6 - 0.5T_8 = 0.302$$

$$-0.5T_6 - T_7 + 2.020T_8 - 0.5T_{10} = 0.302$$

$$-0.5T_8 - T_9 + 2.020T_{10} - 0.5T_{12} = 0.302$$

$$-0.5T_{10} - T_{11} + 2.020T_{12} - 0.5T_{14} = 0.302$$

$$-0.5T_{12} - T_{13} + 2.020T_{14} - 0.5T_{16} = 0.302$$

$$-0.5T_{14} - T_{15} + 2.020T_{16} - 0.5T_{18} = 0.302$$

$$-0.5T_{16} - T_{17} + 2.020T_{18} - 0.5T_{20} = 0.302$$

- Isolated central nodal point (19).

$$2T_{17} + T_{20} + T_{20} - 4T_{19} = 0$$

- For tip fin poits (20). This type of node has not yet been analyzed, so it is necessary to develop an equation that characterizes it.

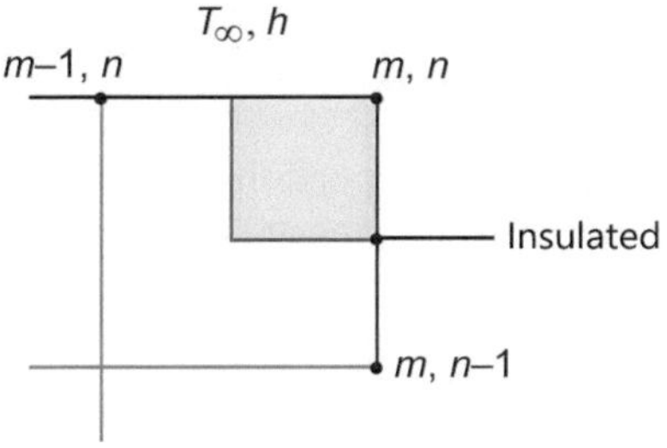

$$k\frac{\Delta y}{2}\frac{\left(T_{m-1,n} - T_{m,n}\right)}{\Delta x} + k\frac{\Delta x}{2}\frac{\left(T_{m,n-1} - T_{m,n}\right)}{\Delta y} + h\frac{\Delta x}{2}\left(T_\infty - T_{m,n}\right) = 0$$

If it is considered $\Delta x = \Delta y$, one obtains:

$$T_{m-1,n} + T_{m,n-1} - T_{m,n}\left(2 + h\frac{\Delta x}{k}\right) + h\frac{\Delta x}{2}T_{\infty} = 0$$

Substituting the values from the example:

$$T_{18} + T_{19} - T_{20}(2 + 150 \times 0.002/14.5) + (150 \times 0.002/14.5) \times 15 = 0$$

$$T_{18} + T_{19} - 2.020T_{20} = -0.302$$

Finally, combining all the equations in matrix form:

$$\begin{pmatrix}
-4 & 2 & 1 & 0 & 0 & 0 & 0 & 0 & 0 & 0 & 0 & 0 & 0 & 0 & 0 & 0 & 0 & 0 & 0 & 0 \\
1 & 0 & -4 & 2 & 1 & 0 & 0 & 0 & 0 & 0 & 0 & 0 & 0 & 0 & 0 & 0 & 0 & 0 & 0 & 0 \\
0 & 0 & 1 & 0 & -4 & 2 & 1 & 0 & 0 & 0 & 0 & 0 & 0 & 0 & 0 & 0 & 0 & 0 & 0 & 0 \\
0 & 0 & 0 & 0 & 1 & 0 & -4 & 2 & 1 & 0 & 0 & 0 & 0 & 0 & 0 & 0 & 0 & 0 & 0 & 0 \\
0 & 0 & 0 & 0 & 0 & 0 & 1 & 0 & -4 & 2 & 1 & 0 & 0 & 0 & 0 & 0 & 0 & 0 & 0 & 0 \\
0 & 0 & 0 & 0 & 0 & 0 & 0 & 0 & 1 & 0 & -4 & 2 & 1 & 0 & 0 & 0 & 0 & 0 & 0 & 0 \\
0 & 0 & 0 & 0 & 0 & 0 & 0 & 0 & 0 & 0 & 1 & 0 & -4 & 2 & 1 & 0 & 0 & 0 & 0 & 0 \\
0 & 0 & 0 & 0 & 0 & 0 & 0 & 0 & 0 & 0 & 0 & 0 & 1 & 0 & -4 & 2 & 1 & 0 & 0 & 0 \\
0 & 0 & 0 & 0 & 0 & 0 & 0 & 0 & 0 & 0 & 0 & 0 & 0 & 0 & 1 & 0 & -4 & 2 & 1 & 0 \\
-1 & 2.02 & 0 & -0.5 & 0 & 0 & 0 & 0 & 0 & 0 & 0 & 0 & 0 & 0 & 0 & 0 & 0 & 0 & 0 & 0 \\
0 & -0.5 & -1 & 2.02 & 0 & -0.5 & 0 & 0 & 0 & 0 & 0 & 0 & 0 & 0 & 0 & 0 & 0 & 0 & 0 & 0 \\
0 & 0 & 0 & -0.5 & -1 & 2.02 & 0 & -0.5 & 0 & 0 & 0 & 0 & 0 & 0 & 0 & 0 & 0 & 0 & 0 & 0 \\
0 & 0 & 0 & 0 & 0 & -0.5 & -1 & 2.02 & 0 & -0.5 & 0 & 0 & 0 & 0 & 0 & 0 & 0 & 0 & 0 & 0 \\
0 & 0 & 0 & 0 & 0 & 0 & 0 & -0.5 & -1 & 2.02 & 0 & -0.5 & 0 & 0 & 0 & 0 & 0 & 0 & 0 & 0 \\
0 & 0 & 0 & 0 & 0 & 0 & 0 & 0 & 0 & -0.5 & -1 & 2.02 & 0 & -0.5 & 0 & 0 & 0 & 0 & 0 & 0 \\
0 & 0 & 0 & 0 & 0 & 0 & 0 & 0 & 0 & 0 & 0 & -0.5 & -1 & 2.02 & 0 & -0.5 & 0 & 0 & 0 & 0 \\
0 & 0 & 0 & 0 & 0 & 0 & 0 & 0 & 0 & 0 & 0 & 0 & 0 & -0.5 & -1 & 2.02 & 0 & -0.5 & 0 & 0 \\
0 & 0 & 0 & 0 & 0 & 0 & 0 & 0 & 0 & 0 & 0 & 0 & 0 & 0 & 0 & -0.5 & -1 & 2.02 & 0 & -0.5 \\
0 & 0 & 0 & 0 & 0 & 0 & 0 & 0 & 0 & 0 & 0 & 0 & 0 & 0 & 0 & 0 & 2 & 0 & -4 & 2 \\
0 & 0 & 0 & 0 & 0 & 0 & 0 & 0 & 0 & 0 & 0 & 0 & 0 & 0 & 0 & 0 & 0 & 1 & 1 & -2.02
\end{pmatrix}
\begin{pmatrix} T_1 \\ T_2 \\ T_3 \\ T_4 \\ T_5 \\ T_6 \\ T_7 \\ T_8 \\ T_9 \\ T_{10} \\ T_{11} \\ T_{12} \\ T_{13} \\ T_{14} \\ T_{15} \\ T_{16} \\ T_{17} \\ T_{18} \\ T_{19} \\ T_{20} \end{pmatrix}
=
\begin{pmatrix} -100 \\ 0 \\ 0 \\ 0 \\ 0 \\ 0 \\ 0 \\ 0 \\ 0 \\ 50.302 \\ 0.302 \\ 0.302 \\ 0.302 \\ 0.302 \\ 0.302 \\ 0.302 \\ 0.302 \\ 0.302 \\ 0 \\ -0.302 \end{pmatrix}$$

The solution of this system of linear equations results in the following temperature distribution (in °C), with their fin positions indicated in parentheses.

100 (b)	89.8 (2)	81.5 (4)	74.5 (6)	68.7 (8)	64.0 (10)	60.3 (12)	57.4 (14)	55.4 (16)	54.3 (18)	53.9 (20)
100 (b)	90.5 (1)	82.1 (3)	75.1 (5)	69.2 (7)	64.5 (9)	60.7 (11)	57.9 (13)	55.9 (15)	54.7 (17)	54.3 (19)

3.5 Conduction Shape Factor

In Sect. 3.4 a numerical technique was presented to solve geometrical complex problems coupled with varied boundary conditions. The finite difference technique is well suited to solve such heat conduction problems. Nevertheless, there are situations of two-dimensional conduction heat transfer between a body and a surrounding medium that allow a simple analysis involving the concept of conduction shape factor, S. Both the body and the medium are isothermal. The rate of heat exchanged is given by

$$\dot{Q} = kS\Delta T_{total} \tag{3.31}$$

Conduction shape factors S per unit of depth is given in Table 3.2 for selected isothermal geometries. It is suggested to consult the work of Sunderland and Johnson (1964) who collected expressions for other conditions including convective boundaries.

Example 3.6 (*Conduction shape factor*)

Obtain the rate of heat lost per unit of length of a long pipe buried underground at a 1 m depth. The 100 mm pipe diameter transports vapor at 110 °C. The ground surface is at 10 °C. Assume the soil conductivity k is equal to 1.2 W/m °C.

Solution

First, let us obtain the shape factor for $L = 1$ m of tube length. For $d = 0.1$ m and $h = 1$ m, the shape factor S is shown in Table 3.2

$$S = \frac{2\pi L}{cosh^{-1}(2h/d)} = \frac{2\pi \times 1}{cosh^{-1}(2 \times 1/0.1)} = 1{,}7\text{m}$$

Therefore, from Eq. (3.31)

$$\dot{Q}' = kS\Delta T_{total} = 1.2 \times 1.7 \times (110 - 10) = 204\text{W/m}$$

Table 3.2 Conduction shape factors for selected isothermal geometries in an infinite medium of uniform conductivity k.

Configuration	Schematics	Shape factor
Sphere to an infinite plane	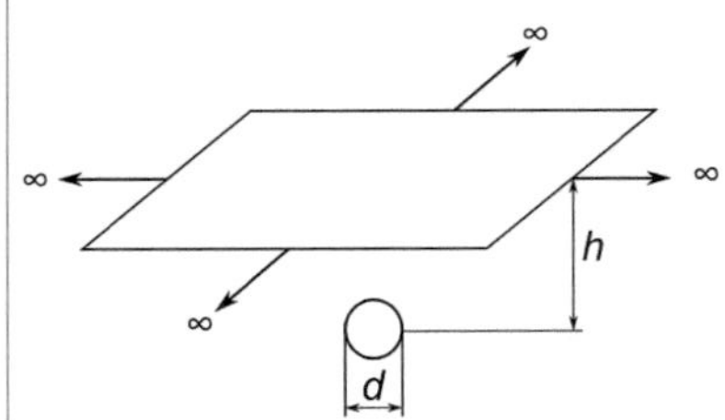	$S = \frac{2\pi d}{1+(d/4h)}$
Cylinder of length L parallel to a plane infinite width and length L ($d << h$; $h << L$)	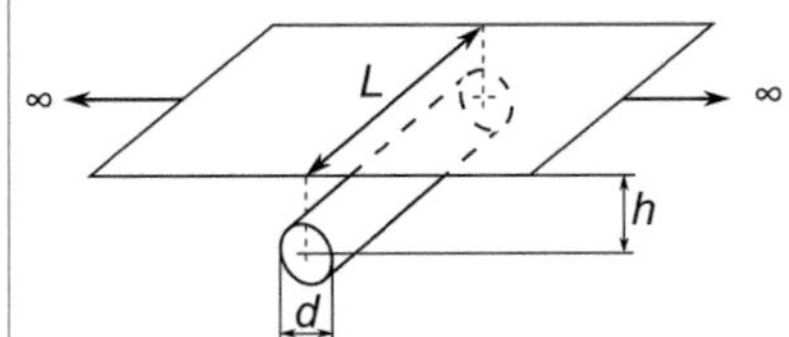	$S = \frac{2\pi L}{cosh^{-1}(2h/d)}$

(continued)

Table 3.2 (continued)

Configuration	Schematics	Shape factor
Cylinder of length L between parallel planes infinite width and length L	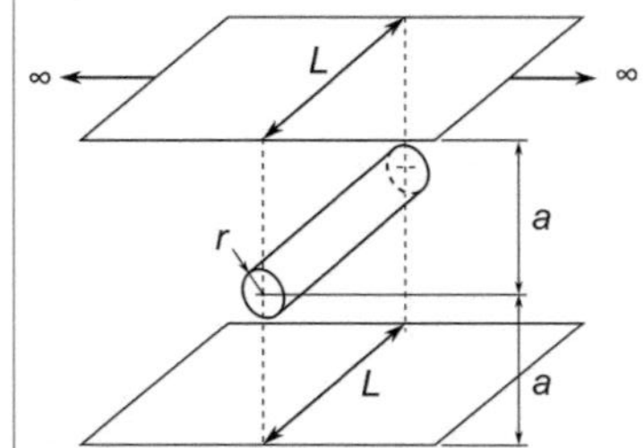	$S = \frac{2\pi L}{ln(4a/r)}$
Cylinder of length L perpendicular to an infinite plane ($d << L$)	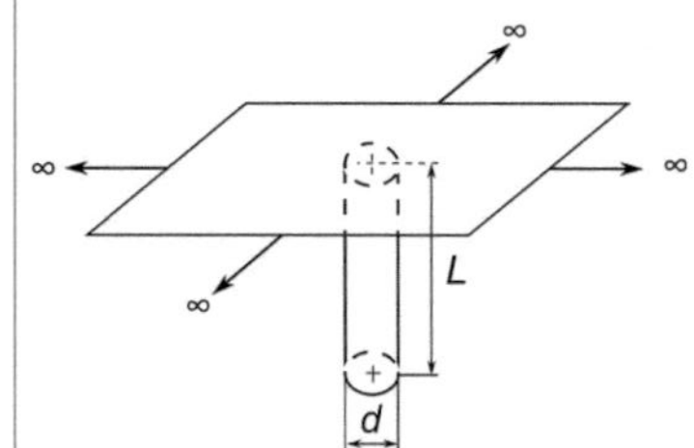	$S = \frac{2\pi L}{ln(4L/d)}$

(continued)

Table 3.2 (continued)

Configuration	Schematics	Shape factor
Circular disk parallel to an infinite plane ($d << h$)	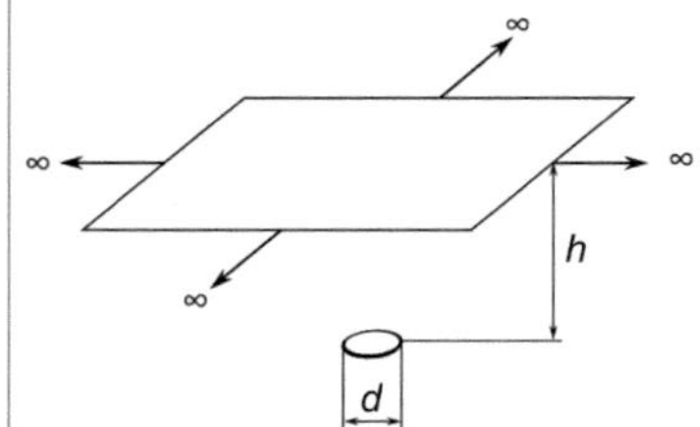	$S = \frac{4.44d}{1-(d/5.66h)}$
Parallel cylinders to an infinite plane ($L >> z$; $L >>$ radii)	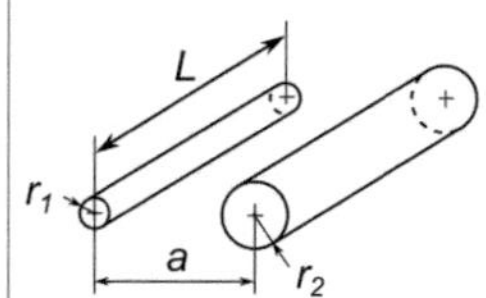	$S = \frac{2\pi L}{cosh^{-1}\frac{\left(a^2-r_1^2-r_2^2\right)}{2r_1r_2}}$
Cylinder of length L inside a square prism of side w ($L >> w$)	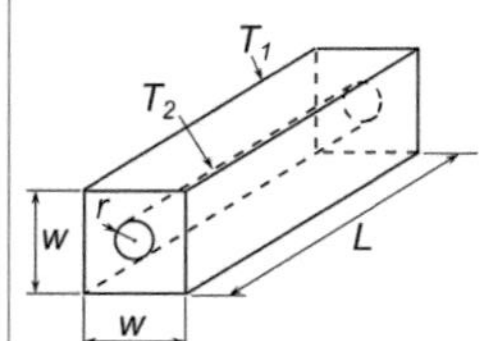	$S = \frac{2\pi L}{ln\left(0.637\frac{w}{r}-0.166\right)}$

Adapted from Sunderland and Johnson (1964)

3.6 Concluding Words

The Laplace equation governing heat conduction is a potential equation, which also represents many physical phenomena, such as the potential velocity in Fluid Mechanics and the electric potential in electricity, among others. Thus, it is possible to solve the problem of heat conduction in complex geometries by using analogy with other phenomena. For example, it used to be in the past a laboratory experiment to obtain the temperature distribution using a special type of aluminized sheet of paper by measuring the distribution of the electric potential on the sheet that was subjected at its edges to equivalent boundary conditions (established electric potential or insulation).

An old technique allows to obtain the shape factor by drawing isothermal lines for some relatively simple geometries, such as circles, squares, rectangles, spheres and triangles. Despite these techniques, numerical methods and the wide access to computers have to some extent made these techniques obsolete. On the other hand, shape factors shown in Table 3.2 are exact solutions to the problems specified in the table. Therefore, whenever possible, exact analytical solutions are preferable because they also become comparison standards for numerical solutions.

3.7 Proposed Problems

3.1 Find the temperature distribution for the rectangular section illustrated below.

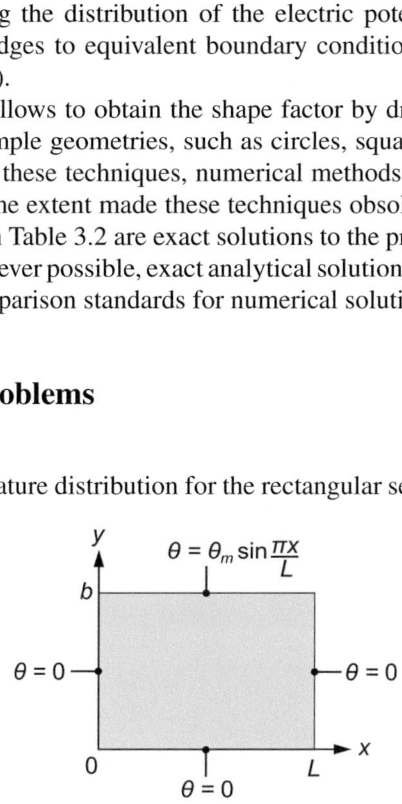

3.2 Determine the temperature at the center of a square plate, i.e., $b = L = 1$ m. The temperatures $T_1 = 100$ °C and $T_2 = 200$ °C as indicated.

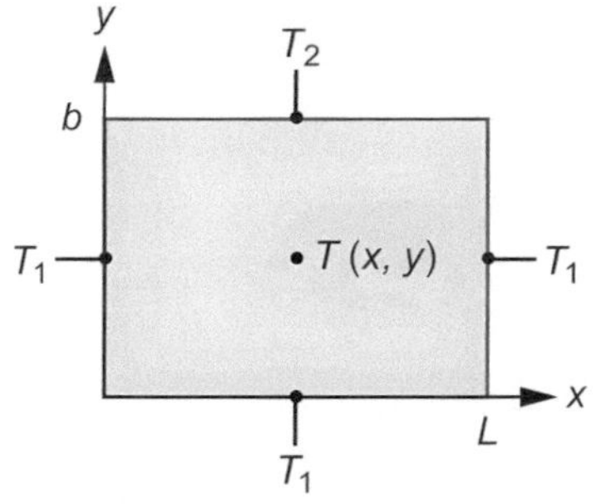

3.3 Determine the temperature at the center of a square plate, i.e., $b = L = 1$ m. The temperatures are $T_1 = 100$ °C and $T_3 = 300$ °C as indicated.

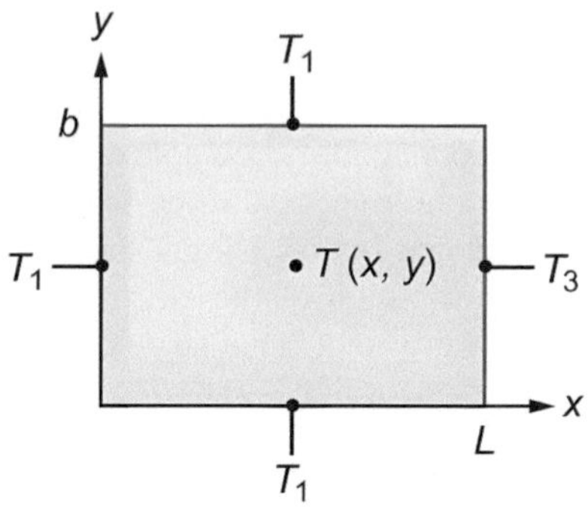

3.4 Determine the temperature at the center of a square plate, i.e., $b = L = 1$ m. The temperatures are $T_1 = 100$ °C, $T_2 = 200$ °C, and $T_3 = 300$ °C as indicated. (*Hint*: use the principle of superposition of solutions with the results of the previous problems—Sect. 3.3).

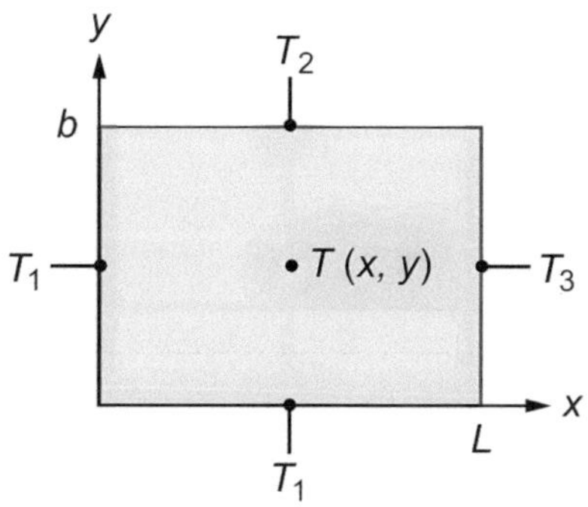

3.5 Redo the Example 3.1 for the point $x = 0.25$ m and $y = 0.5$ m. Compare with the solution of the symmetrical point 5.

3.6 Redo the proposed Problem 3.4 by numerical solution using the finite difference method for a 25 × 25 cm grid. Compare the results.

3.7 Develop the nodal equation for an interior nodal point for the case of a one-dimensional problem, such as a bar or fin, as illustrated. Consider the following cases: (a) thermally isolated surface; (b) surface exposed to heat convection to an environment at T_∞, coefficient h of heat transfer for an exposed area A.

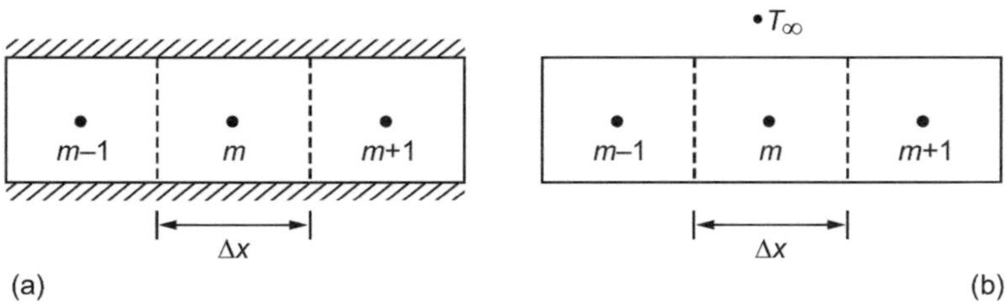

3.8 Redo the proposed problem 2.20 of the circular bar using a one-dimensional numerical solution, according to the discretization of the Problem 3.7, item (b).

3.9 Find the unknown temperatures (1 through 6) in the figure using the finite difference method for a regular mesh ($\Delta x = \Delta y$).

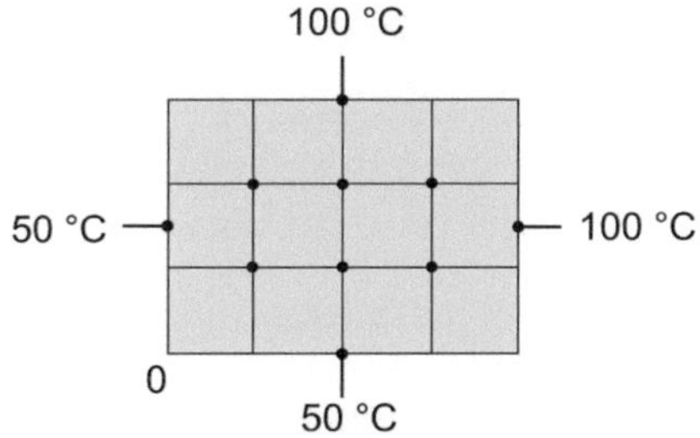

3.10 Derive the nodal equation for an interior nodal point (m, n) for a system with internal energy generation (heat), q'''_G, Equation (3.30).

3.11 Derive the nodal equations in Table 3.1 for a system with internal energy generation (heat), q'''_G.

3.12 Determine the nodal equation for a node (m, n) located on the 45° inclined surface exposed to convection with a fluid that is at T_∞ with a h heat transfer coefficient.

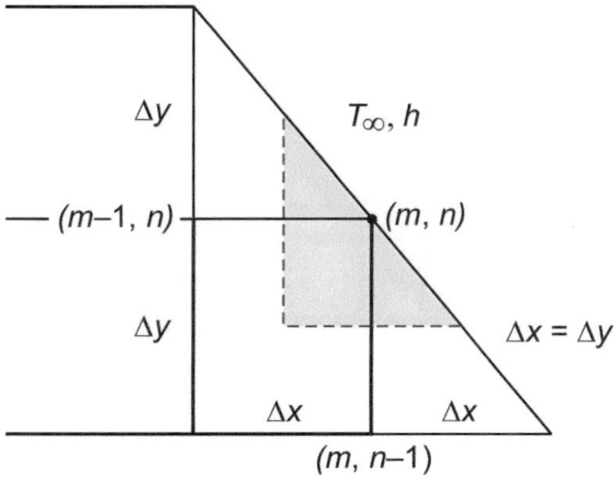

3.13 Determine the nodal equation for a nodal point (m, n) located in the upper right corner isolated at the top and with convective heat exchange to the right.

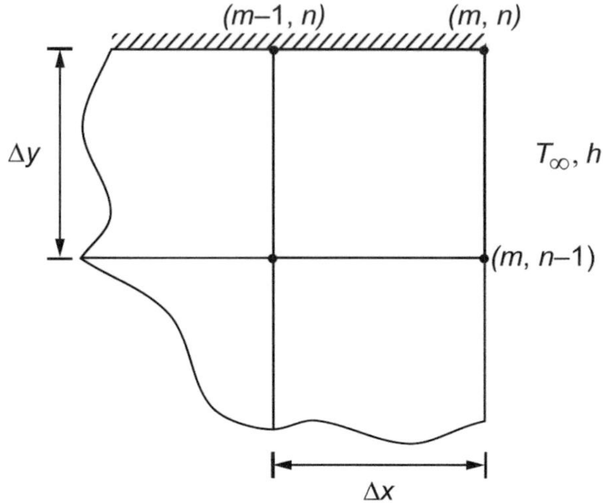

3.14 Find the nodal equation of an internal nodal point (m, n, p) in a three-dimensional system in cartesian coordinates.

3.15 Determine the unknown temperatures T_1, T_2, and T_3 in the middle of the square chimney shown in figure. Symmetric nodal points receive the same label. The external wall is at 50 °C and the internal wall is at 300 °C. The chimney is made of common brick, its dimensions are given in figure.

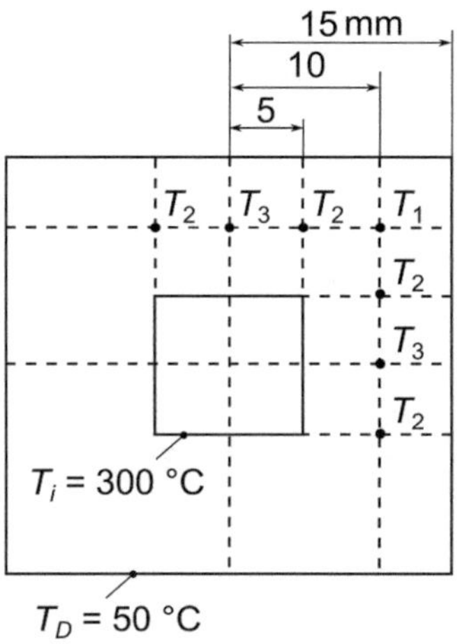

3.16 The boundary conditions have been changed for the chimney from problem 3.15. Now the external wall temperature is no longer known as the wall is exposed to natural convection, $h = 5$ W/m^2 °C, to air at $T_\infty = 20$ °C. Determine the unknown temperatures T_1 … T_6 of the square chimney. The remaining conditions are unchanged.

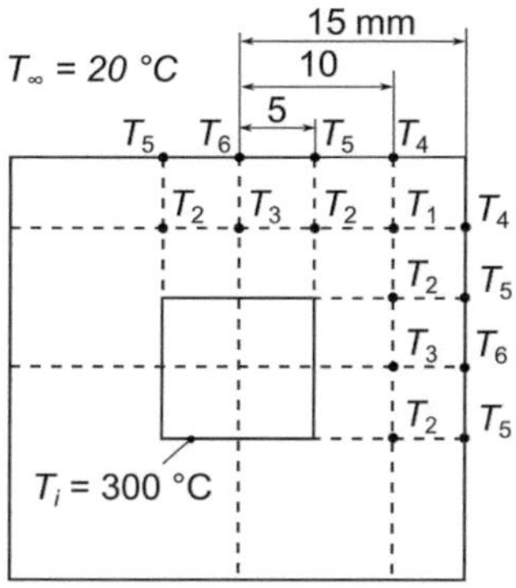

3.17 A steel pipe carrying water heated at 50 °C from a solar water heater is installed inside a square concrete column 40 cm on a side. The pipe is 5 cm in diameter and the surface of the concrete column is at 10 °C. Determine the rate of heat loss per unit length of the pipe.

References

Carslaw, H.S., and J.C. Jaeger. 1959. *Conduction of heat in solids*, 2nd ed. UK: Oxford University Press.

Sunderland, J.E., and K.R. Johnson. 1964. Shape factors for heat conduction through bodies with isothermal or convective boundary conditions. *Ashrae Transactions* 10.

Chapter 4
Transient Heat Conduction

Abstract This chapter is devoted to the study of transient heat conduction. The most elementary transient conduction problem is the case of a lumped system in which the internal resistance to heat conduction is negligible compared to the external convective resistance. An important dimensionless number is the Biot number defined by the ratio between internal and external thermal resistances. A lumped system approach is valid for a low Biot number, more specifically for $Bi < 0.1$. Next, the solution to the one-dimensional transient heat conduction equation is developed for a semi-infinite solid. In this case, three boundary conditions are studied at the exposed face, namely, (a) constant temperature, (b) constant heat flux, and (c) heat convection, which result in analytical solutions based on a similarity variable. For plate, cylinder, and sphere geometries exposed to heat convection, the solutions of transient conduction follow Heisler's diagrams. Finally, the chapter analyzes the finite difference method applied to the transient problems.

Keywords Transient conduction · Concentrated system · Semi-infinite solid · Heisler · Finite differences

As a body is cooled, or heated, it undergoes a heat transfer process in which time becomes an important variable for obtaining the instantaneous temperature as well as the time rate of the transferred heat process. In the previous chapters, the so-called *steady-state* or *stationary* cases were studied resulting in a spatial temperature distribution within the body independent of time. In cases where time becomes an important variable, the heat transfer process is called *transient* or *transitory*, which is the object of analysis in this chapter. To illustrate the analysis, consider a slab that receives thermal radiation $\dot{Q}$ on its left surface as shown in Fig. 4.1a. In a steady state, the slab faces temperature reach T_1 and T_2 having the environment surrounding temperature T_∞. As soon as the radiation incidence ceases, the temperature in each region within the slab decays until the thermal equilibrium with the environment is achieved when the slab reaches a uniform temperature T_∞ in its interior. Until this final equilibrium is reached, the temperature distribution within the slab will

J. R. Simões-Moreira and E. W. Zavaleta-Aguilar, *Heat Transfer*, Heat and Mass Transfer, https://doi.org/10.1007/978-3-031-80318-5_4

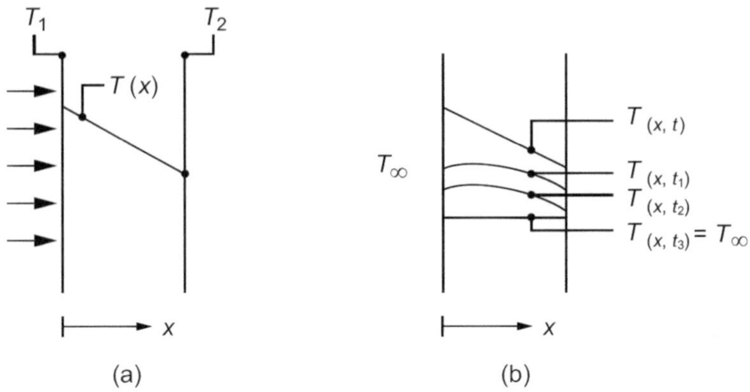

Fig. 4.1 One-dimensional heat transfer process without thermal energy (heat) generation in a slab **a** in steady state and **b** in transient regime

depend on time and, thus, the heat transfer process will occur in a transient regime, as illustrated by the cooling curves for various times t_1, t_2,... t_n in Fig. 4.1b.

In the example Fig. 4.1b, the temperature distribution within the slab varies both in x position and in time t that is, $T(x, t)$ and the analytical solution will be the subject of study in Sect. 4.2. However, there are situations where the temperature dependence on time is more relevant because the instantaneous spatial variation of the temperature within the body is negligible and, therefore, the body's instantaneous temperature is uniform. Therefore, that is the case when time analysis only is sufficient. Such systems are called *lumped systems*, analyzed in the next section.

4.1 Lumped Systems

As a body or system at a given temperature is subjected to a temperature change, for example, by its exposition to a new environment at a different temperature, some time will be necessary until a thermal equilibrium between that body and the environment is established. Practical examples are industrial heating or cooling processes, metals heat treatment, products placed in or removed from a refrigerator, and materials within ovens, among many others. Consider that the body of the scheme in Fig. 4.2 is initially at a uniform temperature T_0 in equilibrium with the environment at $T_{\infty 1}$. Suddenly, the body is removed from that environment and exposed to a second environment at a higher temperature $T_{\infty 2}$ ($T_{\infty 2} > T_{\infty 1} = T_0$). An attempt to illustrate the body heating process is shown in the time graph in Fig. 4.2. Therefore, there will be a heating process and the expected heating curve in that figure is, in a way, intuitive to most people, based on their own personal experience.

An accurate analysis of the body heating process in Fig. 4.2 will indicate that the body heating process may not occur uniformly throughout its interior, with some

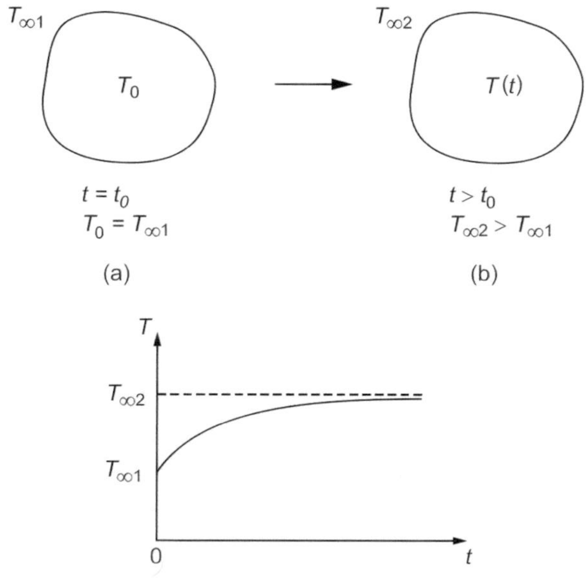

Fig. 4.2 Temperature evolution of a body heated up

regions closer to its surface experiencing a more pronounced temperature change over time, while inner regions experience a less pronounced time change as illustrated in Fig. 4.3, where the central core temperature T_c, and at any position near the surface, T_s vary differently. Therefore, the instantaneous heating process does not occur uniformly. A more precise analysis involves the problem of internal heat conduction and becomes a bit laborious from a mathematical point of view, nevertheless, it can be solved for some cases of simple geometries and boundary conditions, as will be seen in Sect. 4.2. More complex cases must be solved numerically.

In this section, the interest is to solve the transient problem assuming a simplifying hypothesis that works for a large number of practical cases. This hypothesis consists of assuming that the whole body has an instantaneous uniform temperature, being the body temperature dependent only on time, as illustrated by the dashed line in Fig. 4.3.

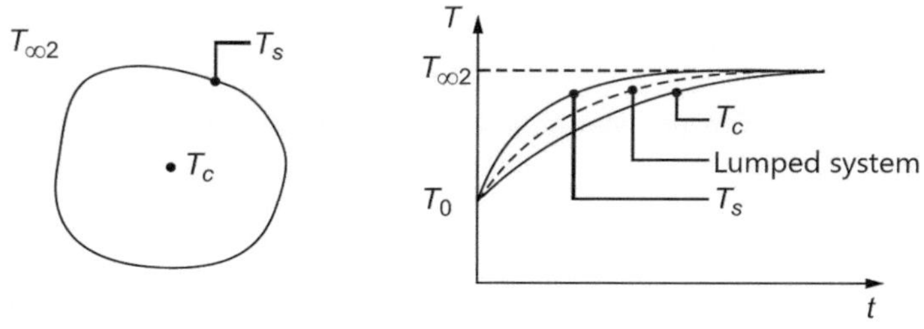

Fig. 4.3 Temperature at the surface and at the body center in the process of heating and its approximation to a lumped system (dotted curve)

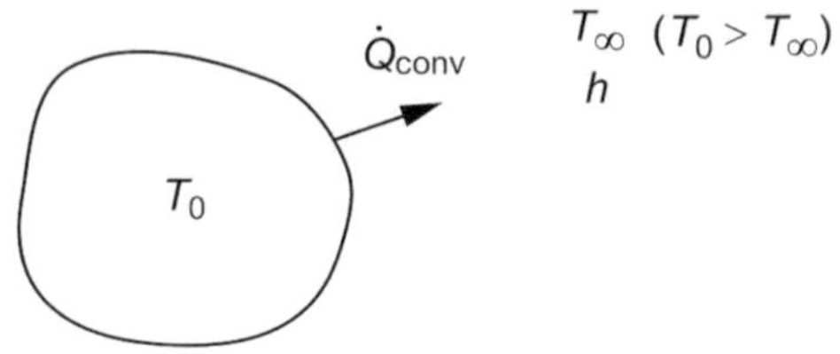

Fig. 4.4 Lumped system heat exchange in a transient regime

That is the *lumped system* hypothesis. Therefore, at any time t the whole body is at the same uniform temperature, $T(t)$. That situation occurs for systems (bodies) having their internal resistance to heat conduction negligible compared to the external heat exchange resistance, which usually occurs by convection; A dimensionless number gives the ratio between those resistances called the *Biot number* (Eq. 4.6) is analyzed below. In the scheme illustrated in Fig. 4.4, a body is at an initial temperature T_0 and is suddenly exposed to a lower temperature environment T_∞ and a convective heat transfer, $\dot{Q}_{\text{conv}}$, will take place on its surface.

The following analysis assumes that there is no thermal energy (heat) generation within the body, neglecting any manifestation of kinetic and potential energy, as well as considering constant transport properties. An energy balance scheme of the body in Fig. 4.4 is the following:

$$\underbrace{\begin{Bmatrix}\text{Rate of change}\\ \text{of body internal energy}\end{Bmatrix}}_{\text{I}} = \underbrace{\begin{Bmatrix}\text{Heat transfer}\\ \text{rate by}\\ \text{convection}\end{Bmatrix}}_{\text{II}}$$

In which:

$$\text{Term (I)}:\ \frac{dU}{dt} = m\frac{du}{dt} = \rho\forall\frac{du}{dt} = \rho C\forall\frac{dT}{dt}.$$

where m is the body mass, U is the body's internal energy, u is the body's specific internal energy, ρ is the body density, $\forall$ is the body volume, and C is the body's specific heat.

$$\text{Term (II)}:\ \dot{Q}_{\text{conv}} = -hA(T - T_\infty).$$

where h is the heat transfer coefficient by convection between the body and the surrounding fluid, A is the body surface area in contact with the fluid, T is the instantaneous and uniform body temperature $T = T(t)$ and T_∞ is the temperature of the far field fluid. Thus, by the above energy balance scheme, we have:

$$\rho C\forall\frac{dT}{dt} = -hA(T - T_\infty) \tag{4.1}$$

This equation is a first-order ordinary differential equation with constant coefficients. Separating the temperature and time variables for integration by parts, we get:

$$\frac{dT}{T - T_\infty} = -\frac{hA}{\rho C\forall}dt$$

For simplicity, consider $\theta = T - T_\infty$ and therefore, $d\theta = dT$, so:

$$\frac{d\theta}{\theta} = -\frac{hA}{\rho C\forall}dt$$

Carrying out the integration from its initial condition, time $t = 0$, when the body temperature is T_0 (or $\theta_0 = T_0 - T_\infty$), up to any time t, when the body will have a temperature $T(t)$ or $\theta(t) = T(t) - T_\infty$, it comes:

$$\int_{\theta_0}^{\theta} \frac{d\theta}{\theta} = -\frac{hA}{\rho C\forall}\int_0^t dt$$

which results in:

$$\ln\left(\frac{\theta}{\theta_0}\right) = -\frac{hA}{\rho C\forall}t \quad \text{or} \quad \frac{\theta}{\theta_0} = e^{-\frac{hA}{\rho C\forall}t}$$

Finally, substituting the original variables, one obtains:

$$\frac{T(t) - T_\infty}{T_0 - T_\infty} = e^{-\frac{hA}{\rho C\forall}t} \tag{4.2}$$

The equation represents a temperature exponential decay. This type of equation occurs in various physical systems, including the field of electricity. There is a perfect analogy between the transient thermal problem under analysis and the case of a charging and discharging a capacitor, as illustrated in Fig. 4.5.

Referring to Fig. 4.5, initially, the capacitor of capacitance C is charged to an electric voltage V_0 (switch *sw* closed). Then, the switch *sw* is opened and the capacitor begins to discharge through the resistance R_e. The solution of this parallel RC circuit is:

$$\frac{V}{V_0} = e^{-\frac{t}{R_e C}} \tag{4.3}$$

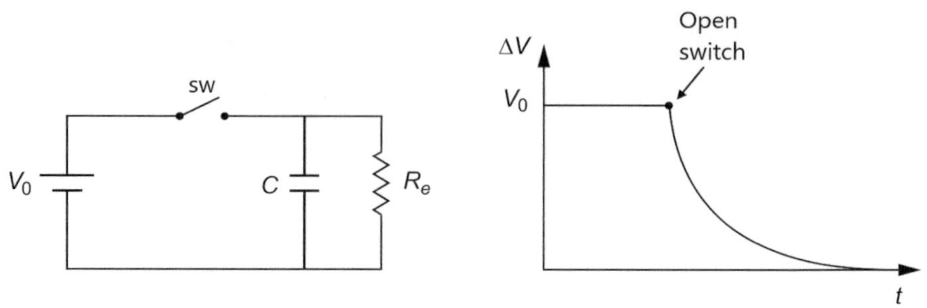

Fig. 4.5 Analogy with a charge/discharge of a capacitor

Table 4.1 Analogous variables in the thermal and electrical case

Electrical	Thermal
Voltage, V	$T - T_\infty$
Capacitance, C	$\rho C \forall$
Ohmic resistance, R_e	$1/hA$

Compare Eqs. (4.2) and (4.3). The analogy between the magnitudes from the transient thermal problem to the electrical ones is indicated in Table 4.1.

The equivalent thermal circuit is indicated in Fig. 4.6.

In electricity, it is usual to define the *time constant*, τ, of the circuit which is given by $\tau = RC$. The time constant is a very practical quantity to indicate how "quickly" the capacitor charges or discharges. The value of time at which $t = \tau$ represents the time when the capacitor voltage reached the value of $e^{-1} \sim 0.368$, as shown below:

$$\frac{V}{V_0} = e^{-\frac{\tau}{\tau}} = e^{-1} = \frac{1}{e} = 0.368$$

With that in mind, a very interesting analysis can be made, as illustrated in the graph in Fig. 4.7, which indicates the discharge of the capacitor for different time constants. The larger the time constant ($\tau_3 > \tau_2 > \tau_1$), the slower the voltage across the capacitor reaches the value of $0.368V_0$.

Fig. 4.6 Analog thermal circuit

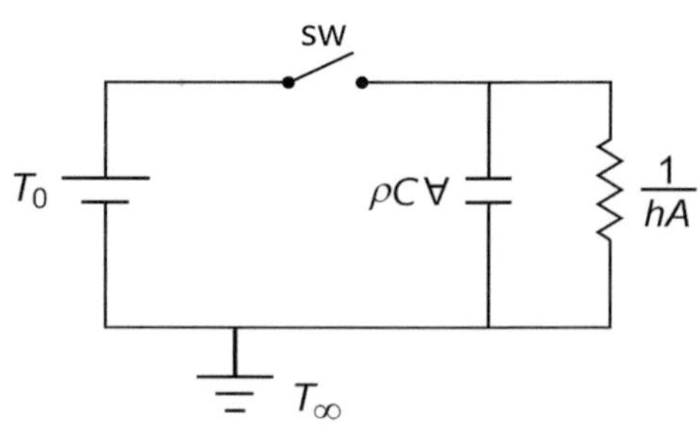

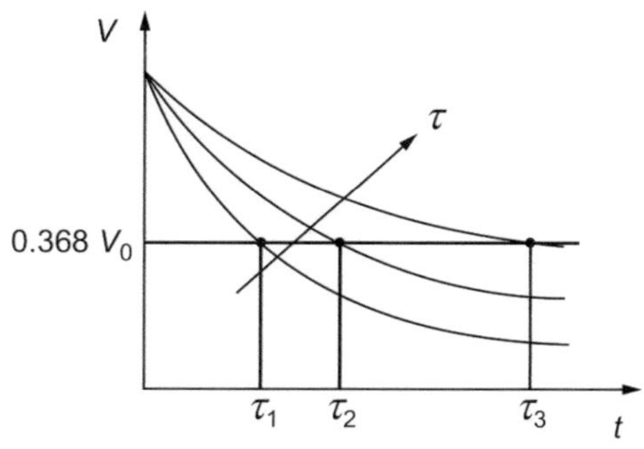

Fig. 4.7 Time constant influence over the discharge of a capacitor

By analogy, the *thermal time constant* τ_t can be defined accordingly

$$\frac{T - T_\infty}{T_0 - T_\infty} = e^{-\frac{hA}{\rho C \forall} t} = e^{-\frac{t}{\tau_t}}$$

From which the thermal time constant is obtained

$$\tau_t = \frac{\rho C \forall}{hA} \tag{4.4}$$

By inspecting the graph in Fig. 4.8, one can see the influence of the thermal time constant over the cooling (or heating) process by analyzing the temperature variation in time of a lumped system.

A relevant example of the thermal time constant is the case of temperature measurement with thermocouple sensors. Thermocouples consist of two wire tips making a soldered or fused junction. This junction is exposed to the environment for measuring the temperature. Illustratively, suppose an environment where the temperature undergoes a square wave variation as illustrated in solid lines in the scheme in Fig. 4.9, that is, its temperature periodically assumes either $T_{\infty 1}$ or $T_{\infty 2}$ (square wave). Ideally, it is demanded to select a sensor that tracks the temperature variation as closely as possible to accurately indicate the temperature-time variation. Graphs

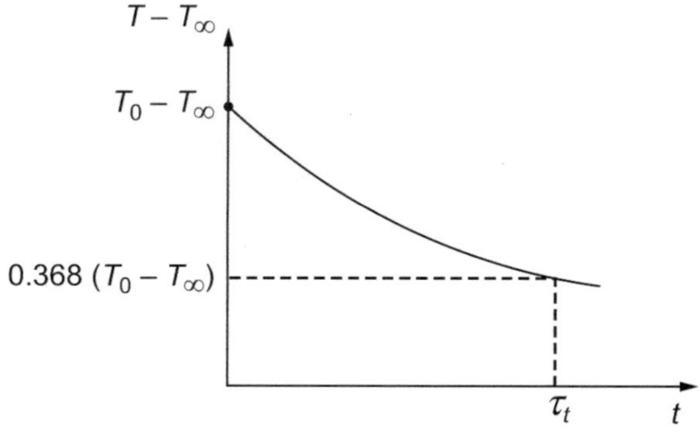

Fig. 4.8 Influence of the thermal time constant on the cooling of a lumped system

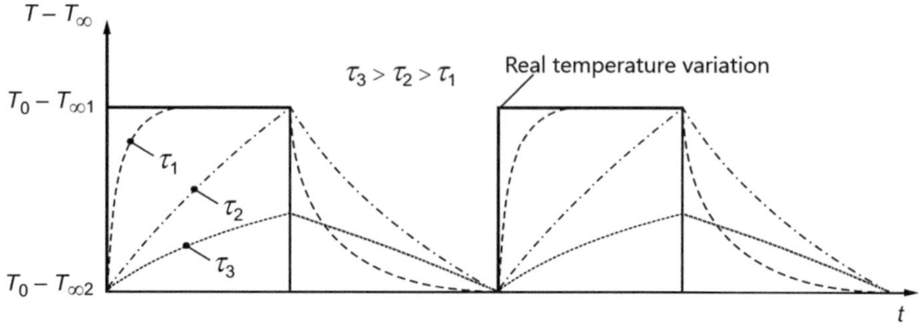

Fig. 4.9 Temperature measurement as a function of the thermal time constant for a square temperature variation

of time responses of three sensors having different thermal time constants are illustrated in the same figure. Note that the sensor with the highest thermal constant, τ_3, practically does not "feel" the temperature variation, whereas the one with the lowest thermal constant, τ_1, follows more closely the temperature variation. This example could be the case of an internal combustion engine, where the chamber temperature cools off due to the admission of air and fuel mixture and heats up due to the combustion process in a continuous cycle. With that simple example, the importance of the thermal constant for temperature measurement is shown: the smaller the thermal constant of the sensor element, the more realistically it will reproduce the temperature variation of the environment. Ideally, the sensor thermal time constant should be much less than the temperature variation period.

The equation that governs the lumped transient regime (Eq. 4.2) can still be rewritten to obtain the following form in dimensionless variables:

$$\frac{\theta}{\theta_0} = \frac{T - T_\infty}{T_0 - T_\infty} = e^{-BiFo} \tag{4.5}$$

where *Bi* is the *Biot number*, defined by

$$Bi = \frac{\frac{L}{k}}{\frac{1}{h}} = \frac{hL}{k} \tag{4.6}$$

h is the convection heat transfer coefficient, *k* is the body thermal conductivity, and *L* is the body characteristic length. The Biot number accounts for the ratio between the internal resistance to heat conduction (L/k) to the external resistance to convection ($1/h$), whereas the characteristic length, *L*, is defined as the ratio between the body volume (∀) and its exposed area (*A*) for convective heat exchange.

$$L = \frac{\forall}{A} \tag{4.7}$$

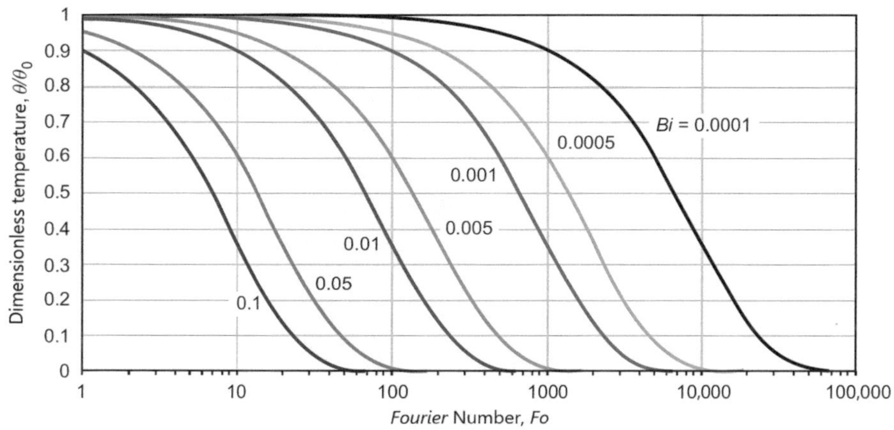

Fig. 4.10 Variation of the dimensionless temperature θ/θ_0 as a function of the Fourier number, *Fo*, for various Biot values, *Bi*, according to Eq. (4.5)

The other dimensionless group *Fo* is the *Fourier number*, defined by

$$Fo = \frac{\alpha t}{L^2} \tag{4.8}$$

α is the thermal diffusivity of the body $(k/\rho C)$, t is time. *Fo* can be seen as a dimensionless "time".

A lower Biot number means that the resistance to internal conduction is less than the resistance to the external convective heat transfer and, in this case, the hypothesis of a lumped system is valid as the instantaneous body temperature varies more uniformly. It is accepted that the limit of the applicability of the lumped system hypothesis occurs for $Bi < 0.1$. The graph in Fig. 4.10 indicates the behavior of the dimensionless temperature θ/θ_0 as a function of the Fourier number, *Fo*, for various Biot values, *Bi*.

Example 4.1 (*Lumped system*)

Watermelons are very juicy and refreshing fruits in warm weather. Consider the case of a watermelon at 25 °C that is placed in the refrigerator, whose internal compartment is at 5 °C. Do you believe that the cooling of the watermelon will occur uniformly, or that, after a few minutes, the same slice will have different temperatures? For estimation purposes, consider the watermelon as a 30 cm sphere with thermal conductivity equal to that of water, $k = 0.59$ W/mK, and thermal diffusivity, $\alpha = 1.4 \times 10^{-7}$ m^2/s. Also, consider that the refrigerator's internal compartment heat transfer coefficient is $h = 5$ W/m^{2}K.

Now redo the problem for a 3 cm grape. How long would it take to reach 10 °C?

iStock.com/GOLFX

iStock.com/alentynVolkov

Solution

Watermelon

$$\text{characteristic length, } L = \frac{D}{6} = \frac{0.3}{6} = 0.05\,\text{m}.$$

The corresponding Biot number is:

$$Bi = \frac{hL}{k} = \frac{5 \times 0.05}{0.59} = 0.42 > 0.1$$

(Therefore, it cannot be analyzed as a lumped system).

Conclusion, the watermelon *will not* cool uniformly. Does that agree with your experience when tasting a watermelon that stayed in the refrigerator for a short time?

Grape

In this case, the characteristic length is $L = \frac{D}{6} = \frac{0.03}{6} = 0.005$ m. The Biot number is $Bi = \frac{hL}{k} = \frac{5 \times 0.005}{0.59} = 0.042 < 0.1$. Therefore, the hypothesis of a lumped system is valid and the grape will cool uniformly.

$$\text{Thus, } \theta = 10 - 5 = 5 \text{ e } \theta_0 = 25 - 5 = 20 \Rightarrow \frac{\theta}{\theta_0} = 0.25$$

With $Bi = 0.042$ and $\theta/\theta_0 = 0.25$, consulting the graph in Fig. 4.10 or using Eq. (4.5), we get $Fo = 33$. Then,

$$t = \frac{FoL^2}{\alpha} = \frac{33 \times 0.005^2}{1.4 \times 10^{-7}} = 5842 \text{ s} \approx 98\,\text{min}.$$

In the case of heating a body by a medium at a higher temperature, T_∞, the same expressions are valid, as discussed in Example 4.2.

The lumped analysis was limited only to the situation where the body is subject to a sudden temperature change due to a convective boundary condition with a new medium at another temperature. However, the same analysis applies if the boundary condition is different, for example, exposure to an intense thermal radiation source. In this case, it is necessary to work the equations from the energy balance (Eq. 4.1) with the newly imposed boundary condition.

Example 4.2 (*Lumped system—temperature measurement*)

A thermocouple is used to measure the temperature of a hot gas at 300 °C from an exhaust system. Consider that the thermocouple tip junction is a small 1 mm sphere and the heat transfer coefficient by convection between the hot gas and thermocouple tip is 300 W/m^2°C. Assuming that the thermocouple junction has properties similar to those of A316 steel and that, initially, it is at 15 °C, evaluate: (a) the time for the thermocouple to reach a temperature of 299.9 °C; (b) redo the previous item considering that the thermocouple junction has diameters of 0.5 and 2 mm; (c) make schematic scaled diagrams showing the temperature variation for the three junction diameters of 0.5, 1, and 2 mm.

Solution

(a) For $d = 1$ mm Characteristic length: $L = \frac{\forall}{A} = \frac{\pi d^3/6}{\pi d^2} = \frac{d}{6} = \frac{1\times10^{-3}}{6} = 1.667 \times 10^{-4}$ m.
Properties of A316 steel (Table A.2): $\rho = 8238$ kg/m^3, $C = 468$ J/kgK, $k = 13.4$ W/mK. Therefore, $\alpha = 3.476 \times 10^{-6}$ m^2/s.
Biot number: $Bi = \frac{hL}{k} = \frac{300\times1.667\times10^{-4}}{13.4} = 0.004 \ll 0.1$, therefore, the thermocouple junction can be analyzed as a lumped system.
Time constant: $\tau_1 = \frac{\rho CL}{h} = \frac{8238\times468\times1.667\times10^{-4}}{300} = 2.14$ s.
From the expression of temperature, Equation (4.2), it follows:

$$\frac{\theta(t)}{\theta_0} = \frac{T(t) - T_\infty}{T_0 - T_\infty} = e^{-\frac{t}{\tau_1}} \Rightarrow t = -\tau_1 \times \ln\left(\frac{299.9 - 300}{15 - 300}\right) \Rightarrow t = 17 \text{ s.}$$

(b) For $d = 0.5$ mm

$$L = 8.3333 \times 10^{-5}\text{m} \Rightarrow \quad Bi = 0.002 \ll 0.1 \quad \Rightarrow \quad t = 8.5 \text{ s.}$$

For $d = 2$ mm,

$$L = 3.3333 \times 10^{-4}\text{m} \Rightarrow \quad Bi = 0.007 \ll 0.1 \quad \Rightarrow \quad t = 34.1 \text{ s.}$$

(c) From the Equation, one can establish the behavior of the temperature variation over time. In addition, one can notice the influence of the thermal time constant $\tau_t = \frac{\rho C \forall}{hA} = \frac{\rho C d}{6h} = \frac{8238 \times 468 \times d}{6 \times 300} = 2141.88d$.

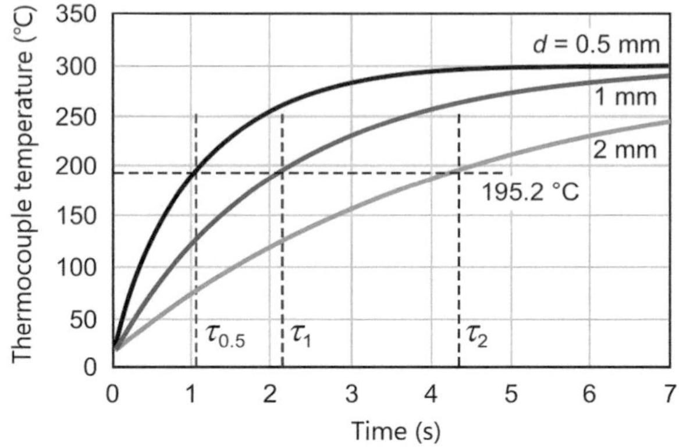

For $d = 0.5\,\text{mm}$, $\tau_{0.5} = 1.07$ s.
For $d = 1.0\,\text{mm}$, $\tau_1 = 2.14$ s.
For $d = 2.0\,\text{mm}$, $\tau_2 = 4.28$ s.

4.2 Heat Conduction in Unidimensional Transient Regime

4.2.1 Semi-infinite Solid

In the previous section, the case of transient heat conduction of a lumped system was studied. This simplified formulation fails when the body has dimensions in combination with transport properties so that the internal resistance to conduction cannot be neglected in view of the external resistance to convection ($Bi > 0.1$). For cases of higher Biot, it is necessary to solve the problem considering the temperature spatial distribution. In this section, the case of only one dimension is studied, for which there are some analytical solutions. Consider the scheme in Fig. 4.11, where a surface (on the left) is exposed to a sudden heat exchange and its dimension extends to the right to infinity (hence the name semi-infinite solid, although it does not seem appropriate), which means an infinite *Bi*. The face exposed on the left undergoes abrupt changes in the boundary condition, represented by (a), (b), and (c), described later. In all cases, the body temperature depends on time t and space x, that is $T = T(x, t)$ and so the heat flux $q = q(x, t)$.

The analysis starts off by establishing the general equation of heat conduction in a new similarity coordinate. The general equation of heat conduction is (Eq. 2.7) is

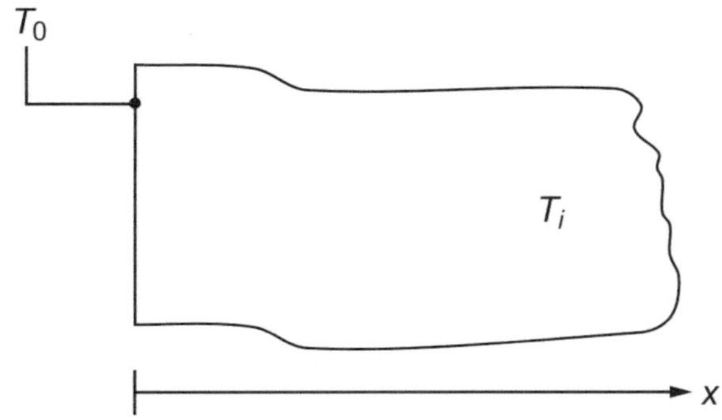

Fig. 4.11 Semi-infinite solid subjected to a constant temperature on the exposed face

$$\nabla^2 T + \frac{q_G'''}{k} = \frac{1}{\alpha}\frac{\partial T}{\partial t} \tag{2.7}$$

As there is no generation of thermal energy (heat) and it is a one-dimensional problem, it follows that:

$$\frac{\partial^2 T}{\partial x^2} = \frac{1}{\alpha}\frac{\partial T}{\partial t} \tag{4.9}$$

The technique to solve the given equation is to combine the two independent variables x and t into a single variable, known as *similarity variable* η; thus, $T(x, t) = T(\eta)$. The similarity variable, in this case, is defined as:

$$\eta = \frac{x}{\sqrt{4\alpha t}} \tag{4.10}$$

The partial derivatives in the independent variables in a function of this new variable are as follows:

$$\frac{\partial T}{\partial x} = \frac{\partial T}{\partial \eta}\frac{\partial \eta}{\partial x} = \frac{\partial T}{\partial \eta}\frac{1}{\sqrt{4\alpha t}};$$

$$\frac{\partial^2 T}{\partial x^2} = \frac{\partial}{\partial \eta}\left(\frac{\partial T}{\partial \eta}\frac{1}{\sqrt{4\alpha t}}\right)\frac{\partial \eta}{\partial x} = \frac{\partial^2 T}{\partial \eta^2}\frac{1}{4\alpha t}; \quad \text{and}$$

$$\frac{\partial T}{\partial t} = \frac{\partial T}{\partial \eta}\frac{\partial \eta}{\partial t} = \frac{\partial T}{\partial \eta}\frac{x}{\sqrt{4\alpha}}\left(-\frac{1}{2t\sqrt{t}}\right) = -\frac{x}{2t\sqrt{4\alpha t}}\frac{\partial T}{\partial \eta}$$

Substituting the partial derivatives into Eq. (4.9), we get:

$$\frac{\partial^2 T}{\partial \eta^2}\frac{1}{4\alpha t} = \frac{1}{\alpha}\left(-\frac{x}{2t\sqrt{4\alpha t}}\frac{\partial T}{\partial \eta}\right)$$

which, after simplification, reduces to:

$$\frac{d^2T}{d\eta^2} = -2\eta\frac{dT}{d\eta} \tag{4.11}$$

Equation (4.11) is expressed in terms of total derivative since T now depends only on η. Changing the variable: $\frac{dT}{d\eta} = \chi$ and solving the equation:

$$\int \frac{d\chi}{\chi} = -2\int \eta d\eta \rightarrow \ln\chi = -\eta^2 + C_1 \rightarrow \chi = e^{-\eta^2+c_1} = C_2 e^{-\eta^2}$$

substituting the derivative:
$\chi = \frac{dT}{d\eta} = C_2 e^{-\eta^2} \rightarrow dT = C_2 e^{-\eta^2} d\eta \rightarrow \int dT = \int C_2 e^{-\eta^2} d\eta$, to finally obtain

$$T = C_2 \int_0^{\eta} e^{-\eta^2} d\eta + C_3$$

The constants of the previous equation are found from the imposed boundary conditions. According to the similarity variable, the conditions are reduced to two, C_2 and C_3, since the third condition is implicit, because whether with $x \rightarrow \infty$ or $t = 0$ results in the same condition $\eta \rightarrow \infty$.

Boundary Conditions

(a) **Constant temperature (T_0) on the exposed face**

In this first boundary condition, the temperature of the left face undergoes a sudden change from the initial temperature, T_i, to a given value, T_0. Thus, the boundary conditions are:

BC$_1$: Boundary condition on the left face ($x = 0$): $T(0, t) = T_0 \Rightarrow \eta = 0$

$$\text{So,}\quad T_0 = C_2 \int_0^0 e^{-\eta^2} d\eta + C_3 \rightarrow C_3 = T_0$$

BC$_2$: Boundary condition, far from the left face: $T(x \rightarrow \infty, t) = T_i \Rightarrow \eta \rightarrow \infty$

$$\text{So,}\quad T_i = C_2 \int_0^{\infty} e^{-\eta^2} d\eta + T_0 \rightarrow C_2 = \frac{T_i - T_0}{\int_0^{\infty} e^{-\eta^2} d\eta} = \frac{T_i - T_0}{\frac{\sqrt{\pi}}{2} erf(\eta)}$$

Define the *Gaussian error function*, $erf(\eta)$, as:

$$erf(\eta) = \frac{2}{\sqrt{\pi}} \int_0^{\eta} e^{-\eta^2} d\eta \tag{4.12}$$

The Gaussian error function, $erf(\eta)$, must be integrated numerically. Some values of this function are presented in Table 4.2. Figure 4.12 shows the general behavior of this function. From Table 4.2, the value of $erf(\eta)$ for $\eta \to \infty$ equals 1, then, the temperature distribution becomes:

$$T(\eta) = \frac{2(T_i - T_0)}{\sqrt{\pi}} \int_0^{\eta} e^{-\eta^2} d\eta + T_0$$

or, substituting the original coordinates:

Table 4.2 Gaussian error function, $erf(\eta) = \frac{2}{\sqrt{\pi}} \int_0^{\eta} e^{-\eta^2} d\eta$

η	erf η	η	erf η	η	erf η	η	erf η	η	erf η	η	erf η
0.00	0.0000	0.38	0.4090	0.76	0.7175	1.14	0.8931	1.52	0.9684	1.90	0.9928
0.02	0.0226	0.40	0.4284	0.78	0.7300	1.16	0.8991	1.54	0.9706	1.92	0.9934
0.04	0.0451	0.42	0.4475	0.80	0.7421	1.18	0.9048	1.56	0.9726	1.94	0.9939
0.06	0.0676	0.44	0.4662	0.82	0.7538	1.20	0.9103	1.58	0.9745	1.96	0.9944
0.08	0.0901	0.46	0.4847	0.84	0.7651	1.22	0.9155	1.60	0.9763	1.98	0.9949
0.10	0.1125	0.48	0.5027	0.86	0.7761	1.24	0.9205	1.62	0.9780	2.00	0.9953
0.12	0.1348	0.50	0.5205	0.88	0.7867	1.26	0.9252	1.64	0.9796	2.10	0.9970
0.14	0.1569	0.52	0.5379	0.90	0.7969	1.28	0.9297	1.66	0.9811	2.20	0.9981
0.16	0.1790	0.54	0.5549	0.92	0.8068	1.30	0.9340	1.68	0.9825	2.30	0.9989
0.18	0.2009	0.56	0.5716	0.94	0.8163	1.32	0.9381	1.70	0.9838	2.40	0.9993
0.20	0.2227	0.58	0.5879	0.96	0.8254	1.34	0.9419	1.72	0.9850	2.50	0.9996
0.22	0.2443	0.60	0.6039	0.98	0.8342	1.36	0.9456	1.74	0.9861	2.60	0.9998
0.24	0.2657	0.62	0.6194	1.00	0.8427	1.38	0.9490	1.76	0.9872	2.70	0.9999
0.26	0.2869	0.64	0.6346	1.02	0.8508	1.40	0.9523	1.78	0.9882	2.80	0.9999
0.28	0.3079	0.66	0.6494	1.04	0.8586	1.42	0.9554	1.80	0.9891	2.90	1.0000
0.30	0.3286	0.68	0.6638	1.06	0.8661	1.44	0.9583	1.82	0.9899	3.00	1.0000
0.32	0.3491	0.70	0.6778	1.08	0.8733	1.46	0.9611	1.84	0.9907	3.10	1.0000
0.34	0.3694	0.72	0.6914	1.10	0.8802	1.48	0.9637	1.86	0.9915	3.20	1.0000
0.36	0.3893	0.74	0.7047	1.12	0.8868	1.50	0.9661	1.88	0.9922	3.30	1.0000

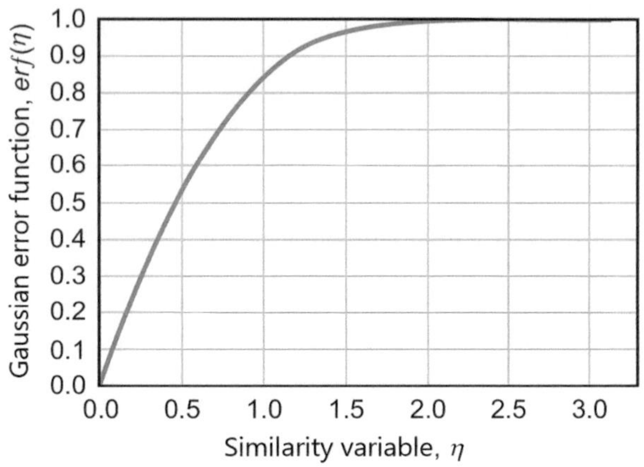

Fig. 4.12 Gaussian error function

$$\frac{T(x,t) - T_0}{T_i - T_0} = \text{erf}\left(\frac{x}{2\sqrt{\alpha t}}\right) \tag{4.13}$$

Once the temperature distribution is determined, the heat transfer rate at a position x and time t, $\dot{Q} = \dot{Q}(x,t)$. To do this, simply apply Fourier's law of conduction to Eq. (4.13), that is:

$$\dot{Q} = -kA\frac{\partial T}{\partial x} = -kA\frac{\partial}{\partial x}\left[T_0 + (T_i - T_0)\text{erf}\left(\frac{x}{2\sqrt{\alpha t}}\right)\right]$$

The partial derivative of the Gauss error function is:

$$\frac{\partial[\text{erf}(\eta)]}{\partial x} = \frac{\partial}{\partial x}\left(\frac{2}{\sqrt{\pi}}\int_0^{\frac{x}{2\sqrt{\alpha t}}} e^{-\eta^2}d\eta\right) \rightarrow \frac{\partial[\text{erf}(\eta)]}{\partial x} = \frac{e^{-\frac{x^2}{4\alpha t}}}{\sqrt{\pi\alpha t}}$$

from which, after substitution, results in:

$$q = \frac{\dot{Q}}{A} = -\frac{k(T_i - T_0)}{\sqrt{\pi\alpha t}}e^{-\frac{x^2}{4\alpha t}} \tag{4.14}$$

(b) **Constant heat flux on the exposed face**

In this second case, we study the situation where the face exposed to the left ($x = 0$) is suddenly subjected to a constant and prescribed heat flux value q_0, as illustrated in Fig. 4.13.

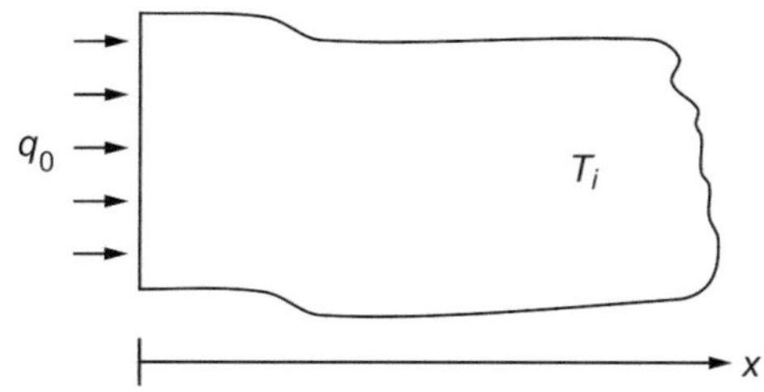

Fig. 4.13 Semi-infinite solid with constant heat flux on the exposed face

Starting from the heat conduction equation: $\frac{\partial^2 T}{\partial x^2} = \frac{1}{\alpha}\frac{\partial T}{\partial t}$.
Subject to the following conditions:
Initial condition (uniform temperature): $T(x, 0) = T_i$.

BC_1: Boundary condition on the left face ($x = 0$): $-k\frac{\partial T}{\partial x}\big|_{x=0} = q_0$.
BC_2: Boundary condition for the very long case: $T(x \to \infty, t) = T_i$.

The solution to this problem is:

$$T(x, t) = \frac{q_0}{k}\left\{e^{-\frac{x^2}{4\alpha t}}\sqrt{\frac{4\alpha t}{\pi}} - x\left[1 - \operatorname{erf}\left(\frac{x}{2\sqrt{\alpha t}}\right)\right]\right\} + T_i \tag{4.15}$$

The heat flux for any position (x) and time (t) is evaluated by Fourier's Law: $q = -k\frac{\partial T}{\partial x}$

$$q = -k\frac{q_0}{k}\frac{\partial}{\partial x}\left\{e^{-\frac{x^2}{4\alpha t}}\sqrt{\frac{4\alpha t}{\pi}} - x\left[1 - \operatorname{erf}\left(\frac{x}{2\sqrt{\alpha t}}\right)\right] + T_i\right\}, \quad \text{or}$$

$$q = q_0\left[1 - \operatorname{erf}\left(\frac{x}{2\sqrt{\alpha t}}\right)\right] \tag{4.16}$$

(c) **Heat convection on the exposed face**

In this last case, we study the situation where heat convection occurs to a fluid of constant temperature T_∞ on the exposed face to the left ($x = 0$) (Fig. 4.14).

Again, starting from the heat conduction equation without thermal energy (heat) generation, we get: $\frac{\partial^2 T}{\partial x^2} = \frac{1}{\alpha}\frac{\partial T}{\partial t}$.
Subject to the following conditions:
Initial condition (uniform temperature): $T(x, 0) = T_i$.

BC_1: Convection on the left face ($x = 0$): $-k\frac{\partial T}{\partial x}\big|_{x=0} = h[T_\infty - T(0, t)]$.

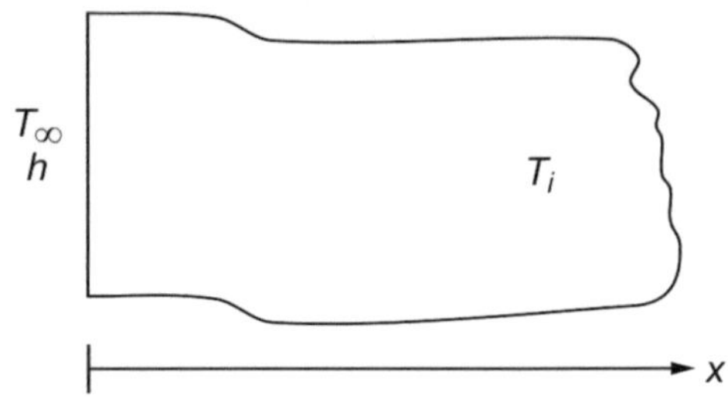

Fig. 4.14 Semi-infinite solid with convection on the exposed face

BC_2: Boundary condition for the very long case: $T(x \to \infty, t) = T_i$.

The solution is given by:

$$\frac{T(x,t) - T_i}{T_\infty - T_i} = 1 - \operatorname{erf}\left(\frac{x}{2\sqrt{\alpha t}}\right) + e^{\left(\frac{hx}{k} + \frac{h^2 \alpha t}{k^2}\right)}\left[\operatorname{erf}\left(\frac{x}{2\sqrt{\alpha t}} + \frac{h\sqrt{\alpha t}}{k}\right) - 1\right] \quad (4.17)$$

The heat flux will be:

$$q = h(T_\infty - T_i)e^{\left(\frac{hx}{k} + \frac{h^2 \alpha t}{k^2}\right)}\left[1 - \operatorname{erf}\left(\frac{x}{2\sqrt{\alpha t}} + \frac{h\sqrt{\alpha t}}{k}\right)\right] \quad (4.18)$$

4.2.2 Other Cases of Interest of Transient Conduction

Slabs, cylinders, and spheres are very common geometries of mechanical parts. For small Biot numbers, the lumped system approach should be used, as it was developed in Sect. 4.1. However, when that is not the case, the general heat conduction equation must be solved, like the solutions of semi-infinite solids from the previous section. For geometries of slabs, cylinders, and spheres, Heisler (1947) developed graphical solutions, which are summarized in Table 4.3.

Table 4.3 Convention for using Heisler's diagrams

Slabs whose thickness is small compared to other dimensions	Cylinders whose diameters are small compared to the length	Spheres
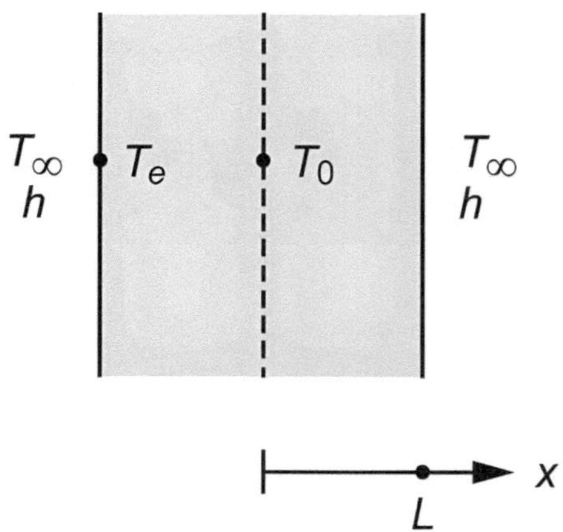	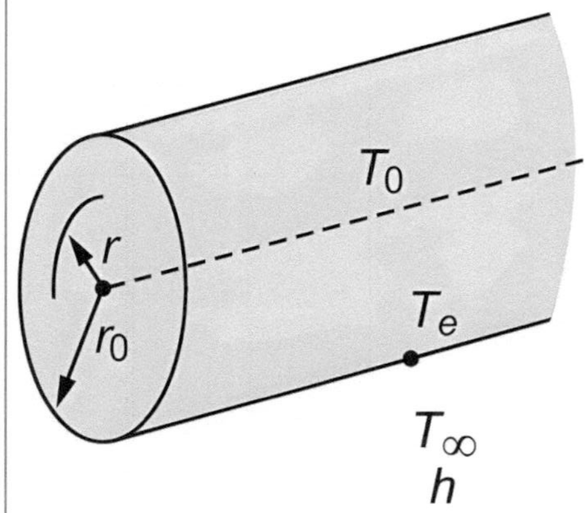	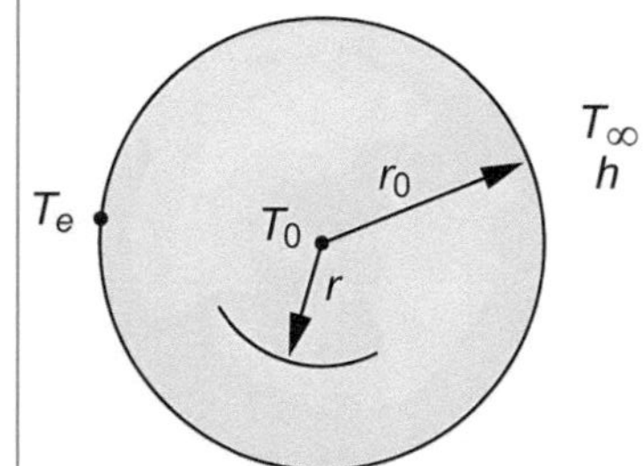
Characteristic length		
L	r_0	r_0

$$\theta = T(x,t) - T_\infty \text{ or } \theta = T(r,t) - T_\infty;\ \theta_i = T_i - T_\infty;\ \theta_0 = T_0 - T_\infty;\ \theta_e = T_e - T_\infty;$$

T_0—temperature in the central region;
T_i—initial temperature.

Biot number: $Bi = \frac{hL}{k}$, $L = r_0$ in the cylindrical or spherical case.
Fourier number: $Fo = \frac{\alpha t}{L^2} = \frac{kt}{\rho C L^2}$ (dimensionless time).
Total heat exchanged by the body: $Q_i = \rho C \forall (T_i - T_\infty) = \rho C \forall \theta_i$.

Example 4.3 (*Use of Heisler's diagrams*)

A 1010 steel plate 5 cm thick is initially at a uniform temperature of 790 °C prepared for a tempering process. Suddenly, the plate is inserted into an oil bath so that both sides of the plate are exposed to the temperature $T_\infty = 65$ °C with $h = 500$ W/m^{2}K. Determine (a) the temperature in the middle plane of the plate, and (b) the temperature 1.25 cm inside it, after 3 min.

Solution

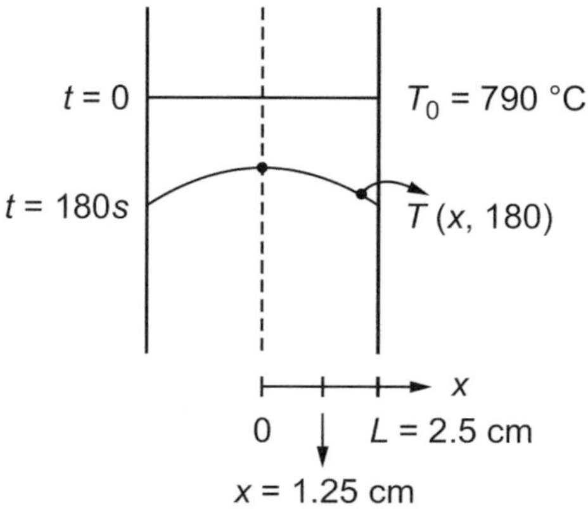

The definitions of the Biot and Fourier numbers should follow the convention pattern (Table 4.3) of Heisler's diagrams.

(a) Steel conductivity: $k = 63.9$ W/mK (Table 4.2)

$$\alpha = 18.8 \times 10^{-6}\,\text{m}^2/\text{s}; \qquad 2L = 0.05\,\text{m} \rightarrow L = 0.025\,\text{m}$$

$$Bi = \frac{hL}{k} = \frac{500 \times 0.025}{43.2} = 0.196$$

Since $Bi > 0.1$. the lumped system solution does not apply. Therefore, the Heisler solution should be used. For this, the parameters for the graphs in Fig. 4.15a are calculated, which are:

$$\frac{1}{Bi} = \frac{1}{0.196} = 5.1 \quad \text{and} \quad Fo = \frac{\alpha t}{L^2} = \frac{18.8 \times 10^{-6} \times 180}{0.025^2} = 5.414.$$

From the Heisler diagram. Figure 4.15a comes: $\frac{\theta_0}{\theta_i} = \frac{T_0 - T_\infty}{T_i - T_\infty} = 0.4$ (approximately). Therefore,

$T_0 = 65 + (790 - 65) \times 0.4 = 355\,°\text{C}$ at the centerline after 3 min.

(a)

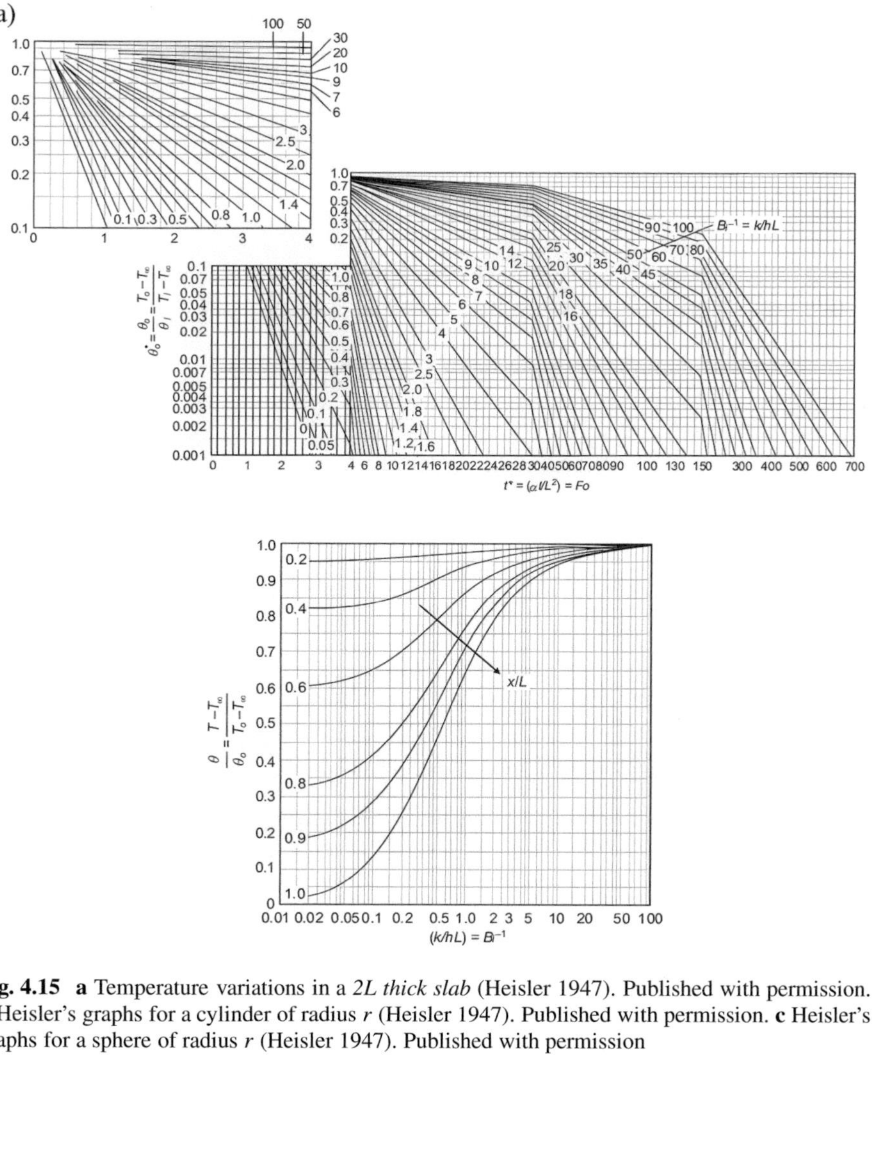

Fig. 4.15 **a** Temperature variations in a *2L thick slab* (Heisler 1947). Published with permission. **b** Heisler's graphs for a cylinder of radius *r* (Heisler 1947). Published with permission. **c** Heisler's graphs for a sphere of radius *r* (Heisler 1947). Published with permission

(b)

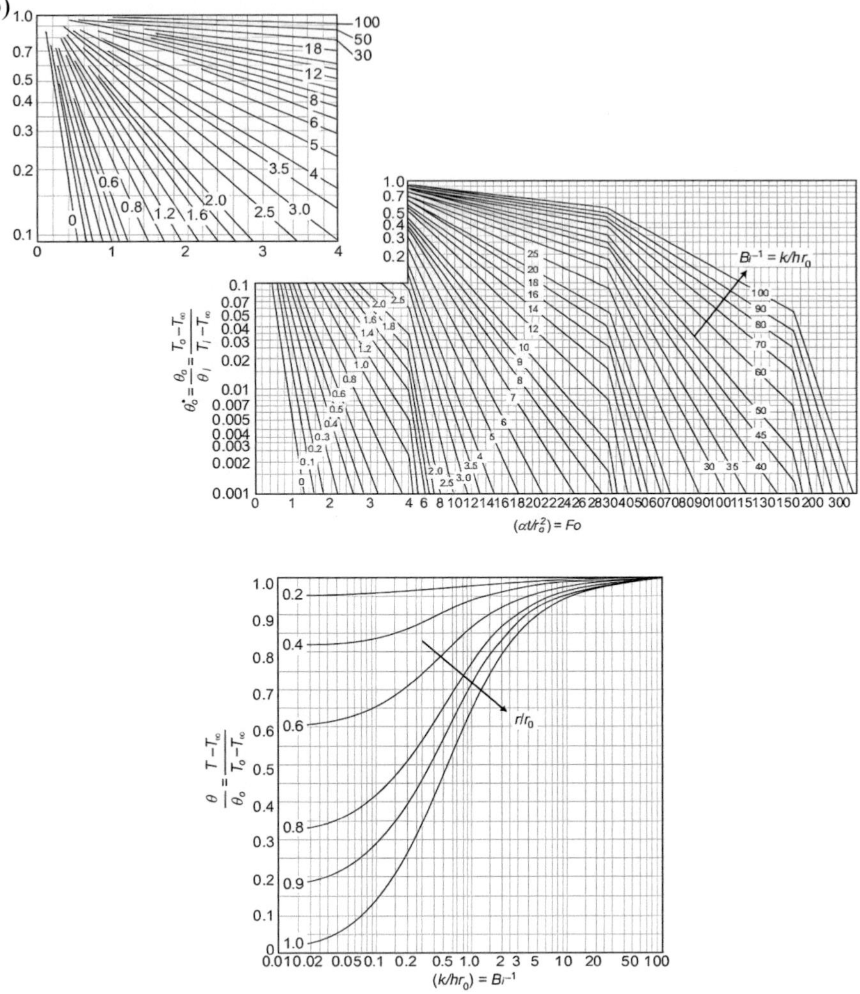

Fig. 4.15 (continued)

(c)

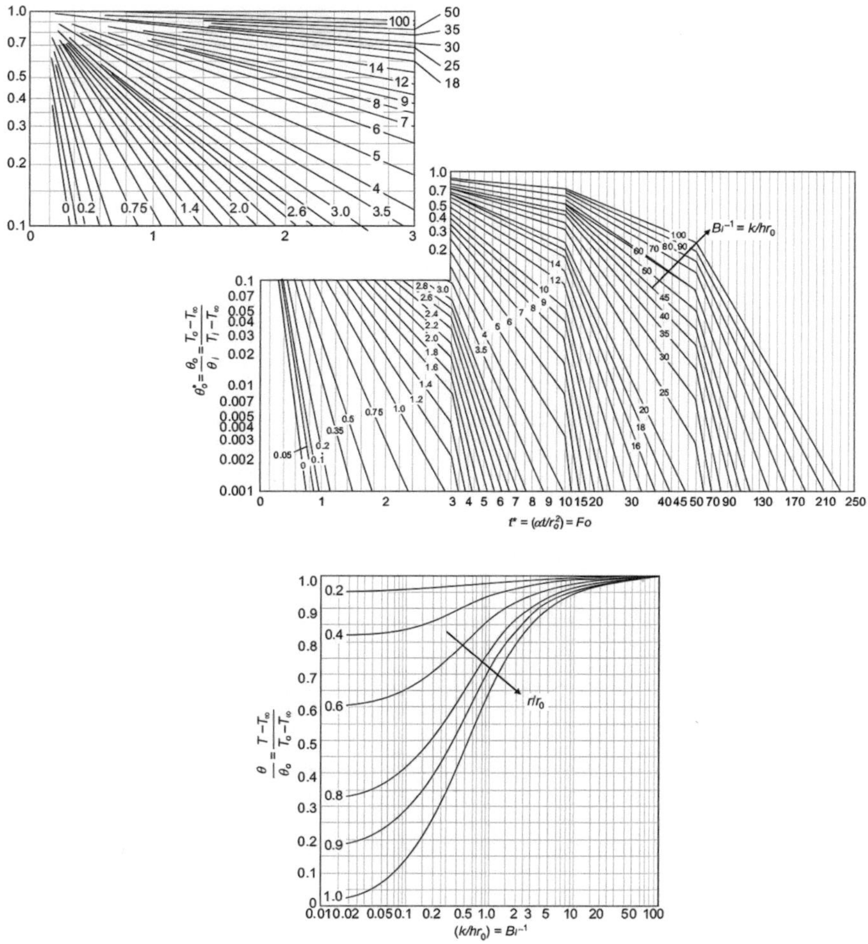

Fig. 4.15 (continued)

(b) From the Heisler diagram. Figure 4.15b. for the position $x = 0.0125$ m: With $\frac{1}{Bi} = 5.1$ and $\frac{x}{L} = \frac{0.0125}{0.025} = 0.5$, it is found that: $\frac{\theta}{\theta_0} \cong 0.96$

$$\frac{T - T_\infty}{T_0 - T_\infty} \approx 0.96 \rightarrow T = 65 + 0.96 \times (355 - 65)$$

$T = 343.4\,°\mathrm{C}$ for the position $\frac{x}{L} = 0.5$ after $t = 3$ min.

Note that it is not such a sharp temperature difference due to the Biot number not being very high. The $Bi = 0.196$ value is close to the validity condition of the lumped system hypothesis ($Bi < 0.1$).

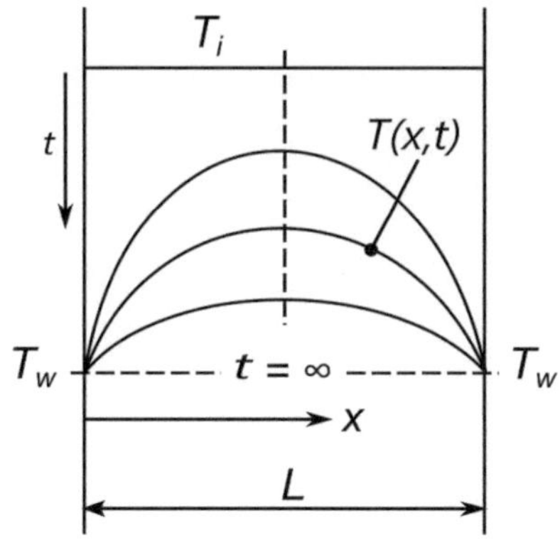

Fig. 4.16 A slab of thickness L at a uniform initial ($t = 0$) temperature T_i subject to a sudden wall temperature T_w ($t > 0$)

4.2.3 Transient Heat Conduction in a Slab

In the previous section, it was shown the final results obtained by Heisler (1947) in diagram forms. Solutions to that class of problems may be reached by using the method of separation of variables used in Chap. 3 in steady-state two-dimensional problems. In this section, it will be solved the case of a transient conduction in a slab.

Consider an infinite slab subject to an initial temperature T_i as shown in Fig. 4.16 in $t = 0$. The two surfaces undergo a sudden temperature change and are held at constant temperature T_w, as shown for t > 0.

Starting off from Eq. (2.6) for a 1-D without internal energy generation (heat), one obtains:

$$\frac{\partial^2 T}{\partial x^2} = \frac{1}{\alpha}\frac{\partial T}{\partial t} \tag{4.19}$$

Consider a new method of separation of variables

$$\frac{\partial^2 \theta}{\partial x^2} = \frac{1}{\alpha}\frac{\partial \theta}{\partial t} \tag{4.20}$$

where $\theta = T - T_w$. Making $\theta = X(x) \times \tau(t)$, one obtains

$$\frac{1}{X}\frac{d^2 X}{\partial x^2} = \frac{1}{\alpha t}\frac{\partial \tau}{\partial t} = -\lambda^2 \tag{4.22}$$

As previously analyzed in Sect. 3.1, λ^2 is a positive constant. There are two ordinary differential equations, one in $X(x)$ and the other in $\tau(t)$, that is

$$\frac{d^2 X}{\partial x^2} + \lambda^2 = 0 \tag{4.23a}$$

$$\frac{\partial \tau}{\partial t} + \alpha\lambda^2\tau = 0 \tag{4.23b}$$

The initial and boundary conditions are:

IC$_1$: $\theta(x, 0) = T_i$ uniform temperature at $t = 0$;
BC$_1$: $T(0, t) = T_w$ temperature on the left face;
BC$_2$: $T(L, t) = T_w$ temperature on the right face.

The solution to the above equations are

$$X(x) = C_1 \cos(\lambda x) + C_2 \sin(\lambda x); \qquad \text{and} \tag{4.24a}$$

$$\tau(t) = C_3 e^{\alpha\lambda^2 t} \tag{4.24b}$$

Finally,

$$\theta(x, t) = C_3[C_1 \cos(\lambda x) + C_2 \sin(\lambda x)]e^{-\alpha\lambda^2 t} \tag{4.25}$$

By applying the boundary conditions, $C_1 = 0$. As explained in detail in Sect. 3.1 this is an eigenvalue problem where there exists a particular solution for every $\lambda_n = n\pi/L$. Therefore, the overall solution is the sum of individual solutions, that is

$$\theta(x, t) = \sum_{n=1}^{\infty} B_n \sin(\lambda_n x) e^{-\alpha\lambda_n^2 t} \tag{4.26}$$

The coefficients B_n are the Fourier coefficients from sine expansion, that is

$$B_n = \int_0^L \sin(\lambda_n x) dx \tag{4.27}$$

Finally, the temperature distribution is

$$\theta(x, t) = \theta_i \frac{4}{\pi} \sum_{n=1,3\ldots}^{\infty} \frac{e^{-(n\pi/L)^2\alpha t}}{n} \sin\left(\frac{n\pi x}{L}\right) \tag{4.28}$$

Note that $\theta_i = T_i - T_w$ is the initial temperature difference. Also, the sum should be carried out only to odd values of n.

The heat flux is given by the Fourier law

$$q = -k\frac{\partial \theta}{\partial x} = -\frac{4k\theta_i}{L} \sum_{n=1,3\ldots}^{\infty} e^{-(n\pi/L)^2\alpha t} \cos\left(\frac{n\pi x}{L}\right) \tag{4.29}$$

4.3 Heat Conduction in Transient Regime—Finite Differences Solution

The solution of the heat conduction equation for the transient multidimensional of any geometry becomes very complex and, in most cases, does not have an analytical solution. In the previous chapter, in Sect. 3.4, a numerical solution method was developed for two-dimensional problems (and that can be extended to three dimensions) in steady state. In this section, the finite difference method is extended to solve two-dimensional transient cases by carrying out the time-dependence term discretization. Consider a node (*m, n*) of interest in a two-dimensional problem, then the discretized transient term of that node can be written as:

$$\left.\frac{\partial T}{\partial t}\right|_{m,n} \approx \frac{\left(T_{m,n}^{i+1} - T_{m,n}^{i}\right)}{\Delta t} \tag{4.30}$$

the superscript i in T is the index that represents the temperature of the node (*m, n*) at the present time, and $i + 1$ is the temperature of the node in the future elapsed time Δt. The past time is given by $t = i\Delta t$. Substituting the time discretization along with the spatial discretization (Eqs. 3.26a and 3.26b) into Eq. (2.6) without thermal energy generation (heat) generation, for uniform grid size, $\Delta x = \Delta y$, come:

$$\frac{T_{m+1,n}^{i} + T_{m-1,n}^{i} + T_{m,n+1}^{i} + T_{m,n-1}^{i} - 4T_{m,n}^{i}}{(\Delta x)^2} = \frac{\left(T_{m,n}^{i+1} - T_{m,n}^{i}\right)}{\alpha \Delta t}$$

After some arrangements come:

$$T_{m,n}^{i+1} = Fo'\left(T_{m+1,n}^{i} + T_{m-1,n}^{i} + T_{m,n+1}^{i} + T_{m,n-1}^{i}\right) + \left(1 - 4Fo'\right)T_{m,n}^{i} \tag{4.31}$$

In this equation, the *numerical Fourier number,Fo'*, is defined by:

$$Fo' = \frac{\alpha \Delta t}{(\Delta x)^2} \tag{4.32}$$

The Eq. (4.31) for obtaining the time evolution of a nodal temperature is explicit in time. That is, the temperature of the nodal point (*m, n*) at the subsequent time step $i + 1$ is completely determined from the temperature distribution at the time present *i*. Evidently, for a long enough time, the steady-state solution will be reached. However, this method has a certain degree of associated numerical instability. The coefficient that multiplies the temperature of the nodal point (*m, n*) at the time present must be positive, that is, $1 - 4Fo' > 0 \rightarrow Fo' < \frac{1}{4}$ to avoid numerical instability. This stability criterion is valid for interior nodes. Therefore, this condition will affect the selection of the time interval Δt to march in time, which *cannot* be dissociated from the distance $\Delta x = \Delta y$ of the nodal cells. That is:

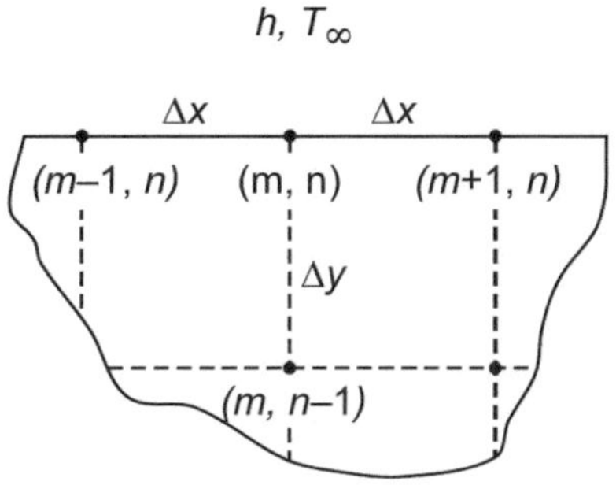

Fig. 4.17 Convective nodal point on the upper surface

$$Fo' = \frac{\alpha \Delta t}{(\Delta x)^2} < \frac{1}{4} \Rightarrow \Delta t < \frac{(\Delta x)^2}{4\alpha} \tag{4.33}$$

For one-dimensional geometries the stability criterion is $Fo' < 1/2$. In the case of a three-dimensional system, the stability criterion is $Fo' < 1/6$. Those criteria apply to an interior nodal point, while the case of a nodal point elsewhere can even be more restrictive, as analyzed next.

Consider a nodal point on an upper surface with convective heat transfer, as shown in Fig. 4.17.

After carrying out the energy balance for the upper surface node (m, n) in Fig. 4.17, the transient nodal equation will be:

$$T_{m,n}^{i+1} = Fo'\left(2T_{m,n-1}^{i} + T_{m-1,n}^{i} + T_{m+1,n}^{i} + 2Bi'T_{\infty}\right) + \left(1 - 4Fo' - 2Bi'Fo'\right)T_{m,n}^{i} \tag{4.34}$$

where the numerical Biot number Bi' is given by:

$$Bi' = \frac{h\Delta x}{k} \tag{4.35}$$

The stability criterion of the temperature coefficient of the nodal point (m, n) of interest must be positive. Therefore, in this case, it must be $1 - 4Fo' - 2Bi'Fo' > 0 \Rightarrow Fo'(2 + Bi') < \frac{1}{2} \Rightarrow Fo' < \frac{1}{2(2+Bi')}$

Table 4.4 summarizes the equations and stability criteria for several cases of interest.

Table 4.4 Finite differential equations in transient regime for nodal points where $\Delta x = \Delta y$

Scheme	Equation	Stability criterion
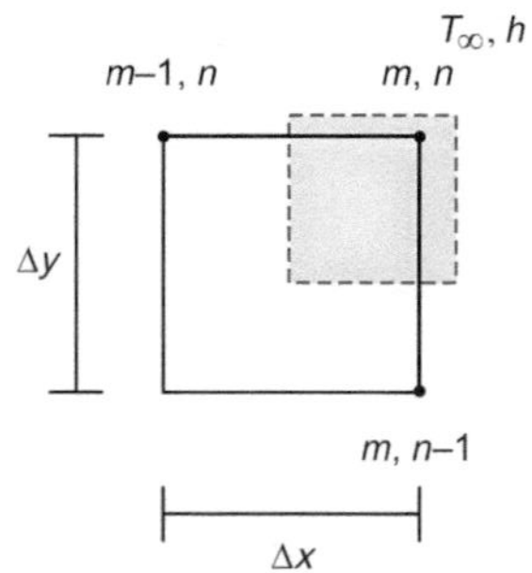	External corner with external convection $T_{m,n}^{i+1} = 2Fo'\left(T_{m-1,n}^{i} + T_{m,n-1}^{i} + 2Bi'T_\infty\right) + (1 - 4Fo' - 4Bi'Fo')T_{m,n}^{i}$	$Fo'(1 + Bi') \le 0.25$
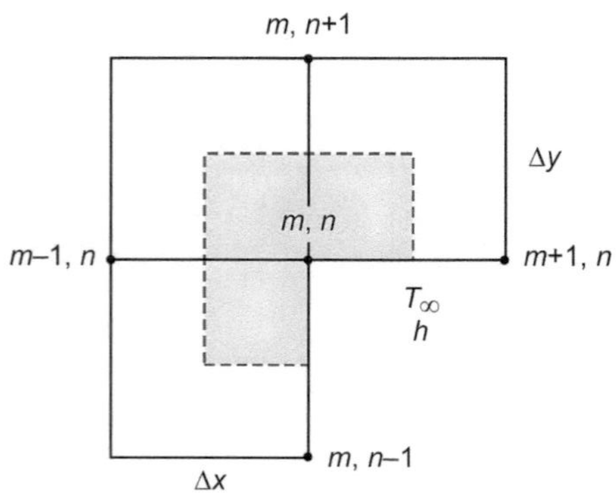	Internal corner with external convection $T_{m,n}^{i+1} = \left(\frac{2Fo'}{3}\right)\left(T_{m+1,n}^{i} + 2T_{m-1,n}^{i} + 2T_{m,n+1}^{i} + T_{m,n-1}^{i} + 2Bi'T_\infty\right) + (1 - 4Fo' - 0.75Bi'Fo')T_{m,n}^{i}$	$Fo'(3 + Bi') \le 0.75$

(continued)

Table 4.4 (continued)

Scheme	Equation	Stability criterion
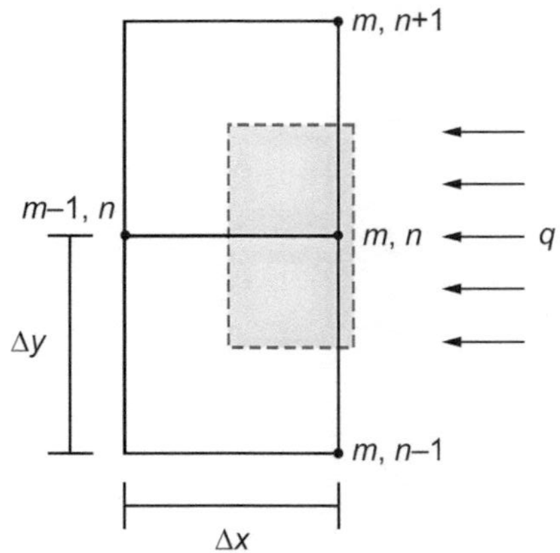	Flat surface with constant heat flux $T_{m,n}^{i+1} = Fo'\left(2T_{m-1,n}^{i} + T_{m,n+1}^{i} + T_{m,n-1}^{i}\right) + 2q\frac{Fo'\Delta x}{k} + \left(1 - 4Fo'\right)T_{m,n}^{i}$	$Fo' \leq 0.25$
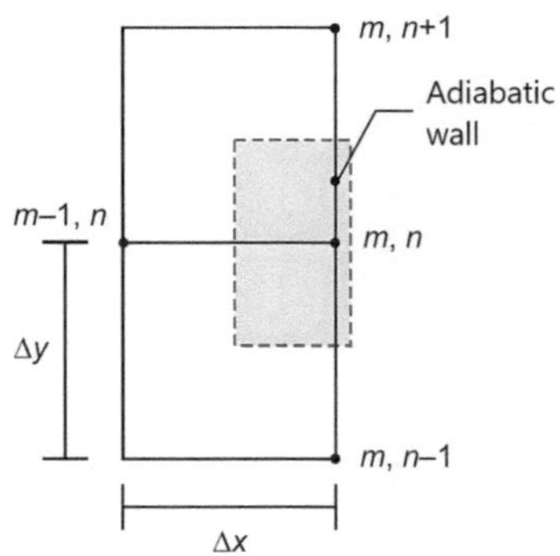	Adiabatic or symmetric flat part $T_{m,n}^{i+1} = Fo'\left(2T_{m-1,n}^{i} + T_{m,n+1}^{i} + T_{m,n-1}^{i}\right) + \left(1 - 4Fo'\right)T_{m,n}^{i}$	$Fo' \leq 0.25$

Example 4.4 (*Two-dimensional transient in a flat plate*)

Consider the transient solution of Example 3.2. A rectangular plate with thermal conductivity $k = 10$ W/mK and $\alpha = 3 \times 10^{-6}$ m^2/s is initially at 100 °C and suddenly exposed to a convective heat exchange on the upper surface to a fluid at 20 °C. The heat transfer coefficient between the fluid and the plate is 200 W/m^{2}K, while the lateral and lower faces are kept at 100 °C, as illustrated in the figure. It is asked to calculate the time evolution of temperature distribution at the indicated nodal points shown in that example. Consider $\Delta x = \Delta y = 0.1$ m.

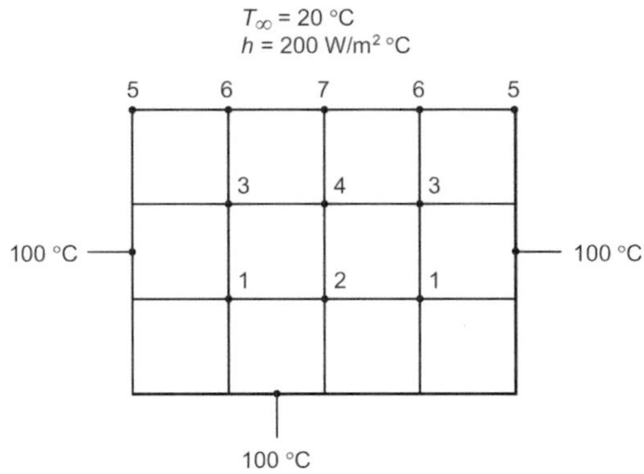

Solution

Calculation of the numerical Biot number (Eq. 4.34): $Bi' = \frac{200 \times 0.1}{10} = 2.$

Applying the stability criterion (more restrictive) of the upper node, we get:

$$Fo' < \frac{1}{2(2 + Bi')} \Rightarrow \frac{1}{2(2 + 2)} = \frac{1}{8} \Rightarrow \Delta t < \frac{(\Delta x)^2}{8\alpha} = \frac{(0.1)^2}{8 \times 3 \times 10^{-6}} = 416.7 \ s$$

Assuming $\Delta t = 60$ s (1 min) $\Rightarrow Fo' = \frac{\alpha \Delta t}{(\Delta x)^2} = \frac{3 \times 10^{-6} \times 60}{(0.1)^2} = 0.018$

For the internal nodal points (1 to 4), Eq. (4.30) is valid:

$T_{m,n}^{i+1} = Fo'\left(T_{m+1,n}^i + T_{m-1,n}^i + T_{m,n+1}^i + T_{m,n-1}^i\right) + \left(1 - 4Fo'\right)T_{m,n}^i$. Therefore.

node 1: $T_1^{i+1} = 0.018 \times \left(T_2^i + T_3^i + 200\right) + 0.928 \times T_1^i$.

node 2: $T_2^{i+1} = 0.018 \times \left(2T_1^i + T_4^i + 100\right) + 0.928 \times T_2^i$.

node 3: $T_3^{i+1} = 0.018 \times \left(T_1^i + T_4^i + T_6^i + 100\right) + 0.928 \times T_3^i$.

node 4: $T_4^{i+1} = 0.018 \times \left(T_2^i + 2T_3^i + T_7^i\right) + 0.928 \times T_4^i$.

node 5: $T_5 = 100$ °C (prescribed temperature).

At the nodal points with convective heat transfer on the upper face (6 and 7), Equation (4.23a) and (4.23b) applies:

$$T_{m,n}^{i+1} = Fo'\left(2T_{m,n-1}^{i} + T_{m-1,n}^{i} + T_{m+1,n}^{i} + 2Bi'T_{\infty}\right) + \left(1 - 4Fo' - 2Bi'Fo'\right)T_{m,n}^{i}.$$

Therefore:

node 6: $T_6^{i+1} = 0.018 \times \left(2T_3^i + T_7^i + 180\right) + 0.856T_6^i.$
node 7: $T_7^{i+1} = 0.018 \times \left(2T_4^i + 2T_6^i + 80\right) + 0.856T_7^i.$

The solution is given by establishing the initial condition of a uniform temperature throughout the plate, for example, $T_0 = 100$ °C at the initial time $t = 0$ s. The problem solution marches up to the next time step ($\Delta t = 60$ s) by solving the above system of nodal points equations at each time step. An extract of the time evolution of the nodal points' temperature every minute is indicated in the following table. As presented in the table, the steady state is reached after about 7 h (420 min) to a convergence of two decimal figures, or an absolute precision less than 0.005. In the case of a numerical convergence to only one decimal figure, or absolute precision less than 0.05, the steady state is reached in about 4 h (240 min). Compare it with the steady-state solution of Example 3.2.

t (s)	t (min)	T_1 (°C)	T_2 (°C)	T_3 (°C)	T_4 (°C)	T_5 (°C)	T_6 (°C)	T_7 (°C)
0	0	100.00	100.00	100.00	100.00	100.00	100.00	100.00
60	1	100.00	100.00	100.00	100.00	100.00	94.24	94.24
120	2	100.00	100.00	99.90	99.90	100.00	89.21	89.10
180	3	100.00	100.00	99.71	99.70	100.00	84.80	84.52
240	4	99.99	99.99	99.45	99.44	100.00	80.94	80.43
...	...	...	...	...	...	...	...	...
1020	17	99.42	99.33	93.80	92.93	100.00	57.71	53.91
1080	18	99.34	99.23	93.34	92.37	100.00	56.98	53.01
1140	19	99.25	99.12	92.90	91.82	100.00	56.33	52.19
1200	20	99.16	99.01	92.46	91.28	100.00	55.74	51.45
1260	21	99.07	98.90	92.04	90.75	100.00	55.21	50.77
...	...	...	...	...	...	...	...	...
4980	83	93.53	91.36	79.89	74.36	100.00	47.76	40.80
5040	84	93.47	91.29	79.82	74.26	100.00	47.74	40.76
5100	85	93.42	91.22	79.75	74.16	100.00	47.71	40.72
5160	86	93.38	91.15	79.68	74.07	100.00	47.68	40.69
5220	87	93.33	91.08	79.62	73.97	100.00	47.66	40.65
...	...	...	...	...	...	...	...	...
14,160	236	91.40	88.36	77.18	70.54	100.00	46.71	39.32
14,220	237	91.40	88.36	77.18	70.54	100.00	46.71	39.32
14,280	238	91.40	88.36	77.18	70.54	100.00	46.71	39.32
14,340	239	91.40	88.36	77.18	70.53	100.00	46.71	39.32
14,400	240	91.40	88.35	77.18	70.53	100.00	46.71	39.32
...	...	...	...	...	...	...	...	...

(continued)

(continued)

t (s)	t (min)	T_1 (°C)	T_2 (°C)	T_3 (°C)	T_4 (°C)	T_5 (°C)	T_6 (°C)	T_7 (°C)
18,000	300	91.36	88.31	77.14	70.47	100.00	46.70	39.29
19,800	330	91.36	88.30	77.13	70.47	100.00	46.69	39.29
21,600	360	91.36	88.29	77.13	70.46	100.00	46.69	39.29
23,400	390	91.36	88.29	77.13	70.46	100.00	46.69	39.29
25,200	420	91.35	88.29	77.13	70.46	100.00	46.69	39.29

Example 4.5 (*One-dimensional transient in wall*)

On a cold night, a 10 cm thick concrete wall reaches a 10 °C uniform temperature. At 9 AM one face reaches 25 °C due to solar irradiation, while the other face is kept at 15 °C because of some environmental heat exchange. Divide the wall into 10 sections ($\Delta x = 1$ cm) and solve the transient temperature distribution along the wall thickness. Of course, the face temperatures will change along the day in a real problem, however, for the sake of analysis suppose the two face temperatures remain constant at the given values. Obtain the temperature distribution within the wall. How long will it take to reach the steady-state solution?

Solution

Let us set the time step to 60 s. Properties from concrete are given in Table A.4 (appendix section). The internal nodal temperature (T_m) equation for the 1-D problem is

$$T_m^{i+1} = Fo'\left(T_{m+1}^i + T_{m-1}^i\right) + \left(1 - 2Fo'\right)T_m^i$$

The numerical Fourier number is $Fo' = \frac{\alpha \Delta t}{(\Delta x)^2} = \frac{6.917 \times 10^{-7} \times 60}{(0.01)^2} = 0.415$, which satisfies the 1-D stability criterion, $Fo' < 1/2$.

Substituting the values, the nodal equation becomes

$$T_m^{i+1} = 0.415 \times \left(T_{m+1}^i + T_{m-1}^i\right) + 0.17T_m^i$$

The following table shows that the convergence to one decimal figure is reached in 135 min. Temperature evolution is shown in figure (curves were smoothed out).

t (min)	Position (cm)										
	0	1	2	3	4	5	6	7	8	9	10
0	10.0	10.0	10.0	10.0	10.0	10.0	10.0	10.0	10.0	10.0	10.0
1	25.0	10.0	10.0	10.0	10.0	10.0	10.0	10.0	10.0	10.0	15.0
5	25.0	19.1	14.6	11.6	10.4	10.0	10.1	10.5	11.5	13.0	15.0
10	25.0	20.9	17.2	14.3	12.4	11.4	11.2	11.6	12.4	13.6	15.0

(continued)

(continued)

t (min)	Position (cm)										
	0	1	2	3	4	5	6	7	8	9	10
25	25.0	22.5	20.2	18.2	16.5	15.3	14.6	14.3	14.3	14.6	15.0
50	25.0	23.5	22.0	20.7	19.4	18.3	17.4	16.7	16.0	15.5	15.0
135	25.0	24.0	23.0	22.0	21.0	20.0	19.0	18.0	17.0	16.0	15.0
300	25.0	24.0	23.0	22.0	21.0	20.0	19.0	18.0	17.0	16.0	15.0

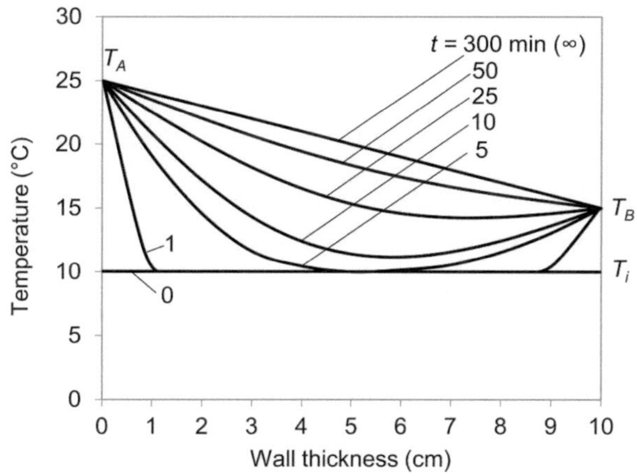

4.4 Proposed Problems

4.1 Determine if the lumped system assumption is valid for a 1010 steel sphere of 10 mm in diameter at 100 °C subjected to (a) natural convection air ($h = 10$ W/m^2°C); (b) forced convection air ($h = 50$ W/m^2°C); (c) forced convection water ($h = 6000$ W/m^2°C).

4.2 Determine the time constant of a type J thermocouple (iron-constantan), whose junction is assumed to be a 0.25 mm diameter sphere. The thermocouple is placed in an ice and liquid water bath at 0 °C, with h equal to 500 W/m^2°C. What do you suggest to do to reduce the time constant by half? (*Hint*: use properties of the junction as the arithmetic mean of the properties of iron—Table A.2—and constantan—specific heat $C = 390$ J/kg°C and $\rho = 8860$ kg/m^3.)

4.3 How long would it take for the thermocouple from proposed problem 4.2 to indicate 9.9 and 0.1 °C, if immersed in water at 10 °C having the same time constant?

4.4 A fuse is used in the protection of an electric circuit to protect it from overloads. An electric fuse is made of a tin wire with a diameter of $d = 0.8$ mm and a length of 10 mm. Normally, the electric circuit operates at 220 V and a 2.5 A nominal current. The fuse is located in an environment where the temperature is $T_\infty = 20$ °C with an average and combined heat transfer coefficient (radiation + convection) equal to $h = 18$ W/m^2°C—disregard heat conduction at its fuse ends. (a) Calculate the steady-state temperature (current of 2.5 A). (b) Suddenly, a 20 A over current happens. Estimate the time for the fuse to break open (electrical protection) as the wire fusion occurs. It is known that the tin electrical resistivity is 1.5×10^{-7} Ωm.

4.5 A full juice box, dimensions 17.5 × 9.5 × 6.4 cm, is in the refrigerator at 5 °C. Joe goes to the refrigerator takes out the box and leaves it on the sink in the environment that is at 20 °C. Obtain the heating curve over time of the juice box, that is, $T(t)$. Use the following main assumptions: (a) assume that the transport properties of the juice are the same as those of water; (b) assume that the box is made of cardboard with a thickness of 1.7 mm. $k_p = 0.05$ W/mK, and does not accumulate internal energy; (c) assume that the internal area is approximately equal to the external one; (d) the heat transfer by convection coefficient is 10 W/m^2°C; (e) the base of the box conducts heat to the sink, but assume that, for calculation purposes, the heat loss through the base is of the same order as the convection on the other faces of the box; (f) the juice heats up evenly (confirm this assumption later).

4.6 A copper sphere, initially at a uniform temperature T_0, is immersed in a fluid. Electric heaters allow temperature control in such a way that the fluid temperature $T_\infty(t)$ presents a periodic variation of the form: $T_\infty(t) - T_m = A\text{sen}(\omega t)$, where:

T_m average fluid temperature over time;
A amplitude of the temperature "wave";
ω angular frequency.

Obtain an expression for the sphere's temperature as a function of time and the heat transfer coefficient between the fluid and the sphere. Assume that the temperatures of the sphere and the fluid are uniform at any moment, thus allowing the use of the lumped analysis method.

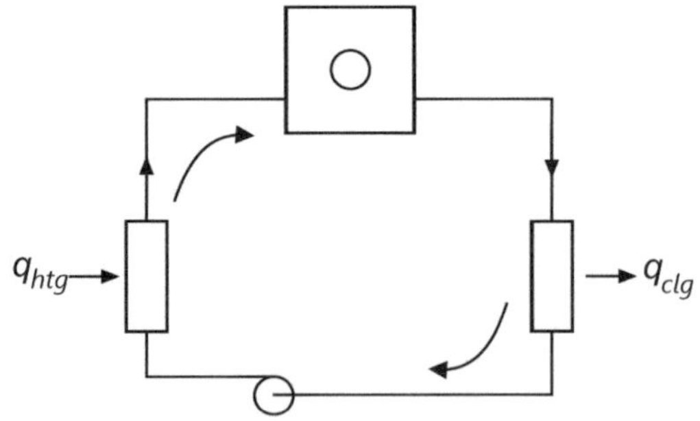

4.7 A 10 cm side aluminum cube at T_1 is placed in contact with a second 10 cm side copper cube at T_2, with $T_1 > T_2$. Assuming that there is only heat transfer by conduction between the contacting two block faces and that the contact resistance between them is negligible, determine the equilibrium temperature of the two blocks and the expression of the temperature difference $\vartheta(t) = T_1(t) - T_2(t)$ as a function of time. Determine the equilibrium temperature and the time required for this difference to be 10% of the initial difference if the aluminum and copper blocks have initial temperatures of 100 and 300 °C, respectively.

4.8 A large solid 20 cm thick brick wall reaches a uniform 5 °C temperature during the night. From 9 am, the air adjacent to the wall is at 18 °C and remains at that value until the beginning of the afternoon. Estimate the brick's temperature at the centerline and on the surface at noon. The convective heat transfer coefficient can be considered constant and equal to 40 W/m^2°C. Consider that both faces of the wall are exposed to the air and neglect thermal radiation incidence.

4.9 A 1010 steel plate 4 cm thick is initially at 40 °C. Suddenly, it is immersed in a hot oil bath at 100 °C and the temperature of the faces is kept at that value. Determine the temperature at the plate center 30 s after the operation. (*Hint*: you may use a Heisler diagram or the analytical solution, Eq. 4.28).

4.10 In a tempering heat treatment of steel-bearing balls $\left(C = 470 \text{ J/kgK}, \rho = 7850 \text{ kg/m}^3, k = 48 \text{ W/mK}\right)$, it is required to raise the surface temperature for a short time without heating significantly the ball interior. This treatment is performed in a melted salt bath at $T_\infty = 1000\,^\circ$C with a convective heat transfer coefficient $h = 5000$ W/m^{2}K. Assume that any location within the sphere whose temperature exceeds 727 °C will be tempered. Estimate the time required to reach the tempering depth of one millimeter in a 20 mm diameter sphere, if its initial temperature is 30 °C.

4.11 Find the transient temperature for an infinite slab subject to an initial temperature T_i in $t = 0$. The two surfaces undergo a sudden temperature change and are held at constant temperatures T_1 and T_2. (*Hint*: repeat the solution in Sect. 4.2.3 applying the given boundary conditions).

4.12 Derive Eq. (4.31).

4.13 Derive an internal nodal temperature for a 3-D transient system with internal energy generation.

4.14 Solve Example 4.5 for the case heat transfer convection occurs rather than fixed face temperatures. Assume the $h = 10$ W/m^2°C for both faces and that the far temperatures are that is, $T_A = 25$ °C and $T_B = 15$ °C (now they are the fluid temperature rather than the wall surface temperatures).

4.15 Solve Example 4.5 by the analytical method of separation of variables (Eq. 4.28) and compare both solutions for the wall centerline temperature (nodal point 5) as a function of time.

References

Gebhart, B. 1993. *Heat conduction and mass diffusion*. McGraw-Hill.
Heisler. B. M. P. 1947. Temperature charts for induction and constant temperature heating. *Transactions of ASME*.

Chapter 5
External Forced Convection

Abstract This chapter analyzes the phenomenon of forced heat transfer over bodies and Newton's cooling law to obtain the heat transfer rate. Firstly, the dimensionless numbers that govern the problem of external forced convection are presented, which are the Nusselt, Reynolds, and Prandtl numbers. The exact and integral or von Karman solutions are developed for the hydrodynamic or fluid dynamic and thermal laminar boundary layers on flat plates or surfaces and the Prandtl number influence on the relative growth between the two boundary layers is discussed. Various working expressions are obtained. The Reynolds-Colburn analogy between surface friction and heat transfer is presented for the laminar regime, which is also valid for the turbulent regime. The instantaneous turbulent conservation equations are presented for forced flow over flat surfaces on which time averages are applied, which give rise to Reynolds's apparent stresses. Working expressions for various situations and geometries of external flow over tubes of circular and other cross-section geometries are presented. The last section of the chapter deals with heat transfer over a tube bank because this is a conventional case of a heat exchanger configuration.

Keywords External flow · Laminar · Turbulent · Tubes · Tube bank

5.1 Introduction to Heat Convection—Newton's Cooling Law

The heat transfer rate by convection from a fluid in contact with a heated (or cooled) surface is governed by *Newton's cooling law* of 1701, defined by:

$$\dot{Q} = Ah(T_w - T_\infty) \tag{5.1}$$

where $\dot{Q}$ is the heat transfer rate, T_w and T_∞ are the temperatures of the surface and the fluid far away, respectively, A is the heat exchange area, that is, the area in contact with the fluid, and h is the convection heat transfer coefficient. The unit of the convection heat transfer coefficient is given by:

J. R. Simões-Moreira and E. W. Zavaleta-Aguilar, *Heat Transfer*, Heat and Mass Transfer, https://doi.org/10.1007/978-3-031-80318-5_5

$$[h] = \frac{[\dot{Q}]}{[A][T]} = \frac{\mathrm{W}}{\mathrm{m}^2\,{}^\circ\mathrm{C}} \text{ or } \frac{\mathrm{W}}{\mathrm{m}^2\mathrm{K}}$$

For a historical perspective on the development of the cooling law, refer to Besson (2012). As can be seen, the expression for calculating heat transfer is considerably simpler than that of heat conduction based on Fourier's differential equation (Eq. 2. 1), as it comprises only a temperature difference and a proportionality coefficient, h. However, the apparent simplicity hides the problem's complexity, because in general, the convection heat transfer coefficient h is a function of a large number of variables, such as the transport properties of the fluid (viscosity, density, and thermal conductivity), relative fluid velocity, geometry, and surface roughness, among others. Thus, it can be broadly stated that the fundamental problem of convection heat transfer is the determination of h specifically valid for the case under analysis. The theoretical value of the heat transfer coefficient by convection can be obtained in the laminar flow regime for some configurations and boundary conditions, such as the case of a flat surface. However, as a general rule for more complex or irregular geometries, the determination of h is usually based on experimental correlations attained in controlled laboratory experiments. Usually, the experimental data are reduced to make up the so-called *dimensionless numbers* that group physical quantities that control the convective heat exchange.

Because of the large number of variables controlling convection heat transfer, a dimensional analysis technique is often used, which is a method of reducing the number of independent variables in a given problem to a smaller set of dimensionless variables or numbers that group together the physical quantities of the phenomenon. Familiar dimensionless numbers are the Reynolds number in Fluid Mechanics and the Biot and Fourier numbers introduced in Chap. 4. The main limitation of dimensional analysis is that it does not provide any additional information about the nature of the phenomenon because all physical quantities that influence the phenomenon *must be known* before any analysis. Therefore, one must have a correct and complete preliminary physical understanding of the problem in advance.

The first step in applying the method consists of establishing the primary quantities. Then, all the physical quantities that influence the problem must be written in terms of those primary quantities, as discussed in the next subsection.

5.1.1 Dimensionless Numbers of Forced Convection Heat Transfer

Forced convective heat transfer is dominated by several physical quantities, such as the transport properties of the fluid, fluid velocity, geometry, dimensional lengths, and surface finishing. Due to the large number of types and nature of fluids (each with its transport property values), velocities, and geometries, it is common to work with groups or dimensionless numbers that allow the establishment of general heat transfer correlations valid in many cases. There are analytical formulations originating from analyses of conservation differential equations and empirical correlations that come from laboratory experiments. This subsection explains how to obtain dimensionless numbers and their interrelation through an example valid for the case of external forced convection on a flat surface. Thus, consider the flow of a fluid with velocity u_∞ and temperature T_∞ that is heated by a flat plate at a constant surface temperature, T_w higher than the fluid temperature, as illustrated in Fig. 5.1. In Table 5.1, all seven quantities that interfere with the heat transfer of this case are listed, including the transport properties of the fluid, its velocity, and the length of the flat plate.

The method of reducing the number of quantities that dominate the phenomenon to dimensionless groups assumes that *all* quantities that interfere with heat transfer are known. In Table 5.1, *M*, *L*, *T*, and *t* represent the quantities mass, length, temperature, and time, respectively, which make up the *primary system of quantities*. The number *N* of quantities that interfere in the phenomenon of convective heat transfer in a flat plate is $N = 7$, that is, 7 physical quantities control the forced heat transfer over flat plates (Table 5.1). On the other hand, there are $P = 4$ *primary quantities* (*MLTt*) that make up the physical dimensions of those 7 quantities. Thus, by the *theorem of* π or of *Buckingham*, there are 3 independent dimensionless groups *I* that govern the phenomenon by reducing the original 7 variables of the problem, that is:

$$I = N - P = 7 - 4 = 3 \text{ (dimensionless groups)}$$

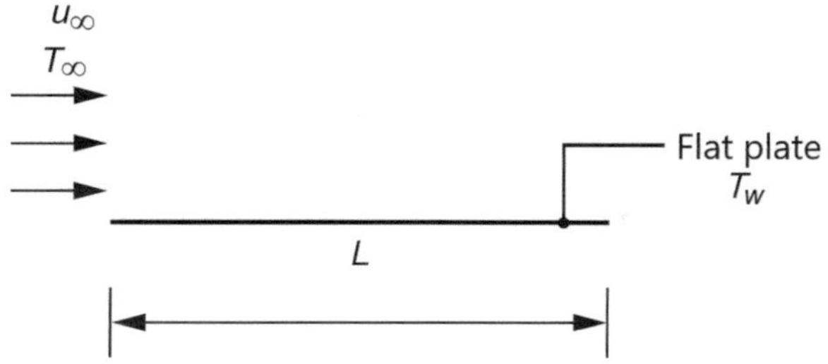

Fig. 5.1 Parallel flow of a fluid over a heat flat plate

Table 5.1 Quantities that interfere with the convective heat transfer of a moving fluid over a flat plate—MLTt system of primary quantities

Quantity		
Symbol	Description	Dimension
ρ	Density	M/L^3
μ	Dynamic viscosity	M/Lt
k	Thermal conductivity	ML/t^3T
C_P	Specific heat at constant pressure	L^2/t^2T
u_∞	Fluid velocity	L/t
L	Plate length	L
h	Convective heat transfer coefficient	M/t^3T

To obtain those 3 dimensionless groups, first a generic dimensionless group π is defined involving the 7 variables of the problem: $\pi = \rho^a \mu^b k^c c_P^d u_\infty^e L^f h^g$, with the sequence a ... g being exponents that make π a dimensionless number.

Now, replace the primary quantities ($MLTt$) of each magnitude, obtained in Table 5.1:

$$\pi = \left(\frac{M}{L^3}\right)^a \left(\frac{M}{Lt}\right)^b \left(\frac{ML}{t^3T}\right)^c \left(\frac{L^2}{t^2T}\right)^d \left(\frac{L}{t}\right)^e L^f \left(\frac{M}{t^3T}\right)^g$$

After grouping the primary quantities, we obtain:

$$\pi = M^{a+b+c+g} L^{-3a-b+c+2d+e+f} T^{-c-d-g} t^{-b-3c-2d-e-3g}$$

As it is a dimensionless π, all the exponents of the primary quantities must vanish, that is:

$$\begin{cases} a+b+c+g=0 \\ -3a-b+c+2d+e+f=0 \\ -c-d-g=0 \\ -b-3c-2d-e-3g=0 \end{cases}$$

Therefore, there is a system of 7 unknowns (a ... g) and only 4 equations. Thus, the system of equations is indeterminate. Next, the method assumes that some values are taken for a set of three exponents so that the system can have a unique solution. That is a critical point of the method, as it requires providing values in a judicious method that, in a certain way, demonstrates some prior knowledge of the problem. Following, the values of the exponents are established and the motivation for the selection is explained.

(a) As h is a quantity of interest and one wants to obtain an explicit relationship of that quantity, then one adopts $g = 1$. Now it remains to assume two more values, for example, assuming a and e equal to 0 to eliminate density and velocity contribution, respectively. So:

$$\begin{cases} g=1 \\ a=e=0 \end{cases}$$

Thus, one can solve the system of equations to obtain the missing exponents, that is: $b = 0$; $c = -1$; $d = 0$; $f = 1$. In this way, the corresponding dimensionless number is:

$$\pi = \rho^0 \mu^0 k^{-1} C_P^0 u_\infty^0 L^1 h^1$$

This first dimensionless group is named the *Nusselt number*, defined by:

$$Nu = \pi_1 = \frac{hL}{k} \tag{5.2}$$

In some texts, Nu is also referred to as a *dimensionless heat transfer coefficient*.

(b) Now, h is eliminated by setting $g = 0$ along with other assumed values:

$$\begin{cases} g = 0 \\ d = 0 \\ f = 1 \end{cases}$$

The solution of the system provides $a = 1; b = -1; c = 0; e = 1$

Resulting in: $\pi = \rho^1 \mu^{-1} k^0 C_P^0 u_\infty^1 L^1 h^0$

From which comes the second relevant dimensionless group of the problem, which is the *Reynolds number*, given by:

$$Re = \pi_2 = \frac{\rho u_\infty L}{\mu} \tag{5.3}$$

The Reynolds number represents the ratio between inertial forces and viscous forces.

(c) Finally, let the following values be for the exponents:

$$b = 1;\ c = -1;\ d = 1$$

This results in the third and last dimensionless number, which is named the *Prandtl number*.

$$Pr = \pi_3 = \frac{C_p \mu}{k} \tag{5.4}$$

As a concluding remark there were 7 variables in the original problem that were reduced to only 3 dimensionless numbers. Thus, those 3 dimensionless numbers can be related through functions (analytical or experimental) of the type:

$$F(\pi_1, \pi_2, \pi_3) = 0 \text{ or } F(Nu, Re, Pr) = 0 \tag{5.5a}$$

Even more, the Nusselt number, Nu, can be isolated to obtain its explicit dependence on the Reynolds, Re, and Prandtl, Pr, numbers that is:

$$Nu = f(Re, Pr) \tag{5.5b}$$

In doing so, experimental data measured in laboratory can provide the unknown function $Nu = f(Re, Pr)$ sought. Also, it is possible to obtain an analytical function itself in some cases, as analyzed in the following sections for laminar regime.

5.2 Laminar Heat Convection Over a Flat Plate

It was shown in the previous section that external forced heat transfer over surfaces results in three dimensionless numbers that control the phenomenon. Those quantities are the Nusselt number, Nu, the Reynolds number, Re, and the Prandtl number, Pr, that combined together form a function expressed by $Nu = f(Re, Pr)$. This section presents the functional relationship in an analytical and exact form for the case of forced flow over a flat surface in laminar regime. For that, the differential equations that govern forced heat transfer are presented.

Consider a fluid flow over a flat plate or surface, as illustrated in Fig. 5.2. Assuming the fluid has a uniform velocity profile (rectangular), u_∞, before hitting the plate. The front and back parts of the plate are known as the *leading* and the *trailing* edges, respectively. After the fluid hits the leading edge, the viscous friction will slow down the fluid portions next to the plate surface, initiating a *laminar hydrodynamic*[1] *boundary layer* (HBL), whose thickness grows continually as the fluid flows along the surface. The boundary layer concept was introduced by Prandtl, in 1904, as he found out that the fluid velocity adjacent to the surface is null, a condition known as *no-slip*, and it increases in the perpendicular direction until it reaches the velocity of the free flow in the far field, u_∞. In Fig. 5.2, the laminar velocity profile is schematized at the x distance from the leading edge. Note that the hydrodynamic laminar boundary layer thickness, δ_H, will continuously grow with x until hydrodynamic instabilities impose a regime transition to start the *turbulent flow regime*, if the plate is lengthy enough. It is assumed that the transition from the laminar flow regime to the turbulent flow regime happens for $Re_{x,\text{trans}} = \rho u_\infty x/\mu = 5 \times 10^5$ (sometimes, 3×10^5 is alternatively used), where x is the distance from the plate leading edge.

In the laminar flow regime, the fluid moves approximately parallel to the surface as if they were sliding "sheets", with the local shear stress, originating from viscous friction, defined by Eq. (5.6).

$$\tau_x = \mu \frac{du}{dy} \tag{5.6}$$

The above equation is valid for a Newtonian fluid (like air, water, and oil). The viscous friction slows down the fluid in the direction perpendicular to the plate, giving rise to a hydrodynamic boundary layer, which undergoes a continuous growth from the leading edge distance, that is, $\delta_H = \delta_H(x)$. The hydrodynamic boundary layer thickness δ_H is defined as the perpendicular distance from the surface $(y = 0)$

[1] In many texts, it is also known as *laminar fluid dynamic boundary layer*

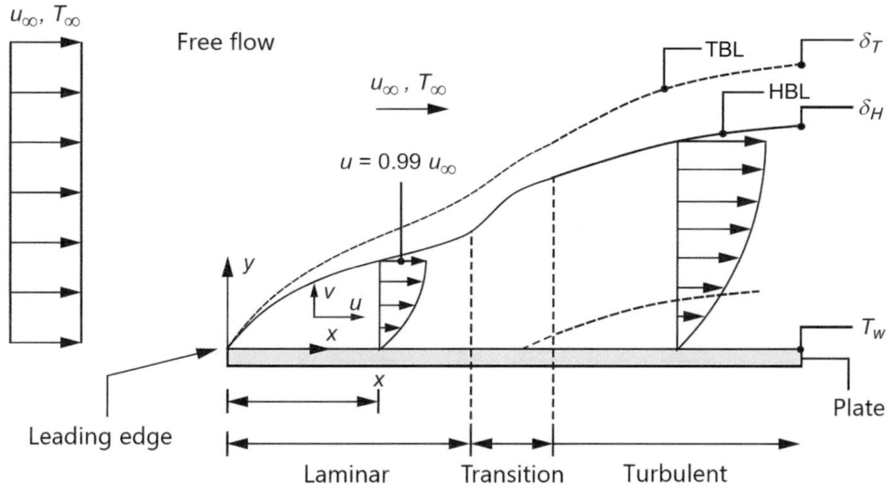

Fig. 5.2 Formation of the hydrodynamic boundary layers, δ_H, and thermal boundary layer, δ_T, on a flat plate

to the position where the fluid velocity reaches 99% of the free flow velocity, that is $u = 0.99u_\infty$.

Before reaching the plate and outside the hydrodynamic boundary layer the flow is known as *free flow*. Similarly, to the hydrodynamic boundary layer, if the fluid and the plate wall are at different temperatures, the *thermal boundary layer* (TBL) arises, as indicated in Fig. 5.2. Assuming that the free flow is at a temperature lower than the surface, the fluid will be gradually heated in the perpendicular direction. The thickness of the TBL, δ_T, is defined as the vertical distance at which the fluid temperature $T = T_{\delta_T}$ (also see Fig. 5.9) satisfies the following relationship:

$$\frac{T_{\delta_T} - T_w}{T_\infty - T_w} = 0.99 \tag{5.7}$$

where T_∞ and T_w are, respectively, the free flow and wall temperature.

5.2.1 *Exact Solution of Blasius (1908)*

To solve the problem of laminar layers, it is necessary to establish the governing differential equations of mass, momentum, and energy. Additionally, the following assumptions are usually made: incompressible fluid, steady state, two-dimensional flow with unit depth, no internal thermal energy (heat) generation, constant pressure in the direction perpendicular to the plate, negligible gravity effects, constant transport properties, and constant force and shear stress in the perpendicular direction. The

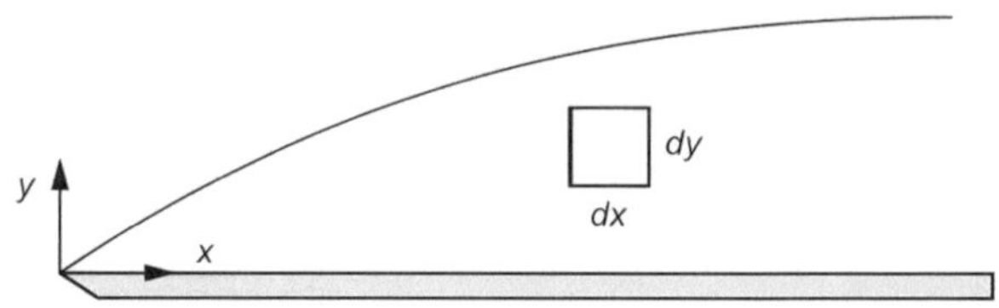

Fig. 5.3 Differential fluid element

balances are carried out in the differential control element (control volume—CV) within the laminar boundary layer (LBL), as indicated in Fig. 5.3.

Continuity or mass conservation equation

In a steady state condition, the mass balance equation for a differential two-dimensional element must obey the fact that the mass flow rates are conserved, that is, $\dot{m}_{\text{out}} = \dot{m}_{\text{in}}$. Keeping Fig. 5.4 in mind, one can write:

$$\rho u dy + \rho v dx = \rho\left(u + \frac{\partial u}{\partial x}dx\right)dy + \rho\left(v + \frac{\partial v}{\partial y}dy\right)dx$$

Simplifying by canceling out common terms and dividing by the elementary area, *dxdy*, we get:

$$\frac{\partial u}{\partial x} + \frac{\partial v}{\partial y} = 0 \tag{5.8}$$

or, in simplified notation,

$$\text{Div}\,\vec{V} = 0 \tag{5.9}$$

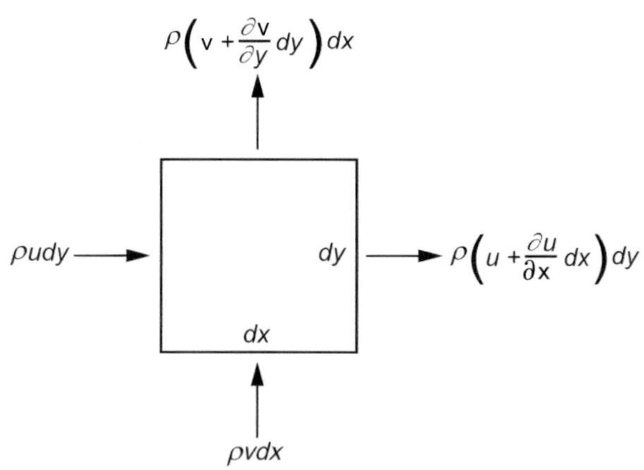

Fig. 5.4 Inputs and outputs of mass flow rates in an elementary 2-D control volume

where the divergent mathematical operator Div in Cartesian two-dimensional form is given by $\mathrm{Div} = \vec{i}\frac{\partial}{\partial x} + \vec{j}\frac{\partial}{\partial y}$ and the velocity vector is given by $\vec{V} = u\vec{i} + v\vec{j}$.

Momentum conservation equation

The next equation corresponds to Newton's Second Law, that is, the sum of all external forces acting on the surface of the fluid element is equal to the rate of change of momentum:

$$\sum \vec{F}_{\text{ext}} = \text{rate of change of momentum}$$

Note that this law is a vector equation, that is, the balance must be carried out in the x and y directions, in the two-dimensional case. Assuming that the fluid is Newtonian and that there are no pressure gradients in the y direction, the balance of forces and momentum in the direction parallel to the plate, that is, the x direction, will be analyzed next. Only the forces resulting from pressure and shear stress are considered.

From Fig. 5.5, it is found that the external forces are given by:

$$\sum F_x = pdy + \left(\tau + \frac{\partial \tau}{\partial y}dy\right)dx - \tau dx - \left(p + \frac{\partial p}{\partial x}dx\right)dy$$

or, simplifying by canceling out common terms, one obtains

$$\sum F_x = \frac{\partial \tau}{\partial y}dxdy - \frac{\partial p}{\partial x}dxdy$$

With the assumption of a Newtonian fluid, it is given that the shear stress is $\tau = \mu\frac{\partial u}{\partial y}$ which, substituting in the previous equation, comes:

$$\sum F_x = \mu\frac{\partial^2 u}{\partial y^2}dxdy - \frac{\partial p}{\partial x}dxdy \tag{5.A}$$

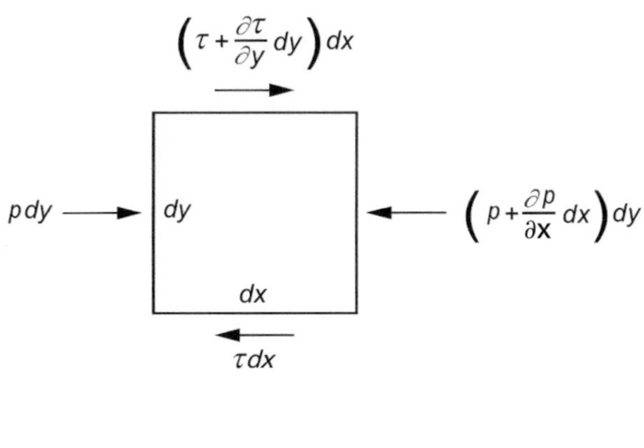

Fig. 5.5 Forces resulting from pressure and shear stress (friction) in the x direction parallel to a surface

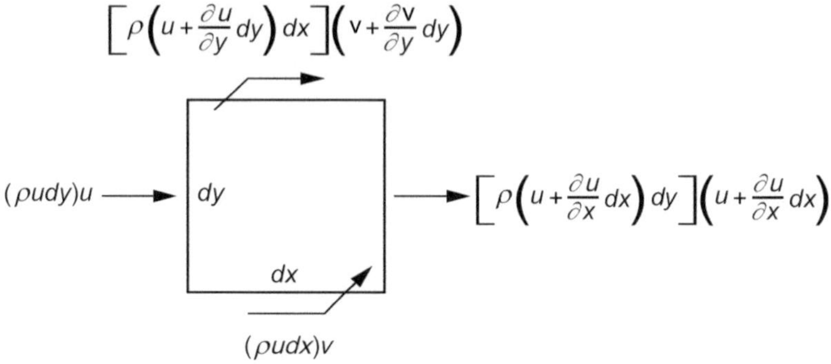

Fig. 5.6 Momentum change rate in the x direction

Now, the rate of momentum change in the x direction must be calculated. That magnitude is defined as the product of the mass flow rate by the velocity. Thus, in this case, the mass flow rate in the x direction on the left face is $\rho u dy$ and the velocity is the one acting on the corresponding face. Likewise, for the other faces. Those considerations are illustrated in Fig. 5.6.

Analyzing the flow crossing the elementary areas indicated in Fig. 5.6, it can be written that the net rate of momentum in the x direction is:

$$\begin{aligned}
\rho\left(u+\frac{\partial u}{\partial x}dx\right)^2 dy - \rho u^2 dy + \rho\left(v+\frac{\partial v}{\partial y}dy\right)\left(u+\frac{\partial u}{\partial y}dy\right)dx - \rho uvdx \\
= \rho u^2 dy + 2\rho u\frac{\partial u}{\partial x}dxdy + \rho\left(\frac{\partial u}{\partial x}dx\right)^2 dy - \rho u^2 dy + \rho vudx \\
+ \rho v\frac{\partial u}{\partial y}dxdy + \rho u\frac{\partial v}{\partial y}dxdy + \rho\frac{\partial v}{\partial y}\frac{\partial u}{\partial y}(dy)^2 dx - \rho uvdx \\
= 2\rho u\frac{\partial u}{\partial x}dxdy + \rho v\frac{\partial u}{\partial y}dxdy + \rho u\frac{\partial v}{\partial y}dxdy + \text{differentials of 3rd order}
\end{aligned}$$

Furthermore, after simplifying terms and some manipulation, it is possible to simplify this equation to obtain:

$$\rho\left(u\frac{\partial u}{\partial x}+v\frac{\partial u}{\partial y}\right)dxdy + \rho u\underbrace{\left(\frac{\partial u}{\partial x}+\frac{\partial v}{\partial y}\right)}_{=0\ (\text{Eq.5.8})}dxdy = \rho\left(u\frac{\partial u}{\partial x}+v\frac{\partial u}{\partial x}\right)dxdy \tag{5.B}$$

where the mass conservation equation (Eq. 5.8) was used. Therefore, it is now possible to equate the terms of the resultant of the external forces (Eq. 5.A) with the rate of momentum change (Eq. 5.B) in steady state, resulting in the momentum conservation equation, Equation (5.10).

$$\rho\left(u\frac{\partial u}{\partial x}+v\frac{\partial u}{\partial y}\right)=\mu\frac{\partial^2 u}{\partial y^2}-\frac{\partial p}{\partial x} \tag{5.10}$$

Energy conservation or the First Law of Thermodynamics equation

In the energy balance, three types of thermal energy exchange on the differential element are considered: the first is due to the enthalpic flow that enters or leaves the element; the second is the heat conduction in the perpendicular direction; and the third is due to the action of viscous forces (friction) that generate thermal energy (heat). It is considered here that this last one is negligible compared to the other terms, so it also neglected the variations of kinetic and potential energy. Figure 5.7 indicates all those forms of energy transfer mentioned.

The energy balance in a steady state without the generation of thermal energy (heat) and negligible work power is expressed by the following equation with the above-mentioned assumptions:

$$\underbrace{\dot{m}_e h_e+\dot{Q}_e}_{(I)}=\underbrace{\dot{m}_s h_s+\dot{Q}_s}_{(II)}$$

Now, each term is analyzed:

(I) *Rate of energy entering the elementary CV*: $\dot{m}_e h_e+\dot{Q}_e$

This rate is composed of the enthalpic flow ($\dot{m}_e h_e$) plus the heat rate transferred by heat conduction ($\dot{Q}_e$), that is:

$$\rho v h dx+\rho u h dy-k dx\frac{\partial T}{\partial y}-k dy\frac{\partial T}{\partial x}$$

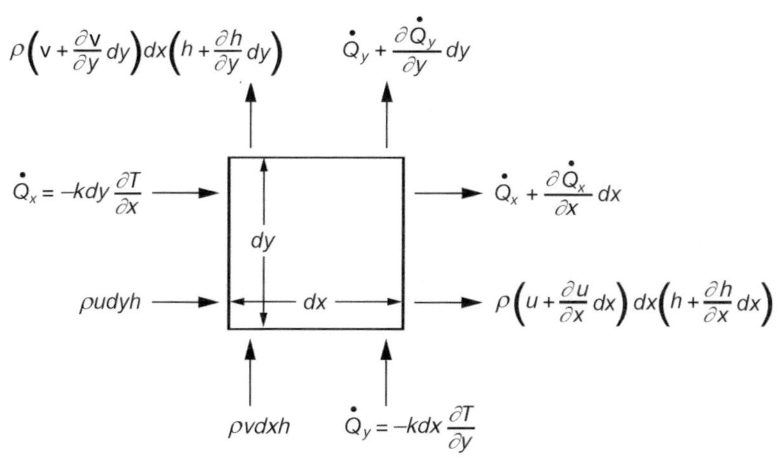

Fig. 5.7 Energy transfer on the surface of the differential element

(II) *Total rate of energy leaving the elementary CV*

$$\rho\left(v+\frac{\partial v}{\partial y}dy\right)\left(h+\frac{\partial h}{\partial y}dy\right)dx+\rho\left(u+\frac{\partial u}{\partial x}dx\right)\left(h+\frac{\partial h}{\partial x}dx\right)dy$$
$$-kdx\left(\frac{\partial T}{\partial y}+\frac{\partial^2 T}{\partial y^2}dy\right)-kdy\left(\frac{\partial T}{\partial x}+\frac{\partial^2 T}{\partial x^2}dx\right)$$

Combining (I) and (II) in the balance equation and disregarding the terms of higher order

$$\rho u\frac{\partial h}{\partial x}dxdy+\rho h\frac{\partial u}{\partial x}dxdy+\rho v\frac{\partial h}{\partial y}dxdy+\rho h\frac{\partial v}{\partial y}dxdy-k\frac{\partial^2 T}{\partial x^2}dxdy-k\frac{\partial^2 T}{\partial y^2}dxdy=0,$$

$$\text{or } \rho\left(u\frac{\partial h}{\partial x}+v\frac{\partial h}{\partial x}\right)+\rho h\underbrace{\left(\frac{\partial u}{\partial x}+\frac{\partial v}{\partial y}\right)}_{=0\ (\text{Eq.5.8})}=k\left(\frac{\partial^2 T}{\partial x^2}+\frac{\partial^2 T}{\partial y^2}\right)$$

Assuming that there is no fluid phase change, one can substitute $\partial h = C_p\partial T$; thus, we have the energy equation in differential form for the laminar boundary layer, given in Eq. (5.11):

$$\rho C_p\left(u\frac{\partial T}{\partial x}+v\frac{\partial T}{\partial y}\right)=k\left(\frac{\partial^2 T}{\partial x^2}+\frac{\partial^2 T}{\partial y^2}\right) \tag{5.11}$$

The work produced by viscous forces is usually negligible compared to the heat conduction term and convective energy transport. This occurs at low velocities. If the flow has high velocity or if its viscosity is high (case of high viscosity oils), on the right side of Eq. (5.11) the viscous dissipation term should be added, $\mu\phi$, where ϕ is the viscous dissipation function defined by Schlichting (1979):

$$\phi=2\left[\left(\frac{\partial u}{\partial x}\right)^2+\left(\frac{\partial v}{\partial y}\right)^2\right]+\left(\frac{\partial u}{\partial y}+\frac{\partial v}{\partial x}\right)^2 \tag{5.12}$$

Thus, the three conservation equations, mass, momentum, and energy, are summarized as:

Mass conservation: $\frac{\partial u}{\partial x}+\frac{\partial v}{\partial y}=0$

Momentum conservation in the x direction: $\rho\left(u\frac{\partial u}{\partial x}+v\frac{\partial u}{\partial y}\right)=\mu\frac{\partial^2 u}{\partial y^2}-\frac{\partial p}{\partial x}$

Energy conservation: $\rho C_p\left(u\frac{\partial T}{\partial x}+v\frac{\partial T}{\partial y}\right)=k\left(\frac{\partial^2 T}{\partial x^2}+\frac{\partial^2 T}{\partial y^2}\right)$

The following approximations are valid for boundary layers where $Re_x=\frac{\rho u_\infty x}{\mu}\gg 1$, given that, in this case, its thickness must be very thin:

$$u\gg v,\ \frac{\partial u}{\partial y}\gg\frac{\partial u}{\partial x},\ \frac{\partial v}{\partial y}\approx 0,\ \frac{\partial v}{\partial x}\approx 0$$

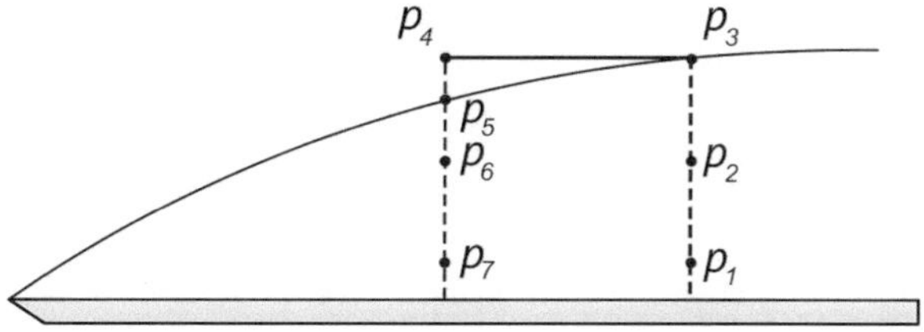

Fig. 5.8 Pressure distribution on a flat plate

It was assumed that there is no pressure gradient in the y direction, thus, $p_1 = p_2 = p_3$ in Fig. 5.8. Therefore, the pressure within the boundary layer can only depend on x, that is, $p = p(x)$. On the other hand, *outside the boundary layer* Eq. (5.10) can also be used and because the flow velocity is constant, it is concluded that $\partial p/\partial x = 0$ ($p_3 = p_4$). An analysis of the conservation of momentum in y outside the boundary layer indicates that $\partial p/\partial y = 0$, due to $v = 0$, therefore $p_4 = p_5$; thus, it is concluded that within the boundary layer $\frac{\partial p}{\partial x} = 0$, which means that the flow pressure is constant inside and outside the boundary layer over a flat plate.

Equation (5.10) can be rewritten in the form of Eq. (5.13), in which the kinematic viscosity, defined by $\nu = \mu/\rho$ is introduced,

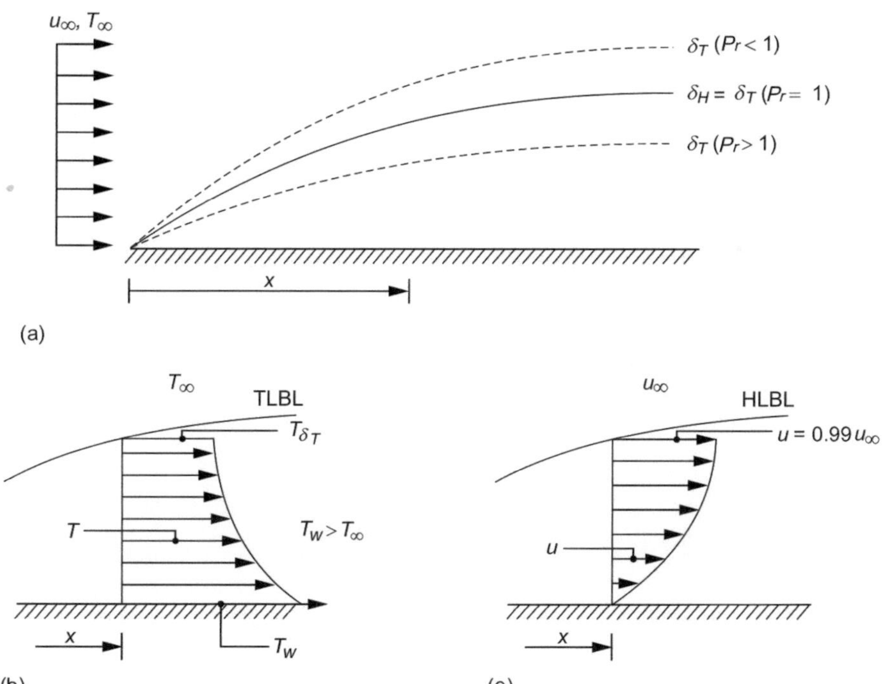

Fig. 5.9 **a** Influence of the Prandtl number on the relative development of boundary layer thicknesses, **b** temperature profile ($T_w > T_\infty$), **c** velocity profile

$$u\frac{\partial u}{\partial x} + v\frac{\partial u}{\partial y} = \nu\frac{\partial^2 u}{\partial y^2} \tag{5.13}$$

Furthermore, in the boundary layer, it is accepted that conduction in the direction x is considered negligible, that is: $\partial T/\partial y \gg \partial T/\partial x$ (conduction in the direction x may be relevant for liquid metals and at very low velocity flow). Thus, the equation of energy becomes:

$$u\frac{\partial T}{\partial x} + v\frac{\partial T}{\partial y} = \alpha\frac{\partial^2 T}{\partial y^2} \tag{5.14}$$

In summary, the three differential equations that govern the heat transfer in the laminar boundary layer are given in Table 5.2.

Comparing the equations of conservation of momentum and energy, an interesting case arises when the kinematic viscosity and thermal diffusivity are equal, $\nu = \alpha$, that is, when $Pr = \nu/\alpha = 1$, which corresponds to the case where the temperature and velocity distribution is similar, as it occurs when the two differential equations are subjected to similar boundary conditions. It is noteworthy to recall that most gases have $0.65 < Pr < 1$ and, therefore, that approach is usually applied to gases.

The three equations shown in Table 5.2 must be solved together with the following boundary conditions:

$$u(x, 0) = 0; T(x, 0) = T_w$$
$$v(x, 0) = 0$$
$$u(x, \infty) = u_\infty; T(x, \infty) = T_\infty$$

In Appendix B, it can be found the classic Blasius solution for the laminar boundary layer and the indicated equations and boundary conditions. The Blasius solution implies the introduction of several parameters of the laminar boundary layer. For this, we define:

The local *Nusselt number*, Nu_x, for a flat plate:

$$Nu_x = \frac{h_x x}{k} \tag{5.15}$$

Table 5.2 Differential equations of conservation of mass, momentum, and energy for a flat plate in a laminar flow regime

Balance	Equation	Number
Mass conservation	$\frac{\partial u}{\partial x} + \frac{\partial v}{\partial y} = 0$	5.8
Momentum conservation in the x direction	$u\frac{\partial u}{\partial x} + v\frac{\partial u}{\partial y} = \nu\frac{\partial^2 u}{\partial y^2}$	5.13
Energy conservation	$u\frac{\partial T}{\partial x} + v\frac{\partial T}{\partial y} = \alpha\frac{\partial^2 T}{\partial y^2}$	5.14

The *Reynolds number*, Re_x, for a flat plate:

$$Re_x = \frac{\rho u x}{\mu} \tag{5.16}$$

The *local friction coefficient*, $C_{f,x}$:

$$C_{f,x} = \frac{\tau_{w,x}}{\rho u_\infty^2/2} \tag{5.17}$$

where $\tau_{w,x}$ is the *local wall shear stress.* A subscript "x" indicates a local value at x distance from the leading edge.

The main results of the Blasius solution (Appendix B) are as follows:

Hydrodynamic laminar boundary layer thickness (HLBL), δ_H:

$$\delta_H = \frac{4.92\,x}{\sqrt{Re_x}} \tag{5.18}$$

Ratio between the thicknesses of the hydrodynamic laminar boundary layer, δ_H, and thermal laminar boundary layer (TLBL) δ_T :

$$\frac{\delta_H}{\delta_T} = Pr^{1/3} \tag{5.19}$$

Local friction coefficient:

$$C_{f,x} = 0.664 Re_x^{-1/2} \tag{5.20}$$

Average friction coefficient from the leading edge: $\overline{C}_{f,L} = \frac{1}{L}\int_0^L C_{f,x} dx$

$$\overline{C}_{f,L} = 1.328 Re_L^{-1/2} = 2C_{f,x=L} \tag{5.21}$$

Drag force imposed by the fluid on the entire plate:

$$F_D = \overline{C}_{f,L} A_p \frac{\rho u_\infty^2}{2} \tag{5.22}$$

Local heat transfer coefficient by convection:

$$h_x = 0.332 \frac{k}{x} Pr^{1/3} Re_x^{1/2} \tag{5.23}$$

Average heat transfer coefficient by convection: $\overline{h}_L = \frac{1}{L}\int_0^L h_x dx$

$$\overline{h}_L = 0.664 \frac{k}{L} Pr^{1/3} Re_L^{1/2} = 2h_{x=L} \tag{5.24}$$

Local Nusselt number:

$$Nu_x = 0.332Re_x^{1/2}Pr^{1/3}; \quad 0.6 \leq Pr \leq 50 \tag{5.25}$$

Average Nusselt number: $\overline{N}u_L = \frac{\overline{h}_L L}{k}$

$$\overline{N}u_L = 0.664Re_L^{1/2}Pr^{1/3} = 2Nu_{x=L} \tag{5.26}$$

Graphs in Fig. 5.9 indicate the relative behavior of the hydrodynamic and thermal laminar boundary layers. Note that the Prandtl number plays a decisive role in the relative thickness growth of the TLBL and HLBL. For $Pr < 1$ (liquid metals), the thermal boundary layer grows less sharply, that is, $\delta_T < \delta_H$; for $Pr \sim 1$ (gases), both boundary layers grow simultaneously; finally, for $Pr > 1$ (common liquids and water), the thermal boundary layer grows more sharply, that is, $\delta_T > \delta_H$.

By analyzing the local equations of friction coefficient (Eq. 5.20) and heat transfer coefficient by convection (Eq. 5.23), it can be inferred that both $C_{f,x}$ and h_x are reciprocal functions the square root of the distance from the leading edge, that is, $C_{f,x} \propto \frac{1}{\sqrt{x}}$, $h_{f,x} \propto \frac{1}{\sqrt{x}}$, whose behavior is indicated in the graphs in Fig. 5.10. It can also be verified, from Eqs. (5.21) and (5.24), that their respective average coefficients up to the distance $x = L$ are twice the local values at position $x = L$. It is concluded that the highest heat transfer (and friction) coefficients occur near the leading edge where the boundary layers begin.

In the design process of surfaces of heat exchangers, the adoption of several grooves spaced in the direction transverse to the flow can be a good practice, as the best heat exchange coefficients occur at the beginning of the flow over the surface and each groove can be seen as the leading edge of a new boundary layer. However, there is a penalty in terms of increased surface friction.

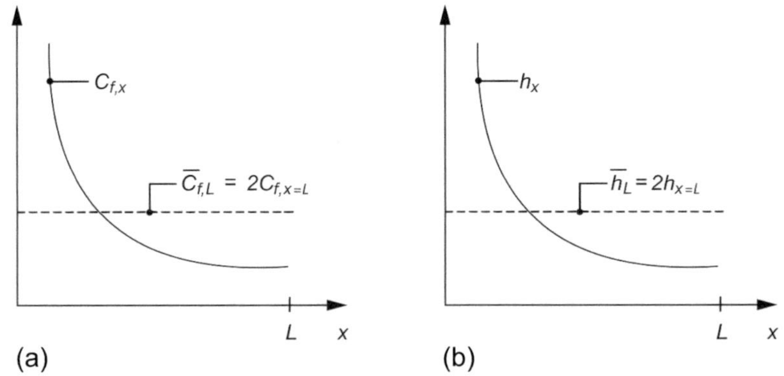

Fig. 5.10 Variation of friction coefficients (**a**) and heat transfer (**b**) by convection with position x from the beginning of the plate

5.2.2 Integral or Approximate Solution of von Kármán

In Sect. 5.2.1, the differential equations of the laminar boundary layer and the results of Blasius´s solution were presented. However, a second solution method of the two boundary layers is the subject of study in this subsection. It is the approximate solution based on the *integral method*, also known as *von Kármán's integral solution.*

In this case, a differential volume is defined only in the x direction parallel to the flow, whose height H extends beyond the boundary layer thickness, that is, $H > \delta_H$, as illustrated in Fig. 5.11. Again, the depth of the plate perpendicular to the page plane is unitary.

The laws of mass and momentum conservation in the differential element 1-2-B-A-1 in Fig. 5.11 in steady state are presented:

Mass balance

Mass flow rate at face 1-A (entering): $\int_0^H \rho u dy$

Mass flow rate at face 2-B (outgoing): $\int_0^H \rho u dy + \frac{d}{dx}\left(\int_0^H \rho u dy\right)dx$

Thus, the net mass flow rate at face A-B is $-\frac{d}{dx}\left(\int_0^H \rho u dy\right)dx$, which is the difference between the mass flow rate entering at 1-A and the one leaving at 2-B.

Momentum flux balance in the x direction

Momentum flux at Face 1-A: $\int_0^H \rho u^2 dy$

Momentum flux on Face 2–B: $\int_0^H \rho u^2 dy + \frac{d}{dx}\left(\int_0^H \rho u^2 dy\right)dx$

Momentum flux on Face A–B: $u_\infty \frac{d}{dx}\left(\int_0^H \rho u dy\right)dx$

The net rate of momentum flux out of the control volume: (face 2 − B) − (face A − B) − (face 1 − A) = $\frac{d}{dx}\left(\int_0^H \rho u^2 dy\right)dx - u_\infty \frac{d}{dx}\left(\int_0^H \rho u dy\right)dx$

Recalling the product differentiation rule, it follows that:

$$d(\alpha\beta) = \alpha d(\beta) + \beta d(\alpha) \text{ or } \alpha d(\beta) = d(\alpha\beta) - \beta d(\alpha)$$

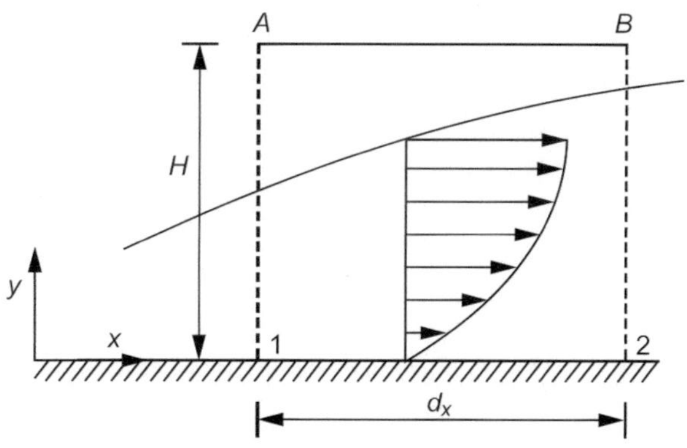

Fig. 5.11 1-D differential control volume for the integral or von Kármán solution

Making $\alpha = u_\infty$ and $\beta = \int_0^H \rho u dy$, we have:

$$u_\infty \frac{d}{dx}\left(\int_0^H \rho u dy\right)dx = \frac{d}{dx}\left(u_\infty \int_0^H \rho u dy\right)dx - \left(\int_0^H \rho u dy\right)\frac{du_\infty}{dx}dx$$
$$= \frac{d}{dx}\left(\int_0^H \rho u_\infty u dy\right)dx - \frac{du_\infty}{dx}\left(\int_0^H \rho u dy\right)dx$$

Now, substituting into the expression of the net flow of momentum flux, we get:

$$\text{momentum flux} = \frac{d}{dx}\left(\int_0^H \rho u^2 dy\right)dx - \frac{d}{dx}\left(\int_0^H \rho u_\infty u dy\right)dx + \frac{du_\infty}{dx}\left(\int_0^H \rho u dy\right)dx$$

The first two terms of the previous equation can be combined to obtain the following more compact form:

$$\text{momentum flux} = \frac{d}{dx}\left(\int_0^H \rho (u - u_\infty) u dy\right)dx + \frac{du_\infty}{dx}\left(\int_0^H \rho u dy\right)dx \quad \text{(A)}$$

Now, the external forces resultant can be obtained. In this case, only pressure and friction forces indicated schematically in Fig. 5.12 must be considered.

- resultant force of pressure: $-H\frac{dp}{dx}dx$
- shear force on the wall: $-\mu \left.\frac{\partial u}{\partial y}\right|_{y=0} dx$

$$\text{Resultant force} = -\mu \left.\frac{\partial u}{\partial y}\right|_{y=0} dx - H\frac{dp}{dx}dx \quad \text{(B)}$$

Finally, the integral equation of the laminar hydrodynamic boundary layer can be written (Second Law of Newton) equating the resultant forces (Eq. B) with the net momentum flux (Eq. A), that is:

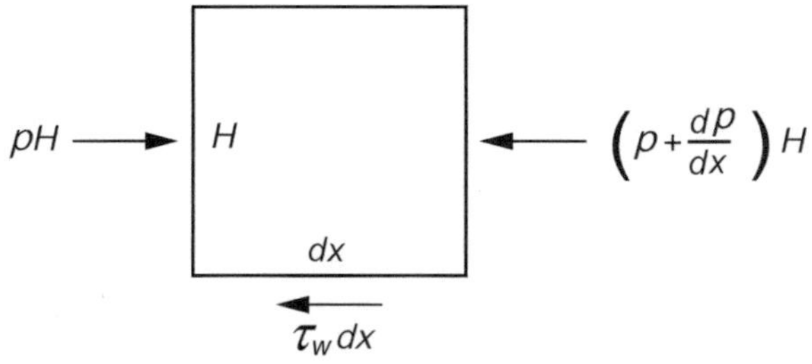

Fig. 5.12 Forces acting on the elementary control volume

$$-\mu \left.\frac{\partial u}{\partial y}\right|_{y=0} dx - H\frac{dp}{dx}dx = \frac{d}{dx}\left(\int_0^H \rho(u - u_\infty)udy\right)dx + \frac{du_\infty}{dx}\left(\int_0^H \rho udy\right)dx \tag{5.27}$$

If the pressure is a constant throughout the flow, as it occurs in a flow over a flat surface, that is, p = cte implies that the far-field velocity is also constant. Thus, Bernoulli's equation is valid outside the boundary layer,

$$\frac{p}{\rho} + \frac{u_\infty}{2} = \text{cte} \tag{5.28}$$

So that, in differential form: $\frac{dp}{\rho} + \frac{2u_\infty du_\infty}{2} = 0 \Rightarrow du_\infty = 0$

Thus, the momentum conservation equation for a free flow over a flat surface is reduced to:

$$-\mu \left.\frac{\partial u}{\partial y}\right|_{y=0} = \frac{d}{dx}\left(\int_0^H \rho(u - u_\infty)udy\right)$$

However, as the differential CV height H is greater than the hydrodynamic boundary layer thickness, that is, $H > \delta_H$, and the velocity is a constant, $u = u_\infty$, outside the boundary layer, then:

$$\rho\frac{d}{dx}\left[\int_0^{\delta_H} (u_\infty - u)udy\right] = \mu \left.\frac{\partial u}{\partial y}\right|_{y=0} \tag{5.29}$$

Equation (5.29) represents the final form of the *momentum conservation equation,* valid for a free laminar flow over a surface or flat plate and it is an exact formulation in integral form. The solution to that equation depends on the knowledge of the velocity profile $u(y)$, which is not known. The approximate solution is given by providing a fitting or adjusted velocity profile, that is, $u(y)$. It is a common technique to curve fit the velocity profile by a polynomial approximation, whose degree depends on the boundary conditions of the problem. Thus, first, consider the four boundary conditions of the laminar hydrodynamic boundary layer, which are:

BC_1: $u = 0$ at $y = 0$: no-slip condition at the wall;

BC_2: $u = u_\infty$ at $y = \delta_H$: outside the HBL the fluid has the free flow velocity $u = u_\infty$;

BC_3: $\frac{\partial u}{\partial y} = 0$ at $y = \delta_H$: the transition from HBL to free flow is smooth;

BC_4: $\frac{\partial^2 u}{\partial y^2} = 0$ at $y = 0$: this condition arises from Eq. (5.10) at the wall.

The first three boundary conditions are simple and immediate. The first informs that the velocity at the plate surface is null (non-slip principle); the second states that outside the HBL the velocity is that of the free flow; and the third states that the transition between the HBL and the free flow is "smooth", hence the derivative is null. The last boundary condition arises from the differential equation of momentum conservation, Eq. (5.10), which requires that this condition be null at the plate surface

($y = 0$), where u and v are null, as well as the pressure gradient. Once these four boundary conditions are established, the polynomial distribution that satisfies those conditions is of 3rd degree, given by:

$$u(y) = C_1 + C_2 y + C_3 y^2 + C_4 y^3 \tag{5.30}$$

Hence, after applying the four boundary conditions, we obtain the approximate velocity profile given by Eq. (5.31):

$$\frac{u(y)}{u_\infty} = \frac{3}{2}\frac{y}{\delta_H} - \frac{1}{2}\left(\frac{y}{\delta_H}\right)^3 \tag{5.31}$$

Introducing this approximate velocity profile into Eq. (5.29), we get:

$$\begin{aligned}\rho u_\infty^2 \frac{d}{dx}\left[\int_0^{\delta_H}\left(1 - \frac{3}{2}\frac{y}{\delta_H} + \frac{1}{2}\left(\frac{y}{\delta_H}\right)^3\right)\left(\frac{3}{2}\frac{y}{\delta_H} - \frac{1}{2}\left(\frac{y}{\delta_H}\right)^3\right)dy\right] \\ = \mu u_\infty \frac{\partial}{\partial y}\left[\frac{3}{2}\frac{y}{\delta_H} - \frac{1}{2}\left(\frac{y}{\delta_H}\right)^3\right]_{y=0}\end{aligned} \tag{5.32}$$

which, after integration, derivation and further simplification, results in:

$$\frac{d}{dx}\left(\frac{39}{280}\rho u_\infty^2 \delta_H\right) = \frac{3}{2}\frac{\mu u_\infty}{\delta_H} \tag{5.33}$$

Equation (5.33) can be integrated by separation of variables, recalling that, for $x = 0 \rightarrow \delta_H = 0$ (the laminar HBL starts at the leading edge), that is: $\frac{78}{840}\frac{\rho u_\infty}{\mu}\int_0^{\delta_H}\delta_H d\delta_H = \int_0^x dx$, to obtain the final form of the approximate hydrodynamic LBL.

$$\delta_H(x) = \frac{4.64x}{\sqrt{Re_x}} \tag{5.34}$$

Considering the assumptions, the approximate result is quite reasonable compared to the exact solution, with the only difference being the constant 4.64, which is 4.92 from the exact solution (Eq. 5.18), signifying a difference of less than—6% between the two solutions.

Thermal laminar boundary layer

Once known the hydrodynamic laminar boundary layer, the approximate therminal laminar boundary layer can be solved in a similar fashion. Heat exchange between the surface and the fluid occurs by heat conduction at $y = 0$. Consequently, the exact definition of the heat transfer coefficient by convection is defined as:

$$h_x(T_w - T_\infty) = -k \left.\frac{\partial T}{\partial y}\right|_{y=0} \tag{5.35}$$

From which the exact definition of the local heat transfer coefficient (local) h_x is:

$$h_x = \frac{-k \left.\frac{\partial T}{\partial y}\right|_{y=0}}{T_w - T_\infty} \tag{5.36}$$

Thus, it is necessary to know the temperature distribution to find the value of h_x, $T = T(y)$. Similarly, to the hydrodynamic case, the following boundary conditions applies to the temperature distribution curve:

BC$_1$: $T = T_w$ at $y = 0$ same surface temperature;

BC$_2$: $\frac{\partial T}{\partial y} = 0$ at $y = \delta_T$ no temperature gradient outside the laminar TBL;

BC$_3$: $T = T_\infty$ at $y = \delta_T$ same temperature as the fluid at the laminar TBL edge;

BC$_4$: $\frac{\partial^2 T}{\partial y^2} = 0$ at $y = 0$ this condition is the energy conservation equation applied to the fluid next to the wall with $u = 0$ and $v = 0$ (see Eq. 5.14).

Consider Fig. 5.13, where the surface heating begins at some distance x_0 from the leading edge and remains at a constant surface temperature, T_w. As before, a 3rd degree polynomial distribution is assumed for the temperature distribution, that is:

$$T(y) = C_1 + C_2 y + C_3 y^2 + C_4 y^3 \tag{5.37}$$

With the introduction of the four boundary conditions, Eq. (5.38) of the approximate temperature distribution inside the laminar TBL is obtained, similarly to what was obtained with the conservation of momentum equation:

$$\frac{\theta(y)}{\theta_\infty} = \frac{T(y) - T_w}{T_\infty - T_w} = \frac{3}{2}\frac{y}{\delta_T} - \frac{1}{2}\left(\frac{y}{\delta_T}\right)^3 \tag{5.38}$$

Analogously to the hydrodynamic case, an energy balance can be developed in a control volume with a thickness H greater than δ_T, as illustrated in Fig. 5.14.

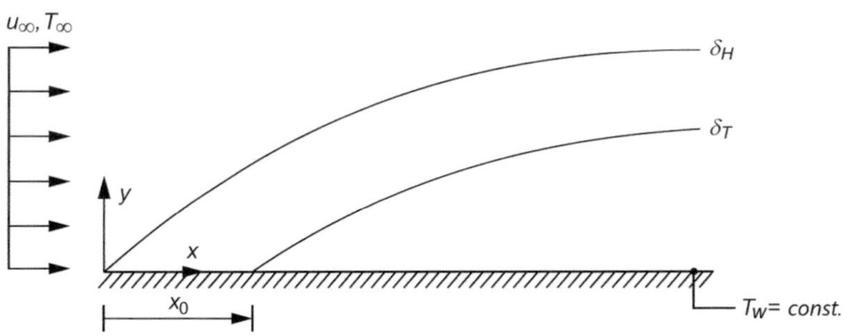

Fig. 5.13 Growth of thermal and hydrodynamic boundary layers

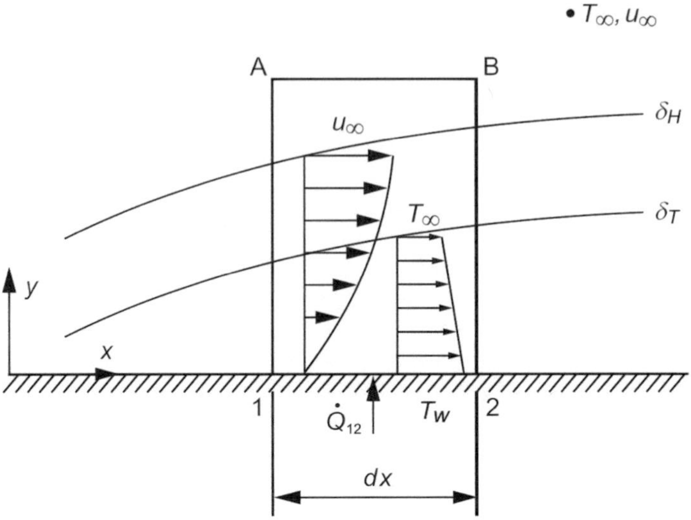

Fig. 5.14 Differential unidimensional control volume considered for the von Kármán solution

Energy balance in the control volume 1-2-A-B

Enthalpic flow at face 1–A: $\int_0^H \rho u C_p T dy$

Enthalpic flow at face 2–B: $\int_0^H \rho u C_p T dy + \frac{d}{dx}\left(\int_0^H \rho u C_p T dy\right)dx$

Enthalpic flow at face A–B: $C_p T_\infty \frac{d}{dx}\left(\int_0^H \rho u dy\right)dx$

Heat rate by conduction on surface face 1–2: $\dot{Q}_{12} = -kdx \left.\frac{\partial T}{\partial y}\right|_{y=0}$

Power of viscous forces inside the element: $\dot{W}_a = \left[\int_0^H \mu\left(\frac{du}{dy}\right)^2 dy\right]dx$

Inserting these quantities into the energy balance, we get:

$$-\int_0^H \rho u C_p T dy - \frac{d}{dx}\left(\int_0^H \rho u C_p T dy\right)dx + \int_0^H \rho u C_p T dy + T_\infty \frac{d}{dx}\left(\int_0^H \rho u C_p dy\right)dx$$
$$+\left[\int_0^H \mu\left(\frac{du}{dy}\right)^2 dy\right]dx - kdx \left.\frac{\partial T}{\partial y}\right|_{y=0} = 0$$

Simplifying and rearranging the terms and introducing the constant properties hypothesis, we obtain:

$$\frac{d}{dx}\left[\int_0^H (T_\infty - T)u dy\right] + \frac{\mu}{\rho C_p}\left[\int_0^H \left(\frac{du}{dy}\right)^2 dy\right] = \alpha \left.\frac{\partial T}{\partial y}\right|_{y=0} \quad (5.39)$$

Compare this result with Eq. (5.29). Generally, in most applications, the thermal power generated by the viscous terms is negligible compared to the enthalpic flow and heat conduction. Therefore, the second term is neglected. Substituting the approximate velocity (Eq. 5.31) and temperature (Eq. 5.38) profiles, the following relationship between the laminar thermal and hydrodynamic boundary layers thickness is obtained:

$$\frac{\delta_T}{\delta_H} = \frac{1}{1.026} Pr^{-1/3} \left[1 - \left(\frac{x_0}{x}\right)^{3/4}\right]^{1/3} \tag{5.40}$$

If heating or cooling starts from the leading edge, $x_0 = 0$, we have:

$$\frac{\delta_T}{\delta_H} = \frac{Pr^{-1/3}}{1.026} \tag{5.41}$$

Comparing this result with the exact solution (Eq. 5.19), a mere difference of 2.6% is found. During the process of deducing the equation, it was assumed $\delta_T < \delta_H$, which encompasses most liquids and many gases, whose Prandtl numbers are on the order of one and higher.

Finally, the local heat transfer coefficient, h_x, can be obtained with the approximate temperature profile (Eq. 5.38), that is:

$$h_x = \frac{-k \left.\frac{\partial T}{\partial y}\right|_{y=0}}{T_w - T_\infty} = -\frac{3}{2} \frac{k(T_\infty - T_w)}{(T_w - T_\infty)\delta_T} = \frac{3}{2} \frac{k}{\delta_T} = \frac{3}{2} \frac{k}{\delta_H} \left(\frac{\delta_H}{\delta_T}\right), \text{ or}$$

substituting the ratio between the thicknesses of the boundary layers (Eq. 5.41), comes:

$$h_x = \frac{3k}{2} \frac{1.026 Pr^{\frac{1}{3}}}{\delta_H} \left[1 - \left(\frac{x_0}{x}\right)^{\frac{3}{4}}\right]^{-\frac{1}{3}}, \text{ or}$$

substituting the thickness of the hydrodynamic LBL (Eq. 5.34), we get:

$$h_x = 0.332 \frac{k}{x} Re_x^{1/2} Pr^{1/3} \left[1 - \left(\frac{x_0}{x}\right)^{3/4}\right]^{-1/3} \tag{5.42}$$

Introducing the definition of Nu_x, we have:

$$Nu_x = 0.332 Re_x^{1/2} Pr^{1/3} \left[1 - \left(\frac{x_0}{x}\right)^{3/4}\right]^{-1/3} \tag{5.43}$$

whose value is remarkedly equal to the exact solution (Eq. 5.23) for $x_0 = 0$. In many engineering calculations, it is often desired to determine the average heat transfer coefficient, $\overline{h}_L$, over the plate length, rather than the local value, h_x. Thus, consider the calculation of the average heat transfer coefficient for a heated plate from its leading edge, that is, $x_0 = 0$, up to a location L distant from the edge. By the definition of average, we have: $\overline{h}_L = \frac{\int_0^L h_x dx}{L}$. Substituting Eq. (5.42), we get:

$$\overline{h}_L = 2\left[0.332\frac{k}{L}Re_L^{1/2}Pr^{1/3}\right] = 2h_{x=L} \tag{5.44}$$

The above equation implies that the average $\overline{h}_L$ is twice the local value, $h_{x=L}$ for an isothermal flat plate. Similarly, it is shown that: $\overline{Nu}_L = \frac{\overline{h}L}{k} = 2Nu_{x=L}$ and, therefore,

$$\overline{Nu}_L = 0.664Re_L^{1/2}Pr^{1/3} = 2Nu_{x=L} \tag{5.45}$$

As already stated, approximate results of the local (Eq. 5.43) and average (Eq. 5.45) Nusselt numbers are the same as the exact solutions, Eqs. (5.25) and (5.26), respectively.

Transport properties of the fluid are to be evaluated at the film temperature, T_f, unless otherwise stated.

$$T_f = \frac{T_w + T_\infty}{2} \tag{5.46}$$

Example 5.1 (*Laminar boundary layer over an isothermal flat plate*)

Hot air at 70 °C and 3 m/s is used to heat a flat surface 1 m long and 0.5 m wide flowing parallel to the plate in the lengthwise direction, as illustrated in the figure. When a steady state is reached, the plate remains at an average temperature of 30 °C. Under these conditions, determine:

(a) the thickness of the hydrodynamic and thermal boundary layer at the end of the plate;
(b) the local and average convection heat transfer coefficient at the end of the plate;
(c) the rate of heat transferred to the plate surface.

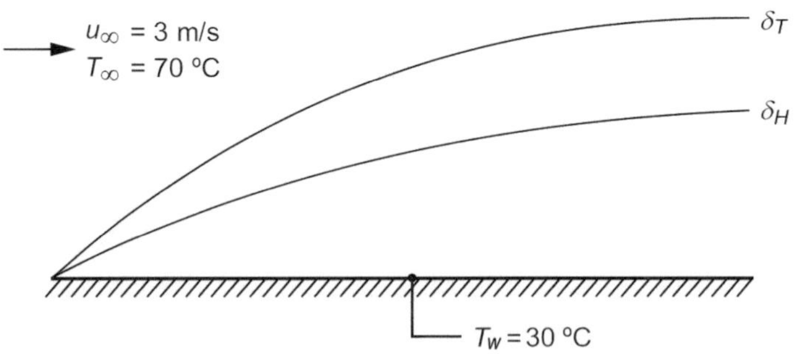

Solution

Properties calculated at film temperature, $T_f = \frac{30+70}{2} = 50\,°\mathrm{C}$

From Appendix A, Table A.5

$$\nu = 1.80 \times 10^{-5}\,\mathrm{m^2/s};\ k = 0.02809\,\mathrm{W/m°C};\ Pr = 0.7044;\ \text{and}\ \rho = 1.092\,\mathrm{kg/m^3}$$

In solving this kind of problems, we first check if the flow is laminar at the plate end, which is done by calculating the Reynolds number.

$Re_L = \frac{u_\infty L}{\nu} = \frac{3\times 1}{1.80\times 10^{-5}} = 1.67 \times 10^5 < 5 \times 10^5$. Therefore, it is laminar.

The expressions of the exact method will be used, although for the calculation of the heat transfer coefficient by convection it is indifferent.

(a) Equation (5.18) Hydrodynamics

$$\delta_H = \frac{4.92x}{\sqrt{Re_x}} = \frac{4.92 \times 1}{\sqrt{1.67 \times 10^5}} \rightarrow \delta_H = 0.012\,\mathrm{m}\ \text{or}\ \delta_H = 12.0\,\mathrm{mm}$$

Equation (5.19) Thermal

$$\delta_T = \frac{\delta_H}{Pr^{1/3}} = \frac{0,012}{0.7044^{1/3}} \rightarrow \delta_T = 0.013\,\mathrm{m}\ \text{or}\ \delta_T = 13.4\,\mathrm{mm}$$

(b) Equation (5.25) Local

$$Nu_x = 0.332 Re_x^{1/2} Pr^{1/3} = 0.332\left(1.67 \times 10^5\right)^{1/2} (0.7044)^{1/3} = 120.7$$

$$h_x = \frac{Nu_x k}{L} = \frac{120.7 \times 0.02809}{1} \rightarrow h_x = 3.39\,\mathrm{W/m^2K}$$

Equation (5.26) Average

$$\overline{Nu_L} = 0.664 Re_L^{1/2} Pr^{1/3} = 0.664 \times 167000^{1/2} \times 0.7044^{1/3} = 241.5$$

$$\bar{h}_L = \frac{\overline{Nu_L} k}{L} = \frac{241.5 \times 0.02809}{1} \rightarrow \bar{h}_L = 6.78\,\mathrm{W/m^2K}$$

(c) $\dot{Q} = A\bar{h}_L(T_\infty - T_w) = 1 \times 0.5 \times 6.78 \times (70 - 30)$

$$\dot{Q} = 135.6\,\mathrm{W}.$$

In the case of a *uniform heat flux* along the plate, $q_w\left[\mathrm{W/m^2}\right]$, as it occurs with electric heating or incidence of thermal and solar or other thermal radiation, the correlation for the local Nusselt number is given by Kays et al. (2005):

$$Nu_x = 0.453 Re_x^{1/2} Pr^{1/3} \tag{5.47}$$

Valid for $Pr > 0.6$ and $Re_x < 5 \times 10^5$. The local heat transfer coefficient for uniform heat flux is:

$$h_x = 0.453kPr^{1/3}\left(\frac{u_\infty}{\upsilon x}\right)^{1/2} \tag{5.48}$$

Notice that *Re* and *Pr* dependencies are the same as those for the isothermal plate case, only varying the multiplying constant (0.453 instead of 0.332). An important quantity in the case of the constant heat flux is the surface or wall temperature since it is now a quantity that varies along the plate, that is, $T_{w,x} = T_w(x)$. For this, we have:

$$q_w = h_x\left(T_{w,x} - T_\infty\right) \rightarrow T_{w,x} = \frac{q_w}{h_x} + T_\infty$$

Substituting the local coefficient from Eq. (5.48), we get:

$$T_{w,x} = \frac{2.208\, q_w}{kPr^{1/3}}\left(\frac{\nu x}{u_\infty}\right)^{1/2} + T_\infty \tag{5.49}$$

Finally, the average temperature on the plate along the length L is given by $\overline{T}_w = \frac{1}{L}\int_0^L T_w dx$, which results in:

$$\overline{T}_w = \frac{1.472\, q_p}{kPr^{1/3}}\left(\frac{\nu L}{u_\infty}\right)^{1/2} + T_\infty \tag{5.50}$$

The transport properties should be evaluated at the temperature $T_f = \frac{\overline{T}_w + T_\infty}{2}$.

Example 5.2 (*Laminar boundary layer over a flat plate—uniform heat flux*)

A thin plate is heated by solar radiation at 500 W/m^2 fully absorbed, while the plate back is thermally insulated. Air at 25 °C and 3 m/s flows parallel 1 m lengthwise and 0.5 m wide, as illustrated in the previous example figure. Under these conditions, determine:

(a) rate of heat exchanged with the air;
(b) the average temperature of the plate;
(c) the temperature of the plate at the trailing edge;
(d) compare the average temperature with the arithmetic mean between the two temperatures at the leading and trailing edge. Are they the same?
(e) make a graph illustrating the temperature variation along the plate, as well as the heat transfer coefficient by convection.

Solution

$$q_w = 500\,\mathrm{W/m^2};\ u_\infty = 3\,\mathrm{m/s};\ T_\infty = 25\ ^\circ\mathrm{C};\ L = 1\,\mathrm{m};\ b = 0.5\,\mathrm{m}$$

(a) In steady state, the absorbed radiation must be transferred to the air by convection, so the rate of heat exchanged with the air is: $\dot{Q} = q_w A = 500 \times 1 \times 0.5 = 250$ W;

(b) The average surface temperature is evaluated by Eq. (5.50). Transport properties should be evaluated at the film temperature $T_f = \frac{\overline{T}_w + T_\infty}{2}$, which is unknown, because the average surface temperature is not known. Thus, an initial wall temperature is assumed $\overline{T}_w = 129\ ^\circ$C, then $T_f = 77\ ^\circ$C. From Table A.5:

$$\nu = 2.071 \times 10^{-5}\,\mathrm{m^2/s};\ k = 0.030\,\mathrm{W/mK};\ Pr = 0.7$$

$$\overline{T}_w = \frac{1.472 \times 500}{0.03 \times 0.7^{1/3}} \left(\frac{2.071 \times 10^{-5} \times 1}{3} \right)^{1/2} + 25 \Rightarrow \overline{T}_w = 98\ ^\circ\mathrm{C}$$

Since $\overline{T}_w = 98\ ^\circ$C is lower than the one initially assumed (129 °C), the calculations must be redone with this new temperature, resulting in $T_f = 61\ ^\circ$C. From Table A.5, the new properties are obtained:

$$\nu = 1.907 \times 10^{-6}\,\mathrm{m^2/s};\ k = 0.02881\,\mathrm{W/mK};\ Pr = 0.703$$

$\overline{T}_w = \frac{1.472 \times 500}{0.02881 \times 0.703^{1/3}} \left(\frac{1.907 \times 10^{-5} \times 1}{3} \right)^{1/2} + 25 \rightarrow \overline{T}_w = 97.4\ ^\circ$C, an acceptable value with what was assumed (98 °C).

(c) The temperature at the end of the plate (trailing edge) is:

$$T_{w,L} = \frac{2.208\ (500)}{0.02881 \times 0.703^{1/3}} \left(\frac{1.907 \times 10^{-5} \times 1}{3} \right)^{1/2} + 25 \rightarrow T_{w,L} = 133.6\ ^\circ\mathrm{C}$$

(d) At the leading edge, the plate temperature is 25 °C, because $L = 0$ in Eq. (5.50). Thus, the arithmetic mean temperature is: (25 + 133.6) / 2 = 79.3 °C, while the actual average temperature is 97.4 °C. Therefore, those temperatures are not equal, because the average temperature is NOT the simple arithmetic mean.

(e) The temperature distribution over the plate is:

$$\begin{aligned} T_{w,x} &= \frac{2.208 \times 500}{0.02881 \times 0.703^{1/3}} \left(\frac{1.907 \times 10^{-5} x}{3} \right)^{1/2} + 25 \rightarrow T_{w,x} \\ &= 108.6\sqrt{x} + 25 \end{aligned}$$

The local coefficient of heat transfer by convection is:

$$h_x = 0.453 \times 0.02881 \times 0.703^{1/3} \left(\frac{3}{1.907 \times 10^{-5} x} \right)^{1/2} \rightarrow h_x = \frac{4.60}{\sqrt{x}}$$

The distribution of the plate temperature and the local heat transfer coefficient are indicated in the following graphs.

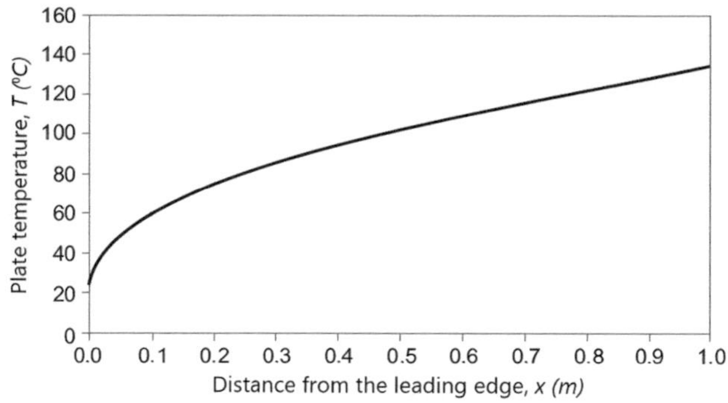

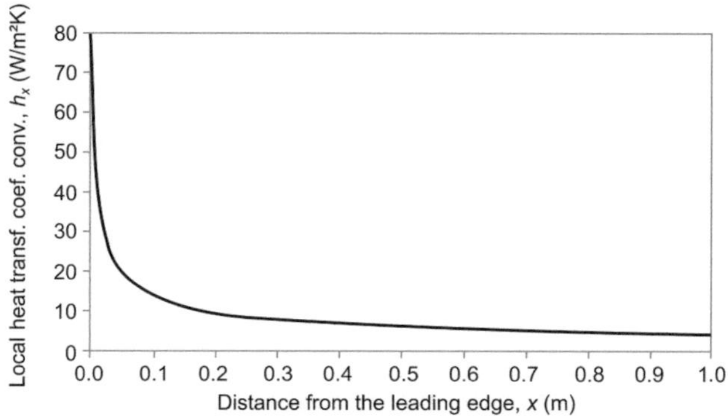

5.2.3 Analogy of Reynolds-Colburn or Analogy Between Heat Transfer and Surface Friction

Heat and momentum transfers are governed by analogous differential equations, as indicated in Table 5.2. The analogy between the two phenomena is investigated in this subsection through the *Reynolds-Colburn analogy*, which correlates surface friction to heat transfer for a laminar flow over a flat plate. What is its utility? The great advantage is that laboratory measurement data on surface friction, which historically are more abundant, can be used to estimate the convection heat transfer coefficient utilizing this analogy.

The local friction coefficient, $C_{f,x}$, was defined by Eq. (5.20), and the local Nusselt number, Nu_x, by Eq. (5.25). Dividing both equations, we obtain Eq. (5.51).

$$\frac{Nu_x}{Re_x Pr} Pr^{2/3} = \frac{C_{f,x}}{2} \tag{5.51}$$

The combination of the three dimensionless numbers outcomes in the definition of the *local Stanton number*, St_x given by:

$$St_x = \frac{Nu_x}{Re_x Pr} = \frac{h_x}{\rho C_p u_\infty} \tag{5.52}$$

So that the analogy between the dimensionless numbers can be written compactly using Eq. (5.53), which results in the analogy:

$$St_x Pr^{2/3} = \frac{C_{f,x}}{2} \tag{5.53}$$

Equation (5.53) connects the surface friction with the heat transfer in laminar flow over a flat plate. Thus, the convective heat transfer coefficient can be determined from the friction drag force measurements on the plate. It can also be applied to turbulent flow (which will be seen in the next section) over a flat plate and modified for turbulent flow inside tubes. The analogy is valid for both local and average values over a length L.

$$\overline{St}_L = \frac{\overline{Nu}_L}{Re_L Pr} = \frac{\overline{h}_L}{\rho C_p u_\infty} \tag{5.54}$$

$$\overline{St}_L Pr^{2/3} = \frac{\overline{C}_{f,L}}{2} \tag{5.55}$$

Example 5.3 (*Analogy between surface friction and heat transfer*)

A thin flat plate is subjected to a parallel water flow, at 1.5 m/s and 80 °C. The dimensions of the plate are indicated in figure. A dynamometer measures the fluid viscous force on the surface of the plate and records 1.74 N. The plate is at 20 °C. Evaluate the average convection heat transfer coefficient between the plate and the water.

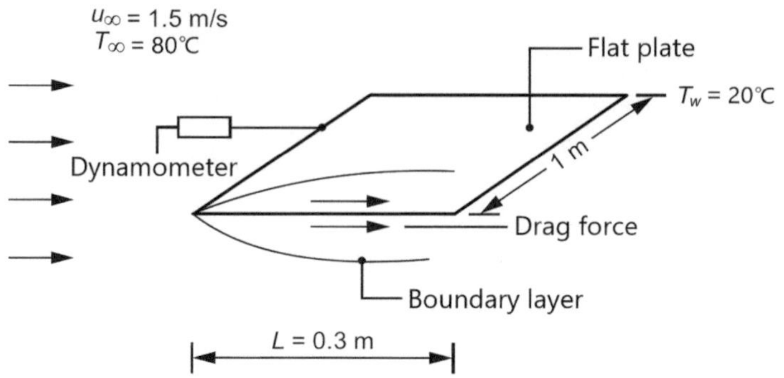

Solution

The viscous drag force on the plate acts on both faces of the plate, thus, on one face it is half, that is, 0.87 N.

The properties are evaluated at the film temperature of 50 °C. From Table A.7:

$$\rho = 989.1\,\text{kg/m}^3;\ \mu = 577 \times 10^{-6}\text{kg/ms};\ Pr = 3.77$$

The drag coefficient is (Eq. 5.20):

$$\overline{C}_{f,L} = \frac{F}{A_s\rho u_\infty/2} = \frac{0.87}{0.3 \times 1 \times 989.1 \times 1.5/2}$$
$$\overline{C}_{f,L} = 0.0026$$

From Eqs. (5.54) and (5.55) of the R-C analogy, it follows:

$$\overline{h}_L = \frac{\overline{C}_{f,L}}{2}\frac{\rho C_p u_\infty}{Pr^{2/3}} = \frac{0.0026}{2}\frac{989.1 \times 4180 \times 1.5}{3.77^{2/3}} \Rightarrow \overline{h}_L = 3328.3\ \text{W/m}^2\text{K}.$$

Note that the Reynolds number of 771.395 > 500.000 indicates that a turbulent flow regime has been reached and consequently, it would not be possible to use Eq. (5.24) throughout the plate because it is valid for laminar flow only. However, we can extend the analogy to the turbulent flow regime as informed above.

5.2.4 Similarity Between Heat Transfer Processes

Let us analyze the momentum equation (Eq. 5.13) in its dimensionless form. For this, the following dimensionless variables are introduced: $x^* = \frac{x}{L}$, $y^* = \frac{y}{L}$, $u^* = \frac{u}{u_\infty}$ in that equation, which results in:

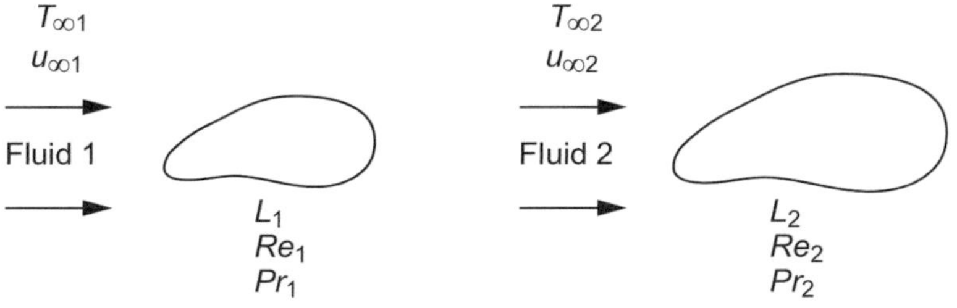

Fig. 5.15 Similarity between two processes with scale geometries

$$u^* \frac{\partial u^*}{\partial x^*} + v^* \frac{\partial u^*}{\partial y^*} = \frac{1}{Re_L} \frac{\partial^2 u^*}{\partial y^{*2}} \tag{5.56}$$

Note that when two geometries are similar, but on a different scale (Fig. 5.15), the dimensionless momentum equation valid for both cases is Eq. (5.56). Thus, it can be stated that if the two cases have the same dimensionless numbers $Re_1 = Re_2$, they will have the same solution and also the friction coefficients will be equal, that is, $C_{f1} = C_{f2}$, within the laminar flow regime.

Similarly, by substituting the dimensionless temperature, $T^* = \frac{(T - T_w)}{(T_\infty - T_w)}$, in the energy equation (Eq. 5.14), we obtain:

$$u^* \frac{\partial T^*}{\partial x^*} + v^* \frac{\partial T^*}{\partial y^*} = \frac{1}{Re_L Pr} \frac{\partial^2 T^*}{\partial y^{*2}} \tag{5.57}$$

Thus, it can be stated that if the product $Re_1 = Re_2$, and $Pr_1 = Pr_2$, therefore it will have that $Nu_1 = Nu_2$, within the laminar flow regime.

5.3 Turbulent Heat Convection on External Surfaces

5.3.1 Turbulent Boundary Layer (TuBL) and the Flat Surface

The convective heat and momentum transfer in the turbulent boundary layer is, phenomenologically, different from those in laminar boundary layers. To understand the mechanism of heat transfer and momentum exchange in the turbulent boundary layer, it is found that the hydrodynamic boundary layer is formed by three sublayers, as illustrated in the scheme in Fig. 5.16.

Those three sublayers are:

- laminar sublayer—similar to laminar flow—molecular action;
- buffer layer—molecular effects are still felt—transition to plain turbulent layer;
- turbulent layer—macroscopic mixtures of fluid portions.

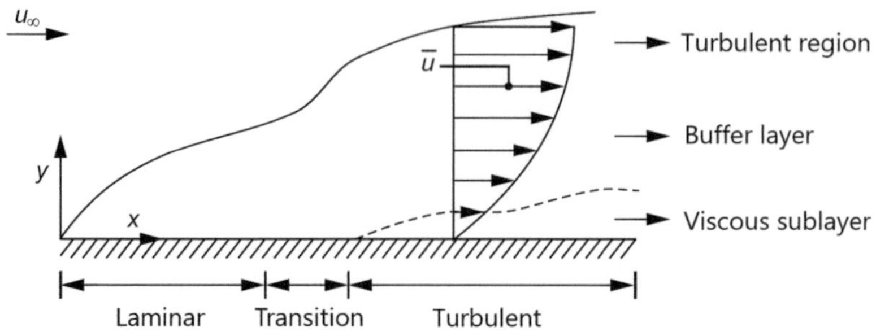

Fig. 5.16 Hydrodynamic turbulent boundary layer

To understand the effect of turbulence, consider an exercise of observing the oscillatory behavior of local velocity at a fixed inner diminute region in the flow, as illustrated in the time graph in Fig. 5.17. It is noted that the instantaneous velocity, u, fluctuates considerably around an average velocity, $\overline{u}$.

Although it may seem irrelevant in a preliminary analysis, velocity fluctuation (u') along with other fluctuation quantities introduces the greatest difficulties in modeling, which gives rise to the "turbulence problem". To analyze the problem, the instantaneous velocity is split into two components: an average value and a fluctuation, as indicated:

– velocity in the parallel direction, $u : u = \overline{u} + u'$
– velocity in the perpendicular direction, $v : v = \overline{v} + v'$.

The same is done with the term for local pressure oscillation:

– pressure: $p = \overline{p} + p'$.

Fig. 5.17 Local (u), average ($\overline{u}$), and fluctuating (u') fluid velocity at a point inside the turbulent flow

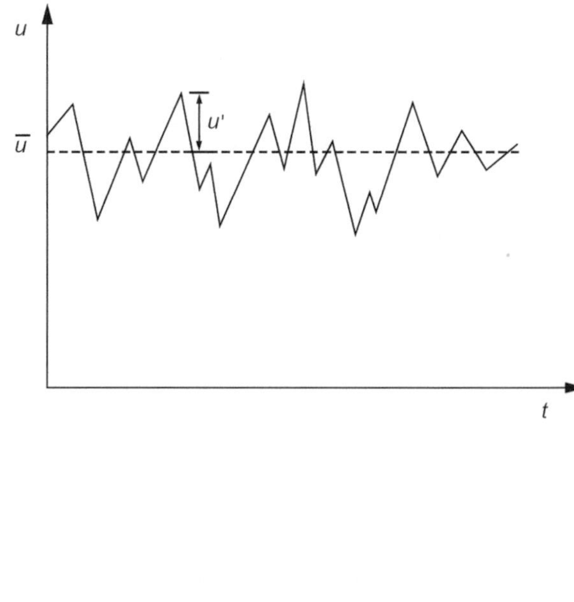

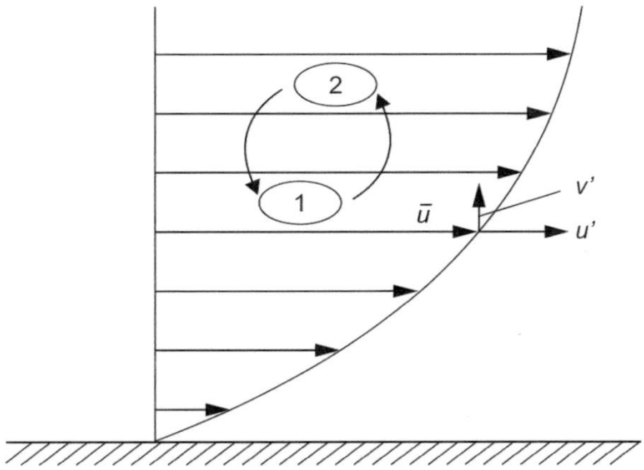

Fig. 5.18 Turbulent movements of the fluid within the boundary layer

The fluctuation terms are accountable for the occurrence of apparent magnitudes, called *Reynolds apparent stresses*, which must be considered in the analysis as seen next.

To have a phenomenological view of the Reynolds apparent stresses, consider the illustration of the turbulent boundary layer in Fig. 5.18. Unlike the laminar case, where the fluid "slides" over itself as if it were thin sheets, there are macroscopic mixtures of "portions" of fluid in the turbulent case. In the example illustrated in Fig. 5.18, a "portion" of fluid (1) is moving upwards, carrying with it its velocity (momentum) and energy (heat transfer). A corresponding "portion" (2) descends to fill the place of the one that ascended. The macroscopic movement of fluid portions produces fluctuations in velocity and other properties. From a mathematical point of view, those fluid movements within the boundary layer give rise to the greatest difficulties in modeling the turbulence problem.

A more detailed analysis of the turbulent boundary layer problem is treated in specific literature, such as Kays et al. (2005) or Bejan (2013). However, the main steps of the modeling are indicated in the following.

The first step is to write down the differential conservation equations. Then, the instantaneous values are replaced by the corresponding average and fluctuation terms, that is, $u = \overline{u} + u'$, $v = \overline{v} + v'$ and $p = \overline{p} + p'$. This procedure is called *Reynolds's time averages.* Starting with the equation of instantaneous momentum conservation (Eq. 5.10), we have:

$$u\frac{\partial u}{\partial x} + v\frac{\partial u}{\partial y} = \nu\frac{\partial^2 u}{\partial y^2} - \frac{1}{\rho}\frac{\partial p}{\partial x} \tag{5.10}$$

Now, substituting the decomposition of the quantities $u = \overline{u} + u'$, $v = \overline{v} + v'$, and $p = \overline{p} + p'$ in this equation, it follows:

$$(\overline{u}+u')\frac{\partial}{\partial x}(\overline{u}+u')+(\overline{v}+v')\frac{\partial}{\partial y}(\overline{u}+u')=\nu\frac{\partial^2}{\partial y^2}(\overline{u}+u')-\frac{1}{\rho}\frac{\partial}{\partial x}(\overline{p}+p')$$

By carrying out the multiplications, the following form of this equation is obtained:

$$\overline{u}\frac{\partial\overline{u}}{\partial x}+\overline{u}\frac{\partial u'}{\partial x}+u'\frac{\partial\overline{u}}{\partial x}+u'\frac{\partial u'}{\partial x}+\overline{v}\frac{\partial\overline{u}}{\partial y}+\overline{v}\frac{\partial u'}{\partial y}+v'\frac{\partial\overline{u}}{\partial y}+v'\frac{\partial u'}{\partial y}$$
$$=\nu\frac{\partial^2\overline{u}}{\partial y^2}+\nu\frac{\partial^2 u'}{\partial y^2}-\frac{1}{\rho}\frac{\partial\overline{p}}{\partial x}-\frac{1}{\rho}\frac{\partial p'}{\partial x}$$

At this point, it is convenient to perform a time average over a time interval between two significant instants, $\Delta t=t_2-t_1$. The time interval Δt must be long enough to capture the relevant flow fluctuation information. For this, the following time average is defined over an instantaneous quantity f any:

$$\overline{f}=\frac{1}{\Delta t}\int_{t_1}^{t_2}f(t)dt \tag{5.58a}$$

The following properties apply to averages:

$$\overline{f_1+f_2}=\overline{f_1}+\overline{f_2}\,,\ \overline{Cf_1}=C\overline{f_1},\ \overline{\overline{f_1}}=\overline{f_1},\ \overline{\frac{\partial f_1}{\partial s}}=\frac{\partial\overline{f_1}}{\partial s}\,,\text{ and }\overline{f'}=0 \tag{5.58b}$$

being C a constant in the time interval and s a spatial coordinate (x, y, and z). Thus, applying the time average over the momentum equation, we get:

$$\overline{\overline{u}\frac{\partial\overline{u}}{\partial x}}+\overline{\overline{u}\frac{\partial u'}{\partial x}}+\overline{u'\frac{\partial\overline{u}}{\partial x}}+\overline{u'\frac{\partial u'}{\partial x}}+\overline{\overline{v}\frac{\partial\overline{u}}{\partial y}}+\overline{\overline{v}\frac{\partial u'}{\partial y}}+\overline{v'\frac{\partial\overline{u}}{\partial y}}+\overline{v'\frac{\partial u'}{\partial y}}$$
$$=\overline{\nu\frac{\partial^2\overline{u}}{\partial y^2}}+\overline{\nu\frac{\partial^2 u'}{\partial y^2}}-\overline{\frac{1}{\rho}\frac{\partial\overline{p}}{\partial x}}-\overline{\frac{1}{\rho}\frac{\partial p'}{\partial x}}$$

Using the properties of time mean (Eq. 5.58b), we have:

$$\overline{u}\frac{\partial\overline{u}}{\partial x}+\overline{u}\underbrace{\overline{\frac{\partial u'}{\partial x}}}_{0}+\underbrace{\overline{u'}}_{0}\frac{\partial\overline{u}}{\partial x}+\overline{u'\frac{\partial u'}{\partial x}}+\overline{v}\frac{\partial\overline{u}}{\partial y}+\overline{v}\overline{\frac{\partial u'}{\partial y}}+\underbrace{\overline{v'}}_{0}\frac{\partial\overline{u}}{\partial y}+\overline{v'\frac{\partial u'}{\partial y}}$$
$$=\nu\frac{\partial^2\overline{u}}{\partial y^2}+\nu\underbrace{\overline{\frac{\partial^2 u'}{\partial y^2}}}_{0}-\frac{1}{\rho}\frac{\partial\overline{p}}{\partial x}-\frac{1}{\rho}\underbrace{\overline{\frac{\partial p'}{\partial x}}}_{0}$$

Rewriting, we have Eq. (5.59a) after simplifications:

$$\overline{u}\frac{\partial \overline{u}}{\partial x} + \overline{v}\frac{\partial \overline{u}}{\partial y} = \nu \frac{\partial^2 \overline{u}}{\partial y^2} - \frac{1}{\rho}\frac{\partial \overline{p}}{\partial x} - \left(\overline{u'\frac{\partial u'}{\partial x}} + \overline{v'\frac{\partial u'}{\partial y}} \right) \tag{5.59a}$$

Furthermore, it is necessary to address the time averages involving fluctuations (terms in parentheses) separately. The following mathematical identities can be written:

$$\overline{u'\frac{\partial u'}{\partial x}} = \overline{\frac{\partial u'u'}{\partial x}} - \overline{u'\frac{\partial u'}{\partial x}} \quad \text{and} \quad \overline{v'\frac{\partial u'}{\partial y}} = \overline{\frac{\partial u'v'}{\partial y}} - \overline{u'\frac{\partial v'}{\partial y}}$$

So that the terms on the right hand side of Eq. (5.59a) can be written as:

$$\overline{u'\frac{\partial u'}{\partial x}} + \overline{v'\frac{\partial u'}{\partial x}} = \frac{\partial \overline{u'u'}}{\partial x} + \frac{\partial \overline{u'v'}}{\partial y} - \overline{u'\underbrace{\left(\frac{\partial u'}{\partial x} + \frac{\partial v'}{\partial y}\right)}_{0}}$$

The term in parentheses on the right side is null by the law of conservation of mass (Eq. 5.8). Substituting the previous equality, we obtain the form of the turbulent differential equation of momentum conservation, which is:

$$\overline{u}\frac{\partial \overline{u}}{\partial x} + \overline{v}\frac{\partial \overline{u}}{\partial y} = \nu \frac{\partial^2 \overline{u}}{\partial y^2} - \frac{1}{\rho}\frac{\partial \overline{p}}{\partial x} - \left(\frac{\partial \overline{u'u'}}{\partial x} + \frac{\partial \overline{u'v'}}{\partial y} \right) \tag{5.59b}$$

In the process of deriving Eq. (5.59a), it was assumed that the time average of the fluctuations and their derivatives are null, which is true considering that those are random fluctuations around an average value and, therefore, their time averages should be null. However, the procedure introduces new terms that involve the time average of the derivative of the fluctuation products that form the terms between parentheses in Eq. (5.59b), which are not null. Finally, there is one last simplification that involves the turbulent boundary layer. For the case of two-dimensional flow, it is found that the gradient of the product of the fluctuations in the main direction x in Eq. (5.59b) (first term in parentheses) is negligible compared to the second term, so the final equation of the conservation of turbulent momentum is:

$$\overline{u}\frac{\partial \overline{u}}{\partial x} + \overline{v}\frac{\partial \overline{u}}{\partial y} = -\frac{1}{\rho}\frac{\partial \overline{p}}{\partial x} + \frac{\partial}{\partial y}\left(\nu \frac{\partial \overline{u}}{\partial y} - \overline{u'v'} \right) \tag{5.60}$$

The modeling of the last term in Eq. (5.60) plays a large role in the turbulence problem, which is precisely to establish models to estimate the gradient of the time average of the product of the fluctuations of the two velocity components, $\overline{u'v'}$. That term originates to the so-called *Reynolds apparent stresses*, which deserve a separate treatment. Advanced literature on the subject is recommended for those interested, such as the book by Kays et al. (2005).

Analogously, the energy equation for the turbulent boundary layer can be established, resulting in:

$$\rho C_p\left(\overline{u}\frac{\partial \overline{T}}{\partial x}+\overline{v}\frac{\partial \overline{T}}{\partial y}\right)=\frac{\partial}{\partial y}\left(k\frac{\partial \overline{T}}{\partial y}-\rho C_p\overline{v'T'}\right) \tag{5.61}$$

In similarity to the laminar case, two turbulent quantities are defined:

- Turbulent viscosity, ε_M:

$$\varepsilon_M\frac{\partial \overline{u}}{\partial y}=-\overline{v'u'} \tag{5.62a}$$

- Turbulent diffusivity, ε_H:

$$\varepsilon_H\frac{\partial \overline{T}}{\partial y}=-\overline{v'T'} \tag{5.62b}$$

Thus, the *total turbulent shear stress,* τ_t, is defined by:

$$\tau_t=\rho(\nu+\varepsilon_M)\frac{\partial \overline{u}}{\partial y} \tag{5.63}$$

and the *total turbulent heat flux,* q_t :

$$q_t=-\rho C_p(\alpha+\varepsilon_H)\frac{\partial \overline{T}}{\partial y} \tag{5.64}$$

Far from the wall, turbulent viscosity and diffusivity govern the phenomenon as corresponding molecular quantities are neglected, that is, $\varepsilon_M \gg \nu$ and $\varepsilon_H \gg \alpha$, so that one can define a *turbulent Prandtl number*, Pr_t, given by:

$$Pr_t=\varepsilon_M/\varepsilon_H \tag{5.65}$$

where Pr_t is approximately unitary, as the energy and momentum transports in this region occur at the same pace due to macroscopic fluid mixtures, which also characterizes the fact that the temperature and velocity average profiles are more uniform in this region.

Next to the wall (laminar sublayer), molecular viscosity and diffusivity dominate and the region behaves similarly to the laminar boundary layer. Finally, a transition from molecular to macroscopic phenomenon takes place in the buffer layer.

The study of turbulent quantities gives rise to universal velocity and temperature profiles. It is important to note that many analyses indicate that the Reynolds-Colburn analogy between surface friction and heat transfer discussed in Sect. 5.2.3 can also be extended to the turbulent region. The object of turbulence studies is to adequately

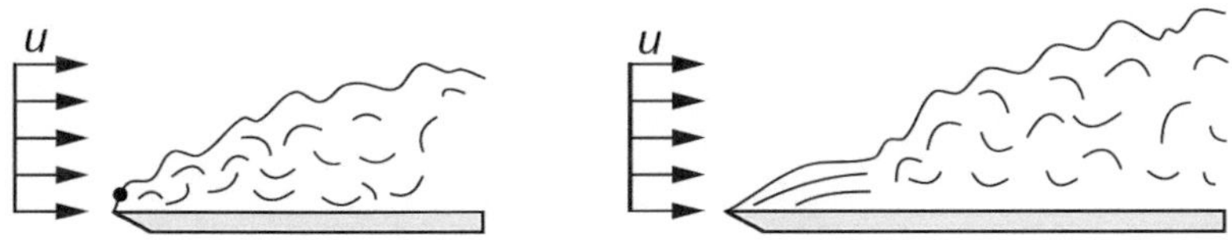

Fig. 5.19 Turbulent boundary layer: only turbulent (on the left) and laminar + turbulent (on the right)

model the effects of instantaneous variations in quantities, which is beyond the scope of this text.

It is relevant to know that there are two heat transfer regimes: laminar and turbulent. There is also a transition region between the two regimes. Note that the onset of turbulence can be achieved in two ways: the first, when a turbulence generator is placed at the leading edge to disturb the flow, such as by installing a thin wire across the flow (*tripping wire*), and the second, when the flow starts as a laminar regime, goes through a regime transition to reach the turbulence regime for $Re_x > 5 \times 10^5$ (other authors use $Re_x > 3 \times 10^5$) (Fig. 5.19).

The main results for the solely turbulent regime

In this section is presented working relationships for turbulent flow onset from the leading edge:

Location:

$$\delta_T \approx \delta_H = \frac{0.37x}{Re_x^{0.2}}; \qquad Re_x \leq 10^7 \tag{5.66}$$

$$C_{f,x} = \frac{0.059}{Re_x^{1/5}}; \qquad Re_x \leq 10^7 \tag{5.67}$$

$$Nu_x = 0.0296 Re_x^{0.8} Pr^{\frac{1}{3}}; \qquad Re_x \leq 10^7; \qquad 0.6 \leq Pr \leq 60 \tag{5.68}$$

Average:

$$\overline{C}_{f,L} = \frac{0.074}{Re_L^{1/5}}; \qquad Re_x \leq 10^7 \tag{5.69}$$

$$\overline{Nu_L} = 0.037 Re_L^{0.8} Pr^{\frac{1}{3}}; \qquad Re_L \leq 10^7; \qquad 0.6 \leq Pr \leq 60 \tag{5.70}$$

The main results for the turbulent regime with a laminar start

It is considered that the transition from a laminar to a turbulent regime occurs at $Re_{cr} = 5 \times 10^5$ for the sake of presenting working equations. In reality, the regime transition occurs within a certain plate length.

Local magnitudes:

$$C_{f,x} = 0.059Re_x^{-1/5}; \qquad 5 \times 10^5 \le Re_x \le 10^7 \tag{5.71}$$

$$Nu_x = 0.0296Re_x^{0.8}Pr^{\frac{1}{3}}; \quad 5 \times 10^5 \le Re_x \le 10^7; 0.6 \le Pr \le 60 \tag{5.72}$$

Average magnitudes:

Average values are evaluated considering the distance from the leading edge, going through the (critical) length at which turbulence begins, $x_{cr} = Re_{cr}\upsilon/u_\infty$, going on up to the total plate length. In the laminar regime, the values of the friction and convective heat transfer coefficients, given by Eqs. (5.20) and (5.23), respectively, are considered. For the turbulent section, Eqs. (5.71) and (5.72) are considered. The average friction and heat transfer coefficients by convection are defined by:

$$\overline{C}_{f,L} = \frac{1}{L}\left(\int_0^{x_{cr}} C_{f,x,\,lam}dx + \int_{x_{cr}}^{L} C_{f,x,turb}dx\right) \text{ and}$$
$$\overline{h}_{f,L} = \frac{1}{L}\left(\int_0^{x_{cr}} h_{x,lam}dx + \int_{x_{cr}}^{L} h_{x,turb}dx\right)$$

With the critical Reynolds number considered $Re_{cr} = 5 \times 10^5$, we have:

$$\overline{C}_{f,L} = \frac{0.074}{Re_L^{1/5}} - \frac{1742}{Re_L}; \qquad 5 \times 10^5 \le Re_L \le 10^8 \tag{5.73}$$

$$\overline{Nu}_L = \left(0.037Re_L^{0.8} - 871\right)Pr^{1/3}; \quad 5 \times 10^5 \le Re_L \le 10^8; \quad 0.6 \le Pr \le 60 \tag{5.74}$$

The transport properties should be evaluated at the film temperature: $T_f = \frac{T_w + T_\infty}{2}$.

If the plate heating occurs at a distance of x_0 from the leading edge, the local Nusselt number for laminar flow is found in Eq. (5.43). If the flow is turbulent from the beginning, the local Nusselt number is (Kays et al., 2005):

$$Nu_x = 0.0296Re_x^{0.8}Pr^{1/3}\left[1 - \left(\frac{x_0}{x}\right)^{9/10}\right]^{-1/9} \tag{5.75}$$

The average Nusselt number ($\overline{Nu}_L = \frac{\overline{h}_L L}{k}$) for the heated section of the plate ($A = b(L - x_0)$), in laminar and solely turbulent flow is expressed by Eqs. (5.76) and (5.77), respectively (Thomas, 1977).

$$\overline{Nu}_L = \frac{0.664Re_L^{0.5}Pr^{1/3}}{1 - \frac{x_0}{L}}\left[1 - \left(\frac{x_0}{L}\right)^{3/4}\right]^{2/3} \tag{5.76}$$

$$\overline{Nu_L} = \frac{0.037Re_L^{0.8}Pr^{1/3}}{1-\frac{x_0}{L}}\left[1-\left(\frac{x_0}{L}\right)^{9/10}\right]^{8/9} \tag{5.77}$$

Example 5.4 (*Turbulent heat transfer over a heated plate*)

Water at 20 °C is heated on an isothermal plate at 80 °C. The fluid flows at 3 km/h over the plate in the 1 m lengthwise direction. Calculate the rate of heat transferred from the plate if it is 80 cm width.

Solution

$$u_\infty = 3\,\text{km/h} = \frac{3\times 1000\,\text{m}}{3600\,\text{s}} = 0.83\,\text{m/s},$$

Properties evaluated at $T_f = \frac{20+80}{2} = 50$ °C (Table A.7)

$$\nu = 5.5359\times 10^{-7}\,\text{m}^2/\text{s};\; k = 0.644\,\text{W/mK};\; Pr = 3.55$$

$$Re_L = \frac{u_\infty L}{\nu} = \frac{0.83\times 1}{5.5359\times 10^{-7}} = 1.50\times 10^6 \text{ (turbulent)}$$

Supposing the regime starts laminar. Then Eq. (5.74) is used

$$\overline{Nu_L} = \left(0.037Re_L^{0.8} - 871\right)Pr^{\frac{1}{3}} = \left[0.037\left(1.50\times 10^6\right)^{0.8} - 871\right]3.55^{\frac{1}{3}}$$
$$= 3597.2$$

$$\bar{h} = \frac{\overline{Nu_L}k}{L} = \frac{3597.2(0.644)}{1} = 2316.6\,\text{W/m}^2\text{K}$$

$$\dot{Q} = \bar{h}A(T_w - T_\infty) = 2316.6\times 1\times 0.8\times(80-20) = 111.2\,\text{W}.$$

5.3.2 *Cross-Flow Over Cylinders and Tubes*

In the case of external cross or perpendicular flow over cylinders and tubes, the analysis of local heat transfer becomes more complex because of the various hydrodynamic effects that can occur in the fluid flow around the tube. In this case, the local Nusselt number is a function of the incidence angle θ, that is, $Nu(\theta)$, which is strongly influenced not only by the formation of boundary layers but also by the effect of boundary layer detachment. Figure 5.20 illustrates the variation of the local Nusselt number for various Reynolds numbers. In the context, the Reynolds number is given as a function of the tube diameter, *D*. It can be inferred from the figure that, for $Re_D < 10^5$, the Nusselt number decreases because of the growth of the laminar

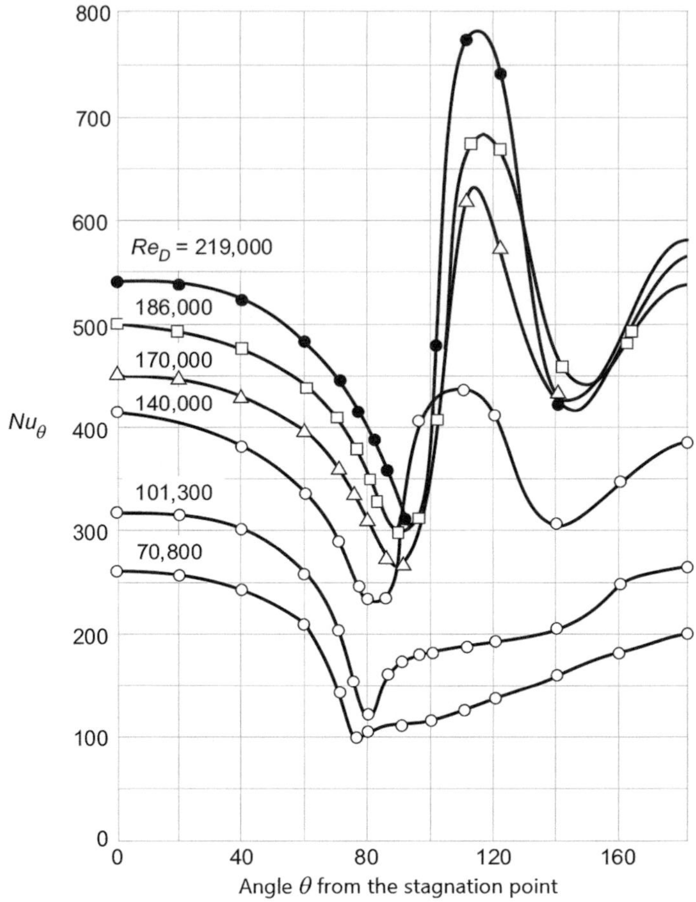

Fig. 5.20 Dependence of the local Nusselt number, Nu_θ, around a cylinder (Giedt 1949) on several Reynolds numbers, Re_D. Published with permission

boundary layer (LBL) up to about 80°. After that angle, the flow detaches from the tube surface destroying (detachment) the LBL, which results in the generation of a vortex system that causes a better mixing of the fluid and, consequently, an increase in the heat transfer coefficient (increase of Nu_θ). For $Re_D > 10^5$, the transition from the laminar boundary layer to the turbulent boundary layer (TuBL) occurs before the detachment phenomenon. In the transitional boundary layer regime, θ from 80 to 100°, there is an improvement in the heat transfer process. Once the TuBL starts, there is again a decrease in the local heat transfer coefficient due to the growth of the TuBL up to around 140 °. After that, the flow detaches from the surface and the destruction of the TuBL happens to then generate vortex and mixing systems that improve heat transfer once again. In the turbulent flow regime case, there are two minimum regions.

Although from the point of view of increasing heat transfer, it may be important to analyze the local effect of the Nusselt number, under the perspective of application and calculation, an expression should be established for the average convective heat transfer coefficient over the circumference of the tube. Thus, an old expression is still being used—the empirical correlation of Hilpert (1933):

$$\overline{Nu_D} = \frac{\overline{h}D}{k} = CRe_D^m Pr^{1/3} \tag{5.78}$$

where D is the tube diameter. The constants C and m are given in Table 5.3 as a function of the Reynolds number. The transport properties should be evaluated at the film temperature.

In the case of air over other cross-sections, Sparrow et al. (2004) gathered several results for some geometries, which are indicated in Table 5.4. Although the given empirical equations in this table have been developed for air, it is an acceptable procedure to multiply the Nusselt number by $Pr^{1/3}$ and use the modified equations for other gases. The transport properties should be evaluated at the film temperature.

For the cross-flow of other fluids over circular cylinders, a widely used expression is due to Zukauskas (1972), given by:

$$\overline{Nu_D} = CRe_D^m Pr^n \left(\frac{Pr}{Pr_w}\right)^{1/4}; \quad 1 < Re_D < 10^6; \quad 0.7 < Pr < 500 \tag{5.79}$$

The constants C and m are obtained from Table 5.5. All properties are evaluated at T_∞, except Pr_w, which is evaluated at the surface temperature (wall). If $Pr \leq 10$, use $n = 0.37$, and, if $Pr > 10$, use $n = 0.36$.

Example 5.5 (*Comparing the heat transfer coefficient between flat surfaces and tubes*)

Consider an airflow at a velocity of 4 m/s at 30 °C. In this flow, a thin flat plate is placed parallel to it, 25 cm lengthwise and 1 m wide. The plate temperature is 60 °C. Subsequently, the plate is rolled up (in lengthwise direction) forming a cylinder over which the airflow will occur crosswise. All other conditions are kept the same. It is asked:

Table 5.3 Constants of the Hilpert correlation for the external flow of a circular cylinder

Geometry	Re_D	C	m
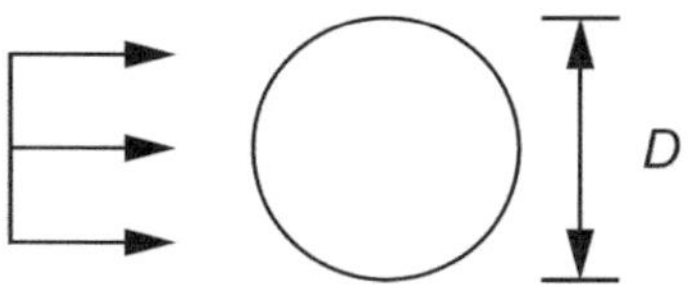	0.4–4	0.989	0.330
	4–40	0.911	0.385
	40–4000	0.683	0.466
	4000–40,000	0.193	0.618
	40,000–400,000	0.027	0.805

Adapted from Hilpert (1933), Knudsen and Katz (1958)

Table 5.4 Nusselt numbers for different cross-flow geometries of air over various cross-sections

Geometry	Re_D	Correlation
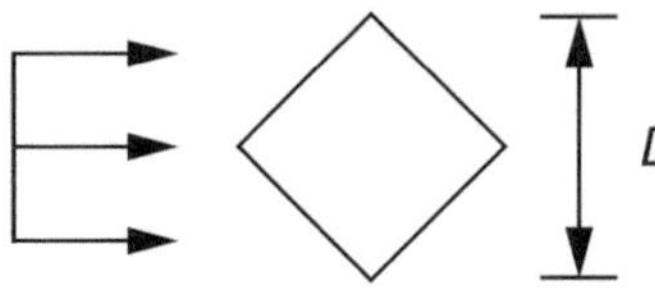	6000–60,000	$\overline{Nu_D} = 0.27Re^{0.59}$
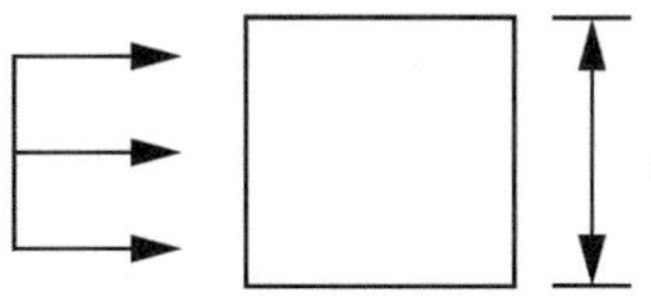	5000–60,000	$\overline{Nu_D} = 0.14Re^{0.66}$
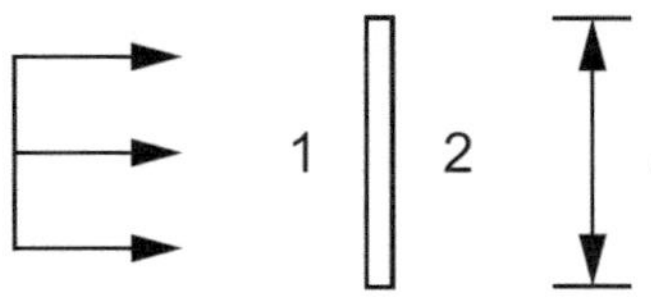	1: 10,000–50,000 2: 7000–80,000 1 and 2: 10,000–50,000	$\overline{Nu_D} = 0.592Re^{0.5}$ $\overline{Nu_D} = 0.17Re^{0.667}$ $\overline{Nu_D} = 0.25Re^{0.61}$
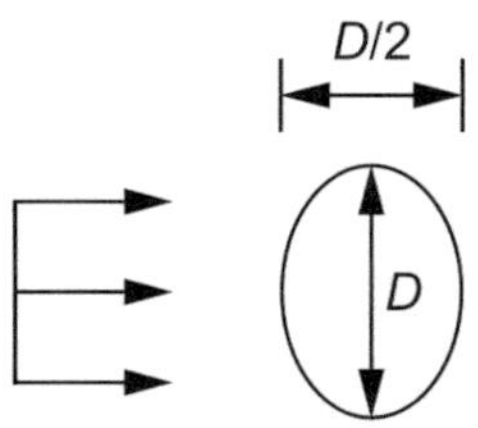	5000–90,000	$\overline{Nu_D} = 0.566Re^{0.545}$
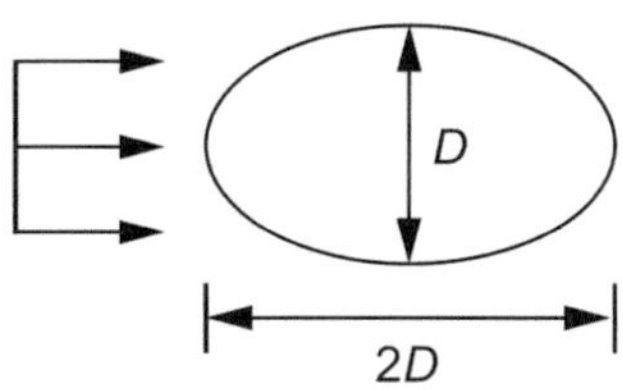	2500–45,000	$\overline{Nu_D} = 0.256Re^{0.573}$

(continued)

Table 5.4 (continued)

Geometry	Re_D	Correlation
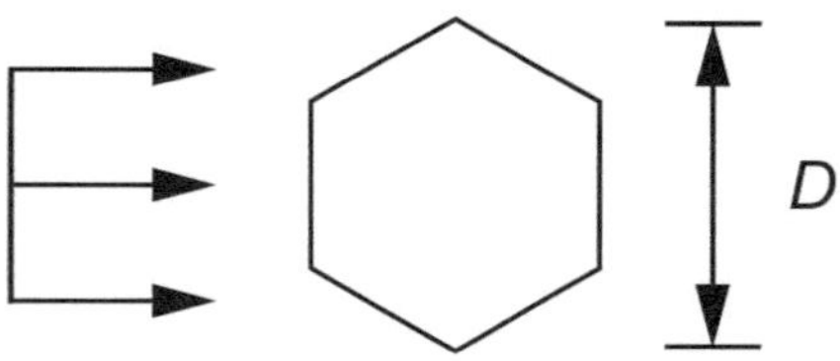	5200–20,400 20,400–105,000	$\overline{Nu_D} = 0.146Re^{0.638}$ $\overline{Nu_D} = 0.035Re^{0.782}$
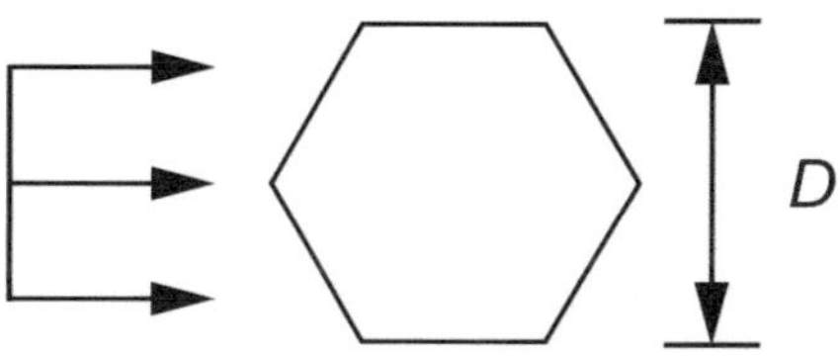	4500–90,700	$\overline{Nu_D} = 0.133Re^{0.638}$

Adapted from Sparrow et al. (2004). For flow of other gases, multiply the right side of the correlation by $Pr^{1/3}$

Table 5.5 Constants to be used in Eq. (5.79)

Geometry	Re_D	C	m
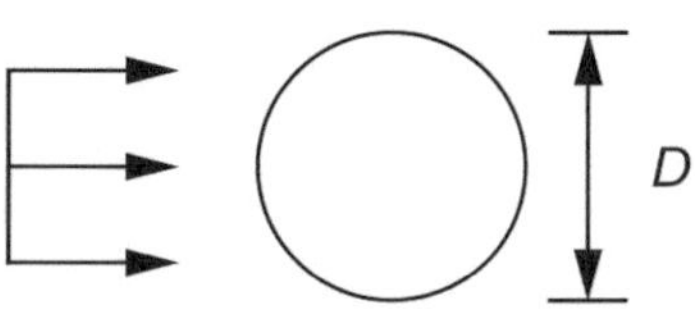	1–40	0.75	0.4
	40–1000	0.51	0.5
	1000–200,000	0.26	0.6
	200,000–1,000,000	0.076	0.7

Adapted from Zukauskas (1972)

(a) In which case is the heat exchange greater?
(b) What is the rate of heat exchanged in both cases?
(c) Analyze whether there is always a greater heat exchange in a given configuration than in the other, regardless of the length and air velocity.

Solution

(a) Air properties calculated at film temperature: $T_f = \frac{T_\infty + T_w}{2} = 45$ °C. From Appendix A, Table A.5: $\nu = 1.68 \times 10^{-5}$ m²/s; $k = 2.69 \times 10^{-2}$ W/mK; $Pr = 0.706$

Plate

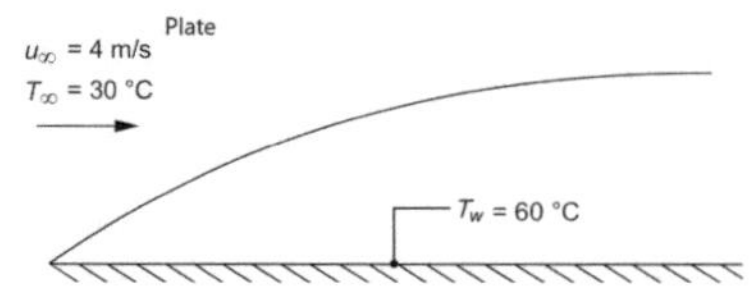

Cylinder

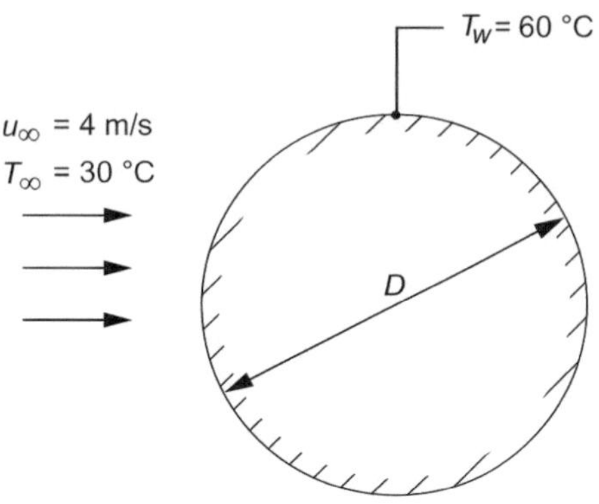

Plate:

$$Re_L = \frac{u_\infty L}{\nu} = \frac{4 \times 0.25}{1.68 \times 10^{-5}} \cong 5.95 \times 10^4 < Re_{crit}$$

$$\overline{Nu}_L = 0.664 Re_L^{1/2} Pr^{1/3} = 0.664 \times \left(5.95 \times 10^4\right)^{1/2} \times (0.706)^{1/3} = 144.2$$

$$\bar{h}_L = \frac{\overline{Nu}k}{L} = \frac{144.2 \times 0.02697}{0.25} = 15.56\,\mathrm{W/m^2 {}^\circ C}$$

Cylinder:

Cylinder diameter after rolling up: $\pi D = L \rightarrow D = 0.25/\pi = 0.0796\,\mathrm{m}$

$$Re_D = \frac{4 \times 0.0796}{1.68 \times 10^{-5}} = 1.895 \times 10^4$$

Using Hilpert's expression (Eq. 5.78): $\overline{Nu}_D = CRe_D^m Pr^{1/3}$
For $Re = 1.895 \times 10^4 \rightarrow C = 0.193,\ m = 0.618$ (Table 5.3)

$$\overline{Nu}_D = 0.193 \times \left(1.895 \times 10^4\right)^{0.618} \times (0.706)^{1/3} = 75.63$$

$$\bar{h}_D = \frac{\overline{Nu}_D k}{D} = \frac{75.63 \times 0.02697}{0.0796} = 25.63\,\mathrm{W/m^2}K$$

The heat transfer is greater in the case of the cylinder, since $\overline{h}_D > \overline{h}_L$ as the heat exchange area is the same and so the temperatures.

(b)

$$Plate: \dot{Q}_{\text{plate}} = \overline{h}_L A_p (T_w - T_\infty) = 15.56 \times 0.25 \times 30 = 116.7 \text{ W}$$
$$Cylinder: \dot{Q}_{\text{cyl}} = \overline{h}_D A_c (T_w - T_\infty) = 25.63 \times 0.25 \times 30 = 192.2 \text{ W}$$

(c)

Laminar portion $Re_{crit,L} = 5 \times 10^5$
Note that: $Re_D = Re_L/\pi \Rightarrow Re_D = 1.59 \times 10^5$

$$\overline{h}_L = \frac{0.664 \times k}{L} Re_L^{1/2} Pr^{1/3} \tag{A}$$

$$\overline{h}_D = \frac{k}{D} Pr^{1/3} C Re_D^m = \pi \frac{k Pr^{1/3}}{L} C Re_D^m \tag{B}$$

Therefore, from (A), $\frac{kPr^{1/3}}{L} = \frac{\overline{h}_L}{0.664 Re_L^{1/2}}$, which can be substituted into (B) to obtain:

$$\overline{h}_D = \frac{\pi C Re_D^m \overline{h}_L}{0.664 \pi^{1/2} Re_D} = 2.669 C Re_D^{m-0.5} \overline{h}_L$$

or $\frac{\overline{h}_D}{\overline{h}_L} = 2.669 C Re_D^{m-0.5}$ for the case of laminar flow on the plate.

Laminar-turbulent portion $Re_L > Re_{crit} = 5 \times 10^5$
$\overline{Nu_L} = \left(0.037 \times Re_L^{0.8} - 871\right) Pr^{1/3}$ (mixed boundary layer)
From which

$$\frac{\overline{h}_L L}{k} = \left(0.037 Re_L^{0.8} - 871\right) Pr^{1/3} \text{ and } \frac{k Pr^{1/3}}{L} = \frac{\overline{h}_L}{0.037 Re_L^{0.8} - 871} \tag{C}$$

substituting into (B), we get $\overline{h}_D = \frac{\pi C Re_D^m \overline{h}_L}{0.037 Re_L^{0.8} - 871}$
substituting $Re_L = \pi Re_D$, we get: $\frac{\overline{h}_D}{\overline{h}_L} = \frac{\pi C Re_D^m}{0.037 Re_L^{0.8} - 871}$.

The various values of C and m from Hilpert's expression (Eq. 5.78) were substituted into the expressions of the ratios between the heat transfer coefficients and appear in the following table and in graphic form in the figure. It is found that the heat transfer will always be greater in the case of the cylinder (within the validity range of the expressions), as indicated in the graph below.

Re_D	C	m	$\frac{\bar{h}_D}{\bar{h}_L}$	Regime
4	0.989	0.33	2.09	Laminar
40	0.911	0.385	1.59	"
4000	0.683	0.466	1.38	"
40,000	0.193	0.618	1.80	"
159,000	0.027	0.805	2.78	"
200,000	0.027	0.805	2.15	Lam-turb
400,000	0.027	0.805	1.43	"

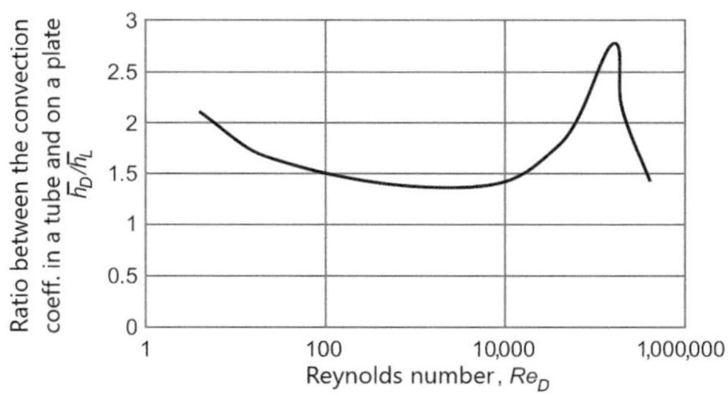

5.4 Cross-Flow Over Tube Bank

Cross-flow of a fluid over a tube bank is a very common configuration in heat exchangers. One of the fluids flows perpendicularly and externally to the tubes, while the other circulates internally, as represented in the scheme of Fig. 5.21. In this section, the interest lies in the analysis of the flow external to the tube bank and not in just one single tube. The next chapter will deal with the heat transfer inside the tubes.

In Fig. 5.22, two typical tube bank arrangements are presented. The first is called "in-line arrangement" and the other, "staggered arrangement". The quantities in the figure are N_L and N_T—number of tubes in the longitudinal and transverse direction; D—external diameter of the tubes; S_L, S_T and S_D—longitudinal, transverse and diametrical pitch (the latter only in staggered configuration); V—average velocity of the external fluid upstream of the tube bank; and T_i and T_o—temperature of the external fluid's input and output.

When the flow occurs around tubes, the tubes installed downstream will be subjected to a wake of vortices produced in previous tubes, so that the fluid upstream of the tube will be disturbed (see illustration of the vortex wakes in Fig. 5.22). This behavior is repeated and intensified throughout the flow to the exit section of the heat

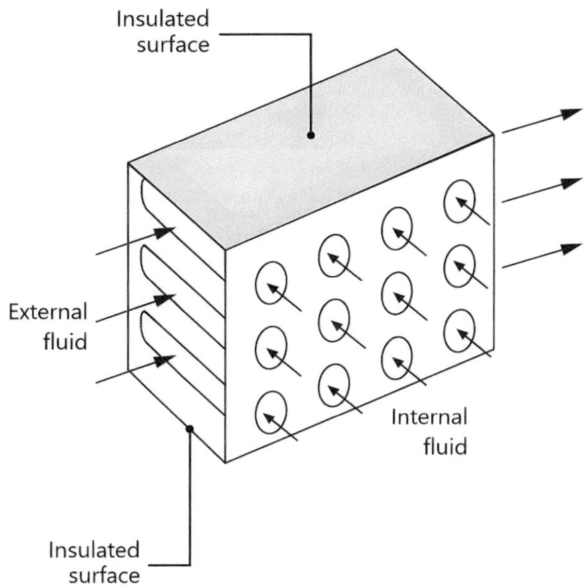

Fig. 5.21 External flow over a tube bank

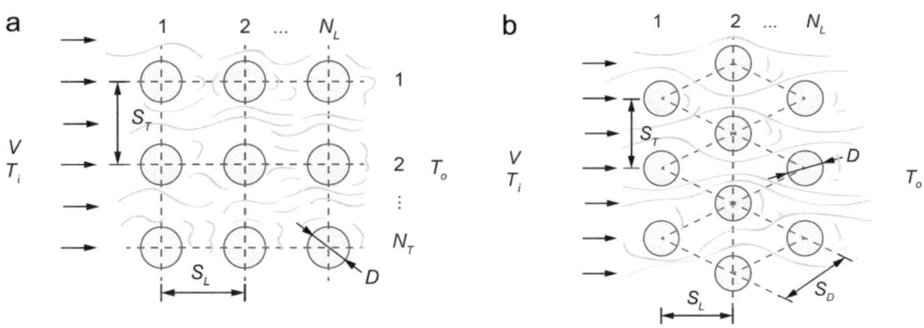

Fig. 5.22 Arrangements of tube banks in **a** line and **b** staggered configuration

exchanger. Therefore, the heat transfer coefficient by convection is different in tube banks compared to an individual tube.

There are several empirical correlations for obtaining heat transfer over tube banks. One can use the expression of Grimison (1937), who analyzed several works and concluded that the heat transfer coefficient by convection for a bank of at least 10 tubes in the direction of the fluid, can be found following the form of Hilpert's equation (Eq. 5.78), which can be used for air and for other fluids. On the other hand, Zukauskas (1972) proposed a correlation, which is based on the maximum Reynolds number, which is a function of the maximum fluid velocity between the tubes:

$$Re_{D,\max} = \frac{V_{\max} D}{\nu} \tag{5.80}$$

In the *line arrangement*, the maximum velocity occurs in the space between the tubes and, therefore, will be:

$$V_{\max} = \frac{S_T}{S_T - D} V \tag{5.81}$$

In the *staggered arrangement*, the maximum velocity can occur in two sections, as illustrated in Fig. 5.22. $V_{\max}$ will occur in the section between two tubes separated by the distance $S_D - D$, if it is satisfied that $2(S_D - D) < (S_T - D)$; after a simple trigonometric analysis, the following equivalent condition is obtained:

$$S_D = \sqrt{S_L^2 + \left(\frac{S_T}{2}\right)^2} < \frac{S_T + D}{2} \tag{5.82}$$

If the inequality of Eq. (5.82) occurs, then:

$$V_{\max} = \frac{S_T}{2(S_D - D)} V \tag{5.83}$$

If the condition of Eq. (5.82) is not met, then, the maximum velocity occurs in the section between two vertically aligned tubes, and, therefore, Eq. (5.81) is used again.

Once obtained $Re_{D,\max}$, then the Nusselt number is evaluated through Eq. (5.84), for $N_L \geq 20$.

$$\overline{Nu}_D = C Re_{D,\max}^m Pr^{0.36} \left(\frac{Pr}{Pr_p}\right)^{0.25}, \text{ valid for } \begin{bmatrix} N_L \geq 20 \\ 0.7 < Pr < 500 \\ Re_{D,\max} < 2 \times 10^6 \end{bmatrix} \tag{5.84}$$

All properties, except Pr_w (which is evaluated at the temperature of the tube walls), are evaluated at the average temperature between the fluid inlet and outlet, $\overline{T} = (T_i + T_o)/2$, and the constants C and m are summarized in Table 5.6.

If the number of tube rows is less than 20, that is, $N_L < 20$, then Eq. (5.84) should be modified, multiplying the result obtained by a constant C_2 as expressed in Eq. (5.85) and values given in Table 5.7.

$$\overline{Nu}_D\big|_{N_L<20} = C_2 \overline{Nu}_D\big|_{N_L \geq 20} \tag{5.85}$$

The heat transfer rate between the tube bank and the external flow can be evaluated according to Newton's cooling law $\dot{Q} = hA\Delta T$, where A is the external surface area of all the tubes and ΔT is a temperature difference. In this section, the temperature of the tube wall is constant, but the fluid changes temperature on its path from T_i to

Table 5.6 Constants C and m from Eq. (5.84) (Zukauskas 1987)

Configuration	$Re_{D,\max}$		C	m
Aligned	16–100		0.90	0.40
	100–1000		0.52	0.50
	1000–200,000[a]		0.27	0.63
	200,000–2,000,000		0.033	0.80
Staggered	1.6–40		1.04	0.40
	40–1000		0.71	0.50
	1000–200,000	$S_T/S_L < 2$ $S_T/S_L > 2$	$0.35\,(S_T/S_L)^{0.2}$ 0.40	0.60 0.60
	200,000–2,000,000		$0.031(S_T/S_L)^{0.2}$	0.80

[a]For $\frac{S_T}{S_L} < 0.7$, heat transfer is inefficient and should be avoided

Table 5.7 Correction factor C_2 in Eq. (5.85) for $N_L < 20$ (adapted from Zukauskas 1987)

Re_D	N_L	1	2	3	4	5	7	10	13	16	20
100–1000	Aligned	1.00	1.00	1.00	1.00	1.00	1.00	1.00	1.00	1.00	1.00
	Alternating	0.83	0.88	0.92	0.94	0.96	0.97	0.98	0.98	0.99	1.00
>1000	Aligned	0.70	0.80	0.86	0.90	0.92	0.95	0.97	0.98	0.99	1.00
	Alternating	0.64	0.76	0.84	0.89	0.92	0.95	0.97	0.98	0.99	1.00

T_o. Thus, the concept of *logarithmic mean temperature difference* (*LMTD*), which will be discussed in detail in Sects. 6.1.5 and 6.3, should be used. As will be seen, the rate of heat transferred is defined by:

$$\dot{Q} = hALMTD \tag{5.86}$$

where

$$LMTD = \frac{\Delta T_1 - \Delta T_2}{\ln\left(\frac{\Delta T_1}{\Delta T_2}\right)} \tag{5.87}$$

ΔT_1 is the temperature difference between the tube wall and the fluid at the entrance and ΔT_2 between the wall and the fluid at the exit. See a schematic representation of the possible fluid and tube wall temperatures in Fig. 5.23.

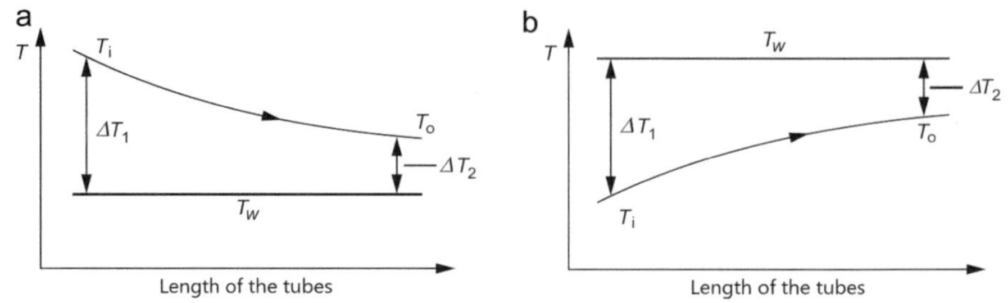

Fig. 5.23 Configuration of the fluid cooling or heating in the tube bank

The heat transfer rate from the First Law of Thermodynamics applied to the fluid is:

$$\dot{Q} = \dot{m}C_p(T_i - T_o) \tag{5.88}$$

Equating Eqs. (5.86) and (5.88):

$$\dot{m}C_p(T_i - T_o) = hA\frac{(T_i - T_w) - (T_o - T_w)}{\ln\left[\frac{(T_i - T_w)}{(T_o - T_w)}\right]}$$

$\ln\left(\frac{T_o - T_w}{T_i - T_w}\right) = -\frac{hA}{\dot{m}C_p}$. From which the outlet temperature of the fluid after the heat exchange is obtained:

$$T_o = T_w + (T_i - T_w)e^{-\frac{hA}{\dot{m}C_p}} \tag{5.89}$$

Note that the mass flow rate can be evaluated either at the inlet or outlet flow.

Example 5.6 (*Cross-flow over tube bank*)

Water at $T_i = 17$ °C flows crosswise over a staggered bundle of tubes inside a rectangular structure of 30 × 30 cm. Combustion gases flow inside the tubes, keeping the tube walls at an average temperature of 280 °C. The water circulates inside the tube of $\phi = 12$ cm, flowing at an average velocity $V = 2.3$ m/s. Estimate the water temperature after passing through the tube bundle.

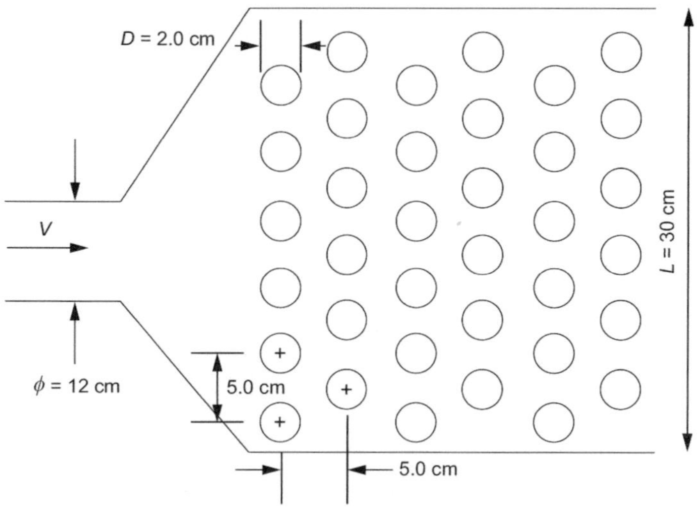

Solution

The $T_e = 17$ °C, $\rho = 998.7$ kg/s (Table A.7). The mass flow rate of the water will be $\dot{m} = \rho V \pi \frac{D^2}{4} = 998.7 \times 2.3\pi \frac{0.12^2}{4} = 25.96$ kg/s.

By mass conservation and assuming that there is no variation in the fluid temperature until it enters the tube bank: $\rho V_e LW = 25.96$, therefore, the fluid entry velocity to the tube bank will be: $998.7 V_e 0.3 \times 0.3 = 25.976 \rightarrow V_e = 0.289$ m/s.

Finding the maximum velocity:

$$\sqrt{S_L^2 + \left(\frac{S_T}{2}\right)^2} < \frac{S_T + D}{2} \rightarrow \sqrt{0.05^2 + 0.025^2} < \frac{0.05 + 0.02}{2}$$

$0.056 < 0.035$, which is not true; thus, the maximum velocity occurs at $S_T - D$:

$$V_{\max} = \frac{S_T}{S_T - D} V = \frac{0.05}{0.05 - 0.02} 0.289 = 0.482 \,\text{m/s}.$$

Transport properties depend on the average temperature between the fluid's inlet and outlet, which, in turn, depends on the outlet temperature, which is unknown. Thus, initially, an outlet temperature is assumed and will be corrected later. Note that this temperature must be between the inlet temperature and the surface temperature of the tubes. Initially, it is assumed that $T_o = 103$ °C.

$$\overline{T} = \frac{(103 + 17)}{2} = 60\,°\text{C}.$$

From Table A.7: $C_p = 4185$ J/kgK; $k = 0.6542$ W/mK; $\nu = 4.744 \times 10^{-7}$ m^2/s, $Pr = 2.983$ and $Pr_w = 0.8346$

$$Re_{D,\max} = \frac{0.482 \times 0.02}{4.744 \times 10^{-7}} = 20{,}298$$

$$\overline{Nu_D} = CRe_{D,\max}^{m} Pr^{0.36}\left(\frac{Pr}{Pr_w}\right)^{0.25}$$

$$\overline{Nu_D} = 0.35\left(\frac{0.05}{0.05}\right)^{0.2} 20{,}298^{0.6} 2.983^{0.36}\left(\frac{2.983}{0.8346}\right)^{0.25} = 274$$

As $N_L = 6$, from Table 5.7, $C_2 = 0.935$

$$\overline{Nu_D} = 0.935 \times 274 = 256.2$$

$$\overline{h} = \frac{\overline{Nu_D} k}{D} = \frac{256.2 \times 0.6542}{0.02} = 8380.3\,\mathrm{W/m^2K}$$

The total number of tubes is $N = 36$, therefore, the total heat exchange area will be: $A = N\pi DL = 36\pi 0.02 \times 0.3 = 0.678$ m^2

The exit temperature:

$$T_o = T_w + (T_i - T_w)e^{-\frac{hA}{\dot{m}C_p}} = 280 + (17 - 280)e^{-\frac{8380.3 \times 0.678}{25.96 \times 4185}}$$

$$T_o = 30.4\ ^\circ\mathrm{C}$$

It is verified that this temperature is different from the one initially assumed at 103 °C and, therefore, it needs to be corrected. For this, the found temperature is used and the new average temperature is $\overline{T} = (17 + 30.4)/2 = 23.7$ °C. Following the procedure already carried out, we find $\overline{h} = 8523.3$ W/m^{2}K and $T_o = 30.63$ °C, which is close enough to the previous value of 30.4 °C; thus, the value of 30.63 °C is accepted. Usually, two to three iterations are needed for the calculation to converge.

5.5 Proposed Problems

5.1 A fluid flows parallel over a 2 m long flat plate. If free flow velocity before reaching the plate is 2 m/s, evaluate the following dimensionless numbers: $Re_L = \frac{\rho u_\infty L}{\mu}$, $Nu_L = \frac{hL}{k}$ and $Pr = \frac{C_p \mu}{k}$, as well as the thermal diffusivity, α, for liquid water, air, liquid ammonia, liquid R134a and Therminol 59. Consider for all cases a film temperature of 25 °C and the heat transfer coefficient by convection is 100 W/m^{2}K.

5.2 From experimental tests find a correlation for the Nusselt number as a function of the Reynolds and Prandtl numbers for a part whose characteristic length is $L = 10$ cm. This part exchanges heat on its external surface when it is at 80 °C. For the first experiment, air at 25 °C and 2.5 m/s is used, which provides a heat flux loss of 630 W/m^2. In a second experiment, the air is at the same temperature, but its velocity is 20 m/s and the heat flux is 3400 W/m^2. A last experiment is carried out with water at 16 °C and 0.55 m/s, whose heat flux is 120,000 W/m^2. Assuming a general correlation to adjust the data collected of the type $Nu_L = CRe_L^a Pr^b$, find the numerical values of the constants *a*, *b*, and c.

5.3 A thin blade at 80 °C with 1 m in length and 0.5 m in width is immersed in an air flow, which moves parallel in the lengthwise direction. If the air velocity is 1.5 m/s and its temperature is 10 °C, evaluate the thickness of the thermal and hydrodynamic boundary layers at the trailing edge and the local and average heat transfer coefficients, as well as the total heat transfer rate.

5.4 An analytical expression for the variation of air temperature in the perpendicular direction over a flat plate was developed, and for a certain distance from the leading edge is $T = 25 + 150e^{-1500y}$ (T in °C and y in m). Evaluate (a) the wall temperature; (b) the free flow temperature ($y \to \infty$); and (c) the local heat transfer coefficient by convection. Is the fluid heating or cooling?

5.5 Water at 0.5 m/s and 10 °C flows over a flat plate of 20 cm × 10 cm kept at 70 °C. Evaluate the rate of heat transferred if the plate is positioned with the flow in the 10 cm and 20 cm lengthwise direction. Which direction would you choose to transfer more heat?

5.6 A fluid at a velocity of 3 m/s and temperature at 80 °C flows over a lengthy flat plate that is at 25 °C and is 0.5 m width. For the fluids glycerin, Therminol VP1, and ethylene glycol, determine (a) the distance from the leading edge to the point of transition from laminar to turbulent; for this distance evaluate (b) the rate of heat transferred to the plate and (c) the ratio between the thickness of the hydrodynamic and thermal boundary layer at the trailing edge.

5.7 Air at 15 °C and 1 m/s flows between two long flat plates that are at 60 °C and 5 cm apart. Evaluate the distance L from the leading edge to where the upper and lower laminar thermal boundary layers meet.

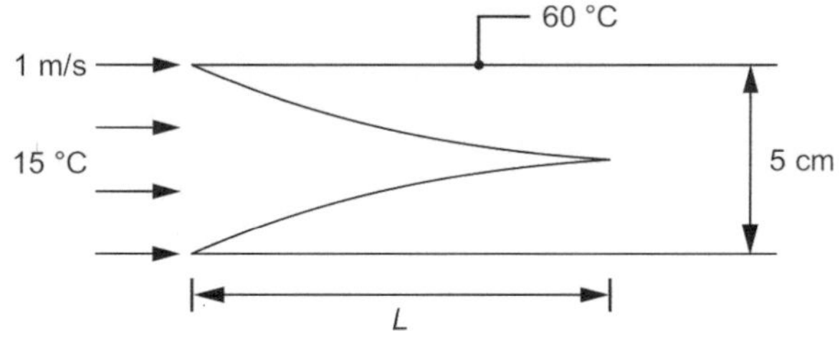

5.8 Liquid refrigerant R32 (HFO-32) at 50 °C flows at 0.2 m/s over a flat 30 cm lengthwise plate whose temperature is 10 °C. Evaluate the velocities $u = u(x, y)$ and $v = v(x, y)$, where $x = 15$ cm is the distance from the leading edge and $y = \delta_T/2$ (for $x = 15$ cm).

5.9 A new thermal solar roof is under test. The roof is composed of several tiles, which have ceramic sides with fittings. To simplify calculations, they can be considered flat. In the center, the tile has a solar panel used for water heating. See the dimensional details of the tile in the figure that follows. Neglecting the thermal exchange in the ceramic part, evaluate the heat loss from the thermal solar panel.

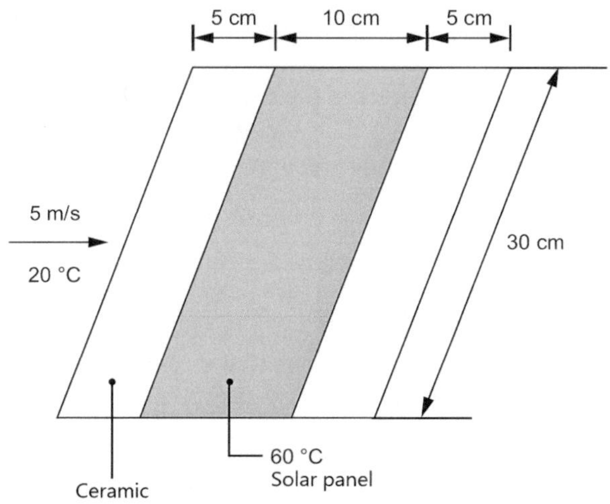

5.10 Air at 20 °C and 2 m/s flows parallel to an iron powered at 1000 W. Assuming that the base of the iron is a 10 cm × 20 cm rectangle, that the air flows in the direction of the 10 cm length, and that the portion of heat transferred by convection (other forms are radiation and conduction to the clothes) is 70 W, (a) find the numerical distribution of temperatures on the iron face in the 10 cm direction; (b) evaluate the temperature in the middle of the iron; and (c) find the average temperature on the iron face.

5.11 Given the small distance (when compared to the radius of the shaft), δ, occupied by the oil in a sliding bearing, the analysis of thermal exchange in the lubricating oil is approximated as if it were between two flat plates (with temperatures T_1 and T_2), where one plate would be the shaft and the other the bearing housing. The oil has dynamic viscosity μ, thermal conductivity k and the shaft rotates with angular velocity ω. From the conservation equations of mass, momentum, and energy, evaluate: (a) the temperature distribution in the oil; (b) the position y of maximum oil temperature; and (c) the heat flux to the shaft and to the bearing.

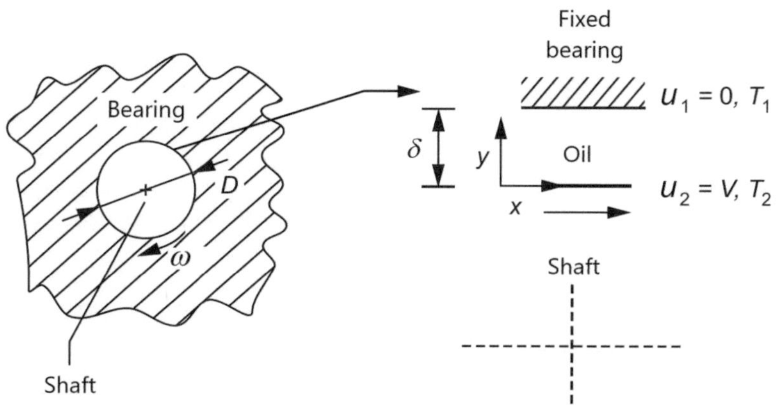

5.12 The figure shows two flat surfaces separated by a distance W. The lower surface is stationary and the upper one moves at a constant velocity V. The velocity profile behaves linearly $u = yV/W$ throughout its course. From a certain distance, the stationary plate provides a constant heat flux q, starting at $x = 0$. Considering that the temperature profile in the thermal boundary layer up to δ_T is linear and using the integral formulation of the thermal boundary layer $\frac{d}{dx}\int_0^{\delta_T} u(T - T_0)dy = \frac{q}{\rho C_p}$ and $q = h_x\left(T_{w,x} - T_0\right) = -k\left.\frac{\partial T}{\partial y}\right|_{y=0}$, evaluate the expressions (a) of the average wall temperature $(\overline{T}_w)$ from 0 to L as a function of T_0, q, k, W, L, C_p, ρ, and V ; and (b) of the average Nusselt number $(\overline{Nu})$ as a function of k, W, L, C_p, ρ, and V.

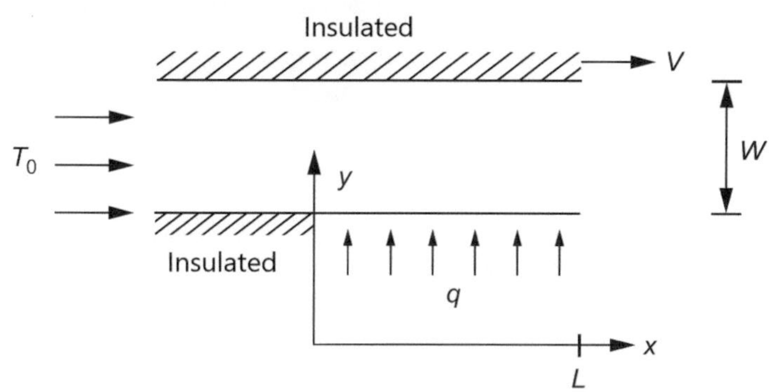

5.13 An airplane wing is tested in a wind tunnel where an average drag coefficient $C_f = 0.002$ was found for a certain Reynolds number. In a real situation where the wing temperature (on a larger scale than the one tested in the wind tunnel) is 7 °C and the air temperature is –20 °C, what will be the heat flux transferred to a characteristic length of the wing of 2.2 m and an airplane velocity of 500 km/h, which gives approximately the same Reynolds number tested in the wind tunnel?

5.14 A flat plate 50 cm long and 20 cm wide at 70 °C is positioned parallel to the water flowing at 1 m/s and 20 °C in the direction of the length. Evaluate (a) the boundary layer thermal and hydrodynamic thicknesses, as well as the local heat transfer coefficients by convection in the middle and at the end of the plate; and (b) the total heat rate transferred to the water.

5.15 Evaluate the heat exchange rate of the proposed problem 5.14, but assuming a plate at 20 °C for the first 5 cm and, after that, 70 °C. A turbulence generator was installed at the leading edge.

5.16 A car travels at 90 km/h in a region where the air temperature is 25 °C. The car's roof has an approximately a square 1,5 m plate geometry and absorbs 300 W/m^2 of solar radiation. In addition, the roof has an emissivity of 0.85. Assuming that the sky temperature for thermal exchange by radiation is 0 °C and that 10% of the absorbed radiation is transferred to the interior of the car, evaluate the external temperature of the car's roof.

Tip: all absorbed radiation must be transferred by convection to the air, by radiation to the sky and to the inside of the car: $q_a = q_{\text{rad}} + q_{\text{conv}} + q_{\text{int}}$.

5.17 Air at 10 °C and 7 m/s flows perpendicularly to a stainless steel AISI 304 electrical resistance wire of 3 m in length, 2.4 mm in diameter, and 1200 W of power. Evaluate the external temperature of the wire, assuming it to be constant.

5.18 Choose the cross-section, within the possibilities shown in Table 5.4 (except the plate), of a cylinder that provides the greatest heat transfer from the cylinder to the air. All cylinders have the same characteristic diameter $D = 20$ mm and the air flows at 15 °C and 8.5 m/s. The temperature of the cylinder wall is 65 °C. Your choice should be based on the evaluation of the heat flux.

5.19 The hot wire anemometer shown in the scheme below has a 10 μm diameter, a length of 5.5 mm and measures the air velocity that passes through it, which is at $T_\infty = 20$ °C. The electrical resistance (R_e) of this hot wire varies with the temperature according to the expression $R_e = 42[1 + 0.0037(T_w - 20)]$, with units R_e [Ω] and T_w [°C]. In a test, the current I that passes through it is 0.028 A and the voltage is 1.80 V, under these conditions evaluate the air velocity and the temperature of the hot wire surface.

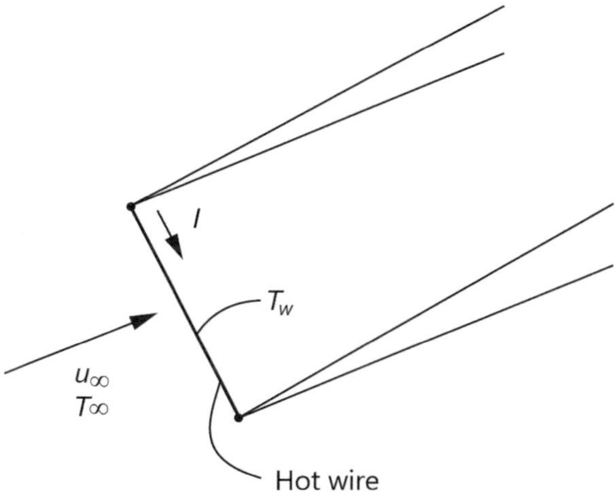

5.20 A computer *chip* is mounted perpendicularly to the air as indicated in figure. Evaluate the total heat dissipated by the *chip* when it is at 90 °C and the air is at 50 °C at a 11 m/s velocity. If it were mounted parallel to the air in the direction of the 25 mm, what would be the rate of heat dissipated? Based on these results, which position would you choose for the *chip* assembly?

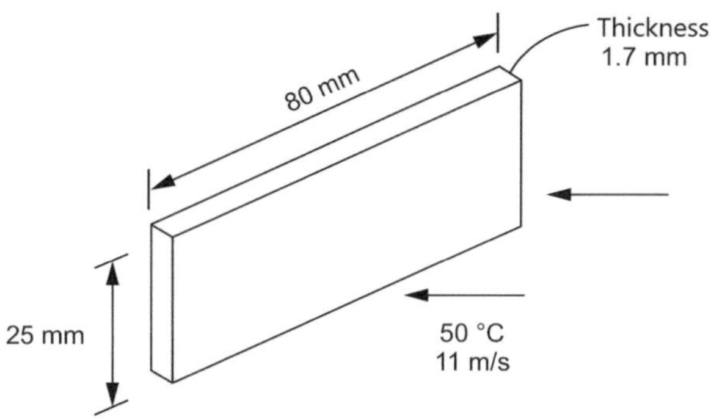

5.21 Condensing steam at 190 °C is used to heat carbon dioxide gas from 25 °C and 1 atm to 70 °C. For this, an in-line tube bank is used where the steam passes inside the tubes and the CO_2 passes on the outside of them in a cross-configuration. The length of each tube is 0.5 m, diameter of 20 mm, $S_T = S_L = 30$ mm, $N_L = 10$ and $N_T = 15$. Evaluate the CO_2 inlet velocity and its mass flow rate.

5.22 A combustion process is more efficient if the air used is preheated. Thus, in an industry, it is planned to heat air with condensing steam at 110 °C in a tube bank with an in-line arrangement. A fan promotes the movement of the ambient air and enters perpendicularly to the tube bank at 18 °C, 1 atm and 6 m/s. The tubes through which the steam passes have an external diameter of 25 mm, with longitudinal and transversal pitches of 5.7 cm. In this tube bank, there are 8 transversal tubes and 9 longitudinal tubes. For a 1 m tube length, evaluate the air outlet temperature and rate of heat exchanged with the steam.

5.23 In a commercial air conditioning application, there is an evaporator whose wall temperature is at −20 °C. It is known that the evaporator is a bank of staggered tubes with longitudinal and transversal dimensions of 16 mm and 9 mm diameter tubes. Air at 25 °C and 1 atm is forced to circulate around across the evaporator which, in the cross-section at the entrance, has a velocity of 5.5 m/s. Evaluate the number of tubes in the longitudinal direction to achieve an air outlet temperature of 5 °C.

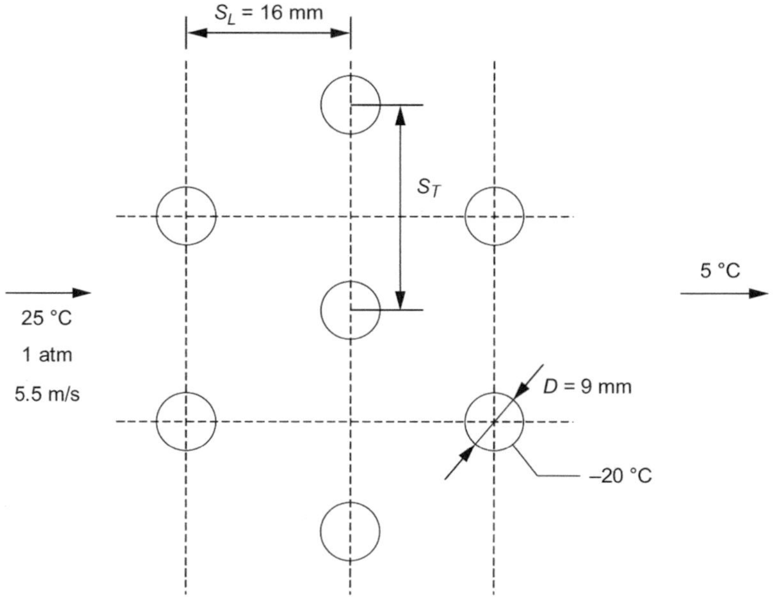

References

Bejan, A. 2013. *Convection heat transfer*, 4th ed. New Jersey: Wiley.

Besson, U. 2012. The history of the cooling law: when the search for simplicity can be an obstacle. *Science and Education* 21: 1085–1110.

Giedt, W.H. 1949. Investigation of variation of point unit heat-transfer coefficient around a cylinder normal to an Air Stream. *ASME Journal of Heat Transfer* 71: 375–81.

Grimison, E.D. 1937. Correlation and utilization of new data on flow resistance and heat transfer cross-flow of gases over tube banks. *Transactions of the ASME* 59: 583–94.

Hilpert, R. 1933. Wärmeabgabe von geheizen Drähten und Rohren im Luftstrom. *Forschung auf dem Gebiet des Ingenieurwesens A* 4: 220.

Kays, W.M., M.E. Crawford, and B. Weigand. 2005. *Convective heat and mass transfer*, 4th ed. Boston: McGraw-Hill Higher Education.

Knudsen, J.D., and D.L. Katz. 1958. *Fluid dynamics and heat transfer*. New York: McGraw-Hill.

Schlichting, H. 1979. *Boundary layer theory*, 7th ed. New York: McGraw-Hill.

Sparrow, E.M., J.P. Abraham, and J.C.K. Tong. 2004. Archival correlations for average heat transfer coefficients for non-circular and circular cylinders and for spheres in cross-flow. *International Journal of Heat and Mass Transfer.* 47: 5285–96.

Thomas, W.C. 1977. Note on the heat transfer equation for forced-convection flow over a flat plate with an unheated starting length. *Mechanical Engineering News.* 9 (1): 361–368.

Zukauskas, A. 1972. Heat transfer from tubes in cross flow. In *Advances in heat transfer*, eds. by J.P. Hartnett and T. F. Jr. Irvine. New York: Academic.

Zukauskas, A. 1987. Heat transfer from tubes in crossflow. In *Advances in Heat transfer*, eds. by J.P. Hartnett and T. F. Jr. Irvine, vol. 18, pp. 87–159.

Chapter 6
Internal Forced Convection

Abstract The forced heat transfer phenomenon inside tubes and ducts is analyzed in this chapter. First, hydrodynamic and thermal boundary layer development in tubes is analyzed. The concept of cup or bulk temperature as the effective cross-section temperature for heat transfer calculation is introduced. The Reynolds number for laminar and turbulent regimes is discussed. The laminar case is studied to analytically obtain the Nusselt number for the constant heat flux at the wall and extend it to constant wall temperature. Values for various cross-sections are presented. Heat transfer in the turbulent regime is analyzed in subsequent sections and working expressions are given for many situations of interest and the friction loss coefficient. In the final section, the logarithmic mean temperature difference, *LMTD*, concept is presented and illustrated by working examples.

Keywords Internal flow · Laminar · Turbulent · Tubes · *LMTD*

6.1 Laminar Heat Transfer Convection Inside Tubes and Ducts

6.1.1 Hydrodynamic Development of the Flow

Figure 6.1 schematizes the development of the velocity profile of a fluid flow inside a pipe, emphasizing the entrance region where the fluid enters the pipe at a uniform velocity u. Unlike external flow over surfaces, studied in Chap. 5, the hydrodynamic[1] boundary layer (HBL) that forms inside the pipe grows from the tube entrance and, depending on the length of the tube, merges at the centerline, as illustrated in Fig. 6.1. The length until this occurs is called the *hydrodynamic entrance length* (L_e), also indicated in the same figure. From this entrance length, it is said that the flow is *hydrodynamically developed* ($x > L_e$).

[1] In many texts, it is also known as *fluid dynamic boundary layer.*

J. R. Simões-Moreira and E. W. Zavaleta-Aguilar, *Heat Transfer*, Heat and Mass Transfer, https://doi.org/10.1007/978-3-031-80318-5_6

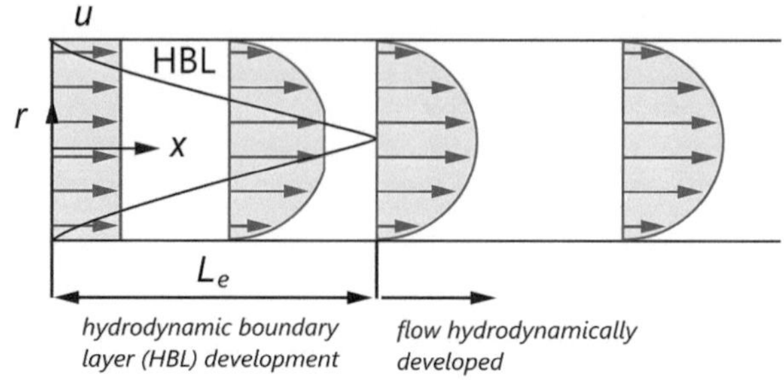

Fig. 6.1 Formation of the hydrodynamic boundary layer inside a tube

Similarly to the hydrodynamic boundary layer, when there is a temperature difference between the inlet fluid and the tube wall, a thermal boundary layer, TBL, growth takes place. The entrance region as the fluid is in thermal development is known as *thermal entrance length*, L_t. After that the nondimensional temperature profile (discussed later) remains unchanged. When the flow is simultaneously hydrodynamic and thermally developed is said to be *fully developed*.

In the case of the external flow studied in the previous chapter, the Reynolds number was defined considering the length L of the region of interest from the leading edge. In the case of flow confined within tubes and pipes, the Reynolds number should now be based on the tube diameter (D), defined by Eq. (6.1).

$$Re_D = \frac{\rho \bar{u} D}{\mu} \tag{6.1}$$

where $\bar{u}$ is the average flow velocity in the tube.

The flow remains in the laminar regime for values of $Re_D < 2300$. Some authors also suggest that the laminar regime is valid for $Re_D < 2100$. In controlled laboratory experiments, higher values of Re_D can also be achieved. The hydrodynamic entry length in the laminar case obeys the following relationship:

$$\frac{L_e}{D} = 0.05 Re_D \tag{6.2}$$

Thus, for the highest Re_D at which the flow remains laminar (2300), it is inferred that the longest possible entry length will be $L_e = 115D$. Thus, in laminar flow, the entry length can extend to tens of diameters, reaching up to 115 diameters, which can occur, for example, with very viscous fluids, such as oils, circulating in small diameter tubes.

The thermal entry length (L_t) is defined by:

$$\frac{L_t}{D} = 0.05 Re_D Pr \tag{6.3}$$

Note that if the Prandtl number is unitary (case of gases where $Pr \sim 1$), the entry lengths, thermal and hydrodynamic will be equal.

6.1.2 Bulk or Cup Temperature

There is difficulty in referencing an effective fluid temperature in the tube cross cross-section to determine the heat transfer rate between the fluid and the wall. To illustrate this difficulty, consider the external and internal flows as illustrated in Fig. 6.2.

In the first case (Fig. 6.2a), by Newton's cooling law, the calculation of heat transfer is given by the difference between the wall, T_w, and the constant fluid, T_∞, temperature far from the wall. The fluid constant temperature at a distance, T_∞, does not occur in the case of a flow confined inside tubes, pipes, and ducts (Fig. 6.2b). Thus, to calculate the heat transfer rate, an effective fluid temperature in the section of interest should be defined, which is called *bulk* or *mixing cup temperature*, T_b. To define the bulk or mixing cup temperature precisely, consider the following temperature and velocity profiles in a fluid inside a heated tube, as illustrated in Fig. 6.3.

Given the velocity and temperature profiles in Fig. 6.3, it can be seen that the highest fluid temperature occurs next to the heated wall, precisely where the fluid has the null velocity due to the no-slip principle. Thus, the simple arithmetic average of the temperature profile does not represent the effective fluid cross-section temperature. Indeed, to obtain the average (effective) mixing temperature of the section,

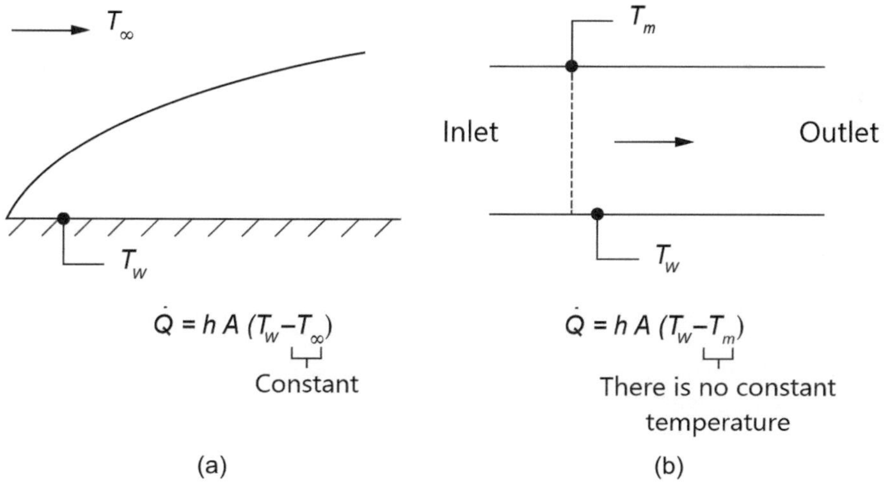

Fig. 6.2 Different ways to evaluate the heat transfer rate in: **a** flat plate—external flow; **b** tube—internal flow

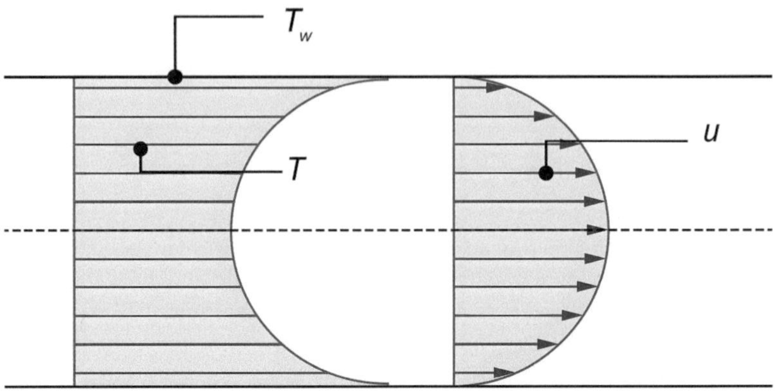

Fig. 6.3 Temperature and velocity laminar profiles in a fluid flowing in a heated tube at wall temperature, T_w

consider the mental exercise in which a tiny "slice" Δx (mathematically, it would be an elementary slice) of the cross-section of the tube with fluid is instantly placed inside a cup (Fig. 6.4). It must be agreed that the effective temperature representing the section is the one from the thermal equilibrium of that fluid portion, right? Yes, this is correct and hence the alternative name of *cup temperature.*

To determine the equilibrium temperature, consider the enthalpic flow, $\dot{E}_h$, in the cross-section given by:

$$\dot{E}_h = \int h d\dot{m} = \int_0^A \rho u h dA \tag{6.4}$$

h, in this context, is the specific enthalpy [J/kg] in the section dA.

Fig. 6.4 Average mixing or bulk flow temperature at the cross-section

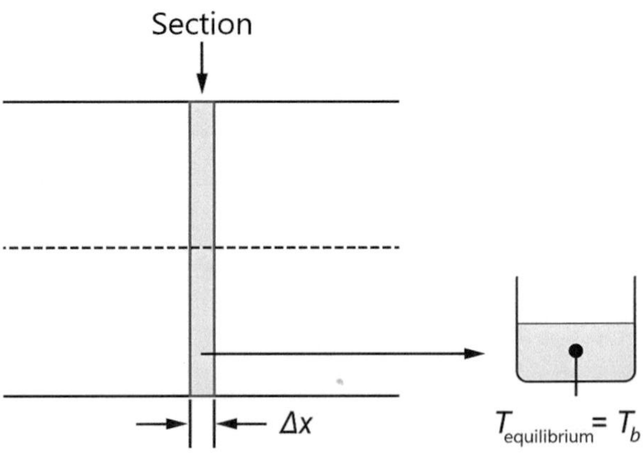

Thus, the average or bulk specific enthalpy, h_b, in the cross-section can be defined by the definition of an average:

$$h_b = \frac{1}{\dot{m}} \int_0^A \rho u h dA = \frac{\int_0^A \rho u h dA}{\int_0^A \rho u dA} \quad (6.5)$$

If there is no phase change, it can be written $h_b = C_p T_b$ and, considering C_p and ρ constants in the section, it follows:

$$T_b = \frac{\int_0^A uTdA}{\int_0^A udA} \quad (6.6)$$

For the case of the circular duct, the cross-sectional area is given by $A = \pi r^2 \Rightarrow dA = 2\pi r dr$, $A = \pi r^2 \Rightarrow dA = 2\pi r dr$, which, substituting into the previous expression, results in:

$$T_b = \frac{\int_0^r uTrdr}{\int_0^r urdr} \quad (6.7)$$

In this way, it is concluded that the bulk temperature is a *weighted average* of the temperature and velocity distributions in the tube cross-section.

6.1.3 Heat Transfer in Laminar Flow Inside Tubes

Once the bulk temperature is known (Eqs. 6.6 or 6.7), it is possible to calculate the heat transfer rate through the expression of Newton's cooling law, that is:

$$\dot{Q} = hA(T_w - T_b) \quad (6.8)$$

Evidently, it is necessary to know the radial velocity, $u(r)$, the temperature, $T(r)$, distribution as defined by Eq. (6.7). In order to obtain T_b, consider the laminar velocity flow profile illustrated in Fig. 6.5 inside a tube. In the diagram on the right, there is a balance of forces acting on the fluid element.

The boundary conditions for the problem indicated in Fig. 6.5 is:

BC_1: $r = 0 \rightarrow \frac{du}{dr} = 0$;
BC_2: $r = R \rightarrow u = 0$.

A balance of forces on the differential fluid element (Fig. 6.5) results in:

$$2\pi r dr dp + 2\pi dx d(r\tau) = 0$$

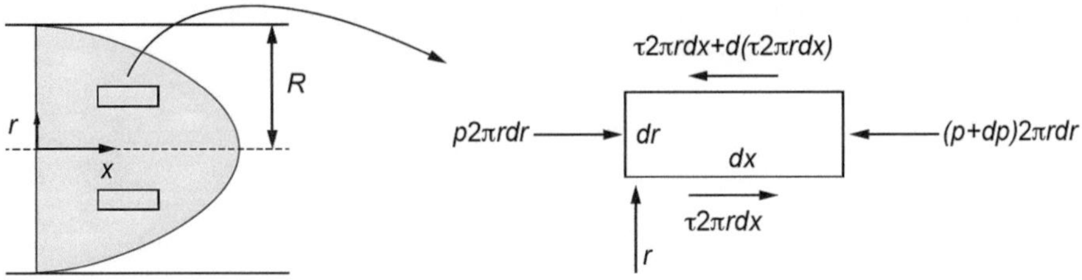

Fig. 6.5 Analysis of forces on a fluid differential element

As it is a Newtonian fluid, $\tau = -\mu \frac{du}{dr}$. Substituting into the previous expression, we get:

$$d\left(r\mu \frac{du}{dr}\right) = r\frac{dp}{dx}dr$$

integrating in the radial direction, r. Note that the static pressure is the same across the entire cross-section:

$$r\mu \frac{du}{dr} = \frac{r^2}{2}\frac{dp}{dx} + C_1$$

Using the BC_1: $0 = \frac{0^2}{2}\frac{dp}{dx} + C_1$, from which we obtain $C_1 = 0$. Integrating again in the radial section (from the center line):

$$u(r) = \frac{r^2}{4\mu}\frac{dp}{dx} + C_2$$

The constant C_2 is determined with BC_2: $\frac{R^2}{4\mu}\frac{dp}{dx} + C_2 = 0$, therefore, $C_2 = -\frac{R^2}{4\mu}\frac{dp}{dx}$. Thus, the velocity distribution is given by Eq. (6.9):

$$u(r) = -\frac{1}{4\mu}\frac{dp}{dx}\left(R^2 - r^2\right) \tag{6.9}$$

The velocity at the tube centerline, u_0, is determined by substituting $r = 0$, which gives

$$u_0 = -\frac{R^2}{4\mu}\frac{dp}{dx} \tag{6.10}$$

Finally, dividing Eq. (6.9) by Eq. (6.10), we get:

$$\frac{u(r)}{u_0} = 1 - \frac{r^2}{R^2} \tag{6.11}$$

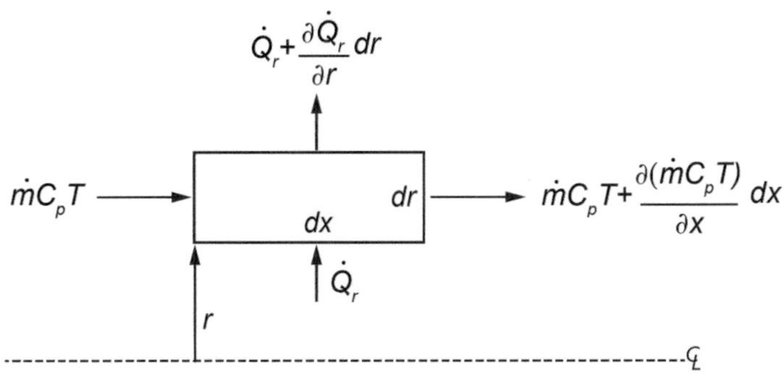

Fig. 6.6 Energy balance in the differential fluid element

Therefore, the velocity profile in laminar flow is a parabolic one of 2nd degree.

It is necessary to carry out an energy balance on a differential fluid element to obtain the temperature distribution, as illustrated in Fig. 6.6. Here, heat conduction is considered in the radial direction and enthalpic flow in the axial direction.

As a result, the energy balance results in

$$\frac{\partial \dot{Q}_r}{\partial r} dr + \frac{\partial}{\partial x}(\dot{m}C_pT)dx = 0,$$

on the other hand, the radial heat conduction is

$$\dot{Q}_r = -k2\pi r dx \frac{\partial T}{\partial r}$$

Also, $\dot{m} = \rho u 2\pi r dr$. Therefore, substituting these quantities into the energy equation, yields

$$\begin{aligned} \frac{1}{ur}\frac{\partial}{\partial r}\left(r\frac{\partial T}{\partial r}\right) &= \frac{\rho C_p}{k}\frac{\partial T}{\partial x}, \quad \text{or} \\ \frac{1}{ur}\frac{\partial}{\partial r}\left(r\frac{\partial T}{\partial r}\right) &= \frac{1}{\alpha}\frac{\partial T}{\partial x} \end{aligned} \tag{6.12}$$

To proceed with the solution, it is necessary to specify the boundary conditions. Assuming constant heat flux at wall, $q_w = constant$, consequently $\frac{dq_w}{dx} = 0$ and the boundary conditions to solve Eq. (6.12) are:

BC$_1$: as the heat flux is constant along the tube wall, then:

$$d\dot{Q} = q_w P \partial x = \dot{m}C_p \partial T, \quad \text{and thus}: \quad \frac{\partial T}{\partial x} = \text{constant};$$

BC$_2$: due to symmetry at the tube center line: $\left.\frac{\partial T}{\partial r}\right|_{r=0} = 0$;

BC$_3$: on the tube wall $k\left.\frac{\partial T}{\partial r}\right|_{r=r_0} = q_w = \text{constant}$.

Introducing these boundary conditions and integrating, results in the following laminar temperature profile:

$$T(r) = T_0 + \frac{1}{\alpha}\frac{\partial T}{\partial x}\frac{u_0 R^2}{4}\left[\left(\frac{r}{R}\right)^2 - \frac{1}{4}\left(\frac{r}{R}\right)^4\right] \tag{6.13}$$

where T_0 is the temperature at the centerline.

Finally, one can introduce the velocity profiles, $u(r)$, from Eq. (6.11) and temperature, $T(r)$, from Eq. (6.13) into the equation of the definition of bulk temperature (Eq. 6.7). After some manipulation, the cup temperature for constant heat flux at the wall is obtained:

$$T_b = T_0 + \frac{7}{96}\frac{u_0 R^2}{\alpha}\frac{\partial T}{\partial x} \tag{6.14}$$

To calculate the heat transfer rate, it is still necessary to obtain the wall temperature ($r = R$). From Eq. (6.13), we get:

$$T_w = T_0 + \frac{1}{\alpha}\frac{3}{16}u_0 R^2\frac{\partial T}{\partial x} \tag{6.15}$$

Now, finally, the heat transfer coefficient in a circular tube with constant properties, constant heat flux on the wall, and fully developed laminar flow can be calculated from its own definition given by Eq. (6.8) by equating it with the conduction inside the fluid next to the wall ($r = R$)—note that the r increases in the opposite direction to the conduction of heat and, therefore, the signal of ($-$) was abolished:

$$\dot{Q} = hA(T_w - T_b) = kA\left.\frac{\partial T}{\partial r}\right|_{r=R} \tag{6.16}$$

From this relationship, the heat transfer coefficient by convection can be obtained:

$$h = \frac{k\left.\frac{\partial T}{\partial r}\right|_{r=R}}{(T_w - T_b)} \tag{6.17}$$

Substituting the derivative of the temperature distribution (Eq. 6.13) into Eq. (6.17), we get:

$$h = \frac{k\frac{\partial}{\partial r}\left\{T_0 + \frac{1}{\alpha}\frac{\partial T}{\partial x}\frac{u_0 R^2}{4}\left[\left(\frac{r}{R}\right)^2 - \frac{1}{4}\left(\frac{r}{R}\right)^4\right]\right\}_{r=R}}{\left(\underbrace{T_0 + \frac{1}{\alpha}\frac{3}{16}u_0 R^2\frac{\partial T}{\partial x}}_{T_w} - \underbrace{T_0 - \frac{7}{96}\frac{u_0 R^2}{\alpha}\frac{\partial T}{\partial x}}_{T_b}\right)} \tag{6.18}$$

After performing the calculations, we arrive at:

$$h = 4.36\frac{k}{D}, \quad \text{or} \quad Nu_D = 4.36 \tag{6.19}$$

This is a remarkable result, as the Nusselt number for fully developed laminar flow, constant properties, subjected to a constant heat flux on the wall *does not depend* on the Reynolds and Prandtl numbers, contrary to the case of heat transfer in an external flow.

If the same calculations are performed for the condition of *constant wall temperature*, we obtain (Kays et al. 2005):

$$Nu_D = 3.66 \tag{6.20}$$

In laminar flow, the effects of surface roughness are negligible. In the development of the previous equations, it was always assumed that transport properties are constant. Thus, transport properties should be obtained at the average between the bulk temperatures at the inlet and outlet of the flow, that is: $\overline{T} = \frac{T_i - T_o}{2}$. However, some properties strongly depend on temperature. As an example, the dynamic viscosity of water at 25 °C is $\mu_{(25\,°\mathrm{C})} = 8.90 \times 10^{-4}$ kg/ms and at 30 °C, $\mu_{(30\,°\mathrm{C})} = 7.98 \times 10^{-4}$ kg/ms. That is, in 5 °C there is a variation of about 10% in the value of dynamic viscosity.

Thus, it follows that the following correction should be used to take this effect into account (Kays et al. 2005).

$$\text{For liquids}: \ \overline{Nu}_{cor} = \overline{Nu}\left(\frac{\mu}{\mu_w}\right)^n \tag{6.21}$$

where μ is the dynamic viscosity at the bulk temperature, T_b, and μ_w is the dynamic viscosity at the wall temperature, T_w. The value $n = 0$ (no correction) applies to gases and $n = 0.14$ for liquids.

The flow is not fully developed for short tubes and thus it is said that the flow is in the *entrance region.* In case the flow is hydrodynamically developed, but thermally developing, the *entrance is thermal*, and when the flow is both hydrodynamically and thermally developing, the *entrance is combined*. Figure 6.7 indicates the variation of the local Nusselt number in the case of thermal entry, starting from the entry until the flow becomes fully developed. Similar behaviors were observed in the case of constant tube wall temperature or when the heat flux on the wall is constant. At high lengths, the Nusselt number approaches 3.66 for constant wall temperature and 4.36 for constant wall flow.

The average Nusselt number for the tube of length L, constant wall temperature and thermal entry is given by Eq. (6.22) (Hausen 1943).

$$\overline{Nu}_D = 3.66 + \frac{0.0668(D/L)Re_D Pr}{1 + 0.04[(D/L)Re_D Pr]^{2/3}} \tag{6.22}$$

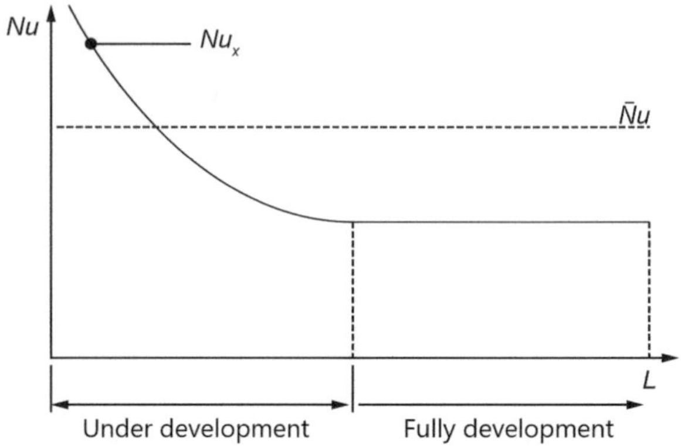

Fig. 6.7 Variation of the local Nusselt number at thermal entry. Adapted from Kays et al. 2005

For the case of constant heat flux at the wall and the flow is fully developed or with thermal entry, the following correlation can be used (Shah and London, 1978).

$$\overline{Nu_D} = 1.953\left(\frac{L}{DRe_DPr}\right)^{-1/3}; \qquad \frac{L}{DRe_DPr} \le 0.03 \tag{6.23}$$

$$\overline{Nu_D} = 4.364 + 0.0722\left(\frac{L}{DRe_DPr}\right)^{-1}; \qquad \frac{L}{DRe_DPr} > 0.03$$

Example 6.1 (*Heat transfer inside tubes in laminar regime*)

A tubular water heater with an internal diameter of 25 mm and is 2 m in length is used to heat water whose average velocity is 0.025 m/s. The heat flux rate at the wall is constant. In a steady state, the temperatures (average) of the water at the inlet and outlet of the heater are 15 and 78.6 °C, respectively. Evaluate (a) the hydrodynamic and thermal entry length, and (b) the average heat transfer coefficients assuming fully developed flow and thermal entry. How much do both evaluations diverge?

Solution

The average fluid temperature is: 46.8 °C (320 K), from Table A.7:

$$\rho = 1/1.011 \times 10^{-3} = 989.1\,\text{kg/m}^3;\ \mu = 577 \times 10^{-6}\,\text{kg/ms};$$
$$k = 0.640\,\text{W/mK};\ Pr = 3.77$$

Initially, it should be evaluated whether the flow is laminar:

$$Re_D = \frac{\rho\,\bar{u}\,D}{\mu} = \frac{989.1 \times 0.025 \times 0.025}{577 \times 10^{-6}} = 1071.4 \quad (< 2300,\ \text{therefore it is laminar})$$

(a)

$$L_e = 0.05Re_D D = 0.05 \times 1071.4 \times 0.025 = 1.34 \text{ m}$$

$$L_t = 0.05Re_D Pr D = 0.05 \times 1071.4 \times 3.77 \times 0.025 = 5 \text{ m}$$

(b)

Assuming fully developed flow:

$$Nu = 4.36$$

$$h = Nu\frac{k}{D} = 4.36\frac{0.640}{0.025} = 111.6 \ \ \text{W/m}^2\text{K}$$

Considering the correlation of Shah and London (1978), Eq. (6.23) to consider the developing part:

$$\frac{L}{DRe_D\text{Pr}} = \frac{2}{0.025 \times 1071.4 \times 3.77} = 0.02$$

$$\overline{Nu_D} = 1.953\left(\frac{L}{DRe_D\text{Pr}}\right)^{-1/3} = 1.953 \times 0.02^{-1/3} = 7.19$$

$$h = \overline{Nu_D}\frac{k}{D} = 7.19\frac{0.640}{0.025} = 184.1 \ \ \text{W/m}^2\text{K}$$

It can be noted that the flow is still developing thermally. The Shah and London's (1978) correlation provides a coefficient 65% higher than the case of fully developed flow and it should preferentially be used for short tubes.

6.1.4 *Heat Transfer in Laminar Flow Inside Ducts of Various Geometries*

Correlations for heat transfer in ducts of other cross-sections are presented in Table 6.1. The *friction factor* f (also known as *Darcy's friction factor*) is also presented, whose definition is given by:

$$f = \frac{8\tau_w}{\rho \bar{u}^2/2} \tag{6.24}$$

Table 6.1 Nusselt numbers and friction factors for fully developed laminar flow

Geometry	Nusselt Number				
		D_h	T_w =const	q_w = const	fRe_D [b]
Circle 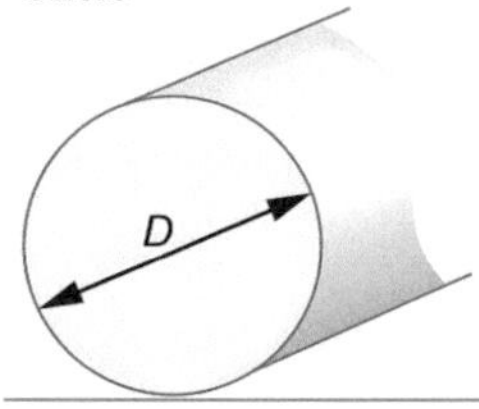					
Rectangle 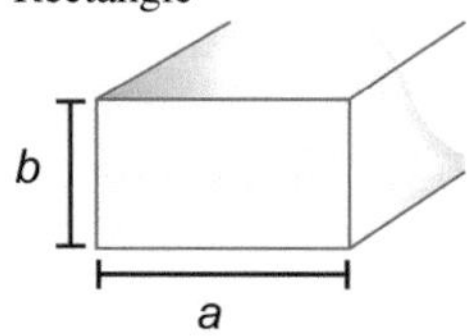	a/b	$\frac{2ab}{a+b}$	2.97	3.61	56.92
	1				
	2		3.38	4.12	62.23
	3		3.94[a]	4.78[a]	68.38
	4		4.39[a]	5.29[a]	72.94
	5		4.80	5.74	76.29
	6		5.01[a]	5.95[a]	78.82
	10		5.86	6.79	84.70
	∞		7.54	8.24	96.00
Triangle isosceles[b] 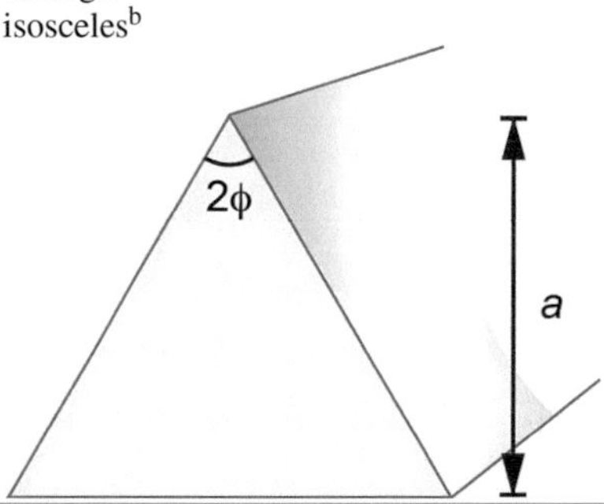	2ϕ	$\frac{2a\sin\phi}{\sin\phi+1}$	1.61	2.45	49.90
	10				
	20		2.00	2.72	51.29
	30		2.26	2.91	52.26
	50		2.45	3.09	53.23
	90		2.34	2.98	52.61
	120		2.00	2.68	50.98
	150		1.50	2.33	48.90

[a]By interpolation; [b]Shah and London (1978)
Adapted from Rohsenow et al. (1998)

In the case of non-circular cross-sections, the *hydraulic diameter*, D_h, should be used in place of tube diameter, D, which is defined as follows:

$$D_h = \frac{4A}{P} \tag{6.25}$$

where A is the cross-sectional area of the duct and P is the so-called *wetted perimeter*, the perimeter encompassing the fluid in contact with the pipe wall.

In the absence of specific expressions for a given cross-section (as it occurs with ventilation and air conditioning duct sections, for example), it is suggested to carry out the calculations replacing the diameter, D, of the working expressions with the hydraulic diameter, D_h (Eq. 6.25).

Example 6.2 (*Heat transfer inside ducts in laminar regime*)

Air at a 0.0004 kg/s mass flow rate flows in a rectangular duct 1 m long. The cross-section can be 4 × 16 mm or 6 × 14 mm. It is known that the duct is uniformly heated at the wall by electronic circuit components, which dissipate 600 W/m^2 of thermal power. It is asked, for each section:

(a) the air exit temperature for each cross-section;
(b) the hydraulic diameters and the Reynolds numbers of these flows;
(c) the average surface temperatures of the ducts;
(d) the pressure drops $\Delta p = f\rho\frac{L}{D}\frac{\bar{u}^2}{2}$;
(e) the ratio between heat exchange rate and pressure drop (W/Pa), and then choose the best cross-section to cool off the circuit.

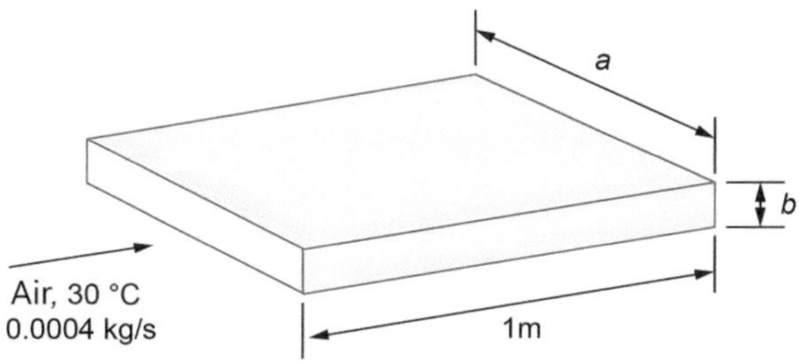

Solution

Initially, let us solve for the 4 mm × 16 mm section.

The lateral area of heat exchange is: $A_L = 2(a + b)L$

$$A_L = 2 \times (0.004 + 0.016) \times 1 = 0.04 \text{ m}^2$$

(a) Initially, assuming an average temperature of 350 K, so $C_p = 1009$ J/kgK:

$qA_L = \dot{m}C_p(T_o - T_i) \rightarrow 600 \times 0.04 = 0.0004 \times 1009 \times (T_o - 30)$, then the exit temperature is $T_o = 89.5$ °C.

In a new interaction, the average temperature is: (30 + 89.5)/2 = 59.8 °C (333 K), therefore $C_p = 1008.3$ J/kgK, so, $600 \times 0.04 = 0.0004 \times 1008.3(T_o - 30)$, then $T_o = 89.5$ °C, which is good enough.

(b) For this average temperature, the following properties are obtained in Appendix A.

$$\rho = 1.052\,\mathrm{kg/m^3};\ \nu = 200.2 \times 10^{-7}\,\mathrm{kg/ms};\ k = 0.029\,\mathrm{W/mk};\ Pr = 0.702$$

$$D_h = \frac{2ab}{a+b} = \frac{2 \times 0.004 \times 0.016}{0.004 + 0.016} = 6.4 \times 10^{-3}\ \mathrm{m}$$

$$\dot{m} = \rho\bar{u}(ab);\quad \bar{u} = \frac{0.0004}{1.05 \times 2 \times 0.004 \times 0.016} = 5.94\ \mathrm{m/s}$$

$$Re_D = \frac{\rho\bar{u}D_h}{\mu} = \frac{1.052 \times 5.94 \times 6.4 \times 10^{-3}}{200.2 \times 10^{-7}} = 1997.6\ \ (< 2300,\ \text{laminar})$$

(c) $\frac{a}{b} = \frac{16}{4} = 4$, therefore, from Table 6.1, $Nu = 5.29$

$$h = Nu\frac{k}{D_h} = 5.29\frac{0.029}{6.4 \times 10^{-3}} = 24.0\ \mathrm{W/m^2K}$$

$$q = h(\overline{T}_w - \overline{T}) \rightarrow \overline{T}_w = \frac{q}{h} + \overline{T} = \frac{600}{24} + 59.8$$

$$\overline{T}_w = 84.8\ ^\circ\mathrm{C}$$

(d) From Table 6.1, $fRe_D = 72.94$, therefore,

$$f = \frac{72.94}{1997.6} = 0.0365$$

$$\Delta p = f\frac{L}{D}\frac{\rho\bar{u}^2}{2} = 0.0365\frac{1}{6.4 \times 10^{-3}}\frac{1.052(5.96^2)}{2}$$

$$\Delta p = 106.6\ \mathrm{Pa}$$

(e) $$\frac{\dot{Q}}{\Delta p} = \frac{qA_L}{\Delta p} = \frac{600 \times 0.04}{106.6} = 0.225\ \mathrm{W/Pa}$$

The values for both geometries are summarized in the following table.

Variable	4 mm × 16 mm	6 mm × 14 mm
T_o (°C)	89.5	89.5
D_h (m)	0.0064	0.0084
Re_D (–)	1997.6	1995.1
$\overline{T}_w$(°C)	84.8	90.3
Δp (Pa)	106.6	41.2
$\dot{Q}/\Delta p$ (W/Pa)	0.225	0.583

Comparing the heat exchanged rate per pressure drop, the 6 mm × 14 mm duct is selected, as it promotes greater thermal exchange with less demand for air ventilation power (which is proportional to the pressure drop).

6.1.5 Bulk Temperature Along the Tube

In this section, the bulk temperature variation along the tube is determined. For this, consider the energy balance in a volume of differential length between two tube sections, as illustrated in Fig. 6.8. Two cases can be analyzed to determine T_b, which depend on boundary conditions imposed on the tube wall: (I) constant heat flux and (II) constant wall temperature.

(I) *Constant heat flux at the wall* ($q_w = \text{cte}$).

Based on the scheme of Fig. 6.8 and considering that, in this analysis, h_x refers to the specific enthalpy in the section, and not to the convective heat coefficient, the following energy balance can be written.

$\dot{m}\left(h_x + \frac{dh_x}{dx}dx\right) = \dot{m}h_x + q_w dA_w$, or simply,

$$\dot{m}\frac{dh_x}{dx}dx = q_w dA_w \tag{6.26}$$

where A_w is the internal tube wall area in contact with the fluid. However, on the other hand, $dA_w = Pdx$, where P is the wetted perimeter. So that:

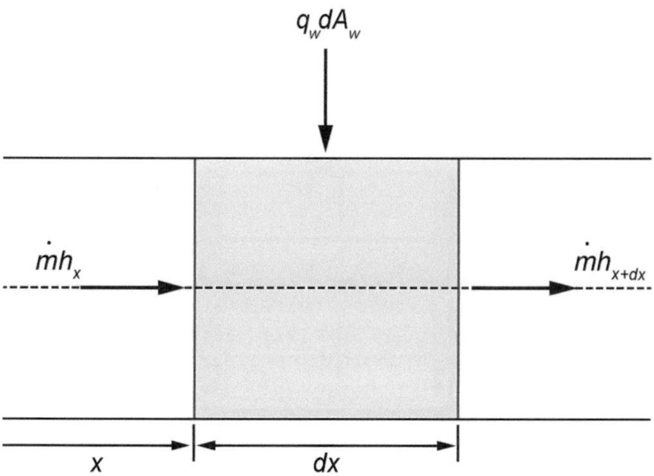

Fig. 6.8 Energy balance between two tube sections separated by *dx*

$$\dot{m}\frac{dh_x}{dx}dx = q_w P dx.$$

or, still, $\dot{m}\frac{dh_x}{dx} = q_w P$. Assuming that the specific enthalpy is $h_x = C_p T_b$ or $dh_x = C_p dT_b$, for $C_p = \text{cte}$, in which $T_b = T_b(x)$, and the mass flow rate being equal to $\dot{m} = \rho\,\bar{u}\,A$, we have:

$$q_w P = \rho C_P\,\bar{u}\,A\frac{dT_b}{dx}$$

Integrating the previous equation, we get: $T_b(x) = \frac{q_w P}{\rho C_P\,\bar{u}\,A}x + \text{const.}$.
For $x = 0$, $T_b = T_i$ (input temperature), so that:

$$T_b(x) = \frac{q_w P}{\rho C_P\,\bar{u}\,A}x + T_i \tag{6.27}$$

On the other hand, the heat flux at wall can be expressed as: $q_w = h(T_w - T_b)$, therefore, the temperature of the inner tube wall is equal to:

$$T_w(x) = \frac{q_w P}{\rho C_P \bar{u} A}x + \frac{q_w}{h} + T_i \tag{6.28}$$

(II) *Constant wall temperature* ($T_w = \text{cte}$).

In this case, applying the conservation of energy,

$$\dot{Q}_x = P dx h(T_w - T_b) = \dot{m}C_p\frac{dT_b}{dx} \quad \text{or,} \quad \frac{dT_b}{T_w - T_b} = \frac{hP}{\dot{m}C_p}dx,$$

where h is the heat transfer coefficient by convection of the internal flow and is considered constant. The integration results in:

$$\ln(T_w - T_m) = \frac{hP}{\dot{m}C_P}x + \text{cte}$$

For the boundary condition at $x = 0$, we have $T_m = T_i$ (entry temperature). Then,

$$\frac{T_w - T_b(x)}{T_w - T_i} = \exp\left[-\frac{hPx}{\dot{m}C_P}\right], \quad \text{or}$$

$$T_b(x) = T_w - (T_w - T_i)\exp\left[-\frac{hPx}{\dot{m}C_P}\right] \tag{6.29}$$

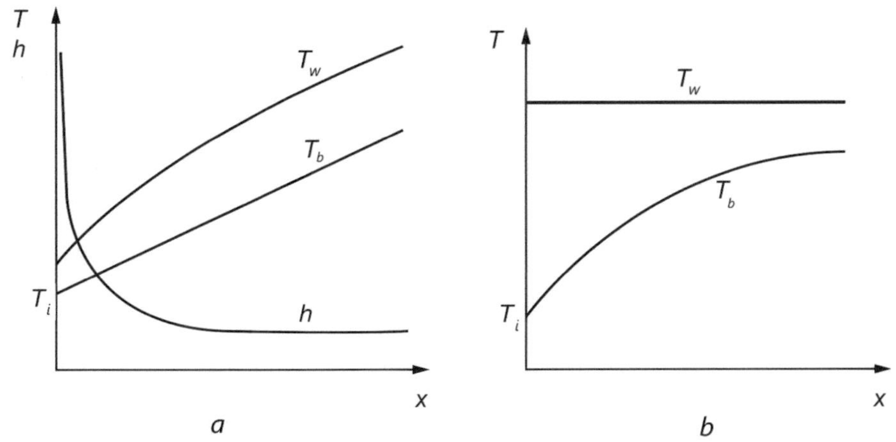

Fig. 6.9 **a** temperatures and heat transfer coefficient variation for constant heat flux at the wall, **b** bulk temperature variation for constant wall temperature

The graph in Fig. 6.9a illustrates the variation of wall temperatures, T_w, and average mixture, T_b, in addition to the value of the local heat transfer coefficient, h, along the length of the tube for the case of constant heat flux on the wall. The graph in Fig. 6.9b shows the increase in the average mixture temperature, T_m, for the case of constant wall temperature, $T_w =$ constant.

6.2 Turbulent Heat Transfer Convection Inside Tubes and Ducts

6.2.1 Development of the Turbulent Boundary Layer

In the case of turbulent flow inside tubes, the boundary layer begins in the laminar regime, goes through a transition to establish a turbulent boundary layer, as illustrated in Fig. 6.10.

The entrance length of the development of the turbulent hydrodynamic and thermal boundary layer is $L_e \approx 10D$. The Reynolds number defines the flow regime; for $Re_D < 2300$, the regime is laminar, and for $Re_D > 4000$, the regime is turbulent. Between 2300 and 4000, a laminar-turbulent transition occurs, being considered fully turbulent at $Re_D > 10,000$.

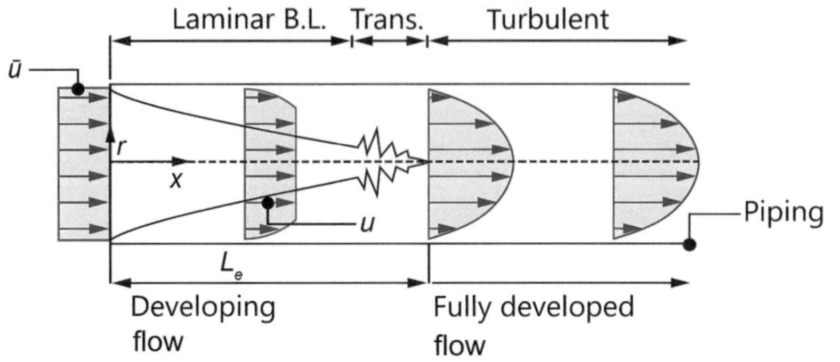

Fig. 6.10 Development of the boundary layer in turbulent flow

6.2.2 Reynolds-Colburn Analogy for Turbulent Flow

The Reynolds-Colburn analogy for laminar flow in a flat plate was discussed in Sect. 5.2.3. In the following analysis of internal fluid, it is assumed that the flow is fully developed. In the laminar case, heat transfer from the wall to the adjacent fluid (or vice versa) occurs by conduction, according to Fourier's Law, that is,

$$\frac{\dot{Q}}{A} = -k\frac{dT}{dr}, \quad \text{or} \quad \frac{\dot{Q}}{A\rho C_p} = -\alpha\frac{dT}{dr}.$$

In the case of turbulent flow, an analogous expression is defined with the following form:

$$\frac{\dot{Q}}{A\rho C_p} = -(\alpha + \varepsilon_H)\frac{dT}{dr} \tag{6.30}$$

in which we introduce the *turbulent thermal diffusivity*, ε_H, already defined in the previous chapter (Eq. 5.59b). Analogous to the shear stress, we have:

$$\tau = \mu\frac{du}{dr} \Rightarrow \frac{\tau}{\rho} = (\nu + \varepsilon_m)\frac{du}{dr} \tag{6.31}$$

in which ε_m is the *turbulent viscosity*, also already defined in the previous chapter (Eq. 5.59a).

As a hypothesis, it is assumed that heat and momentum are transported at the same rate due to macroscopic fluid mixtures, that is, $\varepsilon_H = \varepsilon_m$, then it follows that $Pr_t = 1$. This good assumption is valid because the turbulent mixing mechanisms of the fluid are the same for momentum and thermal energy. With the further assumption of $\alpha = \nu$ or $Pr = 1$ and dividing the previous equations, we get:

$$\frac{\dot{Q}}{AC_p\tau} = -\frac{dT}{du} \quad \text{or} \quad \frac{\dot{Q}}{AC_p\tau}du = -dT$$

Another assumption to be adopted is that the ratio between the heat transfer rate per unit area and the shear is constant in the cross-section, which allows us to write that ratio next to the wall, that is:

$$\frac{\dot{Q}}{C_pA\tau} = \frac{\dot{Q}_w}{C_pA_w\tau_w}, \quad \text{or}$$

$$\frac{\dot{Q}_w}{C_PA_w\tau_w}du = -dT \tag{6.32}$$

The boundary conditions of Eq. (6.32) are:

BC_1: at $r = R \rightarrow u = 0, T = T_w$.

BC_2: for average conditions $\rightarrow u = \overline{u}, T = T_b$.

These conditions, viewed graphically, are illustrated in Fig. 6.11.

So that it is possible to integrate Eq. (6.32) with the established boundary conditions:

$$\frac{\dot{Q}_w}{C_PA_w\tau_w}\int_0^{\overline{u}} du = -\int_{T_w}^{T_b} dT \tag{6.33}$$

Which results in: $\frac{\dot{Q}_w}{C_PA_w\tau_w}\overline{u} = T_w - T_b$

On the other hand, the convective heat transfer rate is given by Eq. (6.8), which, substituting in the previous expression, results in:

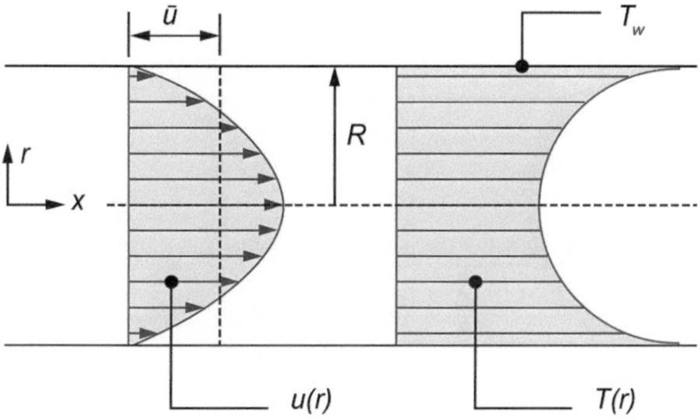

Fig. 6.11 Boundary conditions of Eq. (6.31)

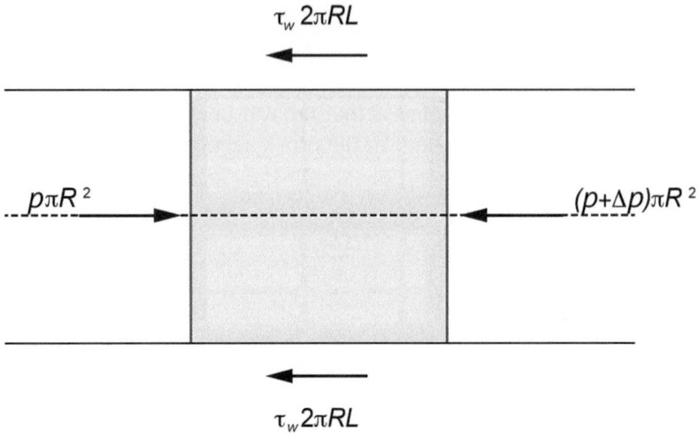

Fig. 6.12 Forces balance in a fluid element

$$\frac{hA_w(T_w - T_b)}{C_p A_w \tau_w}\bar{u} = T_w - T_b$$

and, simplifying, we have:

$$\frac{h\bar{u}}{C_p} = \tau_w \tag{6.34}$$

Forces balance in the fluid element shown in Fig. 6.12 results in $\Delta p \pi R^2 = \tau_w 2\pi RL$, or:

$$\tau_w = \frac{R}{2L}\Delta p \tag{6.35}$$

However, it is known that the distributed pressure loss is given by Eq. (6.36):

$$\Delta p = f\frac{L}{D}\frac{\rho\bar{u}^2}{2} \tag{6.36}$$

being f the Darcy friction factor. Thus, comparing the two Eqs. (6.35) and (6.36), it follows that:

$$\tau_w = \frac{f}{8}\rho\bar{u}^2 \tag{6.37}$$

Finally, the analogy can be concluded by equating Eqs. (6.34) and (6.37). Thus:

$$\frac{h\bar{u}}{C_p} = \frac{f}{8}\rho\bar{u}^2 \tag{6.38}$$

For the dimensionless numbers governing the phenomenon to appear explicitly, additional manipulations are necessary, starting by rearranging Eq. (6.38), to obtain:

$$\frac{h}{\rho C_p \overline{u}} = \frac{f}{8}$$

Now, conveniently, this expression is multiplied and divided by some quantities, as indicated below:

$$\underbrace{\frac{hD}{k}}_{Nu_D} \times \underbrace{\frac{1}{\nu/\alpha}}_{Pr^{-1}} \times \underbrace{\frac{1}{D\overline{u}/\nu}}_{Re^{-1}} = \frac{f}{8}$$

Finally, the dimensionless groups, Nu_D, Pr and Re_D, are substituted:

$$\frac{Nu_D}{\mathrm{Pr}\, Re_D} = \frac{f}{8} \text{ or } St = \frac{f}{8} \tag{6.39}$$

In Eq. (6.39), the first group of dimensionless numbers is the so-called *Stanton number*, *St*, already defined in the previous chapter. This is the Reynolds analogy for turbulent flow inside tubes. It is in accordance with experimental data for gases ($Pr \sim 1$). Based on experimental data, Colburn recommends that the previous ratio be multiplied by $Pr^{2/3}$ to extend the application to other fluids in the range of $0.5 < Pr < 100$, to include most common gases and liquids. Recall that this analogy had already been developed for laminar flow regime in Sect. 5.2.3.

$$\frac{Nu_D}{Pr^{1/3} Re_D} = \frac{f}{8} \quad \text{or} \quad StPr^{2/3} = \frac{f}{8} \tag{6.40}$$

It should be noted that, in the context of external flow, Eq. (5.52), the analogy was made with the friction coefficient (or Fanning's friction factor), $\overline{C}_{f,L}$. In the case of internal flow, Eq. (6.40), the analogy is made with the friction factor, f (or Darcy's friction factor). So that comparing these equations, it is concluded that $f = 4\overline{C}_{f,L}$.

In the Reynolds range between $1 \times 10^4 - 2 \times 10^5$, for *smooth tubes*, the friction factor f can be approximated by the following curve-fitting equation (Colburn 1933):

$$f = 0.184 Re_D^{-0.2} \tag{6.41}$$

Then, substituting the smooth tube friction factor in the analogy (Eq. 6.40), we obtain the Nusselt number:

$$Nu_D = 0.023 Re_D^{0.8} Pr^{1/3} \tag{6.42}$$

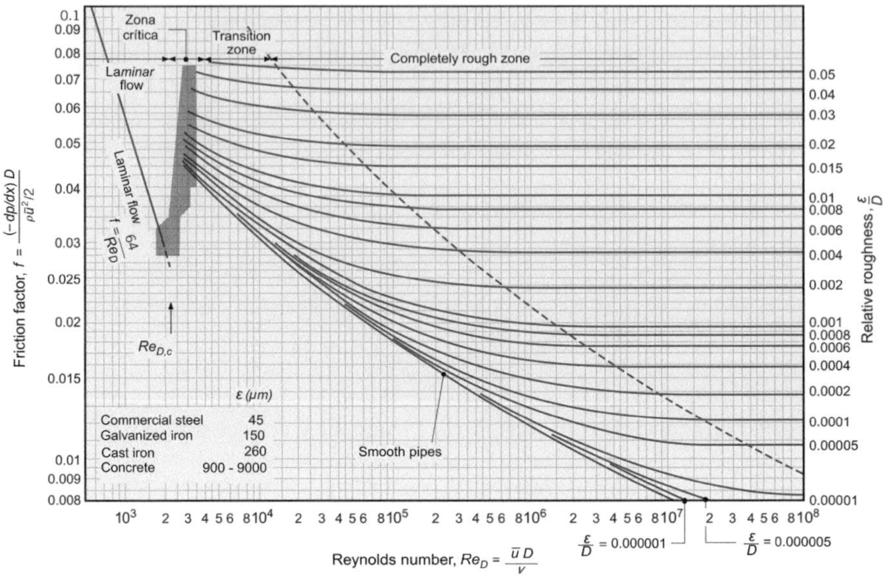

Fig. 6.13 Moody diagram for the friction factor (Moody, 1944). Published with permission

However, Dittus and Boelter[2] (1985) suggest that the exponent of the Prandtl number be $n = 0.3$ if the fluid is being cooled, and $n = 0.4$ if the fluid is being heated. Thus, the equation known as the Dittus-Boelter is given by Eq. (6.43).

$$Nu_D = 0.023 Re_D^{0.8} Pr^n; \quad Re_D \geq 10.0000.6 \leq Pr \leq 160 \tag{6.43}$$

For rough tubes, the Moody diagram should be used to obtain f as seen in Fig. 6.13. In the diagram, ε is the absolute roughness of the surface.

The friction factor of rough tubes can also be obtained through a correlation or fitting equation, for example, the expression of Churchill (1977) given by Eq. (6.44). This expression has the advantage of being explicit and smoothly adjusting the laminar-turbulent transition (2300–4000).

$$f = 8\left[\left(\frac{8}{Re_D}\right)^{12} + \frac{1}{(A+B)^{1.5}}\right]^{\frac{1}{12}}; \quad A = \left\{2.457\ln\left[\frac{1}{\left(\frac{7}{Re_D}\right)^{0.9} + \left(\frac{0.27\varepsilon}{D}\right)}\right]\right\}^{16},$$
$$B = \left(\frac{37,530}{Re_D}\right)^{16} \tag{6.44}$$

[2] Originally, in Ditttus and Boelter´s work (1985), the equation is given for heating: $Nu_D = 0.0243\mathrm{Re}_D^{0.8} Pr^{0.4}$; and for cooling: $Nu_D = 0.0265 Re_D^{0.8} Pr^{0.3}$, it was McAdams (1942) who suggested the expression as indicated in Eq. (6.42).

Another older expression often used is that of Colebrook (1939), given by Eq. (6.45). This equation was used by Moody to construct his diagram (Fig. 6.13). It has the disadvantage of having the friction factor defined implicitly, which requires an iterative numerical solution.

$$\frac{1}{f^{1/2}} = -2.0\log\left(\frac{\varepsilon/D}{3.7} + \frac{2.51}{Re_D f^{1/2}}\right) \tag{6.45}$$

In the case where transport properties vary significantly with temperature, the Nusselt number should be corrected by the ratio of absolute temperatures, as indicated by Eq. (6.46) in the case of gases.

$$Nu_{\text{corr}} = Nu\left(\frac{\overline{T}}{T_w}\right)^n \tag{6.46}$$

where $\overline{T}$ is the average absolute temperature between the inlet and the outlet and T_w is the absolute temperature of the wall, $n = 0$ for gas cooling, $n = 0.45$ for heating ($n = 0.15$ for CO_2) if $0.5 < \frac{\overline{T}}{T_w} < 2.0$.

In the case of liquids, the following expression is suggested to correct the Nusselt number:

$$Nu_{\text{corr}} = Nu\left(\frac{Pr}{Pr_w}\right)^{0.11} \tag{6.47}$$

where Pr is the Prandtl number at the average temperature ($\overline{T}$) and Pr_w is the Prandtl number at the wall temperature. Similar corrections have been proposed for the laminar case (Eq. 6.21).

Gnielinski's (1976) correlation is recommended by several authors and should be used along with the friction factor calculated from Eq. (6.50), which was recommended by Gnielinski (2013) based on the work of Konakov (1946).

$$Nu_D = \frac{(f/8)(Re_D - 1000)Pr}{1 + 12.7(f/8)^{1/2}\left(Pr^{2/3} - 1\right)}\left[1 + \left(\frac{D}{L}\right)^{2/3}\right]K \tag{6.48}$$

$4000 < Re_D < 5 \times 10^6$, $0.5 < Pr < 2000$

$$K = \left(\frac{Pr}{Pr_w}\right)^{0.11} \quad \text{for liquids} \tag{6.49}$$

$$K = \left(\frac{T}{T_w}\right)^n \quad \text{for gases, } n = 0.45 \quad \text{(air)}$$

$$f = (0.782\ln Re_D - 1.5)^{-2} \tag{6.50}$$

In the transition region $2300 < Re_D < 4000$, Gnielinski (2013) proposed a linear interpolation in the Reynolds number between the laminar Nusselt evaluated at $Re = 2300$ and the turbulent evaluated at $Re = 4000$.

Example 6.3 (*Heat transfer inside ducts in a turbulent regime*)

Air flows in a 5 cm diameter smooth tube. The air velocity is 30 m/s and its temperature is 15 °C. The tube heated length is 0.6 m and a wall temperature, $T_w = 38\,°\text{C}$. Assume that the flow is fully developed. Obtain the heat transfer rate and the air outlet temperature.

Solution

Calculate the properties at the average temperature: $\overline{T} = \frac{T_i + T_o}{2}$.

Modified Dittus-Boelter (Eq. 6.43):

$$Nu_D = 0.023 Re_D^{0.8} Pr^{0.4} \tag{6.A}$$

Energy balance:

$$hA_w\left(T_w - \overline{T}\right) = \dot{m}C_p(T_o - T_i) \tag{6.B}$$

Note that there are two Eqs. (6.A) and (6.B) and two unknowns (T_o and h).

This family of problems should normally be solved iteratively, as shown in the following scheme. First, assume an output temperature, calculate all the quantities involved, and then check whether it corresponds to the result of the second equation. If not, take a new output temperature and iterate the solution procedure up to an acceptable convergence.

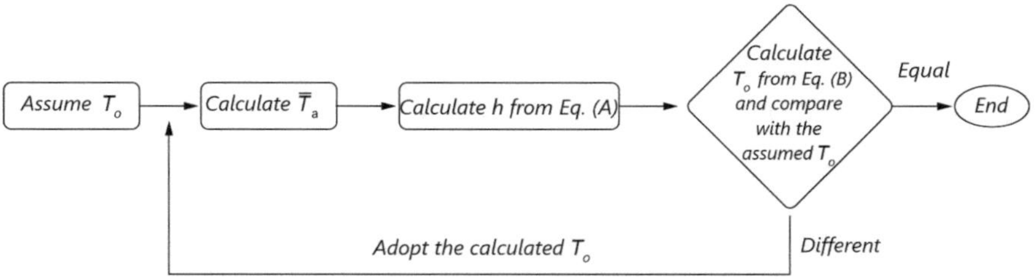

Assuming, $T_o = 21°\text{C} \rightarrow \overline{T} = \frac{15+21}{2} = 18\,°\text{C}$. The transport properties of air (Table A.5) are:

$\rho = 1.2025\,\text{kg/m}^3$; $\nu = 15.11 \times 10^{-6}\,\text{m}^2/\text{s}$; $Pr = 0.709$; $C_p = 1.0068\,\text{kJ/kg°C}$; $k = 0.02560\,\text{W/m°C}$.

Thus, we obtain:

$$Re_D = \frac{\overline{u}D}{\nu} = \frac{30 \times 0.05}{14.11 \times 10^{-6}} = 1.063 \times 10^5 > 4000 \quad \text{(turbulent)}$$

$$Nu_D = 0.023Re_D^{0.8}Pr^{0.4} = 0.023 \times \left(1.063 \times 10^5\right)^{0.8} \times 0.709^{0.4} = 210.49$$

$$h = \frac{Nu_D k}{D} = \frac{210.49 \times 0.02560}{0.05} = 107.8 \frac{\text{W}}{\text{m}^2\,°\text{C}}$$

From (6.B):

$$A_w = \pi DL = \pi \times 0.05 \times 0.6 = 0.094\,\text{m}^2$$

$$\dot{m} = \rho\overline{u}\left(\pi\frac{D^2}{4}\right) = 1.2025 \times 30 \times \left(\pi\frac{0.05^2}{4}\right) = 0.071\ \text{kg/s}$$

$$T_o = T_i + \frac{hA_w}{\dot{m}C_p}\left(T_w - \overline{T}\right) = 15 + \frac{107.8 \times 0.094}{0.071 \times 1006.8}(38 - 18) = 17.8\,°\text{C}$$

It does not match, as the assumed output temperature was 21 °C. Therefore, a new iteration is necessary.

Now assuming $T_o = 17.8\,°\text{C} \rightarrow \overline{T} = 16.4\,°\text{C}$. New properties:

$$\rho = 1.21\,\text{kg/m}^3;\ \nu = 14.96 \times 10^{-6}\,\text{m}^2/\text{s}; Pr = 0.71$$

$$Re_D = 1.0 \times 10^5;\ \dot{m} = 0.071\,\text{kg/s}$$

$$h = 102.3\,\text{W}/\text{m}^{2°}\text{C}$$

$T_o = 17.7\,°\text{C}$, it checks!

Performing the next calculations. By Newton's cooling law,

$$\dot{Q} = hA_w\left(T_w - \overline{T}\right) = 102.3 \times 0.094 \times (38 - 16.4) = 207.7\,\text{W}.$$

Or, by the First Law of Thermodynamics,

$$\dot{Q} = \dot{m}C_p(T_o - T_i) = 0.071 \times 1006.8 \times (17.8 - 15) = 200.2\ \text{W}.$$

The differences are justified due to the approximations used for properties calculation. A new iteration would provide a more accurate result, but the values obtained are already sufficient for most practical cases.

Example 6.4 (*Heat transfer inside ducts in turbulent regime—another example*)

Water flows in a 19 mm diameter tube at a 0.3 kg/s flow rate. The water enters the tube at 20 °C and leaves the tube at 60 °C. The internal surface of the tube is kept at 90 °C. Determine the average heat convection coefficient, knowing that the tube is long. Also, calculate the heat transfer rate per unit of internal tube surface.

Solution

The transport properties of the water are obtained considering the average of the inlet and outlet mixture temperatures, that is, at 40 °C. From Table A.7,

$$\rho = 991.8\,\text{kg/m}^3;\ k = 0.6318\,\text{W}/\,\text{m}^\circ\text{C};\ C_p = 4.179\,\text{kJ}/\,\text{kg}^\circ\text{C};$$
$$\mu = 6.54\times10^{-4}\,\text{kg/ms};\ \ Pr = 4.33;\ \ Pr_w = 1.95\ (90^\circ\text{C})$$

Another way to evaluate the Reynolds number in circular tubes is through the mass flow rate: $\dot{m} = \rho\bar{u}\frac{\pi D^2}{4}$; substituting into the Reynolds number $Re_D = \frac{\rho\bar{u}D}{\mu}$, we obtain:

$$Re_D = \frac{4\dot{m}}{\pi D\mu}$$

Therefore, the Reynolds number is:

$$Re_D = \frac{4\times0.3}{\pi\times0.019\times6.54\times10^{-4}} = 3.076\times10^4 > 4000\ \ \text{(turbulent)}$$

The average Nusselt number is obtained using the Gnielinski´s correlation. Before that, the friction factor is evaluated:

$$f = \left[0.782\times\ln\left(3.076\times10^4\right) - 1.5\right]^{-2} = 0.0231$$

$Nu_D = \frac{(f/8)(Re_D-1000)Pr}{1+12.7(f/8)^{1/2}(Pr^{2/3}-1)}\left[1+\left(\frac{D}{L}\right)^{2/3}\right]\left(\frac{Pr}{Pr_w}\right)^{0.11}$, as the tube is long, we disregard the ratio *D* / *L*.

$$Nu_D = \frac{(0.0231/8)\left(3.076\times10^4 - 1000\right)4.33}{1+12.7(0.0231/8)^{1/2}\left(4.33^{2/3}-1\right)}\left(\frac{4.33}{1.95}\right)^{0.11} = 190.7$$

The average heat transfer coefficient is:

$$h = Nu\frac{k}{D} = 190.7\frac{0.6318}{0.019} = 6340\,\frac{\text{W}}{\text{m}^2\,^\circ\text{C}}$$

$$q = h\left(T_w - \bar{T}\right) = 6340\times(90-40) = 317\times10^3 = 317\text{kW/m}^2.$$

Table 6.2 Summary of friction factor equations (Darcy) in circular tubes

Source	Condition	Friction factor
	Laminar, smooth or rough tube	$f = \frac{64}{Re_D}$
Colburn (1933)	Turbulent, smooth tube, $1 \times 10^4 < Re_D < 2 \times 10^5$	$f = 0.184 Re_D^{-0.2}$
Churchill (1977)	Turbulent, rough tube	$f = 8\left[\left(\frac{8}{Re_D}\right)^{12} + \frac{1}{(A+B)^{1.5}}\right]^{\frac{1}{12}}$ $A = \left\{2.457\ln\left[\frac{1}{\left(\frac{7}{Re_D}\right)^{0.9} + \left(\frac{0.27\varepsilon}{D}\right)}\right]\right\}^{16}$ $B = \left(\frac{37{,}530}{Re_D}\right)^{16}$
Colebrook (1939)	Turbulent, rough tube	$\frac{1}{f^{1/2}} = -2.0\log\left(\frac{\varepsilon/D}{3.7} + \frac{2.51}{Re_D f^{1/2}}\right)$
Konakov (1946)	Turbulent, smooth tube, $4000 < Re_D < 5 \times 10^6$	$f = (0.782\ln Re_D - 1.5)^{-2}$

6.2.3 *Correlations Summary*

Table 6.2 presents a summary of the Darcy friction factor, f, for laminar and turbulent regimes in circular tubes. Other values of f are also provided in Table 6.1 for laminar flow in several cross-sections. Table 6.3 summarizes the expressions of the Nusselt number for circular tubes. Lacking specific expressions for non-circular cross-sections, the concept of hydraulic diameter, D_h, is recommended for calculating f and Nu_D.

6.3 Logarithmic Mean Temperature Difference (*LMTD*)

In Example 6.4, the tube wall temperature is constant and, as the fluid flows, the heating process is not linear. However, the convective heat transfer calculations performed in the solution are approximations, as we use the average of the inlet and outlet bulk temperatures. However, it has been shown that the variation of the bulk temperature varies exponentially between the inlet and the outlet section, whose expression is given by Eq. (6.29), from which we obtain,

$$\dot{m}C_p = \frac{hA}{\ln\left(\frac{T_w - T_i}{T_w - T_o}\right)} \tag{6.51}$$

Table 6.3 Summary of Nusselt number equations in circular tubes

Source	Condition	Nusselt Number
	Laminar, theoretical, fully developed, q_w = constant	$Nu_D = 4.36$
Kays et al. (2005)	Laminar, theoretical, fully developed, T_w = constant	$Nu_D = 3.36$
Hausen (1943)	Laminar, thermal entrance, T_w = constant	$\overline{Nu}_D = 3.66 + \frac{0.0668(D/L)Re_D Pr}{1+0.04[(D/L)Re_D Pr]^{2/3}}$
Shah and London (1978)	Laminar, fully developed or thermal entry, q_w = constant	$\overline{Nu}_D = 1.953\left(\frac{L}{DRe_D Pr}\right)^{-\frac{1}{3}}, \frac{L}{DRe_D Pr} \leq 0.03$ $\overline{Nu}_D = 4.364 + 0.0722\left(\frac{L}{DRe_D Pr}\right)^{-1}, \frac{L}{DRe_D Pr} > 0.03$
Colburn (1933)	Turbulent, fully developed, $0.7 \leq Pr \leq 160$, $Re_D \geq 10,000$	$Nu_D = 0.023 Re_D^{0.8} Pr^{1/3}$
Dittus and Boelter (1985), McAdams (1942)	Turbulent, fully developed, $0.6 \leq Pr \leq 160$, $Re_D \geq 10,000$	$Nu_D = 0.023 Re_D^{0.8} Pr^n$ $n = 0.3$ cooling fluid $n = 0.4$ heating fluid
Gnielinski (1976, 2013)	Turbulent, smooth tube, $0.5 < Pr < 2000$, $4000 < Re_D < 5 \times 10^6$	$Nu_D = \frac{(f/8)(Re_D - 1000)Pr}{1+12.7(f/8)^{1/2}(Pr^{2/3}-1)}\left[1 + \left(\frac{D}{L}\right)^{2/3}\right]K$ $K = \left(\frac{Pr}{Pr_w}\right)^{0.11}$; Liquids $K = \left(\frac{T}{T_w}\right)^{0.11}$; Gases, $n = 0.45$ (ar)

On the other hand, from the First Law of Thermodynamics, it is known that $\dot{Q} = \dot{m}C_p(T_o - T_i)$. Substituting the product $\dot{m}C_p$ into Eq. (6.51), we obtain:

$$\dot{Q} = hA\frac{(T_o - T_i)}{\ln\left(\frac{T_w - T_i}{T_w - T_o}\right)} \tag{6.52}$$

Therefore, the heat transfer rate should be written as:

$$\dot{Q} = hA \times LMTD \tag{6.53}$$

where the logarithmic mean temperature difference (*LMTD*) is defined by:

$$LMTD = \frac{\Delta T_i - \Delta T_o}{\ln\left(\frac{\Delta T_i}{\Delta T_o}\right)} \tag{6.54}$$

with $\Delta T_i = T_w - T_i$ and $\Delta T_o = T_w - T_o$, where T_i is the inlet temperature and T_o, the outlet one. More will be said about *LMTD* in Chap. 9.

Example 6.5 (*Use of LMTD method*)

Recalculate the heat flux rate from Example 6.4 using the *LMTD* concept.

Solution

$$LMTD = \frac{70 - 30}{\ln(70/30)} = 47.2\,°\text{C}$$

Thus, $q = h(LMTD) = 6340 \times 47.2 = 299.3 \times 10^3 = 299.3\,\text{kW/m}^2$.

As a final piece of information, note that if the temperature differences between the inlet and the outlet are not very large, the *LMTD* will tend towards the difference between the wall temperature and the average between the inlet and outlet bulk temperatures.

Example 6.6 (*Tube heating*)

A 50 mm tube diameter consists of a solar heater exposed to solar thermal radiation. The tube can absorb a uniform and constant thermal heat flux of 2500 W/m^2 due to a parabolic trough concentrator. Water enters the tube at $\dot{m} = 0.015$ kg/s and $T_i = 23.6$ °C. (a) What should be the tube length so that the output temperature reaches $T_o = 90$ °C? (b) What is the tube wall temperature at the outlet?

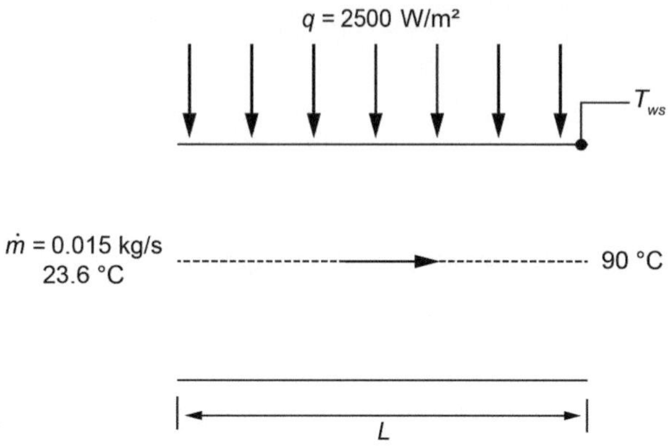

Solution

Water properties are evaluated at 56.8 °C (Table A.7):

$$\rho = 997\,\mathrm{kg/m^3};\; k = 0.650\,\mathrm{W/m°C};\; C_p = 4.184\ \mathrm{kJ/kg°C};\; \mu = 4.89 \times 10^{-4}\ \mathrm{kg/ms}$$

(a) From the First Law of Thermodynamics: $\dot{Q} = \dot{m}C_p(T_o - T_i)$; on the other hand, $\dot{Q} = q\pi DL$, and therefore, $q\pi DL = \dot{m}C_p(T_o - T_i)$, so that:

$$L = \frac{\dot{m}C_p\Delta T}{q\pi D} = \frac{0.015 \times 4184 \times (90 - 23.6)}{2500\pi \times 0.05} = 10.6\ \mathrm{m}$$

(b) $q = h(T_{wo} - T_o)$ or $T_{wo} = \frac{q}{h} + T_o$.
Now, it is necessary to make an estimate of h

$$Re_D = \frac{4\dot{m}}{\pi D\mu} = \frac{4 \times 0.015}{\pi \times 0.05 \times 4.89 \times 10^{-4}} = 781.1 < 2300\ \text{(Laminar)}$$

As it is a case of constant heat flux at the wall, we have $Nu_D = 4.36$.
Thus,$h = Nu_D\frac{k}{D} = 4.36\frac{0.650}{0.05} = 56.7\ \mathrm{W/m^2K}$.
Finally, $T_{wo} = \frac{2500}{56.7} + 90 = 134.1$ °C.

6.4 Proposed Problems

6.1 It is desired to find the heat transfer coefficient of a fluid flowing inside a tube. For this, initially, the hydrodynamic and thermal entries are evaluated. For the cases indicated in the table, considering an average fluid temperature of 30 °C, evaluate: (a) the average flow velocity, (b) the hydrodynamic and thermal entry lengths, and (c) the percentages of these lengths to the length of the tubes.

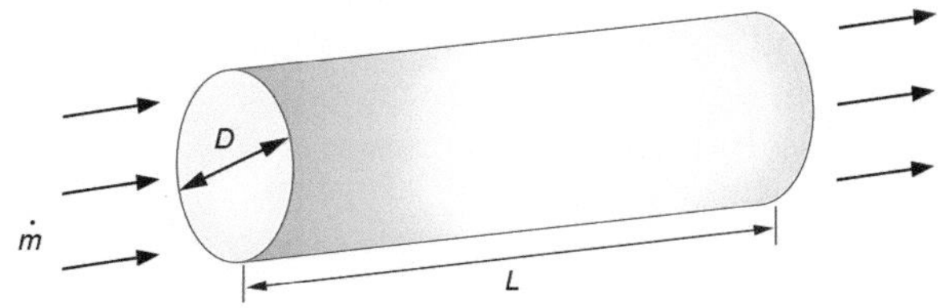

Fluid	$\dot{m}$(kg/s)	D (mm)	L (m)
Water	0.025	20	10
	0.025	20	100
	0.2	20	10
	0.2	20	100

6.2 Measurements of fluid velocity and temperature as a function of radius were made within a circular tube of 25 mm in diameter, namely: $u = 0.08 - 0.08\left(\frac{r}{R}\right)^2$ and $T = 350 + 70\left(\frac{r}{R}\right)^2 - 25\left(\frac{r}{R}\right)^3$, where R is the inner radius of the tube, the velocity is in m/s and the temperature in K. (a) Is the fluid being heated or cooled? Determine (b) the average flow velocity, (c) the average temperature of the flow, (d) the mass flow rate of the fluid, and (e) the local heat transfer coefficient.

6.3 Air at a mass flow rate of 0.28 g/s in fully developed conditions, flows within a 10 mm diameter steel tube, with constant wall temperature. Considering an average air temperature of 25 °C, evaluate (a) the average air velocity, (b) the air velocity at the tube centerline ($r = 0$), (c) the Nusselt number, (d) the heat transfer coefficient by convection, and (e) the pressure drop per meter of the tube. If the tube were made of PVC, would there be a difference in the calculations?

6.4 A fluid flows at an average velocity of 0.8 m/s through a 25.4 mm diameter pipe that is 1 m long. Considering an average fluid temperature of 50 °C, fully developed flow, and constant wall heat flux, evaluate the heat transfer coefficient for (a) carbon dioxide and (b) glycerin. Evaluate the heat transfer coefficient for the same fluids considering the thermal entry. Is a simplified calculation justified in the case of a fully developed flow?

6.5 A 30 mm diameter stainless steel round tube is used to heat Therminol VP1 oil using an electric resistance installed on its wall, which causes a constant heat flux of 4725 W/m^2 on it. Oil enters at a velocity of 0.08 m/s and 50 °C and exits at 110 °C. Considering fully developed flow, evaluate (a) the heater length, (b) the wall temperature at the entrance and exit of the tube, and (c) the fluid pressure drop in the tube.

6.6 A triangular duct is used to transport 0.5 kg/s of ethylene glycol at 40 °C. The external environment imposes a duct wall temperature to be approximately uniform at 100 °C. Check if the temperature variation is significant in 6 m of duct length, as well as the heat gain in this length, assuming fully developed flow.

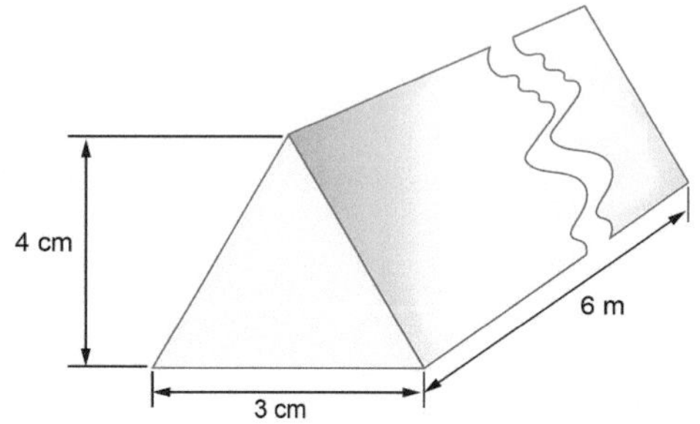

6.7 In a cold region, oil at 17 °C and 80 kg/s mass flow leaves a pumping station and is transported in a carbon steel pipeline of 50 cm in diameter. At a certain distance from the station, the pipeline must cross 100 m in a lake that promotes a nearly constant tube wall temperature at 0 °C. Assuming that the average oil temperature entering the lake is 13 °C, evaluate (a) the oil temperature at the lake exit and (b) the rate of heat lost in the lake. In your calculations, assume the properties of unused (new) motor oil.

6.8 A water-tube boiler is used to vaporize water. In the furnace part, the tubes are arranged vertically. Assuming that in each 3 cm diameter tube, the water flow rate is 1.5 kg/s, evaluate the heat transfer coefficient inside the tube and the friction factor considering an average temperature of 80 °C and a relative roughness of 0.001.

6.9 Saturated steam at 110 °C flows externally to a tube through which a fluid at 0.75 kg/s and 30 °C flows. The tube has an internal diameter of 24 mm. For tube lengths of 1 m and 10 m and for the fluids water and Therminol 59, evaluate (a) the outlet temperature of the fluid and (b) the *LMTD* and the average between the inlet and outlet temperatures of the fluid, comparing these differences. The condensation of the steam outside the tube ensures a temperature of 110 °C on the tube wall.

6.10 A Fresnel solar collector consists of several mirrors arranged on a horizontal plane, which have a tracking system that makes them rotate to reflect the sun's rays and concentrate the radiation on a stainless steel AISI 304 tube receiver (roughness of 0.002 mm) with 30 mm internal diameter. In an industrial application, it is necessary to preheat hot water to 90 °C, using for this the water that enters the receiver at 50 °C and 0.12 kg/s. After the radiation is absorbed by the receiver, consider a constant and uniform heat flux transferred by the receiver tube wall of 4000 W/m^2 to the water. Under these conditions evaluate (a) the collector length, (b) the average temperature of the receiver tube, comparing it with the arithmetic average of the receiver temperature at the inlet and outlet.

(c) What is the pressure drop and the power that the pump should provide regarding this portion of the heating system tube?

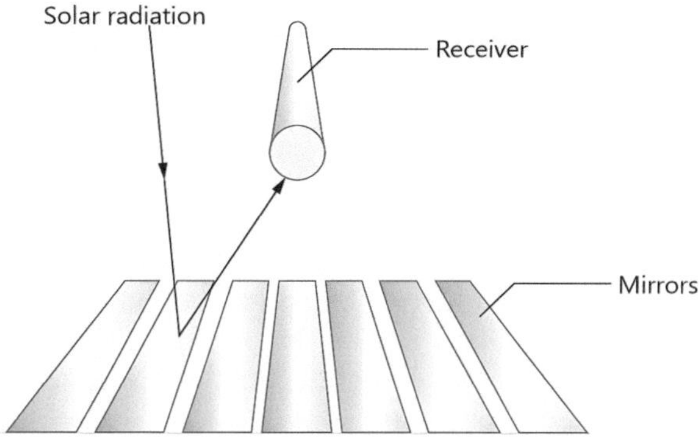

6.11 A fluid at a mass flow rate $\dot{m}$ and temperature T_i enters a pipeline of diameter D being heated on its wall with a linearly varying thermal heat flux $q = Cx$, with C in W/m^3. If the average specific heat of the fluid is C_p, find expressions for (a) the variation of the fluid temperature as a function of x, (b) the length L necessary to heat 50 % of the fluid temperature, and (c) the total heat transferred in this length.

6.12 Therminol VP1 oil at a 0.025 kg/s mass flow rate, and a temperature of 60 °C, should be heated in a 1.5 cm tube diameter to a desired oil outlet temperature of 140 °C. The tube temperature is uniform and equal to 200 °C. Evaluate the necessary length (a) without correction of the Nusselt number by viscosity and (b) with correction of the Nusselt number by viscosity. Compare the results and comment on the differences.

6.13 A fluid at a temperature T_∞ flows externally to a tube, while another fluid with mass flow rate $\dot{m}$ and average specific heat C_p flows internally. From the indicated control volume, demonstrate that the overall heat transfer coefficient in reference to the tube internal area is equal to $U_i = \left[\frac{1}{h_i} + \frac{D_i \ln(D_o/D_i)}{2k} + \frac{D_i}{h_o D_o}\right]^{-1}$

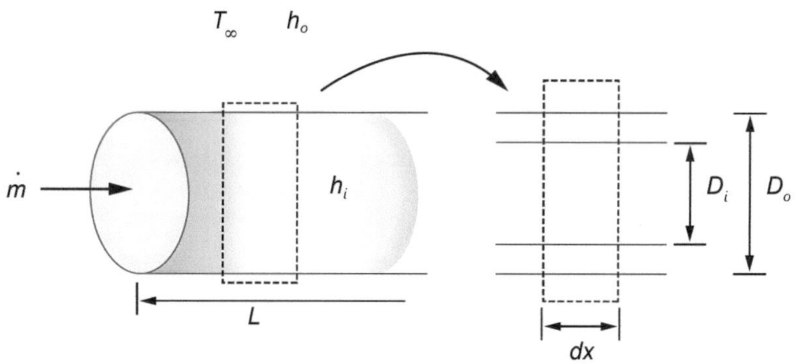

6.14 In a small laboratory boiler, the exhaust gases are expelled into the environment through a carbon steel chimney of 60 cm in diameter, 2 mm in thickness, and 5 m in height. The gases enter the chimney at the bottom with a temperature of 500 °C and 0.4 kg/s. Determine the exit temperature of the gases, if the external ambient temperature to the chimney is 20 °C and the external heat transfer coefficient is 25 W/m^2K. In your calculations, approximate the properties of the combustion gases as those of air.

6.15 A water and ice bath at 0 °C is used for cooling 0.45 kg/s of ethylene glycol-water (50/50) flowing through a 25 mm diameter A304 stainless steel tube with a roughness of 0.0021 mm. The fluid's inlet temperature in the tube is 60 °C and the outlet temperature is 15 °C. Considering fully developed conditions, evaluate (a) the heat transfer rate lost by the fluid, and (b) the length of the tube. (c) If the tube diameter were 120 mm, what would be the heat transfer rate and the necessary length, (d) in addition, obtain the required pumping power for the two tube diameters.

6.16 In a heating application, 5.8×10^{-4} kg/s of air must be heated from 20 °C to 35 °C by flowing inside a 20 mm and 1.8 m long tube. Evaluate the necessary wall temperature in this heating process, assuming the wall temperature is uniform and constant.

6.17 In a thermal recovery process, combustion gases (with properties similar to air at 700 °C and 1.2 kg/s) are redirected to a process within an industry. The transport pipe has a 1.5 m diameter, a length of 50 m, and negligible thickness. Externally, air is at a temperature of 25 °C and 6 m/s. Evaluate the thickness of a layer of ceramic blanket insulation of 128 kg/m^3 that should be installed so that the gases do not decrease to less than 100 °C in their path. Only to evaluate the properties of the external air, consider that the external wall temperature of the insulation is 70 °C.

6.18 A 3 cm diameter pipe is installed in a boiler furnace for water heating. The water enters at 30 °C and 0.5 kg/s. Externally to the tube, there are combustion products at 380 °C and the heat transfer coefficient is 405 W/m^2K, the emissivity of the external surface of the tube is 0.85 and the furnace walls are at 410 °C. If the tube length is 11.5 m and assuming that the tube wall is thin and its temperature constant, evaluate (a) the exit temperature of the water and (b) the temperature on the tube wall.

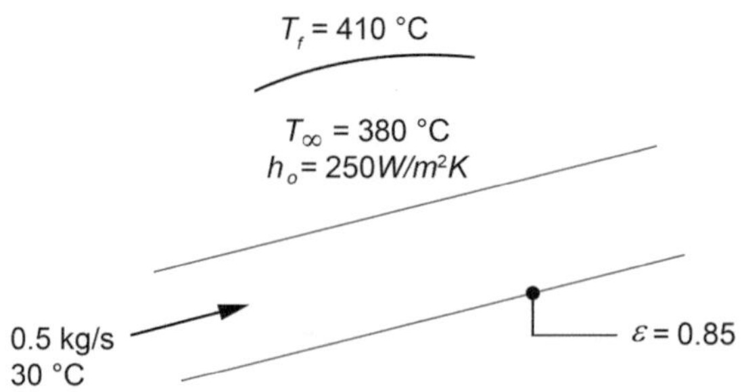

References

Churchill, S.W. 1977. Friction-factor equation spans all fluid-flow regimes. *Chemical Engineering.* 84 (24): 91–92.

Colburn, A.P. 1933. A method of correlating forced convection heat transfer data and a comparison with fluid friction. *International Journal of Heat and Mass Transfer.* 7: 1359–1384.

Colebrook, C.F. 1939. Turbulent flow in pipes with particular reference to the transition region between the smooth and rough pipe laws. *Journal of the Institution of Civil Engineers (London)* 11: 133–156.

Dittus, F.W., and M.K. Boelter. 1985. Heat transfer in automobile radiators of the tubular type. *International Communications in Heat and Mass Transfer* 12: 3–22.

Gnielinski, V. 1976. New equations for heat and mass transfer in turbulent pipe and channel flow. *International Chemical Engineer* 16 (2): 359–368.

Gnielinski, V. 2013. On heat transfer in tubes. *International Journal of Heat and Mass Transfer* 63: 134–140.

Hausen, H. 1943. Darstellung des Wärmeuberganges in Rohren durch verallgemeinerte Potenzbeziehungen. *VDIZ* 4 (91).

Kays, W.M., M.E. Crawford, and B. Weigand. 2005. *Convective heat and mass transfer*, 4th ed. Boston: McGraw-Hill Higher Education.

Konakov, P.K. 1946. A new equation for the friction coefficient for smooth tubes. *Report of the Academic Society for Science of the URSS* 7: 503–506.

McAdams, W.H. 1942. *Heat transmission*, 2nd ed. New York: McGraw-Hill.

Moody, L.F. 1944. Friction factors for pipe flow. *ASME Transactions* 66: 671–684.

Rohsenow, W.M., J.P. Hartnett, and Y.I. Cho. 1998. *Handbook of heat transfer*, 3rd ed. New York: McGraw-Hill.

Shah, R.K., and A.L. London. 1978. *Laminar flow forced convection in ducts.* New York: Academic.

Chapter 7
Natural Convection

Abstract This chapter covers heat transfer by natural or free convection heat transfer driven by gravitational force (buoyancy) due to the fact that the density decreases with increasing temperature. The first situation analyzed is heat transfer by natural laminar convection over isothermal surfaces. For this, differential conservation equations are presented whose solutions are given in Appendix B.2. Dimensionless numbers of natural convection are the Grashof and Rayleigh numbers. Practical working correlations of the Nusselt number as a function of Grashof (or Rayleigh) and Prandtl are presented for several configurations and boundary conditions for both laminar and turbulent flow regimes. Practical working correlations are presented for natural convection confined between two or more vertical, inclined, and horizontal surfaces as well as for a confined space. Various solved problems are presented. At a low fluid velocity the phenomenon of simultaneous natural and forced convection may be relevant and for this purpose criteria for mixed convection are established.

Keywords Natural convection · Grashof · Rayleigh · Mixed convection

7.1 External Natural Convection

7.1.1 Laminar Regime

Forced convection over external surfaces and inside tubes and ducts was studied in the two preceding chapters. Forced convection occurs because of fluid motion over a heated (or cooled) surface caused by an external device, such as a pump, a fan, or another fluid-moving piece of equipment, and, therefore, the gravitational force plays little or no effect on the heat transfer process in these cases. However, for a quiescent fluid in contact with a heated (or cooled) surface, heat transfer between the surface and the fluid will occur by a mechanism other than forced convection. In this situation, the fluid velocity is zero, thus the Reynolds number is also zero. Naturally, the correlations developed for general forced convection $Nu = f(Re, Pr)$ developed in previous chapters do not apply. Consequently, the local motion of the

J. R. Simões-Moreira and E. W. Zavaleta-Aguilar, *Heat Transfer*, Heat and Mass Transfer, https://doi.org/10.1007/978-3-031-80318-5_7

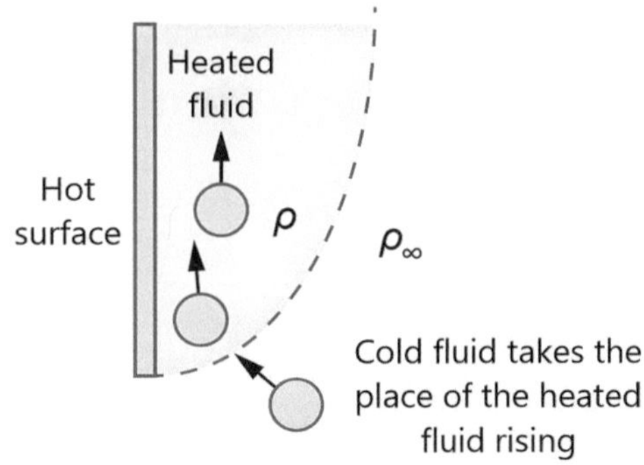

Fig. 7.1 Difference in gravitational buoyancy causes convective motion

fluid next to the surface occurs due to another phenomenon, which originates from the difference in gravitational thrust on the fluid caused by the variation of its density due to temperature gradients. To better understand this aspect, consider a heated vertical surface, as indicated in Fig. 7.1, in contact with a fluid at rest. The region of the fluid in contact with the heated surface naturally heats up and, consequently, its density ρ decreases relative to the density of the distant unheated portions of the fluid, ρ_∞. The buoyancy force per unit volume is given by $f_E = \rho_\infty g$. Near the heated surface, however, the buoyancy force acting on the heated fluid is smaller, because $\rho < \rho_\infty$. Therefore, the net buoyant force per unit volume of fluid will be $(\rho_\infty - \rho)g$, giving rise to an upward fluid motion in which the warmer portions tend to rise, while the cooler ones take their place, giving rise to *natural* or *free convection streams*.

Figure 7.2 illustrates convection streams in water being heated in a pot. The heated portions of fluid at the bottom rise, as the less heated fluid portions descend to replace the fluid portions that rise, which generates the convection currents.

By originating a fluid motion, the hydrodynamic and thermal boundary layers are established next to the surface, as illustrated in Fig. 7.3, for the case of a heated vertical plate. A different boundary condition from the HBL (hydrodynamic boundary layer) in natural convection compared to the forced one, is that the fluid velocity is null both at the surface and at the boundary layer end, where $u_\infty = 0$.

In a similar method to the development of the hydrodynamic laminar boundary layer in Sect. 5.2, it is possible to carry out a differential control volume analysis of the hydrodynamic laminar boundary layer in natural convection, which results in Eq. (7.1). The striking difference between the laminar differential equation of forced (Eq. 5.10) and natural (Eq. 7.1) flow is the buoyancy gravitational term per unit volume, ρg, in the momentum equation in the x (vertical) direction:

$$\rho\left(u\frac{\partial u}{\partial x} + v\frac{\partial u}{\partial y}\right) = -\frac{\partial p}{\partial x} - \rho g + \mu\frac{\partial^2 u}{\partial y^2} \tag{7.1}$$

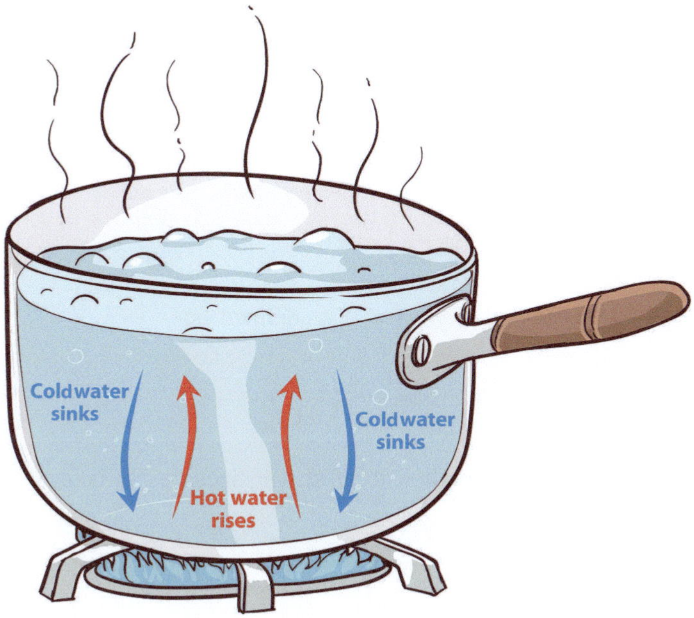

Fig. 7.2 Natural convection in water being heated in a pot. Adapted from iStock.com/corbac40

Fig. 7.3 Thermal (TBL) and hydrodynamic (HBL) boundary layers in natural convection over a heated surface

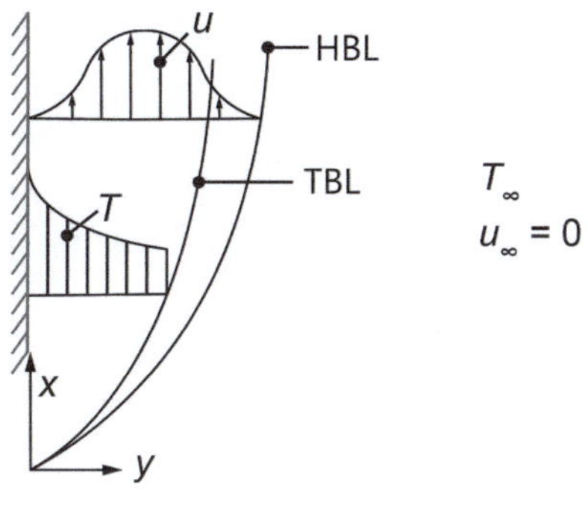

The momentum equation in y (horizontal) does not have the gravitational term:

$$\rho\left(u\frac{\partial v}{\partial x}+v\frac{\partial v}{\partial y}\right)=-\frac{\partial p}{\partial y}+\mu\frac{\partial^2 v}{\partial x^2} \tag{7.2}$$

Applying the momentum equation in y outside the HBL $(u_\infty = 0,\ v_\infty = 0)$.

Applying the momentum equation in y inside the HBL, considering that $u \gg v$ and $\partial v/\partial x \approx \partial v/\partial y \approx 0$, we arrive at: $\frac{\partial p}{\partial y} = 0$.

Therefore, for a given position x, the pressure does not vary in the direction y (horizontal), inside and outside of the HBL.

Now, applying the momentum equation in x outside the HBL $(u_\infty = 0)$: $\frac{\partial p_\infty}{\partial x} = -\rho_\infty g$.

However, as pressure is the same at any position x, so: $p(x) = p_\infty(x)$, and $\frac{\partial p}{\partial x} = \frac{\partial p_\infty}{\partial x}$, thus: $\frac{\partial p}{\partial x} = -\rho_\infty g$.

Substituting this result and rearranging, we obtain the momentum balance in x,

$$\rho\left(u\frac{\partial u}{\partial x} + v\frac{\partial u}{\partial y}\right) = g(\rho_\infty - \rho) + \mu\frac{\partial^2 u}{\partial y^2} \tag{7.3}$$

The *volumetric expansion coefficient*, β, is defined as:

$$\beta = -\frac{1}{\rho}\left(\frac{\partial \rho}{\partial T}\right)_p \tag{7.4}$$

The volumetric expansion coefficient can also be linearized by using the following approximation:

$$\beta = -\frac{1}{\rho}\left(\frac{\partial \rho}{\partial T}\right)_p \cong \frac{1}{\rho}\left(\frac{\rho_\infty - \rho}{T - T_\infty}\right) \tag{7.5}$$

which leads to Eq. (7.6), known as the *Boussinesq approximation.*

$$\rho_\infty - \rho \cong \beta\rho(T - T_\infty) \tag{7.6}$$

Therefore, substituting the Boussinesq approximation into Eq. (7.3), we get:

$$u\frac{\partial u}{\partial x} + v\frac{\partial u}{\partial y} = g\beta(T - T_\infty) + \nu\frac{\partial^2 u}{\partial y^2} \tag{7.7}$$

In the case of the fluid being an *ideal gas*, it is still possible to show that β is the reciprocal of the absolute temperature, T, that is:

$$\beta = -\frac{1}{\rho}\left(\frac{\partial \rho}{\partial T}\right)_p = -\frac{RT}{P}\frac{\partial}{\partial T}\left(\frac{P}{RT}\right)_p = \frac{1}{T} \tag{7.8}$$

The energy balance equation in the differential control volume within the thermal boundary layer is given by Eq. (5.14):

$$u\frac{\partial T}{\partial x} + v\frac{\partial T}{\partial y} = \alpha\frac{\partial^2 T}{\partial y^2} \tag{5.14}$$

The boundary conditions to solve the differential equations are:

$$\mathrm{BC}_1 : y = 0 : u(x, 0) = 0, v(x, 0) = 0, T(x, 0) = T_w$$
$$\mathrm{BC}_2 : y \to \infty : u(x, \infty) \to 0, v(x, \infty) \to 0, T(x, \infty) = T_\infty$$

The differential conservation equations of momentum (7.7) and energy (5.14) of natural convection cannot be solved separately, as the gravitational buoyancy term couples both two equations, which does not occur with the solution of the laminar hydrodynamic and thermal boundary layers of forced convection on a flat plate (Appendix B.1). Details of the coupled equations solution can be seen in Appendix B.2 for an isothermal vertical plate. It appears in the solution the definition of a new dimensionless group called the *Grashof number*, *Gr*, by:

$$Gr_x = g\beta \frac{(T_w - T_\infty)x^3}{\nu^2} \tag{7.9}$$

where x represents the vertical distance from the lower edge. The Grashof number represents the ratio between the *buoyancy* and the *viscous forces* in natural convection. This number is always positive; thus, in case that $T_w < T_\infty$, the term in parentheses in Eq. (7.9) should change to $T_\infty - T_w$. The Grashof number plays a role similar to the Reynolds number in forced convection, which represents the ratio between the *inertial forces* and the *viscous forces*. Thus, the solution of the momentum and energy equations can be written so that the Nusselt number (*Nu*) is a function of *Gr* and *Pr*, that is:

$$Nu = f(Gr, Pr) \tag{7.10}$$

Appendix B.2 indicates the main steps of the exact solution for the laminar case of a vertical isothermal plate. Given that the exact solution only applies to a limited number of geometries and boundary conditions, empirical results are used, which are the subject of the next subsection.

7.1.2 Empirical Correlations

Various natural convection heat transfer conditions can be related as follows:

$$\overline{Nu} = C(Gr_L \mathrm{Pr})^m = CRa_L^m \tag{7.11}$$

The product *GrPr* is called the *Rayleigh number*, *Ra*:

$$Ra_L = Gr_L Pr = \frac{g\beta(T_w - T_\infty)L^3}{\nu\alpha} \tag{7.12}$$

where L is the *characteristic length*. The properties are calculated at the film temperature, T_f, which is the average between the wall temperature, T_w, and the fluid far away, T_∞, as already defined, that is,

$$T_f = \frac{T_w + T_\infty}{2} \tag{7.13}$$

Churchill and Chu (1975a) suggest using the Rayleigh number to characterize whether a fluid is laminar or turbulent, as will be seen next.

(a) **Natural Convection in Isothermal Vertical Plates, T_w = cte**

Besides the exact solution of a vertical isothermal plate in the laminar regime given in Appendix B.2, the correlation of Churchill and Chu (1975a) is generally also accepted for the isothermal vertical plate of height L and width b, Fig. 7.4, in laminar natural convection, which the average Nusselt number value results in:

$$\overline{Nu_L} = 0.68 + \frac{0.67 Ra_L^{\frac{1}{4}}}{\left[1 + (0.492/Pr)^{\frac{9}{16}}\right]^{\frac{4}{9}}}; \quad Ra_L < 10^9, 0 < Pr < \infty \tag{7.14}$$

where $Ra_L > 10^9$ represents the transition from laminar to turbulent boundary layers on the vertical plate. These same researchers suggest Eq. (7.15) for the isothermal vertical plate, which is valid for both regimes, laminar and turbulent. Although, physically, the transition may occur discontinuously, this equation adjusts the data for a continuous and smooth transition.

$$\overline{Nu_L} = \left\{0.825 + \frac{0.387 Ra_L^{\frac{1}{6}}}{\left[1 + (0.492/Pr)^{\frac{9}{16}}\right]^{\frac{8}{27}}}\right\}^2; \quad 0 < Pr < \infty, \ 0 < Ra_L < \infty \tag{7.15}$$

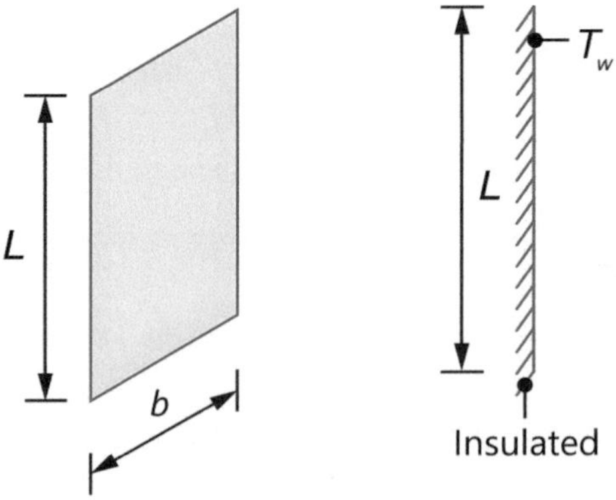

Fig. 7.4 Natural convection on a vertical plate–laminar regime

The transport properties should be evaluated at the film temperature, T_f, given by Eq. (7.13).

(b) **Natural Convection on Horizontal Plates, T_w = cte**

In this configuration, the Rayleigh number is a function of a characteristic length L, defined as the ratio between the surface area, A, and the perimeter, P:

$$L = \frac{A}{P} \tag{7.16}$$

Furthermore, it is assumed that one face of the plate is heated (or cooled), while the other is thermally insulated, as indicated in Fig. 7.5. The correlations are grouped into two situations: the first, when the top surface of the plate is heated or the bottom surface is cooled (Fig. 7.5). In these cases, the Nusselt number (Lloyd and Moran, 1974) is given by Eqs. (7.17).

$$\begin{aligned} \overline{Nu_L} &= 0.27\, {Ra_L}^{1/4} \qquad 10^4 < Ra_L < 10^7 \\ \overline{Nu_L} &= 0.15\, Ra_L^{1/3} \qquad 10^7 < Ra_L < 10^{11} \end{aligned} \tag{7.17}$$

The other situation, Fig. 7.6, is when the plate top surface is cooled or the plate bottom surface is heated (Fishenden and Saunders 1950).

$$\overline{Nu_L} = 0.27 Ra_L^{1/4}; \quad 3 \times 10^5 < Ra_L < 3 \times 10^{10} \tag{7.18}$$

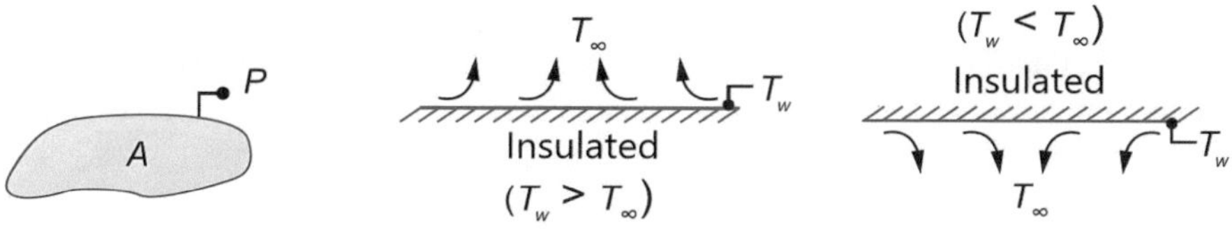

Fig. 7.5 Convection on a horizontal flat plate—up surface heated or down surface cooled

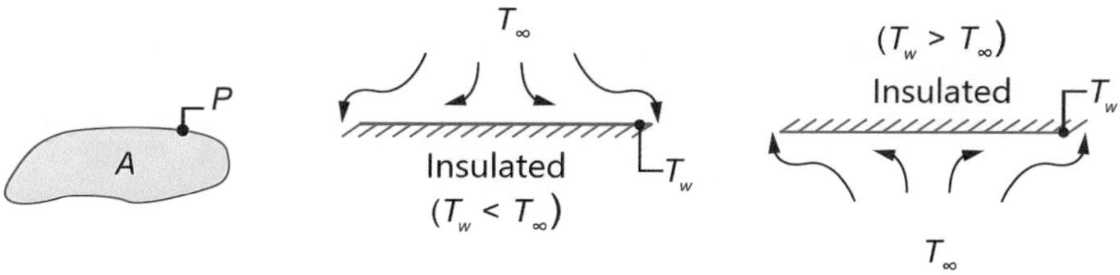

Fig. 7.6 Convection on a flat horizontal plate—up surface cooled or down surface heated

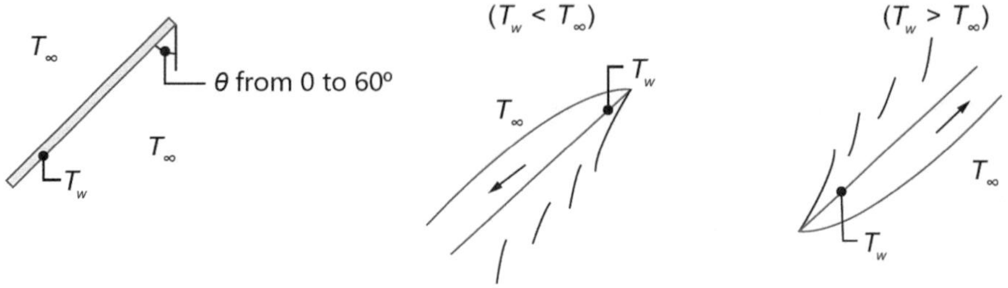

Fig. 7.7 Convection on an inclined flat plate

Fig. 7.8 Natural convection on a horizontal cylinder

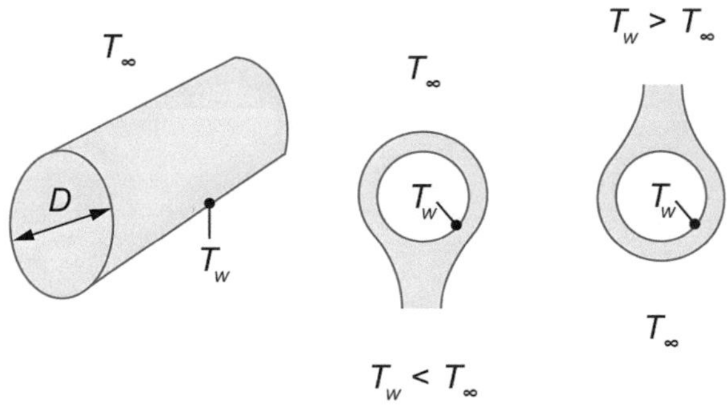

(c) Natural Convection on Inclined Plates, T_w = constant

Inclined surfaces at different temperatures, as illustrated in Fig. 7.7, a boundary layer forms on one side due to the decrease in the gravitational force acting parallel to the surface, and on the other side, the boundary layer breaks up. For the side where the boundary layer is formed (without breaking), Churchill and Chu (1975a) recommend using the same equations as in the vertical case, but one should replace g with $g\cos\theta$ in Eq. (7.14); in Eq. (7.15) no modification is necessary.

(d) Natural Convection on a Horizontal Isothermal Cylinder, T_w = constant

The local natural convection heat transfer coefficient varies along the cylinder circumference. Churchill and Chu (1975b) evaluated the average coefficient for an isothermal wall cylinder, as illustrated in Fig. 7.8.

$$\overline{Nu}_D = \left\{ 0.6 + 0.387 \left[\frac{Ra_D}{\left[1 + (0.559/Pr)^{9/16}\right]^{16/9}} \right]^{\frac{1}{6}} \right\}^2; \quad 10^{-7} < Ra_D < 10^{13} \tag{7.19}$$

Fig. 7.9 Convection in vertical cylinders

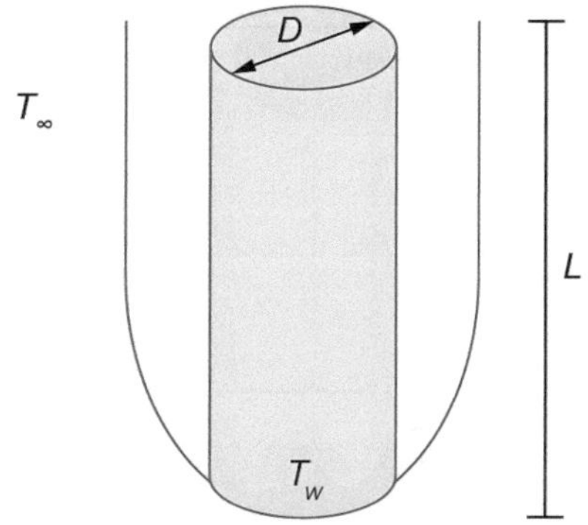

where the Rayleigh number is based on the cylinder diameter D (characteristic length).

(e) Natural Convection in Vertical Cylinder, T_w = constant

Given the ratio between the diameter and the length of the cylinder, the heat transfer coefficient in the cylinder diverges with the flat plate, where the correction given by Popiel et al. (2007) is valid for $0.01 \leq Pr \leq 100$ and $Gr_L \leq 4 \times 10^9$ (laminar boundary layer). See Fig. 7.9.

$$\overline{Nu_L} = \left\{ 0.68 + \frac{0.67 Ra_L^{\frac{1}{4}}}{\left[1 + (0.492/Pr)^{\frac{9}{16}}\right]^{\frac{4}{9}}} \right\} \left\{ 1 + B \left[\frac{5.657}{Gr_L^{0.25}} \left(\frac{L}{D} \right) \right]^C \right\} \tag{7.20}$$

$$B = 0.0571 + 0.2031 Pr^{-0.43}$$

$$C = 0.9165 - 0.0043 Pr^{0.5} + 0.0133 \ln Pr + 0.00048/Pr$$

(f) Natural Convection in Isothermal Spheres, T_w = constant

Spheres correlations can be used to evaluate the heat transfer of suspended particles, droplets, bubbles, and the like. Churchill (1983) suggests his correlations to evaluate the Nusselt number—the first for the laminar case ($Gr_D < 10^9$) (Eq. 7.21) and the other for the turbulent case ($Gr_D > 10^9$) (Eq. 7.22), namely:

$$\overline{Nu_D} = 2 + \frac{0.589 Ra_D^{1/4}}{\left[1 + (0.43/Pr)^{9/16}\right]^{4/9}}; \quad Gr < 10^9;\ 0 < Pr < \infty \tag{7.21}$$

$$\overline{Nu_D} = 2 + \frac{0.589 Ra_D^{1/4}}{\left[1 + (0.43/Pr)^{9/16}\right]^{4/9}} \left\{ 1 + \frac{7.44 \times 10^{-8} Ra}{\left[1 + (0.43/Pr)^{9/16}\right]^{16/9}} \right\}^{1/12} \tag{7.22}$$

$$Gr > 10^9;\ 0 < Pr < \infty$$

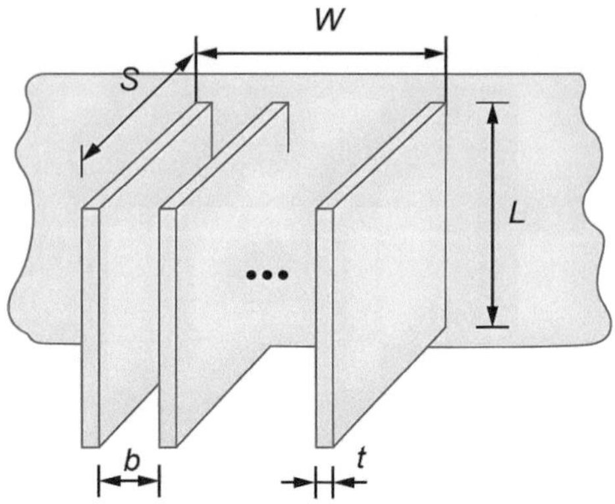

Fig. 7.10 Natural convection in horizontal fins

Equation (7.22) can also be used for laminar, however when $Pr < 0.7$.

(g) **Natural Convection on Isothermal Finned Surfaces, T_w = constant**

Electronic devices commonly dissipate their heat through finned surfaces by natural convection. A typical finned heat sink is illustrated in Fig. 7.10, whose indicated dimensions apply in Eq. (7.23).

The Nusselt number is based on the characteristic length b, which is the spacing between fins (Bar-Cohen and Rohsenow, 1984).

$$Nu_b = \left[\frac{576}{\left(\frac{Ra_b b}{L}\right)^2} + \frac{2.873}{\left(\frac{Ra_b b}{L}\right)^{1/2}} \right]^{-1/2}; \quad 0.1 < Ra_b < 10^5 \tag{7.23}$$

The heat transfer rate dissipated by the fins contemplates the total heat exchange area of all the fins:

$$\dot{Q} = h \times (2NLS) \times (T_w - T_\infty) \tag{7.24}$$

The approximate number of fins N is given by:

$$N = \frac{W}{b + t} \tag{7.25}$$

When the space b is reduced, the number of fins increases, however, the convective flow between the fins decreases, resulting in a decrease in the convective coefficient. In the same work by Bar-Cohen and Rohsenow, they concluded that the optimal spacing, b_{op}, is a function of the Rayleigh number based on the height L:

$$b_{\text{op}} = 2.714 \frac{L}{Ra_L^{1/4}} \tag{7.26}$$

With the optimal spacing, the Nusselt number is:

$$Nu_{b,\text{op}} = 1.31 \tag{7.27}$$

Table 7.1 summarizes empirical correlations for different geometries and situations analyzed so far.

Example 7.1 (*Convection in a small cylinder*)

A 12.5 cm length and 6 cm diameter soda can is cooled in a refrigerator, whose interior air is at 4 °C. The soda is at 26 °C. In what position, vertical or horizontal, the soda will cool quickly? Ignore the thermal exchange at its ends (caps).

Solution

In the vertical position:

Film temperature: $T_f = \frac{26+4}{2} = 15\,°\text{C}\ (288\ \text{K})$

Water transport properties (Table A.7)

$$k = 0.02476\text{W/mK};\ \ \nu = 1.470 \times 10^{-5}\text{m}^2/\text{s};\ \ Pr = 0.7323;$$

$$\beta = \frac{1}{288.15} = 0.00347/\text{K}$$

$$Gr_L = \frac{g\beta(T_w - T_\infty)L^3}{\nu^2} = \frac{9.81 \times 0.00347 \times (26 - 4) \times 0.125^3}{\left(1.470 \times 10^{-5}\right)^2} = 6.77 \times 10^6$$

$$Ra_L = Gr_L Pr = 6.77 \times 10^6 \times 0.7323 = 4.96 \times 10^6\ (\leq 4 \times 10^9)$$

Using Eq. (7.20),

$$B = 0.0571 + 0.2031Pr^{-0.43} = 0.0571 + 0.2031(0.7323)^{-0.43} = 0.2893$$

$$C = 0.9165 - 0.0043\left(0.7323)^{0,5} + 0.0133\ln(0.7323\right) + 0.00048/0.7323 = 0.9093$$

$$\overline{Nu_L} = \left\{0.68 + \frac{0.67\left(4.96 \times 10^6\right)^{0.25}}{\left[1 + \left(\frac{0.492}{0.7323}\right)^{\frac{9}{16}}\right]^{\frac{4}{9}}}\right\}\left\{1 + 0.2893\left[\frac{5.657}{\left(6.77 \times 10^6\right)^{0.25}}\left(\frac{12.5}{6}\right)\right]^{0.9093}\right\}$$

$$\overline{Nu_L} = 26.94$$

$$h = 26.94\frac{0.02476}{0.125} = 5.34\,\text{W/m}^2\text{K}$$

So that the heat transfer rate on the can side is:

$$\dot{Q} = 5.34 \times \pi 0.06 \times 0.125 \times (26 - 4) = 2.8\ \text{W}.$$

Table 7.1 Natural convection correlations on isothermal surfaces

Geometry	Charact length	Limits	Correlation	Equations
Vertical plate 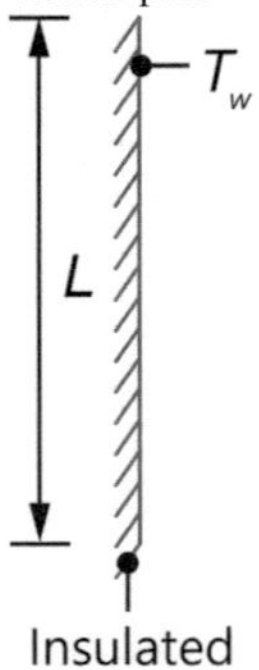	L	$Ra_L < 10^9$ $0 < \mathrm{Pr} < \infty$	$\overline{Nu}_L = 0.68 + \frac{0.67Ra_L^{\frac{1}{4}}}{\left[1+(0.492/Pr)^{\frac{9}{16}}\right]^{\frac{4}{9}}}$	(7.14)
		$0 < Ra_L < \infty$ $0 < \mathrm{Pr} < \infty$	$\overline{Nu}_L = \left\{0.825 + \frac{0.387Ra_L^{\frac{1}{6}}}{\left[1+(0.492/Pr)^{\frac{9}{16}}\right]^{\frac{8}{27}}}\right\}^2$	(7.15)
Horizontal plate 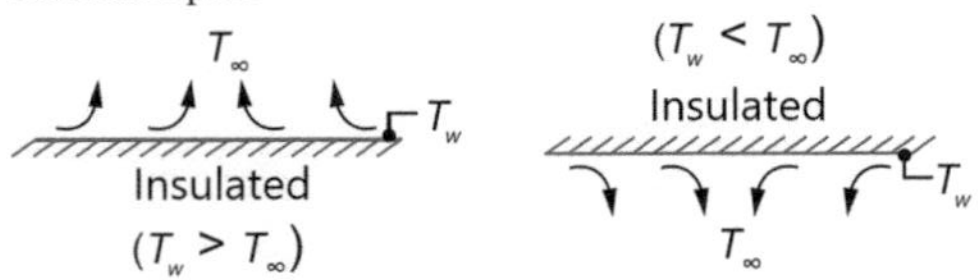	$L = \frac{A}{P}$	$10^4 < Ra_L < 10^7$	$\overline{Nu}_L = 0.54Ra_L^{1/4}$	(7.17)
		$10^7 < Ra_L < 10^{11}$	$\overline{Nu}_L = 0.15Ra_L^{1/3}$	

(continued)

Table 7.1 (continued)

Geometry	Charact length	Limits	Correlation	Equations
Horizontal plate 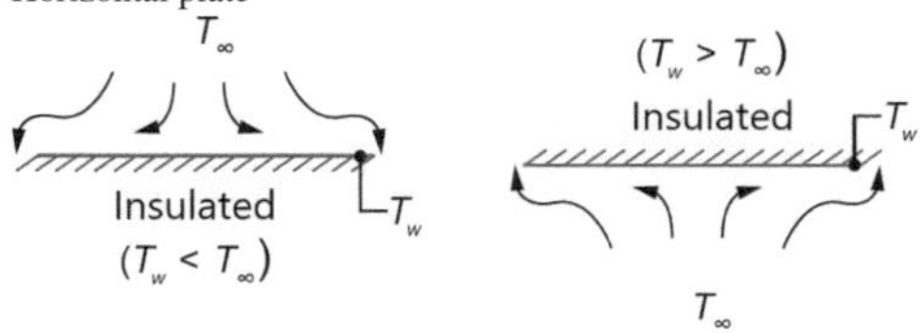	$L = \frac{A}{P}$	$3 \times 10^5 < Ra_L < 3 \times 10^{10}$	$\overline{Nu_L} = 0.27 Ra_L^{1/4}$	(7.18)
Inclined plate $(T_w < T_\infty)$ $(T_w > T_\infty)$ 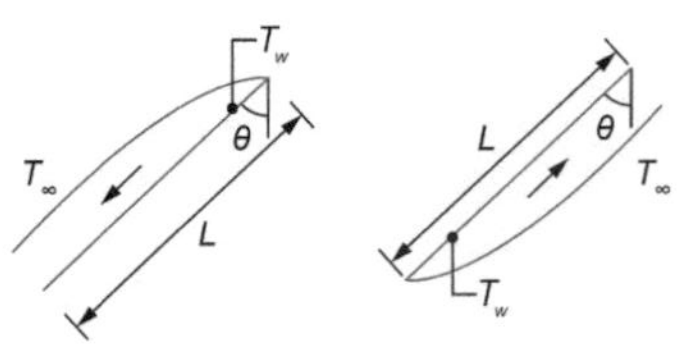	L	$Ra_L < 10^9$ $0 < \Pr < \infty$ $0 < \theta < 60$	$\overline{Nu_L} = 0.68 + \frac{0.67 Ra_L^{\frac{1}{4}}}{\left[1+\left((0.492/Pr)^{\frac{9}{16}}\right)\right]^{\frac{4}{9}}}$ Replace g with $g\cos\theta$	(7.14)
		$0 < Ra_L < \infty$ $0 < \Pr < \infty$ $0 < \theta < 60$	$\overline{Nu_L} = \left\{0.825 + \frac{0.387 Ra_L^{\frac{1}{6}}}{\left[1+(0.492/Pr)^{\frac{9}{16}}\right]^{\frac{8}{27}}}\right\}^2$	(7.15)

(continued)

Table 7.1 (continued)

Geometry	Charact length	Limits	Correlation	Equations
Horizontal cylinder 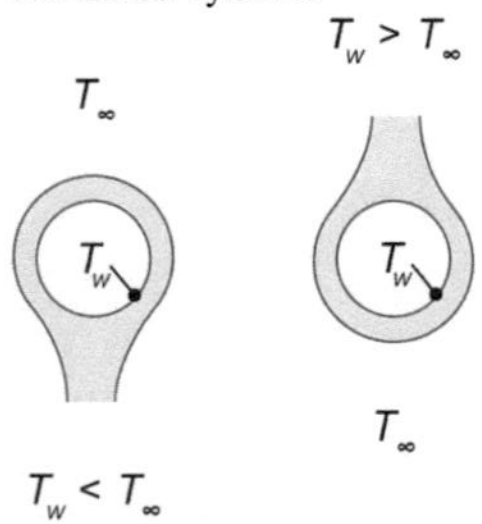	D	$10^{-7} < Ra_D < 10^{13}$	$\overline{Nu}_D = \left\{0.6 + 0.387\left(\frac{Ra_D}{[1+(0.559/Pr)^{9/16}]^{16/9}}\right)^{1/6}\right\}^2$	(7.19)
Vertical cylinder 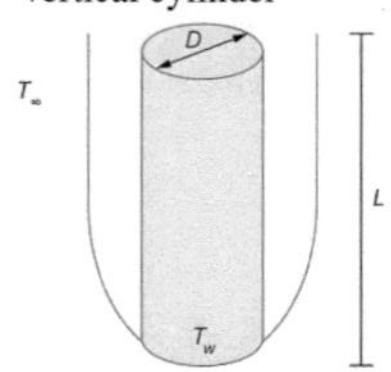	L	$Gr_L \leq 4 \times 10^9$ $0.01 \leq \mathrm{Pr} \leq 100$	$\overline{Nu}_L =$ $\left\{0.68 + \frac{0.67Ra_L^{\frac{1}{4}}}{\left[1+(0.492/Pr)^{\frac{9}{16}}\right]^{\frac{4}{9}}}\right\}\left\{1 + B\left[\frac{5.657}{Gr_L^{0.25}}\left(\frac{L}{D}\right)\right]^C\right\}$ $B = 0.0571 + 0.2031Pr^{-0.43}$ $C = 0.9165 - 0.0043Pr^{0.5} + 0.0133\ln Pr + 0.00048/Pr$	(7.20)

(continued)

Table 7.1 (continued)

Geometry	Charact length	Limits	Correlation	Equations
Sphere 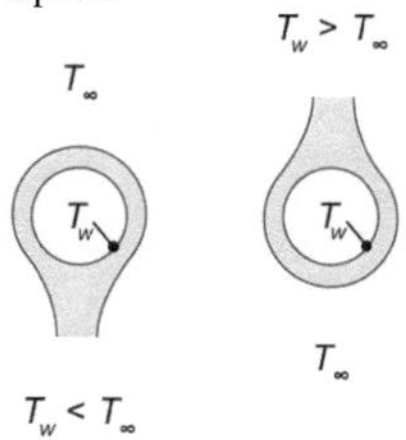	D	$Gr_D < 10^9$ $0 < \mathrm{Pr} < \infty$	$\overline{Nu}_D = 2 + \frac{0.589Ra_D^{1/4}}{[1+(0.43/Pr)^{9/16}]^{4/9}}$	(7.21)
		$Gr_D > 10^9$ $0 < \mathrm{Pr} < \infty$	$\overline{Nu}_D =$ $2 + \frac{0.589Ra_D^{1/4}}{[1+(0.43/Pr)^{9/16}]^{4/9}}\left\{1 + \frac{7.44\times10^{-8}Ra}{[1+(0.43/Pr)^{9/16}]^{16/9}}\right\}^{1/12}$	(7.22)
Finned surfaces 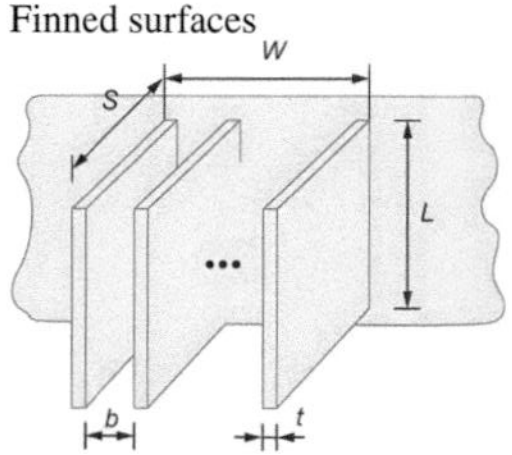	b	$0.1 < Ra_b < 10^5$	$Nu_b = \left[\frac{576}{\left(\frac{Ra_b b}{L}\right)^2} + \frac{2.873}{\left(\frac{Ra_b b}{L}\right)^{1/2}}\right]^{-1/2}$	(7.23)

In the horizontal position:

$$Ra_D = \frac{g\beta(T_w - T_\infty)D^3}{\nu^2}Pr = \frac{9.81 \times 0.00347 \times (26 - 4) \times 0.06^3}{\left(1.470 \times 10^{-5}\right)^2}0.7323$$

$$Ra_D = 5.5 \times 10^5$$

From Eq. (7.19):

$$\overline{Nu_D} = \left\{0.6 + 0.387\left(\frac{5.5 \times 10^5}{\left[1 + (0.559/0.7323)^{9/16}\right]^{16/9}}\right)^{1/6}\right\}^2 = 12.4$$

$$h = Nu\frac{k}{D} = 12.4\frac{0.024760.06}{=}5.11\ \mathrm{W/m^2}$$

So that the heat rate by the side of the horizontal can is:

$$\dot{Q} = 5.11 \times \pi 0.06 \times 0.125 \times (26 - 4) = 2.6\ \mathrm{W}.$$

Although with a small difference, the vertical position is slightly better for faster cooling. An analysis considering the ends should be carried out for a better evaluation. Also, this heat transfer rate would be the initial one, since this is a transient problem where the temperature of the can and the soft drink will decrease over time.

(h) **Natural Convection with Constant Heat Flux on the Wall**

In the case of a *vertical plate* subjected to a uniform heat flux on its surface, $q_w\left[\mathrm{W/m^2}\right]$, the difficulty of establishing the film temperature (to obtain the transport properties) arises, since the wall temperature is not a fixed amount. An alternative to solve these problems would be to adopt the surface temperature at half the height of the plate, that is, $T_{w,L/2}$ (Fuji and Imura 1972) and, in this sense, the Rayleigh number and the heat flux would be a function of the difference in temperatures $T_{w,L/2} - T_\infty$.

Another alternative is to eliminate the temperature dependence $(T_w - T_\infty)$ from the number of Ra_L. For this, the modified Rayleigh number is defined, Ra_L^*, given by:

$$Ra_L^* = \frac{g\beta q_w L^4}{k\nu\alpha} \tag{7.28}$$

With this definition, the Rayleigh number is given by Eq. (7.29), which should be used in Eqs. (7.14) and (7.15) for the case of the vertical plate subjected to a constant heat flux on its surface.

$$Ra_L = \frac{Ra_L^*}{Nu_L} \tag{7.29}$$

(i) **Local Nusselt number on a vertical plate, q_w = cte (Vliet and Ross,** 1975**)**

Modified Grashof:

$$Gr_x^* = Gr_x Nu_x = \frac{g\beta q_w x^4}{k\nu^2} \tag{7.30}$$

Laminar:

$$Nu_x = 0.55\left(Gr_x^* Pr\right)^{1/5}; \quad 10^8 < Gr_x^* Pr < 10^{13} \tag{7.31}$$

Turbulent:

$$Nu_x = 0.17\left(Gr_x^* Pr\right)^{1/4}; \quad 10^{13} < Gr_x^* Pr < 10^{15} \tag{7.32}$$

The local convective heat transfer coefficient is evaluated from: $h_x = \frac{q_w}{(T_{w,x} - T_\infty)}$ and the local Nusselt number from: $Nu_x = \frac{h_x x}{k}$. The transport properties are calculated with the local film temperature: $T_{f,x} = \frac{(T_{w,x} + T_\infty)}{2}$. According to Rohsenow et al. (1998), the laminar regime ends at approximately, $Ra \sim 10^{13}$ and the turbulent regime begins at approximately, $Ra \sim 10^{14}$ however, Eq. (7.32) from Vliet and Ross demonstrated that it can be used from $Ra \sim 10^{13}$.

7.2 Natural Convection in Confined Spaces

Vertical parallel plates

A common case of natural convection is that of two vertical walls maintained at different temperatures T_1 and T_2 (as is the case of windows formed by two glasses, for example) separated by a distance δ, as illustrated in Fig. 7.11. This figure shows the velocity and temperature profiles that may occur, according to MacGregor and Emery (1969).

Temperature and velocity profiles vary according to the Grashof number and, for low Grashof numbers, heat is transferred through conduction through the fluid between the glasses. In this case, the Nusselt number is expressed as a function of the plate distance, δ, that is:

$$Nu_\delta = \frac{h\delta}{k} \tag{7.33}$$

and the Grashof number is based on the distance δ between the plates:

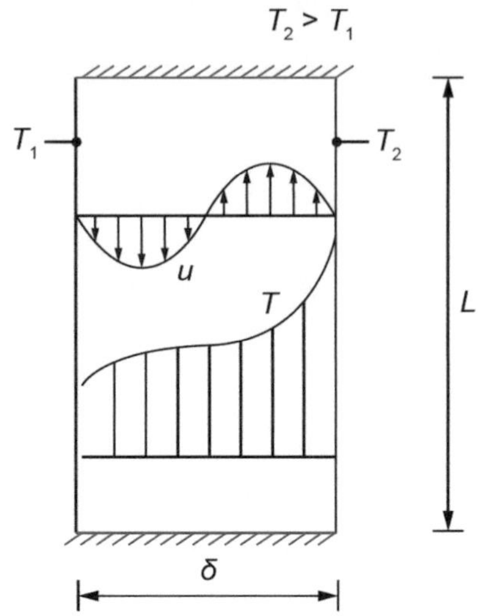

Fig. 7.11 Natural convection between two vertical plates

$$Gr_\delta = g\beta \frac{(T_1 - T_2)\delta^3}{\nu^2} \tag{7.34}$$

MacGregor and Emery (1969) express the Nusselt number for this case in Eqs. (7.35) and (7.36), where L is the plate height.

$$\begin{aligned} &Nu_\delta = 0.42 Ra_\delta^{0.25} Pr^{0.012} \left(\frac{L}{\delta}\right)^{-0.30} \\ &10^4 < Ra_\delta < 10^7;\ 1 < \mathrm{Pr} < 20,000 \\ &10 < L/\delta < 40 \end{aligned} \tag{7.35}$$

$$\begin{aligned} &Nu_\delta = 0.046 Ra_\delta^{1/3} \\ &106 < Gr_\delta Pr < 109;\ 1 < Pr < 20 \\ &1 < L/\delta < 40 \end{aligned} \tag{7.36}$$

The heat flux transferred between the walls, with one being cooled while the other is heated, is calculated as:

$$q = h\,(T_1 - T_2) \tag{7.37}$$

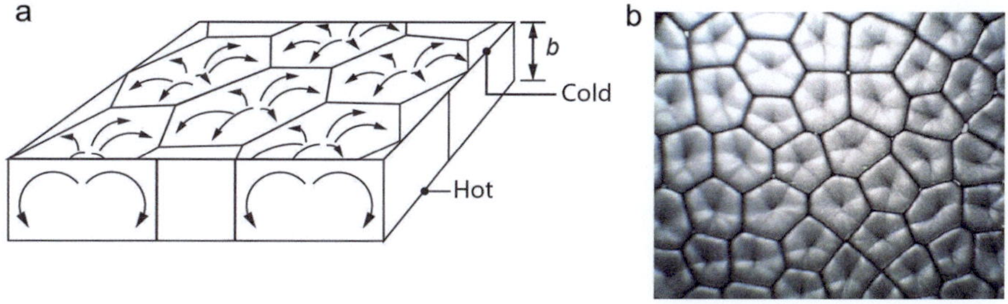

Fig. 7.12 **a** Convection hexagonal cells formed in confined between horizontal plates (Benard 1901); **b** natural convection on a heated surface exposed to air, DOI-10.1088/0143-0807/28/2/016 authorized reproduction @ European Physical Society. Reproduced with permission from IOP Publishing.

Horizontal parallel plates

There are two situations to be considered in the case of horizontal plates. There is no convection if the temperature of the upper plate is higher than that of the lower plate and, in this case, the heat transfer will occur only by simple heat conduction ($Nu_\delta = 1$). In the reciprocal case, that is, the temperature of the lower plate is higher than that of the upper plate, there will be only conduction if $Ra_\delta < 1708$; for Rayleigh numbers greater than 1708 there will be convection with the formation of hexagonal convection cells known as *Bernard cells*, as illustrated in Fig. 7.12a. This pattern of cells becomes unstable with the onset of fluid turbulence, which according to Krishnamurti (1970) varies with the Rayleigh and Prandtl numbers, from $Ra_\delta \sim 5500$ and $Pr = 0.7$ up to $Ra_\delta \sim 55,000$ and $Pr = 8500$, becoming fully turbulent for $Ra_\delta \sim 10^6$. The photograph in Fig. 7.12b illustrates the case of natural convection in contact with atmospheric air, without the upper plate, but which also forms the Bernard convection cells, known as Bernard-Marangoni.

According to Hollands et al. (1975), the Nusselt number for two horizontal and parallel plates is given for air and for water, according to Eqs. (7.38) and (7.39), respectively.

$$Nu_\delta = 1 + 1.44\left[1 - \frac{1708}{Ra_\delta}\right]^* + \left[\left(\frac{Ra_\delta}{5830}\right)^{1/3} - 1\right]^*$$
$$1708 < Ra_\delta < 10^8 \tag{7.38}$$

$$Nu_\delta = 1 + 1.44\left[1 - \frac{1708}{Ra_\delta}\right]^* + \left[\left(\frac{Ra_\delta}{5830}\right)^{1/3} - 1\right] + 2\left(\frac{Ra_\delta^{1/3}}{140}\right)^{\left[1-\ln\left(\frac{Ra_\delta^{1/3}}{140}\right)\right]}$$
$$1708 < Ra_\delta < 4 \times 10^9 \tag{7.39}$$

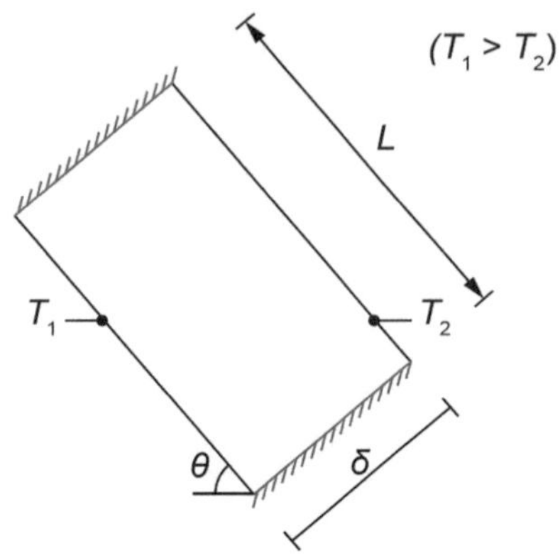

Fig. 7.13 Confined space between two inclined isothermal plates

In the previous equations, the term inside the bracket marked with an asterisk, $[\ ]^*$ is null if the inside term is negative.

Inclined parallel plates

In this case of two isothermal parallel plates at temperatures T_1 and T_2, as schematized in Fig. 7.13. The heat transfer in inclined plates strongly depends on the angle of inclination relative to the horizontal, θ, as expressed in Eq. (7.40) (Hollands et al., 1976), which was tested for air. The Rayleigh number is a function of the temperature difference $(T_1 - T_2)$. Heat transfer occurs solely by conduction ($Nu_\delta = 1$) for $Ra_\delta \cos\theta < 1708$. This expression holds true for an inclination angle up to 60° and provides an error of up to 5%; if the angle is up to 75°, the error will be up to 10%.

$$Nu_\delta = 1 + 1.44\left[1 - \frac{1708}{Ra_\delta \cos\theta}\right]^* \left\{1 - \frac{1708[\sin(1.8\theta)]^{1.6}}{Ra_\delta \cos\theta}\right\} + \left[\frac{(Ra_\delta \cos\theta)^{\frac{1}{3}}}{18} - 1\right]^*$$

$$0 < Ra_\delta \leq 10^5, \quad 0 \leq \theta \leq 60°, \quad L\delta \geq 12\delta > 12 \tag{7.40}$$

The term $[\ ]^*$ in the previous equations means that, if the term evaluated between the brackets, [ ], is negative, the term [ ] should be adopted as zero.

These plates' arrangement occurs in flat solar collectors with air in the confined space where the absorbing plate would be at average temperature T_1 and the glass cover, for example, the surface at T_2, being $T_1 > T_2$.

Concentric cylinders

A common case of parabolic solar trough collector's application is that of concentric tubes, such as the ones mounted at the linear focus. In this case, the convective motion is symmetrical in the annular section, as illustrated in Fig. 7.14.

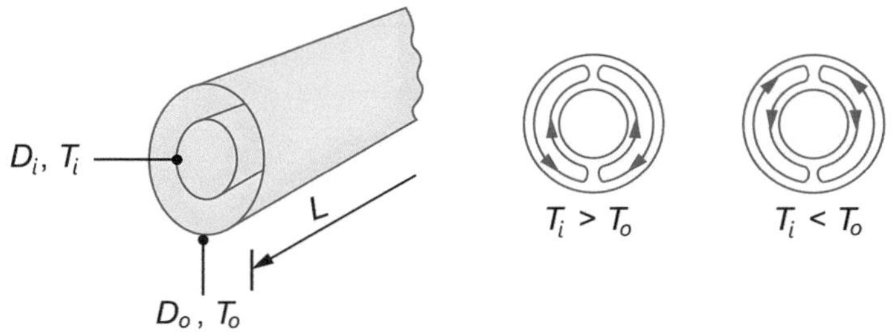

Fig. 7.14 Natural convection in concentric annular section

One way to evaluate the heat transfer between two isothermal tubes is by the relationship of Raithby and Hollands (1975), which evaluates an effective conductivity k_e, and the heat transfer between a surface and the fluid and the fluid and the other surface is evaluated by Eq. (7.41):

$$\begin{aligned} & k_e > k \rightarrow k_e = 0.386k\left(\frac{Pr}{0.861 + Pr}\right)^{\frac{1}{4}} Ra_a^{\frac{1}{4}} \\ & 10^2 < Ra_a < 10^7;\ 0.7 < Pr < 6000 \\ & k_e < k \rightarrow k_e = k \end{aligned} \tag{7.41}$$

The characteristic length for the Rayleigh number is found by:

$$L_a = \frac{[\ln(D_e/D_i)]^{4/3}}{\left(D_i^{-3/5} + D_e^{-3/5}\right)^{5/3}} \tag{7.42}$$

The total heat exchange rate is given by:

$$\dot{Q} = 2\pi k_e L \frac{T_i - T_e}{\ln(D_e/D_i)} \tag{7.43}$$

Example 7.2 (*Convection between two parallel plates*)

Estimate the heat exchange flow between two horizontal parallel flat plates, the lower one being at $T_1 = 100\ °C$ and the upper one at $T_2 = 40\ °C$, at the following conditions:

(a) horizontal position, spacing between plates of 20 mm;
(b) horizontal position, spacing between plates of 50 mm;
(c) inclined position at 45°, spacing between plates of 20 mm.

Solution

The transport properties of the air are evaluated at the average temperature between the isothermal plates:

$$\overline{T} = \frac{(T_1 + T_2)}{2} = \frac{(100 + 40)}{2} = 70\ ^\circ\text{C or } 343.15\ \text{K}.$$

Transport properties from Table A.5 at $\overline{T} = 70\ ^\circ\text{C}$

$$\beta = 1/343.15 = 2.914 \times 10^{-3}\text{K};\ k = 0.02952\ \text{W/mK};$$
$$\nu = 1.999 \times 10^{-5}\ \text{m}^2/\text{s};\ \alpha = 2.845 \times 10^{-5}\ \text{m}^2/\text{s};\ Pr = 0.7025$$

(a) Horizontal and $\delta = 0.02$ m

$$Ra_\delta = \frac{g\beta(T_1 - T_2)\delta^3}{\nu\alpha} = \frac{9.81 \times 2.914 \times 10^{-3}(100 - 40) \times 0.02^3}{1.999 \times 10^{-5} \times 2.845 \times 10^{-5}} = 24,127$$

Using Eq. (7.38):

$$Nu_\delta = 1 + 1.44\left[1 - \frac{1708}{Ra_\delta}\right]^* + \left[\left(\frac{Ra_\delta}{5830}\right)^{1/3} - 1\right]^*$$

$$Nu_\delta = 1 + 1.44\left[1 - \frac{1708}{24,127}\right]^* + \left[\left(\frac{24,127}{5830}\right)^{\frac{1}{3}} - 1\right]^* = 2.943$$

$$h = Nu_\delta \frac{k}{\delta} = 2.943 \frac{0.02952}{0.02} = 4.344\ \text{W/m}^2\text{K}$$
$$q = h(T_1 - T_2) = 4.344(100 - 40) = 260.6\ \text{W/m}^2$$

(b) Horizontal and $\delta = 0.05$ m. Following the same procedure as the previous item,

$$Ra_\delta = 376,986;\ Nu_\delta = 5.447;\ h = 3.216\ \text{W/m}^2\text{K}$$
$$q = 193.0\ \text{W/m}^2$$

(c) Inclined 45° and $\delta = 0.02$ m, from item (a) $Ra_\delta = 24,127$

From Eq. (7.40) for inclined plate:

$$Nu = 1 + 1.44\left[1 - \frac{1708}{Ra_\delta \cos\theta}\right]^* \left\{1 - \frac{1708[\sin(1.8\theta)]^{1.6}}{Ra_\delta \cos\theta}\right\} + \left[\frac{(Ra_\delta \cos\theta)^{1/3}}{18} - 1\right]^*$$

$$Nu = 1 + 1.44\left[1 - \frac{1708}{24127 \times \cos 45^\circ}\right]^* \left\{1 - \frac{1708\left[\sin\left(1.8 \times 45^\circ\right)\right]^{1.6}}{24127 \times \cos 45^\circ}\right\}$$

$$+\left[\frac{\left(24127 \times \cos 45^\circ\right)^{1/3}}{18} - 1\right]^* Nu = 2.599$$

$$h = Nu\frac{k}{\delta} = 2.599\frac{0.02952}{0.02} = 3.835\ \mathrm{W/m^2K}$$

$$q = h(T_1 - T_2) = 3.835(100 - 40) = 230.1\ \mathrm{W/m^2}$$

It can be concluded that increasing the spacing between plates and tilting them causes the decreasing of heat losses.

Example 7.3 (*Convection between two concentric cylinders*)

A parabolic trough solar collector was designed to heat a thermal fluid for industrial use. The collector's receiver consists of an inner metallic tube inside a glass tube. Between the two concentric tubes, there is air in natural convection. Evaluate the heat lost by convection by the receiver, if it is at 180 °C and the glass tube is at 60 °C. The receiver's geometry is indicated in the figure.

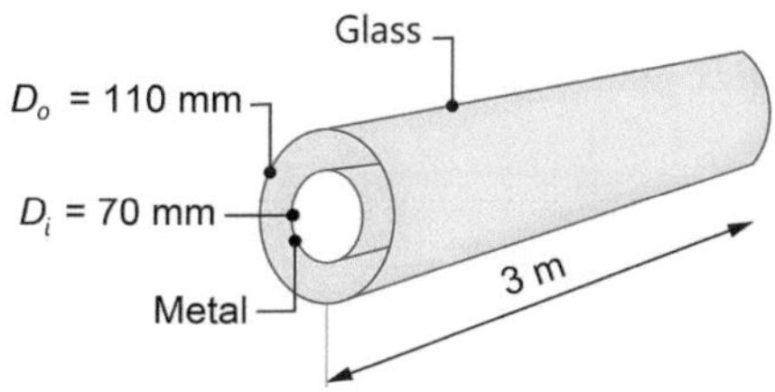

Solution

Transport properties of air should be evaluated at the average temperature of the cylinders: $\overline{T} = \frac{(180+60)}{2} = 120\,°\mathrm{C}$ or 393.15 K.

Transport properties from Table A.5 at $\overline{T} = 120$ °C.

$$\beta = 1/393.15 = 2.543 \times 10^{-3}\ 1/\mathrm{K};\ k = 0.03299\ \mathrm{W/mK};$$
$$\nu = 2.536 \times 10^{-5}\ \mathrm{m^2/s};\ \alpha = 3.627 \times 10^{-5}\ \mathrm{m^2/s};\ Pr = 0.6992$$

From Eq. (7.42), the characteristic length of the Rayleigh number is obtained:

$$L_a = \frac{[\ln(D_e/D_i)]^{4/3}}{\left(D_i^{-3/5} + D_e^{-3/5}\right)^{5/3}} = \frac{[\ln(0.110/0.070)]^{4/3}}{\left(0.070^{-3/5} + 0.110^{-3/5}\right)^{5/3}} = 9.45 \times 10^{-3}\,\text{m}$$

$$Ra_a = \frac{g\beta(T_i - T_o)L_a^3}{\nu\alpha}$$

$$Ra_a = \frac{9.81 \times 2.543 \times 10^{-3} \times (180 - 60) \times \left(9.45 \times 10^{-3}\right)^3}{2.536 \times 10^{-5} \times 3.627 \times 10^{-5}} = 2747$$

From Eq. (7.41):

$$k_e = 0.386k\left(\frac{Pr}{0.861 + Pr}\right)^{1/4} Ra_a^{1/4}$$

$$k_e = 0.386 \times 0.03299\left(\frac{0.6992}{0.861 + 0.6992}\right)^{1/4} 2747^{1/4} = 0.07543\,\text{W/mK}$$

which results greater than k, and thus the value of $k_e = 0.07543$ W/mK is used in Eq. (7.43) to evaluate the heat loss by convection:

$$\dot{Q} = 2\pi k_e L \frac{T_i - T_e}{\ln(D_e/D_i)} = 2\pi \times 0.07543 \times 3 \times \frac{180 - 60}{\ln(0.110/0.070)}$$

$$\dot{Q} = 377\,\text{W}$$

Vacuum is usually made between the receiver and the glass cover, to eliminate the large heat losses by convection. These are called evacuated tube solar collectors.

7.3 Mixed Convection

The cases of natural and forced convection have been treated separately. Of course, nature does not follow our classification, and heat transfer phenomena will occur according to the forces that control them (buoyancy, friction, and inertial forces). So, there are certain situations where both convective effects (natural and forced) are significant, for which the term *mixed convection* is used. Mixed convection is considered to occur when $\frac{Gr_L}{Re_L^2} \approx 1$; in the case of $\frac{Gr_L}{Re_L^2} \ll 1$, the leading convection is the forced one; and when $\frac{Gr_L}{Re_L^2} \gg 1$, the process will be controlled by natural

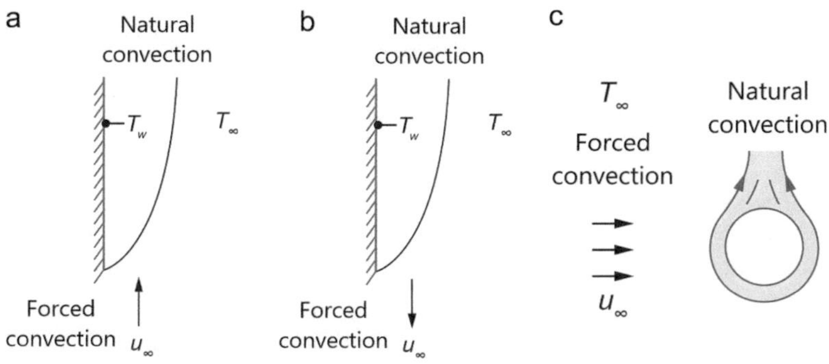

Fig. 7.15 Flow **a** parallel, **b** opposite and parallel, **c** cross

convection. The combined forms of these two types of convection can be grouped into three general categories:

(a) parallel flow occurs when the fluid motion induced by the two forms of convection are in the same direction (example of a heated plate with forced upward movement, Fig. 7.15a);
(b) opposite flow occurs when the movements induced by the two forms of convection are in opposite directions (example of a heated plate with forced downward movement, Fig. 7.15b);
(c) transverse flow is exemplified by the forced cross-movement over a heated cylinder, as shown in Fig. 7.15c.

It is standard to consid, on a first evaluation, that the mixed Nusselt number (Nu) is the result of the combination (Eq. 7.44) of the Nusselt numbers calculated as if there was only forced convection (Nu_F) and only natural convection (Nu_N), according to Eq. (7.44).

$$Nu^n = Nu_F^n \pm Nu_N^n \tag{7.44}$$

In *flat plates*, pure or mixed convection is considered when (Lloyd and Sparrow, 1970):

- Forced convection: $\frac{Gr_L}{Re_L^2} < 0.1$
- Mixed convection: $0.1 < \frac{Gr_L}{Re_L^2} < 10.$
- Natural convection: $\frac{Gr_L}{Re_L^2} > 10$

Example 7.4 (*Natural convection in a sphere*)

In a certain laboratory experiment, a small aluminum 2 cm diameter sphere is kept heated, reaching a constant surface temperature $T_w = 50\,°\text{C}$ and is surrounded by water at $T_\infty = 15\,°\text{C}$. Determine the rate of heat transferred from the sphere when:

(a) the water is at rest;
(b) the water moves at a horizontal velocity of 0.05 m/s;
(c) from what water velocity could natural convection be disregarded?

Note: for item (b), if necessary, use the expression $Nu^3 = Nu_F^3 + Nu_N^3$. For a sphere, the Nusselt number in convection can be evaluated by: $Nu_F = 2 + \left(0.4Re_D^{0.5} + 0.06Re_D^{2/3}\right)Pr^{0.4}\left(\frac{\mu}{\mu_s}\right)^{0.25}$

Solution

As the temperature of the sphere is higher than the environment, there will be an upward natural convective flow, as schematized in the figure on the left. The figure on the right also indicates the influence of forced convection.

(a) Water properties at (Table A.7): $\overline{T} = \frac{50+15}{2} = 32.5$

$$\beta = 0.000326/\text{K};\ \mu = 7.58 \times 10^{-4}\text{kgm/s};\ k = 0.621\,\text{W/mK};$$
$$C_p = 4178\,\text{J/kgK};\ Pr = 5.12;\ \rho = 995\,\text{kg/s},\ \alpha = \nu/Pr = 1.49 \times 10^{-7}\,\text{m}^2/\text{s}$$

$$Gr_D = \frac{g\beta(T_w - T_\infty)D^3}{\nu^2} = \frac{9.81 \times 0.000326 \times (50 - 15) \times 0.02^3}{(7.62 \times 10^{-7})^2} = 1.54 \times 10^6 < 10^9 \text{ laminar}$$

$$Ra_D = Gr_D Pr = 7.9 \times 10^6$$

$$Nu = 2 + \frac{0.589Ra^{1/4}}{\left[1 + \left(\frac{0.43}{Pr}\right)^{9/16}\right]^{4/9}} = 2 + \frac{0.589\left(7,9 \times 10^6\right)^{1/4}}{\left[1 + \left(\frac{0.43}{5,12}\right)^{9/16}\right]^{4/9}} = 30.3\ (Nu_N = 30.3)$$

$$h = Nu\frac{k}{D} = 30.3\frac{0.621}{0.02} = 940.8\,\mathrm{W/m^2}K$$

$$A_p = \pi D^2 = \pi \times 0.02^2 = 0.00126\,\mathrm{m}^2$$

$$\dot{Q} = hA_p(T_w - T_\infty) = 940.8 \times 0.00126 \times (50 - 15) = 41.5\,\mathrm{W}$$

(b) $Re_D = \frac{u_\infty D}{\nu} = \frac{0.05(0.02)}{7.62\times10^{-7}} = 1312.3$

$$\frac{Gr_D}{Re_D^2} = \frac{1.54 \times 10^6}{1312,3^2} = 0.9 \approx 1(\text{mixed})$$

At 50 °C , we have $\mu_w = 546 \times 10^{-6}\,\mathrm{kg/kgms}$

$$Nu_F = 2 + \left(0.4Re_D^{0.5} + 0.06Re_D^{2/3}\right)Pr^{0.4}\left(\frac{\mu}{\mu_w}\right)^{0.25}$$

$$Nu_F = 2 + \left[0.4(1312.3)^{0.5} + 0.06(1312.6)^{2/3}\right]5.12^{0.4}\left(\frac{758}{546}\right)^{0.25} = 47.2$$

$$Nu^3 = Nu_F^3 + Nu_N^3 \Rightarrow Nu = 51$$

$$h = Nu\frac{k}{D} = 51\frac{0.621}{0.02} = 1584.2\,\mathrm{W/m^2K.}$$

$$\dot{Q} = hA_p(T_w - T_\infty) = 1584.2 \times 0.00126 \times (50 - 15) = 69.9\,\mathrm{W}$$

(c) Natural convection is negligible when $\frac{Gr_D}{Re_D^2} \ll 1$ therefore, assuming $\frac{Gr_D}{Re_D^2} < 0.01$.

$u_\infty > \frac{\nu}{D}\sqrt{\frac{Gr_D}{0.01}} = \frac{7.62\times10^{-7}}{0.02}\sqrt{\frac{1.54\times10^6}{0.01}} = 0.47$

Therefore, for a water circulation velocity of $u_\infty > 0.47$ m/s, natural convection is neglegible.

7.4 Proposed Problems

7.1 Evaluate the Grashof and Rayleigh numbers for a flat 20 cm × 10 cm plate in vertical and horizontal positions with a constant wall temperature at 100 °C immersed in a still fluid at 20 °C, when the fluid is (a) air at 1 atm, (b) unused motor oil and (c) water (liquid). Consider only one side of the plate.

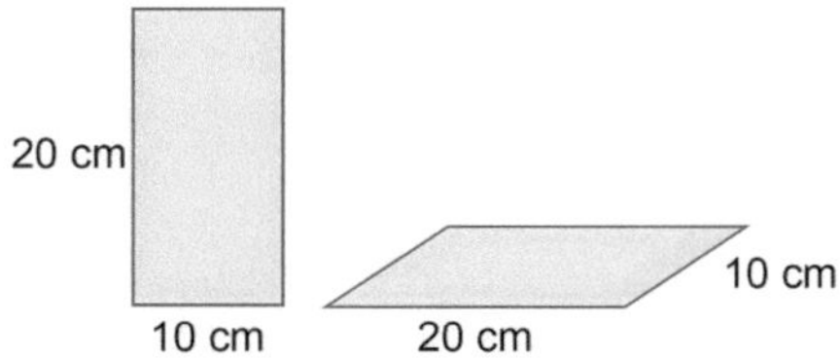

7.2 For the vertical plate of Proposed Problem 7.1 and the fluid being water, evaluate the Grashof and Rayleigh numbers in the following cases: (a) water temperature at 5 °C and plate temperature at 85 °C, (b) water temperature at 50 °C and plate temperature at 70 °C. Compare these results with those of Proposed Problem 7.1 and explain the influence of temperature on these figures.

7.3 A 1 m height vertical plate with a uniform temperature at 120 °C is in a still air environment at 20 °C at 1 atm. From Figure B.2 of Appendix B, evaluate at $x = 20$ cm from the bottom of the plate, (a) the thickness of the hydrodynamic boundary layer, (b) the position where the velocity u is maximum, and (c) the value of the maximum velocity.

7.4 A 20 cm × 10 cm flat plate at 80 °C surface temperature is immersed in still water at 15 °C, having the lengthier side in the vertical position. If the plate is turned by 90°, so that the shorter side is now in the vertical position, it is desired to know whether the new position transfers more heat.

7.5 It is necessary to heat a 15 cm square titanium plate at 10 °C in a still air environment at 50 °C. Evaluate the convection heat transfer coefficient and the rate of heat transferred.

7.6 A horizontal 25 mm diameter tube is used to transport saturated steam. Poor insulation installation causes it to detach from the tube over a 1 m length. The steam is at 150 °C. Neglecting the thermal resistance of the tube, evaluate the heat loss rate in this tube section if it is in a still air environment at 10 °C.

7.7 In Problem 7.6, what would be the heat loss rate if the tube were in the vertical position?

7.8 A horizontal 20 cm × 25 cm plate at 40 °C is suspended in still air at 15 °C. The plate has an emissivity of 0.85 and the environment walls are at 20 °C. Assuming a steady state and that the lower face is thermally insulated, evaluate the heat loss rate from the plate.

7.9 An electrical resistance dissipates 80 W thermal power in still air at 15 °C. The resistance is 3.5 m long, has a diameter of 4.5 mm, and is in a horizontal position. Evaluate its wall temperature, assuming it is uniform over its whole surface.

7.10 Some cities in the Southern Region of Brazil experience cold winters, where the external environment can reach temperatures below 0 °C. The glass window of a house has an internal temperature of 0 °C and an emissivity of 0.9, while the internal still air of the house is at 20 °C. The internal walls of the house are at an average temperature of 18 °C. Evaluate the heat loss rate in 1 m square glass window.

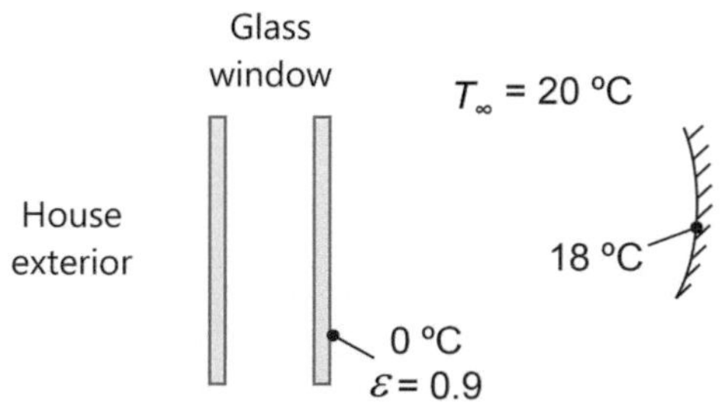

7.11 A 1 m square flat solar panel is tested on the vertical wall of a house. The evaluation test comprises obtaining the panel temperature by placing thermal insulation on its back. Evaluate the temperature in the center of the panel, if it has an emissivity of 0.15, absorbs 300 W/m^2 and the still air is at 20 °C. Consider that the environment temperature for the thermal transfer by radiation (effective sky temperature) is 2 °C.

7.12 A vertical 0.5 m high plate receives a constant 500 W/m^2 heat flux that must be transferred to the quiescent air at 15 °C. Evaluate the maximum temperature of the plate.

7.13 A horizontal circular disc is suspended in still air at 20 °C. The disc diameter is 6 cm. Assuming a uniform disc temperature at 85 °C, evaluate the heat lost rate by the disc (consider both faces).

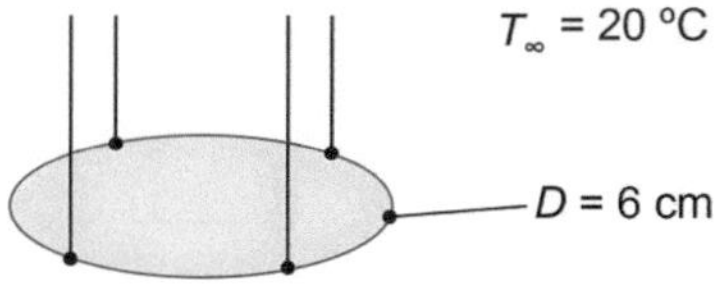

7.14 After a heat treatment up to 180 °C, a steel billet 12 cm diameter and 60 cm long must be slowly cooled in a still oil bath at 120 °C. To slow down the cooling process of the bar, which position (horizontal or vertical) would be most suitable and what is the initial heat loss rate of the bar?

7.15 A vertical fin heat sink was installed to cool a heated surface in still air at 15 °C. Considering the temperature of the fins constant at 80 °C, find (a) the optimal fin spacing, (b) the optimal number of fins, and (c) the heat dissipated under these conditions.

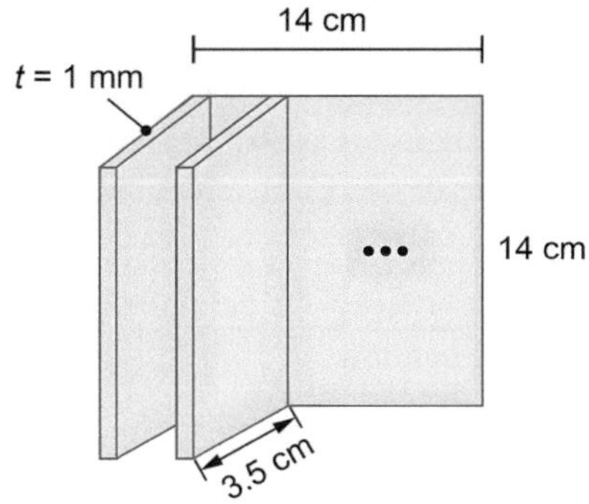

7.16 A mixed solar dryer consists of two parts: the first is a flat solar air heating collector and the second is essentially a box in which the product to be dried is placed. The solar collector is composed of a glass cover and a thermally insulated absorbing plate. The air enters at the bottom and is heated by the plate. Consider a 1 m × 2 m collector tilted 21° equipped with the glass and absorber spaced 2 cm apart and at average glass and absorber temperatures of 42 and 70 °C, respectively. Evaluate the heat loss rate by natural convection between the absorbing plate and the glass cover.

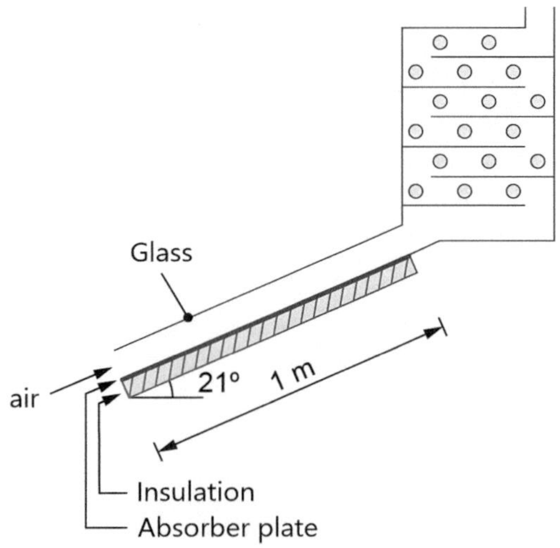

7.17 A large and long water reservoir has two 3 cm tubes inside it, one at 35 °C and the other at 20 °C. Evaluate the heat flux with the water from each tube, assuming the properties evaluated at the average water temperature.

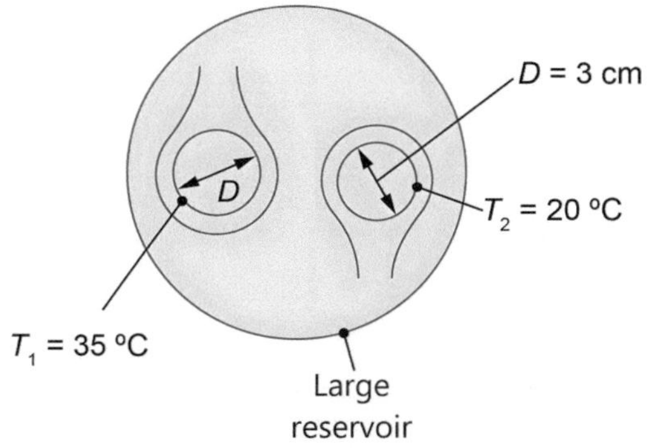

7.18 A laboratory experiment for curve fitting of the Nusselt number in natural convection was carried out. In this, a horizontal 9 cm diameter and 0.5 m length tube surrounded by quiescent air at 18 °C was used. When the tube was at a uniform wall temperature at 110 °C, the heat loss rate was 160 W. For a correlation of the type $Nu_D = CRa_D^m$, where the exponent m depends on the range of the Rayleigh number, according to the table below, it is asked: (a) evaluate the coefficient C, and (b) find the heat loss rate if the air were at 35 °C.

Ra_D	m
10^{-10}–0.01	0.058
0.01–100	0.148
100–10,000	0.188
10,000–10^7	0.250
10^7–10^{12}	0.333

7.19 A foil sheet is hanging, as indicated in the figure below. A strain gauge was placed to measure its weight accurately. The sheet is 40 cm × 30 cm with a mass of 5 g, in a quiescent air environment at 25 °C. When solar radiation hits the sheet, it is heated to a roughly constant temperature at 95 °C. Knowing that the variation of the local shear stress on the plate varies according to the relationship $\tau_w = 1.31 Gr_x^{\frac{3}{4}} \frac{\rho \nu^2}{x^2} \frac{Pr^{\frac{1}{4}}}{\left(Pr + \frac{20}{11}\right)^{\frac{3}{4}}}$, evaluate the weight recorded by the sensor, without the convective effect and with the convective effect. Use $g = 9.81$ m/s^2

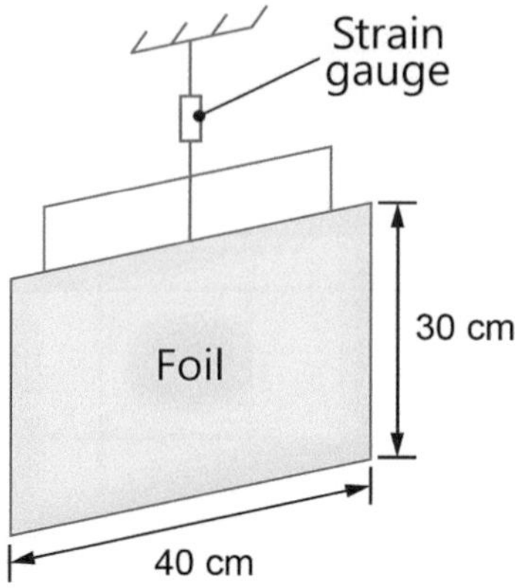

7.20 Saturated ethanol vapor at 120 °C is transported in a horizontal 24.5 mm diameter pipe. The pipe is in a quiescent air environment at 15 °C. Evaluate the heat loss under conditions of (a) pure natural convection, (b) when air flows transversely to the tube at a velocity of 0.15 m/s.

7.21 A vertical electronic plate of 14 cm × 16 cm (14 cm vertically) dissipates approximately 250 W/m^2 from one side, while the other can be considered isolated. The plate is in air environment at 20 °C. Evaluate the temperature of the "isothermal" plate where (a) the air is stagnant, (b) the air has an upward movement of 0.4 m/s and (c) the air has a downward movement of 0.4 m/s. Of these three situations, which one would you choose to avoid overheating problems on the plate?

References

Bar-Cohen, A., and W.M. Rohsenow. 1984. Thermally optimum spacing of vertical, natural convection cooled, parallel plates, pp. 116–123.

Benard, H. 1901. Les Tourbillons cellulaires dans une nappe liquide transportant de la chaleur par convection en r'egime permanent. *Annales de chimie et de physique* 23: 62–144.

Churchill, S.W., and H.H.S. Chu. 1975a. Correlating equations for laminar and turbulent free convection from a vertical plate. *International Journal of Heat and Mass Transfer* 18: 1323–1329.

Churchill, S.W., and H.H.S. Chu. 1975b. Correlating equations for laminar and turbulent free convection from a horizontal cylinder. *International Journal of Heat and Mass Transfer* 18: 1049–1053.

Churchill, S.W. 1983. Comprehensive, theoretically based, correlating equations for free convection from isothermal spheres. *Chemical Engineering Communications* 24: 339–352.

Fishenden, M., and O.A. Saunders. 1950. *An introduction to heat transfer*, Oxford: New York.

Fuji, T., and H. Imura. 1972. Natural-convection heat transfer from a plate with arbitrary inclination. *International Journal of Heat and Mass Transfer* 15: 755–767.

Hollands, K.G.T., G.D. Raithby., and L. Konicek. 1975. Correlation equations for free convection heat transfer in horizontal layers of air and water. *International Journal of Heat and Mass Transfer* 18: 879–884,

Hollands, K.G.T., T.E. Unny., G.D. Raithby., and L. Konicek. 1976. Free convection heat transfer across inclined air layers. *Journal of Heat Transfer* 98: 189–193.

Lloyd, J.R., E.M. Sparrow.1970. Combined forced and free convection flow on vertical surfaces. *International Journal of Heat and Mass Transfer* 13: 434–438.

Lloyd, J.R., and W.R. Moran. 1974. Natural Convection Adjacent to Horizontal Surface of Various Planforms. *Journal of Heat Transfer* 96: 443–447.

MacGregor, R.K., and A.P. Emery. 1969. Free convection through vertical plane layers: moderate and high Prandtl number fluids. *Journal of Heat Transfer* 91: 391–411.

Popiel, C.O., J. Wojtkowiak., and K. Bober. 2007. Laminar free convective heat transfer from isothermal vertical slender cylinder. *Experimental Thermal and Fluid Science* 32: 607–613,

Raithby, G.D., K.G.T. Hollands. 1975. A general method of obtaining approximate solutions to laminar and turbulent free convection problems. In *Advances in Heat Transfer*, eds. by T.F. Jr. Irvine and J.P. Hartnett, vol. 11, pp. 265–315. New York: Academic Press, New York.

Rohsenow, W.M., J.P. Hartnett., and Y.I. Cho. 1998. *Handbook of heat transfer*, 3rd ed. EUA: McGraw-Hill

Vliet, G.C., and D.C. Ross. 1975. Turbulent natural convection on upward and downward facing inclined constant heat flux surfaces. *Journal of Heat Transfer* 97(4): 549–554.

Chapter 8
Condensation and Boiling

Abstract The phenomena of heat transfer with phase change are the subjects of this chapter. The first part of the chapter studies the condensation of vapor on cold surfaces, where the condensation in the form of a laminar descending film on vertical surfaces is modeled following the pioneering work of Nusselt. Practical working expressions are presented for the turbulent and wavy regimes on tubes and tube banks. Droplet condensation is the subject that concludes the study of condensation. The second part of the chapter studies the boiling phenomenon, with the first topic addressed being pool boiling and its boiling curve that displays significant pieces of information such as the onset of nucleate boiling (ONB), characterized by the formation of individual bubbles, followed by the formation of bubble columns and film boiling (of vapor). The subject is presented in detail along with practical working equations and its proper modeling. It also presents the effect of the hysteresis of the boiling curve. The chapter concludes with the presentation of convective heat transfer boiling and the modeling resulting in expressions for calculating the critical heat flux (CHF).

Keywords Condensation · Liquid film · Droplet · Pool boiling · Convective boiling · Critical heat flux

8.1 Introduction

Condensation and boiling are phase change phenomena. In the first case, the vapor phase of a given fluid or a mixture of fluids undergoes the cooling process through heat transfer until the condensation temperature, or dew point temperature in the case of fluid mixtures, is reached, which gives rise to the condensation process, that is, the vapor becomes liquid. In the case of boiling, a given liquid or mixture of liquids undergoes the heating process through heat transfer until the vaporization temperature or bubble temperature, in the case of fluid mixtures, is reached to initiate the vaporization or boiling process. Nature and our daily experience are rich in examples of phase change, such as water boiling in a kettle and the drying process

J. R. Simões-Moreira and E. W. Zavaleta-Aguilar, *Heat Transfer*, Heat and Mass Transfer, https://doi.org/10.1007/978-3-031-80318-5_8

of liquid water from surfaces by vaporization to atmospheric air. There is also the process of water vapor condensation from air on cold surfaces at night, which forms dew. On very cold nights, the water on the surface can freeze. The term *vaporization* refers to the phase change that occurs at the liquid–vapor interface and is due to the difference in partial pressures of the saturated vapor at the interface and the vapor far away. On the other hand, *boiling* is the phenomenon that occurs at the solid–liquid interface with the formation of bubbles on a heated surface. This chapter deals with condensation and boiling. Analysis of the evaporative and condensation processes of water in the air makes up the scope of Psychrometry (Simões-Moreira and Hernandez Neto 2019).

In engineering, there are many examples of condensation processes in industrial equipment, such as condensing units of air conditioning and refrigeration systems and condensers of thermal power cycles. Figure 8.1a shows the scheme of an industrial condenser formed by two passes of tube bundles where water vapor is condensed on the cold tube external surfaces, with cooling water circulating inside the tubes. The vapor condenses on the tube's cold surface establishing a liquid film, as illustrated in Fig. 8.1b, which drains by gravity and is collected at the base of the condenser. Figure 8.1c shows a photograph of this type of heat exchanger.

The boiling phenomenon occurs in many pieces of engineering equipment. Air conditioning evaporating units are good examples where refrigerant fluid boiling happens by removing heat from the environment. Another widely used example is steam generators or boilers (Fig. 8.2). In this piece of equipment, liquid water receives heat in the furnace due to fuel burning and undergoes the boiling process producing the vapor phase. The boiling temperature depends on the liquid pressure.

An example of vaporization occurs in cooling towers where, by evaporative cooling, hot water is cooled in contact with atmospheric air by its partial vaporization. The following sections discuss the phenomena of condensation and boiling that dominate heat transfer in equipment where phase change occurs.

8.2 Condensation

In the case of condensation, vapor comes into contact with a cold surface, whose temperature is at or below the saturation temperature, giving rise to the condensation process on the surface. Depending on the characteristics of the surface and surface finishing, two forms of condensation are possible: by forming a liquid film on the surface or by forming droplets on the surface. Film condensation is treated in the following sections because it is the most common and occurs with larger temperature differences between the saturated vapor and the cold surface temperature, which is named the *degree of sub-cooling*. However, at lower degrees of sub-cooling and surfaces with low wettability (high surface tension) there is a tendency to form condensate droplets, which increases the heat transfer rate by one or more orders of magnitude, thus being more effective.

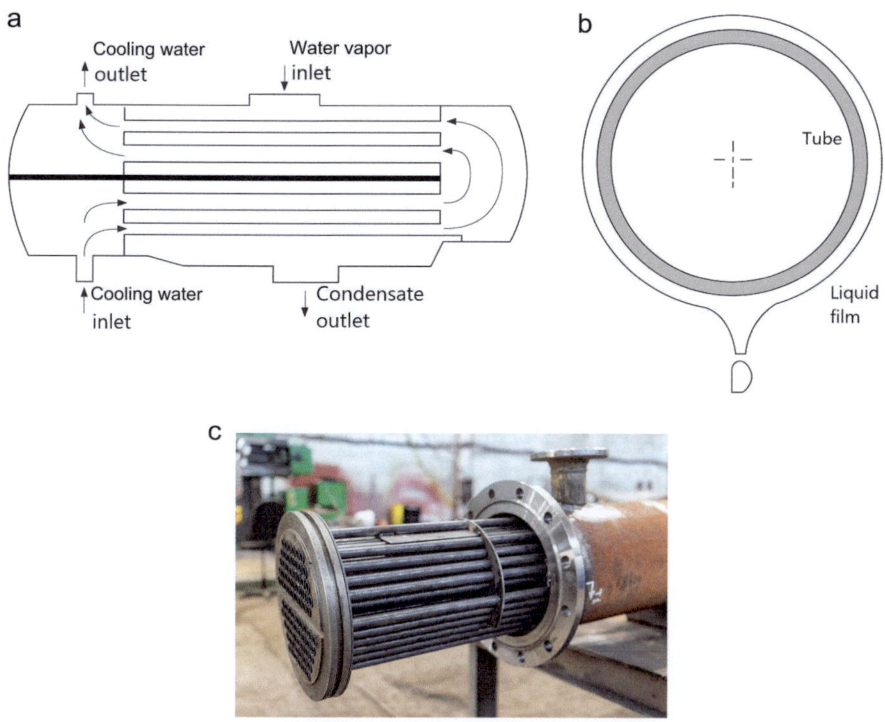

Fig. 8.1 **a** Schematic of an industrial condenser. Water vapor is cooled on the cold surfaces of the tubes. Inside the tubes circulates cooling water; **b** Condensed liquid film formed around the tube cold wall. **c** a photograph of an industrial shell and tube condenser showing the condensing tube bundle. iStock.com/Berkut_34

Fig. 8.2 Natural gas industrial boiler. iStock.com/Huyangshu

8.2.1 Downward Film Condensation on Flat Surfaces

Laminar case

Consider a vertical surface, whose temperature, T_w, is below the saturation temperature, T_{sat}, of the surrounding vapor, as illustrated in Fig. 8.3. A liquid film is formed by vapor condensation from the upper edge, $x = 0$, whose thickness δ grows in the film´s descending direction. In the illustration to the right of the figure, one can see the velocity and temperature profiles. Heat transfer within the laminar film occurs only by conduction. The following classic solution was first obtained by Nusselt (1916).

The velocity profile $u(y)$ can be obtained using a force balance on the differential element of Fig. 8.3. The usual assumptions adopted are: the descending velocity of the film is low, heat transfer within the film is only by conduction, the shear stress at the liquid–vapor interface is negligible and, finally, the vapor is at saturation temperature. With these assumptions, one can write:

$$\mu_L \frac{du}{dy} dx = g(\rho_L - \rho_V)(\delta - y)dx \tag{8.1}$$

Integrating the equation with the boundary condition of $u(0) = 0$ (no-slip condition), we get:

$$u = \frac{g(\rho_L - \rho_V)}{\mu_L}\left(\delta y - \frac{y^2}{2}\right) \tag{8.2}$$

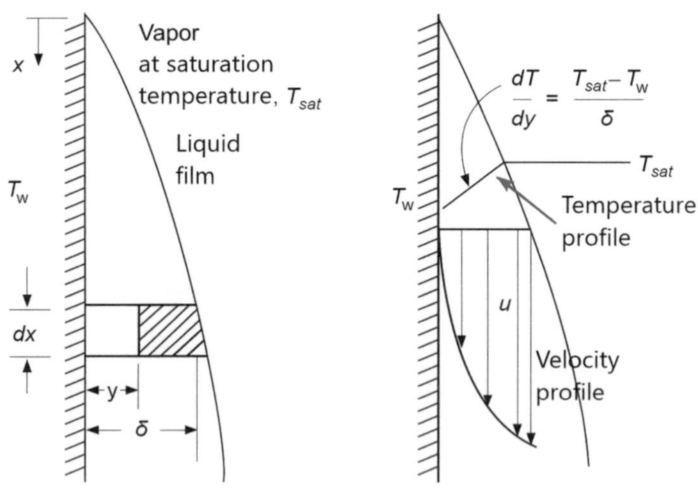

Fig. 8.3 Illustration of the growth of the thickness δ of a descending laminar film and the velocity and temperature profiles

Therefore, it is a 2nd degree parabolic profile. The mass flow rate, Γ_x, of the descending liquid film per unit width of the vertical plate, L_p, is:

$$\Gamma_x = \frac{\dot{m}}{L_p} = \int_0^{\delta} \rho_L u dy = \int_0^{\delta} \rho_L \frac{g(\rho_L - \rho_V)}{\mu_L}\left(\delta y - \frac{y^2}{2}\right)dy = \rho_L \frac{g(\rho_L - \rho_V)}{3\mu_L}\delta^3 \tag{8.3}$$

Therefore, the growth rate of liquid flow in terms of film thickness is:

$$\frac{d\Gamma_x}{d\delta} = \rho_L \frac{g(\rho_L - \rho_V)}{\mu_L}\delta^2 \tag{8.4}$$

On the other hand, the growth rate of liquid flow is due to the vapor condensation. Therefore, from an energy balance, it is possible to write:

$$dq\prime = h_{LV} d\Gamma_x \tag{8.5}$$

where h_{LV} is the specific enthalpy of vaporization of the vapor. The elementary heat rate per unit width of the plate released at the interface due to condensation, dq', is transferred by conduction (linear temperature profile) to the interior of the film to the wall. So that:

$$dq\prime = k_L \frac{T_{sat} - T_w}{\delta} dx \tag{8.6}$$

Finally, by first substituting Eqs. (8.6) into (8.5) and, subsequently, into Eq. (8.4), we get:

$$\frac{k_L(T_{sat} - T_w)}{\delta h_{LV}}\frac{dx}{d\delta} = \rho_L \frac{g(\rho_L - \rho_V)}{\mu_L}\delta^2 \Rightarrow \frac{d\delta}{dx} = \frac{k_L \mu_L (T_{sat} - T_w)}{\delta^3 g \rho_L (\rho_L - \rho_V) h_{LV}}$$

From which we obtain, after integration, the film thickness, δ , growth at a distance x from the initial edge.

$$\delta(x) = \left[\frac{4k_L \mu_L (T_{sat} - T_w)x}{g\rho_L(\rho_L - \rho_V)h_{LV}}\right]^{\frac{1}{4}} \tag{8.7}$$

The local heat transfer coefficient, h_x, can be obtained from the fact that the heat flux by heat convection should be equal to that of conduction in the film, that is:

$$h_x(T_{sat} - T_w) = \frac{k_L(T_{sat} - T_w)}{\delta} \Rightarrow h_x = \frac{k_L}{\delta}$$

$$h_x = \left[\frac{g\rho_L(\rho_L - \rho_V)h_{LV}k_L^3}{4\mu_L(T_{sat} - T_w)x}\right]^{\frac{1}{4}} \tag{8.8}$$

It is important to analyze the behavior of Eqs. (8.7) and (8.8) along the plate. The liquid film thickness increases from the plate's upper edge and grows at a proportion $\delta \propto x^{\frac{1}{4}}$. The local coefficient decreases according to the proportion $h_x \propto x^{-\frac{1}{4}}$, from which it is concluded that $h_x \propto \frac{1}{\delta}$.

The local Nusselt number, Nu_x, based on the distance from the plate edge can be obtained from its own definition:

$$Nu_x = \frac{h_x x}{k_L} = \left[\frac{g\rho_L(\rho_L - \rho_V)h_{LV}x^3}{4k_L\mu_L(T_{\text{sat}} - T_w)}\right]^{\frac{1}{4}} \tag{8.9}$$

The average heat transfer coefficient over the plate length, $\overline{h}_L$, is:

$$\overline{h}_L = \frac{1}{L}\int_0^L h_x dx = \frac{1}{L}\int_0^L \left[\frac{g\rho_L(\rho_L - \rho_V)h_{LV}k_L^3}{4\mu_L(T_{\text{sat}} - T_w)x}\right]^{\frac{1}{4}} dx = \frac{4}{3}h_{x=L} \Rightarrow$$

$$\overline{h}_L = 0.9428\left[\frac{g\rho_L(\rho_L - \rho_V)h_{LV}k_L^3}{\mu_L(T_{\text{sat}} - T_w)L}\right]^{\frac{1}{4}} \tag{8.10}$$

The average Nusselt number is obtained from $\overline{Nu}_L = \frac{\overline{h}_L L}{k_L}$, $\overline{Nu}_L = \frac{\overline{h}_L L}{k_L}$,

$$\overline{Nu}_L = 0.9428\left[\frac{g\rho_L(\rho_L - \rho_V)h_{LV}L^3}{k_L\mu_L(T_{\text{sat}} - T_w)}\right]^{\frac{1}{4}} \tag{8.11}$$

The total condensation flow per unit of plate width, Γ_x, at the position of interest can be obtained from the integration of Eq. (8.5), considering constant properties.

$$\Gamma_x = \frac{q'}{h_{LV}} \tag{8.12}$$

And the total heat transfer rate, $\dot{Q}$, over the total plate area A comes from the law of heat convection:

$$\dot{Q} = \overline{h}_L A(T_{sat} - T_w) \tag{8.13}$$

In the absence of more specific expressions, it is possible to show that, for *inclined plates* at an angle θ tilted vertically, the following expression of the average heat transfer coefficient can be obtained, where g is replaced by $g\cos\theta$. However, it is not valid for $\theta = 90°$ (horizontal plate), or values very close to the horizontal position.

$$\overline{h}_L = 0.9428\left[\frac{g\rho_L(\rho_L - \rho_V)h_{LV}k_L^3\cos\theta}{\mu_L(T_{\text{sat}} - T_w)L}\right]^{\frac{1}{4}} \tag{8.14}$$

The hydraulic diameter, in this case, is $D_h = \frac{4A}{P} = \frac{4L_p\delta}{L_p} = 4\delta$,, thus, the Reynolds number, Re_δ, will be:

$$Re_\delta = \frac{\rho \overline{V} 4\delta}{\mu_L} = \frac{4\dot{m}}{\mu_L L_p} = \frac{4\Gamma_x}{\mu_L} \tag{8.15}$$

Substituting Eq. (8.3) into the previous expression, we get:

$$Re_\delta = \frac{4g\rho_L(\rho_L - \rho_V)\delta^3}{3\mu_L^2} \approx \frac{4}{3}\frac{g\delta^3}{\nu_L^2} \tag{8.16}$$

since $\rho_L \gg \rho_V$, except near critical conditions, and the kinematic viscosity is $\nu_L = \frac{\mu_L}{\rho_L}$.

Finally, substituting the definition of Re_δ into the equations of the local (Eq. 8.8) and average (Eq. 8.10) coefficients, we get:

$$\frac{h_x}{k_L}\left(\frac{\nu_L^2}{g}\right)^{\frac{1}{3}} \approx 1.1\mathrm{Re}_\delta^{-\frac{1}{3}} \tag{8.17}$$

and

$$\frac{\overline{h}_L}{k_L}\left(\frac{\nu_L^2}{g}\right)^{\frac{1}{3}} \approx 1.467\mathrm{Re}_\delta^{-\frac{1}{3}} \tag{8.18}$$

The two previous expressions are valid for the laminar case that covers the range up to $Re_\delta < 30$.

Transport properties vary with temperature. Therefore, it is recommended that the properties of the condensed liquid film be obtained at $T_f = \frac{(T_{\text{sat}} + T_w)}{2}$ and the specific enthalpy of vaporization, h_{LV}, is obtained at T_{sat}.

Improved expressions of the heat transfer coefficient and the previous Nusselt number were obtained by Rohsenow (1956) for $Pr > 0.5$. He proposed the replacement of the specific enthalpy of vaporization, h_{LV}, in the previous expressions by the modified value, h_{LV}^+, including effects related to the subcooling of the film, according to:

$$h_{LV}^+ = h_{LV} + 0.68C_{pL}(T_{\text{sat}} - T_w) \tag{8.19}$$

where C_{pL} is the specific heat at constant pressure of the liquid.

Laminar cases with ripples and turbulent

As the liquid film thickness increases, the film becomes unstable and ripples begin to grow rapidly for $Re_\delta > 30$ and extend up to about $Re_\delta \approx 1600$, when it becomes turbulent. For this transition region, there are several correlations, with Kutadeladze's (1963) being one of the recommended ones that applies to the laminar, wavy, and transition region $0 < Re_\delta < 1600$, given by:

$$\frac{\overline{h}_L}{k_L}\left(\frac{\mu_L^2}{\rho_L(\rho_L-\rho_V)g}\right)^{\frac{1}{3}} = \frac{Re_\delta}{1.08Re_\delta^{1.22}-5.2} \tag{8.20}$$

Based on the work of Labuntsov (1957), Butterworth (1983) recommends the following expression for the turbulent regime, but with laminar, wavy, and transition onset:

$$\frac{\overline{h}_L}{k_L}\left(\frac{\mu_L^2}{\rho_L(\rho_L-\rho_V)g}\right)^{\frac{1}{3}} = \frac{Re_\delta}{8750+58Pr_L^{-\frac{1}{2}}\left(Re_\delta^{\frac{3}{4}}-253\right)} \tag{8.21}$$

Chun and Kim (1991) adjusted the data for the entire laminar region, laminar with waves and turbulent through a semi-empirical expression (with $\rho_L \gg \rho_V$) given by:

$$\frac{\overline{h}_L}{k_L}\left(\frac{\nu_L^2}{g}\right)^{\frac{1}{3}} = 1.33Re_\delta^{-\frac{1}{3}} + 9.56\times10^{-6}Re_\delta^{0.89}Pr_L^{0.94} + 8.22\times10^{-2} \tag{8.22}$$

The validity is in the range $10 < Re_\delta < 3.1\times10^4$ and $1.75 < Pr_L < 5$. The graphics in Fig. 8.4 shows Eq. (8.22) throughout the entire range of validity of the Reynolds number and for various Prandtl numbers.

The heat transfer calculation in condensation on vertical tubes can be performed using the expressions for vertical plates, provided that the film thickness is small compared to the tube diameter, that is, $\delta \ll D$.

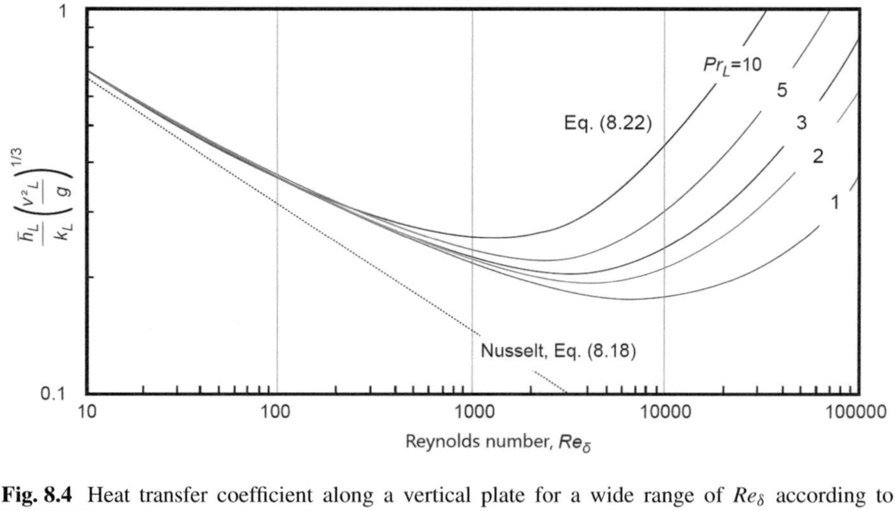

Fig. 8.4 Heat transfer coefficient along a vertical plate for a wide range of Re_δ according to Eq. (8.22)

8.2.2 Downward Film Condensation on Horizontal Tubes

Many condensing industrial equipment are designed with tube banks where the vapor phase condenses on the tube's cold surface. Figure 8.5 illustrates a typical situation where the vapor condenses, forms a liquid film around the tube, and drains to the lower tube.

The recommended expression for the average heat transfer coefficient for a single horizontal tube is similar to that of Eq. (8.10), that is,

$$\overline{h}_D = C\left[\frac{g\rho_L(\rho_L - \rho_V)h_{LV}^{+}k_L^3}{\mu_L(T_{sat} - T_w)D}\right]^{\frac{1}{4}} \tag{8.23}$$

with the constant value $C = 0.725$ for tubes with an external diameter D (Dhir and Lienhard 1971) and $C = 0.815$ in the case of spheres of diameter D (Popiel and Boguslawski 1975).

In the case of a bank of tubes aligned vertically (as illustrated in Fig. 8.5), it has been recommended to multiply the diameter D by the number N of tubes in the same row.

$$\overline{h}_D = C\left[\frac{g\rho_L(\rho_L - \rho_V)h_{LV}^{+}k_L^3}{\mu_L(T_{sat} - T_w)ND}\right]^{\frac{1}{4}} \tag{8.24}$$

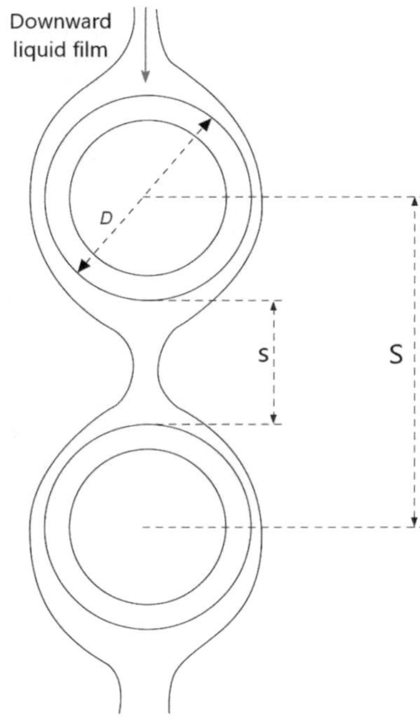

Fig. 8.5 Condensation on tubes (S is the pitch, and s is the distance between tubes)

Example 8.1 (*Condensation on a tube*)

A 50 mm diameter horizontal tube at 30 °C is surrounded by saturated steam at a pressure of 12.35 kPa. For this condition, determine the heat transfer coefficient.

Solution

Problem data: $D = 0.05$ m; $T_w = 30$ °C

From Table A.7, for $P_{\text{sat}} = 12.35$ kPa, we obtain: $T_{sat} = 50$ °C and $h_{LV} = 2382$ kJ/kg. The other properties should be evaluated at $T_f = (30 + 50)/2 = 40\,°\text{C}$;

$k_L = 0.6307$ W/mK; $\rho_L = 992.2$ kg/m^3; $\rho_V = 0.05124$ kg/m^3; $\mu_L = 6.530 \times 10^{-4}$ kg/ms; and $C_{pL} = 4.180$ kJ/kgK.

From Eq. (8.19), we have:

$$h_{LV}^{+} = 2382 + 0.68 \times 4.180 \times (50 - 30) = 2438.8\ \text{kJ/kg} = 2.4388 \times 10^6\ \text{J/kg}$$

Finally, from Eq. (8.23), we get:

$$\begin{aligned}\bar{h}_D &= 0.725 \times \left[\frac{9.81 \times 992.2 \times (992.2 - 0.05124) \times 2.4388 \times 10^6 \times 0.6307^3}{6.53 \times 10^{-4} \times (50 - 30) \times 0.05}\right]^{\frac{1}{4}} \\ &= 7071.0\ \text{W/m}^2\text{K}\end{aligned}$$

8.2.3 *Drop Condensation*

Condensation in the form of droplets presents a heat transfer coefficient much higher than that of the descending film geometry previously studied, being able to reach one or more orders of magnitude. Figure 8.6 illustrates the case of condensation on a cold surface. There is a lot of empiricism in establishing the heat transfer process. Rose (2002) reviewed the theory and results of various experiments resulting in an empirical relationship for heat flux valid for condensation of saturated steam with a pressure between 1 kPa and up to approximately 1 atm on small vertical surfaces.

$$q = T_w^{0.8}\left[5\left(T_{\text{vapor}} - T_w\right) + 0.3\left(T_{\text{vapor}} - T_w\right)^2\right]\ \left[\frac{\text{kW}}{m^2}\right] \tag{8.25}$$

where T_w is in °C.

As it occurs with film condensation, droplet condensation is affected by non-condensable gas presence, such as air in steam. In this case, the respective heat transfer coefficients are reduced.

Fig. 8.6 Steam condensation in droplets on vertical surfaces (glass)

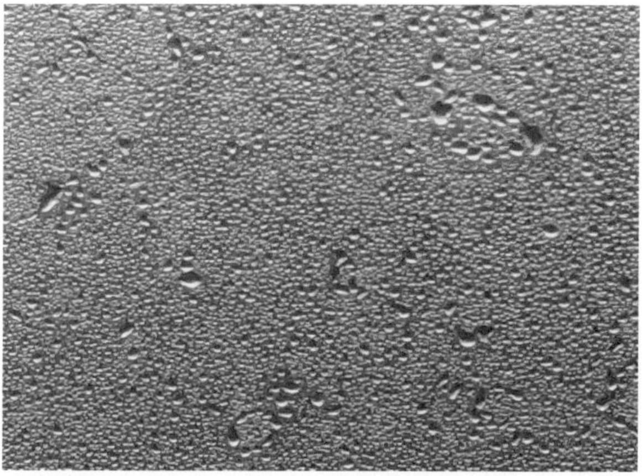

Surface treatment to decrease the surface's wettability, such as a chemical or mechanical treatment, or even surface coating, can increase the heat transfer coefficient. In special cases, the surface can be treated or coated with hydrophobic materials, such as PTFE.

According to some authors, the heat transfer coefficient in droplets can be 10–20 times higher than the liquid film coefficient on copper surfaces in steam condensation (Rohsenow et al., 1998). Other authors report that even higher values can be achieved.

In designing industrial condensers, it is recommended to use the expressions valid for downward film in thermal design, as they present more conservative heat transfer values. Consequently, higher heat transfer areas will be obtained, ensuring the proper operation of the equipment.

8.3 Boiling

The phenomenon opposite to condensation is boiling. Boiling will occur when a liquid medium comes into contact with a sufficiently heated surface. If the surface temperature, T_w, exceeds the saturation temperature of the liquid, T_{sat}, corresponding to the system pressure, the liquid will boil at the surface. In the case of convection without phase change, the parameters controlling the heat transfer phenomenon are well-known based on the dimensionless *Nu, Re,* and *Pr* for forced convection, and *Nu, Gr,* and *Pr* for natural convection. However, in the case of boiling, more parameters are involved in the phenomenon that must be considered. The specific enthalpy of vaporization, h_{LV}, plays an important role in the phase change phenomenon along with other thermodynamic parameters of the substance (saturation properties) and the surface finishing. That dependence has also been pointed out in previous sections in the analysis of the condensation phenomenon. The introduction of these new parameters makes the problem a more complex one. Therefore, the approach is usually empirical, with curve fitting of experimental parameters.

Boiling can be classified into two main classes: *pool boiling* and *convective boiling*. In the first case, studied next, the liquid forms a continuous resting medium surrounding a heated surface. Clearly, the effects of gravitational buoyancy due to the differences in density of the liquid and vapor phases play an important role in conjunction with other thermodynamic and transport properties. In the case of convective boiling, the movement of the liquid over the surface transports the vapor phase formed on the heated surface in addition to the difference in gravitational buoyancy of the phases.

8.3.1 Pool Boiling

Pool boiling is the phenomenon that occurs on a surface surrounded by a quiescent liquid medium. The heated surface could be the one of a tube of a flooded evaporator, for example. It could also be the bottom of a pot full of water heating on the stove. A relevant quantity for analyzing and characterizing boiling is the *excess temperature*, ΔT_e, defined by the difference in temperature of the wall (or surface) and the temperature of the saturated fluid, that is,

$$\Delta T_e = T_w - T_{\text{sat}} \tag{8.26}$$

Laboratory experiments, such as the pioneering one by Nukiyama (1934), indicate that the liquid heating curve, including the phase change region, behaves as illustrated in Fig. 8.7 for water and is characterized by several phenomena related to the excess temperature, ΔT_e. For low values of excess temperature, only natural convection takes place. In the case of water, this value is around 5 °C, and Eqs. (7.17) can be applied for the heat transfer calculation by natural convection. Starting at that point, also called "the onset of nucleated boiling" (ONB), individual bubbles begin to be produced on the heated surface, hence the name *nucleated boiling* for this regime. Following the increase in heat flux, the bubble population increases, as indicated by the photograph in Fig. 8.8, a situation that induces the bubbles to cluster forming *bubble columns*, which coalesce and form larger ascending bubbles until they reach the liquid surface. The heating continues until the *critical heat flux* (CHF) is reached, with the value of $q_{\text{CHF}} \approx 10^3$ kW/m^2 valid for water corresponding to $\Delta T_e \approx 30 - 35\,°\text{C}$. At the moment when the CHF is reached, the bubble population is so intense that a vapor film forms above and adjacent to the surface (or around the surface in the case of tubes and wires), so that a temperature jump ($A \rightarrow B$) occurs in the excess temperature (as indicated by the arrows) due to the rapid deterioration in heat transfer through the film and then continues heating in the regime of *vapor film boiling*. From this point, heat exchange by radiation through the vapor film also becomes relevant. The jump in excess temperature can be in the order of 1000 °C (for water), which will certainly lead to overheating and consequently structural damage to the surface, such as the pipe rupture, which is called DNB (*departure from nucleate boiling*), while others call it a *boiling crisis*, *dryout,* or even *burnout*. Whatever the term used, a local surface *fusion* or *burning* can occur due to the sudden rise in local temperature, above the material melting point. Therefore, knowing the CHF is a very relevant subject in the thermal design of boilers and other boiling equipment for water and other liquids and refrigerants, as well. Recently, the use of boiling techniques for heat dissipation in electronic *chips* has also increased interest, such as the still picture in Fig. 8.8.

The phenomenon of heating beyond the critical heat flux and subsequent cooling has a hysteresis behavior, that is, cooling from film boiling follows another cooling path, as indicated by the opposite arrows in Fig. 8.7. Film cooling continues until the point of *minimum heat flux*, $q_{\min}$, is reached, also called *Leidenfrost point*, when there is a jump in the excess temperature ($C \to D$) and cooling continues in the opposite direction to heating.

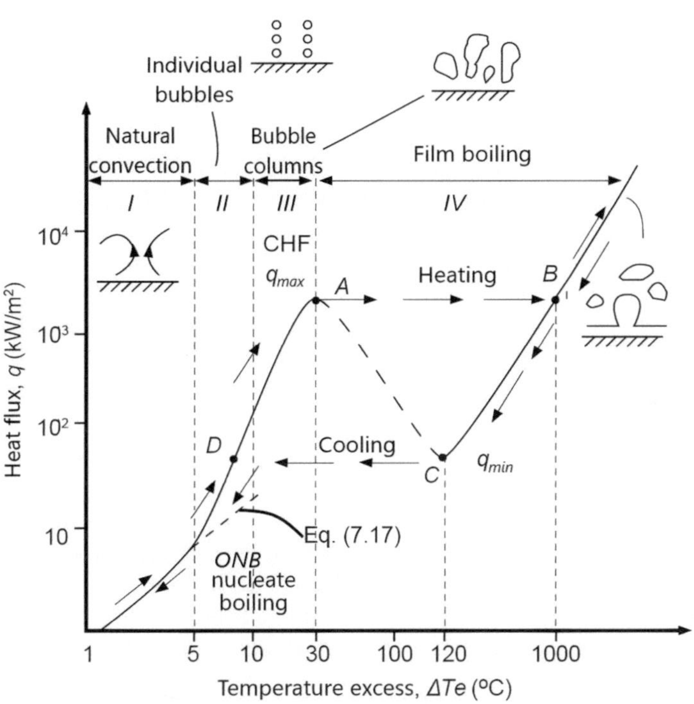

Fig. 8.7 Pool boiling curve (water)

Fig. 8.8 Pool boiling in a bubble column regime of the HFE-7100 fluid (T_{sat} = 60.3 °C at P_{sat} = 98 kPa) on a flat copper surface, for a 250 kW/m^2 heat flux, resulting in an excess temperature, $\Delta T_e = 31.7$ °C, whose study was conducted at the Phase Change Heat Transfer Laboratory-BOILING- at UFSC coordinated by Prof. Julio C. Passos). Courtesy of Prof. Elaine M. Cardoso, UNESP

Practical correlations for engineering calculations

I—Natural convection

The equations valid for natural convection from Chap. 7 can be used for the natural convection regime. If it is a heated horizontal plate, Eqs. (7.17) are valid. For other geometries or conditions, refer to Table 7.1.

II and III—Nucleate boiling—individual and columnar bubbles

Rohsenow (1952) proposed a correlation based on experimental observations, whose final form, modified by Vachon et al. (1968), is given by:

$$\frac{C_{pL}(T_w - T_{sat})}{h_{LV}} = C_{sf}\left[\frac{q}{\mu_L h_{LV}}\sqrt{\frac{\sigma}{g(\rho_L - \rho_V)}}\right]^{0.33} Pr_L^n \tag{8.27}$$

All thermodynamic and transport properties have already been defined and should be obtained at the saturation condition. The liquid Prandtl number, Pr_L, exponent n, is $n = 1$ for water and 1.7 for other fluids, but it can vary depending on the surface conditions. Table 8.1 summarizes the values of the coefficient C_{sf} for various fluid-surface pairs. The surface tension of the surface liquid interface, σ, is a function of surface roughness and can be adjusted for water in the range of $0 \leq T \leq 250\,°\text{C}$ using the following polynomial adjustment equation:

$$\sigma = \left(75.755 - 0.1491T - 0.0002T^2\right) \times 10^{-3} \tag{8.28}$$

with s in N/m^2 and T in °C.

Saiz Jabardo et al. (2004) present a concise table with values obtained by various researchers for different conditions and geometries for the coefficients C_{sf} and exponents of the Prandtl number, n. They also present their experimental results and

Table 8.1 Values of C_{sf} and n for Eq. (8.27)

Fluid-surface pair	C_{sf}
Water—brass	0.0060
Water—scratched copper	0.0068
Water—copper	0.0130
Water—stainless steel PTFE coated	0.0058
Water—ground stainless steel	0.0080
Water—nickel	0.0060
Water—platinum	0.0130
Isopropyl alcohol—copper	0.0022
n-butyl alcohol—copper	0.0030
Benzene—chrome	0.0100
CCl_4—copper	0.0130
n-pentane—polished copper	0.0049
R12—copper	0.0016
R113—copper	0.0022

adjusted values of these coefficients for refrigerant fluids on cylindrical surfaces considering various values of surface roughness. It should be noted that Eq. (8.27) is valid for clean surfaces.

The prediction of the critical heat flux (CHF) is based on one of the oldest models, which is that of Kutateladze (1950), valid for water:

$$q_{\mathrm{CHF}} = kh_{LV}\left[\sigma g \rho_V^2 (\rho_L - \rho_V)\right]^{\frac{1}{4}} \tag{8.29}$$

whose properties have already been defined. Originally, he validated the expression for small disks and heated wires in saturated water for various pressures. The value $k = \pi/24$ was proposed by Zuber (1959) for flat surfaces heated on the upper face in saturated water at atmospheric pressure. The value $k = 0.149$ has also been proposed and is more widely accepted. Many other refined models are available in the literature. Generally, the corrected critical heat flux $q_{\mathrm{CHF-corr}}$ for various geometries and conditions takes the following form from Eq. (8.29), where the indicated function correlates the ratio between the characteristic length of the geometry, L, and the characteristic length of the bubble, L_b:

$$\frac{q_{\mathrm{CHF-corr}}}{q_{\mathrm{CHF}}} = f\left(\frac{L}{L_b}\right) \tag{8.30}$$

L_b is defined by:

$$L_b = \sqrt{\frac{\sigma}{g(\rho_L - \rho_V)}} \tag{8.31}$$

Equation (8.29) is used in the calculation of q_{CHF} with the constant $k = 0.149$. In the case of an infinite heated flat surface, we have $\frac{q_{\mathrm{CHF-corr}}}{q_{\mathrm{CHF}}} = 1.14$. For a large horizontal cylinder, the ratio is $\frac{q_{\mathrm{CHF-corr}}}{q_{\mathrm{CHF}}} = 0.90$, for a large sphere $\frac{q_{\mathrm{CHF-corr}}}{q_{\mathrm{CHF}}} = 0.84$, and, for a large body of unspecified geometry, one can adopt $\frac{q_{\mathrm{CHF-corr}}}{q_{\mathrm{CHF}}} = 0.90$. Therefore, the critical heat flux calculated by Eq. (8.29) overestimates the value of CHF for large cylinders and spheres. In the case of small cylinders (electrical wires), one adopts $\frac{q_{\mathrm{CHF-corr}}}{q_{\mathrm{CHF}}} = 0.94 \times \left(\frac{R}{L_b}\right)^{-1/4}$, for $0.15 < R/L_b < 1.2$ (Lienhard and Dhir 1973) and $\frac{q_{\mathrm{CHF-corr}}}{q_{\mathrm{CHF}}} = 0.90$ if $R/L_b > 1.2$. Values for other geometries can be found in Kreith and Bohn (2001).

Groeneveld et al. (2007) present the CHF for water in pool boiling (zero mass flow) and in convective boiling (subject of the next section) for 24 pressure levels and 20 values of mass flow extracted from a bank of 30,000 experimental points for 8 mm diameter tubes covering the nuclear area as the main area of interest. Values of up to 1.3 MW/m^2 have been reported in the literature for water.

IV—Film boiling

Film boiling is generally avoided, as it results from the combination of high surface temperature with a low heat transfer coefficient. From a phenomenological point of view, the heated surface is enveloped by a vapor film that significantly reduces the heat transfer, with conduction through the film being the dominant effect, except for high surface temperatures, when thermal radiation also becomes relevant. The heat transfer through the film reaches the vapor–liquid interface to form bubbles. Bromley (1950) proposed the following equation to determine the average convective heat coefficient:

$$\bar{h}_c = C\left[\frac{g\rho_V(\rho_L - \rho_V)h_{LV}^{\#}k_V^3}{\mu_V(T_w - T_{\mathrm{sat}})D}\right]^{\frac{1}{4}} \tag{8.32}$$

where the properties of the vapor are calculated at the condition of film temperature $T_f = \frac{(T_{\mathrm{sat}}+T_w)}{2}$. The constant C is 0.62 for cylinders and 0.67 for spheres. The liquid density, ρ_L, and the modified specific enthalpy of vaporization, $h_{LV}^{\#}$, are calculated at the saturation temperature. The modified specific enthalpy is given by:

$$h_{LV}^{\#} = h_{LV} + 0.68C_{pV}(T_w - T_{\mathrm{sat}}) \tag{8.33}$$

Note the similarity of the approach with the condensation studied in the previous sections.

For large-diameter horizontal tubes, Westwater and Breen (1962) recommend the following expression:

$$\overline{h}_c = \left(0.59 + 0.69\frac{\lambda_c}{D}\right)\left[\frac{g\rho_V(\rho_L - \rho_V)h_{LV}^{\#}k_V^3}{\lambda_c\mu_V(T_w - T_{\text{sat}})}\right]^{\frac{1}{4}} \tag{8.34}$$

where

$$\lambda_c = 2\pi\left[\frac{\sigma}{g(\rho_L - \rho_V)}\right]^{\frac{1}{2}} \tag{8.35}$$

In the case of a heated flat surface, Berenson (1960) proposed:

$$\overline{h}_c = 0.425\left[\frac{g\rho_V(\rho_L - \rho_V)h_{LV}^{\#}k_V^3}{\lambda_c\mu_V(T_w - T_{\text{sat}})}\right]^{\frac{1}{4}} \tag{8.36}$$

If a significant difference between the surface and the liquid temperature occurs, the effect of thermal radiation from the surface to the liquid–vapor interface should also be considered. Bromley (1950) suggested the following expression for the combined heat transfer coefficient, $\overline{h}_{comb}$, considering the convective and radiation coefficients. The expression is valid if $\overline{h}_r < \overline{h}_c$:

$$\overline{h}_{\text{comb}} = \overline{h}_c + 0.75\overline{h}_r \tag{8.37}$$

where $\overline{h}_r$ is the linearized coefficient of thermal radiation, defined by:

$$\overline{h}_r = \sigma\varepsilon\left(\frac{T_w^4 - T_{\text{sat}}^4}{T_w - T_{\text{sat}}}\right) \tag{8.38}$$

In the context of this equation, temperatures should be in kelvin, σ is the Stefan-Boltzmann constant, and ε is the emissivity of the surface, quantities defined in Chap. 10.

Minimum heat flux, q_{min}

As the film boiling regime is reached and a stable vapor film is formed around the surface, the cooling process (decrease in heat flux) will follow another curve, as illustrated in Fig. 8.7, characterizing a hysteresis effect. Equation (8.39) shows the value of the minimum heat flux obtained by Zuber (1958) and later experimentally confirmed by Berenson (1961):

$$q_{\text{min}} = 0.09\rho_v h_{LV}\left[\frac{\sigma g(\rho_L - \rho_V)}{(\rho_L + \rho_V)^2}\right]^{\frac{1}{4}} \tag{8.39}$$

Example 8.2 (*Boiling*)

In a PWR-type nuclear reactor (*pressurized water reactor*), the heat flux on the surface of a nuclear fuel rod is 0.5×10^6 W/m^2. The rod is a 1 cm diameter and

1.8 m long which is cooled by water at a pressure of 15 MPa. In the LOCA (*loss of coolant accident*) event, the pressure of the reactor's thermo-hydraulic system can drop rapidly until it reaches atmospheric pressure. It is asked:

a. the rod surface heat transfer rate;
b. the water vaporization rate at operating pressure and atmospheric pressure, after the accident;
c. the ratio R between the vaporization rates before and after the accident.

Solution

a. The heat transfer rate from the rod surface is given by:

$$\dot{Q} = qA = q\pi DL = 0.5 \times 10^6 \times \pi \times 0.01 \times 1.8 = 28.26|\text{kW}.$$

b. The water evaporation rate at 15 MPa pressure is given by:

$$\dot{m}_{evap} = \frac{\dot{Q}}{h_{LV}}$$

For water at a pressure of 15 MPa (Table A.7), whose specific enthalpy of vaporization is $h_{LV} \sim 1 \times 10^6$ J/kg, then we have: $\dot{m}_{evap} = \frac{28.26 \times 10^3}{1 \times 10^6} = 0.02826$ kg/s. After the accident occurs, the water vaporization rate at atmospheric pressure should be calculated with the new specific enthalpy of evaporation (Table A.7), that is, $h_{LV} = 2.257 \times 10^6$ J/kg ($P_{\text{sat}} = 101.325$ kPa). Thus, the new rate of water vaporization is:

$$\dot{m}_{evap} = \frac{28.26 \times 10^3}{2.257 \times 10^6} = 0.01252 \text{ kg/s}$$

c. The ratio R between the vaporization rates before and after the accident is:

$$R = \frac{0.02826}{0.01252} = 2.25$$

Example 8.3 (*Excess temperature*)

Determine the wall temperature, T_w, and the excess temperature, ΔT_e, for a nickel-chrome (nichrome) wire immersed in a water bath at atmospheric pressure. The wire is 1.5 m long and 1 mm in diameter and the total dissipated electric power is 4 kW.

Solution

From Table A.7, for $p_{sat} = 101.325$ kPa, we have: $T_{sat} = 100\,°\text{C}$. Properties evaluated at saturation temperature.

$C_{pL} = 4.216$ kJ/kgK; $\mu_L = 2.817 \times 10^{-4}$ kg/ms; $\rho_L = 958.4$ kg/m^3; $\rho_V = 0.5981$ kg/m^3; $h_{LV} = 2256$ kJ/kg; $Pr_L = 1.74$.

From Eq. (8.28), we have:

$$\sigma = \left(75.755 - 0.1491 \times 100 - 0.0002 \times 100^2\right) \times 10^{-3} = 0.0588\ \mathrm{N/m}$$

Exposed wire surface area: $A = \pi \times 0.001 \times 1.5 = 4.712 \times 10^{-3}\mathrm{m}^2$
Heat flux: $q = \frac{P_{el}}{A} = \frac{4}{4.712 \times 10^{-3}} = 848.8\ \mathrm{kW/m^2}$.
From Table 8.1, $C_{sf} = 0.006$ (assuming only chrome)
Finally, from Eq. (8.27) for $n = 1$ (water), we get:

$$\Delta T_e = T_w - T_{\mathrm{sat}} = C_{sf}\frac{h_{LV}}{C_{pL}}\left[\frac{q}{\mu_L h_{LV}}\sqrt{\frac{\sigma}{g(\rho_L - \rho_V)}}\right]^{0.33} Pr_L^n \Rightarrow$$

$$\Delta T_e = 0.006\frac{2256}{4.216}\left[\frac{848.8}{2.817 \times 10^{-4} \times 2256}\sqrt{\frac{0.0588}{9.81 \times (958.4 - 0.5981)}}\right]^{0.33} \times 1.74$$

$$= 8.3^\circ\mathrm{C}$$

The surface temperature is $T_w = 100 + 8.3 = 108.3$ °C.

Example 8.4 (*Critical heat flux*)

Determine the critical heat flux for Example 8.3 and the maximum electrical power that can be applied without causing damage to the wire.

Solution

Data already obtained or calculated in the previous example.
$h_{LV} = 2256$ kJ/kg; $\rho_L = 958.4$ kg/m^3; $\rho_V = 0.5981$ kg/m^3; $\sigma = 0.0588$ N/m.
Critical heat flux (Eq. 8.29) (using $k = 0.149$):

$$\begin{aligned} q_{\mathrm{CHF}} &= 0.149 \times 2256 \times \left[0.0588 \times 9.81 \times 0.5981^2 \times (958.4 - 0.5981)\right]^{\frac{1}{4}} \\ &= 1260.4\ \mathrm{kW/m^2} \end{aligned}$$

It is recommended to correct this value for small cylinders according to:

$$\frac{q_{CHF-corr}}{q_{CHF}} = 0.94 \times \left(\frac{R}{L_b}\right)^{-1/4}$$

First, the calculation of the characteristic length of the bubble, L_b (Eq. 8.31):

$$L_b = \sqrt{\frac{0.0588}{9.81 \times (958.4 - 0.5981)}} = 0.0025\ \mathrm{m}$$

Therefore, the corrected critical heat flux is:

$$q_{\text{CHF-corr}} = q_{\text{CHF}} \times 0.94 \times \left(\frac{R}{L_b}\right)^{-1/4} = 1260.4 \times 0.94 \times (0.5/2.5)^{-0.25}$$
$$= 1771.6\ \text{kW/m}^2$$

Maximum electric power:

$$P_{el-\max} = q_{\text{CHF-corr}} \times A = 1771.6 \times 4.712 \times 10^{-3} = 8.35\ \text{kW}.$$

8.3.2 Convective Boiling

The case of *convective boiling* differs from pool boiling, as the liquid is forced to flow around the heated surface by some external agent. In the case of pool boiling, gravitational buoyancy due to the differences in density of the liquid and vapor phases plays a central role. Convective boiling will occur when a liquid fluid flows over a surface, whose temperature is higher than the liquid saturation temperature by a certain amount (excess temperature). Furthermore, the flow can occur on the exterior of the surface or inside tubes, ducts, and channels.

Convective nucleate boiling

The case of convective nucleate boiling is similar to the case of pool nucleate boiling. Classic studies identified that the boiling starting point will occur at the intersection of forced convection with fully developed boiling, as illustrated in Fig. 8.9. Several researchers analyzed the problem of convective nucleate boiling, with the expression proposed by Bergles and Rohsenow (1964), which considers the contribution of forced convection with pool boiling, being generally well-accepted, given by:

$$q = q_{\text{conv}}\sqrt{1 + \left[\frac{q_B}{q_{\text{conv}}}\left(1 - \frac{q_{Bi}}{q_B}\right)\right]^2} \tag{8.40}$$

where: q is the heat flux in convective nucleate boiling; q_{conv}, the heat flux due to forced convection calculated according to the geometry by the corresponding expression given in Chap. 5 or 6; q_B, the heat flux calculated in pool boiling, given by Eq. (8.27)—note that here the index "B" was added to the heat flux for clarity; q_{Bi}, the heat flux calculated by the intersection of the pool boiling curve at the point where boiling begins at $\Delta T_e = T_w - T_{\text{sat}}$.

Developed external convective nucleate boiling

In the case of fully developed external boiling, the heat flux is dominated by boiling making the forced convection less significant. An old formulation proposed by

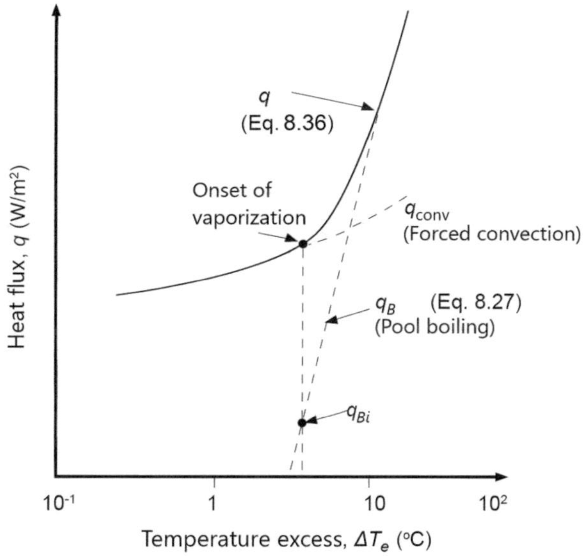

Fig. 8.9 Convective nucleate boiling

Rohsenow (1952) is generally still accepted by many researchers. It involves overlaying the effects of pool boiling with forced convection to obtain the following expression for fully developed convective heat transfer:

$$q = q_{\text{conv}} + q_{\text{ebul}} \tag{8.41}$$

given the terms $q_{\text{conv}} = \overline{h}(T_w - T_b)$, where T_b is the average mixture (or cup) temperature as defined in subsection 6.1.2 and the average heat transfer coefficient $\overline{h}$ is calculated according to the modified Dittus-Boelter correlation (Eq. 6.43) using the properties of the liquid phase. The term q_{ebul} should be calculated by Eq. (8.27).

Maximum heat flux

The maximum heat flux, q_{max}, is greatly increased compared to the CHF pool boiling. Laboratory measurements indicate that the value can reach up to 35 MW/m^2, much higher than 1.3 MW/m^2 for pool boiling water. Below, are two correlations developed by Lienhard and Eichhorn (1976) at low and high velocity for cross-flow over cylinders as a function of the *Weber number*, We_D, based on the cylinder diameter, defined by:

$$We_D = \frac{\rho_V V^2 D}{\sigma} \tag{8.42}$$

The Weber number represents the ratio between the inertia and the surface tension forces. The properties that appear in the number of We_D and the following equations should be evaluated in the saturation state.

Low velocity:

$$\frac{\pi q_{\max}}{\rho_V h_{LV} V} = 1 + \left(\frac{4}{We_D}\right)^{\frac{1}{3}} \tag{8.43}$$

High velocity:

$$\frac{\pi q_{\max}}{\rho_V h_{LV} V} = \frac{\left(\frac{\rho_L}{\rho_V}\right)^{0.75}}{169} + \frac{\left(\frac{\rho_L}{\rho_V}\right)^{0.5}}{19.2 We_D^{\frac{1}{3}}} \tag{8.44}$$

Low velocity is considered if $\frac{\pi q_{\max}}{\rho_V h_{LV} V} > \left[1 + 0.275\left(\frac{\rho_L}{\rho_V}\right)^{0.5}\right]$. Otherwise, it is high velocity.

Example 8.5 (*Convective nucleate boiling*)

Consider the case of the wire from Example 8.3 for water (100 °C) flowing perpendicular to the wire at a velocity of 3 m/s. What is the total heat transfer rate?

Solution

Using Eq. (8.41), it is necessary to calculate the convective heat transfer. For this, the properties at the film temperature $T_f = (T_w + T_\infty)/2 = (108.3 + 100)/2 \sim 104\,°\text{C}$.

Interpolated properties from Table A.7.

$$C_{pL} = 4.22 \text{ kJ/kgK};\ \mu_L = 2.709 \times 10^{-4} \text{ kg/m} \cdot \text{s};$$
$$\rho_L = 955.4 \text{ kg/m}^3;\ Pr_L = 1.67;\ k = 0.6843 \text{ W/mK}$$

Calculation of Re_D:

$$Re_D = \frac{955.4 \times 3 \times 0.001}{2.709 \times 10^{-4}} = 10,580$$

From Eq. (5.78), we have:

$$\overline{Nu_D} = \frac{\overline{h}D}{k} = C\text{Re}_D^m \text{Pr}^{1/3}$$

where $C = 0.193$ and $m = 0.618$ (Table 5.3). Then:

$$\overline{Nu_D} = \frac{\overline{h}D}{k} = 0.193 \times 10,580^{0.618} \times 1.67^{\frac{1}{3}} = 70.3 \Rightarrow$$

$$\overline{h} = \frac{70.3 \times 0.6843}{0.001} = 48,106$$

Thus,

$$q_{conv} = \overline{h}(T_w - T_\infty) = 48,106 \times (108.3 - 100) = 399.28\,\text{W/m}^2$$

Finally, from Eq. (8.41), we get:

$$q = 399.28 + 848.8 = 1248,1\ \text{kW/m}^2.$$

Flow boiling inside tubes and pipes is very relevant, as it is the form that occurs in tube boilers and steam generators of the water-tube type, among other vaporization devices, including evaporator units. In upward flow inside a vertical tube, the flow assumes distinct flow patterns depending, mainly, on the vapor quality x. The flow patterns are indicated in Fig. 8.10a and are commonly known as two-phase flow regimes (for example, see Hewitt 1998). Figure 8.10b shows the qualitative behavior of the heat transfer coefficient as a function of the flow regimes. Thus, the vertical flow patterns are characterized and named as:

- Bubble regime—in this regime bubbles form on the heated wall. Initially, they are dispersed bubbles, whose population continuously increases, forcing the bubbles to cluster.

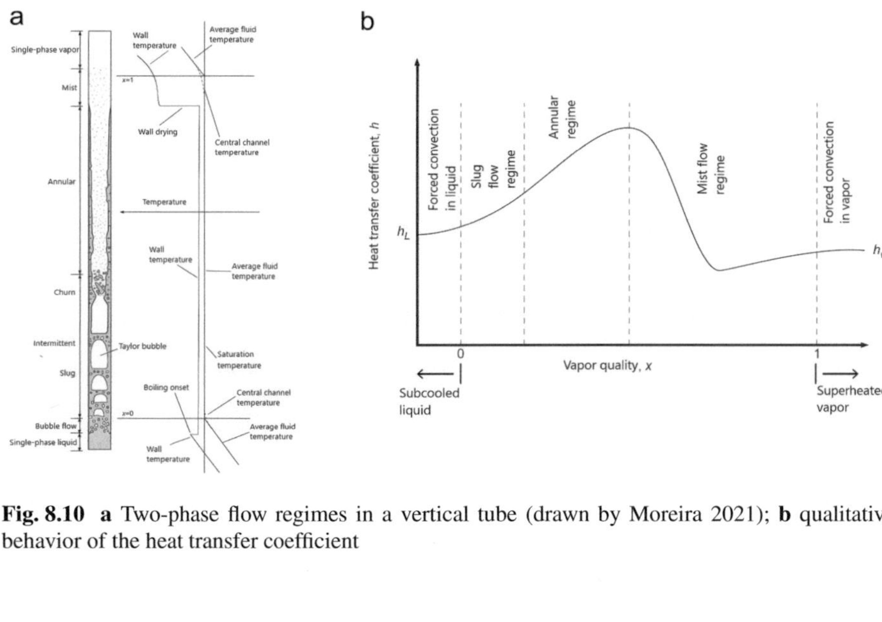

Fig. 8.10 **a** Two-phase flow regimes in a vertical tube (drawn by Moreira 2021); **b** qualitative behavior of the heat transfer coefficient

- Slug regime—the structures of clustered bubbles grow and coalesce generating larger bubbles, called "Taylor bubbles" that are surrounded by a film of liquid next to the wall. Taylor bubbles are alternated by *slugs* of liquid (hence the name of the regime).
- Churn regime—the Taylor bubbles coalesce and the flow becomes somewhat chaotic avoiding the identification of characteristic structures. Hence the name churn regime.
- Annular and mist regime—continuing the heat flux at the wall, more liquid vaporizes so that the vapor begins to flow through the central region of the tube, while the liquid phase flows along the heated wall of the tube. This regime initially has dispersed droplets in its midst forming a structure of the *mist* type, that is, droplets flowing with the vapor phase. Later, the wall *dries*, then, the flow transforms only into vapor ($x = 1$). Continuing the heating, the fluid becomes superheated single-phase vapor.

Kandlikar (1990, 1991, 1999) proposed that the ratio between the average heat transfer coefficient $\overline{h}_c$ two-phase and the monophasic heat transfer coefficient for only liquid $\overline{h}_{c-Lo}$ is given according to the following test:

$$\overline{h}_c = \text{highest value of} \begin{cases} \overline{h}_c \text{ calculated by Eq.(8.45)} \\ \overline{h}_c \text{ calculated by Eq.(8.46)} \end{cases}$$

with:

$$\frac{\overline{h}_c}{\overline{h}_{c-Lo}} = \frac{0.6683}{Co^{0.2}}(1-x)^{0.8} f(Fr_{L0}) + 1058.0 Bo^{0.7}(1-x)^{0.8} F_{FL} \tag{8.45}$$

$$\frac{\overline{h}_c}{\overline{h}_{c-Lo}} = \frac{1.136}{Co^{0.9}}(1-x)^{0.8} f(Fr_{L0}) + 667.2 Bo^{0.7}(1-x)^{0.8} F_{FL} \tag{8.46}$$

where:

$$\text{Convection number} : Co = \left(\frac{\rho_V}{\rho_L}\right)^{0.5}\left(\frac{1-x}{x}\right)^{0.8} \tag{8.47}$$

$$\text{Boiling number} : Bo = \frac{qA_T}{\dot{m}h_{LV}} \tag{8.48}$$

A_T is the internal lateral area of the tube.

Froude number (Fr) of the same total flow liquid phase:

$$Fr_{L0} = \frac{\dot{m}^2}{g\rho_L D} \tag{8.49}$$

The function $f(Fr_{L0})$ is 1 for vertical tubes and horizontal tubes with $Fr_{L0} > 0.4$. For $Fr_{L0} \leq 0.4$, it is given by the following correlation:

$$f(Fr_{L0}) = \begin{cases} (25Fr_{L0})^{0.3} \text{for } Fr_{L0} < 0.04 \text{(horizontal tubes)} \\ 1 \text{(vertical tubes and for) } Fr_{L0} \geq 0.04 \text{(horizontal tubes)} \end{cases} \tag{8.50}$$

Dittus-Bolter equation only for the liquid phase,

$$\overline{h}_{c-Lo} = 0.023 Re_D^{0.8} Pr_L^{0.4} \left(\frac{k_L}{D} \right) \tag{8.51}$$

Kandlikar (1999) recommends that the single-phase heat transfer coefficient for liquid only $\overline{h}_{c-Lo}$ be obtained from the Gnielinski (1976) correlation, given by Eq. (6.45), a form updated by Gnielinski himself (2013). Use the Reynolds number for the saturated liquid phase flowing with the total mass flow rate of the two-phase mixture.

The parameter F_{FL} represents the characteristics of the surface and the fluid in contact. Table 8.2 (Kandlikar 1999) indicates recommended values for various fluids inside copper tubes. For stainless steel tubes, a unit value for this quantity is suggested.

There are other alternative methods to Kandlikar's for obtaining the average two-phase heat transfer coefficient, such as the Liu and Winterton (1991) correlation, adopted by some researchers.

Table 8.2 Values of F_{FL} for Eqs. (8.45) and (8.46) in copper tubes (adapted from Kandlikar 1999)

Fluid	F_{FL}
Water	1.00
Butane	1.50
Kerosene	0.488
Propane	2.15
R-11	1.30
R-12	1.50
R-13B1	1.31
R-22	2.20
R-113	1.30
R-114	1.24
R-134a	1.63
R-152a	1.10
R-32(60%)—R-132(40%)	3.30

8.4 Proposed Problems

8.1 A vertical plate 1 m high by 1 m wide is kept at 50 °C inside a saturated steam environment at atmospheric pressure. Obtain the heat transfer rate and the rate of condensed liquid obtained on each face.

8.2 Recalculate the case of Problem 8.1 for a plate 2 m high by 1 m wide, keeping the other conditions the same.

8.3 Regarding the boiling curve below, it can be stated:

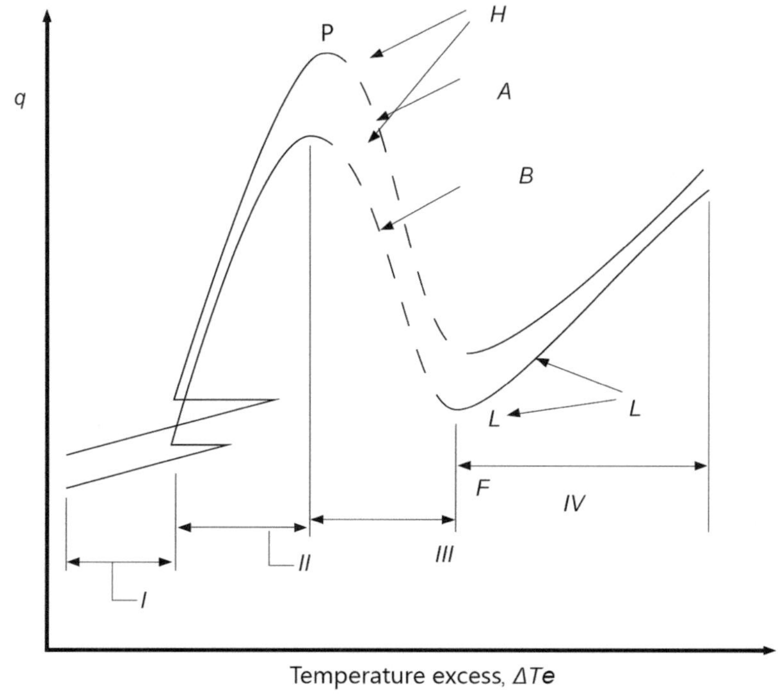

1. Curve "A" represents the pool boiling process and curve "B" the convective boiling process.
2. Curve "B" represents the pool boiling process and curve "A" the convective boiling process.
3. Segments I, II, III, and IV of the curve correspond, respectively, to: single-phase convection; nucleate boiling; transition from nucleate boiling to film boiling; and film boiling.
4. Curves I, II, III, and IV of the curve correspond, respectively, to: the transition from nucleate boiling to film boiling; film boiling; single-phase convection; and nucleate boiling.
5. Point "P" on the curves represents the boiling crisis (DNB), and point "L" is the maximum heat flux.

6. Point "P" on the curves represents the onset of nucleate boiling (ONB), and point "L' the average heat flux.
7. Point "P" on the curves represents the critical heat flux (CHF) and point "L" is the minimum heat flux.

8.4 Consider the case of a large diameter vertical tube (therefore, it is possible to use the vertical plate correlations) designed to operate with condensation of steam on its external side by circulating water inside it. However, in the installation, it was mounted horizontally. Assuming that all operating conditions are maintained, what is the heat transfer ratio between the two configurations as a function of its length and diameter? Consider laminar film condensation.

8.5 An electrical resistance is encapsulated with a stainless steel tube and used to heat water (atmospheric pressure) in a storage tank of a solar heating system. When heated, the temperature of the resistance's surface reaches 106 °C for a 3 kW electrical power. Using the modified Rohsenow model (Eq. 8.27), determine the heat flux in W/m^2, and the length of the resistance if its diameter is 8 mm.

8.6 Determine the Critical Heat Flux for Problem 8.5. (*Hint*: Use the Correction Factor.)

References

Berenson, P. 1960. Transition boiling heat transfer from a horizontal surface. AIChE paper n. 18, *ASME-AIChE Heat Transfer Conference*. NY: Buffalo.

Berenson, P. 1961. Film boiling heat transfer from a horizontal surface. *Journal of Heat Transfer* 83: 351–358.

Bergles, A.E., and W.M. Rohsenow. 1964. The determination of forced-convection surface-boiling heat transfer. *Journal of Heat Transfer* 86: 365–372.

Bromley, L.A. 1950. Heat transfer in stable film boiling. *Chemical Engineering Progress* 46: 221–227.

Butterworth, D. 1983. Film condensation of a pure vapour. In *Heat exchanger design handbook*, ed. by E.U. Schlünder. New York: Hemisphere.

Chun, M.H., and K.T.A. Kim. 1991 Natural convection heat transfer correlation for laminar and turbulent film condensation on a vertical surface. In *Proceedings of ASME/JSME thermal engineering conference*, vol. 2, pp. 459–464. Reno.

Dhir, V.K., and J.H. Lienhard. 1971. Laminar film condensation on plane and axisymmetric bodies in nonuniform gravity. *Journal of Heat Transfer* 93 (1): 97–100.

Gnielinski, V. 1976. New equations for heat and mass transfer in turbulent pipe and channel flow. *International Chemical Engineer* 16: 359–368.

Gnielinski, V. 2013. On heat transfer in tubes. *International Journal of Heat and Mass Transfer* 63: 134–140.

Groeneveld, D.C., J.Q. Shan, A.Z. Vasić, L.K.H. Leung, A. Durmayaz, J. Yang, S.C. Cheng, and A. Tanase. 2007. The 2006 CHF look-up table. *Nuclear Engineering and Design* 237: 1909–1922.

Hewitt, G. Boiling in W. Rohsenow, J. Hartnett, and Y. Cho. 1998. *Handbook of heat transfer*, 3rd ed. McGraw-Hill.

Kandlikar, S.G. 1990. A general correlation for two-phase flow boiling heat transfer coefficient inside horizontal and vertical tubes. *Journal of Heat Transfer* 112: 219–228.

Kandlikar, S.G. 1999. *A handbook of phase change: Boiling and condensation*. CRC Press: Taylor & Francis.

Kandlikar, S.G. 1991. Development of a flow boiling map for subcooled and saturated flow boiling of different fluids in circular tubes. *Transactions of ASME—Journal of Heat Transfer* 113 (1): 190–200.

Kreith, F., and M.S. Bohn. 2001. *Principles of heat transfer*, 6th ed. New York: Brooks/Cole.

Kutateladze, S.S. 1963. *Fundamentals of heat transfer*. New York: Academic.

Kutateladze, S.S. 1950. Hydromechanical model of the crisis of boiling under conditions of free convection. *Journal of Technical Physics, USSR* 20 (11): 1389–1392.

Labuntsov, D.A. 1957. Heat transfer in film condensation of pure steam on vertical surfaces and horizontal tubes. *Teploenergetika* 4 (7): 72–80.

Lienhard, J.H., and V.K. Dhir. 1973. Hydrodynamic predictions of peak pool-boiling heat fluxes from finite bodies. *ASME Journal of Heat Transfer* 95: 152–158.

Lienhard, J.H., and R. Eichhorn. 1976. Peak boiling heat flux on cylinders in a cross flow. *International Journal of Heat and Mass Transfer* 19 (10): 1135–1142.

Liu, Z., and R.H.S. Winterton. 1991. A general correlation for saturated and subcooled flow boiling in tubes and annuli, based on a nucleate pool boiling equation. *International Journal of Heat and Mass Transfer* 34 (11): 2759–2766.

Moreira, T.A. 2021. Doctoral student, EESC-USP. *Personal Communication.*

Nukiyama, S. 1934. The maximum and minimum values of the heat transmitted from metal to boiling water under atmospheric pressure. *Journal Japan Society Mechanical Engineering* 37: 367.

Nusselt, W. 1916. The Condensation of steam on cooled surfaces. *VDIZ* 60 (541–546): 569–575.

Popiel, C.O., and L. Boguslawski. 1975. Local heat-transfer coefficients on the rotating disk in still air. *International Journal of Heat and Mass Transfer* 18 (1).

Rohsenow, W.M. 1952. A method of correlating heat transfer data for surface boiling liquids. *Transaction of ASME* 74: 969–975.

Rohsenow, W.M. 1956. Heat transfer and temperature distribution in laminar-flow condensation. *Transaction of ASME* 78: 1645–1648.

Rohsenow, W.M., J. Hartnet, and Y. Cho. 1998. *Handbook of heat transfer*, 3rd ed. New York: McGraw-Hill.

Rose, J.W. 2002. Dropwise condensation theory and experiment: A review. *Proceedings of the Institution of Mechanical Engineers, Part a: Journal of Power and Energy* 216: 115.

Saiz Jabardo, J.M., E.F. Silva, G. Ribatski, and S.F. Barros. 2004. Evaluation of the Rohsenow correlation through experimental pool boiling of halocarbon refrigerants on cylindrical surfaces. *Journal of the Brazilian Society of Mechanical Sciences and Engineering* 26 (2): 218–230.

Simões-Moreira, J.R., and A. Hernandez Neto. 2019. *Fundaments and applications of psychrometry (Fundamentos e Aplicações da Psicrometria, in Portuguese)*, 2nd ed. São Paulo: Blucher.

Vachon, R. I., G.H. Nix, and G.E. Tanger. 1968. Evaluation of constants for the Rohsenow pool-boiling correlation. *Transaction of ASME, Serie C, Journal of Heat Transfer*. 90: 239–247.

Westwater, J.W., and B.P. Breen. 1962. Effect of diameter of horizontal tubes on film boiling heat transfer. *Chemical Engineering Progress* 58: 67–72.

Zuber, N. 1959. *Hydrodynamic aspects of boiling heat transfer* (Ph.D. Thesis). Los Angeles: University of California

Zuber, N. 1958. On the stability of boiling heat transfer. *Journal of Heat Transfer—Transactions of ASME* 80.

Chapter 9
Heat Exchangers

Abstract This chapter presents the theory and the main types of liquid-liquid, liquid-gas, and gas-gas heat exchangers. The modeling of the most basic double-tube heat exchanger is presented, and its working equations are based on the logarithmic mean temperature difference, *LMTD*, for parallel flow and counterflow heat exchanger configurations. In parallel flow, the two fluids circulate in the same direction, while in the counterflow configuration, they circulate in opposite directions. To solve the cases of more complex heat exchangers than the double tube, the *F* solution method is presented as a parameter that corrects the *LMTD* for more complex configurations. Graphs and equations for various configurations are presented. In the case of the inlet and outlet temperatures not being specific or known, the most common method used is the effectiveness-number of transfer units, ε—*NTU*. The theoretical basis of the method is developed in terms of the ratio between the thermal capacities of the two fluids. Graphs for various heat exchanger configurations are presented. The chapter concludes with an appropriate discussion on irreversibily in heat exchangers.

Keywords Heat exchanger · *LMTD* · *F* method · Effectiveness-*NTU* method · Incrustation · Irreversibility

9.1 Types of Heat Exchangers

One of the most immediate applications of the convective heat transfer theory is the thermal design of heat exchangers. Heat exchangers are pieces of equipment designed to transfer heat from one fluid to another. A familiar example of these devices is the automotive radiator, where the engine cooling fluid is pumped to circulate in existing channels inside the engine block to cool it. In this process, the cooling fluid is heated by removing heat from the engine block and directing it to the radiator, which is intended to transfer the thermal energy of the heated fluid to the surrounding air. The radiator is, therefore, a liquid-air heat exchanger type. The main objectives of the thermal design of heat exchangers are the determination of the heat exchange area and the configuration necessary to transfer the total heat rate, also called *thermal*

J. R. Simões-Moreira and E. W. Zavaleta-Aguilar, *Heat Transfer*, Heat and Mass Transfer, https://doi.org/10.1007/978-3-031-80318-5_9

load or *thermal capacity*, from a heated (or cooled) fluid to another fluid, at flow rates and temperatures established by operational conditions.

A wide variety of heat exchangers exist, whose specific configurations depend on the thermal load, area available for their installation, flow rates, temperatures, and the transport properties of the involved fluids. Figure 9.1 presents a set of pictures of various types of heat exchangers selected to exemplify the vastness of applications and configurations. Figure 9.1a–c show equipment suitable for transferring a liquid to the atmospheric air, generically classified as liquid-gas, notably, they are equipped with fins on the air side; Fig. 9.1d–g exemplifies industrial heat exchangers; finally, Fig. 9.1h shows an example taken from nature, where the large ears of the elephant irrigated by warm blood due the animal's metabolism provide a large heat transfer area to the air. In addition, by waving its ears, the stunning animal allows a relative movement to the surrounding air, which provides higher heat transfer coefficients.

In previous chapters, correlations and techniques for obtaining the heat transfer coefficient for various configurations and flow regimes were presented. In Sect. 2.7, the overall heat transfer coefficient, U, was defined. This quantity allows one to obtain the thermal load of the heat exchanger from Eq. (9.1), already defined in Chap. 2.

$$\dot{Q} = UA\overline{\Delta T} \tag{9.1}$$

where $\overline{\Delta T}$ is an *effective* average temperature difference between the two working fluids temperatures valid throughout the heat exchanger; U was already defined as the overall heat transfer coefficient (Sect. 2.7); and A, the total heat exchange area of the equipment.

Table 9.1 provides approximate values of U for some commonly used fluids in heat exchangers. The relatively wide ranges of U result from the properties of the working fluids and the flow conditions, as well as the constructive configuration of the heat exchanger.

9.2 Double Tube Heat Exchangers

One of the most elementary types of heat exchangers is the *double tube* or *tube-in-tube* heat exchanger type. A still picture of a small device of this type can be seen in Fig. 9.2. As seen in the image, this type of heat exchanger consists of a tube inserted inside another larger diameter tube. In the annular space, one of the heat exchange fluids flows, while inside the inner tube, the other fluid flows. Depending on the tube connections, the two fluids circulate in the same direction, which configures , or circulate in opposite directions, configuring the *parallel flow*.

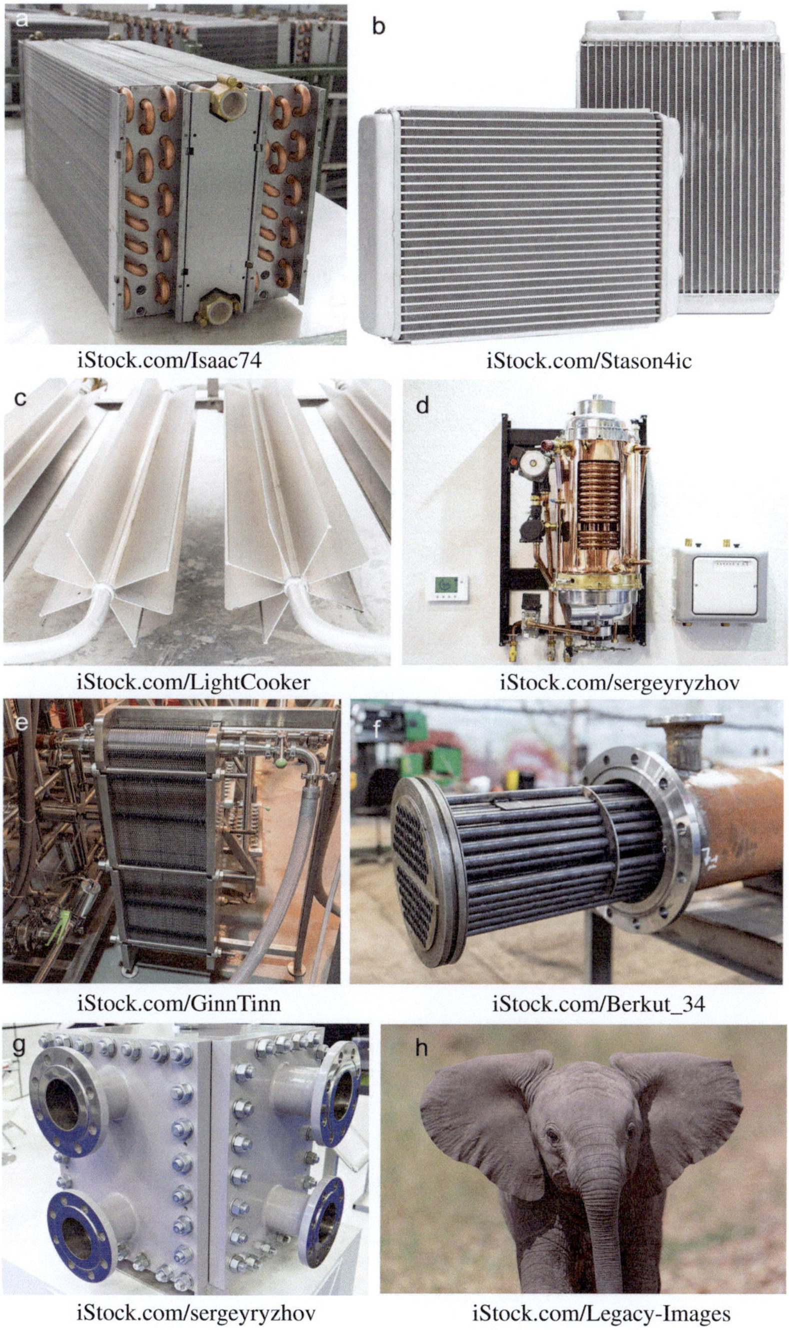

Fig. 9.1 Examples of types of heat exchangers: **a** compact heat exchanger; **b** automotive radiator; **c** vaporizer for cryogenic liquids or LPG; **d** helical heat exchanger; **e** plate heat exchanger; **f** shell and tube type heat exchanger; **g** oil and gas heat exchanger from the oil industry; **h** the elephant's ear acts as a heat exchanger to cool the blood of the beautiful animal

Table 9.1 Representative values of the overall heat transfer coefficient. Published with permission from *The Engineering Toolbox*

Type	Application	Overall heat transfer coeff.—U (W/m^{2}K)
Tubular, Heating and cooling	Gases at atmospheric pressure inside and outside the tubes	5–35
	Gases at high pressure inside and outside the tubes	150–500
	Liquid outside (inside) and gas, at atmospheric pressure, inside (outside) the tubes	15–70
	High pressure gas inside and liquid outside the tubes	200–400
	Liquids inside and outside tubes	150–1200
	Steam outside and liquid inside the tubes	300–1200
Tubular, Condensation	Steam outside and cooling water inside the tubes	1500–4000
	Organic or ammonia vapors outside and cooling water inside the tubes	300–1200
Tubular, Vaporization	Steam outside and highly viscous liquid inside the tube, natural convection	300–900
	Steam outside and low viscosity liquid inside the tube, natural convection	600–1700
	Steam outside and liquid inside the tube, forced convection	900–3000
Cross flow cooling to air	Water cooling	600–750
	Cooling of light hydrocarbon liquids	400–550
	Tar cooling	30–60
	Cooling of air or combustion gases	60–180
	Cooling of gaseous hydrocarbons	200–450
	Condensation of low pressure steam	700–850
	Condensation of organic vapors	350–500
Plates	Liquid—liquid	1000–4000
Spiral	Liquid—liquid	700–2500
	Condensing steam—liquid	900–3500

Fig. 9.2 Still picture of a double tube heat exchanger

9.2.1 Parallel Flow Double Tube Heat Exchangers

It is necessary to precisely define the effective temperature difference $\overline{\Delta T}$ from Eq. (9.1) to proceed with the heat transfer calculations in heat exchangers. To achieve that, consider the parallel flow double tube heat exchanger scheme, as shown in Fig. 9.3. The following terminology is used to designate the temperatures of the *hot fluid*, T_h, and the *cold fluid*, T_c. Therefore, at the entrance of the hot fluid, we have, T_{hi}, and at the exit, T_{ho}. At the entrance of the cold fluid, we have T_{ci}, and at the exit, T_{co}. The figure upper part shows the temperature distributions of the hot and cold fluids.

To solve the thermal problem of the heat exchanger, consider the following assumptions:

(a) The overall heat transfer coefficient, U, is constant throughout the whole heat exchanger;
(b) The heat exchanger is thermally insulated so that heat exchange occurs only between the two fluids;
(c) The temperature of each fluid in the cross section is the average mixing temperature;
(d) The transport properties of each fluid are constant and evaluated at the average temperatures of each fluid;
(e) The entire process occurs in a steady state.

Based on the above assumptions, the rate of elementary heat exchange between the hot and cold fluids in an infinitesimal heat exchange area, *dA*, illustrated in Fig. 9.3 is:

$$d\dot{Q} = U(T_h - T_c)dA \tag{9.2}$$

However, the rate of heat received by the cold fluid is, necessarily, equal in magnitude to that transferred by the hot fluid, which, by the law of conservation of energy, neglecting the variations of kinetic and potential energy in the fluids, results in:

$$d\dot{Q} = \dot{m}_c C_{pc} dT_c = -\dot{m}_h C_{ph} dT_h \tag{9.3}$$

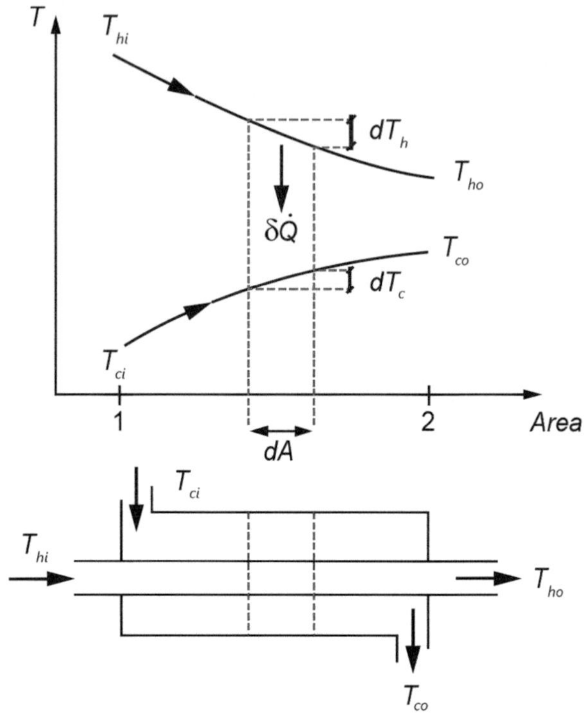

Fig. 9.3 Temperature distribution in a parallel flow double tube heat exchanger

where $\dot{m}$ is the mass flow rate and C_p is the specific heat at constant pressure. By rearranging Eq. (9.3), we can obtain:

$$d(T_h - T_c) = -\left(\frac{1}{\dot{m}_h C_{ph}} + \frac{1}{\dot{m}_c C_{pc}}\right) d\dot{Q} \tag{9.4}$$

Substituting $d\dot{Q}$ from Eqs. (9.2) into (9.4), we get:

$$\frac{d(T_h - T_c)}{(T_h - T_c)} = -U\left(\frac{1}{\dot{m}_h C_{ph}} + \frac{1}{\dot{m}_c C_{pc}}\right) dA \tag{9.5}$$

whose integration results in:

$$\ln\frac{\Delta T_2}{\Delta T_1} = -UA\left(\frac{1}{\dot{m}_h C_{ph}} + \frac{1}{\dot{m}_c C_{pc}}\right) \tag{9.6}$$

with $\Delta T_1 = T_{hi} - T_{ci}$ and $\Delta T_2 = T_{ho} - T_{co}$, as indicated in the temperature distribution graph in Fig. 9.3. Through a global energy balance in each fluid, we have:

$$\dot{m}_h C_{ph} = \frac{\dot{Q}}{(T_{hi} - T_{ho})} \tag{9.7}$$

and

$$\dot{m}_c C_{pc} = \frac{\dot{Q}}{(T_{co} - T_{ci})} \tag{9.8}$$

Substituting these two expressions into Eq. (9.6), we get:

$$\ln \frac{\Delta T_2}{\Delta T_1} = -UA \frac{(T_{hi} - T_{ho}) + (T_{co} - T_{ci})}{\dot{Q}} \tag{9.9}$$

or:

$$\dot{Q} = UA \frac{\Delta T_2 - \Delta T_1}{\ln\left(\frac{\Delta T_2}{\Delta T_1}\right)} \tag{9.10}$$

Comparing this result with Eq. (9.1), it is concluded that the "effective or representative average difference of temperatures", $\overline{\Delta T}$, sought is given by:

$$\overline{\Delta T} = \frac{\Delta T_2 - \Delta T_1}{\ln\left(\frac{\Delta T_2}{\Delta T_1}\right)} \equiv LMTD \tag{9.11}$$

LMTD is known as the *logarithmic mean temperature difference*, already defined in Sect. 6.3.

Thus, the final expression for the total heat transfer rate, or thermal load of the heat exchanger, is:

$$\dot{Q} = UA\,LMTD \tag{9.12}$$

9.2.2 *Counterflow Double Tube Heat Exchangers*

In the case of the counterflow double tube heat exchanger, the two fluids flow in opposite directions, as indicated in Fig. 9.4. The temperature distributions in the hot and cold fluids along the tubes' length are also illustrated. A particularity of this configuration is that the cold fluid outlet temperature (T_{co}) can be higher than the hot fluid outlet temperature (T_{ho}).

Based on the same assumptions as the previous case, the elementary heat exchange rate between the hot and cold fluids for the infinitesimal heat exchange area, dA, illustrated in Fig. 9.4, is:

$$d\dot{Q} = U(T_h - T_c)dA \tag{9.13}$$

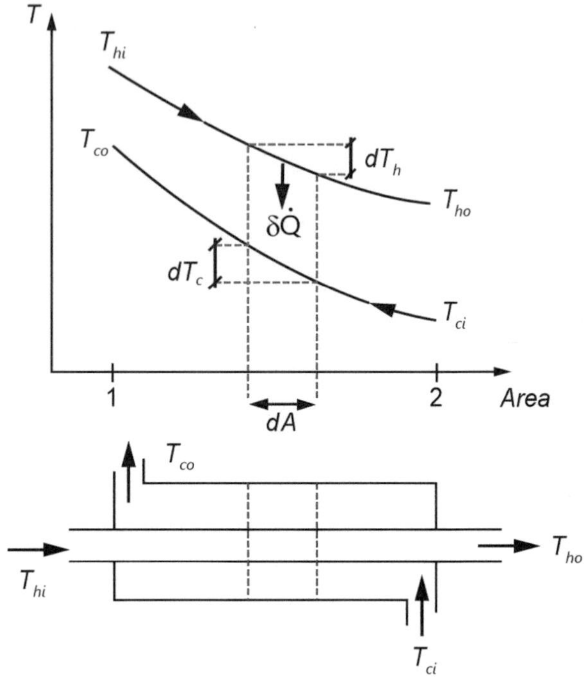

Fig. 9.4 Temperature distribution in a counterflow double tube heat exchanger

By the law of conservation of energy, the heat rate received by the cold fluid is necessarily equal to that transferred from the hot fluid and, also neglecting the variations of kinetic and potential energy in the fluids as in the previous case, we obtain:

$$d\dot{Q} = \dot{m}_c C_{pc} dT_c = -\dot{m}_h C_{ph} dT_h \tag{9.14}$$

By rearranging Eq. (9.14), we obtain:

$$d(T_h - T_c) = -\left(\frac{1}{\dot{m}_h C_{ph}} + \frac{1}{\dot{m}_c C_{pc}}\right) d\dot{Q} \tag{9.15}$$

By substituting $d\dot{Q}$ from Eqs. (9.13) into (9.15), the result is:

$$\frac{d(T_h - T_c)}{T_h - T_c} = -U\left(\frac{1}{\dot{m}_h C_{ph}} + \frac{1}{\dot{m}_c C_{pc}}\right)dA \tag{9.16}$$

whose integration is equivalent to:

$$\ln\frac{\Delta T_2}{\Delta T_1} = -UA\left(\frac{1}{\dot{m}_h C_{ph}} + \frac{1}{\dot{m}_c C_{pc}}\right)dA \tag{9.17}$$

Compare the similarity of this equation with Eq. (9.6). In the case of Eq. (9.17), the definitions of temperature differences diverge from the previous case (in parallel), since $\Delta T_1 = T_{hi} - T_{co}$ and $\Delta T_2 = T_{ho} - T_{ci}$. In order to avoid difficulties and confusion with the definitions of these temperature differences, note that, regardless of whether the configuration is parallel flow or counterflow, ΔT_1 can be interpreted as the temperature difference of the fluids at their connections positioned to the left of the heat exchanger, and ΔT_2 as the temperature difference of the fluids at their connections positioned to the right of the heat exchanger, as illustrated in Fig. 9.5.

The total heat transfer rate is obtained from the application of the First Law of Thermodynamics to the hot and cold streams of the entire heat exchanger, resulting in:

$$\dot{Q} = \dot{m}_h C_{ph}(T_{hi} - T_{ho}) \tag{9.18}$$

and

$$\dot{Q} = \dot{m}_c C_{pc}(T_{co} - T_{ci}) \tag{9.19}$$

Substituting these two expressions into Eq. (9.17), we get:

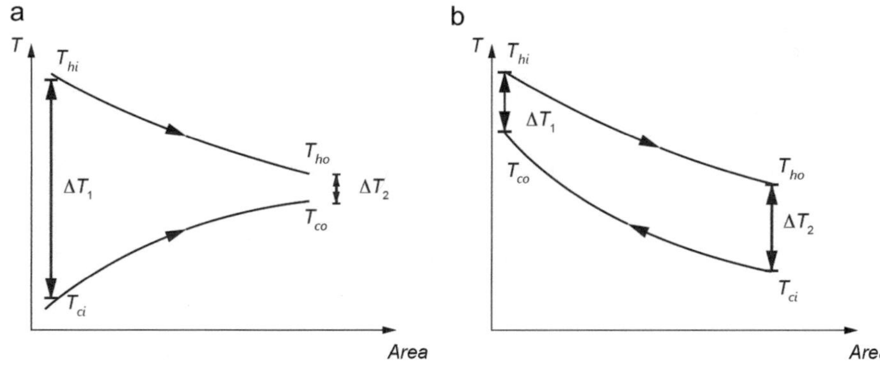

Fig. 9.5 Temperature difference ΔT_1 and ΔT_2 in heat exchanger **a** parallel and **b** counterflow

$$\ln\frac{\Delta T_2}{\Delta T_1} = -UA\frac{(T_{hi} - T_{ho}) + (T_{co} - T_{ci})}{\dot{Q}} \tag{9.20}$$

or:

$$\dot{Q} = UA\frac{\Delta T_2 - \Delta T_1}{\ln\left(\frac{\Delta T_2}{\Delta T_1}\right)} \tag{9.21}$$

From the above equation, we obtain the same expression as in the case of parallel current (Eq. 9.10).

$$\dot{Q} = UA\ LMTD \tag{9.22}$$

As demonstrated, the expression for calculating the heat exchange rate (thermal load) of the heat exchanger is the same (compared with Eq. 9.12), regardless of the configuration, with the warning of using the correct definitions of *LMTD* for parallel or counterflow accordingly.

For the same inlet and outlet temperatures of the two fluids, the *LMTD* in counterflow is greater than for parallel flow. Therefore, for the same overall heat transfer coefficient, U, the area required for a given heat transfer rate is smaller for a counterflow exchanger.

Example 9.1 (*Counterflow double tube heat exchanger*)

After a condensation process, (pressurized) liquid water at 120 °C at a 2.3 kg/s mass flow rate is used to heat another stream of liquid water at 15 °C and 1.0 kg/s up to 40 °C, which will be stored for later use. If a counterflow double tube heat exchanger with an internal copper tube and diameters $d_i = 24$ mm and $d_o = 25$ mm, evaluate the necessary length of the heat exchanger considering a heat transfer coefficient in the annular section of 550 W/m^{2}K, where the hot water passes.

Solution

The configuration of the heat exchanger and the temperature distribution scheme are indicated in the figures.

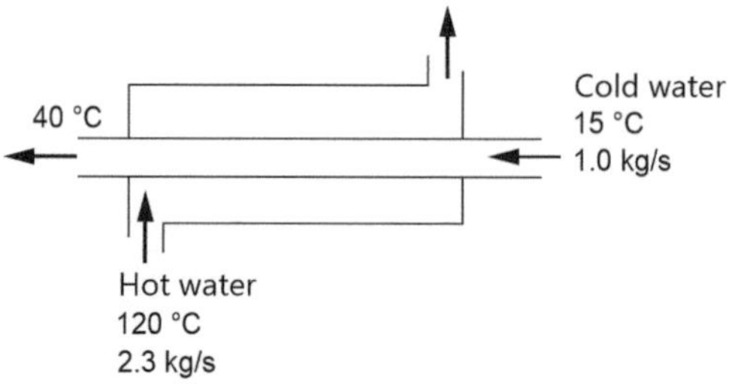

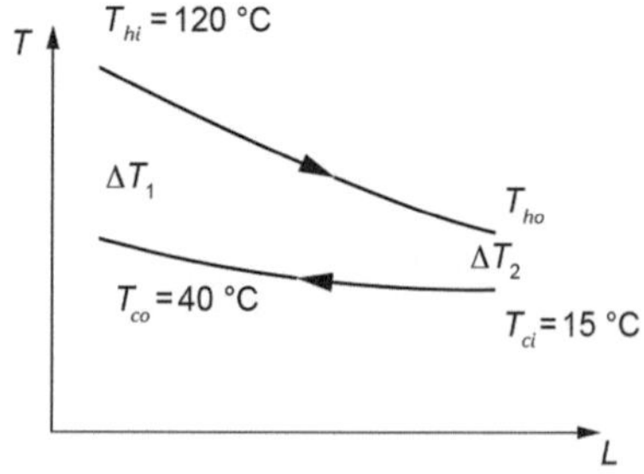

In the global energy balance (1st Law of Thermodynamics) of the entire exchanger, the heat supplied by the hot fluid must be equal to the heat received by the cold fluid. Notice that C_{ph} cannot be accurately evaluated because T_{ho} (and $\bar{T}_h$) is not known. Therefore, an initial value is assumed $T_{ho} = 95$ °C: $\overline{T}_h = 107.5$ °C and $C_{ph} = 4227$ J/kgK. For the cold water, you can adopt the average value $C_{pc} = 4179$ J/kgK.

$$1 \times 4179 \times (40 - 15) = 2.3 \times 4227 \times (120 - T_{ho}) \rightarrow T_{ho} = 109.3\,°\mathrm{C}$$

With this new temperature, a new average temperature is evaluated: $\overline{T}_h = 114.7°C$ and $C_{ph} = 4236$ J/kgK $\rightarrow T_{ho} = 109.3$ °C, which is the correct temperature.

The logarithmic mean temperature difference can be evaluated with $\Delta T_1 = 80$ °C and $\Delta T_2 = 94.3$ °C.

$$LMTD = \frac{80 - 94.3}{\ln(80/94.3)} = 86.9\ °\mathrm{C}$$

The rate of heat transfer is: $\dot{Q} = 1 \times 4179 \times (40 - 15) = 104,475$ W
Heat transfer calculations—flow inside tubes
Internal flow: $\overline{T}_c = 27.5°$C (from the properties of water, Table A.7)

$$\rho = 996.7\,\mathrm{kg/m^3};\ k = 0.614\,\mathrm{W/mK};\ Pr = 5.74;\ \mu = 843 \times 10^{-6}\ \mathrm{kg/ms};$$
$$C_p = 4179\,\mathrm{J/kgK};\ \nu = 843 \times 10^{-6}/996.7 = 8.46 \times 10^{-7}\,\mathrm{m^2/s}$$

$$Re = \frac{4\dot{m}}{\pi d_i \mu} = \frac{4 \times 1}{\pi 0.024 \times 843 \times 10^{-6}} = 62,964\ \text{(turbulent)}$$

Using the Dittus-Boelter expression (Eq. 6.43):

$$Nu = 0.023 Re^{0.8} Pr^{0.4} = 0.023 \times 62964^{0.8} \times 5.74^{0.4} = 319.6$$

$$h_i = Nu\frac{k}{d_i} = 319.6\frac{0.614}{0.024} = 8175.7\ \mathrm{W/m^2}K$$

External flow: $h_o = 550$ W/m^{2}K
Conductivity of the inner copper tube wall: $k = 396.5$ W/mK (Table A.2)
Overall heat transfer coefficient as a function of the external area of the inner tube:

$$\frac{1}{U_o A_o} = \frac{1}{h_i A_i} + \frac{\ln(d_o/d_i)}{2\pi k L} + \frac{1}{h_o A_o}$$

$$\frac{1}{U_o \pi \times 0.025 \times L} = \frac{1}{8175.7 \times \pi \times 0.024 \times L} + \frac{\ln(0.025/0.024)}{2\pi \times 396.5 \times L} + \frac{1}{550 \times \pi \times 0.025 \times L}$$

$$U_o = 513.6\ \text{W/m}^2\text{K}$$

Therefore, the length is evaluated from:

$$\dot{Q} = U_o A_o DMLT \rightarrow 513.6 \times \pi \times 0.025 \times L \times 86.9 = 104,475$$

$$L = 29.8\ \text{m}$$

Note: realize that the heat transfer rate in the hot fluid is: $\dot{Q} = 2.3 \times 4236 \times (120 - 109.3) = 104{,}247$ W. An approximation with more decimal places would provide a more accurate result.

9.3 *F*-Method

Industrial heat exchangers present more complex geometries and configurations than the fundamental double tube configuration, such as multitubular ones, which can have several passes in the shell or the cross-flow configuration. In these cases, determining the average temperature difference is a difficult task and, therefore, the usual calculation procedure is based on modifying the basic equations by introducing a *correction factor F*. Thus, the equation for calculating the total thermal load of the heat rate becomes the following formulation:

$$\dot{Q} = UAF\ LMTD_{cc} \tag{9.23}$$

where $LMTD_{cc}$ is that for a double tube counterflow heat exchanger with the same inlet and outlet temperatures as those of the complex configuration. The sequence of Figs. 9.6, 9.7, 9.8 and 9.9 provides the correction factors F for various configurations. The notation (T, t) represents the temperatures of the two fluid streams, regardless of whether the hot fluid flows inside the tubes or in the shell. In this temperature convention, index 1 indicates inlet and 2, outlet. Care should be taken in Fig. 9.9 where T indicates the mixed fluid and t the unmixed fluid. The graphs correspond to the equations indicated in the same figures, in which the parameters P and R are defined as:

$$P = \frac{t_2 - t_1}{T_1 - t_1} \tag{9.24}$$

$$R = \frac{T_1 - T_2}{t_2 - t_1} \tag{9.25}$$

where P varies from 0 to 1 and R, from 0 to infinity.

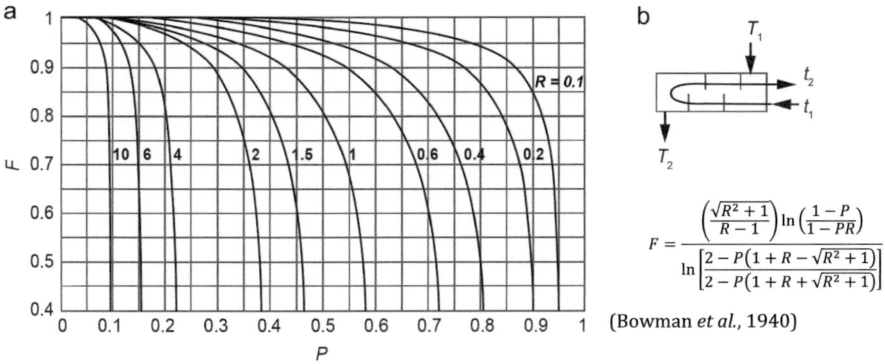

$$F = \frac{\left(\frac{\sqrt{R^2+1}}{R-1}\right)\ln\left(\frac{1-P}{1-PR}\right)}{\ln\left[\frac{2-P\left(1+R-\sqrt{R^2+1}\right)}{2-P\left(1+R+\sqrt{R^2+1}\right)}\right]}$$

(Bowman *et al.*, 1940)

Fig. 9.6 Correction factor in shell and tube exchanger with one pass in the shell and multiple of 2 in the tubes

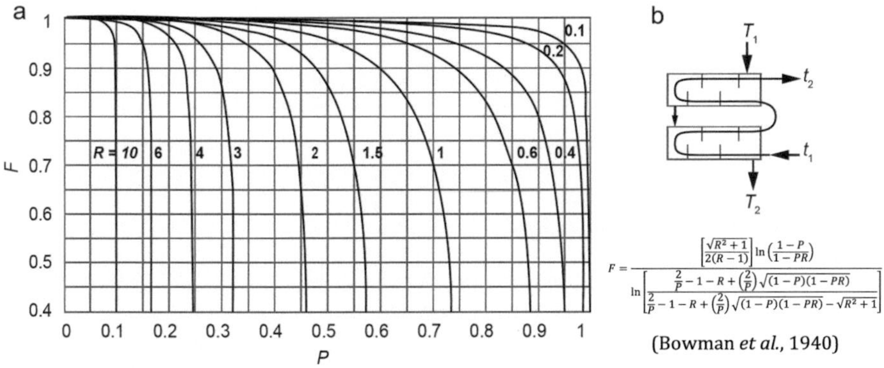

$$F = \frac{\left[\frac{\sqrt{R^2+1}}{2(R-1)}\right]\ln\left(\frac{1-P}{1-PR}\right)}{\ln\left[\frac{\frac{2}{P}-1-R+\left(\frac{2}{P}\right)\sqrt{(1-P)(1-PR)}}{\frac{2}{P}-1-R+\left(\frac{2}{P}\right)\sqrt{(1-P)(1-PR)}-\sqrt{R^2+1}}\right]}$$

(Bowman *et al.*, 1940)

Fig. 9.7 Correction factor in shell and tube heat exchanger with two passes in the shell and multiple of 4 in the tubes

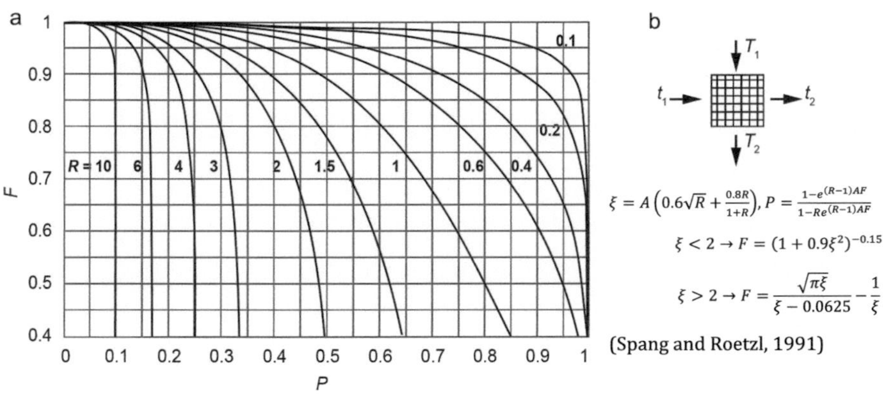

$$\xi = A\left(0.6\sqrt{R}+\frac{0.8R}{1+R}\right),\ P = \frac{1-e^{(R-1)AF}}{1-Re^{(R-1)AF}}$$

$$\xi < 2 \rightarrow F = (1+0.9\xi^2)^{-0.15}$$

$$\xi > 2 \rightarrow F = \frac{\sqrt{\pi\xi}}{\xi-0.0625}-\frac{1}{\xi}$$

(Spang and Roetzl, 1991)

Fig. 9.8 Correction factor in cross-flow heat exchanger, single pass, both fluids unmixed

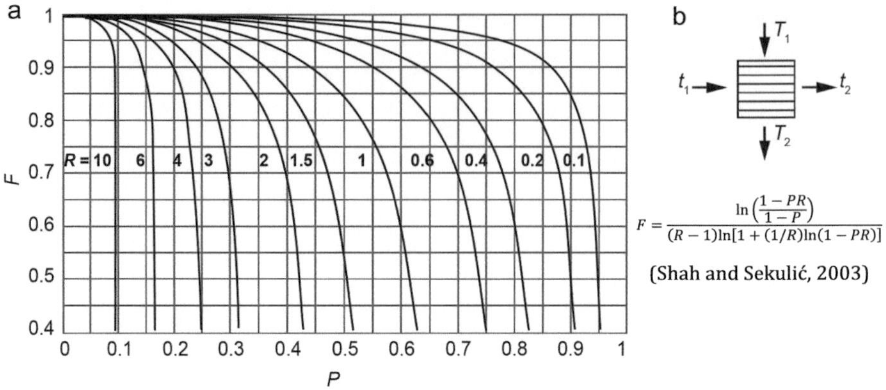

Fig. 9.9 Correction factor in cross-flow heat exchanger, single pass, one fluid unmixed (t) and the other mixed (T)

Other configurations can be found in specific works on heat exchangers, such as in Shah and Sekulić's work (2003). In case of phase change (evaporation or condensation), it may happen that $R = 0$ or $R \to \infty$ and $P = 0$; in both situations, we have $F = 1$, regardless of the side where the fluid is changing phase.

Example 9.2 (*Shell and tube heat exchanger*)

A 2-pass shell and 4-pass tube heat exchanger (2 passes per shell) are to be designed to cool oil off. The oil is at 120 °C and 2 kg/s and should be cooled to 50 °C, while water is available at 15 °C and 1.3 kg/s. Knowing that the oil passes in the shell and the water in the tubes and that the overall heat transfer coefficient is 500 W/m^{2}K, for a 2 cm tube diameter and 8 m long, evaluate the number of tubes needed to achieve the demanded thermal load.

Solution

The heat exchanger configuration and the temperature distribution are indicated in the figures.

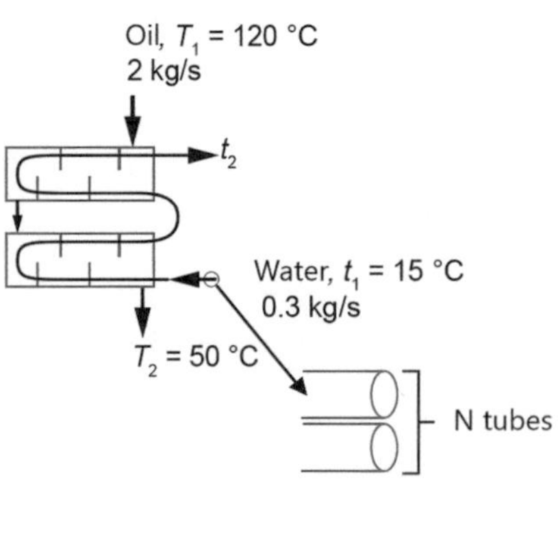

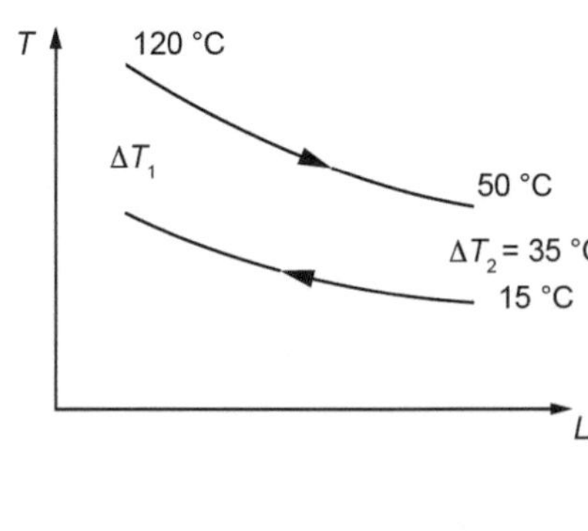

Apparently, this problem cannot be solved, as the exiting water temperature T_2 is unknown. However, that magnitude can be evaluated using the energy balance. For that, consider the following average transport properties:

Oil at $\overline{T}_h = (120 + 50)/2 = 85$ °C, $C_{ph} = 2152$ J/kgK (Table A.8)

For water, an initial value of $C_{pc} = 4179$ J/kgK (Table A.7). From an energy balance, we obtain

$$\dot{m}_h C_{ph}(T_{hi} - T_{ho}) = \dot{m}_c C_{pc}(T_{co} - T_{ci})$$

$2 \times 2152 \times (120 - 50) = 1.3 \times 4179 \times (T_{co} - 15) \rightarrow T_{co} = 70.5\,°\text{C}$. Therefore, $\overline{T}_c = 42.7\,°\text{C}$, and $C_{pc} = 4179$ J/kgK—check!

$$\Delta T_1 = 49.5°C, \, \Delta T_2 = 35\,°\text{C}$$

$$DMLT_{cc} = (49.5 - 35)/\ln(49.5/35) = 41.8\,°\text{C}$$

Also note that the T corresponds to the fluid (oil) that flows the shell side and the t, to the fluid (water)flowing inside the tubes.

$$P = \frac{t_2 - t_1}{T_1 - t_1} = \frac{70.5 - 15}{120 - 15} = 0.53$$

$$R = \frac{T_1 - T_2}{t_2 - t_1} = \frac{120 - 50}{70.5 - 15} = 1.26$$

From Fig. 9.7, $F \approx 0.91$

$$\dot{Q} = \dot{m}_h C_{ph}(T_{hi} - T_{ho}) = 2 \times 2152 \times (120 - 50) = 301,280 \text{ W}$$

$$\dot{Q} = UAF \;\; LMTD_{cc} \rightarrow 301,280 = 500 \times A \times 0.91 \times 41.8 \Rightarrow A = 15.84\,\text{m}^2$$

That is the total area. To obtain the number or tubes, N, one has

$$A = N\pi dL \rightarrow 15.84 = N \times \pi \times 0.020 \times 8$$

$N = 31.5$, which should be rounded up to 32 tubes.

Example 9.3 (*Cross-flow heat exchanger*)

A finless radiator is tested to cool down hot water with forced air. The water enters at 90 °C and exits at 70 °C, while air flows around the tubes entering at 20 °C and exiting at 30 °C. Given the overall heat transfer coefficient of 450 W/m^{2}K and the heat exchange area of 0.5 m^2, evaluate the heat transfer rate and the mass flow rates of the working fluids.

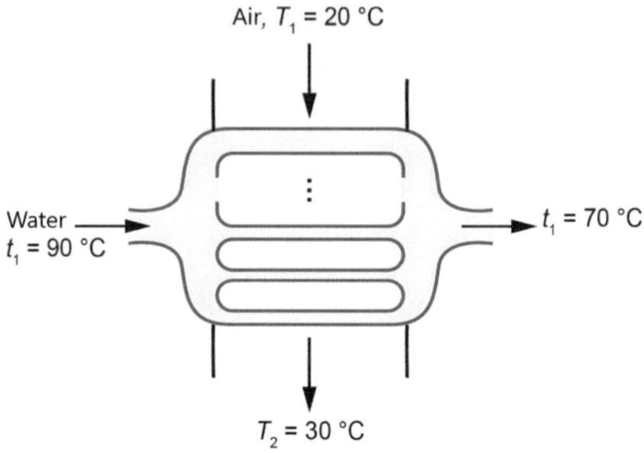

Solution

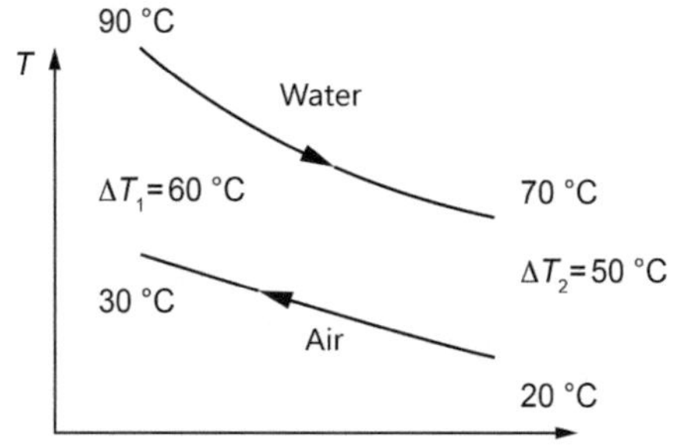

From the schematic figure, we have:

$$P = \frac{t_2 - t_1}{T_1 - t_1} = \frac{70 - 90}{20 - 90} = 0.29$$

$$R = \frac{T_1 - T_2}{t_2 - t_1} = \frac{20 - 30}{70 - 90} = 0.5$$

$F \approx 0.98$ (Fig. 9.9)
$\Delta T_1 = 60$ °C, $\Delta T_2 = 50$ °C

$$LMDT_{cc} = (60 - 50)/\ln(60/50) = 54.8\,°\mathrm{C}$$

$$\dot{Q} = UAF\ \ LMTD_{cc} = 450 \times 0.5 \times 0.98 \times 54.8 = 12{,}083\,\mathrm{W}$$

The mass flow rates can be obtained from an energy balance:
Water at $\overline{T}_h = 80$ °C (Table A.7): $C_{ph} = 4193$ J/kgK
At a $\overline{T}_c = 25$ °C (Table A.5): $C_{pc} = 1007$ J/kgK

$$\dot{Q} = \dot{m}_h C_{ph}(T_{hi} - T_{ho}) \rightarrow 12083 = \dot{m}_h 4193 \times (90 - 70) \rightarrow \dot{m}_h = 0.14 \ \text{kg/s}$$

$$\dot{Q} = \dot{m}_c C_{pc}(T_{co} - T_{ci}) \rightarrow 12083 = \dot{m}_c 1007 \times (30 - 20) \rightarrow \dot{m}_c = 0.2 \ \text{kg/s}$$

9.4 Effectiveness-NTU Method

The F method analyzed in the previous section is useful when all the inlet and outlet temperatures of the cold and hot fluids are known, or can be immediately evaluated from an energy balance. In thermal projects, however, it is common for one or more inlet or outlet temperatures of the heat exchanger to be unknown. In this case, the F method becomes laborious and requires an iterative trial and error method in which temperatures are assumed and the values updated until convergence occurs. With many *software* and numerical solution techniques, this does not represent an insuperable hurdle. However, there is a method that does not require the knowledge of all the involved temperatures known as the *effectiveness* (ε) and *number of transfer units* (*NTU*) method. For that, consider the following definition of effectiveness, ε:

$$\varepsilon \equiv \frac{\text{Actual heat exchange rate}}{\text{Maximum possible heat exchange rate}} = \frac{\dot{Q}_{\text{actual}}}{\dot{Q}_{\text{max}}} \tag{9.26}$$

The maximum possible heat exchange load is the one that would result if *one of the fluids* underwent a temperature change equal to the maximum possible temperature difference in the heat exchanger, that is, the inlet hot fluid temperature subtracted from the inlet cold fluid temperature. This method uses the effectiveness to eliminate the unknown outlet temperature and provides the solution for effectiveness in terms of other known operational parameters $\dot{m}$, C_p, A, and U.

Let the *thermal capacity*, C, be defined as the product of the mass flow rate by the specific heat at constant pressure, that is:

$$C = \dot{m}C_p \tag{9.27}$$

Then, by carrying out an energy balance for the fluids, we have:

$$\dot{Q}_{real} = C_h(T_{hi} - T_{ho}) = C_c(T_{co} - T_{ci}) \tag{9.28}$$

which indicates that the heat rate transferred from the hot fluid is the same as that received by the cold fluid. Therefore, the maximum possible theoretical heat rate exchange, $\dot{Q}_{\text{max}}$, occurs when the fluid with the lowest thermal capacity, C, undergoes the greatest possible temperature variation, which is exactly the difference between the hot inlet temperature, T_{hi}, and cold inlet temperature, T_{ci}, that is,

$$\dot{Q}_{\max} = C_{\min}(T_{hi} - T_{ci}) \tag{9.29}$$

Therefore, $C_{\min}$ can occur on either the cold side or the hot side.

$$C_{\min} = \min \begin{cases} C_c = C_{pc}\dot{m}_c \\ \quad \text{or} \\ C_h = C_{ph}\dot{m}_h \end{cases}$$

This maximum heat rate exchange would theoretically occur in a counterflow heat exchanger with infinite area. By combining Eqs. (9.26) and (9.29), we get:

$$\dot{Q}_{\text{real}} = \varepsilon C_{\min}(T_{hi} - T_{ci}) \tag{9.30}$$

To find a relationship between ε and $C_{\min}$, consider a *parallel flow heat exchanger*, like the one in Fig. 9.3, using the same simplifying assumptions indicated in Sect. 9.2.1.

By combining Eqs. (9.28) and (9.29), we obtain the expressions for the effectiveness of the hot and cold sides

$$\varepsilon = \frac{C_h(T_{hi} - T_{ho})}{C_{\min}(T_{hi} - T_{ci})} = \frac{C_c(T_{co} - T_{ci})}{C_{\min}(T_{hi} - T_{ci})} \tag{9.31}$$

As the minimum value of C, that is, $C_{\min}$, can occur for both the hot and the cold fluid, there are two possible values for the effectiveness:

$$\text{If} \quad C_h < C_c \Rightarrow \varepsilon_h = \frac{T_{hi} - T_{ho}}{T_{hi} - T_{ci}} \tag{9.32}$$

$$\text{If} \quad C_h > C_c \Rightarrow \varepsilon_c = \frac{T_{co} - T_{ci}}{T_{hi} - T_{ci}} \tag{9.33}$$

The effectiveness ε_c and ε_h indicate which fluid has the minimum value of C. Substituting into Eq. (9.6), it can be written in terms of C in the following way:

$$\ln\frac{T_{ho} - T_{co}}{T_{hi} - T_{ci}} = -UA\left(\frac{1}{C_h} + \frac{1}{C_c}\right) \tag{9.34}$$

or

$$\frac{T_{ho} - T_{co}}{T_{hi} - T_{ci}} = e^{-\frac{UA}{C_h}\left(1 + \frac{C_h}{C_c}\right)} \tag{9.35}$$

Assuming that the hot fluid has the $C_{\min} = C_h$

Then, from the effectiveness Eq. (9.31), the following relationship can be obtained:

$$\varepsilon_h = \frac{T_{hi} - T_{ho}}{T_{hi} - T_{ci}} = \frac{T_{hi} - T_{ci} + T_{ci} + T_{co} - T_{co} - T_{ho}}{T_{hi} - T_{ci}}$$

$$\varepsilon_h = \frac{T_{hi} - T_{ci}}{T_{hi} - T_{ci}} - \frac{T_{ho} - T_{co}}{T_{hi} - T_{ci}} - \frac{T_{co} - T_{ci}}{T_{hi} - T_{ci}}, \text{ or,}$$

$$\varepsilon_h = 1 - \frac{T_{ho} - T_{co}}{T_{hi} - T_{ci}} - \frac{T_{co} - T_{ci}}{T_{hi} - T_{ci}} = 1 - \frac{T_{ho} - T_{co}}{T_{hi} - T_{ci}} - \frac{\frac{\dot{Q}_{\text{real}}}{C_c}}{\frac{\dot{Q}_{\text{real}}}{\varepsilon_h C_h}}$$

By rearranging, we get

$$\varepsilon_h\left(1 + \frac{C_h}{C_c}\right) = 1 - \frac{T_{ho} - T_{co}}{T_{hi} - T_{ci}}$$

By substituting Eq. (9.17)

$$\varepsilon_h = \frac{1 - e^{-\frac{UA}{C_h}\left(1 + \frac{C_h}{C_c}\right)}}{1 + C_h/C_c} \quad \text{if} \quad C_{\min} = C_h \tag{9.36a}$$

Assuming that the cold fluid has the $C_{\min} = C_c$

A similar analysis provided the following expression:

$$\varepsilon_c = \frac{1 - e^{-\frac{UA}{C_c}\left(1 + \frac{C_c}{C_h}\right)}}{1 + C_c/C_h} \quad \text{if} \quad C_{\min} = C_c \tag{9.36b}$$

By comparing Eqs. (9.36a) and (9.36b), it is possible to generalize to encompass both situations, that is:

$$\varepsilon = \frac{1 - e^{-\frac{UA}{C_{\min}}\left(1 + \frac{C_{\min}}{C_{\max}}\right)}}{1 + C_{\min}/C_{\max}} \tag{9.37}$$

In addition, the following group of quantities is called the *number of transfer units* (*NTU*):

$$NTU = \frac{UA}{C_{\min}} \tag{9.38}$$

Thus, for a parallel flow double tube heat exchanger, the final expression is given by:

$$\varepsilon = \frac{1 - e^{-NTU\left(1+\frac{C_{min}}{C_{max}}\right)}}{1 + C_{min}/C_{max}} \tag{9.39}$$

Note that for heat exchangers where one of the fluids is changing phase, as it occurs in evaporators or condensers, it is found that $C_{min}/C_{max} = 0$, because one of the fluids remains at a constant temperature, making its (apparent) specific heat infinite.

By carrying out a similar analysis for the double tube counterflow heat exchanger, we obtain:

$$\varepsilon = \frac{1 - e^{-NTU\left(1-\frac{C_{min}}{C_{max}}\right)}}{1 - \frac{C_{min}}{C_{max}} e^{-NTU\left(1-\frac{C_{min}}{C_{max}}\right)}} \tag{9.40}$$

Other Configurations

Expressions for the effectiveness of other configurations are given in Table 9.2, where $C = C_{min}/C_{max}$. The expressions listed in the table are graphically represented in Fig. 9.10.

The effectiveness of heat exchangers for a certain *NTU* in shell and tube configuration with 2 passes in the shell and 4 (6, 8...) in the tubes is greater than one pass in the shell and multiples of 2 in the tubes. In addition, it is noted that the effectiveness is greater for 3 passes in the shells and 6 (9, 12...) in the tubes than in the case of 2 passes in the shells and 4 (6, 8...) in the tubes.

Figure 9.10f shows that, for a specific *NTU*, the effectiveness of a cross-flow exchanger when C_{min} is unmixed and C_{max} with mixture is less than in the case where C_{min} is with mixture and C_{max} is without mixture, except when $C = 0$ or $C = 1$, in which the effectivenesses are equal.

Example 9.4 (*Effectiveness in counterflow and parallel heat exchanger*)

Ethylene glycol at 90 °C and 1 kg/s needs to be cooled in a double tube heat exchanger, and for this water at 15 °C and 1 kg/s is available. By assuming that the overall heat transfer coefficient is 900 W/m^{2}K and the thermal exchange area is 9 m^2, evaluate the effectivenesses, the heat transferred (thermal load), and the outlet temperatures of the ethylene glycol for a double tube exchanger in parallel and counterflow. Assume that the overall coefficient remains constant in both cases.

Table 9.2 Effectiveness and *NTU* for heat exchangers (Kays and London, 1984)

Type	Effectiveness	*NTU*	Equations
Parallel flow double tube	$\varepsilon = \frac{1-e^{-\mathrm{NTU}(1+C)}}{1+C}$	$NTU = -\frac{\ln[1-\varepsilon(1+C)]}{1+C}$	(9.39)
Counterflow double tube	$C < 1, \varepsilon = \frac{1-e^{-\mathrm{NTU}(1-C)}}{1-Ce^{-\mathrm{NTU}(1-C)}}$ $C = 1, \varepsilon = \frac{\mathrm{NTU}}{1+\mathrm{NTU}}$	$NTU = \frac{1}{C-1}\ln\left(\frac{\varepsilon-1}{\varepsilon C-1}\right)$ $NTU = \frac{\varepsilon}{1-\varepsilon}$	(9.40)
Shell and tube with 1 shell pass and 2, 4, 6... tube passes	$\varepsilon =$ $2\left\{1 + C + \sqrt{1+C^2}\,\frac{1+e^{-\mathrm{NTU}\sqrt{1+C^2}}}{1-e^{-\mathrm{NTU}\sqrt{1+C^2}}}\right\}^{-1}$	$NTU =$ $-\frac{1}{\sqrt{1+C^2}}\ln\left(\frac{\frac{2/\varepsilon-1-C}{\sqrt{1+C^2}}-1}{\frac{2/\varepsilon-1-C}{\sqrt{1+C^2}}+1}\right)$	(9.41)
Shell and tube with *n* passes in the shell and 2*n*, 4*n*, 6*n*... passes in the tubes	$\varepsilon = \frac{\left(\frac{1-\varepsilon_1 C}{1-\varepsilon_1}\right)^n-1}{\left(\frac{1-\varepsilon_1 C}{1-\varepsilon_1}\right)^n-C}$ ε_1 evaluated as 1 pass in the shell and 2, 4, 6... in the tubes	$\mathrm{NTU} =$ $-\frac{n}{\sqrt{1+C^2}}\ln\left(\frac{\frac{\frac{2}{A-1-C}}{\sqrt{1+C^2}}-1}{\frac{\frac{2}{A-1-C}}{\sqrt{1+C^2}}+1}\right)$ $A =$ $\left(\sqrt[n]{\frac{\varepsilon C-1}{\varepsilon-1}}-1\right)\left(\sqrt[n]{\frac{\varepsilon C-1}{\varepsilon-1}}-C\right)^{-1}$	(9.42)
Cross flow, single pass, both fluids unmixed	$\varepsilon = 1 - e^{\frac{\mathrm{NTU}^{0.22}}{C}e^{-C\mathrm{NTU}^{0.78}}-1}$		(9.43)
Cross-flow, single pass: $C_{\min}$ without mix $C_{\max}$ with mix	$\varepsilon = \frac{1-e^{C\left(e^{-\mathrm{NTU}}-1\right)}}{C}$	$NTU = -\ln\left[1 + \frac{ln(1-\varepsilon C)}{C}\right]$	(9.44)
Cross-flow, single pass: $C_{\min}$ with mix $C_{\max}$ without mix	$\varepsilon = 1 - e^{\frac{e^{-C\mathrm{NTU}}-1}{C}}$	$NTU = -\frac{\ln[C\ln(1-\varepsilon C)+1]}{C}$	(9.45)
Any exchanger with: $C=0$	$\varepsilon = 1 - e^{-NTU}$	$NTU = -\ln(1-\varepsilon C)$	(9.46)

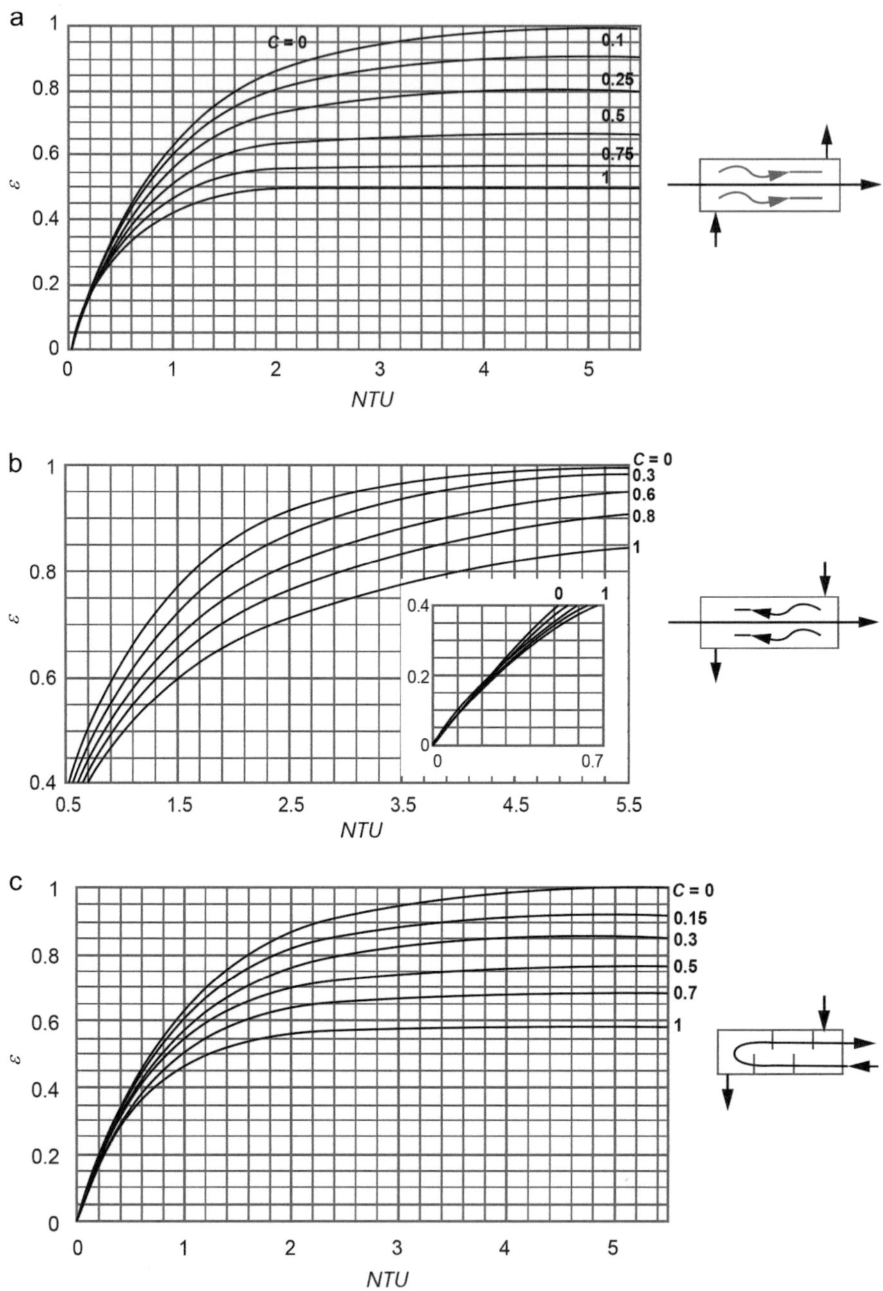

Fig. 9.10 **a** Effectiveness of a parallel flow double tube heat exchanger. **b** Effectiveness of a counterflow double tube heat exchanger. **c** Effectiveness of a shell and tube heat exchanger with one pass in the shell and multiple of 2 in the tubes. **d** Effectiveness of a shell and tube heat exchanger with n passes in the shell and $2n$, $4n$, $6n$... passes in the tubes ($n = 2$ and $n = 3$). **e** Effectiveness of a one-pass cross-flow heat exchanger with both fluids unmixed. **f** Effectiveness of a cross-flow heat exchanger with one pass with one unmixed fluid and another mixed

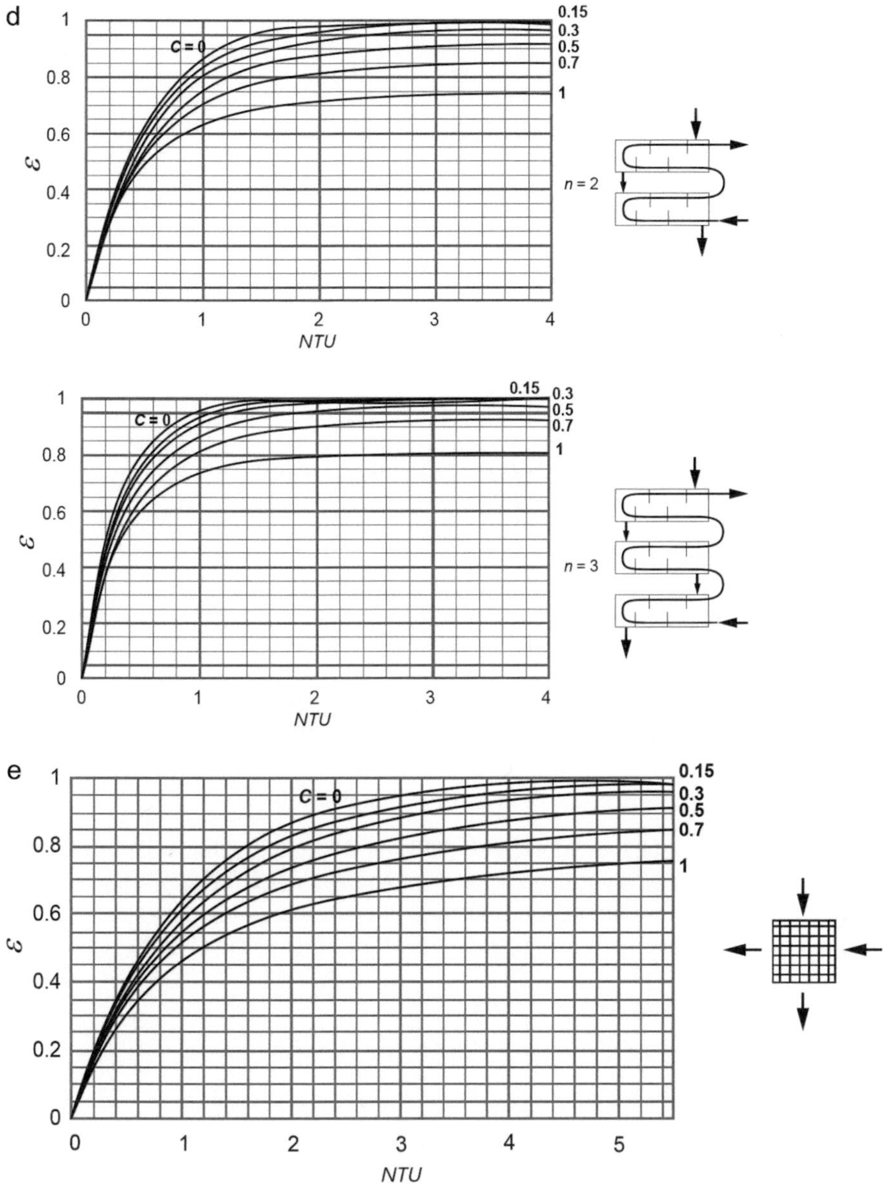

Fig. 9.10 (continued)

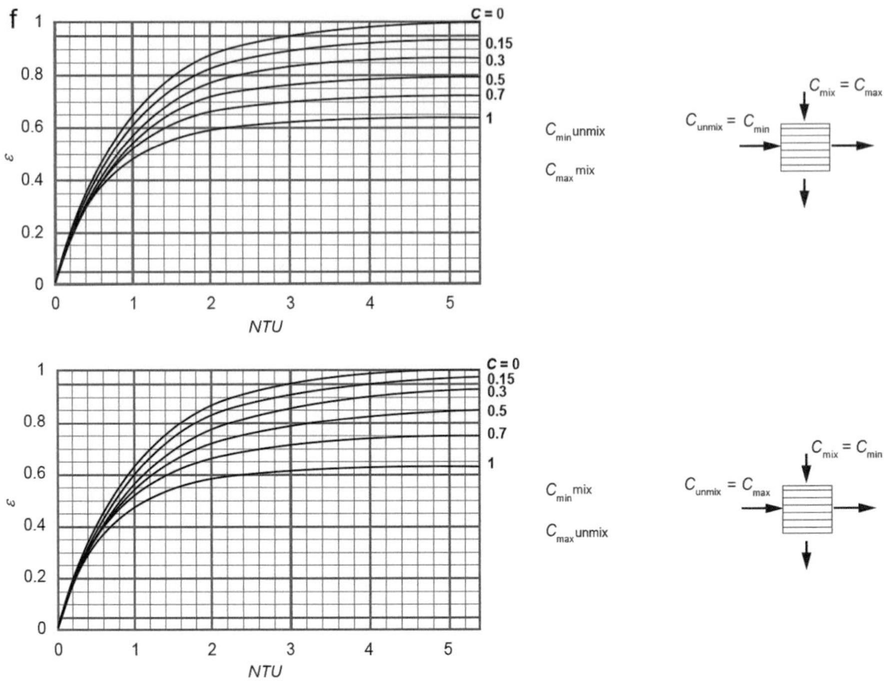

Fig. 9.10 (continued)

Solution

In this problem, only the two inlet temperatures of the fluids are known, and the other two outlet temperatures are unknown and cannot be evaluated directly. Therefore, the effectiveness method should be used. The specific heats are assumed to be according to the average temperature of the initial guest, which, if necessary, will be reevaluated by reiteration.

$$\overline{T}_h = 77\ °\text{C}$$

$C_{ph} = 2637$ J/kgK (Table A.8) and $C_h = \dot{m}_h C_{ph} = 1 \times 2637 = 2637$ W/K $= C_{\min}$

$$\bar{T}_c = 37\ °\text{C}$$

$C_{pc} = 4178$ J/kgK (Table A.7) and $C_c = \dot{m}_c C_{pc} = 1 \times 4178 = 4178$ W/K $= C_{\max}$
Therefore, $C = C_{\min}/C_{\max} = 2637/4178 = 0.63$ and

$$\text{NTU} = UA/C_{\min} = 900 \times 9/2637 = 3.07$$

Parallel flow:

$$\dot{Q}_{\max} = C_{\min}\Delta T_{\max} = 2637 \times (90 - 15) = 197,775\ \text{W}$$

$$C = 0.63,\ \text{NTU} = 3.07 \text{ from Fig. 9.10} \Rightarrow \quad \varepsilon = 0.61$$

$$\dot{Q} = \varepsilon\dot{Q}_{\max} = 0.61 \times 197,775 = 120,643\ \text{W}$$

$$120,643 = 2637 \times (90 - T_{ho}) \rightarrow T_{ho} = 44.3\ ^{\circ}\text{C}$$

$$120,643 = 4178 \times (T_{co} - 15) \rightarrow T_{co} = 43.9\ ^{\circ}\text{C}$$

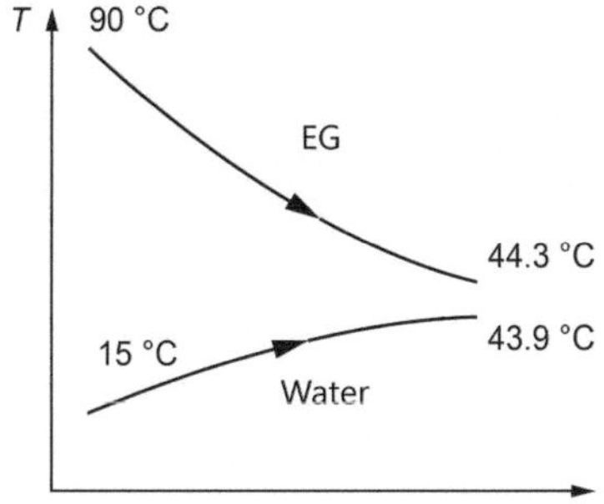

Counterflow:

$$\dot{Q}_{\max} = C_{\min}\Delta T_{\max} = 2637 \times (90 - 15) = 197{,}775\ \text{W}$$
$$C = 0.63,\ \text{NTU} = 3.07 \text{ from Fig. 9.10b} \quad \Rightarrow \quad \varepsilon = 0.85$$

$$\dot{Q} = \varepsilon\dot{Q}_{\max} = 0.85 \times 197,775 = 168{,}109\ \text{W}$$

$$168{,}109 = 2637 \times (90 - T_{ho}) \rightarrow T_{ho} = 26.3\ ^{\circ}\text{C}$$

$$168{,}109 = 4178 \times (T_{co} - 15) \rightarrow T_{co} = 55.2\ ^{\circ}\text{C}$$

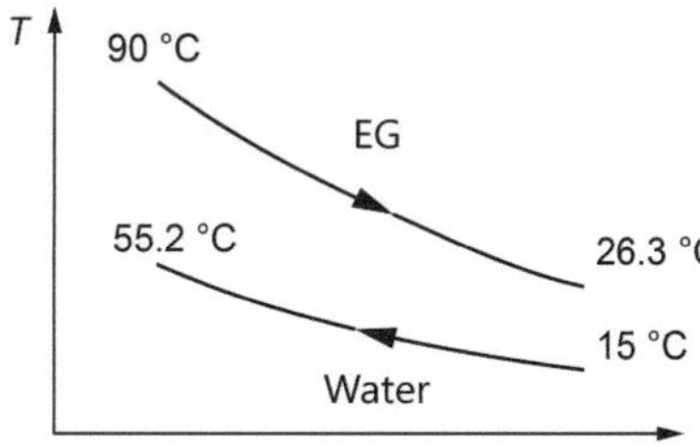

Note 1: notice that the effectiveness and the heat exchange rate in the counterflow heat exchanger are higher than the parallel one, thus the outlet temperature of the hot fluid is lower.

Note 2: the average temperatures for both heat exchangers (parallel: $\overline{T}_h = 67.2$ °C, $\overline{T}_c = 29.5$ °C, counterflow: $\overline{T}_h = 58.2$ °C, $\overline{T}_c = 35.1$ °C) are different from that initial assumption (parallel and counterflow: $\overline{T}_h = 77$ °C, $\overline{T}_c = 37$ °C); if more precise values are wanted, iterative calculations should be performed with the average temperatures found for each heat exchanger. Performing these calculations, it is found that the output temperatures are: parallel: $T_{ho} = 45.3$ °C, $T_{co} = 43.9$ °C; and countercurrent: $T_{ho} = 26.6$ °C, $T_{co} = 55$ °C, which are close to the previous calculations and, therefore, accepted.

9.5 The Problem of Incrustations

During the operation of heat exchangers, deposits of incrustations often occur, which are present in the fluid or produced by chemical reactions of the fluid with the equipment material. These deposits can be oxides, sludge, soot, and mineral salts. Those deposits will degrade the heat transfer rate, as they present an additional thermal resistance. Thus, it must be added an additional thermal resistance R_d that occurs on the "dirty" fluid side to the original overall heat transfer coefficient, U. Thus, the new overall coefficient, U_d, becomes:

$$\frac{1}{U_d} = R_d + \frac{1}{U} \tag{9.47}$$

The North American Association Tubular Exchanger Manufacturers Association (Tubular Exchanger Manufacturers Association, 1999) published several values of deposit thermal resistances associated with various industrial applications, as seen in Table 9.3.

Table 9.3 Deposit thermal resistances

Fluid	R_d (m^2°C/W)
Treated boiler feed water	0.0002
Sea water—below 50 °C	0.00009
Sea water—above 50 °C	0.0002
River water	0.0004
Fuel oil	0.0009
Oils (hydraulic, lubricant and transformer)	0.0002
Ethylene glycol	0.00035
Steam	0.0001
Air	0.0004
Engine exhaust gases	0.0018
Refrigerants—liquid	0.0002
Refrigerants—vapor	0.0004

Adapted from Tubular Exchanger Manufacturers Association (1999)

9.6 Irreversibility in Heat Exchangers

Heat exchangers are pieces of equipment highly irreversible as the heat exchange between the two fluid streams occur at a high temperature difference. In this section it is evaluated the entropy generation rate, $\dot{S}_{\text{gen}}$, in heat exchanges as a function of operational parameters and fluid properties. Consider an externally insulated (adiabatic) heat exchanger as depicted in Fig. 9.11.

The Second Law can be written for the heat exchanger (a control volume)

$$\frac{dS_{\text{HE}}}{dt} = \sum \dot{m}_{\text{in}} s_{\text{in}} - \sum \dot{m}_{\text{out}} s_{\text{out}} + \frac{\dot{Q}_{\text{HE}}}{T} + \dot{S}_{\text{gen}} \tag{9.48}$$

where, S_{HE} is the heat exchanger total entropy; the first and second terms in the right-hand side are the entropy inflow and outflow associated with the mass flow into and out of the heat exchanger, respectively. Therefore, they refer to the cold and hot mass flow rate and the related specific entropy. The third term is the heat exchanged to the environment, which is null for an adiabatic device. Finally, the last term is the entropy generation rate due to the internal irreversibility. Considering a steady-state operation, it yields.

$$\dot{S}_{\text{gen}} = \sum \dot{m}_o s_o - \sum \dot{m}_i s_i = \dot{m}_c (s_{co} - s_{ci}) + \dot{m}_h (s_{ho} - s_{hi}) \tag{9.49}$$

The specific entropy variation between entrance and exit from the heat exchanger for each fluid stream (without phase change) at a constant pressure process is

$$Tds = dh - vdp = C_p dT \tag{9.50}$$

$$\Delta s = \int_{\text{in}}^{\text{out}} ds = \int_{T_i}^{T_o} \frac{C_p dT}{T} = C_p \ln\left(\frac{T_o}{T_i}\right) \tag{9.51}$$

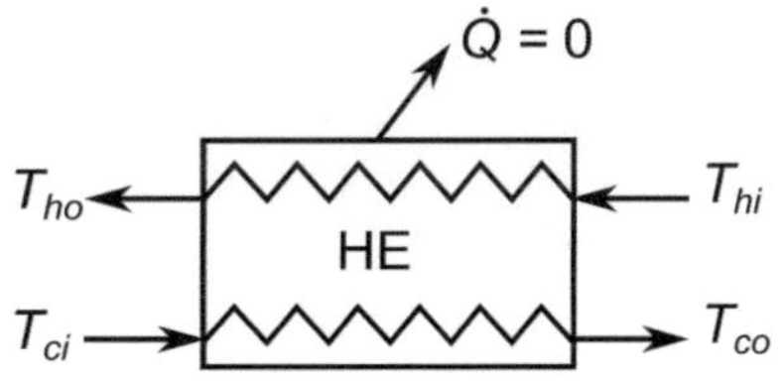

Fig. 9.11 Heat exchanger illustration

Applying to cold and hot sides, one obtains

$$s_{co} - s_{ci} = C_{pc} \ln\left(\frac{T_{co}}{T_{ci}}\right) \tag{9.52a}$$

and

$$s_{ho} - s_{hi} = C_{ph} \ln\left(\frac{T_{ho}}{T_{hi}}\right) \tag{9.52b}$$

Substituting above equations into Eq. (9.49) along with the heat exchanger thermal capacity definition given by Eq. (9.27), it yields,

$$\dot{S}_{\text{gen}} = C_c \ln\left(\frac{T_{co}}{T_{ci}}\right) + C_h \ln\left(\frac{T_{ho}}{T_{hi}}\right) \tag{9.53}$$

Also, introducing the energy balance for each fluid stream following Eq. (9.28), one obtains

$$\dot{S}_{\text{gen}} = C_c \ln\left(1 + \frac{\dot{Q}_{\text{real}}}{C_c T_{ci}}\right) + C_h \ln\left(1 - \frac{\dot{Q}_{\text{real}}}{C_h T_{hi}}\right) \tag{9.54}$$

Introducing the definition of the effectiveness, ε, given by Equation (9.30)

$$\dot{S}_{\text{gen}} = C_c \ln\left(1 + \frac{\varepsilon C_{\min}(T_{hi} - T_{ci})}{C_c T_{ci}}\right) + C_h \ln\left(1 - \frac{\varepsilon C_{\min}(T_{hi} - T_{ci})}{C_h T_{hi}}\right) \tag{9.55a}$$

Let $\theta_i = T_{hi} - T_{ci}$ the difference between the hot and cold entrance temperatures. After rearranging the above equation, one obtains

$$\dot{S}_{\text{gen}} = C_c \ln\left(1 + \frac{\varepsilon C_{\min}\theta_i}{C_c T_{ci}}\right) + C_h \ln\left(1 - \frac{\varepsilon C_{\min}\theta_i}{C_h T_{hi}}\right) \tag{9.55b}$$

Assuming that the hot fluid has the $C_{\min} = C_h$, then

$$\frac{\dot{S}_{\text{gen}}}{C_h} = \ln\left[\left(1 + \frac{\varepsilon C_h \theta_i}{C_c T_{ci}}\right)^{\frac{C_c}{C_h}} \times \left(1 - \frac{\varepsilon \theta_i}{T_{hi}}\right)\right] \tag{9.56a}$$

Assuming that the cold fluid has the $C_{\min} = C_c$, then

$$\frac{\dot{S}_{\text{gen}}}{C_c} = \ln\left[\left(1 + \frac{\varepsilon \theta_i}{T_{ci}}\right) \times \left(1 - \frac{\varepsilon C_c \theta_i}{C_h T_{hi}}\right)^{\frac{C_h}{C_c}}\right] \tag{9.56b}$$

It is important to note that effectiveness and the thermal capacity ratio are not independent variables; they are interconnected based on the heat exchanger configuration, as given in Table 9.2 and the set of graphs in Fig. 9.10. Further analysis and optimization of specific heat exchangers can be carried out by integrating the relationship between these parameters.

Example 9.5 (*Irreversibility in counterflow and parallel heat exchanger*)
Calculate the entropy generation rate for the two heat exchangers from Example 9.4. Which one generates more irreversible?

Solution
the hot fluid has the $C_{\min} = C_h = 2637\ \mathrm{W/K}$. Other data from the Example 9.4 are: $C_c = 4178\ \mathrm{W/K}$, $T_{hi} = 90\ °\mathrm{C} = 363.15\ \mathrm{K}$; $T_{ci} = 15\ °\mathrm{C} = 288.15\ \mathrm{K}$; $\theta_i = 75\ \mathrm{K}$.

Parallel flow:

$\varepsilon = 0.61 \Rightarrow$ from Eq. (9.56a) $\dot{S}_{\text{gen}} = 43.9\ \mathrm{W/K}$

Counterflow:

$\varepsilon = 0.85 \Rightarrow$ from Eq. (9.56a) $\dot{S}_{\text{gen}} = 37.1\ \mathrm{W/K}$

Considering the previous example, a few comments are worthy to be drawn. The entropy generation rate is higher in the parallel flow configuration in this example. The average temperature of the hot and cold fluids (see data in Example 9.4) for both heat exchangers are parallel flow: $\overline{T}_h = 67.2$ °C, $\overline{T}_c = 29.5$ °C; counterflow: $\overline{T}_h = 58.2$ °C, $\overline{T}_c = 35.1$ °C, which gives average temperature differences between the two fluids in parallel: $\Delta T_{pf} = \overline{T}_h - \overline{T}_c = 67.2 - 29.5 = 37.7$ °C; counterflow: $\Delta T_{cc} = \overline{T}_h - \overline{T}_c = 58.2 - 35.1 = 23.1$ °C. Therefore, the heat transfer in counterflow heat exchanger configuration occurs at an average temperature difference lower than the one in parallel flow, which may indicate a lower irreversibility as well. Finally, the counterflow heat exchanger besides being less irreversible, it also transfers more heat (higher heat duty).

9.7 Proposed Problems

9.1 An ethylene glycol solution at –10 °C exchanges heat with water in a double tube heat exchanger and exits at 0 °C. The water enters the heat exchanger at 25 °C and leaves it at 15 °C at a volumetric flow rate of 0.54 m^3/h. The heat transfer coefficient is 900 W/m^2°C. Determine the necessary heat exchange area for parallel and counterflow configurations.

9.2 A counterflow double tube heat exchanger is used to cool an industrial oil (C_p = 1.5 kJ/kg°C) from 90 to 60 °C. The cooling is carried out with water at 20 °C and a 0.5 kg/s mass flow rate, which is heated to 35 °C. The overall heat transfer coefficient was estimated as 300 W/m^2°C. Calculate the oil flow rate that can be cooled and the heat transfer area.

9.3 Using the same data from the Proposed Problem 9.2, what would be the new area if a shell and tube heat exchanger were used with water in one pass in the shell and two passes in the oil tubes?

9.4 Using the same data from the Proposed Problem 9.2, what would be the new area if a shell and tube heat exchanger were used with water in two passes in the shell and four passes in the oil tubes?

9.5 A single pass cross-flow heat exchanger is used to heat air with hot water. Water enters the heat exchanger tubes at a mass flow rate of 0.02 kg/s and 70 °C. Atmospheric air at 20 °C and a 0.07 m^3/s volumetric flow rate circulates in cross configuration with the tubes. The product *UA* is 56 W/K. Determine the exit temperatures of the water and the air. Both flows are unmixed.

9.6 The heat exchanger from the Proposed Problem 9.5 is formed by 12 parallel 0.1 cm diameter tubes. By installing fins, a balance of the heat transfer coefficients was achieved so that the internal water coefficient, h_1, was equal to the external air coefficient, h_2. Neglecting the wall thickness and assuming fully developed water flow inside the tubes, determine the length of the tubes. For calculation purposes, consider that the tube wall temperature is constant and equal to the average water temperature between the inlet and the outlet.

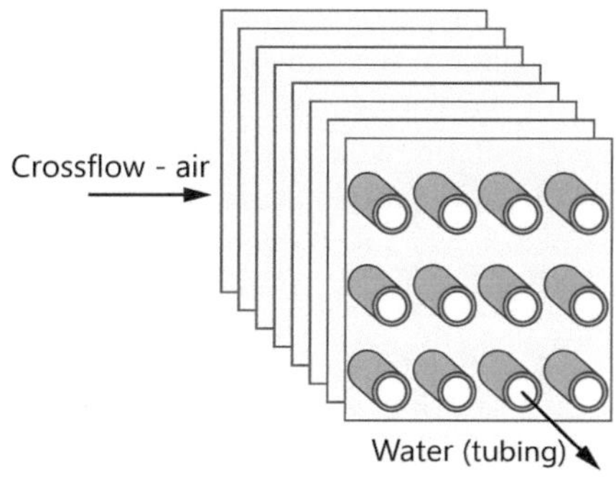

9.7 A unmixed crossflow heat exchanger is used to heat water for a restaurant from 30 °C to 75 °C at a flow rate of 3 m^3/h. The water is heated with therminol 59 available at 100 °C flowing at a mass flow rate of 2.5 kg/s. What is the NTU for this heat exchanger if the heat transfer area is 20 m^2? What is the overall heat transfer coefficient, *U*? What is its heat load?

9.8 Unused motor oil is heated from 10 °C to 30 °C in a shell and tube exchanger with one pass in the shell (hot water) and 2 passes in the tubes (oil). The water is available at 60 °C and it discharges at 40 °C at a flow rate of 1 m^3/h. The overall heat transfer coefficient $U = 800$ W/m^2 °C. What is the total heat transfer area? Considering the tubes are 3, 4, and 5 m long, calculate the corresponding number of tubes of 25 mm in diameter.

9.9 Prove that for a condenser and a boiler, the effectivity is given by Eq. 9.46. (*hint*: $C = C_{min}/C_{max} = 0$)

9.10 An evaporator unit is installed in a cold chamber to keep the internal temperature constant. Ammonia is used as the refrigeration circuit removing the cooling load by evaporation at −30 °C. Considering a cooling load of 50 TR, estimate the area of a cross-flow heat exchanger, single pass, both fluids unmixed. Cold chamber air enters the heat exchanger (evaporator) at −15 °C and leaves it at −20 °C as saturated ammonia is completely evaporated. (*hint*: I TR = 3,517 kW and vaporization enthalpy $h_{LV} = 1359{,}8$ kW/kg for ammonia at −30 °C)

9.11 In a process, a shell and tube type condenser is used, where steam, initially saturated, completely condenses in the shell side at 95 °C. In this case, the convection heat transfer coefficient is given by $h_{\text{vapor}} = 18{,}000$ W/m^{2}K, the heat exchange rate $\dot{Q} = 2250\ \text{kW}$, and the overall heat transfer coefficient $U = 5000$ W/m^{2}K. Water at 27 °C enters the tubes in two passes in a single shell and exits at 42 °C ($C_p = 4.180$ kJ/kgK). It is asked:

a. the flow rates of the hot and cold fluids;
b. plot the temperature distribution graphs $T - x$;
c. the necessary tube area and effectiveness of this exchanger.

9.12 A heat exchanger must be designed to cool 2 kg/s of oil from 120 to 40 °C. After initial considerations, a one-pass shell and six-pass tube type were selected. Each tube pass consists of 25 thin-walled tubes, with a diameter of 2 cm connected in parallel. The oil should be cooled using water that enters the heat exchanger at 15 °C and exits at 45 °C. A scheme of the unit can be seen in the following figure. The overall heat transfer coefficient is 300 W/m^2°C. Determine:

a. the mass flow rate of water;
b. the total heat transfer area;
c. the tubing length.

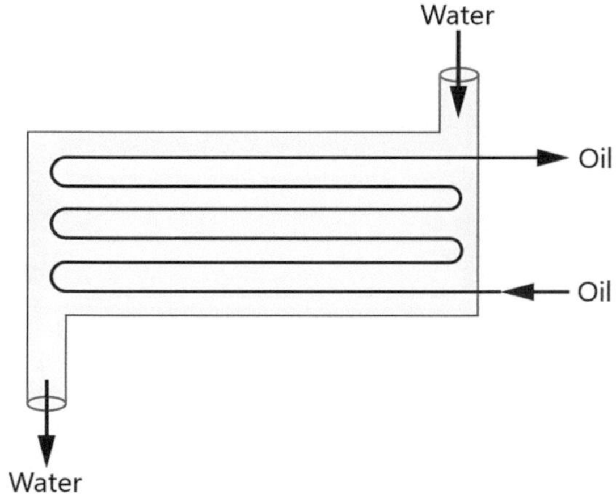

9.13 Water at a volumetric flow rate of 2.7 m^3/h enters a double tube counterflow heat exchanger at 35 °C and is heated by an industrial oil ($C_p = 1.5$ kJ/kg °C) at a 1.5 kg/s mass flow rate and 120 °C. The heat exchange area is 14 m^2 and the overall coefficient is 350 W/m^2°C. Determine the total heat transfer rate and the outlet temperatures of the water and the oil.

9.14 Fouling occurs in the heat exchange from proposed Problem 9.9, which deteriorates the heat transfer rate by 10%, that is, the heat transfer rate becomes 90 % of the designed original value. Determine the new outlet temperatures of the water and oil. What is the overall heat transfer coefficient and the value of the fouling thermal resistance after fouling?

References

Bowman, R.A., A.C. Mueller, and W.M. Nagle. 1940. Me temperature difference in design. *Transaction of the ASME* 62: 283–294.

Kays, W.M., and A.L. London. 1984. *Compact heat exchangers*, 3rd ed. New York: McGraw-Hill.

Shah, R.K., and D.P. Sekulić. 2003. *Fundamentals of heat exchanger design.* New Jersey: Wiley.

Spang, B., and W. Roetzl. 1991. Approximate equations for the design of cross-flow heat exchangers. In *Proceedings of the Eurotherm Seminar No. 18, Design and Operation of Heat Exchangers*, 125–134. Berlin: Hamburg, Springer.

Tubular Exchanger Manufacturers Association. 1999. *Standard of the tubular exchanger manufacturers association*, 8th ed. New York: TEMA.

Chapter 10
Fundamentals of Thermal Radiation

Abstract This chapter aims to study the fundamentals of thermal radiation. Thermal radiation is the radiation emitted by bodies as a function of the fourth power of their absolute temperature. Planck's black body radiation law is presented, which establishes the emissive power of the black body as a function of temperature and wavelength. The integration of this law for all wavelengths results in the Stefan-Boltzmann law. The properties of thermal radiation are presented: absorptivity, transmissivity, emissivity, and reflectivity. The concept of black cavity radiation is discussed, as well as Kirchhoff's law between emissivity and absorptivity. Next, the approach of gray bodies as radiation models for real bodies is presented. The following section deals with the calculations of thermal radiation flux on surfaces. Finally, the chapter concludes with the presentation of concepts of environmental thermal radiation.

Keywords Thermal radiation · Planck's law · Stefan-Boltzmann law · Radiation properties · Kirchhoff's law · Environmental radiation

10.1 Introduction to Thermal Radiation

Thermal radiation is the third form of heat transfer. Out of the forms of heat transfer, it is the most interesting and intriguing, as all bodies emit thermal radiation because thermal radiation emission depends on the body's absolute temperature, more precisely on the fourth power of its absolute temperature. Thus, everything around us emits thermal radiation, including ourselves. In addition to bodies emitting, they also absorb part of the incident thermal radiation. Unlike heat transfer by conduction and convection, thermal radiation does not need a material medium to occur. This is how thermal radiation, as well as other forms of radiation emitted by the Sun, runs across space to reach the Earth bringing heat and light to sustain all life.

The physical mechanism of energy transport by thermal radiation, and by electromagnetic radiation in a broader sense, has yet to be definitively elucidated. In certain

J. R. Simões-Moreira and E. W. Zavaleta-Aguilar, *Heat Transfer*, Heat and Mass Transfer, https://doi.org/10.1007/978-3-031-80318-5_10

laboratory experiments, radiant energy is considered to be transported by electromagnetic waves, while in others, it is considered to be transported by photons. These two physical models give rise to what is known as *wave-particle duality*. However, it is known that radiation travels at the speed-of-light regardless of the physical model considered. The energy associated with each photon is $h\nu$, where h is the *Planck constant*, whose value is $h = 6.625 \times 10^{-34}$ Js. The frequency, ν, is related to the wavelength, λ, through

$$c = \lambda \nu \tag{10.1}$$

where c is the speed of light, having the value $c = 3 \times 10^8$ m/s in vacuum. The unit of wavelength is the micrometer, which is 1 μm $= 10^{-6}$ m. The angstrom unit is also used, 1 Å $= 10^{-10}$ m, which corresponds to the order of magnitude of the size of an atom.

Radiation phenomena are classified by their wavelength (or frequency), as shown in Fig. 10.1. The type of radiation characterized by its wavelength depends on the source of the radiation. High-frequency electrons bombarding a metallic surface produce X-rays, while certain crystals can be excited to produce radio waves at longer wavelengths. Visible light comprises a small portion of the radiation spectrum, ranging from 0.35 to 0.7 μm. The deadly gamma and cosmic rays are forms of radiation with very short wavelengths, or high frequency and high energy. Gamma and X rays and, more generally, cosmic rays from space and the Sun are deflected or interact with the *Van Allen radiation belt*, which forms a protective electromagnetic shield around the planet, safeguarding life on Earth from these forms of radiation, which are destructive to living beings. The space outside this electromagnetic shield is deadly to human life and shielding from cosmic radiation in spacecraft is one of the major challenges for interplanetary travel or even for a visit to our natural satellite (Chancellor et al. 2018). FM and AM radio waves have frequencies below 10^8 Hz. Microwaves fall within the range of approximately 10^8 to 10^{11} Hz. *Thermal radiation* is produced by a body solely due to the fourth power of its absolute temperature and is continuous throughout the entire wavelength spectrum, as illustrated in Fig. 10.1 and will be discussed further next.

10.2 Laws of Thermal Radiation

10.2.1 Black Body Thermal Radiation

Thermal radiation is the form of radiation emitted by a body solely due to the fourth power of its absolute temperature. Nevertheless, different bodies at the same temperature do not emit the same amount of thermal radiation, what introduces the reader to the following natural question: is there a body that, at a given temperature, emits

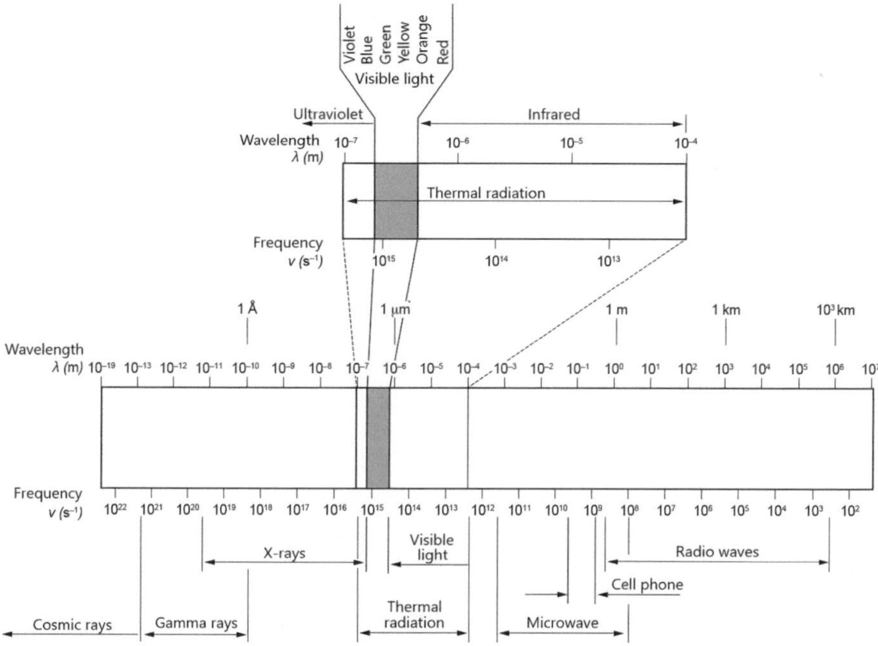

Fig. 10.1 Electromagnetic radiation spectrum and the various names according to its range

the maximum possible amount of thermal radiation? The answer is: yes! This idealized body is called a *black body*. It should be noted that the adjective *black* is not associated with the color (or absence of color) that our eyes perceive. For example, the bright Sun emits thermal radiation resembling a black body. A black body, also called an *ideal radiator*, is a body that emits and absorbs, at a given temperature, the maximum possible amount of thermal radiation at any wavelength, what makes the black body an idealization and a standard of comparison for the radiation emitted and absorbed by real bodies.

In the end of the nineteenth century, several studies in the area of thermal radiation were conducted when, in 1900, Max Planck presented his expression refined from earlier analyses, being later confirmed both experimentally and theoretically. *Planck's Law*, enables the calculation of the thermal radiation emitted by a black body at a given temperature and wavelength per unit area of the body's surface. This "amount of radiation emitted" is defined as the *black body spectral emissive power* or *monochromatic black body emissive power*, $E_{b\lambda}$, where the index b refers to the fact that it is a black body and λ refers to its spectral nature, that is, for a single wavelength of the spectrum. The units of $E_{b\lambda}$ are $W/m^2\mu m$. Planck demonstrated that the spectral emissive power of the black body is distributed according to the following expression:

$$E_{b\lambda} = \frac{C_1}{\lambda^5\left(e^{C_2/\lambda T} - 1\right)} \tag{10.2}$$

where

$$C_1 = 8\pi hc = 3.7415 \times 10^8\ \text{W}\mu\text{m}^4/\text{m}^2 = 3.7415 \times 10^{-16}\ \text{Wm}^2;$$

$$C_2 = \frac{hc}{k_B} = 1.4388 \times 10^4 \mu\text{mK} = 1.4388 \times 10^{-2}\text{mK};$$

$k_B = 1.381 \times 10^{-23}\ \text{J/K}$ is the Boltzmann constant;
h, c are the Planck constant and the speed-of-light already defined

The spectral emissive power of the black body, $E_{b\lambda}$, is therefore a function of the wavelength, λ, and the absolute temperature of the body, T, that is, $E_{b\lambda} = f(\lambda, T)$. To highlight the dependence of these two quantities, various isothermal curves are shown in the graph in Fig. 10.2 as a function of the wavelength. Emissions from bodies at certain temperatures are also shown in the graphics for the sake of comparison.

Based on Planck's law, shown in its graphical representation in Fig. 10.2, some relevant information about the thermal radiation of a black body can be extracted, highlighting the following:

- The thermal radiation emitted by a black body is a continuous magnitude that spans for all wavelengths. That is, it is a quantity distributed continuously from $\lambda = 0$ to the longest possible wavelength ($\lambda = \infty$);
- At a given wavelength, λ, the black body spectral emissive power, $E_{b\lambda}$, increases with temperature;
- The thermal radiation at a given temperature tends to concentrate at shorter wavelengths—the distribution is not a symmetric one;
- Thermal radiation has a maximum emission point as a function of wavelength for a given temperature, with the maximum point given by the *Wien's displacement law*, presented next;
- A substantial fraction of the thermal radiation emitted by the Sun falls within the visible region (0.35–0.7 μm). Sun thermal radiation spectrum is similar to the spectrum of a black body at 5800 K (or more precisely 5778 K).

The spectral emissive power distribution reaches a maximum for a given temperature. From Calculus, the maximum takes place where the partial derivative of Planck's law with respect to λ for an isothermal is null, given by:

$$\left(\frac{\partial E_{b\lambda}}{\partial \lambda}\right)_T = \frac{\partial}{\partial \lambda}\left(\frac{C_1}{\lambda^5\left(e^{C_2/\lambda T} - 1\right)}\right)_T = 0 \Rightarrow 1 - e^{\frac{C_2}{\lambda_{\max} T}} - \frac{C_2}{5\lambda_{\max} T} = 0,$$

whose numerical solution results in $\lambda_{\max}[\mu\text{m}]$.

$$\lambda_{\max} = \frac{2897.8}{T} \tag{10.3}$$

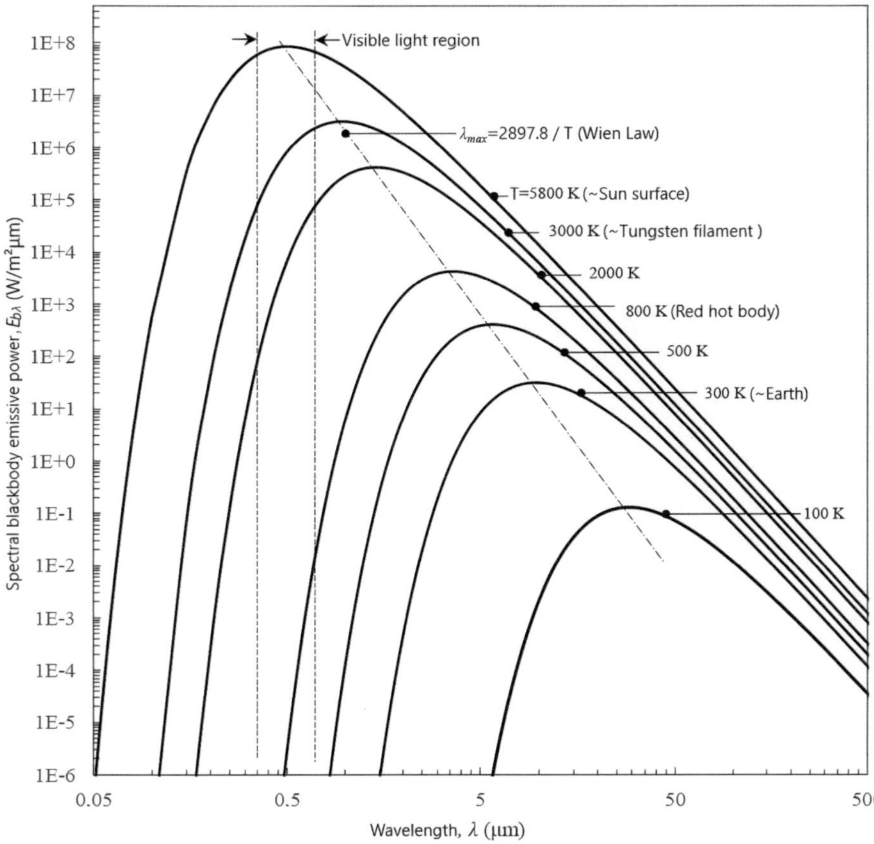

Fig. 10.2 Spectral emissive power of a black body as a function of temperature and wavelength

This is *Wien's displacement law*, which is shown in Fig. 10.2 as the dashed line connecting the maximum points of various isothermal curves.

Example 10.1 (*Maximum emissive power of the sun*)

Considering the Sun as a black body with a 5800 K surface temperature, it is asked: (a) determine the wavelength at which its spectral emissive power is maximum and its value; (b) suppose a giant red star, what is the approximate temperature of its surface?

Solution

(a) According to Wien's Law (Eq. 10.3): $\lambda_{\max} = \frac{2897.8}{5800} = 0.5\ \mu\text{m}$, therefore

$$E_{b\lambda} = \frac{3.7415 \times 10^8}{0,5^5\left[e^{1.4388\times10^4/(0.5\times5800)} - 1\right]} = 8 \times 10^7\ \text{W/m}^2\mu\text{m}$$

The actual color of our Sun is white. However, as the Sun's radiation crosses the atmosphere, the blue or violet wavelength is scattered (Rayleigh scattering effect), which gives the sky its blue color. On the other hand, the orange and yellow colors wavelengths pass through giving the impression that this wavelength combination is the color of our Sun.

(b) By analyzing Figs. 10.1 and 10.2, it is noted that bodies, like red giant stars with surface temperatures around 2000–3000 K, emit maximum radiation at wavelengths in the infrared region, along with visible red. Conversely, blue stars have higher surface temperatures (20,000–30,000 K).

The spectral emissive power of a black body reveals its wavelength dependence. However, in most practical situations, it is necessary to obtain the total radiation emitted at all wavelengths at a given temperature. This *total emissive power of the black body*, E_b, can be determined by integrating the spectral emissive power, (Eq. 10.2), for all wavelengths λ, from 0 to ∞. Thus:

$$E_b = \int_0^\infty E_{b\lambda} d\lambda = \int_0^\infty \frac{C_1}{\lambda^5 \left(e^{C_2/\lambda T} - 1\right)} d\lambda$$

let the following change of variable: $\xi = \frac{C_2}{\lambda T} \Rightarrow d\lambda = -\frac{C_2}{T\xi^2} d\xi$, substituting the new variable into the integral, we get:

$$E_b = -\int_\infty^0 \frac{C_1}{\left(\frac{C_2}{\xi T}\right)^5 \left(e^\xi - 1\right)} \frac{C_2}{T\xi^2} d\xi = \frac{C_1 T^4}{C_2^4} \underbrace{\int_0^\infty \frac{\xi^3}{\left(e^\xi - 1\right)} d\xi}_{\frac{\pi^4}{15}} = \underbrace{\frac{\pi^4 C_1}{15 C_2^4}}_{\sigma} T^4 \Rightarrow$$

$$E_b = \sigma T^4 \tag{10.4}$$

where $\sigma = 5.669 \times 10^{-8}\ \text{W/m}^2\text{K}^4$ is the *Stefan-Boltzmann constant* (S-B).

This is known as the *Stefan-Boltzmann Law* of black-body thermal radiation. Note that σ is related to the constants C_1 and C_2 in the Planck equation. Stefan derived this expression in 1879 from the analysis of Tyndall's experimental data. Later, Boltzmann, who had been his student, derived it theoretically. Thus, the Stefan-Boltzmann Law precedes chronologically the Planck Law.

10.2.2 Fraction of Black Body Thermal Radiation

In certain cases, it is necessary to determine the thermal radiation emitted within a specific band or range of the total black body radiation spectrum. To achieve this, the *fraction of thermal radiation*, $F_{[0-\lambda_1]}$, emitted by a black body in the wavelength range $[0 - \lambda_1]$ is defined by Eq. (10.5),

$$F_{[0-\lambda_1]} = \frac{E_{b,0-\lambda_1}}{E_b} = \frac{\int_0^{\lambda_1} \frac{C_1}{\lambda^5 (e^{C_2/\lambda T} - 1)} d\lambda}{\sigma T^4} \tag{10.5}$$

Figure 10.3a shows a graph of the fraction of thermal radiation in the range 0–λ as a function of the product λT. The tabulated values of this quantity can be found in Table 10.1. Figure 10.3b illustrates the case where one wants to determine the total emissive power within the spectral band between the wavelengths λ_1 and λ_2. To achieve this, one can integrate Planck's Law (Eq. 10.2) between the two specified wavelengths, as indicated in the figure. Alternatively, one can use the concept of the fraction of thermal radiation within the spectral band $\lambda_1 - \lambda_2$, which is:

$$F_{[\lambda_1-\lambda_2]} = F_{[0-\lambda_2]} - F_{[0-\lambda_1]} \tag{10.6}$$

So the black body emissive power in the specified interval is given by:

$$E_{b,\,\lambda_1-\lambda_2} = F_{[\lambda_1-\lambda_2]} E_b \tag{10.7}$$

Table 10.1 also shows the *normalized spectral emissive power* by the total black body emissive power as a function of the product λT, according to the following expression from Planck's Law (Eq. 10.2):

$$\frac{E_{b\lambda}}{\sigma T^5} = \frac{C_1}{\sigma (\lambda T)^5 \left(e^{C_2/\lambda T} - 1\right)} \tag{10.8}$$

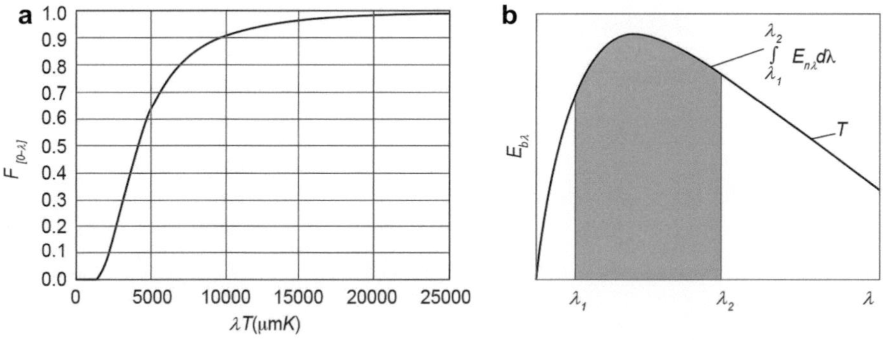

Fig. 10.3 Black body thermal radiation: **a** fraction of thermal radiation as a function of the product λT; **b** thermal radiation between wavelengths λ_1 and λ_2

Table 10.1 Fraction of thermal radiation and spectral emissive power as a function of λT

λT (μmK)	$F_{[0-\lambda]}$	$\frac{E_{b\lambda}}{\sigma T^5}$ (μmK)$^{-1}$	λT (μmK)	$F_{[0-\lambda]}$	$\frac{E_{b\lambda}}{\sigma T^5}$ (μmK)$^{-1}$
200	0.0000	1.18×10^{-27}	6200	0.7542	7.85×10^{-5}
400	0.0000	1.54×10^{-13}	6400	0.7692	7.26×10^{-5}
600	0.0000	3.27×10^{-9}	6600	0.7832	6.72×10^{-5}
800	0.0000	3.11×10^{-7}	6800	0.7962	6.22×10^{-5}
1000	0.0003	3.72×10^{-6}	7000	0.8082	5.77×10^{-5}
1200	0.0021	1.65×10^{-5}	7200	0.8193	5.35×10^{-5}
1400	0.0078	4.22×10^{-5}	7400	0.8296	4.97×10^{-5}
1600	0.0197	7.83×10^{-5}	7600	0.8392	4.61×10^{-5}
1800	0.0393	1.18×10^{-4}	7800	0.8481	4.29×10^{-5}
2000	0.0667	1.55×10^{-4}	8000	0.8563	4.00×10^{-5}
2200	0.1009	1.85×10^{-4}	8500	0.8747	3.35×10^{-5}
2400	0.1402	2.07×10^{-4}	9000	0.8901	2.83×10^{-5}
2600	0.1831	2.20×10^{-4}	9500	0.9031	2.40×10^{-5}
2800	0.2279	2.26×10^{-4}	10,000	0.9143	2.05×10^{-5}
3000	0.2732	2.26×10^{-4}	10,500	0.9238	1.76×10^{-5}
3200	0.3181	2.22×10^{-4}	11,000	0.9320	1.52×10^{-5}
3400	0.3617	2.14×10^{-4}	11,500	0.9390	1.32×10^{-5}
3600	0.4036	2.04×10^{-4}	12,000	0.9452	1.14×10^{-5}
3800	0.4434	1.93×10^{-4}	13,000	0.9552	8.78×10^{-6}
4000	0.4809	1.82×10^{-4}	14,000	0.9630	6.84×10^{-6}
4200	0.5160	1.70×10^{-4}	15,000	0.9691	5.40×10^{-6}
4400	0.5488	1.58×10^{-4}	16,000	0.9739	4.32×10^{-6}
4600	0.5793	1.47×10^{-4}	18,000	0.9809	2.85×10^{-6}
4800	0.6076	1.36×10^{-4}	20,000	0.9857	1.96×10^{-6}
5000	0.6338	1.26×10^{-4}	25,000	0.9923	8.69×10^{-7}
5200	0.6580	1.16×10^{-4}	30,000	0.9954	4.41×10^{-7}
5400	0.6804	1.08×10^{-4}	40,000	0.9981	1.49×10^{-7}
5600	0.7011	9.94×10^{-5}	50,000	0.9990	6.33×10^{-8}
5800	0.7202	9.18×10^{-5}	75,000	0.9998	1.32×10^{-8}
6000	0.7379	8.49×10^{-5}	100,000	1.0000	4.27×10^{-9}

Example 10.2 (*Solar radiation within the visible range*)

Solar radiation has approximately the same spectral distribution as an ideal black body radiator at 5800 K. Find the amount of solar radiation (W/m^2) emitted within the visible wavelength band (0.35 to 0.7 μm).

Solution

From Table 10.1 (by interpolation):

$$0 \leq \lambda \leq 0.35 \rightarrow \lambda_1 T = 0.35 \times 5800 = 2030\ \mu\text{mK} \Rightarrow F_{[0-0.35]} = 0.0718$$

$$0 \leq \lambda \leq 0.7 \rightarrow \lambda_2 T = 0.7 \times 5800 = 4060\ \mu\text{mK} \Rightarrow F_{[0-0.7]} = 0.4914$$

Therefore, the fraction of radiation (Eq. 10.6) within the visible band (0.35–0.7 μm) is:

$$F_{[0.35-0.7]} = F_{[0-0.7]} - F_{[0-0.35]}$$

$$F_{[0.35-0.7]} = 0.4914 - 0.0718 = 0.4196$$

Thus, using Eq. (10.7), we have:

$$E_b = 0.4196 \times 5.669 \times 10^{-8} \times 5800^4 = 2.7 \times 10^7\ \text{W/m}^2$$

Comment: about 42% of solar thermal radiation is emitted within the visible band. A superficial inspection of the di-log graph in the figure on the left below may lead to the wrong conclusion that the amount of solar radiation within the visible band is small. However, the graph on a linear scale in the figure on the right-hand side would correctly show the fraction of visible radiation within the visible band.

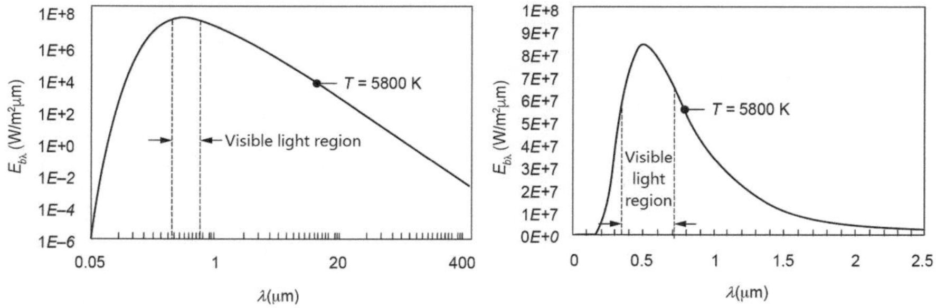

Example 10.3 (*Incandescent lamp*)

Incandescent lamps were used until recently for lighting, but they were banned for their high electrical consumption. This occurred because their brightness comes from the thermal radiation emitted by the tungsten filament, which typically reaches a temperature of approximately 2850 K. Under these conditions, consider the following: (a) determine the wavelength at which the thermal emission is maximum; (b) is this wavelength within the visible spectrum? (c) make a graph of $E_{b\lambda}$ as a function of λ for this temperature, using linear scales and estimate the fraction of the total lamp emission that lies within the visible region of the spectrum, i.e., between 0.35 and 0.7 μm; and (d) if the bulb consumes 60 W of electrical power, what is the maximum amount of visible light it produces, in watts? Assume that the lamp emits thermal radiation as a black body.

Solution

(a) From Wien's displacement law, Eq. (10.3): $\lambda_{max} \times 2850 = 2897.8$, thus, the wavelength at which the radiation is maximum will be: $\lambda_{max} = 1.02\,\mu m$, within the infrared band;

(b) The visible range lies within the 0.35 and 0.7 μm band, thus the wavelength at which the maximum spectral emissive power occurs, $\lambda_{max} = 1.02\,\mu m$, is outside the visible range;

(c) Graph of emissive power as a function of wavelength:

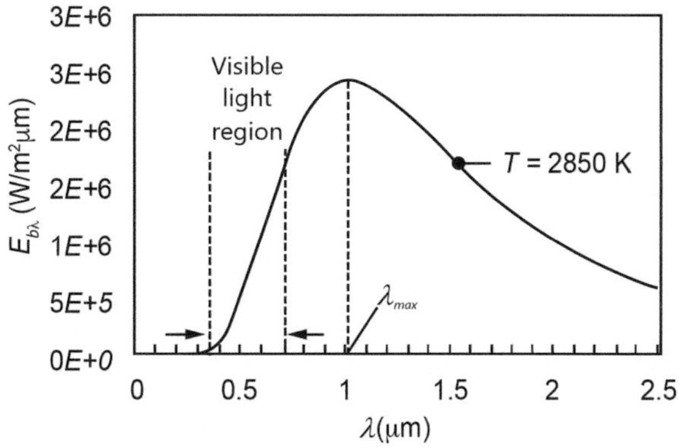

From Table 10.1 (by interpolation):

$$0 \le \lambda \le 0.35 \rightarrow \lambda_1 T = 0.35 \times 2850 = 998\ \mu mK \Rightarrow F_{[0-0.35]} = 0.0003$$

$$0 \le \lambda \le 0.7 \rightarrow \lambda_2 T = 0.7 \times 2850 = 1995\ \mu mK \Rightarrow F_{[0-0.7]} = 0.0660$$

$$F_{[0.35-0.7]} = 0.0660 - 0.0003 = 0.0657$$

(d) At a temperature of 2850 K, the incandescent lamp emits only 6.6% of the thermal radiation in the visible range. Thus, at most, it produces $0.0657 \times 60 = 4$ W of luminous power in this range.

10.2.3 Intensity of Thermal Radiation

The thermal radiation emitted by a surface, or its emissive power, outspreads in all directions. A relevant question is to determine the amount of thermal radiation in a given angular direction emitted from a surface element and that will reach a second surface. For this, it is necessary to define the *solid angle*, ω, which represents the angle formed by an area element on the surface of a sphere of radius r. To understand the definition and function of this quantity, an analogy is usually established between a differential plane angle $d\alpha$ and the differential solid angle $d\omega$, as well illustrated in Fig. 10.4.

The differential solid angle is given by:

$$d\omega = \frac{dA_n}{r^2} \tag{10.9}$$

where dA_n is the area element of the sphere of radius r. The unit of the solid angle is *steradian*, whose symbol is sr, and varies between 0 and 4π sr on a unit radius sphere. A hemispherical surface of unit radius has 2π sr.

Consider a surface element dA_1 that emits thermal radiation in a direction θ, measured from the straight line normal to the surface (zenith angle), and azimuthal direction ϕ, as illustrated in Fig. 10.5a. The *intensity of radiation* $I(\theta, \phi)$ is defined as the rate of radiant energy emitted by the area element dA_1 in the elementary solid angle $d\omega$, whose unit is W/m^{2}sr. This emitted radiant energy crosses the elementary area dA_n located on the surface of the enveloping hemisphere of radius r, as illustrated in Fig. 10.5a.

Figure 10.5b illustrates the calculation of the elementary area element dA_n on the surface of the hemisphere, given by:

$$dA_n = r^2 \sin\theta d\theta d\phi \tag{10.10}$$

Substituting in the expression of the solid angle (Eq. 10.9), it comes that:

$$d\omega = \sin\theta d\theta d\phi \tag{10.11}$$

Let $d\dot{Q}$ be the rate of radiant energy emitted by the surface element dA_1 that crosses the area element dA_n on the hemisphere. The projection of the area element is $dA_1 \cos\theta$ in the direction θ. Therefore, it can be shown that the relationship between

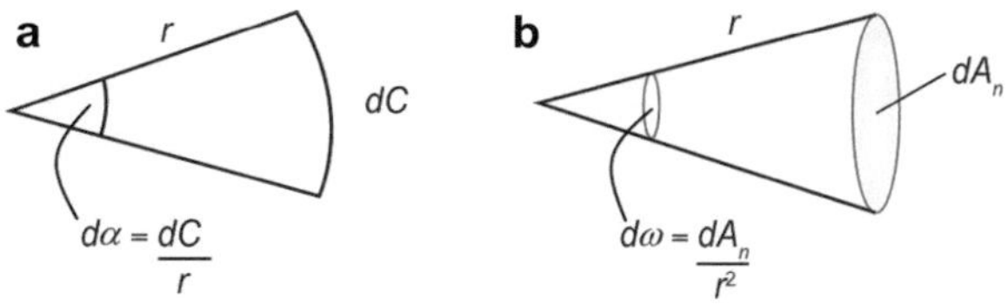

Fig. 10.4 **a** Differential plane angle $d\alpha$; **b** differential solid angle $d\omega$

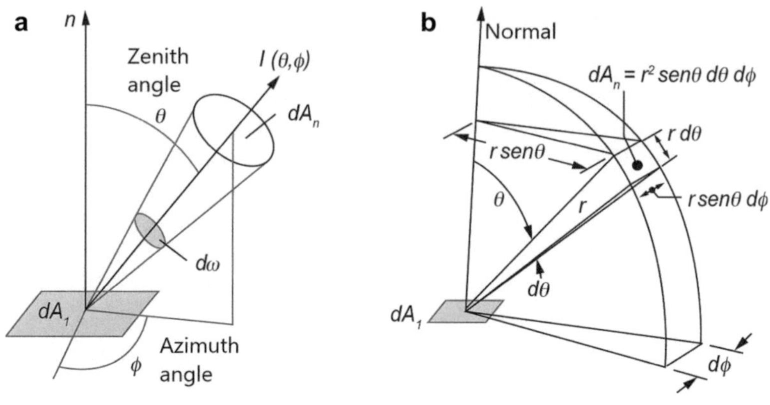

Fig. 10.5 **a** Radiation intensity $I(\theta, \phi)$ emitted by an elementary surface dA_1 in the direction (θ, ϕ) and the elementary solid angle in that direction $d\omega$ that crosses the elementary area of the hemispherical surface dA_n; **b** area element dA_n on the surface and its calculation

$d\dot{Q}$ and the radiation intensity is given by:

$$\frac{d\dot{Q}}{dA_1} = I(\theta, \phi) \cos\theta d\omega \tag{10.12}$$

On the other hand, the elementary solid angle $d\omega$ is given by Eq. (10.11). Thus, substituting this quantity into the previous equation, we get:

$$\frac{\dot{Q}}{A_1} = \int_0^{2\pi} \int_0^{\frac{\pi}{2}} I(\theta, \phi) \cos\theta \sin\theta d\theta \tag{10.13}$$

The thermal radiation emitted by black bodies is generally considered *diffuse*, that is, it is distributed uniformly in all directions, which means that the emission obeys *Lambert's cosine law*. In this case, the intensity of black body radiation, I_b, is independent of the directions (θ, ϕ) and can come out of the integration signal of Eq. (10.13). On the other hand, the ratio $\frac{\dot{Q}}{A_1}$ is the very black body emissive power E_b of the surface, which results from the double integration of Eq. (10.13) into the value π. So that:

$$E_b = \pi I_b \tag{10.14}$$

The same applies to the spectral emissive power of a black body, $E_{b\lambda}$ (whose value can be found by Planck's Law, Eq. (10.2), that is:

$$E_{b\lambda} = \pi I_{b\lambda} \tag{10.15}$$

10.3 Surface Properties for Thermal Radiation

When radiant energy hits a surface, part of the thermal radiation can be reflected, part can be absorbed, and part can be transmitted, as schematically shown in the diagram in Fig. 10.6.

Based on the scheme in Fig. 10.6, the following properties of incident radiation on a surface are defined:

α *absorptivity*—the fraction of incident radiation absorbed by the surface;
ρ *reflectivity*—the fraction of incident radiation reflected from the surface; and
τ *transmissivity*—the fraction of incident radiation transmitted through the surface.

Thus, by energy conservation, the sum of the three fractions must be one, that is:

$$\alpha + \rho + \tau = 1 \tag{10.16}$$

Many solid bodies do not transmit thermal radiation, having their transmissivity null, $\tau = 0$. For these cases of *opaque bodies* to thermal radiation, we have:

$$\alpha + \rho = 1 \tag{10.17}$$

These properties can also be spectral or monochromatic, that is:

$$\alpha_\lambda + \rho_\lambda + \tau_\lambda = 1 \tag{10.18}$$

in which α_λ is the *spectral* or *monochromatic absorptivity*, or fraction of incident radiation absorbed by the surface at the wavelength λ; ρ_λ, the *spectral* or *monochromatic reflectivity*, or fraction of the incident radiation reflected from the surface at the wavelength λ; and τ_λ is the *spectral* or *monochromatic transmissivity* or fraction of the incident radiation transmitted through the surface at the wavelength λ.

As an example, consider the spectral transmissivity of a commercial glass widely used in household tackles, whose wavelength dependence curve is illustrated in

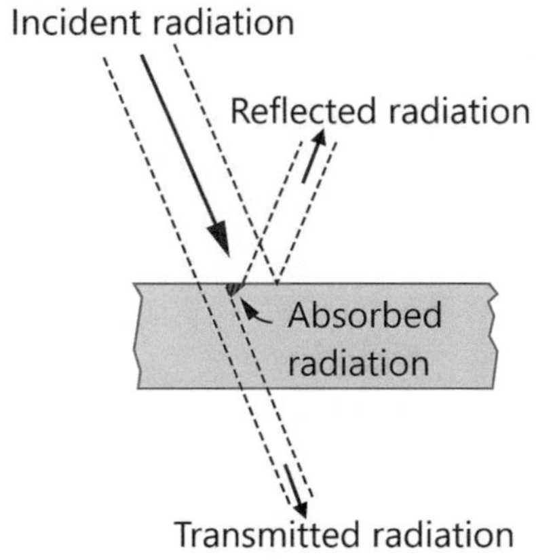

Fig. 10.6 Incident, reflected, absorbed, and transmitted radiation

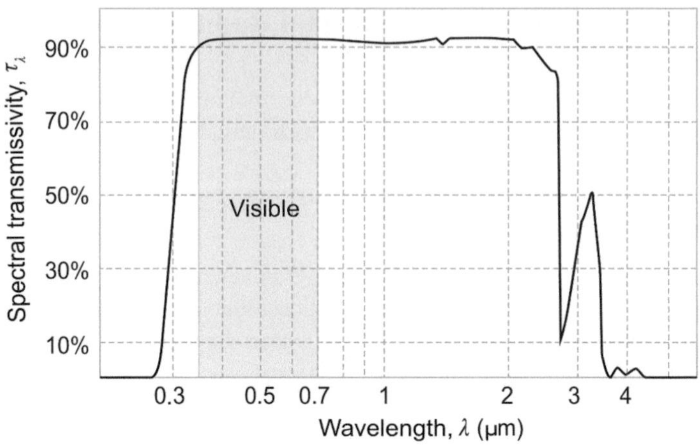

Fig. 10.7 Transmissivity of a commercial borosilicate glass 2 mm thick. Courtesy of Praezisions Glas and Optik, Germany—www.pgo-online.com

Fig. 10.7. The transmissivity is virtually zero in the ultraviolet region and above the near-infrared region, which means the glass is opaque in these regions. However, in the band between these two regions, including the entire visible and near-infrared region, the transmissivity is around 92%.

The thermal radiation emitted by a surface varies according to the surface material and finishing and view direction along with the temperature. The dependence of these surface characteristics results in the definition of the *emissivity* discussed next.

Example 10.4 (*Transmissivity in commercial glass*)

Someone stands behind a glass window whose transmissivity curve is indicated in figure. Assuming that solar radiation incidence is perpendicular to the glass, determine the spectrum of the solar thermal radiation ($T_{sun} = 5800$ K) transmitted through the glass, and the radiation flux as well. Assume that the solar radiation power reaching the glass is 1000 W/m^2.

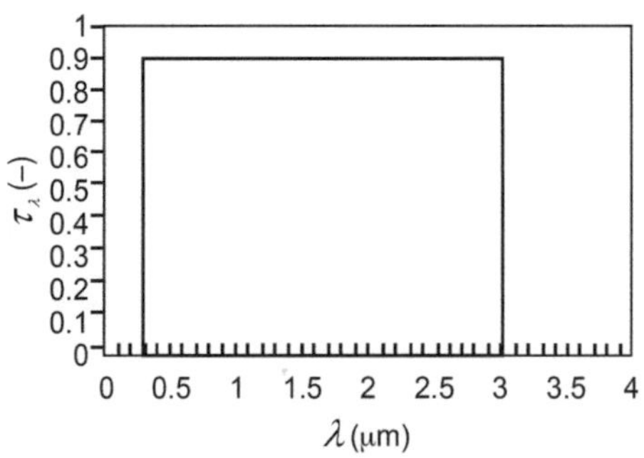

Solution

Figures below show the spectral emissive power of a black body at 5800 K and the glass transmissivity curve, whose non-zero value is between 0.3 and 3 μm and zero everywhere else.

Of the irradiation of 1000 W/m^2 that hits the glass, 90% are transmitted in the wavelength range of 0.3–3 μm. Outside this range, the glass is opaque.

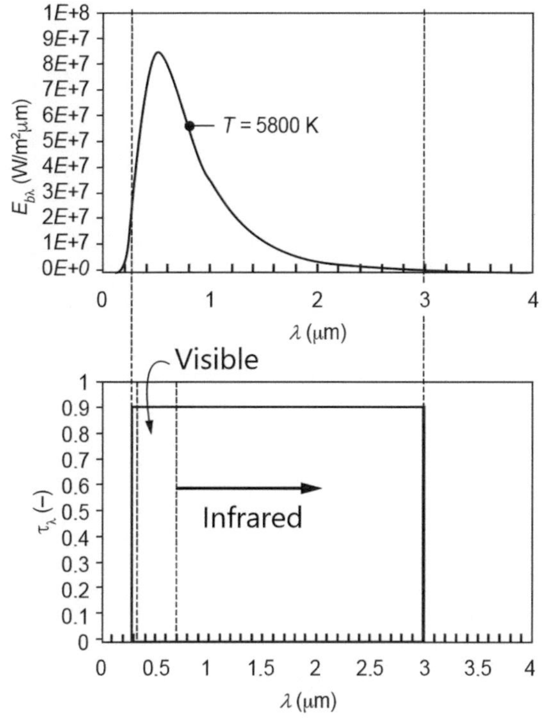

Irradiation of the Sun that reaches the glass: 1000 W/m^2.
From Table 10.1, we have:

$$0 \le \lambda \le 0.3 \rightarrow \lambda_1 T = 0.3 \times 5800 = 1740\,\mu\text{mK} \rightarrow F_{[0-0.3]} = 0.0335$$

$$0 \le \lambda \le 3 \rightarrow \lambda_2 T = 3 \times 5800 = 17400\,\mu\text{mK} \rightarrow F_{[0-3]} = 0.9788$$

Therefore, $F_{[0.3-3]} = F_{[0-3]} - F_{[0-0.3]} = 0.9788 - 0.0335 = 0.9453$.
Radiation flux transmitted through the glass:

$$E = \tau \times F_{[0.3-3]} \times E_{sun} = 0.9 \times 0.9788 \times 1000 = 881\,\text{W/m}^2$$

10.3.1 Emissivity, ε, and Irradiation, G

The *actual emissive power* of a body's surface is the rate of radiation emitted by the body, which can be spectral, E_λ, or total, E. The *spectral emissivity* or *monochromatic*, ε_λ, is defined as the ratio between the actual spectral emissive power, E_λ, and the spectral emissive power of a black body at the same temperature, $E_{b\lambda}$. That is:

$$\varepsilon_\lambda = \frac{E_\lambda}{E_{b\lambda}} \tag{10.19}$$

It should be noted that the emissivity of real bodies may not be uniform in all directions, wavelengths, and temperature. For example, consider Fig. 10.8, where polar diagrams of the directional emissivity ε_θ for various materials. On the left side of Fig. 10.8, the values of ε_θ for different non-metallic materials are shown, while on the right side, the values for metallic materials are presented. When comparing the two classes of materials, it can be inferred that the directional emissivity of non-metallic materials decreases with an increase in the zenith angle θ. In contrast, the opposite behavior is observed for metallic materials.

The *spectral emissivity*, ε_λ, can be obtained from the *spectral directional emissivity*, ε_θ, by integration, given by:

$$\varepsilon_\lambda = \int_0^{\frac{\pi}{2}} \varepsilon_\theta \sin 2\theta d\theta \tag{10.20}$$

Figure 10.9 shows the *normal spectral emissivity*, $\varepsilon_{\lambda n}$, for aluminum and copper as a function of wavelength for room temperature.

The *total hemispherical emissivity*, or *average*, ε, is defined as the ratio between the emissive power of the body, E, and the emissive power of a black body, E_b, at the same temperature, as given by Eq. (10.21):

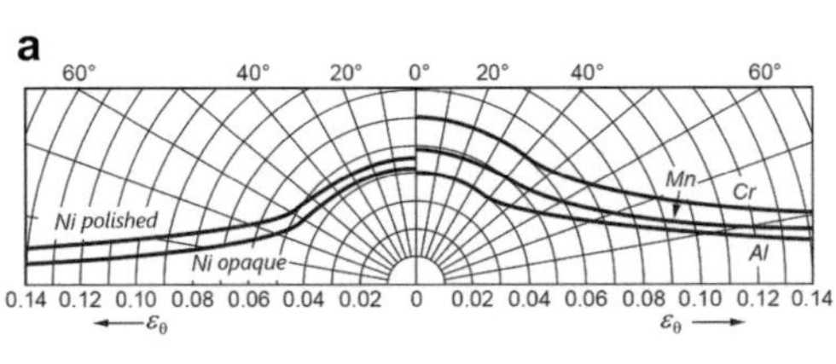

Fig. 10.8 Directional emissivity for various materials. On the left side, non-conductive materials: **a** ice; **b** wood; **c** glass; **d** paper; **e** clay; **f** copper oxide; **g** aluminum oxide. On the right side, conductive materials (Schmidt and Eckert, 1935)

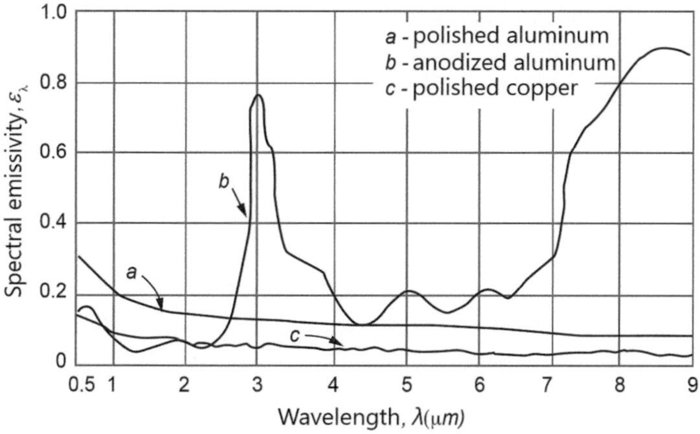

Fig. 10.9 Normal spectral emissivity ($\varepsilon_{\lambda n}$) of some metals at room temperature. Adapted from Kreith and Bohn (2001)

$$\varepsilon = \frac{E}{E_b} = \frac{\int_0^\infty \varepsilon_\lambda E_{b\lambda} d\lambda}{\int_0^\infty E_{b\lambda} d\lambda} = \frac{\int_0^\infty \varepsilon_\lambda E_{b\lambda} d\lambda}{\sigma T^4} \tag{10.21}$$

If it is possible to approximate the spectral emissivity in n wavelength intervals where it can be assumed constant, as illustrated in Fig. 10.10, the corresponding total emissivity would be:

$$\varepsilon = \varepsilon_{\lambda_1} \frac{\int_0^{\lambda_1} E_{b\lambda} d\lambda}{\sigma T^4} + \varepsilon_{\lambda_2} \frac{\int_{\lambda_1}^{\lambda_2} E_{b\lambda} d\lambda}{\sigma T^4} + \cdots \varepsilon_{\lambda_n} \frac{\int_{\lambda_n}^{\infty} E_{b\lambda} d\lambda}{\sigma T^4}$$

$$\varepsilon = \varepsilon_{\lambda_1} F_{[\lambda_1 - \lambda_2]} + \cdots \varepsilon_{\lambda_n} F_{[\lambda_{n-1} - \infty]}$$

$$\varepsilon = \varepsilon_{\lambda_1} F_{[0 - \lambda_1]} + \varepsilon_{\lambda_2} F_{[\lambda_1 - \lambda_2]} + \cdots \varepsilon_{\lambda_n} \left(1 - F_{[0 - \lambda_{n-1}]}\right) \tag{10.22}$$

The total hemispherical emissivity is a property of the material and its surface finishing and temperature. For example, at 300 K, the total emissivity for a highly polished aluminum surface is 0.04, for aluminum foil it is 0.07, and for anodized aluminum it is 0.82. A set of emissivity data can be found in Table 10.2. Figure 10.11 shows the variation of emissivity as a function of temperature and surface finishing for copper. Note that the surface finishing or treatment is a determining factor in establishing the emissivity of the surface.

Table 10.2 presents the normal emissivity, ε_n, of various metallic and non-metallic surfaces, as well as construction materials.

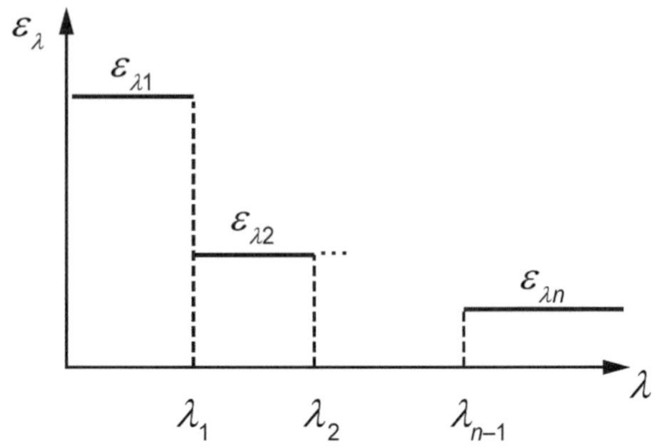

Fig. 10.10 Approximation of spectral emissivity within bands

As a general rule, the following remarks apply to total emissivity, ε, and normal emissivity, ε_n, on the surface:

- emissivity of highly polished surfaces is low;
- emissivity of metallic surfaces increases with temperature;
- oxidized and rough surfaces have a higher emissivity than those of the base material;
- the total emissivity of metallic surfaces, ε, is generally about 20% higher than the normal total emissivity on the surface, ε_n, that is, $\varepsilon \sim 1.2\ \varepsilon_n$. Conversely, in the case of non-metallic surfaces, the opposite is observed, but it tends to have a minor variation, i.e., $\varepsilon \sim 0.96\ \varepsilon_n$;
- The emissivity of non-metallic surfaces generally is higher than those of metallic surfaces. In addition, they tend to decrease with temperature, contrary to metallic surfaces tendency.

Example 10.5 (*Emissivity of paint*)

Suppose the spectral emissivity of an experimental high-temperature coating is approximated by the curve indicated below, which, ideally, is independent of temperature. Determine the total emissivity at 27 and 327 °C and the emissive power of the surface.

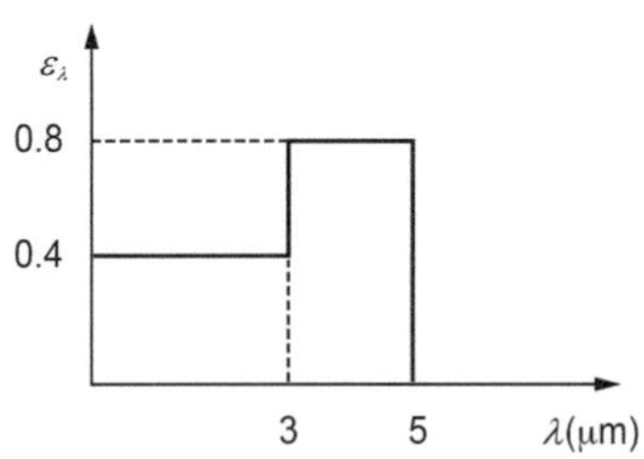

Table 10.2 Normal emissivity, ε_n, of various metallic and non-metallic surfaces, as well as construction materials

Surface	T (°C)	ε_n
Metals and their oxides		
Stainless steels		
Polished	100	0.074
Type 310 (25% Cr, 20% Ni) brown, stained, oxidized due to service in ovens	216–527	0.90–0.97
Aluminum		
Highly polished plate, 98.3% pure	277–577	0.039–0.057
Commercial sheet	100	0.09
Oxidized at 600 °C	199–599	0.11–0.19
Highly oxidized	93–504	0.20–0.31
Lead		
Pure (99.96%), not oxidized	127–200	0.057–0.075
Gray oxidized	24	0.28
Copper		
Polished	100	0.052
Plate heated to 600 °C	200–600	0.57
Cuprous oxide	800–1100	0.66–0.54
Cast Copper	1077–1277	0.16–0.13
Chrome (see nickel alloys for Ni–Cr steels)		
Polished	38–1093	0.08–0.36
Tin		
Bright Tin-plated Iron	24	0.043–0.064
Bright	50	0.06
Commercial Tin-plated Iron Sheet	100	0.07–0.08
Iron and Steel (excluding stainless)		
Metallic Surfaces (or very thin oxide layer)		
Iron, Polished	427–1027	0.14–0.38
Cast Iron, Polished	200	0.21
Wrought Iron, Highly Polished	38–427	0.28
Oxidized Surfaces		
Iron Plate, Completely Oxidized	19	0.69
Steel Plate, Rough	38–371	0.94–0.97
Cast Surfaces		
Cast Iron	1299–1399	0.29
Mild Steel	1599–1799	0.28

(continued)

Table 10.2 (continued)

Surface	T (°C)	ε_n
Metals and their oxides		
Brass		
Polished	38–316	0.1
Oxidized by heating at 600 °C	200–600	0.61–0.59
Nickel alloys		
Nickel-chrome	52–1034	0.64–0.76
Nickel-copper, polished	100	0.059
Bright nickel-chrome wire	50–1000	0.65–0.79
Oxidized nickel-chrome wire	50–1000	0.95–0.98
Gold		
Pure, highly polished	227–627	0.018–0.035
Platinum		
Pure, polished plate	227–627	0.054–0.104
Strip	927–1627	0.12–0.17
Filament	27–1227	0.036–0.192
Wire	227–1377	0.073–0.182
Silver		
Polished, pure	227–627	0.020–0.032
Polished	38–371	0.022–0.031
Tungsten		
Aged filament	27–3300	0.032–0.35
Filament	3300	0.39
Polished coating	100	0.066
Zinc		
Commercial, 99.1% pure, polished	227–327	0.045–0.053
Oxidized by heating to 400 °C	339	0.11
Refractories, building materials, paints and miscellaneous		
Water	0–100	0.95–0.963
Mortar, rough lime	10–88	0.91
Asbestos		
Plate	23	0.96
Paper	38–371	0.93–0.94
Rubber		
Hard and glossy plate	23	0.94
Soft, gray, rough (regenerated)	24	0.86

(continued)

Table 10.2 (continued)

Surface	T (°C)	ε_n
Metals and their oxides		
Carbon		
Filament	1040–1400	0.526
Carbon black coating for water glass	100–230	0.96–0.95
Thin layer of carbon on an iron plate	21	0.927
Planed oak	21	0.9
Plaster, 0.02″ thickness on a smooth or blackened plate	21	0.903
Marble, light gray, polished	22	0.93
Cover paper	21	0.91
Brick		
Red, rough, but no serious irregularity	21	0.93
Glazed brick	1100	0.75
Building brick	1000	0.45
Refractory brick	1000	0.75
Magnesite refractory brick	1000	0.38
Paints, lacquers, varnishes		
Very white enamel varnish on wrought iron plate	23	0.906
Glossy black lacquer, sprayed on iron	24	0.875
Glossy black shellac on tinned iron laminate	21	0.821
Matte black shellac	77–146	0.91
Black or white lacquer	38–93	0.80–0.95
Flat black lacquer	38–93	0.96–0.98
Oil-based paints, 16 different, all colors	100	0.92–0.96
Al paint, after heating to 325 °C	150–316	0.35
Glass		
Smooth	22	0.94
Pyrex, lead and sodium carbonate	260–538	0.95–0.85

Aadapted from McAdams (1954)—table compiled by H. C. Hottel

Solution

At 27 °C (~300 K), from graphics $\varepsilon_{\lambda_1} = 0.4$, $\varepsilon_{\lambda_2} = 0.8$, $\varepsilon_{\lambda_3} = 0$

$$\lambda_1 T = 3 \times 300 = 900\,\mu\text{mK} \rightarrow F_{[0-\lambda_1]} = 0.0002\ \text{(Table 10.1)}$$

$$\lambda_2 T = 5 \times 300 = 1500\,\mu\text{mK} \rightarrow F_{[0-\lambda_2]} = 0.0138\ \text{(Table 10.1)}$$

$$\varepsilon = 0.4 \times 0.0002 + 0.8 \times (0.0138 - 0.0002) + 0 \times (1 - 0.0138) = 0.0110$$

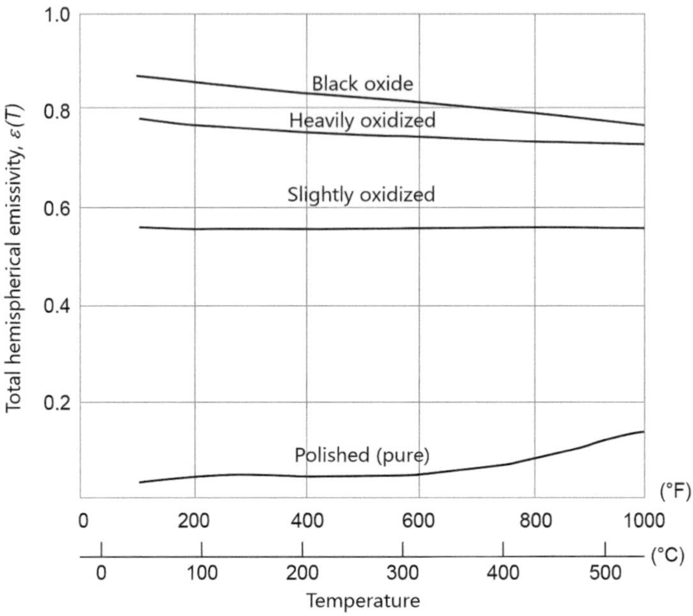

Fig. 10.11 Influence of temperature on the total hemispherical emissivity for copper considering various types of surface finish. Adapted from Kreith and Bohn (2001)

$$E = \varepsilon E_b = \varepsilon \sigma T^4 = 0.011 \times 5.67 \times 10^{-8} \times 300^4 = 5.1\,\mathrm{W/m^2}$$

At 327 °C (~600 K), from graphics (same values) $\varepsilon_{\lambda_1} = 0.4$, $\varepsilon_{\lambda_2} = 0.8$, $\varepsilon_{\lambda_3} = 0\varepsilon_{\lambda_3} = 0$

$$\lambda_1 T = 3 \times 600 = 1800\,\mu\mathrm{mK} \rightarrow F_{[0-\lambda_1]} = 0.0393\ \text{(Table 10.1)}$$

$$\lambda_2 T = 5 \times 600 = 3000\,\mu\mathrm{mK} \rightarrow F_{[0-\lambda_2]} = 0.2732\ \text{(Table 10.1)}$$

$$\varepsilon = 0.4 \times 0.0393 + 0.8 \times (0.2732 - 0.0393) + 0 \times (1 - 0.2732) = 0.2028$$

$$E = \varepsilon E_b = \varepsilon \sigma T^4 = 0.2028 \times 5.67 \times 10^{-8} \times 300^4 = 1490\,\mathrm{W/m^2}.$$

Another quantity of interest is the *spectral* or *monochromatic irradiance, G_λ*. This quantity can be understood as the rate of thermal radiation that reaches a given surface at a certain wavelength λ and has units of W/m^{2}μm. The incident thermal radiation was generated by one or more sources external to the surface. Just like emissivity, the irradiance has directional characteristics, as illustrated in Fig. 10.12 for the incident radiation rate, $I_{\lambda i}$ $(\lambda, \theta, \varphi)$ on the area element $dA_{1,}$ at the zenith angle

Fig. 10.12 Irradiation hitting an area element

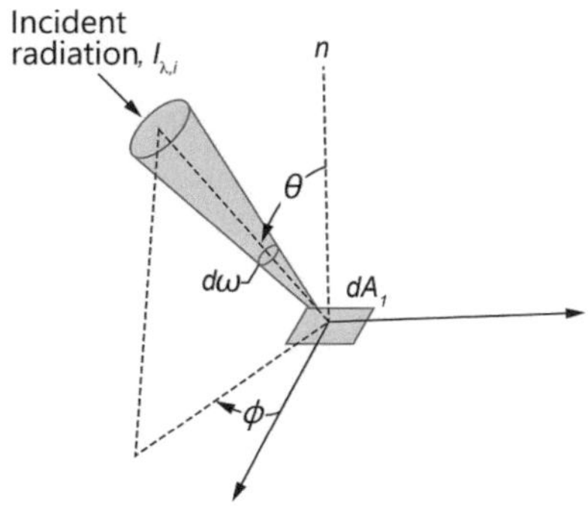

θ, azimuthal angle ϕ and differential solid angle $d\omega$.

$$G_\lambda = \int_0^{2\pi} \int_0^{\frac{\pi}{2}} I_{\lambda i}(\lambda, \theta, \phi) \cos\theta \ \sin\theta d\theta d\phi \tag{10.23}$$

Evidently, the *total irradiance*, *G*, is the integral of the spectral irradiance and has units of W/m^2, that is:

$$G = \int_0^\infty G_\lambda d\lambda \tag{10.24}$$

If the incident radiation rate is diffuse, the directional dependence is disregarded, resulting in:

$$G_\lambda = \pi I_{\lambda i} \ and \ G = \pi I_i \tag{10.25}$$

10.3.2 Kirchhoff's Law

Consider a large cavity, with a surface temperature T_c inside it. Regardless of the internal surface properties of the cavity, it will behave as a black body, since it will emit and absorb all its radiation internally. Thus, the isothermal cavity behaves like a black cavity (Sect. 10.3.3) with emissive power $E_b = \sigma T_c^4$. Now, a body is placed inside it that is in thermal equilibrium with the cavity, that is, it has the same temperature. This body receives total irradiance, $G = E_b$, as illustrated in Fig. 10.13.

On the other hand, the body placed inside the cavity has an emissive power, *E*. So that an energy balance in the body results in:

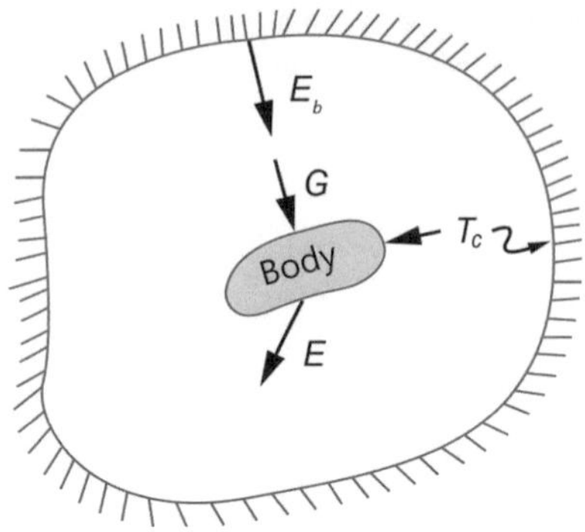

Fig. 10.13 Thermal equilibrium in a cavity and Kirchhoff's Law

$$\alpha G = E = \varepsilon E_b \tag{10.26}$$

As the irradiance G is equal to the black body emissive power of the cavity E_b, then, $\alpha E_b = \varepsilon E_b$, or:

$$\alpha = \varepsilon \tag{10.27}$$

The equality between the two properties reproduces the so-called *Kirchhoff's Law*. However, it is important to note that the total emissivity and absorptivity properties are temperature-dependent, as discussed in the previous section. Although it was derived for the case where the cavity and the body temperatures are equal, this law is still valid if the temperature difference between them is not too high.

The validity also extends to spectral or monochromatic properties. However, in this case, the only condition is that either the irradiation source or the surface that receives it is diffuse.

$$\alpha_\lambda = \varepsilon_\lambda \tag{10.28}$$

10.3.3 Grey Body

A body for which the surface emissivity and absorptivity are independent of wavelength and direction is called a *grey body*. In practice, real bodies are often approximated as grey bodies. The fundamental characteristic of the grey body approximation is that the emissivity and absorptivity are constant and independent of wavelength. Thus, for the grey body, the following relationship is valid:

$$\varepsilon = \varepsilon_\lambda = \text{constant} \tag{10.29}$$

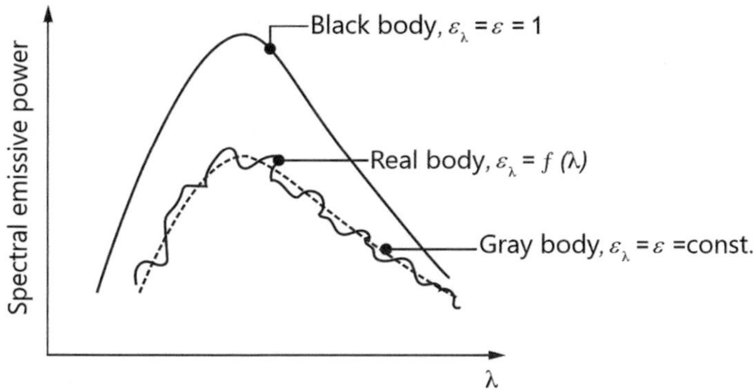

Fig. 10.14 Spectral emissive power for black, real, and grey bodies

$$\alpha = \alpha_\lambda = \text{constant}$$

The graphs in Fig. 10.14 illustrate the behavior of the emissive power of three bodies at a given temperature. As already studied in previous sections, a real body has a spectral emissivity dependency on several factors, such as the material and surface finishing, exhibiting a general pattern that depends on the wavelength, as illustrated in the figure. The black body is the one that has a unit emissivity (total and spectral) and its spectral emissive power follows Planck's Law. A grey body, on the other hand, has constant spectral and total emissivity for all wavelengths (illustrated by the dashed line).

10.3.4 Radiation Fluxes on a Surface

When thermal radiation hits a *surface*, three phenomena can occur: it can be absorbed, reflected, or transmitted by the surface as already seen. Figure 10.15 shows the reflected portion of the incident irradiation (irradiance) and the surface's emission (emissive power), which make the total radiation leaving the surface. These quantities can be spectral or total.

The irradiance, G, already defined by Eq. (10.23), is the incident radiation flux on a given surface. It can be spectral or monochromatic G_λ, or total, G.

Fig. 10.15 Surface radiation fluxes

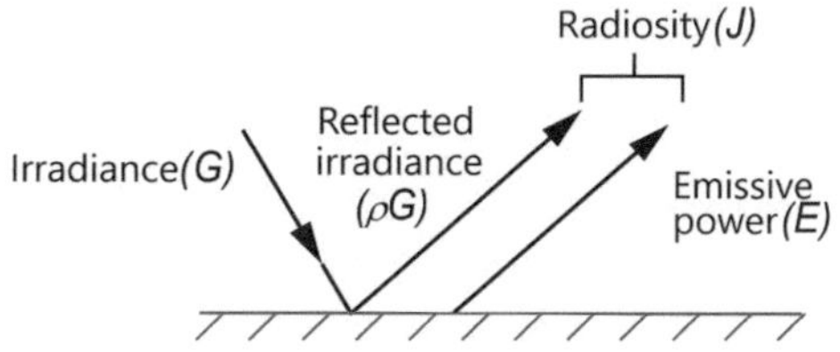

The *emissive power*, already defined in Sect. 10.3.1, constitutes the radiation flux emitted from a surface. It can be spectral or monochromatic, E_λ, or total, E.

The *radiosity*, J, is *the total radiation flux* leaving a surface. It consists of two parts: the reflected incident irradiation, ρG, and the own emission from the surface (emissive power), $E = \varepsilon E_b$. Therefore,

$$J = \rho G + \varepsilon E_b \tag{10.30}$$

On the other hand, if a surface receives the spectral irradiance G_λ, the total properties of this surface, α, ρ, and τ can be evaluated using the following equations:

$$\alpha = \frac{\int_0^\infty \alpha_\lambda G_\lambda d\lambda}{\int_0^\infty G_\lambda d\lambda} \tag{10.31}$$

$$\rho = \frac{\int_0^\infty \rho_\lambda G_\lambda d\lambda}{\int_0^\infty G_\lambda d\lambda} \tag{10.32}$$

$$\tau = \frac{\int_0^\infty \tau_\lambda G_\lambda d\lambda}{\int_0^\infty G_\lambda d\lambda} \tag{10.33}$$

If the emission source is a black body, then $G_\lambda = E_{b\lambda}$ and if also the spectral radiation properties on the surface are constant in certain wavelength bands, as is the case, for example, where there are three wavelength bands in which it is constant, then:

$$\alpha = \alpha_{\lambda_1} F_{[0-\lambda_1]} + \alpha_{\lambda_2} F_{[\lambda_1-\lambda_2]} + \alpha_{\lambda_3} F_{[\lambda_2-\infty]} \tag{10.34}$$

$$\rho = \rho_{\lambda_1} F_{[0-\lambda_1]} + \rho_{\lambda_2} F_{[\lambda_1-\lambda_2]} + \rho_{\lambda_3} F_{[\lambda_2-\infty]} \tag{10.35}$$

$$\tau = \tau_{\lambda_1} F_{[0-\lambda_1]} + \tau_{\lambda_2} F_{[\lambda_1-\lambda_2]} + \tau_{\lambda_3} F_{[\lambda_2-\infty]} \tag{10.36}$$

The bands of wavelength properties considered constant do not necessarily coincide with each other property.

Example 10.6 (*Equilibrium temperature*)

On a hot, cloudless summer day, an asphaltic road pavement receives 700 W/m^2 of solar irradiance, G_{sun}. The effective sky temperature is 290 K. Heat is transferred from the surface to the air at 27 °C by natural convection with a $h = 8\ \mathrm{W/m^{2\circ}C}$. Assume that no heat is conducted to the ground, $q_{ground} = 0$. The asphalt's absorptivity for solar radiation is $\alpha_{sun} = 0.92$, while the average emissivity of the road surface is $\varepsilon = 0.15$ for low temperature. Determine the temperature of equilibrium, T, of the asphalt pavement.

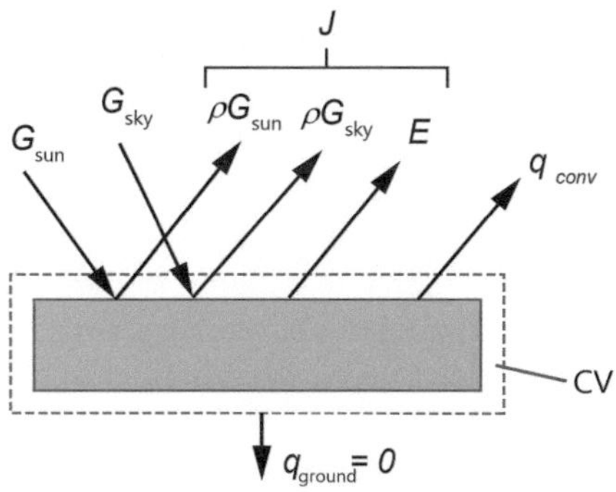

Solution

Let the control volume, CV, indicated in the figure that encompasses a piece of the road pavement surface. An energy balance can be performed at steady state:

$$\text{Energy balance:}\ \begin{bmatrix}\text{Energy rate}\\ \text{entering the CV}\end{bmatrix} = \begin{bmatrix}\text{Energy rate}\\ \text{leaving the CV}\end{bmatrix}.$$

Knowing that the area is the same for all heat fluxes, the previous equation can be expressed in terms of all heat fluxes, in this case, only heat and thermal radiation.

$$G_{sun} + G_{sky} = J + q_{conv} + \underbrace{q_{ground}}_{=0} \Rightarrow$$

$$G_{sun} + G_{sky} = \underbrace{\rho_{sun} G_{sun} + \rho_{sky} G_{sky} + \varepsilon E_b}_{J} + q_{conv} \Rightarrow$$

$$\underbrace{(1 - \rho_{sun})}_{\alpha_{sun}} G_{sol} + \underbrace{(1 - \rho_{sky})}_{\alpha_{sky}} G_{sky} = q_{conv} + \varepsilon E_b \Rightarrow$$

$$\alpha_{sun} G_{sun} + \varepsilon \sigma T_{sky}^4 - h(T - T_\infty) - \varepsilon \sigma T^4 = 0$$

It was assumed that the surface's absorptivity for sky radiation is equal to the pavement's surface emissivity, $\alpha_{sky} = \varepsilon$, since both are properties at low temperature, which are very distinct from the high-temperature absorptivity for solar radiation, α_{sun}. Substituting the values, we get:

$$0.92 \times 700 + 0.15 \times 5.67 \times 10^{-8} \times 290^4 - 8 \times (T - 300.15)$$
$$-0.15 \times 5.67 \times 10^{-8} \times T^4 = 0$$

After solving this fourth-degree polynomial equation, the surface temperature is 368.2 K or 95,4 °C. (Enough to burn your foot sole and to fry an egg.)

10.4 Solar and Environmental Radiation

Solar radiation behaves as a black body at about 5800 K (or precisely 5778 K). Nevertheless, the temperatures inside our star are on the order of millions of kelvins, thanks to nuclear fusion reactions, and the black body value represents the equivalent temperature on its surface. The solar radiation spectrum that reaches our planet outside the atmosphere and the spectrum at sea level are shown in Fig. 10.16. The figure also indicates the absorption bands of various gases present in the atmosphere, such as H_2O, CO_2, O_2, and O_3.

As illustrated in Fig. 10.16, the absorption spectra of solar radiation show that ozone (O_3) absorbs a significant part of the ultraviolet (UV) radiation below 400 nm and completely attenuates below 300 nm. A large fraction of the visible radiation is scattered by clouds and reflected back into space, with about 70% of visible radiation reaching sea level. Water vapor (H_2O) and carbon dioxide (CO_2) in the atmosphere have broad absorption bands in the infrared range. In contrast, nitrogen (N_2), oxygen (O_2), and argon (Ar), which constitute approximately almost 100% of the dry air composition, do not absorb solar radiation. The blue color of the sky is a result of *Rayleigh scattering*, where molecules smaller than a given wavelength of light scatter radiation in all directions. This phenomenon occurs in our atmosphere, which is primarily composed of nitrogen and oxygen gases, both of which have molecular diameters smaller than the wavelength of blue light. Consequently, the scattering of the blue spectrum of solar radiation by atmospheric gases provides a magnificent view of the blue sky.

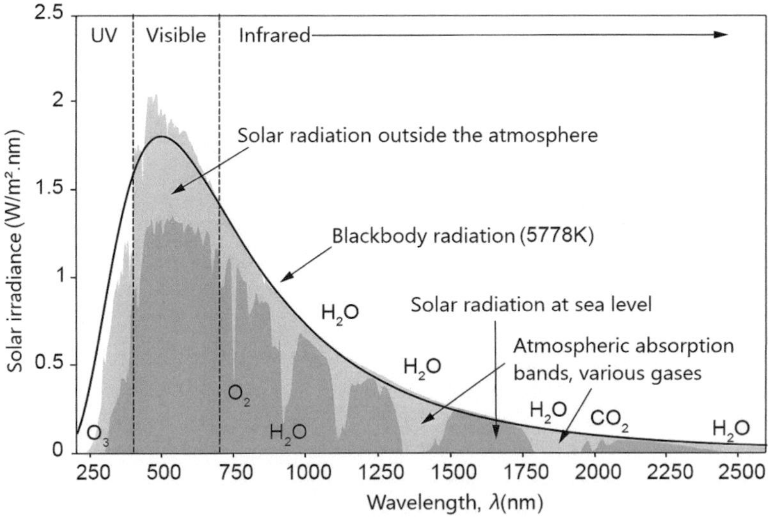

Fig. 10.16 Extraterrestrial solar spectrum and at sea level, as well as the absorption bands of the gases present in the atmosphere. *Credits* Adapted from Nick 84(*CC BY-SA 3.0*), via Wikimedia Commons. Available at: https://commons.wikimedia.org/wiki/File:Solar_spectrum_en.svg. Accessed on: April 14, 2022

Example 10.7 (*Solar radiation and solar constant, G_{cs}*)

Assuming that the Sun is a black body at 5800 K, its total emissive power is $E_{sun} = \sigma T^4 = 5.669 \times 10^{-8} \times 5800^4 = 64\ \mathrm{MW/m^2}$. Therefore, the Sun throws into space an unimaginable amount of 64 MW per unit area of its surface. The question is: how much of this solar thermal radiation reaches the Earth?

Solution

In this case, the Sun's total emission into space is $\dot{Q}_{\mathrm{sun}} = E_{\mathrm{sun}}A_{\mathrm{sun}}$, where A_{sun} is the area of the solar globe surface. This amount is radiated into space in radial direction. Thus, the total radiation emitted should reach an imaginary surface, approximately spherical, that contains the average orbit of the Earth with an area A_{Earth}. *Note*: this is not the area of the Earth's surface, but of the imaginary spherical surface (approximate) that encompasses the orbit of the Earth's translational motion around the Sun. Thus, as the total solar radiation emitted by the Sun´s surface must cross the spherical orbital surface, then one can write

$$\dot{Q}_{sun} = const = E_{sun}A_{sun,} = q_{Earth}A_{Earth} \Rightarrow q_{Earth} = E_{sun} \times \left(\frac{R_{sun}}{R_{Earth}}\right)^2$$

in which R_{sun} is the Sun's radius $\left(6.96 \times 10^5\ \mathrm{km}\right)$; R_{Earth} is the radius of the approximate imaginary sphere that contains the average orbit of the Earth (~149.6×10^6 km), and q_{Earth} is the heat flux in the form of solar thermal radiation that arrives per unit area on the Earth's sphere orbit. Thus,

$$q_{\mathrm{Earth}} = E_{\mathrm{sun}} \times \left(\frac{R_{\mathrm{sun}}}{R_{\mathrm{Earth}}}\right)^2 = 64.000.000 \times \left(\frac{0.696}{149.6}\right)^2 \approx 1385\ \mathrm{W/m^2}$$

So, about 1385 W/m^2 of solar irradiation reaches the Earth's orbit on a normal plane of incidence. The most well-accepted values of this magnitude, called the solar constant G_{sc}, vary between 1321 and 1411 W/m^2 depending on the month due to the planet's elliptical orbit.

The *solar constant*, G_{cs}, presented in the solution of Example 10.7 represents the average solar irradiation incident on a plane perpendicular to solar radiation outside the atmosphere (extraterrestrial). The value of $G_{sc} = 1366.1$ W/m^2 was proposed by the ASTM E-490 standard, which was established from satellite and high-altitude flights data, solarimetric base measurements, among other estimation and measuring techniques. To consider the variation of incident solar radiation throughout the year, the following expression (Duffie and Beckman, 2013) has been used:

$$G_{0N} = G_{sc}\left[1 + 0.033\cos\left(\frac{360\,\mathrm{N}}{365}\right)\right] \tag{10.36}$$

where G_{0N} is the normal extraterrestrial radiation outside the atmosphere on the N-th day since the beginning of the year.

As solar radiation crosses the Earth's atmosphere, it is attenuated by the atmospheric gases, particulate material, and aerosols in suspension. Thus, the lengthier the path to be traveled by the Sun's rays in the atmosphere, the greater will be the attenuation by all combined effects. This path to be traveled is given by the thickness of the *mass of air*, *MA*, as illustrated in Fig. 10.17. The mass of air is given by Eq. (10.38), and it is essentially a trigonometric relationship, involving the zenith angle, θ, and the distances to be traveled by the Sun's rays (without considering the planet´s curvature). At sea level, with the Sun at its zenith, the zenith angle $\theta = 0°$ establishes the standard condition of $MA = 1$. With $\theta = 60°$, $MA = 2$. Outside the atmosphere, $MA = 0$.

$$MA = \frac{BC}{AB} = \frac{1}{\cos\theta} \tag{10.38}$$

Before reaching our planet's surface, solar radiation is attenuated by the atmosphere and clouds. So, the incident solar radiation is divided into parts, namely *direct radiation* and *diffuse radiation*, as illustrated in Fig. 10.18. Direct radiation, as the name suggests, is the irradiation coming directly from the incident Sun's rays, with a portion that can be reflected, depending on the surfaces. Diffuse radiation is the emission by clouds, soil, surrounding objects, and the atmosphere itself, as well. Direct radiation at ground level is less than that extraterrestrial which provides the solar constant G_{sc}.

There are several maps of solar incidence radiation at sea and ground level. Many of these maps are based on satellite measurements and solarimetric measuring stations. Figure 10.19 shows one of these maps.

In thermal radiation calculations, it is necessary to establish an equivalent sky temperature, called the *effective sky temperature*, T_{sky}. There is not a single fixed value of this magnitude. However, some equations to estimate the sky temperature for a clear and cloudless sky depend on some psychrometric properties of the air and day hour at the most elaborate versions. However, a relatively simple equation for an initial estimate is that of Swinbank (1963), given by:

$$T_{sky} = 0.0552 T_{air}^{1.5} \tag{10.39}$$

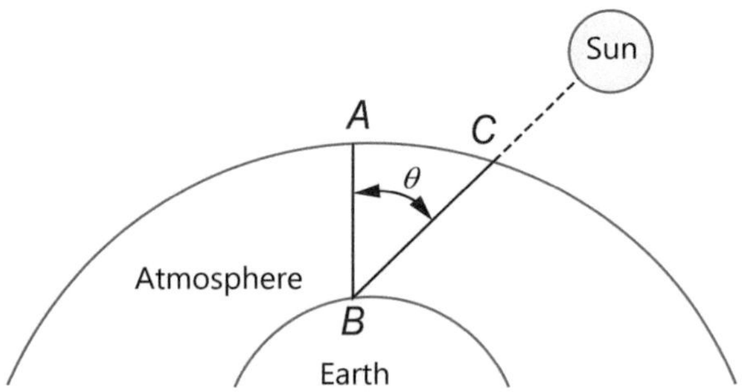

Fig. 10.17 Effect of mass of air, MA, on the absorption of solar radiation

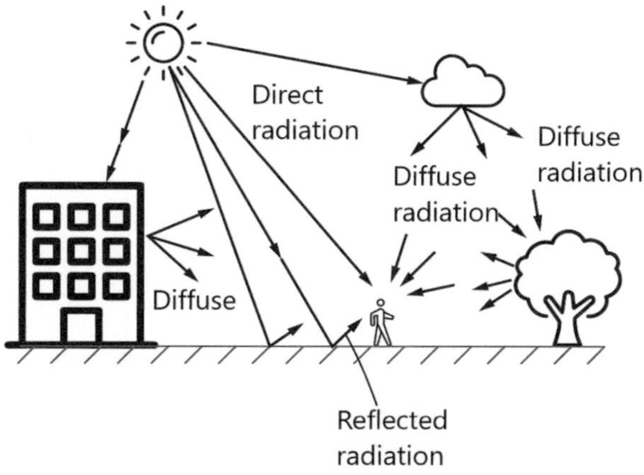

Fig. 10.18 Solar radiation and its incidence on bodies that can be direct or diffuse. The diffuse occurs through clouds and surrounding objects

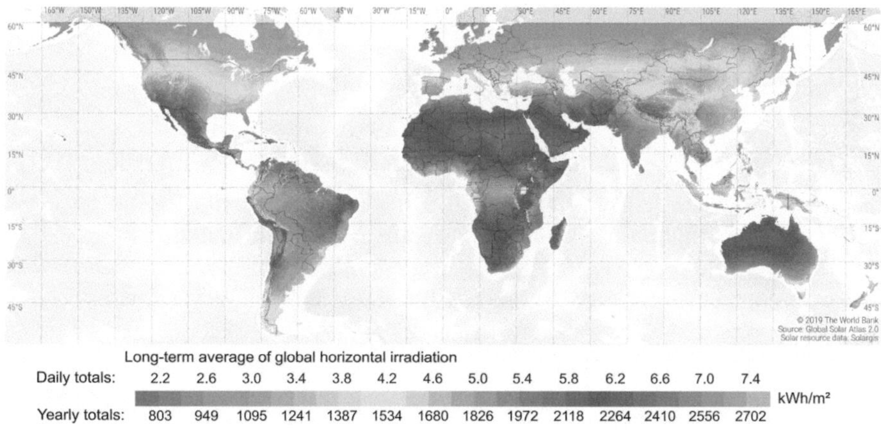

Fig. 10.19 World map of incidence of horizontal global radiation. This map was published by the World Bank Group, funded by ESMAP and prepared by Solargis. With permission from Creative Commons 4.0 Attribution International, CC BY 4.0. Adapted from https://solargis.com/maps-and-gis-data/download/world. Accessed on: June 7, 2022

where T_{air} is the air temperature in K and T_{sky} is the effective sky temperature in K. Clouds tend to increase the effective sky temperature. Finally, essentially the atmosphere is transparent in the radiation band from 8 to 14 μm, which covers the infrared region, allowing Earth's radiation to escape into space.

Example 10.8 (*Extraterrestrial radiation*)

Evaluate the extraterrestrial solar radiation on the southern summer solstice (December 21th) on a surface that is normal to radiation and on a horizontal surface, knowing that it has a zenith angle of 45°.

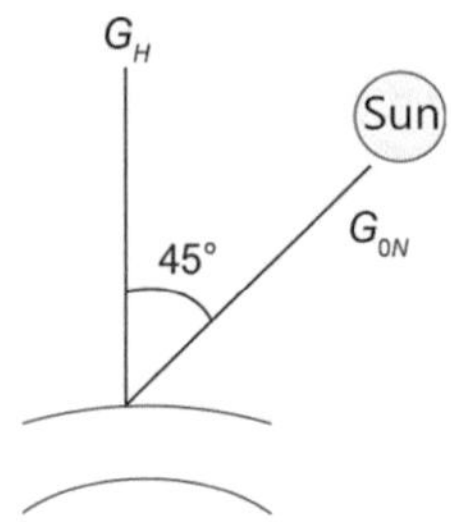

Solution

From Eq. (10.37), the extraterrestrial radiation on a surface that is normal to radiation is obtained:

$$G_{0N} = 1366.1\left[1 + 0.033\cos\left(\frac{360\,\mathrm{N}}{365}\right)\right]$$

December 21th is day $N = 355$ of the year (from January 1st), so:

$$G_{0N} = 1366.1\left[1 + 0.033\cos\left(\frac{360 \times 355}{365}\right)\right] = 1410.5\,\mathrm{W/m^2}$$

The radiation on the surface that has a zenith angle of 45° will be:

$$G_H = G_{0N}\cos\theta \Rightarrow |G_H = 1410.5 \times \cos 45° = 997.4\,\mathrm{W/m^2}.$$

10.5 Proposed Problems

10.1 Evaluate the wavelength of the electromagnetic wave of electricity from a power grid, 60 Hz frequency, considering the speed of light is equal to that of a vacuum.

10.2 In an astronomical observatory, the spectrum of a star was observed, whose peak emission occurred at 0.25 μm. Estimate its surface temperature considering that its emission follows the black body radiation. Research in proper literature to find to what class of stars it belongs.

10.3 A black body at 400 K emits thermal radiation. (a) Make a graph of spectral emissive power as a function of wavelength. Determine (b) the total emissive power; (c) the wavelength of maximum spectral emissive power; and (d) the maximum spectral emissive power.

10.4 The brightest star in the Orion constellation is Rigel, a bluish supergiant, with a temperature of approximately 12,100 K, distant close to 860 light years from Earth. For this star, evaluate (a) the wavelength corresponding to the maximum spectral emissive power; (b) its maximum spectral emissive power, comparing it with that of the Sun; and (c) the radiation flux in the visible range, also comparing it with that of the Sun.

10.5 A parallelepiped of 10 cm × 10 cm × 5 cm records a temperature of 500 °C. Evaluate the maximum rate of thermal radiation that the overall surface can emit.

10.6 Assuming that the Sun is a black body at 5800 K, its total emissive power is: $E_{sun} = \sigma T^4 = 5.669 \times 10^{-8} \times 5800^4 = 64\text{MW/m}^2$. Therefore, the Sun throws into space an unimaginable amount of 64 MW per unit area of its surface. The question is: how much of this solar thermal radiation per unit area reaches the planets Mercury, Venus, Earth, and Mars? (*Hint*: see Example 10.7).

10.7 A nearly black body is at 1500 K. Determine (a) the spectral emissive intensity at $\lambda = 0.6\ \mu$m; (b) the total emissive intensity of the surface.

10.8 Consider that a flat surface of 1.5 cm^2 behaves like a black body at 2000 K. Evaluate the radiation rate emitted in the range delimited by (a) $\phi = 0-180°$ and $\theta = 0-45°$ (see the following figure); and (b) $\phi = 0-180°$ and $\theta = 45-90°$.

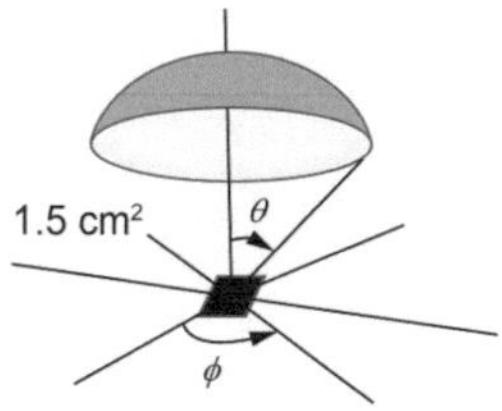

10.9 A flat plate of 1.5 cm^2 behaves like a black body at 1500 K. Evaluate the rate of radiation emitted by the plate and that exits through a hole of 1.5 cm in diameter positioned above the flat plate in a sphere of diameter of (a) 1 m, (b) 2 m, and (c) 4 m.

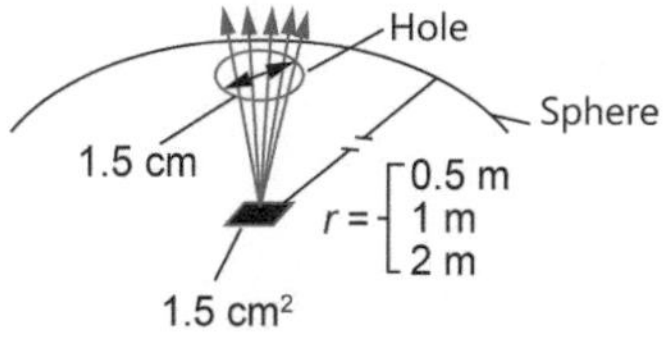

10.10 A flat body of 2 cm^2 behaves like a black body at 500 °C. It is placed 25 cm in front of a radiation flux sensor. Evaluate the radiant flux detected by the sensor, if the angle θ is 0° and 20°.

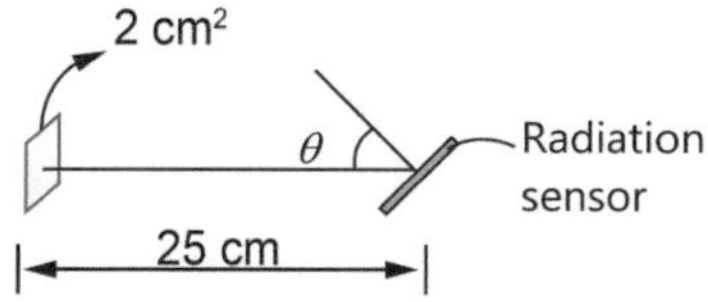

10.11 For the angle $\theta = 0°$ in Problem 10.10, evaluate the radiation flux detected by the sensor if an optical filter is placed between the flat body and the sensor, with a spectral transmissivity equal to 0 in the 0–3 μm band and 0.6 for wavelength greater than 3 μm.

10.12 A small opaque object with a diffuse surface is at 400 K. The object is then placed in an oven whose walls are at 2000 K. The walls are diffuse and gray and have an emissivity equal to 0.20. For the surface of the object, the spectral emissivities are: $\varepsilon_\lambda = 0$ for $0 < \lambda < 1.5\,\mu\text{m}$, $\varepsilon_\lambda = 0.6$ for $1.5 \leq \lambda < 3.0\,\mu\text{m}$ and $\varepsilon_\lambda = 0.4$ for $\lambda \geq 3.0\,\mu\text{m}$. Calculate (a) the total emissivity and absorptivity of this surface when its temperature is 400 K, (b) the radiosity on this surface and the net radiation flux to the surface at the moment when the surface temperature reaches 800 K, (c) the spectral emissive power for $\lambda = 2.5\,\mu\text{m}$, (d) the initial heat transfer rate to the object.

10.13 A new transparent material was subjected to a radiation flux of 2000 W/m^2. Through instrumentation, it was determined that 400 W/m^2 were reflected and 850 W/m^2 were absorbed. The question is: what is the transmissivity of this material?

10.14 In cold cities, it is common to place a wood stove to heat the interior of a house. Evaluate the rate of radiation heat transmitted by a stove having a 50 cm × 40 cm glass window. The stove can be approximated as a black body at 900 °C and the window glass has a spectral transmissivity as indicated in the figure.

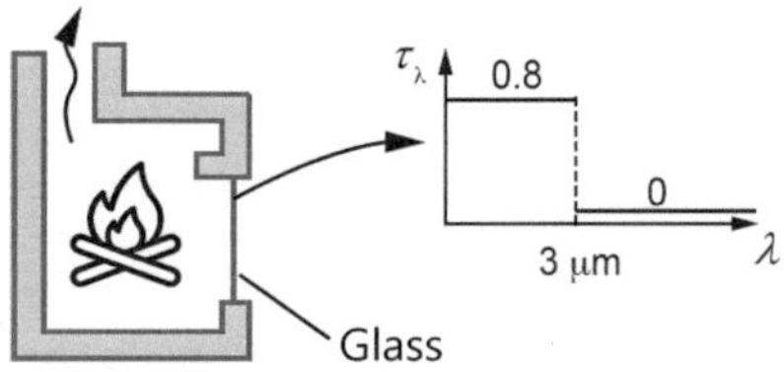

10.15 An opaque surface presents a spectral absorptivity of 0.2 for $\lambda \leq 2.5\,\mu\text{m}$ and 0.7 for $\lambda > 2.5$ for a black body radiation source at 2000 K. Evaluate (a) the absorptivity and the reflectivity of the surface at this temperature.

10.16 Irradiation hits a surface whose spectral reflectivity behavior is shown in the following figure. Evaluate (a) the total irradiation hitting the surface, (b) the absorbed radiation, and (c) the average absorptivity of the surface.

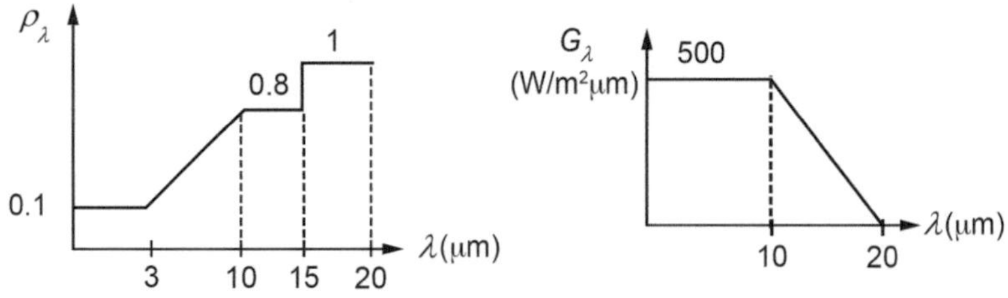

10.17 Determine the radiosity of a surface at 100 °C, with an emissivity of 0.9 and reflectivity of 0.5, subjected to a solar irradiation (1000 W/m^2).

10.18 A plate receives radiation on the upper side, while other sides are perfectly thermally insulated. The plate is opaque and the incident irradiation is 2000 W/m^2, with 1350 W/m^2 being absorbed; its emissive power is 1000 W/m^2 when the plate has a temperature of 200 °C. If the surrounding air records a temperature of 110 °C and the heat transfer coefficient by convection is 5 W/m^{2}K, evaluate the following magnitudes of the plate (a) absorptivity, (b) reflectivity, (c) transmissivity, (d) emissivity, and (e) the net heat flux transferred on the plate by thermal radiation.

10.19 A plate receives irradiation of 2200 W/m^2 on the lower part, while the upper and the sides are insulated. The plate is at 130 °C and the air below it is at 25 °C with $h = 50\,\text{W/m}^2\text{K}$. The plate has an area of 20 cm × 20 cm with a thickness of 4 mm, $\alpha = 0.8$, $\varepsilon = 0.6$, density of 2300 kg/m^3, and specific heat $C = 920$ J/kgK. Evaluate the temperature variation (K/s) of the plate at this moment.

10.20 A large closed cavity has a circular opening of 5 cm diameter through which 6 W of thermal radiation escapes. Evaluate the internal surface cavity temperature when its surface is (a) a black body and (b) when made of stainless steel, but has become oxidized.

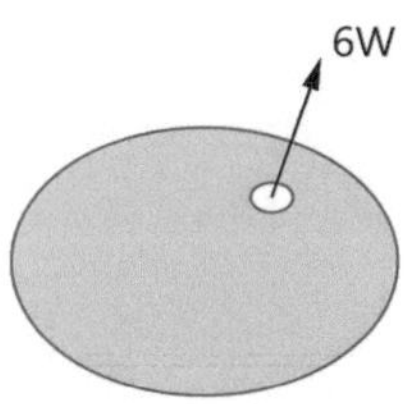

10.21 A flat 1 m × 1 m square solar collector for water heating receives 650 W/m^2 of solar radiation. The absorbing plate is at 65 °C, its absorptivity for the Sun is 0.90 and its emissivity is 0.12. Considering that the ambient air above the plate is at 15 °C, with $h = 8$ W/m^{2}K, at a sky temperature equal to 0 °C, evaluate (a) the thermal energy flux transferred to the water. (b) When the water stops circulating and assuming that everything under the solar plate is thermally isolated, evaluate the equilibrium temperature that the plate will reach.

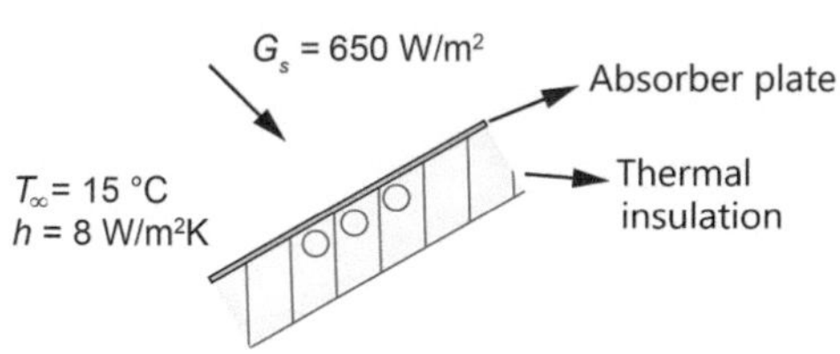

10.22 A plate (isolated on its bottom side) has an electric heater at the bottom. It is exposed to air at 20 °C and $h = 15$ W/m^{2}K, to solar radiation of 550 W/m^2. The "sky" effective temperature is 0 °C. If its absorptivity for solar radiation is 0.35, its emissivity is 0.3, and it is at 70 °C, evaluate the heat flux from the heater necessary to maintain that steady state condition.

References

Chancellor, J.C., R.S. Blue, K.A. Cengel, S.M. Auñón-Chancellor, K.H. Rubins, H.G. Katzgraber, A.R. Kennedy. 2018. Limitations in predicting the space radiation health risk for exploration astronauts. *NPJ Microgravity* 4 (8).

Duffie, J. A., and W.A. Beckman. 2013. *Solar engineering of thermal processes*, 4th ed. Wiley.

Kreith, F., and M.S. Bohn. 2001. *Principles of heat transfer*, 6th ed. New York: Brooks/Cole.

McAdams, W.H. 1954. *Heat transmission*, 3rd ed. New York: McGraw-Hill.

Schmidt, E., and E. Eckert. 1935. Über die Richtungsverteilung der Wärmestrahlung von Oberflächen. *Forsch Ing-Wes* 6: 175–183.

Swinbank, W.C. 1963. Long-wave radiation from clear skies. *Quarterly Journal of the Royal Meteorological Society* 89: 339–348.

Chapter 11
Applied Thermal Radiation

Abstract This chapter presents the techniques for solving thermal radiation problems in various cases of engineering interest. First, elementary heat transfer cases of thermal radiation are studied, which are the cases of two black and grey isothermal parallel and infinite surfaces in a vacuum. In this first analysis, the technique of electrical analogy is presented to solve problems of thermal radiation. The concept of thermal radiation shape factors is introduced to solve more complex cases. Shape factors for various situations and geometries are presented next. From this theory, several cases of heat exchange by radiation are solved, starting with three enclosed surfaces and extending the solution to *n* surfaces forming a closed envelope. The concept of thermal radiation shielding is presented with illustrations of practical cases, such as temperature readings from a thermocouple. Chapter presents an analysis of the combined heat transfer of thermal radiation and convection. A black cavity modelling is also analized. Finally, the chapter concludes with subject of thermal radiation in gases.

Keywords Combined radiation · Gas radiation · Radiation shielding · Shape factors · Thermal radiation

In the previous chapter, the thermal radiation fundamentals were introduced, based on Planck's law. The properties of surfaces for thermal radiation were presented and analyzed, namely absorptivity, transmissivity, reflectivity, and emissivity. This chapter presents the application aspects of thermal radiation in engineering problems and their solution methods.

J. R. Simões-Moreira and E. W. Zavaleta-Aguilar, *Heat Transfer*, Heat and Mass Transfer, https://doi.org/10.1007/978-3-031-80318-5_11

11.1 Heat Exchange by Thermal Radiation Between Two Parallel and Infinite Surfaces

Consider the definition of radiosity, J, as the rate of total thermal radiation leaving a surface per unit area as seen in Sect. 10.3.4. As indicated in Fig. 11.1a, the radiosity is a compound of two components, namely the reflected incident irradiation, ρG, and the surface's emission or the emissive power, E, as given by the following equation:

$$J = \rho G + E \tag{11.1}$$

An *opaque body* has a null transmissivity, $\tau = 0$, and if the body is also a *gray* one, the properties of emissivity, ε, reflectivity, ρ, are independent of the wavelength λ (Sect. 10.3.3). Therefore, the radiosity can be written as:

$$J = \rho G + E = (1 - \alpha)G + \varepsilon E_b \tag{11.2}$$

In addition, consider that $\alpha = \varepsilon$, according to Kirchhoff's Law (Sect. 10.3.2). Thus, one can simplify Eq. (11.2) according to $\rho = 1 - \alpha = 1 - \varepsilon$, obtaining:

$$J = (1 - \varepsilon)G + \varepsilon E_b, \text{ or}$$

$$G = \frac{J - \varepsilon E_b}{1 - \varepsilon} \tag{11.3}$$

On the other hand, the net rate of heat by radiation, $\dot{Q}$, transferred from the surface of area A (Fig. 11.1b) is the difference between its radiosity, J, and irradiation, G, that is:

$$\dot{Q} = A(J - G) \tag{11.4}$$

Substituting the irradiation G from Eq. (11.3), we get:

$$\dot{Q} = A\left(J - \frac{J - \varepsilon E_b}{1 - \varepsilon}\right) = A\left(\frac{J - \varepsilon J - J + \varepsilon E_b}{1 - \varepsilon}\right), \textit{or } \dot{Q} = \frac{E_b - J}{(1 - \varepsilon)/\varepsilon A} \tag{11.5}$$

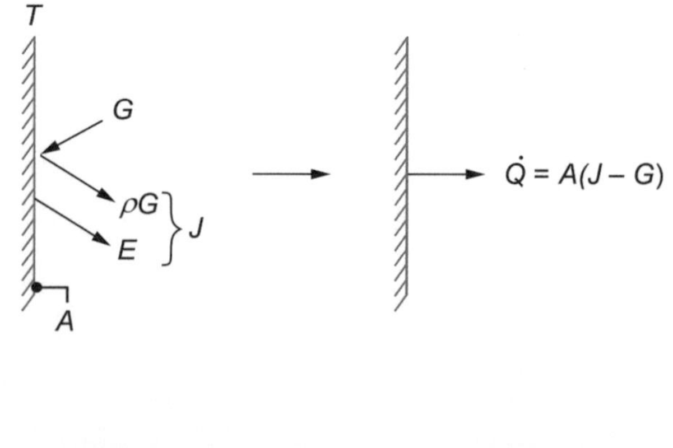

Fig. 11.1 **a** Radiosity, J, and **b** net radiation rate leaving a surface, $\dot{Q}$

Relevant comments regarding Eq. (11.5):

1. A black body by definition has $\varepsilon = 1$. In this case, the radiosity and the black body emissive power are equal and $J = E_b = \sigma T^4$ (Eq. 11.3). Thus, the net rate of heat transferred from the surface is evaluated by: $\dot{Q} = A(E_b - G)$.
2. An *adiabatic—re-radiating* surface is a one where there is no net heat transfer from the surface, which means that $\dot{Q} = 0$. In this case, the radiosity and the black body emissive power are equal, that is, $J = E_b = \sigma T^4$.

11.1.1 Net Rate of Thermal Radiation Exchanged Between Two Parallel and Infinite Isothermal Surfaces

One of the most elementary cases of heat exchange by thermal radiation to be analyzed is that of two isothermal and parallel flat surfaces in vacuum. Consider the situation of two large and parallel surfaces at temperatures T_1 and T_2, as illustrated in Fig. 11.2. The idea of large surfaces is that the edge effects do not affect the solution.

The net radiation heat rate, $\dot{Q}_{1-2}$, exchanged between surfaces 1 and 2 is the difference in radiosity between them since the radiosity of surface 1 becomes the irradiation of surface 2 and vice-versa. So that:

$$\dot{Q}_{1-2} = J_1 A_1 - J_2 A_2 = A(J_1 - J_2) \tag{11.6}$$

Note the order of indexes 1 and 2 in $\dot{Q}_{1-2}$. They represent and should be read as the rate of radiation leaving surface 1 and hitting surface 2 subtracted from the rate of radiation leaving surface 2 and hitting surface 1.

In the case of the surfaces being black (black body), it is given that: $\varepsilon_1 = \varepsilon_2 = 1$ and $\rho_1 = \rho_2 = 0$. Thus, the emissive power of a surface, *E*, becomes its black body emissive power E_b in Eq. (11.1). So that, substituting into Eq. (11.6),

$$\dot{Q}_{1-2} = \sigma A\left(T_1^4 - T_2^4\right) \tag{11.7}$$

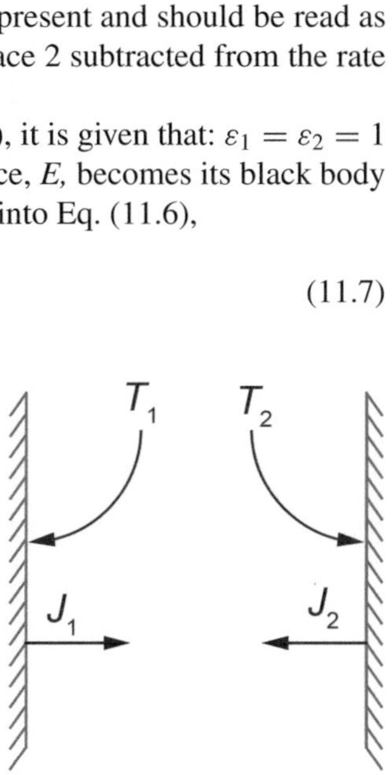

Fig. 11.2 Radiosities on surfaces 1 and 2

In the case of both surfaces being gray and opaque, the radiosity of either of the two surfaces (Eq. 11.1) is given by:

$$J = \rho G + \varepsilon E_b = (1 - \varepsilon)G + \varepsilon E_b \tag{11.8}$$

The next step is to isolate the irradiation term G from Eq. (11.8), that is,

$$G = \frac{J - \varepsilon E_b}{1 - \varepsilon} \tag{11.9}$$

The rate of heat transferred by radiation from a surface is given by the difference between its black body emissive power and its radiosity, as defined by Eq. (11.5). On the other hand, in the case of two infinite parallel surfaces, the rate of heat transferred from one surface necessarily reaches the other, that is, the radiosity of surface 1, J_1, becomes the irradiation of surface 2, G_2. Therefore, the net heat rate $\dot{Q}_{1-2}$ by radiation exchanged between the two surfaces is:

$$\dot{Q}_{1-2} = \frac{E_{b1} - J_1}{(1 - \varepsilon_1)/\varepsilon_1 A} = -\frac{E_{b2} - J_2}{(1 - \varepsilon_2)/\varepsilon_2 A} \tag{11.10}$$

Equation (11.6) and the two Eq. (11.10) have three unknowns $\left(\dot{Q}_{1-2}, J_1, \text{ and } J_2\right)$, since the temperatures of the surfaces T_1 and T_2 and therefore, their black body emissive powers, E_{b1} and E_{b2}, along with their respective emissivities and total area, are known quantities. Therefore, the radiosities and the heat rate can be obtained by solving those three equations simultaneously. However, before proceeding in this line, it is possible to establish an electrical analogy with Ohm's Law, where the black body emissive powers E_{b1} and E_{b2} represent the "electrical potentials" established by the temperatures of the surfaces. J_1 and J_2 are "electrical potentials" associated with surfaces 1 and 2, respectively. The quantities involving the emissivities and area are "resistances", as illustrated in the analog thermal circuit of Fig. 11.3.

So that the net heat rate, $\dot{Q}_{1-2}$, which "flows" from E_{b1} to E_{b2}, is given by:

$$\dot{Q}_{1-2} = \frac{E_{b1} - E_{b2}}{\frac{1-\varepsilon_1}{\varepsilon_1 A} + \frac{1}{A} + \frac{1-\varepsilon_2}{\varepsilon_2 A}} = \frac{\sigma\left(T_1^4 - T_2^4\right)}{R_1 + R_{1-2} + R_2} \tag{11.11}$$

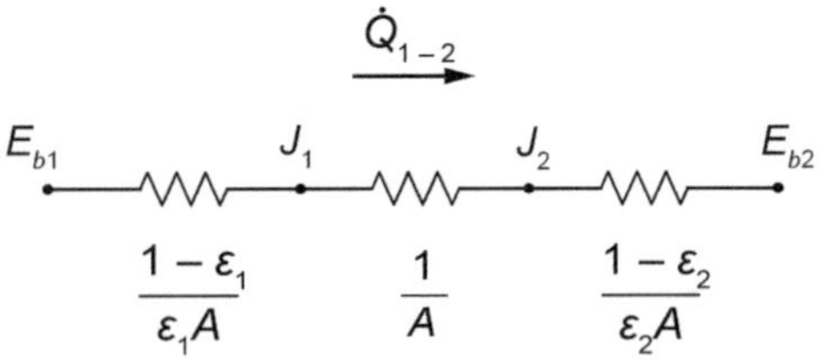

Fig. 11.3 Thermal circuit between the grey surfaces 1 and 2

Note that there are two *resistances* R_1 and R_2 between the potentials E_b and J of each surface, which depend on their emissivity (besides the area) and, therefore, are resistances associated only with the surface, hence they are called *surface resistances*, that is:

$$R_i = \frac{1 - \varepsilon_i}{\varepsilon_i A_i} \tag{11.12}$$

where $i = 1$ or 2, stands for each surface.

The resistance between the radiosity potentials J_1 and J_2 forms a resistance that depends on the relative position and sizes of the plates, referred to as *spatial resistance* $R_{1\text{-}2}$. More will be said about this type of resistance associated with the concept of shape factor in the next section (see Eq. 11.21). For the case of two infinite parallel plates, the spatial resistance is simply the inverse of the area of the surfaces.

$$R_{1-2} = \frac{1}{A} \tag{11.13}$$

Example 11.1 (*Net heat exchange by radiation between two surfaces*)

Find the radiosity and the heat flux exchanged between two parallel, gray, and opaque surfaces maintained at $T_1 = 400$ K and $T_2 = 300$ K, respectively. The surfaces' emissivity is 0.5 and they are in vacuum.

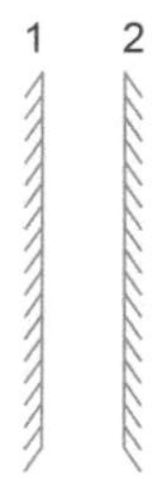

Solution

Equivalent thermal circuit:

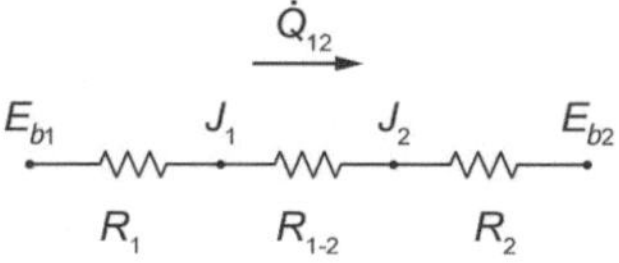

Calculation of the surface resistances R_1 and R_2 (Eq. 11.12), recalling that in this case $\varepsilon_1 = \varepsilon_2 = \varepsilon$, then,

$$R_1 = R_2 = \frac{1 - \varepsilon}{A\varepsilon} = \frac{1}{A}\frac{0.5}{0.5} = \frac{1}{A}$$

Spatial resistance, $R_{1\text{-}2}$ (Eq. 11.13), $R_{1-2} = \frac{1}{A}$, therefore, the total resistance, R_T, is given by: $R_T = \frac{1}{A} + \frac{1}{A} + \frac{1}{A} = \frac{3}{A}$.

Finally, the net rate of heat exchanged between the two surfaces is:
$\dot{Q}_{1-2} = \frac{E_{b1}-E_{b2}}{R_T} = \frac{\sigma\left(T_1^4-T_2^4\right)}{3/A}$, and the heat flux exchanged is:

$$q_{1-2} = \frac{\dot{Q}_{1-2}}{A} = \frac{5.67\times10^{-8}\times\left(400^4-300^4\right)}{3}$$

$$q_{1-2} = 330.75\,\mathrm{W/m^2}$$

Once the net heat flux exchanged is calculated, one can calculate the radiances, which are the intermediate "potentials" of the analogous thermal circuit:

$$q_{1-2} = \frac{E_{b1}-J_1}{AR_1} \Rightarrow J_1 = E_{b1} - A\times R_1\times q_{1-2} = \sigma 400^2 - 330.75 \Rightarrow$$

$$J_1 = 1120.77\,\mathrm{W/m^2}$$

$$J_2 = E_{b2} - A\times R_2\times q_{1-2} = \sigma 300^2 + 330.75 \Rightarrow$$

$$J_2 = 790.62\,\mathrm{W/m^2}.$$

Comment: note that the radiosity calculation comprises the initial assessment of the heat flux that "circulates" through the circuit of series resistances. It is the same electrical problem for obtaining intermediate potentials, or voltage drops over the resistances. One must first obtain the electrical current through the whole circuit and thus it may be calculated the intermediate voltage drops on each resistance.

11.2 Thermal Radiation Exchange Between Any Surfaces

11.2.1 Radiation Shape Factors—Definition

In the previous section, the case of net heat exchange between two parallel and infinite isothermal surfaces was analyzed. This is a particular situation, and its usefulness was for introducing the concepts of surface and spatial radiation resistances. Nevertheless, in most general cases, the net radiation heat rate exchange between two surfaces at any relative configurations and any dimensions must be found. To solve this problem, consider two surfaces A_1 and A_2 at any spatial orientation, as schematized in Fig. 11.4. The angles θ_1 and θ_2 are those that the normal lines make with the line connecting the surfaces in the area elements dA_1 and dA_2, respectively, distant from r each other.

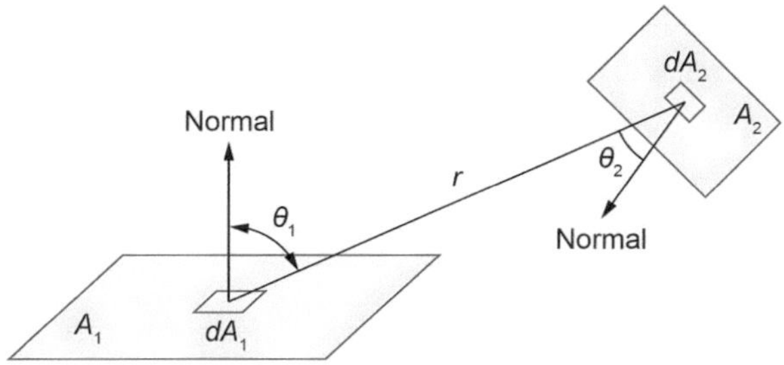

Fig. 11.4 General arrangement of two surfaces for determining the radiation shape factor

The concept of radiation intensity from Sect. 10.2.3 and the solid angle ω allows one to determine the rate of heat by radiation that leaves the elementary surface area dA_1 hitting the elementary surface area dA_2 as $d\dot{Q}_{dA_1-dA_2} = I_1 dA_1 \cos\theta_1 d\omega_1$. In this case, the differential solid angle is $d\omega_1 = dA_2 \cos\theta_2/r^2$, therefore:

$$d\dot{Q}_{dA_1-dA_2} = \frac{I_1 \cos\theta_1 \cos\theta_2}{r^2} dA_1 dA_2$$

The total radiation heat rate transferred from surface A_1 to the surface A_2 is obtained by integrating the previous equation over the two areas, resulting in the following expression:

$$\dot{Q}_{1-2} = \int\limits_{A_2}\int\limits_{A_1} \frac{I_1 \cos\theta_1 \cos\theta_2}{r^2} dA_1 dA_2$$

The *radiation shape factor* F_{1-2} is defined as the fraction of radiation that leaves surface 1 and reaches surface 2, that is:

$$F_{1-2} = \frac{\dot{Q}_{1-2}}{\dot{Q}_1} = \frac{\int_{A_2}\int_{A_1} \frac{I_1 \cos\theta_1 \cos\theta_2}{r^2} dA_1 dA_2}{A_1 J_1} = \frac{\int_{A_2}\int_{A_1} \frac{I_1 \cos\theta_1 \cos\theta_2}{r^2} dA_1 dA_2}{A_1 \pi I_1} \Rightarrow$$

$$F_{1-2} = \frac{1}{A_1 \pi} \int\limits_{A_2}\int\limits_{A_1} \frac{\cos\theta_1 \cos\theta_2}{r^2} dA_1 dA_2 \qquad (11.14a)$$

When performing an analogous analysis from surface 2 with respect to surface 1, one obtains the *radiation shape factor* F_{2-1} as the fraction of radiation that leaves surface 2 and hits surface 1, resulting in the following equation:

$$F_{2-1} = \frac{1}{A_2 \pi} \int\limits_{A_1}\int\limits_{A_2} \frac{\cos\theta_1 \cos\theta_2}{r^2} dA_2 dA_1 \qquad (11.14b)$$

Comparing Eqs. (11.14a), and (11.14b), one can conclude that there is a relationship between the two shape factors, that is:

$$A_1F_{1-2} = A_2F_{2-1} \tag{11.15}$$

Generalizing, for any two surfaces i and j, it can be established that:

$$A_iF_{i-j} = A_jF_{j-i} \tag{11.16}$$

The previous equation is important for the calculation of radiation shape factors and it is known as the *law of reciprocity*. It is useful, because, in many practical situations, one of the shape factors is known or easily computed, while obtaining the other can be more complex. Thus, by using the law of reciprocity the more complex shape factor can be obtained as exemplified in Example 11.2. The shape factor is also known in the literature as *configuration factor*, and *view factor*. For simplicity, the radiation shape factor will be referred to as shape factor in this chapter without mentioning the radiation term.

Consider the case of a *closed surface* formed by n surfaced enclosured, as well illustrated in Fig. 11.5a. Surface 1 exchanges heat by radiation with all the other surfaces of the enclosure. So that surface 1 "sees" all the other n surrounding surfaces, which allows to write the following relation of the shape factors of surface 1 in relation to the others:

$$F_{1-1} + F_{1-2} + \cdots + F_{1-n} = 1 \Rightarrow \sum_{j=1}^{n} F_{1-j} = 1$$

Generalizing, for any surface i of interest, we have:

$$\sum_{j=1}^{n} F_{i-j} = 1 \tag{11.17}$$

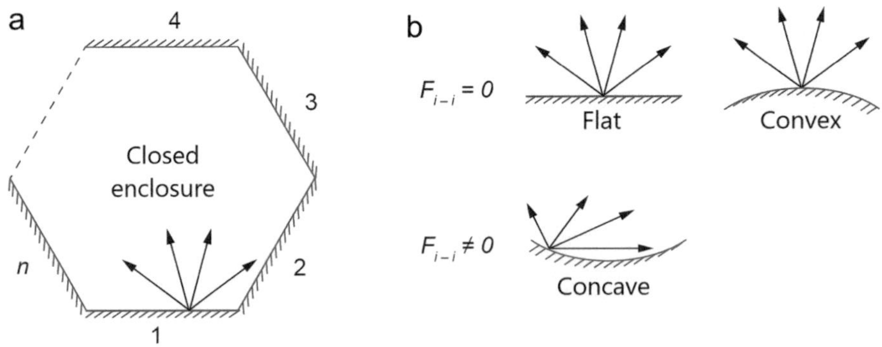

Fig. 11.5 **a** Closed surface and the closure law; **b** shape factor of flat, convex, and concave surfaces

This is the so-called *closure law* or *summation law* of shape factors. This law, in combination with the law of reciprocity (Eq. 11.16) form the *shape factor algebra* that are used to solve many problems of thermal radiation between surfaces. It is worth noting that the shape factor of the surface i to itself, that is, F_{i-i}, is only zero if the surface is flat or convex (Fig. 11.15b); in the case of a concave surface, $F_{i-i} \neq 0$, because part of the radiation emitted from surface i hits itself. The use of the shape factor algebra is best illustrated by Example 11.2.

Example 11.2 (*Shape factor*)

Determine the shape factor F_{2-2} of the shown configuration, knowing that surface 2 completely encloses surface 1.

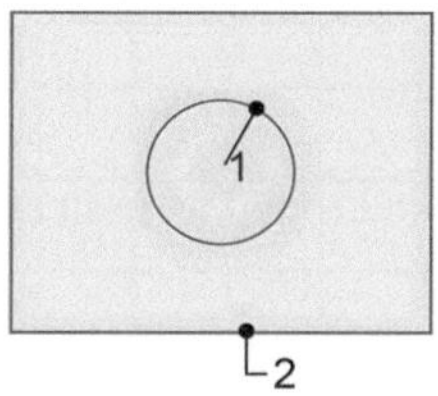

Solution

By the closure law, Eq. (11.17), we have that $F_{2-1} + F_{2-2} = 1$. Therefore, $F_{2-2} = 1 - F_{2-1}$.

The problem now lies on obtaining F_{2-1}, which is not found immediately. However, it is possible to use the law of reciprocity, Eq. (11.16), to write:

$$A_1 F_{1-2} = A_2 F_{2-1} \Rightarrow F_{2-1} = \frac{A_1}{A_2} \underbrace{F_{1-2}}_{1} \Rightarrow F_{2-1} = \frac{A_1}{A_2}$$

knowing that $F_{1-2} = 1$, since the surface 1 is completely enveloped by the surface 2. Finally,

$$F_{2-2} = 1 - \frac{A_1}{A_2}$$

Comment: needless to say, the result obtained is always valid for the case of a surface that completely envelopes the other regardless of the geometrical surface shape; such would be the cases of long double tubes and concentric spheres, among others. In addition, if the inner surface is very small compared to the enveloping surface, that is, $A_1 \ll A_2$, then $F_{2-2} \to 1$.

11.2.2 Shape Factors—Various Situations

As seen, the shape factor from Eqs. (11.14a) and (11.14b) refers only to a spatial trigonometric problem that considers the relative position between the two surfaces, as well as their dimensions and forms. Shape factor analytical solution is limited to some geometries and configurations that allow to carry out the double integration. In the broader case, it is necessary to resort to numerical methods to solve the shape factor equation for a more complex case. Howell (2010) compiled hundreds of geometries and configurations and made the material available for free access over the *Internet*. Table 11.1 indicates a collection of expressions of shape factors from numerous geometries and configurations of interest adapted from his work. The sequence of graphs in Fig. 11.6 shows selected shape factors.

Example 11.3 (*Shape factor—perpendicular planes*)

Three flat plates 1, 2, and 3 are positioned as indicated. Determine the shape factors F_{1-2}, F_{2-1} and F_{1-3}.

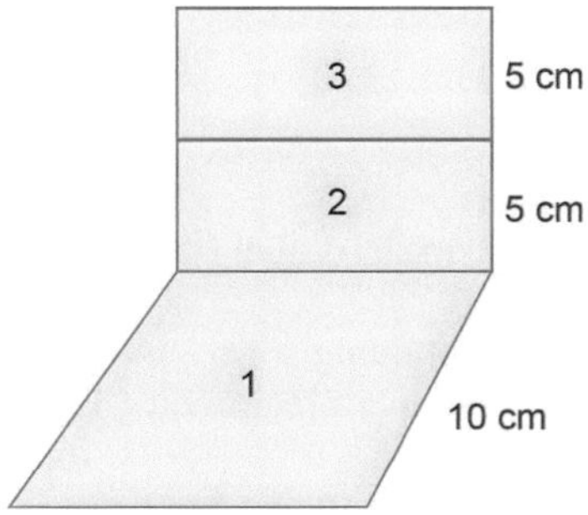

Solution

Using the graph from Fig. 11.6b, we get:

$$F_{1-2}: \frac{h}{a} = \frac{5}{10} = 0.5; \ \frac{w}{a} = \frac{10}{10} = 1 \rightarrow F_{1-2} = 0.15$$

However, $F_{1-2,3} = F_{1-2} + F_{1-3} \rightarrow F_{1-3} = \underbrace{F_{1-2,3} - F_{1-2}}_{graph}$

$$F_{1-2,3}: \frac{h}{a} = \frac{10}{10} = 1; \ \frac{w}{a} = \frac{10}{10} = 1 \rightarrow F_{1-2,3} = 0.2$$

Which results in $F_{1-3} = 0.2 - 0.15 = 0.05$
Using the law of reciprocity (Eq. 11.16), we have:

$$A_1F_{1-2} = A_2F_{2-1} \rightarrow F_{2-1} = \frac{A_1}{A_2}F_{1-2} \Rightarrow$$

$$F_{2-1} = \frac{10 \times 10}{10 \times 5} \times 0.15 \rightarrow F_{2-1} = 0.30$$

Table 11.1 Shape factors for various situations (adapted from Howell, 2010)

Geometry	Equation
Infinitely long parallel plates 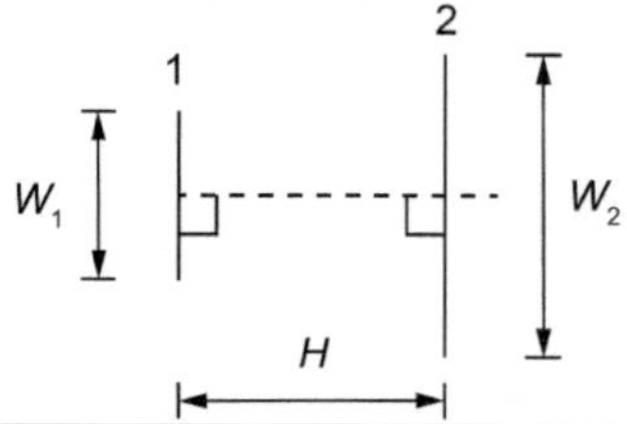	$B = W_1/H,\ C = W_2/H$ $F_{1-2} = \frac{1}{2B}\left[\sqrt{(B+C)^2+4} - \sqrt{(C-B)^2+4}\right]$
Infinitely long inclined plates 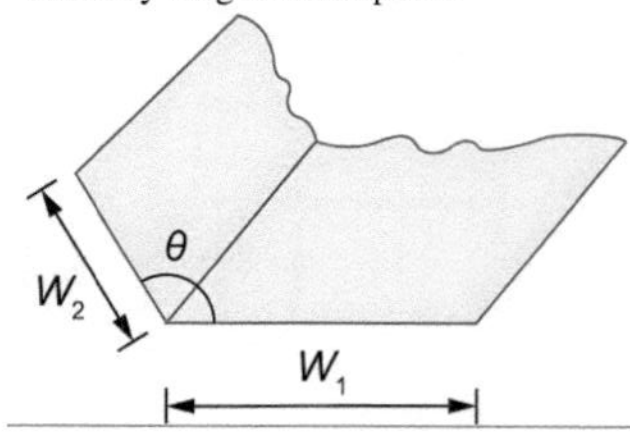	$A = W_2/W_1$ $F_{1-2} = \frac{A+1-\sqrt{A^2+1-2\cos\theta}}{2}$
Enclosure with three infinitely long sides 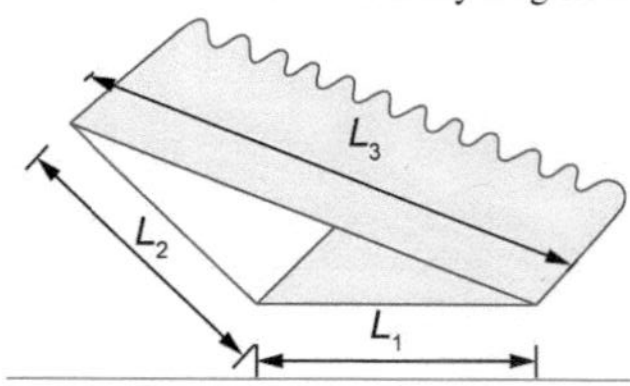	$F_{1-2} = \frac{L_1+L_2-L_3}{2L_1}$

(continued)

Table 11.1 (continued)

Geometry	Equation
Plate and 1… n infinitely long cylinders 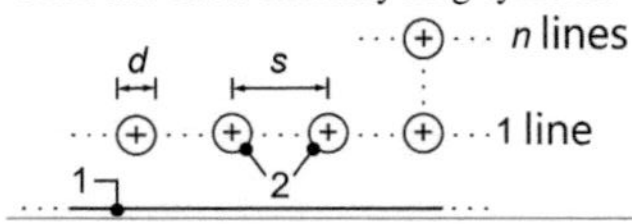	$A = \frac{d}{s}$ $F_{1-2} = 1 - \sqrt{1 - A^2} + A\arctan\sqrt{\frac{1-A^2}{A^2}}$ $F_{1-n} = 1 - (1 - F_{1-2})^n$
Finite plate inclined on infinite plate 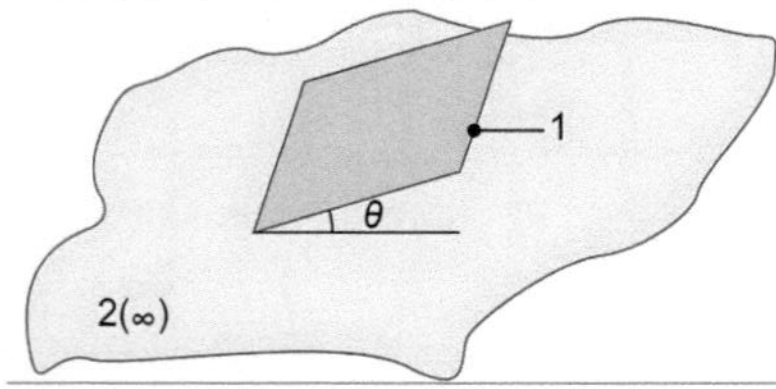	
Equal parallel plates 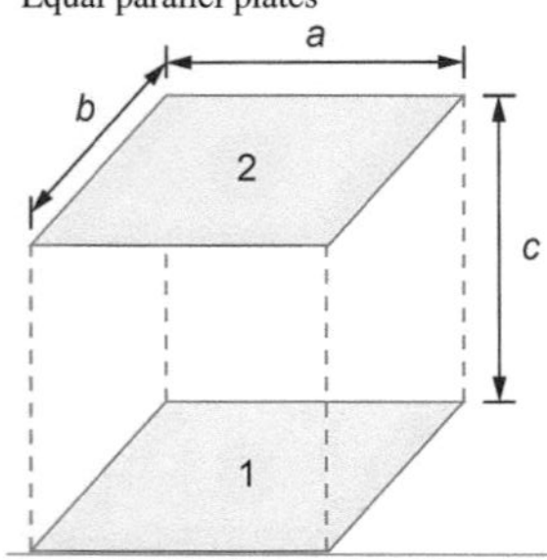	$X = \frac{a}{c},\ Y = \frac{b}{c}$ $F_{1-2} = \frac{2}{\pi XY}\left\{\ln\left[\frac{(1+X^2)(1+Y^2)}{1+X^2+Y^2}\right]^{1/2} + X\sqrt{1+Y^2}\arctan\left(\frac{X}{\sqrt{1+Y^2}}\right)\right.$ $\left. +Y\sqrt{1+X^2}\arctan\left(\frac{Y}{\sqrt{1+X^2}}\right) - X\arctan X - Y\arctan Y\right\}$

(continued)

Table 11.1 (continued)

Geometry	Equation
Perpendicular plates 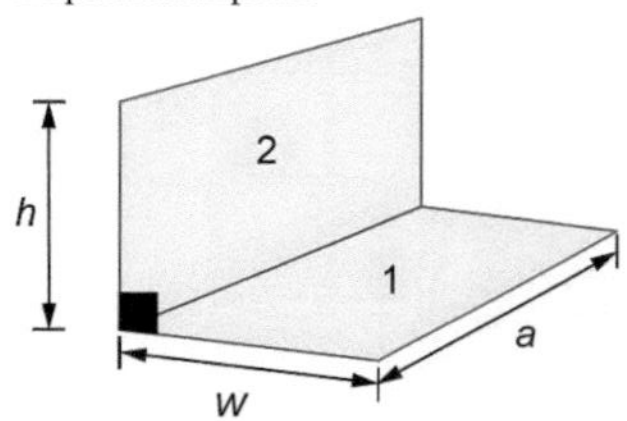	$H = \frac{h}{a},\ W = \frac{w}{a}$ $F_{1-2} =$ $\frac{1}{W\pi}\left\{W \arctan\frac{1}{W} + H \arctan\frac{1}{H} - \sqrt{H^2+W^2} \times \arctan\sqrt{\frac{1}{H^2+W^2}}\right.$ $\left. + \frac{1}{4}\ln\left\{\frac{\left(1+W^2\right)\left(1+H^2\right)}{1+W^2+H^2} \times \left[\frac{W^2\left(1+W^2+H^2\right)}{(1+W^2)(W^2+H^2)}\right]^{W^2}\left[\frac{H^2\left(1+H^2+W^2\right)}{(1+H^2)(H^2+W^2)}\right]^{H^2}\right\}\right\}$
Isosceles triangles 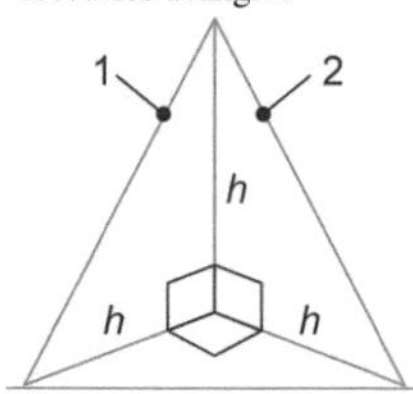	$F_{1-2} = 0.2113$
Coaxial parallel discs 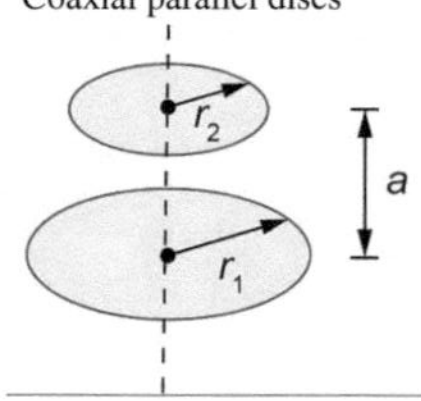	$R_1 = \frac{r_1}{a},\ R_2 = \frac{r_2}{a},\ X = 1 + \frac{1+R_2^2}{R_1^2}$ $F_{1-2} = \frac{1}{2}\left\{X - \left[X^2 - 4\left(\frac{R_2}{R_1}\right)^2\right]^{\frac{1}{2}}\right\}$

(continued)

Table 11.1 (continued)

Geometry	Equation
Coaxial parallel rings 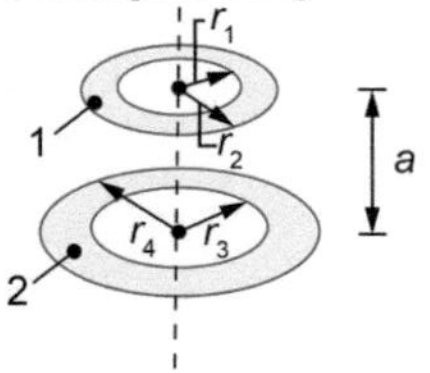	$H = \frac{a}{r_1}, R_2 = \frac{r_2}{r_1}, R_3 = \frac{r_3}{r_1}, R_4 = \frac{r_4}{r_1}$ $F_{1-2} = \frac{1}{2\left(R_2^2 - 1\right)} \left\{ \left[\left(R_2^2 + R_3^2 + H^2\right)^2 - (2R_3R_2)^2 \right]^{\frac{1}{2}} - \left[\left(R_2^2 + R_4^2 + H^2\right)^2 - (2R_2R_4)^2 \right]^{\frac{1}{2}} \right.$ $\left. + \left[\left(1 + R_4^2 + H^2\right)^2 - (2R_4)^2 \right]^{\frac{1}{2}} - \left[\left(1 + R_3^2 + H^2\right)^2 - (2R_3)^2 \right]^{\frac{1}{2}} \right\}$
Coaxial cylinders and ring 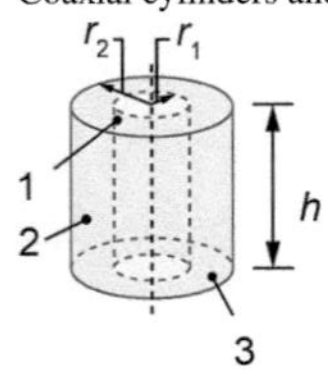 1. inner cylinder 2. outer cylinder 3. lower ring	$R = \frac{r_2}{r_1}, H = \frac{h}{r_1}$ $F_{1-2} = 1 - \frac{H^2 + R^2 - 1}{4H}$ $- \frac{1}{\pi} \left[\arccos \frac{H^2 - R^2 + 1}{H^2 + R^2 - 1} - \frac{\sqrt{\left(H^2 + R^2 + 1\right)^2 - 4R^2}}{2H} \right.$ $\left. \arccos \frac{H^2 - R^2 + 1}{R\left(H^2 + R^2 - 1\right)} - \frac{H^2 - R^2 + 1}{2H} \arcsin \frac{1}{R} \right]$ $h \to \infty : F_{1-2} = 1, F_{2-1} = \frac{d_1}{d_2}, F_{2-2} = 1 - \frac{d_1}{d_2}$
	$R = \frac{r_1}{r_2}, H = \frac{h}{r_2}, X = \sqrt{1 - R^2}, Y = \frac{R\left(1 - R^2 - H^2\right)}{1 - R^2 + H^2}$ $F_{2-3} = \frac{1}{\pi} \left\{ R\left(\arctan \frac{X}{H} - \arctan \frac{2X}{H} \right) + \frac{H}{4} \left[\arcsin\left(2R^2 - 1\right) - \arcsin R \right] \right.$ $+ \frac{X^2}{4H}\left(\frac{\pi}{2} + \arcsin R\right) - \frac{\sqrt{\left(1 + H^2 + R^2\right)^2 - 4R^2}}{4H}$ $\left. \times \left(\frac{\pi}{2} + \arcsin Y\right) + \frac{\sqrt{4 + H^2}}{4} \left[\frac{\pi}{2} + \arcsin\left(1 - \frac{2R^2H^2}{4X^2 + H^2}\right) \right] \right\}$

(continued)

Table 11.1 (continued)

Geometry	Equation
Ring to hemisphere 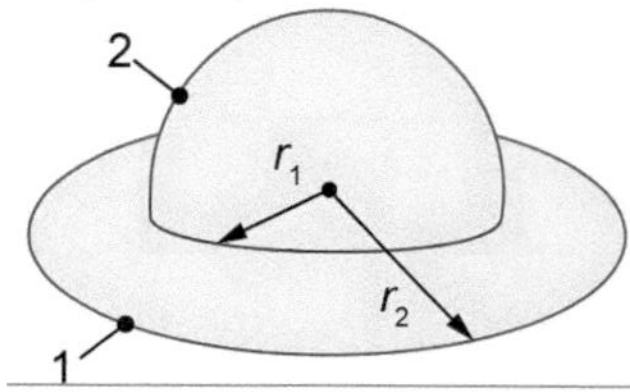	$R = \frac{r_2}{r_1}$, $Z = \sqrt{R^2 - 1}$ $F_{1-2} = \frac{1}{\pi Z^2}\left(-\frac{\pi}{2} - Z + R^2 \tan^{-1}\frac{1}{Z} + 2\tan^{-1} Z\right)$
Coaxial cylindrical external side and disk 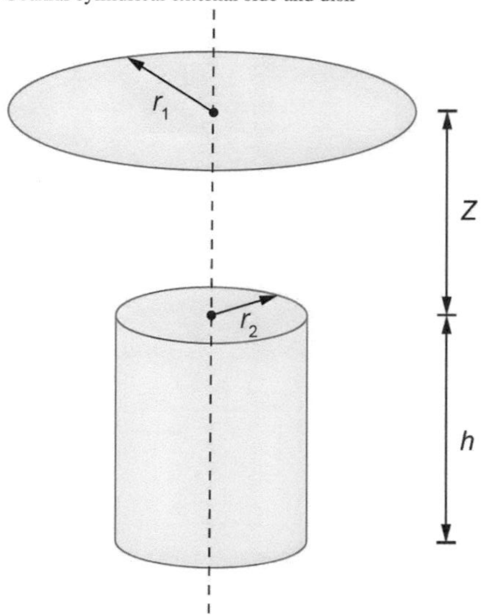	$r_1 > r_2$ $R_1 = \frac{r_1}{z}$, $R_2 = \frac{r_2}{z}$, $H = \frac{h}{z}$ $F_{1-2} = \left(\frac{R_2 H}{\pi R_1^2}\right)\left\{\arccos\left[\frac{(1+H)^2 - R_1^2 + R_2^2}{(1+H)^2 + R_1^2 - R_2^2}\right] - \arccos\left[\frac{1 - R_1^2 + R_2^2}{1 + R_1^2 - R_2^2}\right]\right\}$ $- \left(\frac{H}{\pi R_1^2}\right)\left\{\frac{\left\{\left[(1+H)^2 + R_1^2 + R_2^2\right]^2 - 4R_1^2 R_2^2\right\}^{1/2}}{2(1+H)}\right.$ $\left.\times \arccos\left\{\left(\frac{R_2}{R_1}\right)\left[\frac{(1+H)^2 - R_1^2 + R_2^2}{1 + R_1^2 - R_2^2}\right]\right\}\right\}$ $+ \left(\frac{H}{\pi R_1^2}\right) \times \left\{\frac{\left[\left(1^2 + R_1^2 + R_2^2\right)^2 - 4R_1^2 R_2^2\right]^{1/2}}{2} \arccos\left[\left(\frac{R_2}{R_1}\right)\left(\frac{1 - R_1^2 + R_2^2}{1 + R_1^2 - R_2^2}\right)\right]\right\}$ $+ \left(\frac{H^2}{2\pi R_1^2}\right)\left\{\arccos\left(\frac{R_2}{R_1}\right) - \left(\frac{\left(R_1^2 - R_2^2\right)}{1+H}\right)\left[\frac{\pi}{2} + \arcsin\left(\frac{R_2}{R_1}\right)\right]\right\}$

(continued)

Table 11.1 (continued)

Geometry	Equation
Coaxial cylindrical internal side and disk 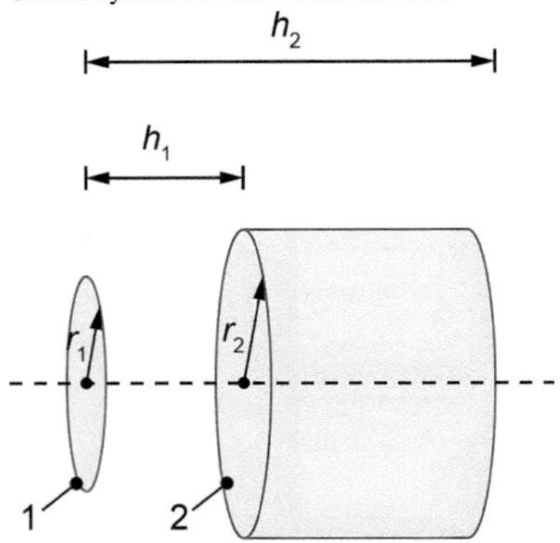	$r_1 \le r_2$ $R = \frac{r_2}{r_1},\ H_1 = \frac{h_1}{r_1},\ H_2 = \frac{h_2}{r_1}$ $F_{1-2} = \frac{1}{2}\left[H_1^2 - H_2^2 + \sqrt{\left(H_2^2 + R^2 + 1\right)^2 - 4R^2} - \sqrt{\left(H_1^2 + R^2 + 1\right)^2 - 4R^2}\right]$
Cylinder to rectangle, both infinite 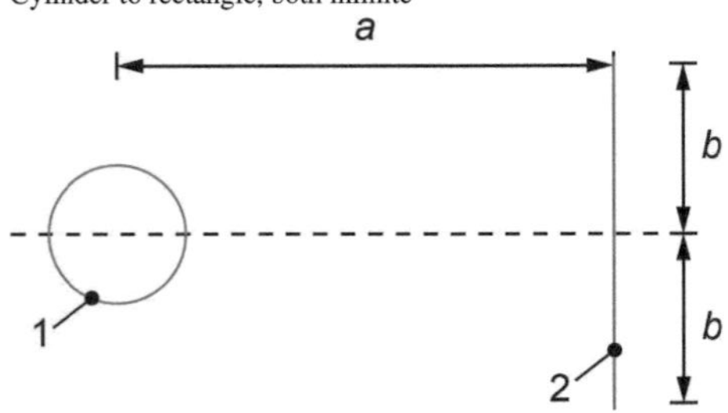	$F_{1-2} = \frac{1}{\pi}\arctan(b/a)$

(continued)

Table 11.1 (continued)

Geometry	Equation
Cylinder to semicylinder 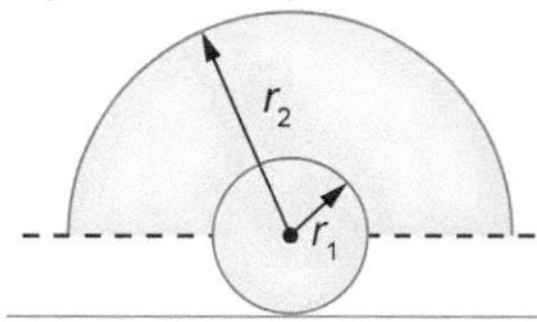	$R = \frac{r_1}{r_2}$ $F_{1-2} = 0.5$ $F_{2-2} = 1 - \frac{2}{\pi}\left[\left(1 - R^2\right)^{\frac{1}{2}} + R\sin^{-1}R\right]$
Sphere to rectangle 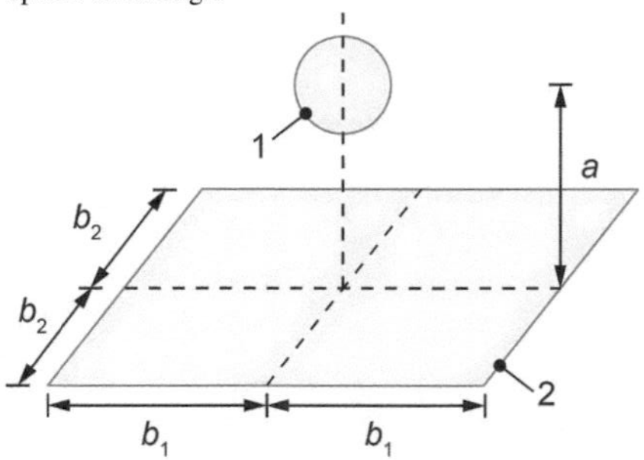	$B_1 = \frac{b_1}{a}, B_2 = \frac{b_2}{a}$ $F_{1-2} = \frac{1}{2\pi}\left\{\arcsin\left[\frac{2B_1^2-\left(1-B_1^2\right)\left(B_1^2+B_2^2\right)}{\left(1+B_1^2\right)\left(B_1^2+B_2^2\right)}\right] + \arcsin\left[\frac{2B_2^2-\left(1-B_2^2\right)\left(B_1^2+B_2^2\right)}{\left(1+B_2^2\right)\left(B_1^2+B_2^2\right)}\right]\right\}$

(continued)

Table 11.1 (continued)

Geometry	Equation
Sphere to a coaxial disk sector 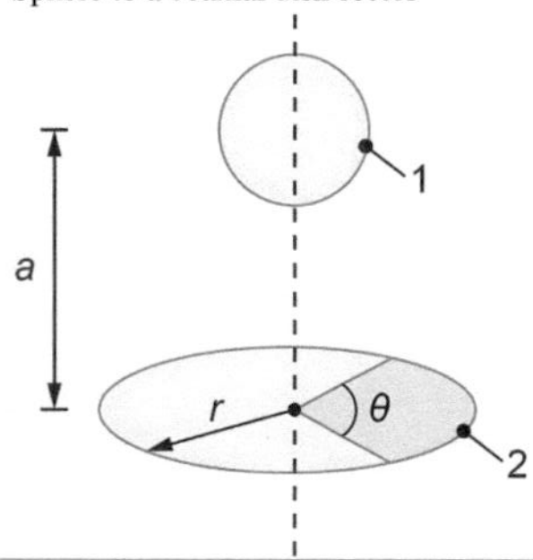	$R = \frac{r}{a}$ $F_{1-2} = \frac{\theta}{4\pi}\left[1 - \frac{1}{\left(1+R^2\right)^{\frac{1}{2}}}\right]$
Two spheres 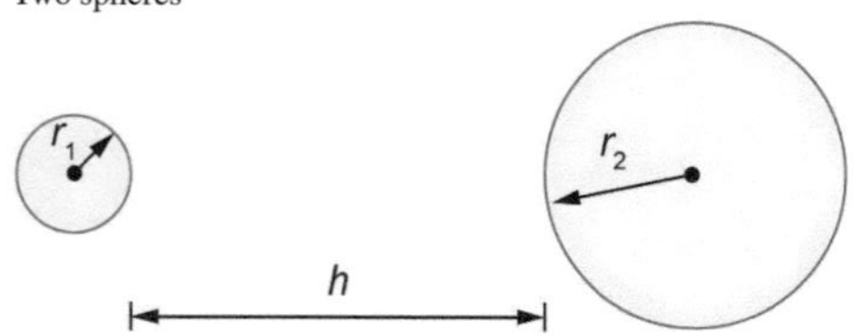	$R = \frac{r_1}{r_2}, S = \frac{h}{r_2}$ (up to 5.8% accuracy) $F_{1-2} = \left\{1 - \left[1 - \left(\frac{R}{S+R+1}\right)^2\right]^{\frac{1}{2}}\right\}\left\{1 - \left[1 - \left(\frac{1}{S+R+1}\right)^2\right]^{\frac{1}{2}}\right\} \times \left(\frac{S+R+1}{R}\right)^2$
Two concentric parallel square plates 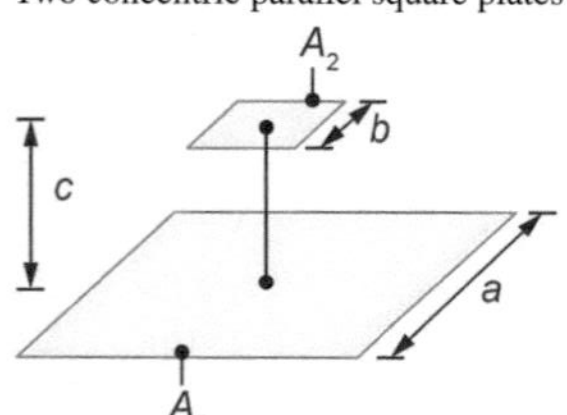	$A = \frac{a}{c}, B = \frac{b}{a}, X = A(1+B), Y = A(1-B)$ $F_{1-2} = \frac{1}{\pi A^2}\left\{\left\{\ln \frac{\left[A^2\left(1+B^2\right)+2\right]^2}{(Y^2+2)(X^2+2)} + \left(Y^2+4\right)^{\frac{1}{2}}\left[Y \tan^{-1}\frac{Y}{(Y^2+4)^{\frac{1}{2}}} - X \tan^{-1}\frac{X}{(Y^2+4)^{\frac{1}{2}}}\right]\right.\right.$ $\left.\left. + \left(X^2+4\right)^{\frac{1}{2}}\left[X \tan^{-1}\frac{X}{(X^2+4)^{\frac{1}{2}}} - Y \tan^{-1}\frac{Y}{(X^2+4)^{\frac{1}{2}}}\right]\right\}\right\}$ *For* $A < 0.2: \; F_{1-2} = \frac{(AB)^2}{\pi}$

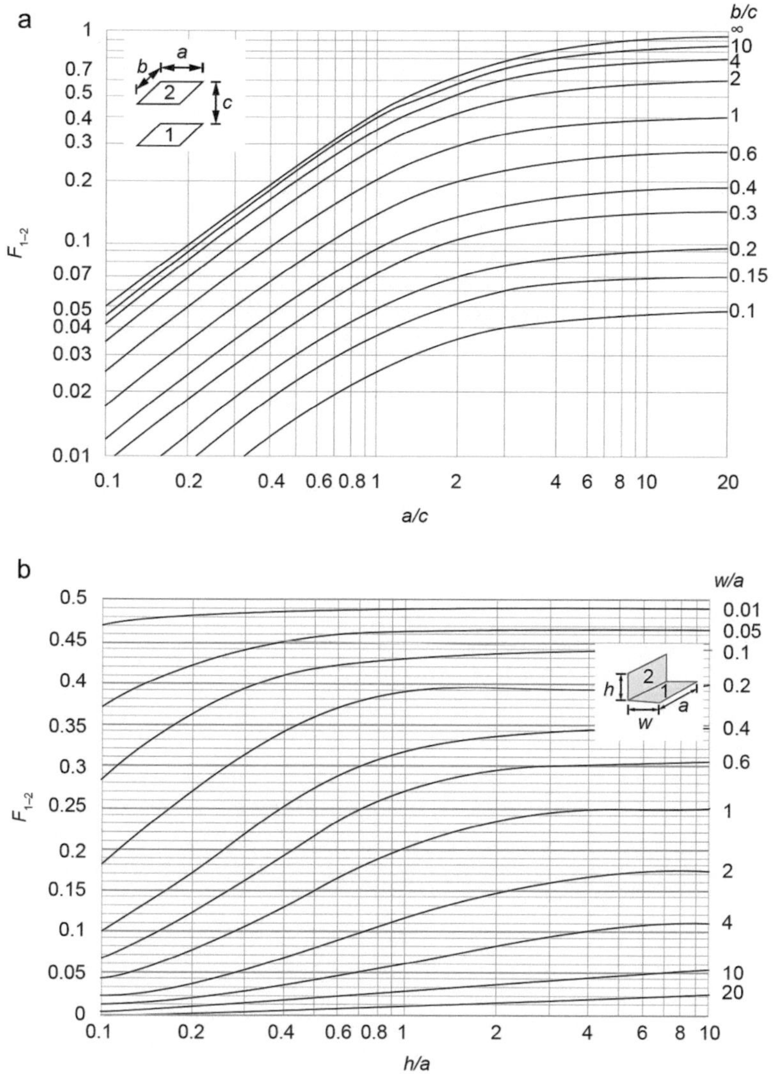

Fig. 11.6 Shape factor graphs for various configurations (the analytical expressions are in Table 11.1): **a** equal parallel plates, **b** perpendicular plates, **c** coaxial parallel discs, **d** coaxial cylinders, **e** outer cylinder and ring, **f** two spheres, and **g** two concentric parallel square plates

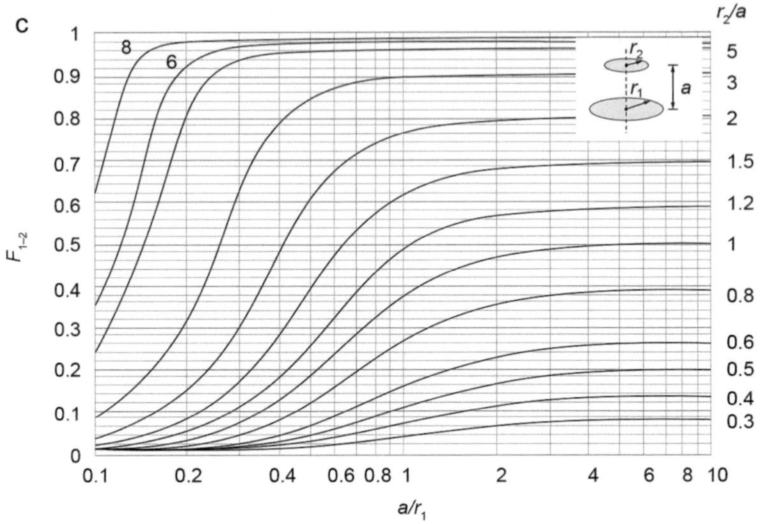

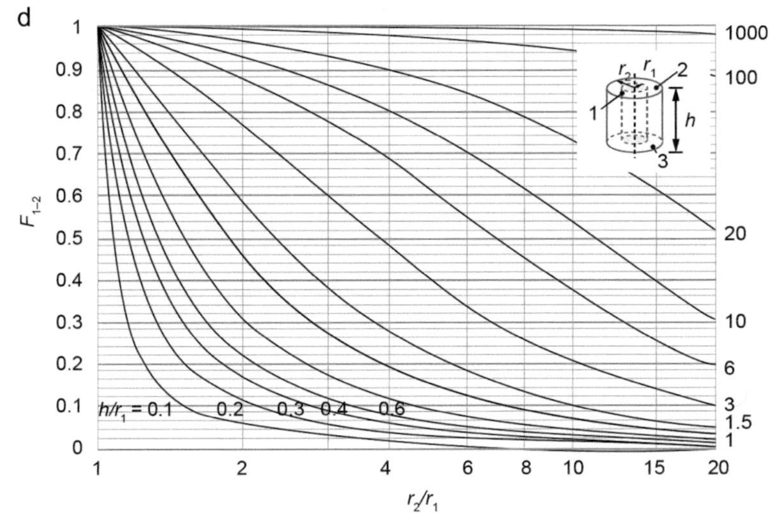

Fig. 11.6 (continued)

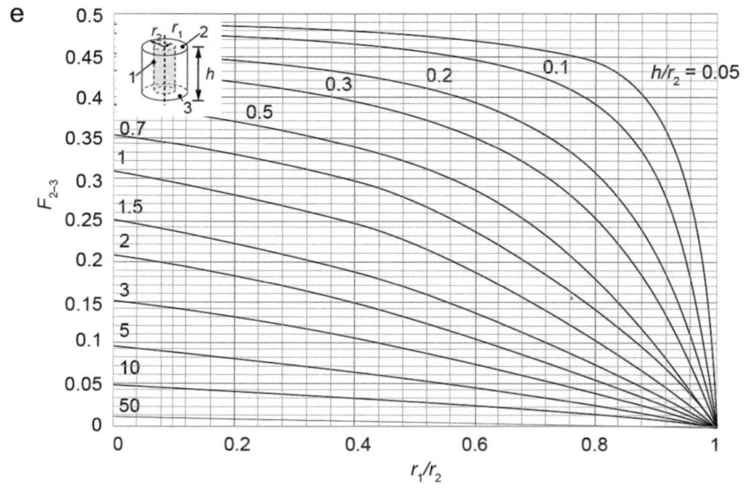

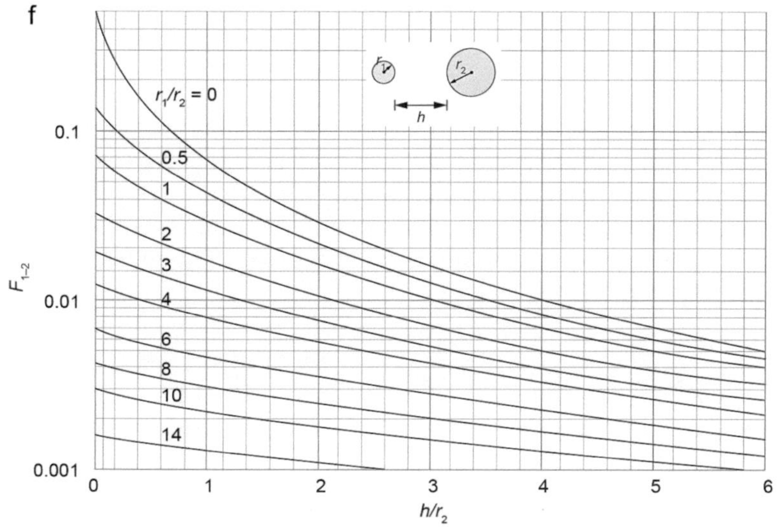

Fig. 11.6 (continued)

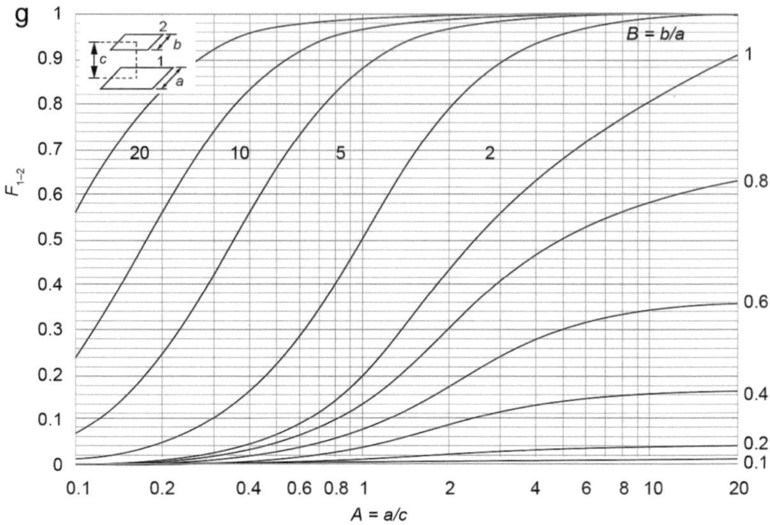

Fig. 11.6 (continued)

Comments: although the areas of surfaces 2 and 3 are equal, the shape factors F_{1-2} and F_{1-3} are different. Indeed, the radiation rate leaving surface 1 and hitting 2 is 15%, while from surface 1 to 3 it is only 5%.

Based on the results of the shape factors, it can be stated that the radiation rate leaving 1 and hitting 2 and 3 combined totals 20%, and therefore, the remaining 80% goes to the surrounding environment.

Example 11.4 (*Shape factor—parallel discs*)

A small truncated cone is formed by two parallel discs that are connected by a conical surface as shown in the figure. Determine the fraction of radiant energy that leaves the cylindrical surface (3) and hits its own.

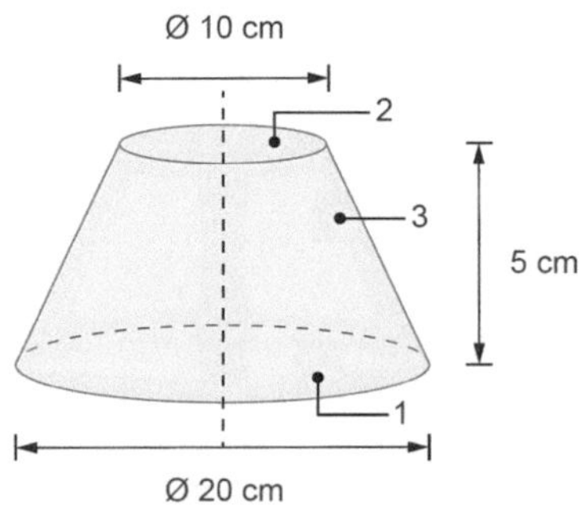

Solution

$$A_1 = \pi \frac{D^2}{4} = \pi \frac{0.2^2}{4} = 0.0314\,\text{m}^2$$

$$A_2 = \pi \frac{0.1^2}{4} = 7.85 \times 10^{-3}\ \text{m}^2$$

$$g = \sqrt{h^2 + (R - r)^2} = \sqrt{0.05^2 + (0.1 - 0.05)^2} = 0.0707$$

$$A_3 = \pi g(r + R) = \pi 0.0707 \times (0.05 + 0.1) = 0.0333$$

Closure law for surface 1: $F_{1-1} + F_{1-2} + F_{1-3} = 1,\ F_{1-1} = 0$
$\frac{a}{r_1} = \frac{5}{10} = 0.5;\ \frac{r_2}{a} = \frac{5}{5} = 1 \rightarrow F_{1-2} = 0.19$ (Fig. 11.6c)

$$F_{1-3} = 1 - F_{1-2} = 1 - 0.19 = 0.81$$

$$A_1F_{1-3} = A_3F_{3-1} \rightarrow F_{3-1} = F_{1-3}\frac{A_1}{A_3} = 0.81 \times \frac{0.0314}{0.0333} = 0.7638$$

Closure law for surface 2: $F_{2-1} + F_{2-3} + F_{2-2} = 1$, but $F_{2-2} = 0$

$$A_1F_{1-2} = A_2F_{2-1} \rightarrow F_{2-1} = F_{1-2}\frac{A_1}{A_2} = 0.19 \times \frac{0.0314}{7.85 \times 10^{-3}} = 0.76$$

$$F_{2-3} = 1 - F_{2-1} = 1 - 0.76 = 0.24$$

$$A_2F_{2-3} = A_3F_{3-2} \rightarrow F_{3-2} = F_{2-3}\frac{A_2}{A_3} = 0.24 \times \frac{7.85 \times 10^{-3}}{0.0333} = 0.0566$$

Finally, by the closure law for surface 3: $F_{3-1} + F_{3-2} + F_{3-3} = 1$

$$F_{3-3} = 1 - F_{3-1} - F_{3-2} = 1 - 0.7638 - 0.0566$$

$$F_{3-3} = 0.1796.$$

11.2.3 Shape Factors—Hottel's Chord Method

A technique credited to Hottel (1954), called the *chord method* allows solving geometries of constant cross-section tubes and channels lengthy enough so that the edges do not interfere with the calculation of the shape factor. Thus, these geometries can be considered as two-dimensional problem geometries. The technique is quite simple and consists of the following equation:

$$F_{i-j} = \frac{\sum \text{chords that cross} - \sum \text{chords that do not cross}}{2 \times \text{chord of the surface } i} \quad (11.18)$$

The "chords" are the sizes of the chords identified in Fig. 11.7. To illustrate, consider the heat exchange by radiation between surfaces 1 and 2 long in the perpendicular direction to the paper illustrated in Fig. 11.7. The two surfaces can have any shape, concave, flat, or convex. The shape factors F_{1-2} and F_{2-1} are indicated.

Shape factors

$$F_{1-2} = \frac{(L_5 + L_6) - (L_3 + L_4)}{2 \times L_1}$$

$$F_{2-1} = \frac{(L_5 + L_6) - (L_3 + L_4)}{2 \times L_2}$$

Example 11.5 (*Shape factor—chord method*)

Two long and parallel plates of width $L_1 = 12$ cm and $L_2 = 7$ cm are spaced 5 cm apart, as shown in the figure. Find the shape factors $F_{1\text{-}2}$ and F_{2-1}.

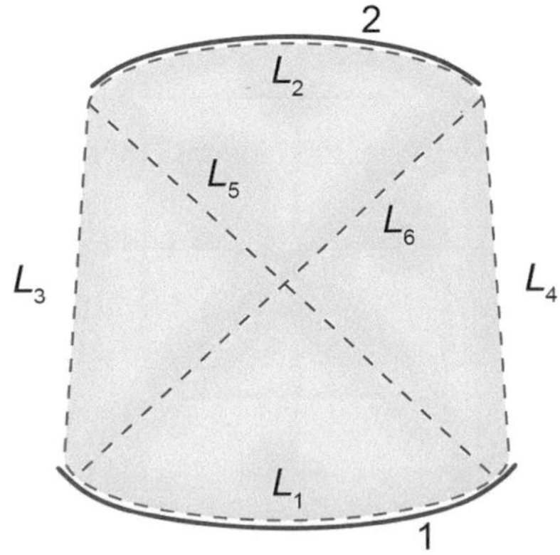

Fig. 11.7 Hottel's chord method for obtaining shape factors for long tubes and channels

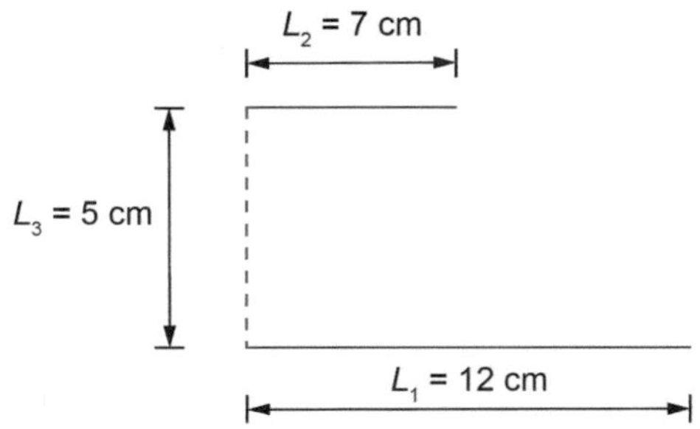

Solution

It is necessary to determine the lengths of the chords. Using a basic geometric relationship, one obtains

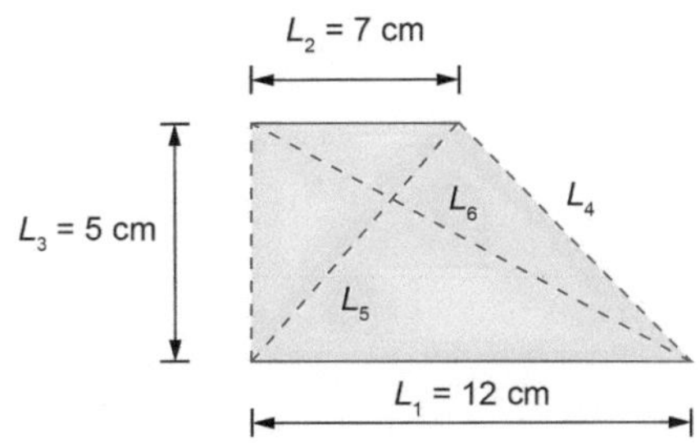

$$L_4 = \sqrt{(12-7)^2 + 5^2} = 7.07 \text{ cm}, L_5 = \sqrt{7^2 + 5^2} = 8.60 \text{ cm and}$$
$$L_6 = \sqrt{12^2 + 5^2} = 13 \text{ cm}$$

Thus, the shape factors are:

$$F_{1-2} = \frac{8.6 + 13 - (7.07 + 5)}{2 \times 12} = \frac{9.53}{24} = 0.397$$

$$F_{2-1} = \frac{8.6 + 13 - (7.07 + 5)}{2 \times 7} = \frac{9.53}{14} = 0.681$$

Compare the solution with the law of reciprocity (Eq. 11.16).

11.3 Heat Exchange by Radiation Between Surfaces

11.3.1 Thermal Radiation Exchanged Between Two Surfaces

The net rate of heat exchanged by radiation between two or more surfaces can now be calculated considering the shape factors between involved surfaces. For that, consider two isothermal gray and opaque surfaces at temperatures T_1 and T_2. The net radiation heat rate exchange between surfaces 1 and 2, $\dot{Q}_{1-2}$, forming an enclosure depends on their radiosity and shape factors, as indicated in Eq. (11.19).

$$\dot{Q}_{1-2} = A_1 F_{1-2} J_1 - A_2 F_{2-1} J_2 \tag{11.19}$$

Using the law of reciprocity, Eq. (11.16), that is, $A_1 F_{1-2} = A_2 F_{2-1}$, it follows that:

$\dot{Q}_{1-2} = -\dot{Q}_{2-1} = A_1 F_{1-2}(J_1 - J_2) = A_2 F_{2-1}(J_1 - J_2)$, or

$$\dot{Q}_{1-2} = \frac{J_1 - J_2}{R_{1-2}} \tag{11.20a}$$

where

$$R_{1-2} = \frac{1}{A_1 F_{1-2}} = \frac{1}{A_2 F_{2-1}} \tag{11.20b}$$

where R_{1-2} is known as the *spatial resistance* between the two surfaces, which depends only on the shape factors and areas involved. More generally, the spatial resistance R_{i-j} of surface i in relation to surface j can be written as:

$$R_{i-j} = \frac{1}{A_i F_{i-j}} = \frac{1}{A_j F_{j-i}} \tag{11.21}$$

This relationship is a generalization of the spatial resistance in Eq. (11.13), which was only valid for two parallel and infinite plates where the shape factor F_{1-2} is unitary. Finally, the equivalent thermal radiation circuit is indicated in Fig. 11.8 with the insertion of the surface resistances (Eq. 11.12) and spatial (Eq. 11.21), which form the total equivalent resistance, R_T, of the circuit:

$$R_T = \frac{1}{R_1} + \frac{1}{A_1 F_{1-2}} + \frac{1}{R_2} \tag{11.22}$$

and the net rate of heat exchanged, $\dot{Q}_{1-2}$, between the two surfaces is:

$$\dot{Q}_{1-2} = \frac{E_{b1} - E_{b2}}{R_T} = \frac{\sigma\left(T_1^4 - T_2^4\right)}{R_T} \tag{11.23}$$

Example 11.6 (*Heat exchange between two surfaces*)

Find the radiation heat transfer rate of a heated sphere inside a closed cylindrical surface kept in a vacuum, as indicated in figure. The 10 cm sphere diameter has a 0.8 emissivity and is kept at 300 °C. The inner surface of the cylinder, whose area is 2 m^2, has 0.2 emissivity and is kept at a 20 °C temperature.

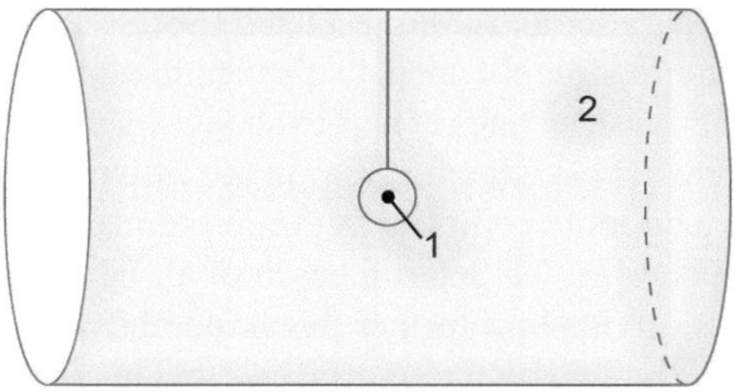

Solution

The equivalent radiation circuit is equal to that indicated in Fig. 11.8. The radiation shape factor from the sphere to the cylindrical enclosure is $F_{1-2} = 1$, since the thorough radiation emitted by the sphere hits only the cylindrical surface. The net rate of heat transfer, $\dot{Q}_{1-2}$, is obtained by:

$$\dot{Q}_{1-2} = \frac{\sigma\left(T_1^4 - T_2^4\right)}{\frac{1-\varepsilon_1}{\varepsilon_1 A_1} + \frac{1}{A_1 F_{1-2}} + \frac{1-\varepsilon_2}{\varepsilon_2 A_2}}$$

where $A_1 = \pi \frac{d^2}{4} = 7.85 \times 10^{-3}\,\text{m}^2$

$$T_1 = 300\,°\text{C} = 573.2\,\text{K and } T_2 = 20\,°\text{C} = 293.2\,\text{K}$$

$$\dot{Q}_{1-2} = \frac{5.67 \times 10^{-8}\left(573,2^4 - 293,2^4\right)}{(1 - 0.8)/\left(0.8 \times 7.85 \times 10^{-3}\right) + 1/\left(7.85 \times 10^{-3} \times 1\right) + (1 - 0.2)/(0.2 \times 2)}$$

$$\dot{Q}_{1-2} = 35.4\,\text{W}$$

Fig. 11.8 Thermal circuit for the case of two surfaces

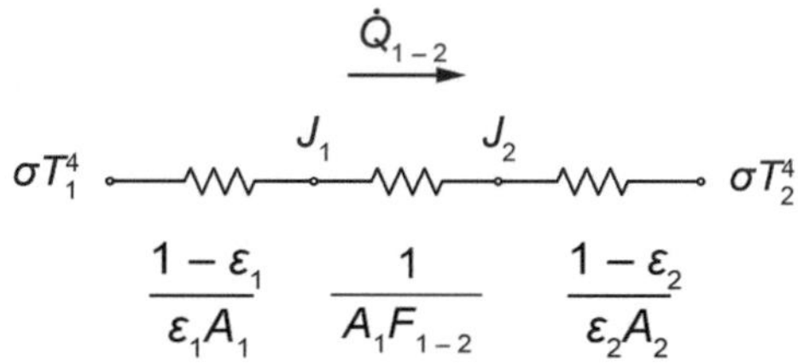

Note that this radiation heat rate of 35.4 W must be transferred from the cylindrical outer surface to the external environment by some heat transfer mechanism: conduction, convection, radiation, or a combination of convection and radiation. The sphere also needs a source of thermal energy to be kept at 300 °C. The radiation problem deals only with the net radiation exchange between the two surfaces, regardless of the way to keep the surface temperatures.

In the case of two parallel and infinite plates (Sect. 11.1), the radiosity of the surfaces (J_1 and J_2) are readily obtained, as the irradiation of one surface becomes the radiosity of the other, and the rates of heat exchange by radiation from the surfaces are also the net heat rate between them, $\dot{Q}_{1-2}$. In the case of *more than two surfaces*, the problem becomes a bit more complex, as the determination of the radiosity of the surface of interest depends on the determination of all heat exchanges by radiation with the other surfaces. To explore further this issue, the next subsection examines the case of an enclosure formed by three surfaces. Next in the following subsections, the solution technique is expanded to situations involving any number of enclosed surfaces.

11.3.2 Thermal Radiation Exchanged Between Three Surfaces

Consider the scheme of Fig. 11.9, where three isothermal surfaces maintained at temperatures T_1, T_2, and T_3 are shown. Although the figure shows an enclosure of a triangular section, the analysis applies to any closed enclosure formed by three surfaces. The temperatures of the surfaces are maintained by the heat rates that are transferred to or from the external environment, which represent the rates of heat $\dot{Q}_1$, $\dot{Q}_2$, and $\dot{Q}_3$, respectively.

The net radiation heat exchanges between these three surfaces, taken two by two, can be obtained through Eq. (11.20a), which results in:

Fig. 11.9 Closed envelope formed by three surfaces

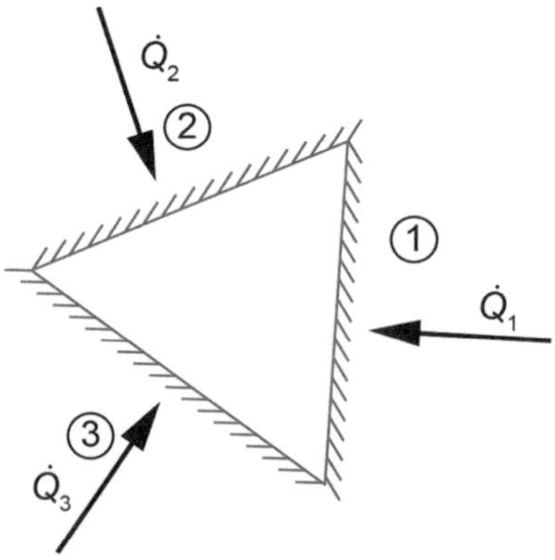

$$\begin{aligned}
&\text{surfaces } (1-2): \dot{Q}_{1-2} = -\dot{Q}_{2-1} = \frac{(J_1 - J_2)}{R_{1-2}} \\
&\text{surfaces } (1-3): \dot{Q}_{1-3} = -\dot{Q}_{3-1} = \frac{(J_1 - J_3)}{R_{1-3}} \\
&\text{surfaces } (2-3): \dot{Q}_{2-3} = -\dot{Q}_{3-2} = \frac{(J_2 - J_3)}{R_{2-3}}
\end{aligned} \tag{11.24}$$

where the *spatial resistances* (Eq. 11.20b) between the surfaces are given by:

$$\begin{aligned}
&\text{spatial resistance}(1-2): R_{1-2} = \frac{1}{A_1 F_{1-2}} = \frac{1}{A_2 F_{2-1}} \\
&\text{spatial resistance}(1-3): R_{1-3} = \frac{1}{A_1 F_{1-3}} = \frac{1}{A_3 F_{3-1}} \\
&\text{spatial resistance } (2-3): R_{2-3} = \frac{1}{A_2 F_{2-3}} = \frac{1}{A_3 F_{3-2}}
\end{aligned} \tag{11.25}$$

Therefore, in order to determine the net radiation heat exchanges between the surfaces, it is necessary that, first, the associated radiosities, J_1, J_2, and J_3, of each surface are determined. On the other hand, the radiation heat rate of each surface is given by Eq. (11.5), which allows us to write:

$$\begin{aligned}
&\text{surface 1}: \dot{Q}_1 = \frac{E_{b1} - J_1}{R_1} \\
&\text{surface 2}: \dot{Q}_2 = \frac{E_{b2} - J_2}{R_2} \\
&\text{surface 3}: \dot{Q}_3 = \frac{E_{b3} - J_3}{R_3}
\end{aligned} \tag{11.26}$$

where the *surface resistances* of the surfaces are given by:

$$\begin{aligned}
&\text{surface resistance 1}: R_1 = \frac{1-\varepsilon_1}{\varepsilon_1 A_1} \\
&\text{surface resistance 2}: R_2 = \frac{1-\varepsilon_2}{\varepsilon_2 A_2} \\
&\text{surface resistance 3}: R_3 = \frac{1-\varepsilon_3}{\varepsilon_3 A_3}
\end{aligned} \tag{11.27}$$

Now, it is important once again to use the electrical analogy to establish a relationship between the radiation heat rates of each surface, $\dot{Q}_i$ (Eq. 11.26), with the net radiation heat rates exchanged between the surfaces, $\dot{Q}_{i-j}$ (Eq. 11.24), and the other quantities involved. In the present case of three surfaces, the equivalent radiation circuit is indicated in Fig. 11.10. Observe the surface resistances R_{i-j}, the surface resistances R_i and the rates of heat exchange by radiation in the analogous circuit. The

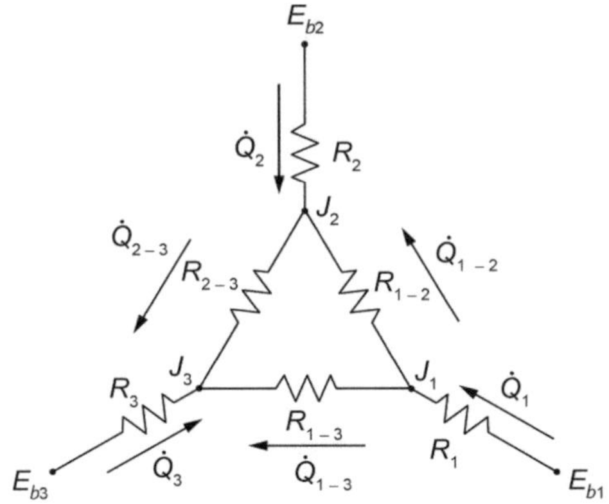

Fig. 11.10 Equivalent radiation circuit of three enclosed surfaces with arbitrary heat radiation rate directions randomly assumed

solution to these thermal problems requires that we impose at first arbitrary directions of the radiation heat rates at each node 1, 2, and 3. Therefore, the directions of the radiation heat rates indicated in the figure are *arbitrary* in order to establish the solution of the circuit. In the end, if the direction of the heat rate originally assumed results in a negative signal (−), then, it will be known that the correct direction is in opposition to what was assumed.

The solution to electrical circuits (or in this case, equivalent thermal circuits) of this nature involves the use of *Kirchhoff's Law* (which differs from discussed in Chap. 10), which states that "*the sum of the currents entering a node is equal to the sum of the currents leaving the node*". Therefore, the heat rates of the surfaces, $\dot{Q}_1$, $\dot{Q}_2$, and $\dot{Q}_3$ are related to the net rates of radiation heat exchange between surfaces, $\dot{Q}_{1-2}$, $\dot{Q}_{1-3}$, and $\dot{Q}_{2-3}$, through the energy balance at each node. Again, the signals that appear are associated with the arbitrary directions of the heat rates already adopted previously in Fig. 11.10 by consistency:

$$\begin{aligned} \dot{Q}_1 &= \dot{Q}_{1-2} + \dot{Q}_{1-3} \\ \dot{Q}_2 &= -\dot{Q}_{1-2} + \dot{Q}_{2-3} \\ \dot{Q}_3 &= -\dot{Q}_{1-3} - \dot{Q}_{2-3} \end{aligned} \tag{11.28}$$

Substituting these radiation heat exchange rates from each surface into the set of Eqs. (11.26) results in a linear system of six equations and six unknowns, namely, $\dot{Q}_{1-2}$, $\dot{Q}_{1-3}$, $\dot{Q}_{2-3}$, J_1, J_2, and J_3 whose solution can be achieved through some numerical method of solving linear systems of equations, such as Gaussian elimination. The advantage of this approach is that the radiation net rates of heat exchanged between surfaces result directly from the solution.

Example 11.7 (*Radiation heat exchange between three surfaces*)

The truncated cone analyzed in Example 11.4 is in vacuum and the surfaces have the following temperatures and emissivity: $T_1 = 1500$ K, $T_2 = 300$ K, $T_3 = 500$ K,

$\varepsilon_1 = 0.9$, $\varepsilon_2 = 0.2$, and $\varepsilon_3 = 0.3$. Evaluate the heat rate transferred by radiation for each surface.

Solution

Using the signal convention from Fig. 11.10

$$\text{Node 1}: \frac{E_{b1} - J_1}{\frac{(1-\varepsilon_1)}{A_1\varepsilon_1}} = \frac{J_1 - J_2}{\frac{1}{A_1F_{1-2}}} + \frac{J_1 - J_3}{\frac{1}{A_1F_{1-3}}}$$

$$\text{Node 2}: \frac{E_{b2} - J_2}{\frac{(1-\varepsilon_2)}{A_2\varepsilon_2}} = -\frac{J_1 - J_2}{\frac{1}{A_1F_{1-2}}} + \frac{J_2 - J_3}{\frac{1}{A_2F_{2-3}}}$$

$$\text{Node 3}: \frac{E_{b3} - J_3}{\frac{(1-\varepsilon_3)}{A_3\varepsilon_3}} = -\frac{J_1 - J_3}{\frac{1}{A_1F_{1-3}}} - \frac{J_2 - J_3}{\frac{1}{A_2F_{2-3}}}$$

$$E_{b1} = 5.67 \times 10^{-8} \times 1500^4 = 287,044\,\mathrm{w/m^2}, A_1 = 0.0314\,\mathrm{m^2}$$

$$E_{b2} = 5.67 \times 10^{-8} \times 300^4 = 459\,\mathrm{W/m^2}, A_2 = 7.85 \times 10^{-3}\,\mathrm{m^2}$$

$$E_{b3} = 5.67 \times 10^{-8} \times 500^4 = 3544\,\mathrm{W/m^2}, A_3 = 0.0333\,\mathrm{m^2}$$

$$\text{Shape factors}: F_{1-2} = 0.19, F_{1-3} = 0.81, F_{2-3} = 0.24$$

$$\text{Node 1}: \frac{287,044 - J_1}{\frac{1-0.9}{0.9\times0.0314}} = \frac{J_1 - J_2}{\frac{1}{0.0314\times0.19}} + \frac{J_1 - J_3}{\frac{1}{0.0314\times0.81}}$$

$$\text{Node 2}: \frac{459 - J_2}{\frac{1-0.2}{0.2\times7.85\times10^{-3}}} = -\frac{J_1 - J_2}{\frac{1}{0.0314\times0.19}} + \frac{J_2 - J_3}{\frac{1}{7.85\times10^{-3}\times0.24}}$$

$$\text{Node 3}: \frac{3544 - J_3}{\frac{1-0.3}{0.3\times0.033}} = -\frac{J_1 - J_3}{\frac{1}{0.0314\times0.81}} - \frac{J_2 - J_3}{\frac{1}{7.85\times10^{-3}\times0.24}}$$

Solving the system of three equations, we get:

$$J_1 = 276,795\,\mathrm{W/m^2};\ J_2 = 202,988\,\mathrm{W/m^2};\ J_3 = 180,232\,\mathrm{W/m^2}$$

$$\dot{Q}_1 = \frac{287,044 - 276,795}{\frac{1-0.9}{0.9\times0.0314}} = 2896.4\,\mathrm{W}$$

$$\dot{Q}_2 = \frac{459 - 202,988}{\frac{1-0.2}{0.2\times7.85\times10^{-3}}} = -397.5\,\mathrm{W}$$

$$\dot{Q}_3 = \frac{3544 - 180,232}{\frac{1-0.3}{0.3\times0.0333}} = -2498.9$$

Negative signals indicate that the directions of the heat transfer rates assumed at the beginning were opposite. The correct directions of these rates are represented in the figure below. Notice that $\dot{Q}_1 = \dot{Q}_2 + \dot{Q}_3$, as required by the First Law of Thermodynamics.

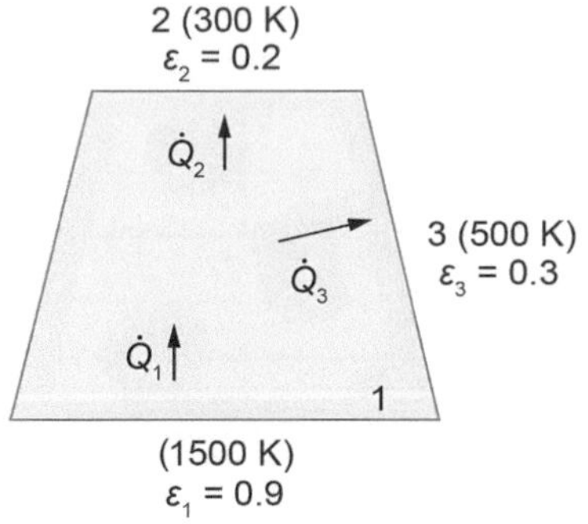

Adiabatic-reradiating surfaces

An adiabatic-reradiating surface is one that fully reflects the incident radiation. Thus, no net heat rate is transferred outside the system, which in, practical terms the surface is adiabatic. Example 11.8 well illustrated such a surface.

Example 11.8 (*Adiabatic-reradiating and black surfaces*)

A small oven has a vertical cylindrical shape. The base, considered black, is kept at 600 K and the upper surface at 300 K. If the adiabatic and reradiating cylindrical surface and the emissivity of the surfaces are: $\varepsilon_2 = 0.5$ and $\varepsilon_3 = 0.8$, evaluate the temperature of the cylindrical surface and the portions of the heat rate that reach the upper surface coming from the lower and from the cylindrical surface.

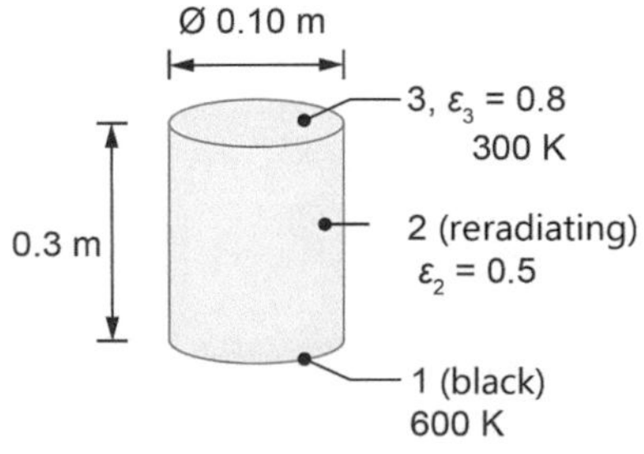

Solution

The figure below shows the equivalent radiation circuit, in which, initially, the directions of the radiation rates are freely assumed; In the end, the heat flux signal indicates the right direction.

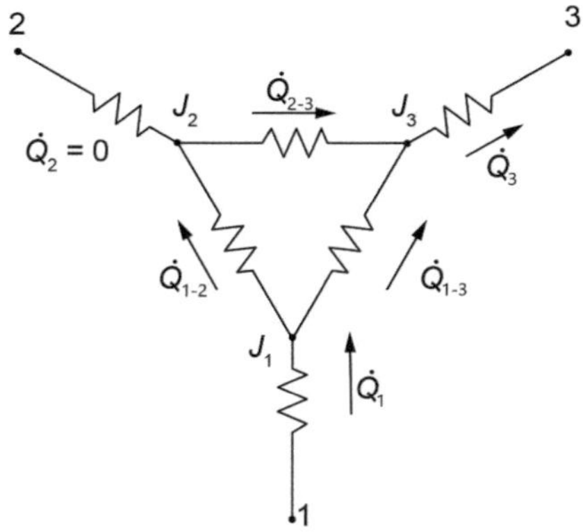

Node 1: black surface, so: $J_1 = E_{b1} = 5.67 \times 10^{-8} \times 600^4 = 7348\,\text{W/m}^2$,
Node 2: adiabatic-reradiating surface, that is: $\dot{Q}_{1-2} = \dot{Q}_{2-3}$: $\frac{J_1 - J_2}{\frac{1}{A_1 F_{1-2}}} = \frac{J_2 - J_3}{\frac{1}{A_2 F_{2-3}}}$.
Node 3: energy balance: $\dot{Q}_3 = \dot{Q}_{1-3} + \dot{Q}_{2-3}$, so, $\frac{J_3 - E_{b3}}{\frac{(1-\varepsilon_3)}{A_3 \varepsilon_3}} = \frac{J_1 - J_3}{\frac{1}{A_1 F_{1-3}}} + \frac{J_2 - J_3}{\frac{1}{A_2 F_{2-3}}}$.
$E_{b3} = 5.67 \times 10^{-8} \times 300^4 = 459\,\text{W/m}^2$,
Shape factors:

$$A_1 = A_3 = \pi 0.18^2/4 = 0.0254\,\text{m}^2$$

$$A_2 = \pi dL = \pi 0.18 \times 0.3 = 0.1696\,\text{m}^2$$

$\frac{r_3}{a} = \frac{0.09}{0.3} = 0.3$, $\frac{a}{r_1} = \frac{0.3}{0.09} = 3.3 \rightarrow F_{1-3} = 0.08$ (Fig. 11.6c)

$$F_{1-1} + F_{1-2} + F_{1-3} = 1, \text{but}\, F_{1-1} = 0 \rightarrow F_{1-2} = 0.92$$

$$F_{3-1} = F_{1-3} = 0.08\,\text{(symmetry)}$$

$$F_{3-1} + F_{3-2} + F_{3-3} = 1, \text{but}\, F_{3-3} = 0 \rightarrow F_{3-2} = 0.92$$

$$F_{2-3} A_2 = F_{3-2} A_3 \rightarrow F_{2-3} = F_{3-2} \frac{A_3}{A_2} = 0.92 \frac{0.0254}{0.1696} \rightarrow F_{2-3} = 0.1378$$

$$F_{1-3} = 0.19, F_{2-3} = 0.0566$$

$$\text{Node 2}: \frac{7348 - J_2}{\frac{1}{(0.0254 \times 0.92)}} = \frac{J_2 - J_3}{\frac{1}{(0.1696 \times 0.1378)}}$$

$$\text{Node 3}: \frac{J_3 - 459}{\frac{(1-0.8)}{(0.0254 \times 0.8)}} = \frac{7348 - J_3}{\frac{1}{(0.0254 \times 0.08)}} + \frac{J_2 - J_3}{\frac{1}{(0.1696 \times 0.1378)}}$$

The previous system of two equations provides the solution of:

$$J_2 = 4313\,\mathrm{W/m^2}\ \text{and}\ J_3 = 1278\,\mathrm{W/m^2}$$

As surface 2 is adiabatic-reradiating:

$$J_2 = E_{b2} = 4313 = 5.67 \times 10^{-8} T_2^4 \rightarrow T_2 = 525\,\mathrm{k}$$

Surface 3 receives a heat rate of:

$$\underbrace{\frac{1278 - 459}{\frac{(1-0.8)}{(0.0254\times 0.8)}}}_{\dot{Q}_3} = \underbrace{\frac{7348 - 1278}{\frac{1}{(0.0254\times 0.08)}}}_{\dot{Q}_{13}} + \underbrace{\frac{4313 - 1278}{\frac{1}{(0.1696\times 0.1378)}}}_{\dot{Q}_{23}}$$

$$\dot{Q}_3 = 83.2\,W, \rightarrow \dot{Q}_{13} = 12.3\,W \rightarrow \dot{Q}_{23} = 70.9\,\mathrm{W}$$

Although surface 1 has the highest temperature, the fact that surface 2 is reradiating and has a larger area means that, of all the heat received by surface 3, 15% comes directly from surface 1 and 85% from surface 2.

Example 11.9 (*Two surfaces and the surrounding environment*)

A new selective paint ($\varepsilon_2 = 0.2$) for medium-temperature solar receivers is being tested. The test consists of raising the temperature to 500 °C in a small oven several times. The paint was deposited on a flat, circular stainless steel substrate 40 mm in diameter. To heat the system, another circular plate ($\varepsilon_1 = 0.9$) whose electrical resistance is heated to reach 1100 K is used. The two circular plates are thermally insulated on the back and placed in an environment where, once a steady state is reached, the temperature reaches 30 °C. Neglecting convective heat losses, evaluate the resistance electrical power required to heat the plate, and the radiation heat rate exchanged between the surfaces in a steady state. Assume that the surrounding environment behaves like a black body.

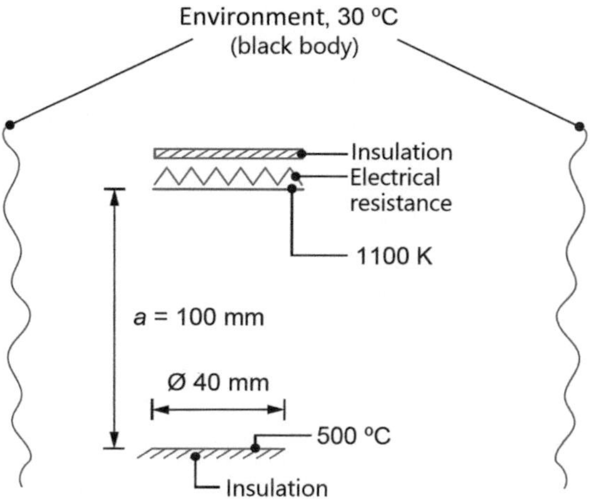

Solution

To solve this problem, we suppose the environment is an imaginary surface (3) surrounding the two surfaces, as illustrated in the figure below. To obtain the shape factors involved, consider the following geometric ratios:

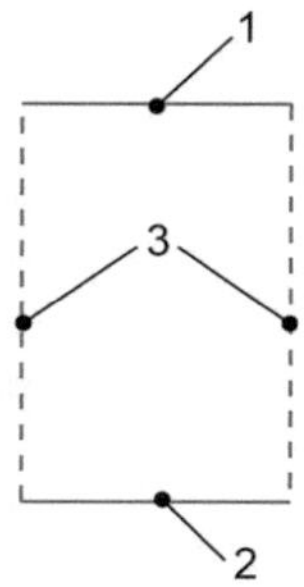

$\frac{a}{r_1} = 0.1/0.04 = 2.5$, $\frac{r_2}{a} = 0.04/0.1 = 0.4$; from Fig. 11.6, $F_{1-2} = 0.12$.
By symmetry: $F_{2-1} = F_{1-2} = 0.12$.
By the closure law, Eq. (11.17): $F_{2-1} + \underbrace{F_{2-2}}_{0} + F_{2-3} = 1$

$$F_{2-3} = 1 - F_{2-1} = 1 - 0.12 = 0.88$$

By symmetry: $F_{13} = F_{23} = 0.88F_{13} = F_{23} = 0.88$

$$A_1 = A_2 = \frac{\pi d_1^2}{4} = \frac{\pi 0.04^2}{4} = 1.256 \times 10^{-3}\,\text{m}^2$$

$$A_3 = \pi da = \pi 0.04 \times 0.10 = 0.0126\,\text{m}^2$$

$$E_{b1} = \sigma T_1^4 = 5.67 \times 10^{-8} \times 1100^4 = 83{,}014.5\,\mathrm{W/m^2}$$

$$E_{b2} = \sigma T_2^4 = 5.67 \times 10^{-8} \times 773,2^4 = 20{,}265.2\,\mathrm{W/m^2}$$

$$E_{b3} = \sigma T_3^4 = 5.67 \times 10^{-8} \times 303,2^4 = 479.2\,\mathrm{W/m^2}$$

The radiation circuit for this case can be represented by:

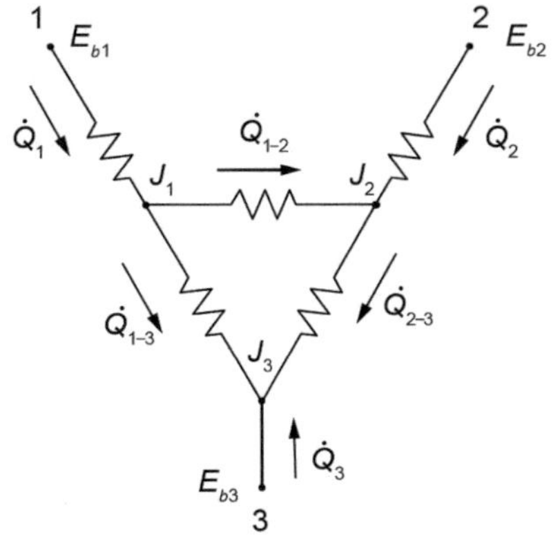

From Kirchhoff's Law:

$$\text{Node 1}: \frac{E_{n1} - J_1}{\frac{(1-\varepsilon_1)}{A_1\varepsilon_1}} = \frac{J_1 - J_2}{\frac{1}{A_1F_{1-2}}} + \frac{J_1 - J_3}{\frac{1}{A_1F_{1-3}}} \rightarrow \frac{83,014.5 - J_1}{\frac{(1-0.9)}{0.9}} = \frac{J_1 - J_2}{\frac{1}{0.12}} + \frac{J_1 - 479.2}{\frac{1}{0.88}}$$

$$\text{Node 2}: \frac{E_{n2} - J_2}{\frac{(1-\varepsilon_2)}{A_2\varepsilon_2}} = -\frac{J_1 - J_2}{\frac{1}{A_1F_{1-2}}} + \frac{J_2 - J_3}{\frac{1}{A_2F_{2-3}}} \rightarrow \frac{20,265.2 - J_2}{\frac{(1-0.2)}{0.2}} = -\frac{J_1 - J_2}{\frac{1}{0.12}} + \frac{J_2 - 479.2}{\frac{1}{0.88}}$$

Solving the system of two equations, we get:

$$J_1 = 74{,}894.2\,\mathrm{W/m^2}\ \text{and}\ J_2 = 11{,}580.2\,\mathrm{W/m^2}$$

Finding the heat rates on surfaces 1 and 2,

$$\dot{Q}_1 = \frac{E_{b1} - J_1}{\frac{(1-\varepsilon_1)}{A_1\varepsilon_1}} = \frac{83{,}014.5 - 74{,}894.2}{\frac{(1-0.9)}{(1.256\times10^{-3}\times0.9)}}$$

$$\dot{Q}_1 = 91.8\,\mathrm{W}$$

$$\dot{Q}_2 = \frac{E_{b2} - J_2}{\frac{(1-\varepsilon_2)}{A_2\varepsilon_2}} = \frac{20{,}265.2 - 11{,}580.2}{\frac{(1-0.2)}{(1.256\times10^{-3}\times0.2)}}$$

$$\dot{Q}_2 = 2.7\,\mathrm{W}$$

$$Node\,3: \frac{J_1 - J_3}{\frac{1}{A_1 F_{1-3}}} + \frac{J_2 - J_3}{\frac{1}{A_2 F_{2-3}}} + \dot{Q}_3 = 0$$

$$\frac{74,894.2 - 479.2}{\frac{1}{(1.256\times10^{-3}\times0.88)}} + \frac{11,580.2 - 479.2}{\frac{1}{(1.256\times10^{-3}\times0.88)}} + \dot{Q}_3 = 0$$

$$\dot{Q}_3 = -94.5\,\mathrm{W}$$

It can be noted that, in a steady state, the heater resistance power should be 91.8 W. The plate with the paint transfers 2.7 W of heat rate and the environment receives 94.5 W. In the actual case, convection will also have a relevant role, as discussed in Sect. 11.5. Finally, this is a particular situation when the lower plate reaches 500 °C.

So far, radiation heat exchanges from isothermal surfaces have been considered. However, if the radiation heat rates on the surfaces are known, that is, $\dot{Q}_1, \dot{Q}_2,$ and $\dot{Q}_3$ rather than the temperatures, then one should proceed to solve the set of Eqs. (11.28) to obtain the net heat rates between the surfaces and then solve Eqs. (11.24) and define the radiosities of the surfaces, $J_1, J_2,$ and J_3. Finally, one obtains the black body emissive powers (Eq. 11.26) and, consequently, the temperatures of the surfaces, $T_1, T_2,$ and T_3. If there are mixed boundary conditions, that is, known temperature and radiation heat rates, then the problem should be treated separately by combining the provided equations.

11.3.3 Thermal Radiation Exchanged Between Multiple Surfaces

In many real situations, the number of enclosed surfaces that exchange radiant heat can be many, which requires solving the corresponding system of equations. To consider the case of N surfaces that form a *closed system*, as illustrated in Fig. 11.11, it is necessary first to obtain the net heat rate of each surface, $\dot{Q}_i$.

By Kirchhoff's Law, which is also an energy balance, one can write:

$$\dot{Q}_i = \sum_{j=1}^{N} \dot{Q}_{i-j} \tag{11.29}$$

By signal convention, $\dot{Q}_i > 0$ indicates heat transfer from outside the system to surface i in accordance with the convention of Fig. 11.11. The net heat exchange by radiation from surface i can also be written according to Eq. (11.26), that is:

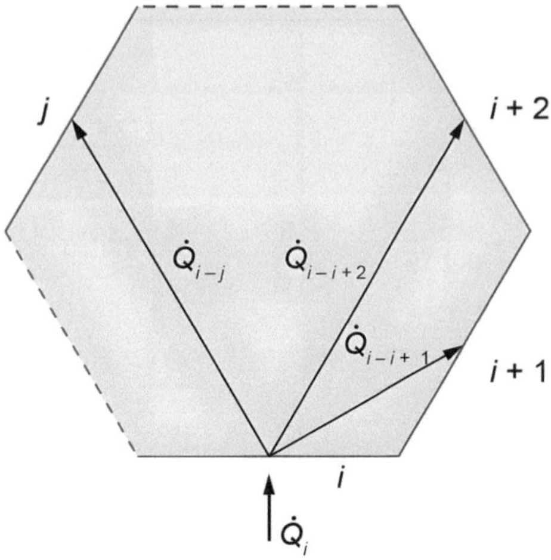

Fig. 11.11 Net heat rate, $\dot{Q}_i$ from surface i and the net exchange with the other surfaces that form a closed system

$$\dot{Q}_i = \frac{E_{bi} - J_i}{R_i} \tag{11.30}$$

and the surface resistances (Eq. 11.27),

$$R_i = \frac{1 - \varepsilon_i}{\varepsilon_i A_i} \tag{11.31}$$

The relevant cases are:

1.1 Black body—this case needs to be analyzed separately, because, from Eq. (11.30), it can induce to a null value of the net heat flux from the surface i, $\dot{Q}_i$, which is false, because the surface may or may not exchange heat with the external environment. The black body condition is the one that $E_{bi} = J_i$. One should solve Eq. (11.29) with the inclusion of the external heat rate (convection, conduction, or radiation out of the system), if necessary.

1.2 Adiabatic-reradiating surface—in this case, the net heat rate from the surface i, $\dot{Q}_i$ is zero.

The net heat rate between two surfaces i and j is given by:

$$\dot{Q}_{i-j} = \frac{J_i - J_j}{R_{i-j}} \tag{11.32}$$

Recalling that the assumed heat rate signals are just a convention adopted initially to solve the problem and that $\dot{Q}_{i-j} = -\dot{Q}_{j-i}$. Substituting the heat rates from Eqs. (11.30) into (11.29) and after rearrangement, we obtain the following set of equations:

$$
\begin{aligned}
\left(1 - F_{1-1} + \tfrac{\varepsilon_1}{1-\varepsilon_1}\right)J_1 - F_{1-2}J_2 - F_{1-3}J_3 \cdots - F_{1-N}J_N &= \tfrac{\varepsilon_1}{1-\varepsilon_1} \\
E_{b1} - F_{2-1}J_1 + \left(1 - F_{2-2} + \tfrac{\varepsilon_2}{1-\varepsilon_2}\right)J_2 - F_{2-3}J_3 \cdots - F_{2-N}J_N &= \tfrac{\varepsilon_2}{1-\varepsilon_2} \\
E_{b2} - F_{3-1}J_1 - F_{3-2}J_2 + \left(1 - F_{3-3} + \tfrac{\varepsilon_3}{1-\varepsilon_3}\right)J_3 \cdots - F_{3-N}J_N &= \tfrac{\varepsilon_3}{1-\varepsilon_3}E_{b3} \\
\vdots \quad \vdots \\
-F_{n-1}J_1 - F_{n-2}J_2 - F_{n-3}J_3 \cdots + \left(1 - F_{N-N} + \tfrac{\varepsilon_N}{1-\varepsilon_N}\right)J_N &= \tfrac{\varepsilon_N}{1-\varepsilon_N}E_{bN}
\end{aligned} \tag{11.33}
$$

In matrix terms, the coefficient matrix A, the vector of unknowns (radiosities) B, and the vector of constants C can be written as:

$$
A = \begin{bmatrix} a_{11} & a_{12} & \dots & a_{1n} \\ a_{21} & a_{22} & \dots & a_{21} \\ \vdots & \vdots & & \vdots \\ a_{N1} & a_{N1} & \dots & a_{NN} \end{bmatrix} \quad B = \begin{bmatrix} J_1 \\ J_2 \\ \vdots \\ J_N \end{bmatrix} \tag{11.34a}
$$

$$
C = \begin{bmatrix} \frac{\varepsilon_1}{1-\varepsilon_1}E_{b1} \\ \frac{\varepsilon_2}{1-\varepsilon_2}E_{b2} \\ \vdots \\ \frac{\varepsilon_N}{1-\varepsilon_N}E_{bN} \end{bmatrix} \tag{11.34b}
$$

where

$$
\begin{aligned}
&\text{diagonal elements of A}: \ a_{ii} = \left(1 - F_{i-i} + \frac{\varepsilon_i}{1-\varepsilon_i}\right) \\
&\text{elements off the diagonal of A}: \ a_{ij} = -F_{i-j} \text{ with } i \neq j \\
&\text{elements of the constant vector C}: C_i = \frac{\varepsilon_i}{1-\varepsilon_i}E_{bi}
\end{aligned} \tag{11.34c}
$$

The peculiarities of black body must be fulfilled. In this case, if the surface i is a black body, then $a_{i-j} = 0 \quad (i \neq j)$, $a_{i-i} = 1$ and $C_i = E_{bi} = \sigma T_i^4$. A linear system of N equations by N unknowns can be solved to obtain the radiosity J_i. From this point, the net heat exchange rates between the surfaces, $\dot{Q}_{i-j}$, are obtained through Eqs. (11.32) to, finally, determine the net heat rates of each surface, $\dot{Q}_i$, by Eq. (11.30).

11.3.4 Thermal Radiation Exchanged Between a Small Surface Surrounded by a Much Larger One

A simplification can be adopted when a small surface A_1 is completely surrounded by another surface A_2, much larger. In this case, the total resistance, R_T (Eq. 11.22), can be simplified. Recalling that surface 1 is completely surrounded by surface 2, the shape factor is a unit, i. e. $F_{1-2} = 1$.

$$R_T = \frac{(1-\varepsilon_1)}{\varepsilon_1 A_1} + \frac{1}{A_1 F_{1-2}} + \frac{(1-\varepsilon_2)}{\varepsilon_2 A_2}$$

$$= \frac{1}{A_1}\left[\frac{(1-\varepsilon_1)}{\varepsilon_1} + 1 + \underbrace{\frac{A_1}{A_2}}_{\approx 0}\frac{(1-\varepsilon_2)}{\varepsilon_2}\right] \approx \frac{1}{\varepsilon_1 A_1}$$

So that, in this case, the heat exchange by thermal radiation will be, approximately:

$$\dot{Q}_{1-2} = \frac{E_{b1} - E_{b2}}{R_T} \approx \varepsilon_1 \sigma A_1 \left(T_1^4 - T_2^4\right) \tag{11.35}$$

Example 11.10 (*Enclosing surface*)

Evaluate this heat transfer rate from Example 11.6 considering the simplification of Eq. (11.3) and compare these results.

Solution

$$\dot{Q}_{12} = 0.8 \times 5.67 \times 10^{-8} \times 7.85 \times 10^{-3} \times \left(573.2^4 - 293.2^4\right)$$

$$\dot{Q}_{12} = 35.9\,\text{W}$$

The difference between the actual and the simplified solution is around 1.4%.

11.4 Radiation Shielding

A technique to reduce the radiation heat transfer rate between surfaces requires using highly reflective materials, that is, of low emissivity, ε. This is easily verifiable by analyzing the surface resistance (Eq. 11.12)—when $\varepsilon \rightarrow 0$ (which denotes a highly reflective surface), the surface resistance tends to infinity and, consequently, the radiation heat rate decreases. Thus, it is a practical procedure that surfaces exposed to intense radiation sources (such as solar or from ovens) be covered with layers of highly reflective materials. Of course, if the goal is to absorb radiation, then the emissivity should be as high as possible, $\varepsilon \rightarrow 1$.

Alternatively, another method of reducing heat transfer by radiation is the installation of *radiation shielding* between the heat transfer surfaces. These shields introduce extra resistances in the radiation circuit reducing the heat rate. To illustrate the technique, consider two *parallel* and *infinite* surfaces with equal emissivity $\varepsilon_1 = \varepsilon_2 = \varepsilon$, as indicated in Fig. 11.12a. Then, it follows that the rate of heat by radiation between these two surfaces ($\dot{Q}_{wo}$, without shielding) is given by:

$$\dot{Q}_{wo} = \frac{\sigma\left(T_1^4 - T_2^4\right)}{\frac{1-\varepsilon}{A\varepsilon} + \frac{1}{AF_{1-2}} + \frac{1-\varepsilon}{A\varepsilon}}, \text{ with } F_{1-2} = 1 \text{ results in } \frac{\dot{Q}_{wo}}{A} = \frac{\sigma\left(T_1^4 - T_2^4\right)}{2\frac{(1-\varepsilon)}{\varepsilon} + 1}$$

Now, consider the installation of a third surface 3 between the original surfaces, as indicated in Fig. 11.12b. Thus, the radiation heat transfer rate in this new configuration must take into account the new resistances associated with the third surface. The equivalent thermal circuit is represented in Fig. 11.13.

By energy conservation on surface 3 (shielding), all the incident radiation heat rates must be equal to the one that leaves, therefore:

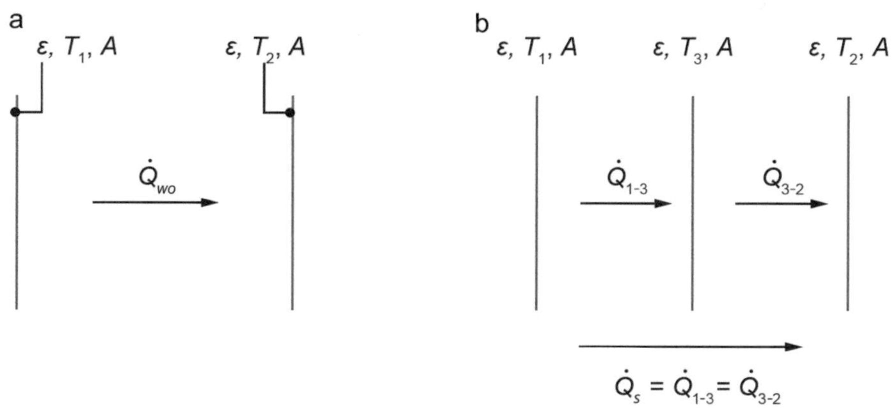

Fig. 11.12 **a** Radiation transfer rate between two parallel and infinite surfaces, without shielding (*wo*); **b** radiation transfer rate between two surfaces with an intermediate shielding surface (*ws*)

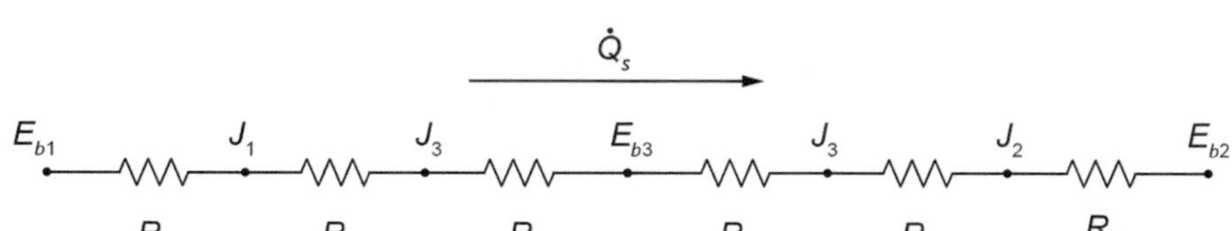

Fig. 11.13 Radiation circuit for two parallel surfaces separated by an intermediate shielding surface

$$\frac{\dot{Q}_{1-3}}{A}=\frac{\dot{Q}_{3-2}}{A}=\frac{\dot{Q}_{ws}}{A}\text{ or }\frac{\dot{Q}_{ws}}{A}=\frac{\sigma\left(T_1^4-T_3^4\right)}{2\frac{(1-\varepsilon)}{\varepsilon}+1}=\frac{\sigma\left(T_3^4-T_2^4\right)}{2\frac{(1-\varepsilon)}{\varepsilon}+1}$$

Obtaining the temperature of the intermediate plate $T_3^4=\frac{T_1^4+T_2^4}{2}$. Finally, substituting T_3 into the previous equation,

$$\frac{\dot{Q}_{ws}}{A}=\frac{1}{2}\left[\frac{\sigma\left(T_1^4-T_2^4\right)}{2\frac{(1-\varepsilon)}{\varepsilon}+1}\right]=\frac{1}{2}\frac{\dot{Q}_{wo}}{A} \tag{11.36}$$

Comment: the addition of a shielding surface halves the initial heat rate (provided all surfaces have the same emissivity ε).

When n shielding surfaces are introduced between the original surfaces 1 and 2 (all with the same emissivity), the heat rate is reduced to:

$$\left.\frac{\dot{Q}}{A}\right|_{with\ n\ shields}=\frac{1}{1+n}\left.\frac{\dot{Q}}{A}\right|_{without\ shield} \tag{11.37}$$

Example 11.11 (*Thermal radiation shielding*)

Prove Eq. (11.37) for surfaces n with the same emissivity.

Solution

The heat flux without shielding is given by:

$$\left.\frac{\dot{Q}}{A}\right|_{wo}=\frac{\sigma\left(T_1^4-T_2^4\right)}{2\frac{(1-\varepsilon)}{\varepsilon}+1}$$

The successive introduction of surfaces 1, 2, … n of thermal shielding results in the following heat rate equation:

$$\dot{Q}_{ws}=\frac{\sigma\left(T_1^4-T_2^4\right)}{\frac{1-\varepsilon}{A\varepsilon}+\frac{1}{A}+\underbrace{\frac{1-\varepsilon}{A\varepsilon}+\frac{1-\varepsilon}{A\varepsilon}+\frac{1}{A}}_{1}+\underbrace{\frac{1-\varepsilon}{A\varepsilon}+\frac{1-\varepsilon}{A\varepsilon}+\frac{1}{A}}_{2}\cdots+\underbrace{\frac{1-\varepsilon}{A\varepsilon}+\frac{1-\varepsilon}{A\varepsilon}+\frac{1}{A}}_{3}+\frac{1-\varepsilon}{A\varepsilon}}$$

Each new surface introduces a set of resistances 1, 2, 3,… n. Thus by a process of induction

$$\frac{\dot{Q}_{ws}}{A}=\frac{1}{(n+1)}\underbrace{\frac{\sigma\left(T_1^4-T_2^4\right)}{2\frac{1-\varepsilon}{A\varepsilon}+1}}_{\frac{\dot{Q}_{wo}}{A}}$$

Finally,

$$\left.\frac{\dot{Q}}{A}\right|_{ws} = \frac{1}{(n+1)}\left.\frac{\dot{Q}}{A}\right|_{wo}$$

Example 11.12 (*Radiation shielding in thermal insulation*)

The average temperature of the cryogenic liquid transport pipe is 80 K, with an external diameter of 15 mm and emissivity of 0.04. To prevent the fluid from heating, a concentric pipe with an internal diameter of 40 mm and an emissivity of 0.03 is installed, and the annular space is evacuated. The average temperature of the outer tube is 30 °C. (a) Calculate the heat gain of the cryogenic fluid per meter of pipe length; (b) If a shielding pipe with a diameter of 30 mm in diameter and emissivity of 0.03 is installed, what would be the heat gain of the cryogenic fluid per meter of pipe length, and by what percentage would the gain be reduced? What would be the temperature of the shielding pipe?

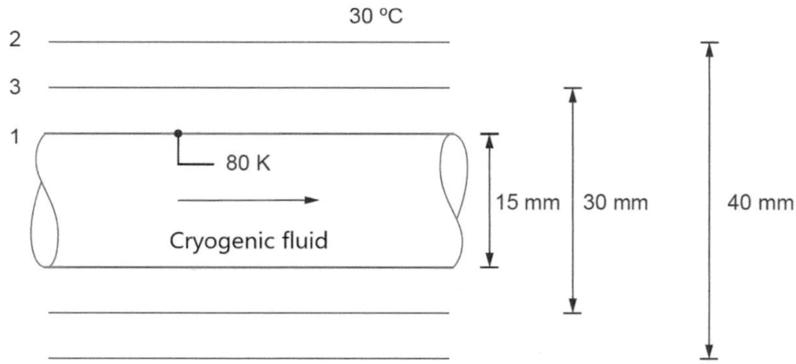

Solution

(a) *Without shielding*—radiation circuit with only the two pipes 1 and 2

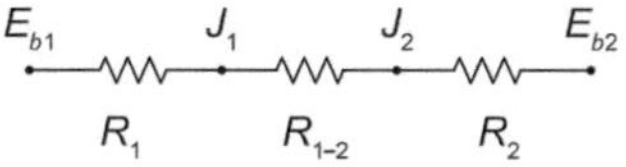

$$F_{12} = 1$$

$$R_T = \frac{1-\varepsilon_1}{\varepsilon_1 A_1} + \frac{1}{A_1 F_{1-2}} + \frac{1-\varepsilon_2}{\varepsilon_2 A_2} = \frac{1}{A_1}\left[\frac{1-\varepsilon_1}{\varepsilon_1} + \frac{1}{1} + \frac{1-\varepsilon_2}{\varepsilon_2}\frac{D_1}{D_2}\right]$$

$$R_T = \frac{1}{\pi D_1 L}\left[\frac{1-\varepsilon_1}{\varepsilon_1} + 1 + \frac{1-\varepsilon_2}{\varepsilon_2}\frac{D_1}{D_2}\right]$$

$$R_T = \frac{1}{\pi \times 0.015 \times 1}\left[\frac{1 - 0.04}{0.04} + 1 + \frac{1 - 0.03}{0.03} \times \frac{15}{40}\right] = 787.8/\text{m}^2$$

$$\dot{Q}_{wo} = \frac{\sigma\left(T_2^4 - T_1^4\right)}{R_T}$$

$$\dot{Q}_{wo} = \frac{5.67 \times 10^{-8}\left(303, 2^4 - 80^4\right)}{787.8} = 0.605\ \text{W (1 meter)}$$

(b) *With shielding*—radiation circuit with the introduction of the third shielding pipe

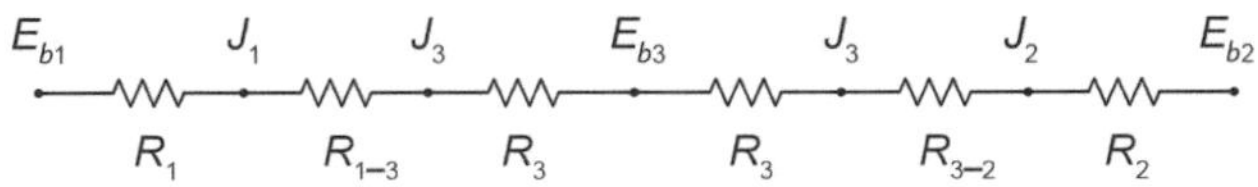

$$F_{1-3} = 1;\ F_{3-2} = 1$$

$$\begin{aligned} R_T &= \frac{1}{A_1}\left[\frac{1 - \varepsilon_1}{\varepsilon_1} + \frac{1}{1} + 2\frac{(1 - \varepsilon_3)}{\varepsilon_3}\frac{D_1}{D_3} + 1\frac{D_1}{D_3} + \frac{1 - \varepsilon_2}{\varepsilon_2}\frac{D_1}{D_2}\right] \\ &= \frac{1}{\pi \times 0.015 \times 1}\left[\frac{1 - 0.04}{0.04} + 1 + 2\frac{(1 - 0.03)}{0.03} \times \frac{15}{30} + \frac{1 - 0.03}{0.03} \times \frac{15}{40}\right] \\ &= 1474.0\ 1/\text{m}^2 \end{aligned}$$

Thus,

$$\dot{Q}_{ws} = \frac{\sigma\left(T_2^4 - T_1^4\right)}{R_T}$$

$$\dot{Q}_{ws} = \frac{5.67 \times 10^{-8}\left(303.2^4 - 80^4\right)}{1474.0} = 0.3241\ \text{W (1 meter)}$$

the radiation heat rate loss decreases by about $\frac{0.605 - 0.3241}{0.605}\ 100\% = 46.4\%$

The shielding temperature can be evaluated from:

$$\dot{Q}_{cb} = \frac{\sigma\left(T_2^4 - T_3^4\right)}{\frac{1-\varepsilon_2}{A_2\varepsilon_2} + \frac{1}{A_2F_{2-3}} + \frac{1-\varepsilon_3}{A_3\varepsilon_3}},\ A_2F_{2-3} = A_3F_{3-2},\ F_{3-2} = 1$$

$$0.3241 = \frac{5.67 \times 10^{-8}\left(303.2^4 - T_3^4\right)}{\frac{1-0.03}{\pi \times 0.040 \times 1 \times 0.03} + \frac{1}{\pi \times 0.030 \times 1} + \frac{1-0.03}{\pi \times 0.030 \times 1 \times 0.03}}$$

$$T_3 = 265\ \text{K}$$

11.5 Combined Heat Transfer

Actual heat transfer problems, of course, can occur in a combined manner, that is, all three forms of heat transfer may and normally are present. In the case of heat transfer between two fluids, the overall heat transfer coefficient, U, was defined to capture the effects of combined conduction and convection. In the case of tube exposure to high (or very cold) temperature sources, such as inside a boiler furnace, or even exposure to solar radiation, heat transfer by thermal radiation cannot be disregarded. Thus, all heat transfer effects need to be analyzed jointly or combined. In this section, a technique for solving these problems through equivalent thermal circuits will be seen. To illustrate the technique, some examples will be presented in the following subsections, starting with the case of temperature measurement.

11.5.1 Effect of Radiation on Temperature Measurement

When a thermometer or other temperature sensor element is installed in a gas stream to measure its temperature, the temperature value indicated by the sensor is determined by the overall energy balance on the sensor bulb. A considerable error may arise between the registered temperature and the actual gas temperature if the effects of radiation are not properly addressed. The error tends to increase for significant differences in the temperatures involved, as seen next.

Consider the sensor shown in Fig. 11.14a. The (unknown) gas temperature is T_∞, the surrounding surface temperature is T_w and the temperature indicated by the thermometer, T_t.

Assuming that the gas is hot, then the heat rate is transferred by convection, $\dot{Q}_{cv}$, from the hot gas to the thermometer bulb and, from this, to the surrounding medium is in the form of thermal radiation, $\dot{Q}_{rd}$. Therefore, the equivalent thermal circuit can be established as indicated in Fig. 11.14b. Note now that the "potentials" of thermal exchanges are not the same. The "potential" of thermal radiation is the black body emissive power, E_{bt}. The "potential" of convective heat transfer is the temperature, T_t. However, the two potentials are interconnected by a dashed line to illustrate that

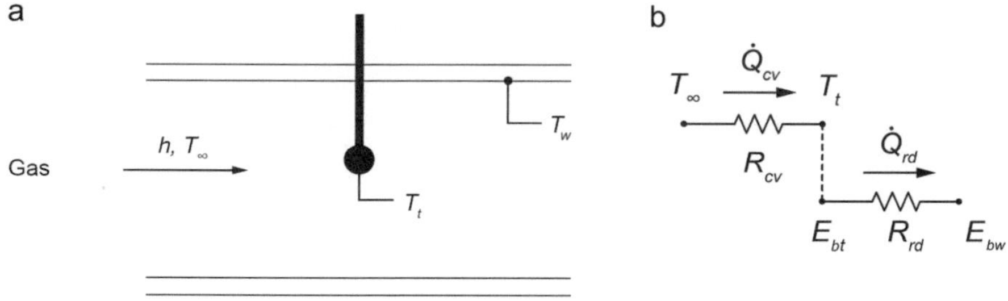

Fig. 11.14 **a** Temperature sensor element inserted in a flow; **b** equivalent thermal circuit

it is the same body that, in this case, is the thermometer bulb. Thus, in equilibrium, the heat rate received by the thermometer bulb must be equal to the radiation heat loss, this is:

$$\dot{Q}_{cv} = \dot{Q}_{rd} \Rightarrow \frac{T_\infty - T_t}{R_{cv}} = \frac{E_{bt} - E_{bw}}{R_{rd}} \Rightarrow$$

$$\frac{T_\infty - T_t}{\frac{1}{hA_t}} = \frac{\sigma\left(T_t^4 - T_w^4\right)}{\frac{1}{\varepsilon_t A_t}} \tag{11.38}$$

where A_t is the bulb surface area, and ε_t its emissivity. For the calculation of thermal radiation resistance, it has already been considered that the thermometer bulb is completely surrounded by the tube wall, as discussed in Sect. 11.3.4. Solving this equation, it is obtained that the fluid temperature is given by:

$$T_\infty = T_t + \underbrace{\frac{\varepsilon_t \sigma\left(T_t^4 - T_w^4\right)}{h}}_{\text{Correction due to radiation}} \tag{11.39}$$

Thus, the actual unknown gas temperature, T_∞, is not the temperature indicated by the thermometer, T_t. The temperature indicated by the sensor must be corrected due to the effect of combined thermal radiation and convection by the term indicated in Eq. (11.39). The relevant observation from this analysis is that the lower the heat transfer coefficient by convection and the higher the sensor's emissivity, the greater the correction to be made. Therefore, in temperature measurements, it is important to use a sensor made of material of low emissivity and the moving gas must have a high heat transfer coefficient. However, it is not always possible to act on these quantities, so thermal shielding of the element should be provided to reduce the radiation losses, as discussed in Sect. 11.4.

Example 11.13 (*Gas temperature reading*)

It is desired to measure the temperature of a gas inside a duct using an exposed junction thermocouple whose emissivity is 0.82. The duct wall temperature is 250 °C. It is asked: (a) the actual gas temperature, if the thermocouple records a temperature of 400 °C, assuming the heat transfer coefficient by convection between the gas and the sensor is 120 W/m^{2}K; (b) to improve the accuracy of the temperature measurement, it is decided to place a highly reflective metal shield ($\rho_s = 0.85$) in a cylindrical shape around the thermocouple, assuming that the area of the small cylinder is 10 times the area of the thermocouple and that, being very thin, its temperature can be considered constant. Considering that the heat transfer coefficient by convection between the gas and the shield is 100 W/m^{2}K, evaluate the new reading of the thermocouple.

Solution

(a) *Without shielding*:

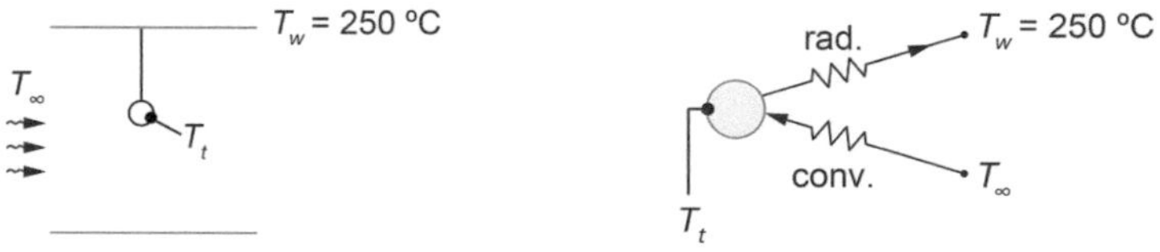

Just use the established solution above (Eq. 11.39).
$T_w = 250$ °C (523.2 K), $T_t = 400$ °C (673.2 K)

$$T_\infty = 673.2 + \frac{0.82 \times 5.67 \times 10^{-8}}{120}\left(673.2^4 - 523.2^4\right)$$

$$T_\infty = 723.7\,\text{K}(450.5\,°\text{C})$$

(b) *With shielding*:

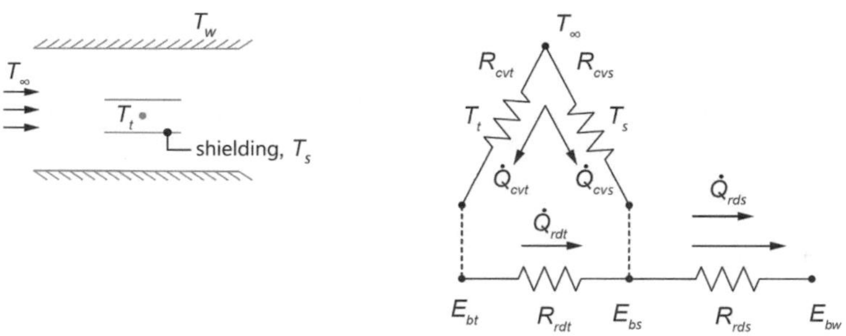

In this configuration, it is noticed that the gas transfers heat to the thermocouple and also to both sides of the shield. The energy balances in the thermocouple and the shield are as follows:
Thermocouple: $h_{gt}A_t(T_\infty - T_t) = \frac{\sigma\left(T_t^4 - T_s^4\right)}{\frac{1-\varepsilon_t}{A_t\varepsilon_t} + \frac{1}{A_tF_{t-s}} + \frac{1-\varepsilon_S}{A_S\varepsilon_S}}$.
Shielding (two sides): $2h_{gs}A_s(T_\infty - T_s) + \frac{\sigma\left(T_t^4 - T_s^4\right)}{\frac{1-\varepsilon_t}{A_t\varepsilon_t} + \frac{1}{A_tF_{t-s}} + \frac{1-\varepsilon_S}{A_S\varepsilon_S}} = \frac{\sigma\left(T_s^4 - T_w^4\right)}{\frac{1-\varepsilon_S}{A_S\varepsilon_S} + \frac{1}{A_SF_{S-w}} + \frac{1-\varepsilon_W}{A_W\varepsilon_W}}$.
Substituting values and considering that $\frac{A_s}{A_t} = 10$, $\frac{A_s}{A_w} \sim 0$, $F_{t-s} = 1$, $F_{s-w} = 1$:
Thermocouple: first divide both sides by A_t:

$$120 \times (723.7 - T_t) = \frac{5.67 \times 10^{-8}\left(T_t^4 - T_s^4\right)}{\frac{1-0.82}{0.82} + \frac{1}{1} + \frac{1-0.15}{10\times0.15}} \Rightarrow$$

$$86{,}844 - 120T_t = 3.174 \times 10^{-8}\left(T_t^4 - T_s^4\right)$$

Shielding: first divide both sides by A_s:

$$2 \times 100 \times (723.7 - T_s) + \frac{5.67 \times 10^{-8}\left(T_t^4 - T_s^4\right)}{\frac{10(1-0.82)}{0.82} + \frac{10}{1} + \frac{1-0.15}{0.15}} = \frac{5.67 \times 10^{-8}\left(T_s^4 - 523.2^4\right)}{\frac{1-0.15}{0.15} + \frac{1}{1} + \underbrace{\frac{A_s}{A_w}}_{\approx 0} \times \frac{1-\varepsilon_w}{\varepsilon_w}} \Rightarrow$$

$$1.447 \times 10^5 - 200T_s + 0.317 \times 10^{-8}\left(T_t^4 - T_s^4\right) = 0.8505 \times 10^{-8}\left(T_s^4 - 523.2^4\right)$$

Solving the system, we get:
$T_t = 721.5$ K (448.3 °C), and $T_s = 715.9$ K (442.7 °C).

Note that the correction of the thermocouple reading in the unshielded condition is $\Delta T_{\text{cor},wo} = 50.5$ °C and with shielding, of $\Delta T_{\text{cor},ws} = 2.2$ °C.

11.5.2 Combined Heat Transfer in Pipelines

In this section, a case of combined heat transfer involving conduction, convection, and thermal radiation is analyzed using illustrated examples. Consider a metallic pipe that transports a cold fluid. To prevent heat gain from environment to the fluid circulating inside the pipe, thermal insulation is installed. If the internal temperature of the insulation (1) is known or if the pipe wall has high thermal conductivity, the resistance to conduction through the pipe wall is negligible, and the internal temperature of the insulation can be assumed to be the same as the fluid temperature. Heat transfer occurs through the insulation by conduction, while the external surface of the insulation (2) is subjected to heat transfer by both convection and thermal radiation. Figure 11.5 illustrates the case of a cold pipe gaining heat from the environment.

The energy balance on the external wall of the insulation (2) for the case of Fig. 11.15 is given by:

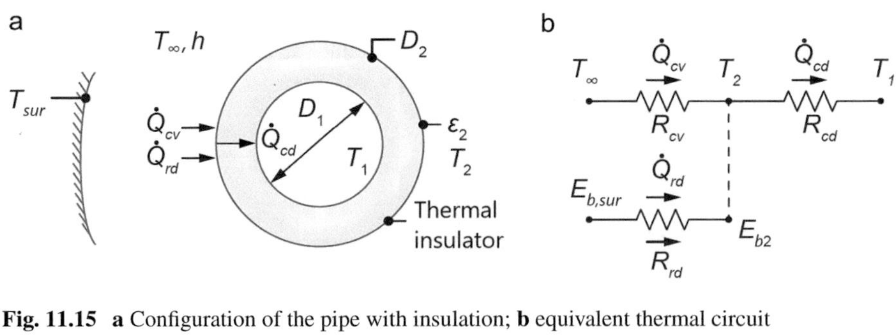

Fig. 11.15 **a** Configuration of the pipe with insulation; **b** equivalent thermal circuit

$$\dot{Q}_{cd} = \dot{Q}_{cv} + \dot{Q}_{rd} \Rightarrow \qquad \frac{T_2 - T_1}{\frac{\ln[D_2/D_1]}{2\pi k_i L}} = \frac{T_\infty - T_2}{\frac{1}{hA_2}} + \frac{E_{b,env} - E_{b2}}{\frac{1}{\varepsilon_2 A_2}} \Rightarrow$$
$$\frac{T_2-T_1}{R_{cd}} = \frac{T_\infty-T_2}{R_{cv}} + \frac{E_{b,env}-E_{b2}}{R_{rd}} \Rightarrow$$
$$\frac{T_2 - T_1}{\frac{\ln[D_2/D_1]}{2\pi k_i L}} = \frac{T_\infty - T_2}{\frac{1}{h\pi D_2 L}} + \frac{E_{b,env} - E_{b2}}{\frac{1}{\varepsilon_2 \pi D_2 L}}$$

$$\frac{2k_i(T_2 - T_1)}{\ln\left(\frac{D_2}{D_1}\right)} = hD_2(T_\infty - T_2) + \varepsilon_2 \sigma D_2\left(T_{env}^4 - T_2^4\right) \qquad (11.40)$$

Example 11.14 (*Combined heat transfer*)

Consider a 10 cm diameter pipe used for the transport of liquefied natural gas (LNG) at 111 K (−162 °C). The problem of heat loss is critical, as it results in the vaporization of the LNG to the environment and consequent loss of NG vapor which, in technical terms, is called *boil off*. Thus, the outer wall of the tube is wrapped in insulating material with a thickness of 5 cm and conductivity $k_i = 0.06$ W/mK. Considering that the inner wall of the transport tube remains at 111 K and that the pipe wall has excellent thermal conductivity and, therefore, does not represent any thermal resistance to conduction. In addition, the environment air temperature is 300 K and the external coefficient of heat transfer by convection between the insulating material and the air is 5 W/m^{2}K; the ground and the sky behave as a single black body at an average temperature of 290 K. Evaluate (a) the external temperature of the insulation assuming that its surface emissivity is 0.5 and (b) plot the sensitivity of the external temperature of the insulation and the heat gained by the pipe per meter of pipe length as a function of emissivity varying from 0.1 to 0.9.

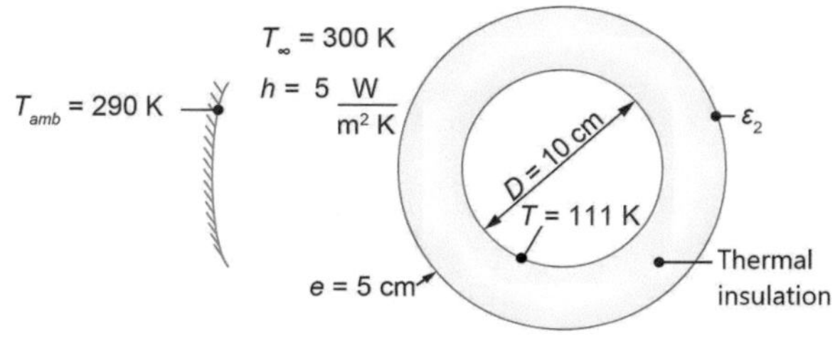

Solution

The external diameter of the insulation is: $D_2 = D_1 + 2e$

$$D_2 = D_1 + 2e = 0.1 + 2 \times 0.05 = 0.2\,\text{m}$$

Substituting values into Eq. (11.40):

$$\frac{2 \times 0.06 \times (T_2 - 111)}{\ln\left(\frac{0.2}{0.1}\right)} = 5 \times 0.2 \times (300 - T_2) + \varepsilon_2 \times 5.67 \times 10^{-8} \times 0.2 \times \left(290^4 - T_2^4\right)$$

(a) Solving the previous polynomial equation for $\varepsilon_2 = 0.5$, we obtain:

$$T_2 = 277.6 \text{ K or } 4.4\ ^\circ\text{C}.$$

(b) The rate of heat gain per unit length of the pipe is:

$$q' = \frac{\dot{Q}}{L} = \frac{2\pi k_i (T_2 - T_1)}{\ln\left(\frac{D_2}{D_1}\right)}$$

Solving the equations for various emissivity values,

ε_2	T_2 (°C)	q' (W/m)
0.1	0.4	88.4
0.2	1.6	89.0
0.3	2.7	89.6
0.4	3.6	90.1
0.5	4.4	90.6
0.6	5.2	91.0
0.7	5.8	91.3
0.8	6.4	91.6
0.9	6.9	91.9

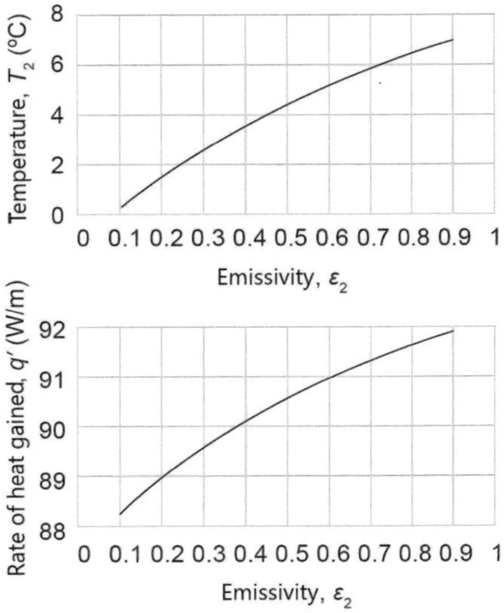

It can be noted in the graphs that the higher the emissivity of the external surface of the insulation, the higher the temperature and the heat gain by the pipe (and the fluid). For example, at 10 m there will be a thermal power gain of 920 W. If the emissivity is 0.9, this thermal power gain will be significant depending on the LNG flow being transported, which will indicate the formation of methane vapor, which will need to be purged (*boil off*) into the environment.

11.5.3 Radiation Heat Transfer Coefficient

In situations of combined heat transfer, it is usual to define the linearized radiation heat transfer coefficient, h_r so that the total heat transfer is the sum of convection and thermal radiation

$$\dot{Q}_T = \dot{Q}_c + \dot{Q}_r$$

For a small body of emissivity ε at a surface temperature T_0 enclosed in a large environment at T_{env} that exchange heat with a fluid at T_∞, the following expression can be obtained for the total heat flux

$$q = \frac{\dot{Q}_T}{A} = hA(T_0 - T_\infty) + h_r A(T_0 - T_{env}) \quad (11.41)$$

If the environment and the fluid are at the same temperature $T_{env} = T_\infty$, then

$$q = \frac{\dot{Q}_T}{A} = h_{comb} A(T_0 - T_{env})$$

$h_{comb} = h + h_r$, and the linearized radiation heat transfer coefficient is

$$h_r = \varepsilon\sigma\left(T_0^2 + T_{env}^2\right)(T_0 + T_{env}) \quad (11.42)$$

Example 11.15 (*Radiation heat transfer coefficient*)

Redo the solved Example 11.6 considering the system is filled with still air with a convection heat transfer of $h = 5\text{W/m}^2\text{K}$.

Solution

From Eq. (11.42)

$$h_r = \varepsilon\sigma\left(T_0^2 + T_{\text{env}}^2\right)(T_0 + T_{\text{env}}) =$$
$$0.8 \times 5.67 \times 10^{-8}\left(573.2^2 + 293.2^2\right)(573.2 + 293.2) = 16.3\,\text{W/mK}$$

Therefore, the combined heat transfer coefficient is

$$h_{\text{comb}} = h + h_r = 5 + 16.3 = 21.3\,\text{W/mK}$$

Finally, the combined rate of heat transfer from the sphere is

$$\dot{Q}_{1-2} = h_{\text{comb}} A_1 (T_0 - T_{\text{env}}) = 21.3 \times 0.0314 \times 280 = 187.3\,\text{W}$$

11.6 Black Cavity Modelling

The concept of a black cavity was presented in Sect. 10.3.2 in connection with Kirchhoff's law. A black cavity is an adiabatic device that behaves as a black body which completely absorbs the impinging thermal radiation at its open window. The thermal radiation undergoes an absorption-reflection process several times within the cavity until it is completely absorbed by the cavity wall. In this process, an equilibrium (adiabatic) cavity temperature is established. The effective cavity absorptivity, α_{eff}, tends to unity, regardless of the internal material of the cavity, as long as the ratio between the opening window area, A_w, and the internal cavity area, A_c, is very small. That is, the cavity behaves like a black body.

Consider the adiabatic cavity depicted in Fig. 11.16a, where a radiation heat flux, q, impinges the opening window area, A_w. The radiation heat flux is absorbed and reflected several times on the internal cavity surface of area A_c; being α_c the absorptivity of internal cavity surface and ε_c, the emissivity of internal cavity surface. According to Kirchhoff's law, $\alpha_c = \varepsilon_c$. The thermal circuit is shown in Fig. 11.16b, where the *env* stands for environment.

In thermal equilibrium the heat fluxes are equal, this is

$$qA_w = \dot{Q}_{c-env} \tag{11.43}$$

From the thermal circuit in Fig. 11.16b, one can write

$$q = \frac{\dot{Q}_{c-env}}{A_w} = \frac{1}{A_w}\frac{E_c - E_{env}}{R_T}$$

where, $R_T = \frac{1-\varepsilon_c}{\varepsilon_c A_c} + \frac{1}{A_w F_{w-c}}$

After substitution, the final expression is obtained considering $F_{w-c} = 1$ and $\alpha_c = \varepsilon_c$

$$q = \frac{\sigma\left(T_c^4 - T_{env}^4\right)}{1 + \frac{A_w}{A_c}\left(\frac{1-\alpha_c}{\alpha_c}\right)} = \alpha_{eff}\,\sigma\left(T_c^4 - T_{env}^4\right) \tag{11.44}$$

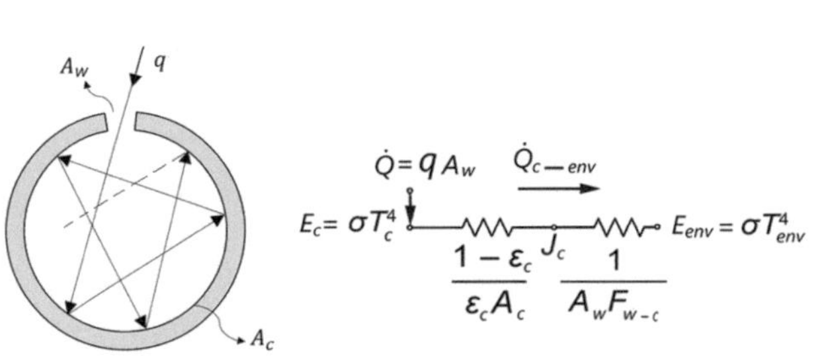

Fig. 11.16 **a** Black cavity schematics; **b** thermal circuit

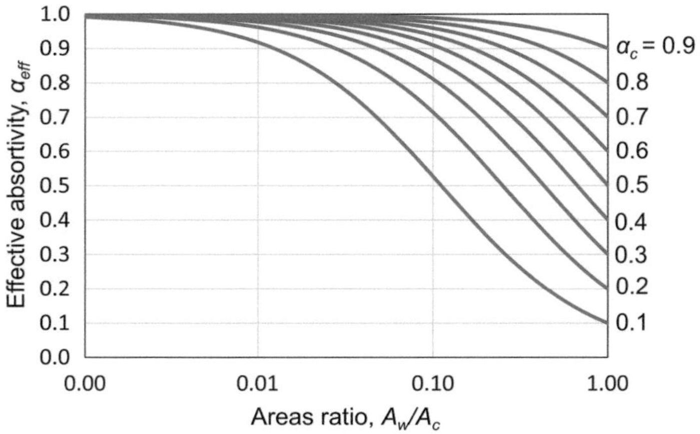

Fig. 11.17 Black cavity α_{eff} as a function of window to cavity areas ratio and the cavity absorptivity, α_{c}.

where, the effective cavity absorptivity α_{eff} is given by

$$\alpha_{eff} = \frac{1}{1 + \frac{A_w}{A_c}\left(\frac{1-\alpha_c}{\alpha_c}\right)} \tag{11.45}$$

Regardless of the absorptivity of the cavity material, α_{c}, if the opening to the cavity area ratio $A_{\text{w}}/A_{\text{c}}$ is very small, then $\alpha_{eff} \approx 1$, as can be seen by analyzing the previous equation through the graph in Fig. 11.17. This means that the cavity behaves like a black body with all the thermal radiation incident on the opening area being absorbed.

Example 11.16 (*Cavity absorptivity*)

For a cylindrical cavity having a 3 cm round aperture window opening, 10 cm i.d., and 30 cm long, find the cavity effective absorptivity α_{eff} as a function of the cavity emissivity α_{c}.

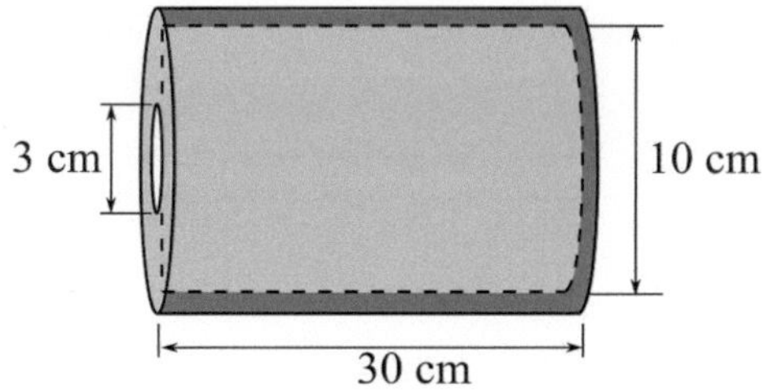

Solution

The opening to the cavity areas A_w/A_c ratio is

$$\frac{A_w}{A_c} = \frac{\pi D_w^2/4}{\pi D_c L + \pi D_c^2/4 + \pi \left(D_c^2 - D_w^2\right)/4} == 0.0065$$

By solving Eq. (11.45) for several emissivity α_c yields

α_c (cavity absorptivity)	α_{eff} (effective absorptivity)
0.9	0.999
0.5	0.994
0.2	0.975
0.1	0.945

For this area ratio, a cavity emissivity above 0.2 would be ideal. Alternatively, one could build a larger cavity as well.

Adiabatic cavity equilibrium temperature, T_{ac}

The adiabatic equilibrium temperature of the cavity, T_{ac}, is the maximum temperature a cavity can reach, which means the cavity is adiabatic without any heat loss. This temperature can be obtained for a given radiation heat flux, q, by solving Eq. (11.44)

$$T_{ac} = \sqrt[4]{\frac{q}{\sigma \alpha_{eff}} + T_{env}^4} = \sqrt[4]{\frac{NK_{sun}}{\sigma \alpha_{eff}} + T_{env}^4} \tag{11.46}$$

where, K_{sun} is a standard solar irradiation equal to 1000 W/m^2, known as *one Sun*, and N is the *number* of suns. Therefore, $q = N \times K_{sun} = 1000 \times N$.

The graph in Fig. 11.18 shows the high increase of the adiabatic cavity equilibrium temperature, T_{ac}, as a function of the number of suns. As seen very high temperature can be reached within a black cavity. Naturally, that theoretical temperature is reached in vacuum, where no other heat losses occur. In real black cavities, a glass window usually covers the cavity, which introduces losses that must be properly modelled. Black cavities have been used in connection with thermochemical reactions at a high temperature, been the solar radiation the source of a clean thermal energy. Indoor solar simulators have been built for studying such reactions in black cavities such as the studies carried out by Varón et al. (2023).

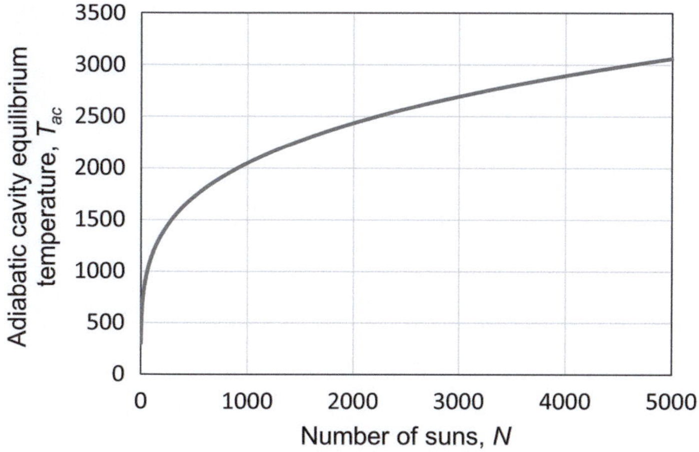

Fig. 11.18 Adiabatic cavity equilibrium temperature, T_{ac} (K), as a function of the number of suns, N_s, for $\alpha_{eff} = 1$, and $T_{env} = 300$ K.

11.7 Thermal Radiation in Gases

Radiation through gases is essentially volumetric in the bulk of the medium, making the problem more complex than radiation exchange between surfaces. As a general rule, symmetric non-polar gases are transparent to thermal radiation. Most of the gases surrounding us fall into this category, including N_2 and O_2, which make up the large portion of dry air. On the other hand, water vapor, CO_2, CH_4, and many hydrocarbons absorb and emit thermal radiation at specific wavelength bands, as seen in Fig. 11.19.

Gas absorption and emission occur in a volumetric manner. Consider the schematics shown in Fig. 11.20 in which a monochromatic radiation beam, $I_{\lambda 0}$, reaches a layer of gas of thickness L at $x = 0$. As the beam radiation crosses the absorbing medium of thickness dx, it is attenuated according to Eq. (11.47)

$$dI_\lambda = -\kappa_\lambda I_\lambda dx \tag{11.47}$$

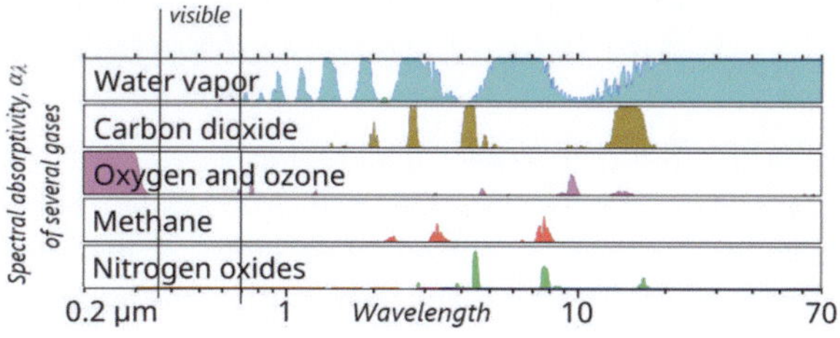

Fig. 11.19 Spectral absorptivity of several gases. Adapted, public domain

Fig. 11.20 Absorption in an elementary gas layer.

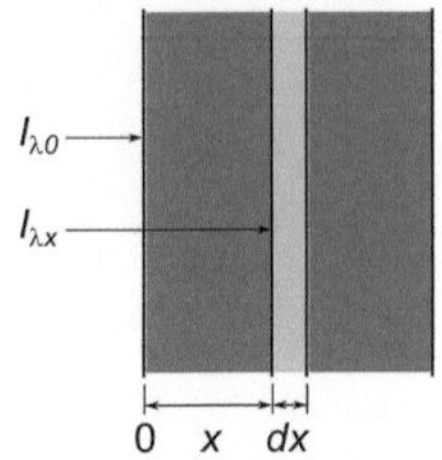

where, κ_λ is the monochromatic or spectral absorption coefficient, whose SI unit is m^{-1}. Integrating this equation yields

$$\int_{I_{\lambda 0}}^{I_{\lambda x}} \frac{dI_\lambda}{I_\lambda} = -\int_0^x \kappa_\lambda dx \tag{11.48}$$

Finally, Beer's law is obtained after carrying out the integration for κ_λ constant.

$$\frac{I_{\lambda x}}{I_{\lambda 0}} = e^{-\kappa_\lambda x} \tag{11.49}$$

The above equation is exactly the transmissivity τ_λ. For non-reflecting media, then it is possible to obtain the monochromatic or spectral absorptivity, α_λ, recalling that

$$\alpha_\lambda + \tau_\lambda = 1$$

Therefore,

$$\alpha_\lambda = 1 - \tau_\lambda = 1 - e^{-\kappa_\lambda x} \tag{11.50}$$

From Kirchhoff´s law, it is known the absorptivity is also the emissivity. Therefore, $\varepsilon_\lambda = \alpha_\lambda$ as long as the law holds.

11.7.1 Radiation Exchange of an Absorbing and Transmitting Medium Between Two Parallel And Infinite Isothermal Surfaces

In Sect. 11.1.1, the case of radiation exchange between two parallel and infinite isothermal surfaces was analyzed. The system was in a vacuum, so only thermal radiation exchange occurred. Now, consider that the space between the surfaces is filled with an absorbing and transmitting gas as depicted in Fig. 11.21. The index m stands for the medium properties.

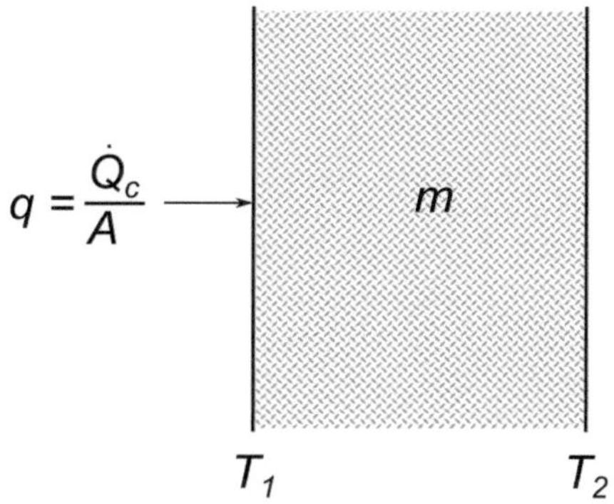

Fig. 11.21 Radiation exchange between two parallel and infinite isothermal surfaces filled with an absorbing and transmitting gas.

The net rate of radiation heat, $\dot{Q}_{1-2}$, exchanged between surfaces 1 and 2 is the difference in radiosity transmitted through the medium between the surfaces. So that:

$$\dot{Q}_{1-2} = J_1 A_1 \tau_m - J_2 A_2 \tau_m = \frac{J_1 - J_2}{1/A\tau_m} = \frac{J_1 - J_2}{1/A(1-\varepsilon_m)} \tag{11.51}$$

The order of indexes 1 and 2 in $\dot{Q}_{1-2}$ represents the rate of radiation leaving surface 1 and hitting surface 2 transmitted through the medium m subtracted from the rate of radiation leaving surface 2 transmitted through the medium m hitting surface 1. Also, note that $A = A_1 = A_2$ and that Eq. (11.50) was used along with Kirchhoff´s law, $\varepsilon_m = \alpha_m$.

The next step is to find the net rate of heat radiation from the medium m to surfaces 1 and 2. First, let us consider the radiosity J_m from medium m

$$J_m = \varepsilon_m E_{bm} \tag{11.52}$$

The rate of heat radiation from the surface 1 absorbed by the medium m is

$$J_1 A_1 \alpha_m = J_1 A_1 \varepsilon_m \tag{11.53}$$

Finally, the net rate of heat radiation, $\dot{Q}_{m-1}$, from medium m to surface 1 is

$$\dot{Q}_{m-1} = A_m J_m - J_1 A_1 \varepsilon_m = A_m \varepsilon_m E_{bm} - J_1 A_1 \varepsilon_m = \frac{E_{bm} - J_1}{1/A_1 \varepsilon_m} \tag{11.54}$$

The gas emission area, A_m, was taken as the projected area over surface 1, A_1.

In an analogous way the net rate of heat radiation from the medium m to surface 2 is

$$\dot{Q}_{m-2} = \frac{E_{bm} - J_2}{1/A_2 \varepsilon_m} \tag{11.55}$$

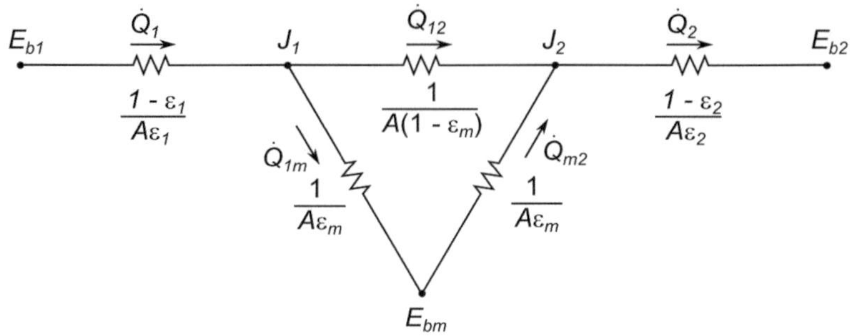

Fig. 11.22 Thermal circuit medium between two infinite parallel isothermal surfaces with a participating medium

Finally, the overall net rate of heat by radiation, $\dot{Q}_t$, is due to the transmitted part (Eq. 11.51) between surfaces and the exchange with the medium (Eqs. 11.54 and 11.55). A better understanding is achieved by examining the analog thermal circuit in Fig. 11.22.

The solution to the thermal circuit depends on whether the medium temperature is a known fixed amount or not. In the unknown gas temperature case, the gas temperature results from the circuit solution given by Eq. (11.56).

$$\dot{Q}_t = \frac{E_{b1} - E_{b2}}{R_1 + R_m + R_2} = \frac{\sigma\left(T_1^4 - T_2^4\right)}{\frac{1-\varepsilon_1}{\varepsilon_1 A} + R_m + \frac{1-\varepsilon_2}{\varepsilon_2 A}} \tag{11.56}$$

where, R_m is equivalent resistance of the resistances from the gas participation, which is given by $\frac{1}{R_m} = \frac{A\varepsilon_m}{2} + A(1 - \varepsilon_m)$. R_1 and R_2 are the surface resistances.

If the gas temperature is a known quantity, the thermal circuit should have the gas's blackbody emissive power fixed and be solved as if the system consisted of three surfaces in a similar fashion is analyzed in Sect. 11.3.2 resulting in a linear system of equations to be solved.

The theory developed here can be extended to glass and other transparent media. However, caution is needed regarding the simplification that reflectivity may not be zero in such media, as assumed for a gas medium. In the lack of the knowing the medium reflectivity a value of 0.1 may be considered.

As the number of involved surfaces increases, the problem becomes more complex, requiring the use of shape factors. Also, convection or even heat conduction should play a role making the solution even harder.

Example 11.17 (*Radiation heat exchange with a participating gas media*)

Two large parallel isothermal surfaces at temperatures T_1 =700 K and T_2 =300 K are separated by a gas absorbing radiation. The radiation properties are $\varepsilon_1 = 0.1$, $\varepsilon_2 = 0.8$, and $\varepsilon_m = 0.15$. Find the gas temperature and the rate of heat transfer between the surfaces.

Solution

The thermal circuit is given in Fig. 11.22, which leads to the thermal Eq. (11.56) to obtain the rate of radiation heat transfer. First, let us calculate the thermal resistances by unit of area A.

$$R_1 = \frac{1-\varepsilon_1}{\varepsilon_1 A} = \frac{1-0.1}{0.1A} = \frac{9}{A};$$

$$R_2 = \frac{1-\varepsilon_2}{\varepsilon_2 A} = \frac{1-0.8}{0.8A} = \frac{0.25}{A};$$

and

$$R_{\mathrm{m}} = \frac{1}{\frac{A\varepsilon_m}{2} + A(1-\varepsilon_m)} = \frac{1}{A\left(\frac{0.15}{2} + 0.85\right)} = \frac{1.081}{A}$$

The overall thermal resistance is

$$R_{\mathrm{t}} = R_1 + R_{\mathrm{m}} + R_2 = 10.331$$

Substituting into Equation (11.50) yields

$$q_t = \frac{\dot{Q}_t}{A} = \frac{5.67 \times 10^{-8}\left(700^4 - 300^4\right)}{10.331} = 1273.3\,\mathrm{W/m}$$

To obtain the gas temperature, first let us calculate the radiosities

$$q_t A = \frac{E_{b1} - J_1}{R_1} = \frac{\sigma T_1^4 - J_1}{R_1} \rightarrow$$
$$J_1 = \sigma T_1^4 - q_t A \times R_1 = 5.67 \times 10^{-8} \times 700^4 - 1273.3 \times 9 = 2153.97\,\mathrm{W/m};$$

The radiosity J_2 is

$$q_t A = \frac{J_2 - E_{b2}}{R_2} = \frac{J_2 - \sigma T_2^4}{R_2} \rightarrow$$
$$J_2 = \sigma T_2^4 + q_t A \times R_2 = 5.67 \times 10^{-8} \times 300^4 + 1273.3 \times 0.25 = 777.60\,\mathrm{W/m}$$

The direct radiation heat flux between surfaces 1 and 2, q_{1-2} is

$$q_{1-2} = \frac{J_1 - J_2}{\frac{1}{(1-\varepsilon_m)}} = 0.85 \times (2153.97 - 777.60) = 1169.9\,\mathrm{W/m}$$

Finally, the gas temperature can be found recalling that

$$q_{1-m} = q_{m-2} = q_t - q_{1-2} = 1273.3 - 1169.9 = 103.4\,\mathrm{W/m}$$

and

$$q_{1-m} = \frac{J_1 - E_{bm}}{\frac{1}{\varepsilon_m}} \rightarrow E_{bm} = J_1 - \frac{q_{1-m}}{\varepsilon_m} = 2153.97 - \frac{103.4}{0.15} \rightarrow$$

$$E_{bm} = 1464.64 \rightarrow T_m = \sqrt[4]{\frac{1464,64}{\sigma}} = 400.9\,\mathrm{K}$$

In this example, only the radiation heat transfer exchange was considered. Evidently, it is necessary to consider the convective heat transfer for a more real analysis.

11.7.2 CO_2 And Water Vapor Absorption

Hottel (1954) solved the problem of radiation heat exchange between a hemispherical gas mass of radius L at a temperature T_G containing CO_2 at a partial pressure p_C and an infinitesimal black body located at the hemisphere center base. The solution is displayed in Fig. 11.23 for a total gas pressure of 1 atm. He also presented the solution for water vapor emissivity in a similar geometry and condition for a total gas pressure of 1 atm as seen in Fig. 11.24.

Figures 11.23 and 11.24 are valid for a 1 atm total gas pressure. Pressure correction factors C_c and C_w for the emissivity for a gas pressure other than the atmospheric are presented in Fig. 11.25a for CO_2 and in Fig. 11.25b for water, respectively. The pressure correction factor is the ratio between the emissivity at a given pressure and the emissivity at 1 atm.

Emission ranges of CO_2 and water vapor overlap in certain emission spectra. Therefore, a correction is necessary to take away the overestimation, according to Eq. (11.57). The overlapping correction, $\Delta\varepsilon$, is given in Fig. 11.26.

$$\varepsilon_G = \varepsilon_C + \varepsilon_w - \Delta\varepsilon \tag{11.57}$$

Emissions from geometries other hemispherical shapes can be obtained in a similar fashion by using the same previous graphics considering the mean beam length L displayed in Table 11.2 for selected geometries. In case the geometry of interest is not available in that table, it is suggested to use $L = 4V/A$, where V is the enclosed volume and A is the surface area.

Example 11.18 (*Radiation heat exchange with a participating gas media*)

A sample of combustion gases at 560 °C has a composition of 7.3% of CO_2, 6.5% of water vapor, and the remaining is formed by N_2 and O_2. The gas is within a pressurized (2 atm) cube shaped furnace whose walls are at 400 °C. The furnace side

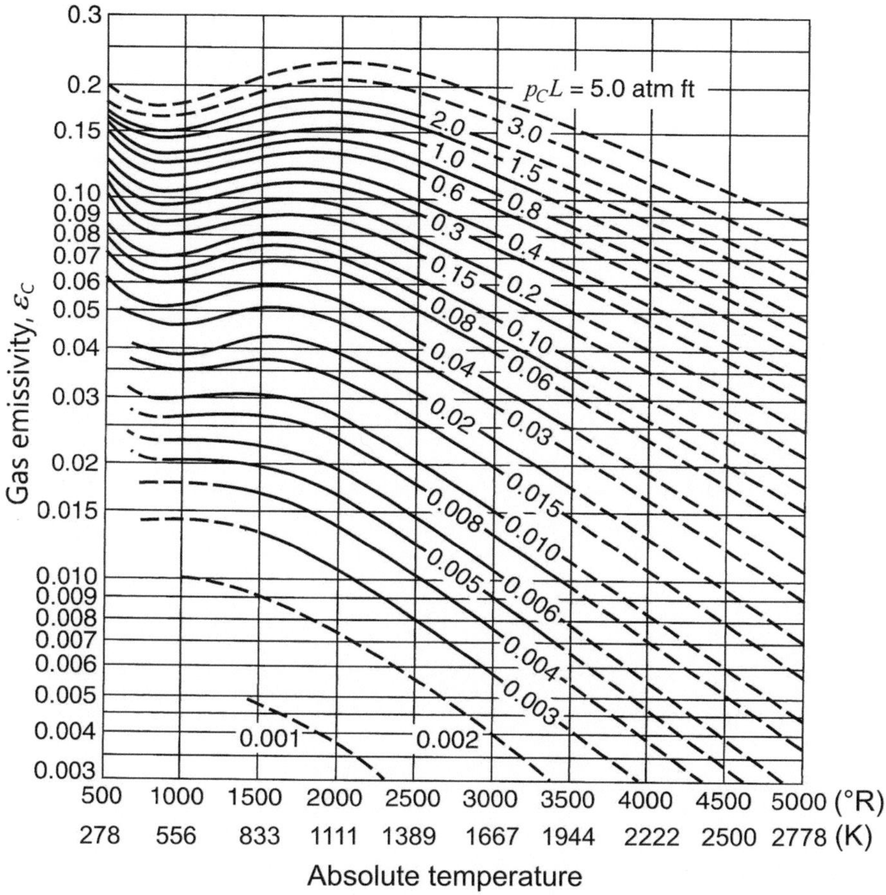

Fig. 11.23 Emissivity of a low carbon dioxide vapor pressure in a 1 atm gas mass. Adapted from Hottel (1954)

is 1 m long and walls can be considered as a black body. Obtain the initial radiation rate of heat transfer between the gases and the furnace wall.

Solution

Let us obain ε_G. First let us calculate the partial pressures:

At 1 atm

CO_2: $p_c = 0.073 \times 1 = 0.073$ atm

H_2O: $p_w = 0.065 \times 1 = 0.065$ atm

From Table 11.2, $L = 2 \times \frac{1}{3} = 0.667$ m $= 2.188$ ft, and $T_G = 560°\text{C} \cong 1500°\text{R}$.Therefore,

CO_2: $p_c L = 0.073 \times 2.188 = 0.16$ atm ft. From Fig. 11.23, $\varepsilon_c^0 = 0.095$

H_2O:$p_w L = 0.065 \times 2.188 = 0.14$ atm ft.From Fig. 11.24, $\varepsilon_w^0 = 0.085$

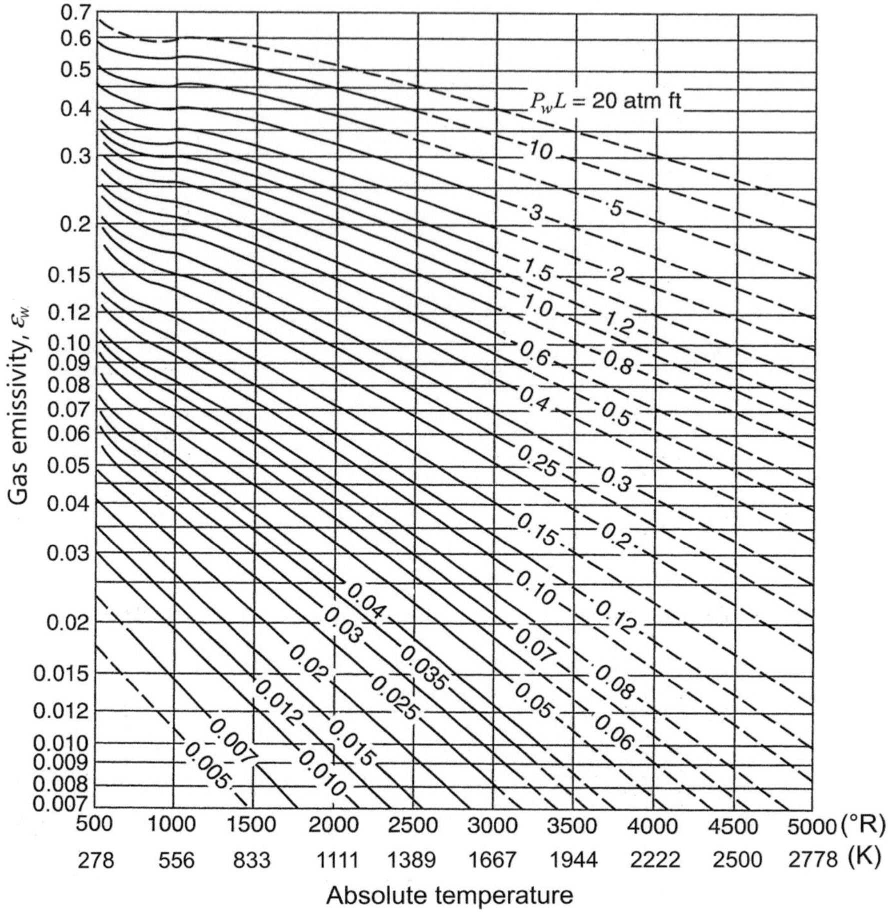

Fig. 11.24 Emissivity of a low vapor pressure water in a 1 atm gas mass. Adapted from Hottel (1954)

Correction for pressure

CO_2: $p_cL = 0.16$ atm ft and $p = 2$ atm. From Fig. 11.25a, $C_c = 1.13\varepsilon_c = C_c \times \varepsilon_c^0 = 1.13 \times 0.095 = 0.107$

H_2O: $p_wL = 0.14$ atm ft and $(p + p_w)/2 = (2 + 0.065)/2 = 1.03$ atm. From Fig. 11.25b, $C_w = 1.58$; $\varepsilon_w = C_w \times \varepsilon_c^0 = 1.58 \times 0.085 = 0.134$

Now, let us calculate de overlapping correction, $\Delta\varepsilon$, for that

$p_cL + p_wL = 0.16 + 0.14 = 0.30$ and $p_w/(p_c + p_w) = 0.065/(0.073 + 0.065) = 0.47$

Using the middle graph (1700 °F) in Fig. 11.26 for those values, one obtains $\Delta\varepsilon = 0.004$

Finally, from Eq. (11.57)

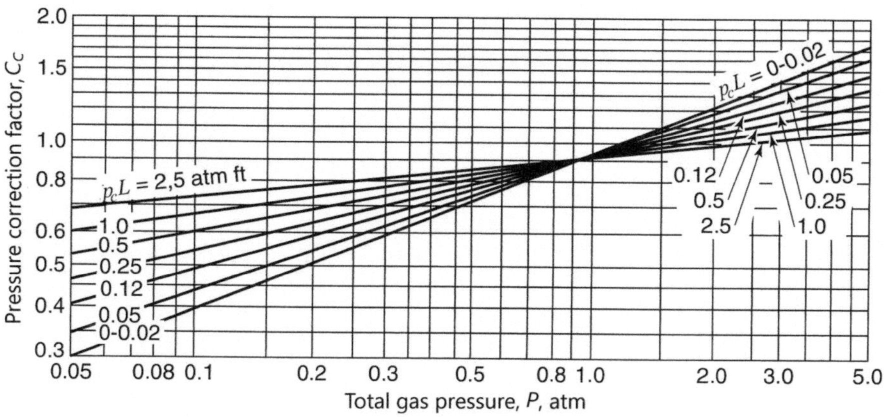

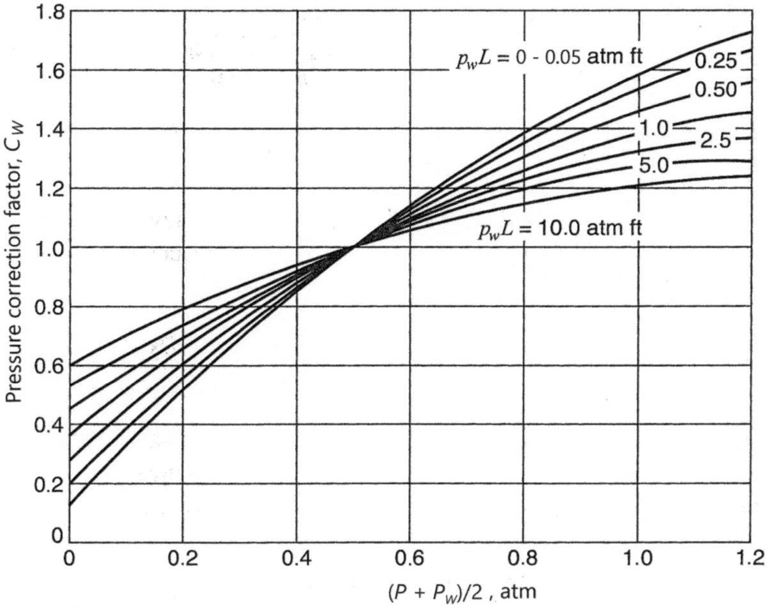

Fig. 11.25 **a** Pressure correction factor for carbon dioxide, C_c. **b** pressure correction factor for water, C_w. Adapted from Hottel (1954)

$$\varepsilon_G = \varepsilon_C + \varepsilon_w - \Delta\varepsilon = 0.107 + 0.134 - 0.004 = 0.237$$

From Eq. (11.55) for $J = E_{bw}$ (wall black body emissivity)

$$\dot{Q}_{G-W} = \frac{E_{bG} - E_{bw}}{\frac{1}{A\varepsilon_G}} = \sigma A\varepsilon_G\left(T_G^4 - T_w^4\right) = 5.67\,\mathrm{W}$$

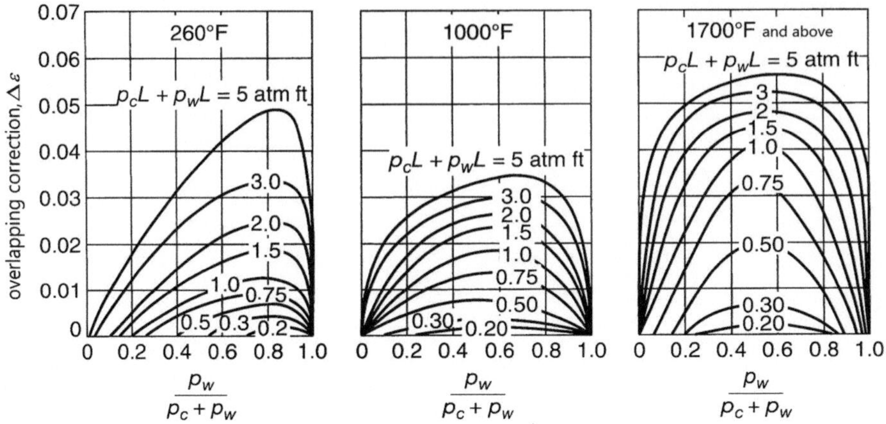

Fig. 11.26 Correction due to emission overlap of carbon dioxide and water vapor. Adapted from Hottel (1954)

Table 11.2 Beam length L for several geometries

Shape	Beam length L
Sphere	$2D/3$
Infinite cylinder	$1\ D$
Infinite parallel planes	$2 \times$ distance between planes
Cube of side a	$2a/3$
Any geometry	$4V/A$

Adapted from Hottel (1954)

11.8 Proposed Problems

11.1 Two large and equal flat parallel surfaces have the following temperatures: $T_1 = 15$ °C and $T_2 = 100$ °C. Evaluate the net radiation flux between the surfaces and their radiosities, if they (a) are black, and (b) have emissivity $\varepsilon_1 = 0.1$, and $\varepsilon_2 = 0.9$.

11.2 A net radiation flux of 400 W/m^2 is transferred from a surface. If its radiosity is 800 W/m^2 and its emissivity is 0.4, evaluate the temperature of the surface.

11.3 Determine the shape factor $F_{1\text{–}2}$ of the following figure, knowing that each flat region 1, 2, 3, and 4 has the same square area A and that 1 and 3 are perpendicular to 2 and 4.

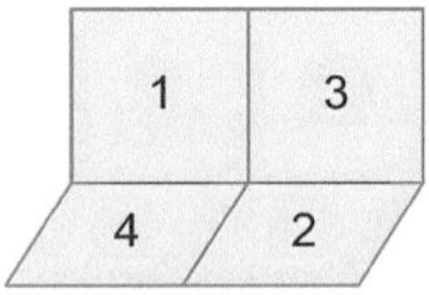

11.4 Evaluate the shape factors $F_{1\text{–}2}$ and $F_{2\text{–}1}$.

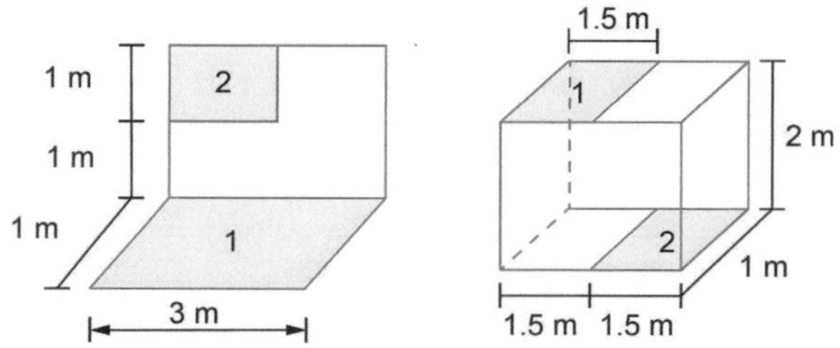

11.5 Evaluate the shape factor $F_{1\text{–}2}$ from the disc to the inner part of the cylinder, which are coaxial.

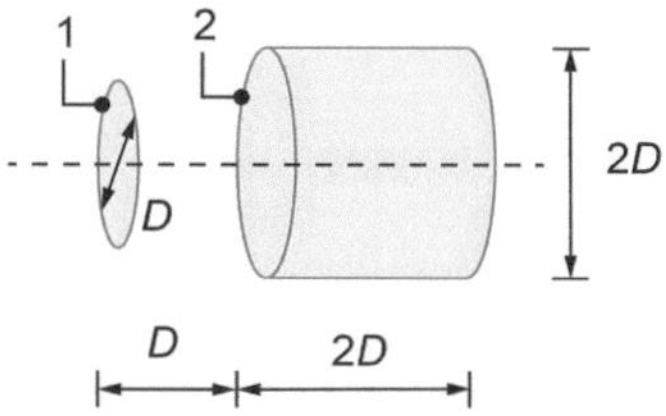

11.6 Two parallel coaxial discs with diameters equal to 1 m are black bodies and spaced 2 m apart; their temperatures are $T_1 = 400$ K and $T_2 = 300$ K. Evaluate the net heat variation by radiation transferred between them when the distance separating them is (a) 1 m and (b) 4 m.

11.7 In a mall, a long and 10 m wide ice skating rink with a semicircular duct cover was installed. Assuming that the rink and the cover behave like black bodies at 0 and 20 °C, evaluate the rate of heat transferred to the ice by radiation per meter of rink length.

11.8 What is the shape factor between a sphere 1 of diameter 0.03 m and an infinite surface 2?

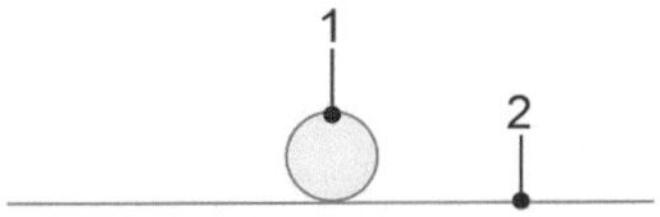

11.9 Using Hottel's expression (Eq. 11.18), determine the shape factor of two plates of widths W_1 and W_2 long (infinite) and spaced by H. Compare the result with the expression from Table 11.1.

11.10 A metallic bar 1 m long, 8 mm in diameter and emissivity of 0.1 dissipates 10 W of heat rate. Evaluate its temperature if it is inserted inside a cylinder with a diameter of 80 mm of the same length, with an emissivity of 0.90 and a temperature of 0 °C.

11.11 Saturated liquefied methane at –100 °C is stored in a 1.5 m diameter spherical tank. To avoid heat gain, the sphere is placed inside another 2 m diameter sphere and the space between the spheres is evacuated. If the outer sphere is at

5 °C, and the emissivity of all surfaces is 0.015, evaluate the rate of methane vaporization (*boil off*) per day (kg/day) if its enthalpy of vaporization is 300 kJ/kg.

11.12 Two parallel plates (1) and (2) of dimensions 2 × 2 m are maintained at 350 °C and 25 °C, respectively, and there is a vacuum between them. The emissivity of the plates is 0.1 and they are spaced 0.1 m apart. Under these conditions, it is asked: (a) calculate the total net radiation heat rate (W) transferred between the plates in their original configuration. Wishing to reduce the heat flux between the plates to 1/3 of the previous value, a third plate is inserted exactly in the middle of the two previous ones (dashed plate in the figure). It is asked (b) to determine the emissivity of this new plate to achieve the desired effect, being the same on both sides of the plate; (c) What material(s) could be used? (d) Calculate the equilibrium temperature of the third plate.

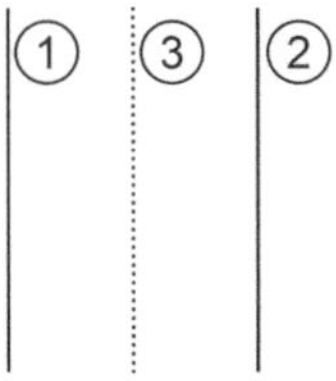

11.13 Water, at atmospheric pressure, flows through the interior of a circular tube, as indicated in the figure, and is heated from 10 to 30 °C. The temperature of the inner surface of this tube is constant at 40 °C, while the water flow rate is 60 kg/h. This tube is surrounded by a second tube and, in the space between the two is evacuated. Other data are indicated in the following table. Evaluate (a) the length of the tubes, and (b) the temperature of the inner surface of the outer tube (T_3).

Feature	Inner tube	Outer tube
Material	Plastic	Carbon steel
Conductivity (W/m°C)	0.23	70
Specific heat (kJ/kg°C)	1.70	0.40
Thermal diffusivity (m^2/s)	1.10×10^{-7}	1.94×10^{-6}
Inner diameter (mm)	20	60
Outer diameter (mm)	25	65
Emissivity	$\varepsilon_1 = 0.9$ $\varepsilon_2 = 0.8$	$\varepsilon_3 = 0.5$ $\varepsilon_4 = 0.7$

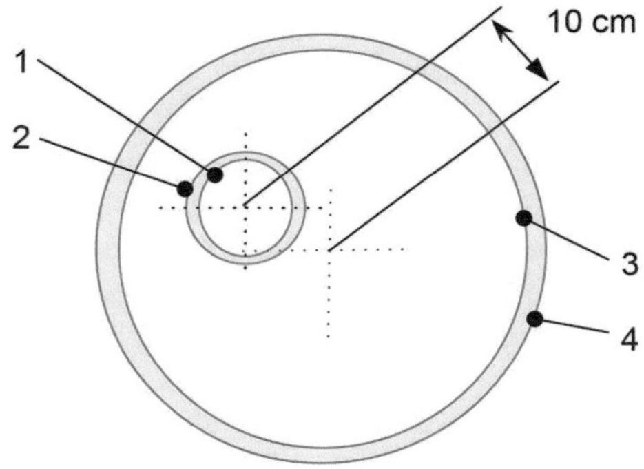

11.14 In a certain industrial installation, liquid nitrogen is used as a coolant. This coolant at 80 K flows inside a stainless steel tube ($\varepsilon = 0.2$), with an external diameter of 6.35 mm, inserted into a large vacuum chamber, whose walls are at 230 K. Determine the rate of heat gain by the tube per unit length. If a second stainless steel tube with a diameter of 12.7 mm is placed around the original tube, determine the new rate of heat gain. Find the temperature of this second tube that acts as a radiation shield.

11.15 A satellite of diameter, $D_S = 1.5$ m, orbits the planet at an altitude $H = 500$ km. The Earth is a sphere of diameter, D_T, equal to approximately 12,700 km. Calculate (a) the equilibrium temperature of the satellite when it is in the "dark" zone of the Earth (without solar radiation incidence). Consider the uniform temperature of the satellite and that the Earth is a black body at 15 °C, and the outer space is a black body at 0 K. When the satellite is on the "bright" side of the Earth (facing the Sun), it receives a solar heat flux of 1366 W/m^2. (b) Recalculate the new equilibrium temperature of the satellite for these new conditions. Use the projected area of the satellite, i.e. $A_S = \pi D_S^2/4$ for calculation purposes, and the coating material has an emissivity of $\varepsilon = 0.06$. If necessary, use in your calculations the shape factor of the satellite face facing the Earth: $F_{S,T} = \frac{D_T^2}{D_T^2+4L^2}$.

11.16 A horizontal tube 6 m long and 10 cm in diameter is maintained at a temperature of 150 °C in a large room where the air is at 20 °C. The walls of the room are at 38 °C. Consider the tube's emissivity $\varepsilon = 0.7$. Calculate (a) the rate of heat lost by the tube by convection, thermal radiation, and combined effect. Subsequently, it was decided to apply a highly reflective paint on the external surface of the tube with $\varepsilon = 0.05$; (b) recalculate the heat losses in this new condition.

11.17 A metal plate is located horizontally on the top of a house roof. The air temperature is 20 °C and flows parallel to the surface of the plate at a speed of 5 m/s. Assuming average values for the heat transfer coefficient, calculate its equilibrium temperature, knowing that the plate only exchanges heat by the upper surface. The plate length is 1 m in the direction that the air flows and 0.5 m in the perpendicular direction, the surface emissivity is 0.9 and the effective sky temperature (black body) for radiation effects is −50 °C.

11.18 The length of the external piping connecting the boiler to a building is 120 m. The boiler generates saturated steam at 120 °C. The piping is made of stainless steel ($k_a = 35$ W/mK, $D_i = 254$ mm, $D_e = 280$ mm) and thermally insulated (magnesia duct). The insulation thickness is $e = 50$ mm. The heat transfer coefficient in the internal flow is 5000 W/m^{2}K and the piping is exposed to a cross-wind at a temperature of 25 °C, resulting in a heat convection coefficient of 8 W/m^{2}K. Knowing that the sky behaves as a black body at −40 °C, the surroundings and soil as a black body at 20 °C, and that the piping is covered with a coating that presents a total hemispherical emissivity equal to 0.2, determine the mass flow of the condensate at the end of the piping.

11.19 Consider a pipeline for transporting liquefied natural gas (LNG) at 111 K (−162 °C). The problem of heat gain from the environment was critical and, therefore, three alternatives to decrease this heat gain were studied. In all cases, the following assumptions were considered:

(a) The inner wall of the pipe is at 111 K;
(b) The involved piping has excellent thermal conductivity and, thus, does not represent any thermal resistance to conduction. Their thicknesses are also negligible;
(c) The external coefficient of heat transfer by convection between the insulating material and the environment air, in all cases, was assumed to be 5 W/m^{2}K;
(d) The environment air temperature is 300 K for all cases. The ground and the sky behave as a single black body at an average temperature of 290 K;
(e) The thermal insulator used was always the same, with a conductivity of 0.06 W/mK;
(f) The diameter of the LNG pipeline is $D = 10$ cm and the thickness of the thermal insulation, $e = 5$ cm.

Make your calculations of the heat transfer rate to the cryogenic fluid for 1 m of pipeline and consider all types of heat transfer (conduction, convection, and thermal radiation), if necessary. Calculate the gain of heat transfer rate in each of the following alternatives, as well as the temperature of the external surface of the thermal insulator. In all cases, start the problem by indicating the thermal circuit.

Alternative 1 The LNG transport pipe was wrapped with thermal insulation (figure below), which was externally "painted" with a paint with an emissivity of 0.98 (black color).

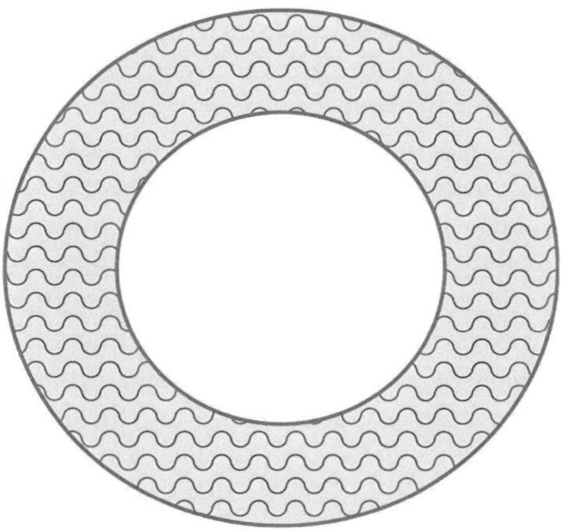

Alternative 2 Someone on the team suggested that the same design as alternative 1 be maintained, but the external surface of the pipe be painted with a paint of lower emissivity, for example, with an emissivity of 0.2.

Alternative 3 A newly graduated engineer, recalling his lessons in *Heat Transfer*, suggested that the inner tube be inserted into a second pipe and that a vacuum be made in the annular space (figure below), in addition to reducing the thickness of the thermal insulation by half, to maintain the total external diameter of the pipeline unchanged and provide space to accommodate the second tube. The inner pipe remained unchanged. Analyze the case with the emissivity of the insulator of 0.98 and 0.2. The emissivity of the pipes is 0.4.

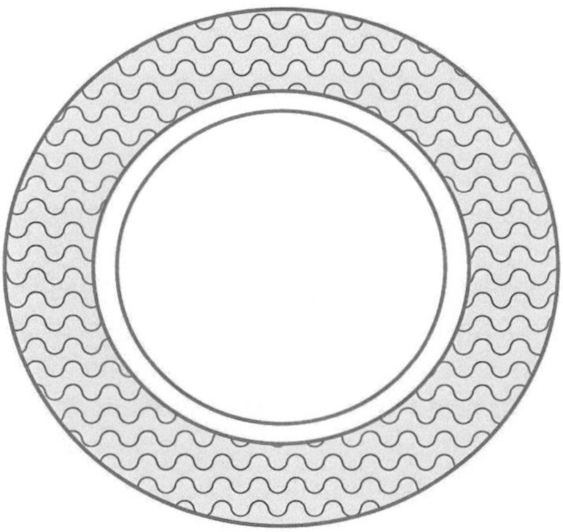

11.20 A 1 m square "hot plate" of a large snack bar is kept at 127 °C using a gas combustion. For radiation properties, the plate has an emissivity of 0.7, and the walls of the snack bar ($T_w = 27$ °C), and $\varepsilon = 0.9$. In addition to thermal radiation, the surface of the plate loses heat to the environment air (27 °C) by natural convection ($h = 6$ W/m^{2}K). Determine (a) what percentage of heat transfer rate is lost by thermal radiation to the walls of the snack bar; (b) what is the minimum flow rate, in kg/h, of gas necessary to keep the plate hot at 127 °C, knowing that the heat transfer rate resulting from combustion can be estimated by $\dot{Q} = \dot{m}_{gas} \times \text{LHV}$, where *LHV* is the lower heating value of the gas and is 50,000 kJ/kg. For calculation purposes, consider only the upper surface of the plate.

11.21 The surfaces of the two concentric cylinders behave like black bodies. The inner cylinder (1) has a temperature and diameter equal to 480 K and 0.2 m, respectively. The outer cylinder (2) has a diameter equal to 0.4 m. The following figure shows a scheme of the indicated configuration. Note that the set is positioned in an environment (enclosure) that has a much larger exposed area than that of the cylinder (2) and with a uniform surface temperature equal to 300 K. Air, also at 300 K, flows over the outer cylinder (2) and provides a heat transfer coefficient equal to 12 W/m^{2}K. Considering that the space formed between the cylinders was evacuated and neglecting the thermal resistance of cylinder (2), it is asked to present: (a) all your simplifying hypotheses and equivalent thermal circuits; (b) all relevant shape factors; (c) the net heat transfer rate between surfaces 1 and 2 per meter of the cylinder; and (d) the temperature of the outer surface of the cylinder (2).

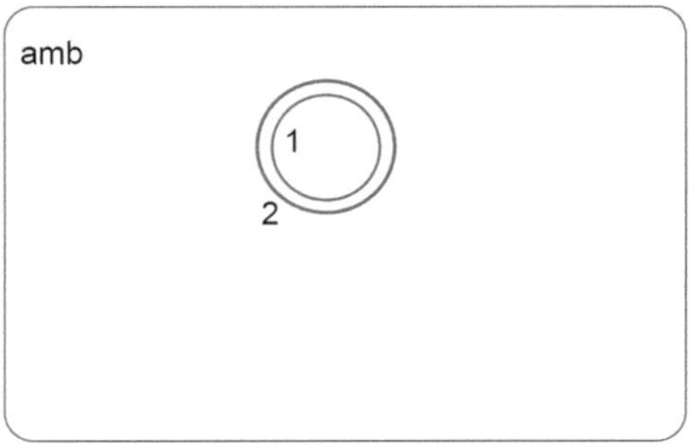

11.22 A small 1 m square (1) metal plate rests in the center of an open square shed of 10 m × 10 m, whose height is 4 m (see figure below). The roof of this shed can be approximated by a flat surface (3) that reaches a temperature of 70 °C due to intense solar radiation. The sides of the shed are open. The small metal plate exchanges convective heat with the atmospheric air that is at 20 °C, with the heat transfer coefficient being $h = 10$ W/m^2°C. There is no conduction of heat to the ground. Under these conditions, it is asked: (a) the temperature of the metal plate when its emissivity $\varepsilon_1 = 1$; and (2) and when its emissivity $\varepsilon_1 = 0.1$ due to the painting of the surface with a highly reflective paint. Consider that all remaining surfaces are black.

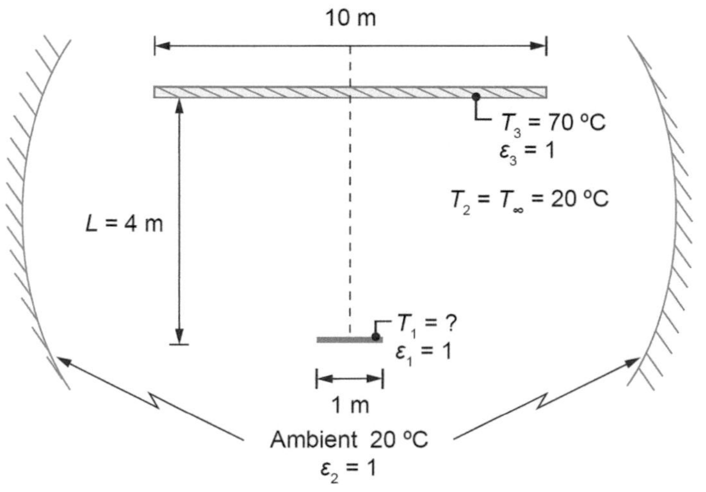

References

Hottel, H.C. 1954. Radiant heat transmission. In *Heat transmission*, ed. by W.H. McAdams, 3rd ed. New York: McGraw-Hill.

Howell, J.R. 2010. *A catalog of radiation configuration factors*. 3rd ed. Available in: http://www.thermalradiation.net. Accessed 18 Aug 2024.

Holman, J.P. 2021. *Heat transfer*, 10th ed. Mcgraw-Hill.

Hottel, H. 1954. *Radiant-heat transmission in McAdams, W. Heat Transmission*, 3rd ed. McGraw-Hill B. C.

Varón, L.M., B. Narváez-Romo, L. Costa-Sobral, G. Barreto and J.R. Simões-Moreira. 2023. Novel high-flux indoor solar simulator for high temperature thermal processes. *Applied Thermal Engineering* **234**: 121188.

Chapter 12
Mass Transfer

Abstract This chapter focuses on mass transfer. Mass transfer can occur within bodies via Fick's law between a material and a fluid medium through convection. Fick's law is introduced, along with the variables involved in calculating mass diffusion. Diffusion in gases, liquids, non-porous materials, and porous materials are analyzed, with guidance on obtaining the respective diffusion coefficients. The differential equation governing mass diffusion is derived, along with the corresponding boundary conditions. Both steady-state and transient analyses are carried out. The next topic addressed in the chapter is convective mass transfer, introducing the theory of concentration boundary layers and flow regimes. Dimensionless mass transfer numbers are presented, as well as the Chilton-Colburn analogy between heat and mass transfer. Working equations for convective mass transfer are provided, with examples of application to typical mass transfer conditions.

Keywords Mass transfer · Mass diffusion · Fick's law · Mass convection · Chilton-Colburn analogy

The mass transfer process comprises the flux or transport of matter and occurs when there is a concentration gradient of a species within a material or in contact with a surrounding medium. This transfer can happen in gases, liquids, or solids. Mass transfer always occurs from regions of higher concentration to regions of lower concentration. The transported material is referred to as the *species* or *solute.*

Similarly to heat transfer, mass transfer can occur via diffusion or convection. Molecular diffusion within materials is generally slow, whereas convection typically progresses in a more rapid rate because of the mixing process, which enhances the transport of matter. To illustrate this difference between those two transport mechanisms, consider the examples in Fig. 12.1a, b, where a bottle of perfume (A) is broken in a large room with stagnant air (B). Slowly, the perfume molecules diffuse through the still air, and, in this case, people would take a while to notice the fragrance due to the mass transfer by diffusion. Now, imagine that a fan turned on the room promoting a convective movement of the perfume in the air. In this case, people will quickly notice the fragrance because this mechanism involves mass

J. R. Simões-Moreira and E. W. Zavaleta-Aguilar, *Heat Transfer*, Heat and Mass Transfer, https://doi.org/10.1007/978-3-031-80318-5_12

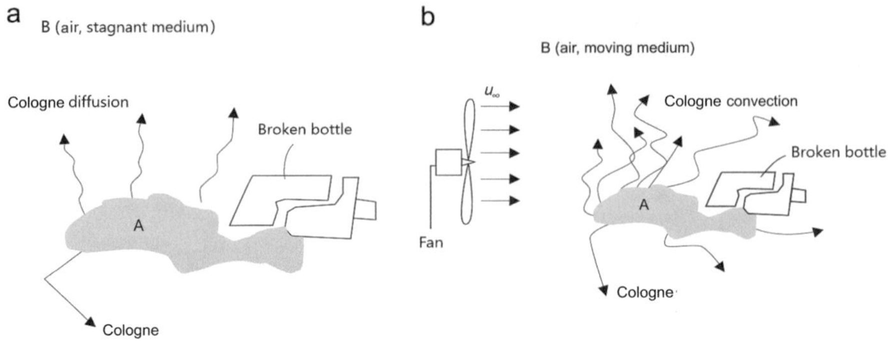

Fig. 12.1 Mass transfer by diffusion (**a**) and convection (**b**)

transfer by convection, where the macroscopic mixing of mass is predominant and more effective.

There is a direct analogy between mass transfer and heat transfer, as both phenomena involve *potential* difference that drives and controls the transfer process. In heat transfer, the potential is the temperature difference, while in mass transfer, it is the concentration difference of a species. In both cases, the transfer direction is from a higher potential (temperature or concentration) to the lower one. Additionally, the constitutive laws governing both phenomena are similar.

12.1 Mass Transfer by Diffusion

The law governing mass transfer by diffusion is *Fick's Law* of 1855, which is analogous to Fourier's Law of heat conduction discussed in Chap. 2. Fick's Law is expressed as:

$$j_A = -D_{AB}\frac{d\rho_A}{dz} \tag{12.1}$$

where j_A [kg/m^{2}s] is the mass flux of species A diffusing into B, D_{AB} is the *mass diffusion coefficient* or *binary diffusivity* of A diffusing into B [m^2/s] (for gases in binary mixture, $D_{AB} = D_{BA}$). Tipically, the order of magnitude of the diffusion coefficient is less than 10^{-4} m^2/s; ρ_A is the mass concentration of A per unit volume of the mixture $(A + B)$ [kg/m^3]; and z is the distance in the direction of diffusion [m].

The minus signal in Eq. (12.1) indicates that the concentration gradient $\frac{d\rho_A}{dx}$ is negative in the direction of diffusion. Figure 12.2 illustrates schematically the mass transfer process where the dots represent species A at a higher concentration in region 1 than in region 2, resulting in mass transfer from 1 to 2.

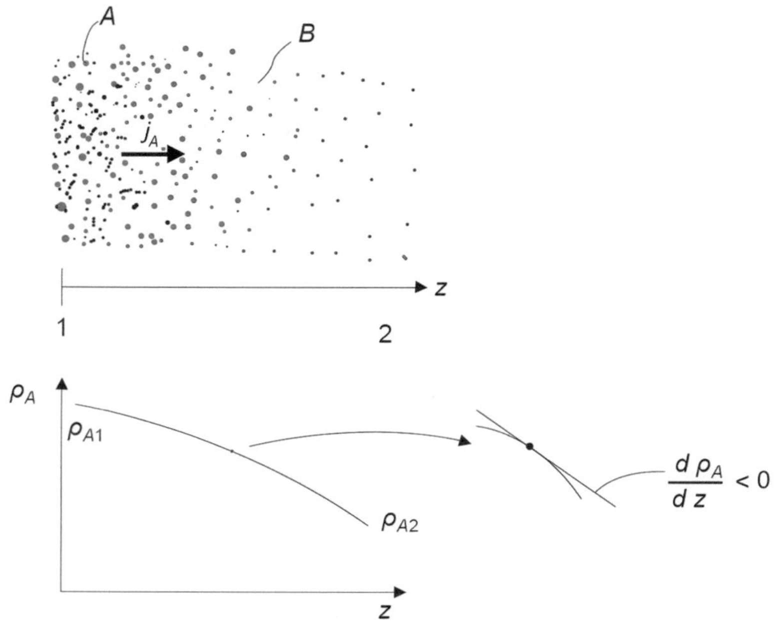

Fig. 12.2 Species A diffusing from a region of higher concentration to a region of lower concentration along the z-direction

Note that the behavior of the concentration curves for species A in Fig. 12.2 is illustrative only. This behavior depends on some variables, as will be discussed throughout this chapter; it can be linear or exhibit other forms.

12.1.1 Definition of Variables

The concentration, Fick's Law, and other variables of interest can be expressed on either a *mass* or *molar* basis. Next, relevant variables are defined.

Mass concentration of species A [kg/m^3] is defined as the ratio of the mass of A (m_A) present in the solution volume ($\forall$).

$$\rho_A = \frac{m_A}{\forall} \tag{12.2}$$

Number of moles of species A [mol] is defined as the ratio between the mass of A (m_A) and the molar mass of A (M_A).

$$n_A = \frac{m_A}{M_A} \tag{12.3}$$

Molar concentration of species A [mol/m^3] is defined as the number of moles of A (n_A) per volume of solution ($\forall$).

$$C_A = \frac{n_A}{\forall} \tag{12.4}$$

Mass fraction of species A [–] w_A, is defined as the ratio between the mass (or mass concentration) of A, m_A, and the mass (or mass concentration) of the mixture, m. For the case of a mixture of only two components A and B, its total mass is:

$$m = m_A + m_B \tag{12.5}$$

The mass fraction of A is:

$$w_A = \frac{m_A}{m} \tag{12.6}$$

or,

$$w_A = \frac{\rho_A}{\rho} \tag{12.7}$$

The sum of the mass fractions of all the n components of a mixture is unity, that is,

$$\sum_{i=1}^{n} w_i = 1 \tag{12.8}$$

Molar fraction of A, y_A, is defined as the ratio between the number of moles (or molar concentration) of A, n_A, and the number of moles (or molar concentration) of the mixture, n. Again, assuming that the mixture is formed by only two components A and B, the total number of moles, n, is:

$$n = n_A + n_B \tag{12.9}$$

The molar fraction of A is:

$$y_A = \frac{n_A}{n} \tag{12.10}$$

or,

$$y_A = \frac{C_A}{C} \tag{12.11}$$

The sum of the molar fractions of all the n components of a mixture is unity.

$$\sum_{i=1}^{n} y_i = 1 \tag{12.12}$$

Molar mass of a mixture of two components, M [kg/mol]:

$$m = m_A + m_B \rightarrow nM = n_A M_A + n_B M_B \rightarrow M = \frac{n_A}{n} M_A + \frac{n_B}{n} M_B$$

For n components, we have:

$$M = \sum_{i=1}^{n} y_i M_i \tag{12.13}$$

or, in terms of mass for two components,

$$n = n_A + n_B \rightarrow \frac{m}{M} = \frac{m_A}{M_A} + \frac{m_B}{M_B} \rightarrow \frac{1}{M} = \frac{1}{M_A}\frac{m_A}{m} + \frac{1}{M_A}\frac{m_B}{m}$$

For n components, we have:

$$\frac{1}{M} = \sum_{i=1}^{n} \frac{w_i}{M_i} \tag{12.14}$$

Relationship between mass and molar fraction,

$$w_A = \frac{m_A}{m} = \frac{n_A M_A}{nM}, \quad \text{or}$$

$$w_A = y_A \frac{M_A}{M} \tag{12.15}$$

Relationship between mass and molar concentration,

$$C_A = \frac{n_A}{\forall} = \frac{m_A}{M_A \forall}, \quad \text{or}$$

$$C_A = \frac{\rho_A}{M_A} \tag{12.16}$$

Fick's law in molar basis [mol/m^{2}s]; in this case, the molar flux of A, J_A, is defined as:

$$J_A = -D_{AB} \frac{dC_A}{dz} \tag{12.17}$$

Table 12.1 summarizes the definitions presented above. Examples 12.1 and 12.2 illustrate their use.

Example 12.1 (*Mass and molar fractions*)

Atmospheric air is a mixture of several gases whose composition can vary according to local and weather conditions. One of the components is water vapor and, in some situations, it is convenient to perform calculations considering dry air, that is, free of water vapor. Assuming that in a given situation the air composition in molar basis is: 76.913 % of N_2; 20.632 % of O_2; 1.5 % of H_2O; 0.920 % of Ar, and 0.037 % of CO_2 and given that their molar masses are 28; 32; 18; 39.9; 44 g/mol, respectively, find:

(a) the molar mass of the mixture and the mass fractions of the air components;
(b) the molar mass of the mixture and the molar and mass fractions of the components for dry air.

Solution

(a) Applying Eq. (12.13), the molar mass of the mixture is:

$$M = \sum_{i=1} y_i M_i = 0.76913 \times 28 + 0.20632 \times 32$$
$$+ 0.015 \times 18 + 0.0092 \times 39.9 + 0.00037 \times 44$$

Table 12.1 Quantities related to diffusive fluxes in mass and molar basis

Quantity	Mass base		Molar base	
	Unit	Equation	Unit	Equation
Diffusive flux of *A* (Fick's Law)	kg/m²s	$j_A = -D_{AB}\frac{d\rho_A}{dz}$	mol/ m²s	$J_A = -D_{AB}\frac{dC_A}{dz}$
Concentration of *A*	kg/m³	$\rho_A = \frac{m_A}{V}$	mol/m³	$C_A = \frac{n_A}{V}$
Fraction	–	$w_A = \frac{m_A}{m} = \frac{\rho_A}{\rho}$	–	$y_A^* = \frac{n_A}{n} = \frac{C_A}{C}$
Sum of fractions	–	$\sum_{i=1}^{n} w_i = 1$	–	$\sum_{i=1}^{n} y_i = 1$
Molar mass of mixture	g/mol	$\frac{1}{M} = \sum_{i=1}^{n} \frac{w_i}{M_i}$	g/mol	$M = \sum_{i=1}^{n} y_i M_i$
Number of moles of *A*	–	–	mol	$n_A = \frac{m_A}{M_A}$
Relationship between mass and molar concentrations	–	–	mol/m3	$C_A = \frac{\rho_A}{M_A}$
Relationship between mass and molar fractions	–	$w_A = y_A \frac{M_A}{M}$		

*The molar fraction in gases will be assigned the letter *y* and in liquids *x*

$$M = 28.791 \text{ g/mol}$$

Thus, the mass fractions are evaluated using Eq. (12.15):

$$w_{N_2} = y_{N_2}\frac{M_{N_2}}{M} = 76.913\frac{28}{28.791} = 74.8\%;$$
$$w_{O_2} = 20.632\frac{32}{28.791} = 22.932\%;$$
$$w_{H_2O} = 1.5\frac{18}{28.791} = 0.938\%;$$
$$w_{Ar} = 0.92\frac{39.9}{28.791} = 1.275\%;$$
$$w_{CO_2} = 0.037\frac{44}{28.791} = 0.057\%.$$

(b) For the case of dry air, one should divide the molar fractions by the term $(1 - y_{H_2O})$ in order to normalize the mixture free of water vapor.

$$y_{N_2} = 76.913\frac{1}{1 - 0.015} = 78.084\%,$$

similarly for the other gases

$$y_{O_2} = 20.946\%, \; y_{Ar} = 0.934\%, \quad \text{and} \quad y_{CO_2} = 0.038\ \%$$

$$M = \sum_{i=1} y_i M_i = 0.78084 \times 28 + 0.20946 \times 32 + 0.00934 \times 39.9 + 0.00038 \times 44$$

$$M = 28.956 \text{ g/mol}$$

$$w_{N_2} = 78.084\frac{28}{28.956} = 75.506\%$$

$$w_{O_2} = 23.148\%, \; w_{Ar} = 1.287\%, \; w_{CO_2} = 0.058\%$$

Example 12.2 (*Fick's law*)

Benzene is a product that can cause dizziness in humans and, for long exposure periods and considerable concentrations, leukemia. Benzene was detected in the air at a location and its concentrations were measured at two different points, separated by 0.5 m, which were 2.58×10^{-5} kg/m^3, and 1.29×10^{-5} kg/m^3. Assuming that the concentration variation is linear and the diffusion coefficient of benzene in still air is $D_{AB} = 9 \times 10^{-6}$ m^2/s, evaluate the mass flux of benzene.

Solution

The mass transfer process in still air is through molecular diffusion. If the distribution of the benzene concentration is linear, then its mass concentration gradient between these measurement points is: $\frac{d\rho_A}{dx} = \frac{\rho_{A,2}-\rho_{A,1}}{x_2-x_1}$, thus,

$$j_A = -D_{AB}\frac{\rho_{A,2} - \rho_{A,1}}{x_2 - x_1} = -9 \times 10^{-6}\frac{1.29 \times 10^{-5} - 2.58 \times 10^{-5}}{0.5 - 0}$$

$$j_A = 2.3 \times 10^{-7}\ \text{g/m}^2\text{s}$$

Assuming that benzene is diluted in the air, the concentration of the mixture is essentially the concentration of the air. At 101.3 kPa and 25 °C, the air concentration is 1.1764 kg/m^3. Thus, the mass fraction of benzene in the air is $2.58 \times 10^{-5}/1.1764 = 22 \times 10^{-6}$ or 22 ppm (parts per million), which, although it seems a negligible value, has already exceeded the limit allowed in many countries (in Brazil the limit is 1 ppm, NR 15, 2021).

Fick's Law (Eq. 12.1) depends on the binary (mass) diffusion coefficient and the concentration gradient of species. The diffusion coefficient varies strongly with the medium in which the mass transfer occurs, that is, they differ for gases, liquids, and solids, as seen next.

12.1.2 Diffusivity in Gases

Based on the kinetic theory of gases, where it is assumed that the molecules are spherical, without intermolecular forces, and undergo elastic collisions—known as the *hard-ball* model—it is possible to derive a theoretical expression for the mass diffusivity between two gases. However, this expression shows some variation compared to experimental values (Foust et al. 2006). More accurate expressions for determining binary diffusivity in non-polar gases were developed by Fuller et al. (1966), with their correlation given by Eq. (12.18). Note that diffusivity increases with temperature and decreases with pressure, molecular size, and the molar masses of the species.

$$D_{AB} = 0.001\frac{T^{1.75}}{p\delta_{AB}^2}\left(\frac{1}{M_A} + \frac{1}{M_B}\right)^{1/2} \tag{12.18}$$

where the corresponding units in Eq. (12.18) are D_{AB} [cm^2/s], T is the absolute temperature [K], M_A and M_B are the molar masses of A and B, respectively [g/mol], p is the absolute pressure of the system [atm] and δ_{AB} (without a specific name) [Å] is defined as:

$$\delta_{AB} = \left(\sum_{i=1}^{n} k_i v_i\right)_A^{1/3} + \left(\sum_{i=1}^{n} k_i v_i\right)_B^{1/3} \tag{12.19}$$

where v_i is the volume associated with diffusion, experimentally obtained through curve fitting to experimental data (see Table 12.2). k_i is the atomic index, and n is the number of atoms that make up the molecule. There are atoms and molecules for which the values of $\sum_{i=1}^{n} k_i v_i$ have been experimentally evaluated, as shown in Table 12.3.

Table 12.2 Increments of atomic diffusion volumes (Fuller et al. 1966)

Atom	v_i [Å^3]
C	16.5
H	1.98
O	5.48
N	5.69
Cl	19.5
S	17.0
Aromatic ring or heterocyclic	–20.2

Table 12.3 Diffusion volumes of simple atoms and molecules (Å^3), Fuller et al. (1966)

Gas	$\sum_{i=1}^{n} k_i v_i$ [Å^3]
H_2	7.07
He	2.88
N_2	17.9
O_2	16.6
Ar	20.1
Ne	5.59
Ar (argon)	16.1
Kr	22.8
Xe	37.9
SO_2	41.1
CO	18.9
CO_2	26.9
N_2O	35.9
NH_3	14.9
H_2O	12.7
CCl_2F_2	114.8
SF_6	69.7
Cl_2	37.7
Br_2	67.2

Example 12.3 (*Diffusion coefficient (mass)*)

Evaluate the diffusion coefficient (diffusivity) of ethane (C_2H_6) in air at 50 °C and 1 atm pressure.

Solution

A = ethane and B = air

$$M_A = 2C + 6H = 2 \times 12 + 6 \times 1 = 30\,\text{g/mol}$$

$$M_B = 29\text{g/mol (Example 12.1)}$$

$$\left(\sum_{i=1}^{n} k_i v_i\right)_A = 2v_C + 6v_H = 2 \times 16.5 + 6 \times 1.98 = 44.88\ \mathring{A}^3 \text{(Table 12.2)}$$

$$\left(\sum_{i=1}^{n} k_i v_i\right)_B = 20.1\ \mathring{A}^3 \text{(Table 12.3)}$$

$$\delta_{AB} = 44.88^{1/3} + 20,1^{1/3} = 6.272\ \mathring{A}$$

$$D_{AB} = 0.001\frac{T^{1.75}}{pv_{AB}^2}\left(\frac{1}{M_A} + \frac{1}{M_B}\right)^{1/2} = 0.001\frac{(50 + 273.2)^{1.75}}{1 \times 6.272^2}\left(\frac{1}{30} + \frac{1}{29}\right)^{1/2}$$

$$D_{AB} = 0.163\,\text{cm}^2/\text{s}$$

It is possible to find in the literature some experimental values of binary diffusion coefficients in gases as in Reid et al. (1987), who collected diffusion coefficients for various substances, which are indicated in Table 12.4.

Once the diffusivity for a gas under certain conditions is known, it is possible to estimate it for different temperatures and pressures using the Fuller et al.'s model (1966),

$$\frac{D_{AB,1}}{D_{AB,2}} = \frac{0.001\frac{T_1^{1.75}}{p_1 v_{AB}^2}\left(\frac{1}{M_A} + \frac{1}{M_B}\right)^{1/2}}{0.001\frac{T_2^{1.75}}{p_2 v_{AB}^2}\left(\frac{1}{M_A} + \frac{1}{M_B}\right)^{1/2}}$$

$$D_{AB,2} = D_{AB,1}\left(\frac{T_2}{T_1}\right)^{1.75}\left(\frac{p_1}{p_2}\right) \tag{12.20}$$

As a gaseous species (species 1) diffuses in a mixture of other still gases (mixture M of species 2, 3,...), an effective diffusion coefficient of 1 in the mixture M can be

Table 12.4 Binary diffusion coefficients in gases at 1 atm for various temperatures

Gases	T [K]	D_{AB} [cm^2/s]
Ar/CO_2	276 317	0.144 0.179
Ar/ethanol	313	0.147
Ar/He	276 346	0.632 0.914
Ar/n-hexane	294 328	0.081 0.094
Ar/2-methylfuran	334	0.107
Ar/naphthalene	303	0.087
Ar/H_2O	313	0.292
NH_3/diethyl ether	288 337	0.101 0.139
Argon/benzene	323 373	0.085 0.112
Argon/helium	276 418	0.655 1.417
Argon/ hexafluorobenzene	323 373	0.082 0.095
Argon/H_2	295 628 1068	0.840 3.25 8.21
Argon/krypton	273	0.121
Argon/CH_4	298	0.205
Argon/SO_2	263	0.078
Argon/xenon	195 378	0.052 0.180
CO_2/He	298 498	0.620 1.433
CO_2/N_2	298	0.169
CO_2/N_2O	313	0.130
CO_2/SO_2	473	0.198
CO_2 /tetrafluoromethane	298 673	0.087 0.385
CO_2/H_2O	307	0.201
CO/N_2	373	0.322
Ethylene/H_2O	328	0.236
He/benzene	423	0.618
He/bromobenzene	427	0.550
H e/2-chlorobutane	429	0.568
He/n-butanol	423	0.595
He/1-iodobutane	428	0.524

(continued)

Table 12.4 (continued)

Gases	T [K]	D_{AB} [cm^2/s]
He/methanol	432	1.046
He/N_2	298	0.696
He/H_2O	352	1.136
H_2/acetone	296	0.430
H_2/NH_3	263 358 473	0.580 1.110 1.890
H_2/cyclohexane	289	0.323
H_2/naphthalene	303	0.305
H_2/nitrobenzene	493	0.831
H_2/N_2	294 573	0.773 2.449
H_2/pyridine	318	0.443
H_2/H_2O	307	0.927
CH_4/H_2O	352	0.361
N_2/NH_3	298 358	0.233 0.332
N_2/aniline	473	0.182
N_2/SO_2	263	0.105
N_2/H_2O	308 352	0.259 0.364
N_2/SF_6	378	0.146
O_2/benzene	311	0.102
O_2/CCl_4	296	0.076
O_2/cyclohexane	289	0.076
O_2/H_2O	352	0.357

Adapted from Reid et al. (1987)

obtained, to be used in Eq. (12.1). The coefficient is defined as follows (Fairbanks and Wilke 1950):

$$D_{1M} = \frac{1 - y_1}{\frac{y_2}{D_{12}} + \frac{y_3}{D_{13}} + \cdots} \tag{12.21}$$

12.1.3 Diffusivity in Liquids

Liquids can mix to form a solution. For molecular diffusion to occur, the solvent (medium) must be still, and this mixture can involve either non-electrolytic or electrolytic species. Electrolytes are species that, when in contact with a liquid, dissociate into ions; for example, table salt (NaCl) in water. Unlike gas mixing, which is based on the kinetic theory of gases, there is no established theory for diffusion in liquids. Two theories have been proposed to study the phenomenon of diffusion in non-electrolytic solutions: the energy jump theory and the hydrodynamic theory. The former, also known as Eyring's theory, proposes that an atom (or molecule) vibrates in its position and, upon gaining sufficient energy, moves to a new position (diffuses). The second theory, the hydrodynamic theory, is more widely accepted. It postulates that diffusion is influenced by both a force caused by the chemical potential difference of the solute and a hydrodynamic drag force, acting on the solute (modeled as a sphere) in infinite dilution, opposite to the direction of movement. By equating these two forces under uniform motion, the diffusion coefficient in an ideal solution, known as the Stokes–Einstein equation, is obtained:

$$D_{AB}^{o} = \frac{kT}{6\pi \mu_B R_A} \tag{12.22}$$

where k is the Boltzmann constant, T is the absolute temperature, μ_B is the dynamic viscosity of the solvent, and R_A is the molecular radius of A. Although Eq. (12.22) is theoretical, it shows that diffusivity increases with temperature and decreases with the viscosity of the medium and the size of the solute.

Diffusivity in liquids, in contrast to diffusion in gases, is strongly dependent on concentration. For non-electrolytic solutions where the diffusing species is diluted, Wilke and Chang (1955) proposed the following correlation:

$$D_{AB}^{o} = \frac{7.4 \times 10^{-8} (\phi M_B)^{0.5} T}{\mu \, \overline{v}_A^{0.6}} \tag{12.23}$$

The superscript (°) of D [cm^2/s] indicates that species A is diluted in B, ϕ [–] is the *association factor* of the solvent correcting its molar mass, M_B [g/mol] is the molar mass of the solvent, T [K] is the temperature, μ [cP] is the dynamic viscosity of the solution and $\bar{v}_A$ [cm^3/mol] is the molar volume of A as a saturated liquid at 1 atm pressure. Table 12.5 provides association factor values for various liquids.

Wilke and Chang (1955) collected many experimental diffusivity values for dilute solutions, some of which are indicated in Table 12.6.

The molar fraction of the solute (x_A) or of the solvent (x_B) in liquids is expressed in a similar way as in gases (Eq. 12.10), according to Eq. (12.24).

$$x_i = \frac{n_i}{n} \tag{12.24}$$

Table 12.5 Association factor (Wilke and Chang, 1955)

Solvent	ϕ [–]
Water	2.6
Methanol	1.9
Ethanol	1.5
Benzene	1.0
Ether	1.0
Heptane	1.0

Table 12.6 Binary diffusion coefficient of diluted non-electrolytes (Wilke and Chang 1955)

Solute/Solvent	T (°C)	D^o_{AB} [cm^2/s]
Acetic acid/acetone	25	3.309×10^{-5}
Acetic acid/benzene	25	2.081×10^{-5}
Acetic acid/CCl_4	25	1.490×10^{-5}
Acetic acid/toluene	25	2.265×10^{-5}
Aniline/ethanol	18.5	2.7×10^{-5}
Benzene/chloroform	15	3.70×10^{-5}
Benzoic acid/acetone	25	2.522×10^{-5}
Benzoic acid/CCl_4	25	0.908×10^{-5}
Bromobenzene/benzene	7.3	1.41×10^{-5}
CCl_4/benzene	25	2.0×10^{-5}
CCl_4/n-heptane	25	3.17×10^{-5}
CCl_4/kerosene	25	0.961×10^{-5}
Formic acid/acetone	25	3.768×10^{-5}
Formic acid/toluene	25	2.646×10^{-5}
Toluene/n-decane	25	2.090×10^{-5}
Toluene/n-dodecane	25	1.380×10^{-5}
Toluene/n-heptane	25	3.720×10^{-5}
Toluene/n-hexane	25	4.210×10^{-5}
Toluene/n-tetradecane	25	1.020×10^{-5}

When the solution is concentrated, the molar fraction of the solute influences the diffusivity. Leffler and Cullinan (1970) proposed the following relationship to evaluate the diffusivity in concentrated solutions:

$$D_{AB} = \alpha \frac{\left(\mu_A D^o_{BA}\right)^{x_A} \left(\mu_B D^o_{AB}\right)^{x_B}}{\mu} \tag{12.25}$$

where α [–] is the *thermodynamic factor* and corrects the non-ideality of the solution. Studies for electrolytic solutions can be found in specialized literature, such as Reid et al. (1988).

12.1.4 Diffusivity in Solids

As a general rule, a solid can be categorized either as non-porous or porous. First, the diffusivity in non-porous solids is studied, and later, in porous ones.

Diffusivity in non-porous solids

The theory to explain diffusion in non-porous solids is due to Eyring or the energy jump theory, already discussed above in diffusion in liquids. This theory assumes that a species is vibrating until it has enough energy to jump to a new equilibrium state; subsequently, this energy decreases, initiating a new cycle of increasing–decreasing.

The diffusion coefficient in non-porous solids is generally lower than in liquids and gases, because the spaces available for diffusion in solids are smaller. These spaces can be *structural flaws* (or vacancies) or *interstices*, schematized in Fig. 12.3.

Diffusion by vacancy (self-diffusion can occur) occurs in spaces within structural flaws in solids, whereas in interstitial diffusion, species A is transferred through the interstitial spaces of the solid's network B. In general, the activation energy Q (or energy barrier—see Eq. 12.26) required for interstitial diffusion is lower than in the case of vacancy. For both cases, the diffusion coefficient of species A in solid B can be expressed by:

$$D_{AB} = D_0 e^{-\frac{Q}{RT}} \tag{12.26}$$

where D_0 [cm^2/s] is the pre-exponential constant that does not depend on temperature, Q [J/mol], the diffusive activation energy, $R = 8.314$ J/mol K is the universal gas constant, and T [K], the absolute temperature.

Diffusion in solids is a common process, for example, it occurs in steel hardening in the process of *carburizing*, increasing the concentration of carbon atoms in a certain thickness of the metal surface. Another example corresponds to doping with boron or phosphorus giving electronic properties to silicon in photovoltaic cells. Parameters of Eq. (12.26) for some cases are indicated in Table 12.7.

Recent applications consider hydrogen for vehicle use, either directly by combustion (combustion engines) or to feed fuel cells to generate electrical energy. Either way, the hydrogen gas has to be pressurized to be stored in cylinders and metal tanks both in the production process and the final use in vehicles. There will be a concentration gradient of hydrogen between the inner surface of the container wall and the external environment air. Thus, a diffusive mass transfer of hydrogen from the inside

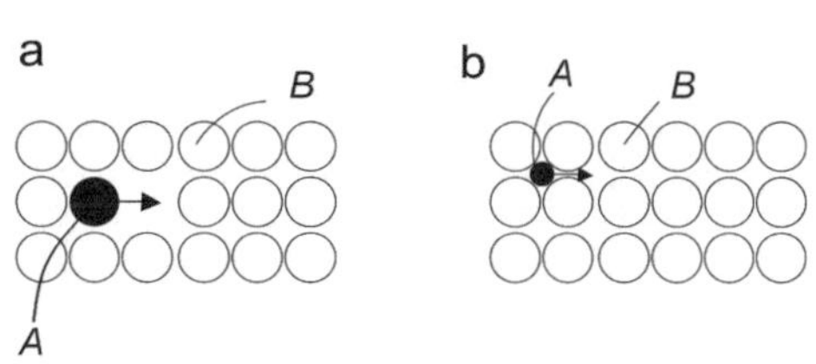

Fig. 12.3 Diffusion of species A in solid B **a** by vacancy, and **b** interstitial

Table 12.7 Parameters of Eq. (12.26) for diffusion on several materials (Vlack, 1964)

Species	Solid	D_0 [cm²/s]	Q [kJ/mol]
C	α-Fe (BCC)[a]	0.0079	75.7
C	γ-Fe (FCC)[b]	0.21	141.4
C	Ti (HCP)[c]	2.24	174.1
Fe	α-Fe (BCC)	5.8	249.8
Fe	γ-Fe (FCC)	0.58	284.1
Ni	γ-Fe (FCC)	0.5	276.1
Mn	γ-Fe (FCC)	0.35	282.4
Zn	Cu	0.033	159.0
Cu	Al	2.0	141.8
Cu	Cu	11.0	239.3
Ag	Ag (crystal)	0.72	188.3
Ag	Ag (grain boundary)	0.14	90.0

[a]BCC: body-centered cubic structure; [b]FCC: face-centered cubic structure; [c]HC: hexagonal compound

to the outside through the wall will be established. According to Wang et al. (2020), when hydrogen is transferred in a steel metallic wall, it is embrittled. The brittleness depends on the composition of the steel, mainly of martensite steels. The diffusion coefficient in these cases can vary between 2.8×10^{-7} and 3.7×10^{-7} cm²/s.

Example 12.4 (*Stored hydrogen diffusion*)

Hydrogen has been seen as the fuel of the future since in the combustion process it results in the production of water vapor. Consider the possibility of transporting pressurized hydrogen gas in a pipeline of $e = 10$ cm thickness used, originally, for natural gas transport. The concentration on the metallic surface on the hydrogen gas side is 0.02 kg/m³. Assume that the diffusivity of hydrogen in steel is 3×10^{-11} m²/s and that on the other side of the pipe, in contact with the external environment, its concentration is negligible. Also, assume a linear concentration gradient in the steel wall and that its radius is large compared to its thickness, to simplify the problem. Evaluate the mass flow of hydrogen lost to the environment. How long will it take to lose 1 mg of H_2 in 1 m² of piping wall area?

Solution

The flow of hydrogen in steel is evaluated by Fick's Law (Eq. 12.1). Assuming a linear concentration gradient in steel, the flow of hydrogen will be:

$$j_A = -D_{AB}\frac{\rho_{Ao} - \rho_{Ai}}{e}$$

Substituting values,

$$j_A = -3 \times 10^{-11} \frac{0 - 0.02}{0.1} = 6 \times 10^{-12} \text{ kg/m}^2\text{s}$$

Time for leaking $m'' = 1 \frac{\text{mg}}{\text{m}^2} = 10^{-6} \frac{\text{kg}}{\text{m}^{2'}}$

$$\Delta t = \frac{m''}{j_A} = \frac{10^{-6}}{6 \times 10^{-12}} = 1,67 \times 10^5 \text{ s} \cong 1 \text{ day and 22 hours}$$

Diffusivity in porous solids

In porous solids, the species (gaseous or liquid) is transported through the tortuous networks of pores. Thus, this diffusion type strongly depends on the geometry of the pores. Diffusivity in pores has applications in adsorption, catalysis, and drying processes. In macropores, diffusion is similar to "normal" molecular diffusion and the walls of the pores do not have a significant influence; however, when the pores are smaller, the diffusivity takes the name of *Knudsen diffusivity*, where the walls interfere with diffusion. More details about this type of diffusion can be found in Brakel (1975), Epstein (1989).

12.2 Differential Analysis of Mass Transfer

Diffusive mass transfer may be modeled in differential form, which for some cases of simple geometries results advantageous, as the analysis can be carried out for a differential volume and then be generalized to finite control volumes. The average velocity of each species are $\vec{v}_A$ and $\vec{v}_B$, as indicated in Fig. 12.4, in which two species can be seen in transport A and B moving along the z-axis.

The average *mass* velocity of the mixture of k components is defined as:

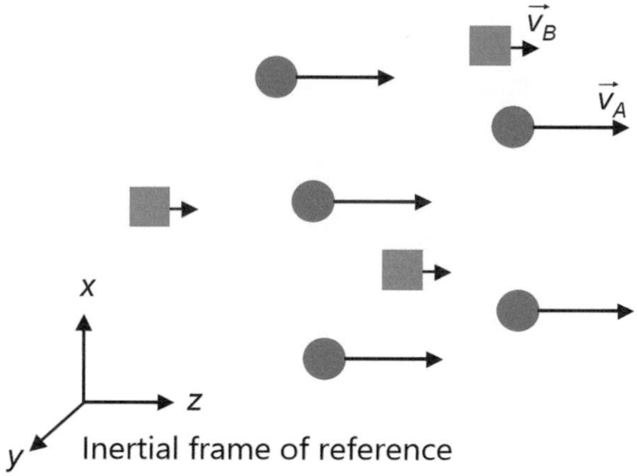

Fig. 12.4 Velocity of the species

$$\vec{v} = \frac{\sum_{i=1}^{k} \rho_i \vec{v}_i}{\rho} = \sum_{i=1}^{k} w_i \vec{v}_i \tag{12.27}$$

On a *molar basis*, the average velocity of the mixture is:

$$\vec{V} = \frac{\sum_{i=1}^{k} C_i \vec{v}_i}{C} = \sum_{i=1}^{k} y_i \vec{v}_i \tag{12.28}$$

Each species, for example A, has its velocity expressed as:

$$\vec{v}_A = u_A \vec{i} + v_A \vec{j} + w_A \vec{k} \tag{12.29}$$

The diffusion velocity (relative velocity), on a mass basis, of species i concerning the mixture is defined as:

$$\vec{v}_i - \vec{v} \tag{12.30}$$

Thus, the diffusive flux, on a mass basis, will be:

$$\vec{j}_A = \rho_A (\vec{v}_i - \vec{v}) \tag{12.31}$$

The diffusive flux from Eq. (12.31) is the same as that defined in Fick's Law (Eq. 12.1), which, expanded to the other coordinate axes, is:

$$\vec{j}_A = -D_{AB} \nabla \rho_A \tag{12.32}$$

Considering an inertial reference, the absolute mass flow of species A is defined by:

$$\vec{n}_A = \rho_A \vec{v}_A \tag{12.33}$$

Other formulations of the mass flow are:

$$\vec{n}_A = \rho_A (\vec{v}_A - \vec{v}) + \rho_A \vec{v} = \vec{j}_A + \rho_A \left(\frac{\vec{n}_A + \vec{n}_B}{\rho} \right)$$

$$\vec{n}_A = \vec{j}_A + \rho_A \vec{v} \vec{n}_A = \vec{j}_A + w_A (\vec{n}_A + \vec{n}_B)$$

$$\vec{n}_A = -\rho D_{AB} \frac{dw_A}{dz} + w_A (\vec{n}_A + \vec{n}_B) \tag{12.34}$$

And in molar basis:

$$\text{Fick's Law}: \quad \vec{J}_A = C_A \left(\vec{v}_i - \vec{V} \right) = -D_{AB} \nabla C_A \tag{12.35}$$

$$\text{Absolute flux}: \vec{N}_A = C_A \vec{v}_i \tag{12.36}$$

$$\vec{N}_A = \vec{J}_A + C_A \vec{v}$$

$$\vec{N}_A = -CD_{AB}\frac{dy_A}{dz} + y_A\left(\vec{N}_A + \vec{N}_B\right)$$

$$\vec{N}_A = -CD_{AB}\frac{dy_A}{dz} + y_A\left(\vec{N}_A + \vec{N}_B\right) \tag{12.37}$$

The absolute molar and mass fluxes of species A are related as follows:

$$\vec{n}_A = M_A \vec{N}_A \tag{12.38}$$

The absolute mass and molar fluxes of the mixture are:

$$\vec{n} = \rho \vec{v} = \vec{n}_A + \vec{n}_B$$
$$\vec{N} = C\vec{V} = \vec{N}_A + \vec{N}_B \tag{12.39}$$

Once these variables are defined, we proceed to analyze the mass balance in a differential control volume. Figure 12.5 shows a mass balance scheme of species A, with the arrows indicating the mass flows of A entering and leaving the differential control volume on the three coordinate axes. Only the equations of the mass flow on the x-axis are indicated to avoid overloading the figure. For the other axes, replace the coordinate x with the coordinate of interest (y or z) and the corresponding cross-sectional areas for each flow.

It may also be an accumulation of A and a chemical reaction in the control volume, being the species A produced or consumed. These quantities are considered in a more generalized form in Eq. (12.40):

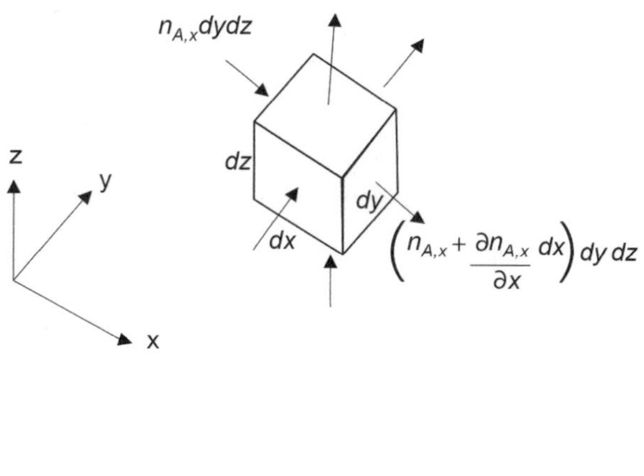

Fig. 12.5 Differential control volume for the mass balance of species A in the x direction only

$$\begin{bmatrix}\text{Rate of}\\ \text{accumulation}\\ \text{of } A\end{bmatrix} = \begin{bmatrix}\text{Entering mass}\\ \text{flowrate of } A\end{bmatrix} - \begin{bmatrix}\text{Exiting mass}\\ \text{flowrate of } A\end{bmatrix} + \begin{bmatrix}\text{Rate of}\\ \text{production or}\\ \text{consumption of } A \text{ by}\\ \text{chemical reaction}\end{bmatrix} \tag{12.40}$$

Developing this equation,

$$\frac{\partial \rho_A}{\partial t}dxdydz = \left(n_{A,x}dydz + n_{A,y}dxdz + n_{A,z}dxdy\right) - \left[\left(n_{A,x} + \frac{\partial n_{A,x}}{\partial x}dx\right)dydz + \left(n_{A,y} + \frac{\partial n_{A,y}}{\partial y}dy\right)dxdz + \left(n_{A,z} + \frac{\partial n_{A,z}}{\partial z}dz\right)dxdy\right] + r_A'''dxdydz$$

and, simplifying it, we get:

$$\frac{\partial \rho_A}{\partial t} + \frac{\partial n_{A,x}}{\partial x} + \frac{\partial n_{A,y}}{\partial y} + \frac{\partial n_{A,z}}{\partial z} = r_A''' \tag{12.41a}$$

which constitutes the *general mass conservation equation of A*. (Compare with the general equation of heat conduction for constant properties, Eq. 2.6). Knowing that the molar flux can be expressed in its vector form, in Cartesian coordinates, as $\vec{n}_A = n_{A,x}\vec{i} + n_{A,y}\vec{j} + n_{A,z}\vec{k}$, Eq. (12.41a) can also be represented in a compact form:

$$\frac{\partial \rho_A}{\partial t} + \nabla \cdot \vec{n}_A = r_A''' \tag{12.41b}$$

where the term $\nabla \cdot \vec{n}_A$ is the *divergence* of $\vec{n}_A$. In the two previous equations, r_A''' [kg/m^{3}s] is the rate of production or consumption of A per unit of volume, which may come from a homogeneous chemical reaction in the control volume.

The divergence of $\vec{n}_A$, $(\nabla \cdot \vec{n}_A)$, in cylindrical and spherical coordinates, is represented, respectively, as:

$$\nabla \cdot \vec{n}_A = \frac{1}{r}\frac{\partial}{\partial r}\left(rn_{A,r}\right) + \frac{1}{r}\frac{\partial n_{A,\theta}}{\partial \theta} + \frac{\partial n_{A,z}}{\partial z} \tag{12.42a}$$

$$\nabla \cdot \vec{n}_A = \frac{1}{r^2}\frac{\partial}{\partial r}\left(r^2 n_{A,r}\right) + \frac{1}{r\sin\theta}\frac{\partial}{\partial \theta}\left(\sin\theta\, n_{A,\theta}\right) + \frac{1}{r\sin\theta}\frac{\partial n_{A,\phi}}{\partial \phi} \tag{12.42b}$$

Note that Eqs. (12.41a), (12.41b), (12.42a) and (12.42b) can also be expressed on a molar basis. Note the similarity of these equations with the heat transfer, Eqs. (2.11) and (2.12). The difference is that, in the heat transfer equations, it refers to a scalar term (temperature, T) and, the Eqs. (12.42a) and (12.42b) refer to a vector magnitude, $\vec{n}_A$.

Some simplifications can be assumed, as is the case of an isothermal and constant pressure process with a negligible average mixture mass velocity ($\vec{v} = 0$) and without chemical reaction. From Eq. (12.34), $\vec{n}_A = \vec{j}_A + \rho_A\vec{v}$, and (12.41b), comes:

$$\frac{\partial \rho_A}{\partial t} + \nabla \cdot \left(\vec{j}_A + \rho_A \vec{v} \right) = r_A''' \rightarrow \frac{\partial \rho_A}{\partial t} + \nabla \cdot (\rho_A \vec{v}) = -\nabla \cdot \vec{j}_A + r_A'''$$

$$\frac{\partial \rho_A}{\partial t} + \nabla \cdot (\rho_A \vec{v}) = -\nabla \cdot (-D_{AB} \nabla \rho_A) + r_A'''$$

As $\vec{v} = 0, T$ and p are constant, therefore D_{AB} and ρ are constants and no chemical reaction, $r_A''' = 0$, thus:

$$\frac{\partial \rho_A}{\partial t} = D_{AB} \nabla^2 \rho_A \tag{12.43a}$$

or, in molar terms,

$$\frac{\partial C_A}{\partial t} = D_{AB} \nabla^2 C_A \tag{12.43b}$$

The Eq. (12.43a) and its variant (Eq. 12.43b) are known as the *Second Law of Fick*. See the similarity with Eq. (2.8).

Example 12.5 (*Mass conservation of A in cylindrical coordinates*)

From the control volume in cylindrical coordinates indicated in the following figure, find the expression for the mass conservation equation of *A* in these coordinates.

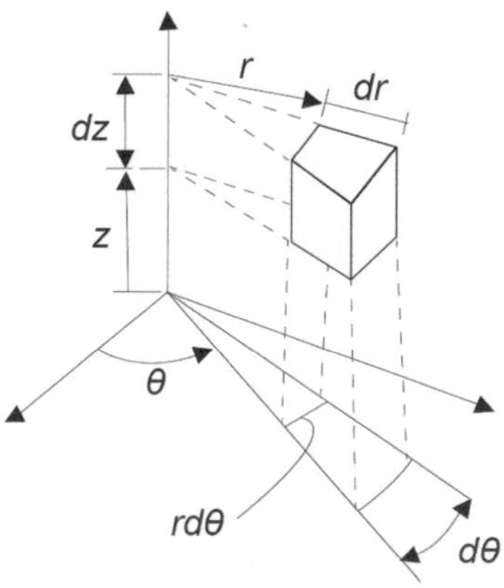

Solution

The cross-sectional areas corresponding to the mass flow in the coordinates (r, θ, z) and the volume of the differential element are:

$$A_r = rd\theta dz, A_\theta = drdz, A_z = rd\theta dr, \forall = rd\theta drdz$$

Substituting into Eq. (12.40), we get:

$$\frac{\partial \rho_A}{\partial t} r dr d\theta dz = \left(n_{A,r} r d\theta dz + n_{A,\theta} dr dz + n_{A,z} r dr d\theta\right) - \left[n_{A,} r d\theta dz + \Delta \frac{\partial(n_{A,r} r d\theta dz)}{\partial r} dr + n_{A,\theta} dr dz + \frac{\partial(n_{A,\theta} dr dz)}{r\partial\theta} r d\theta + n_{A,z} r dr d\theta + \Delta \frac{\partial(n_{A,z} r dr d\theta)}{\partial z} dz\right] + r_A''' r dr d\theta dz$$

Simplifying:

$$\frac{\partial \rho_A}{\partial t} r dr d\theta dz = -\frac{\partial\left(r n_{A,r}\right)}{\partial r} dr d\theta dz - \frac{\partial n_{A,\theta}}{\partial \theta} dr d\theta dz - \frac{\partial n_{A,z}}{\partial z} r dr d\theta dz + r_A''' r dr d\theta dz$$

Dividing by the differential volume, $r dr d\theta dz$, we obtain:

$$\frac{\partial \rho_A}{\partial t} + \frac{1}{r}\frac{\partial}{\partial r}\left(r n_{A,r}\right) + \frac{1}{r}\frac{\partial n_{A,\theta}}{\partial \theta} + \frac{\partial n_{A,z}}{\partial z} = r_A'''$$

12.2.1 *Boundary Conditions*

Some boundary conditions in mass transfer correspond to gas–liquid, and fluid–solid interfaces, where the fluid can be gas or liquid. The term "vapor" is usually used in situations where the liquid and gaseous phases coexist. In many situations, it is assumed that the interface is at thermodynamic equilibrium and there is no accumulation of matter.

Being at thermodynamic equilibrium, for example, as in Fig. 12.6, means that at the interface (boundary) the gaseous and liquid matter have the same pressure and temperature, which does not mean that the molar fractions of A at each phase at the interface are equal. Assuming that the mixtures are ideal, then, at the interface of each phase, we have:

$$\text{Dalton's Law} \quad - \quad \text{ideal gas}: p_{A,s} = y_{A,s} p \tag{12.44}$$

$$\text{Raoult's Law} \quad - \quad \text{ideal solution}: p_{A,s} = x_{A,s} p_A^{\text{vap}} \tag{12.45}$$

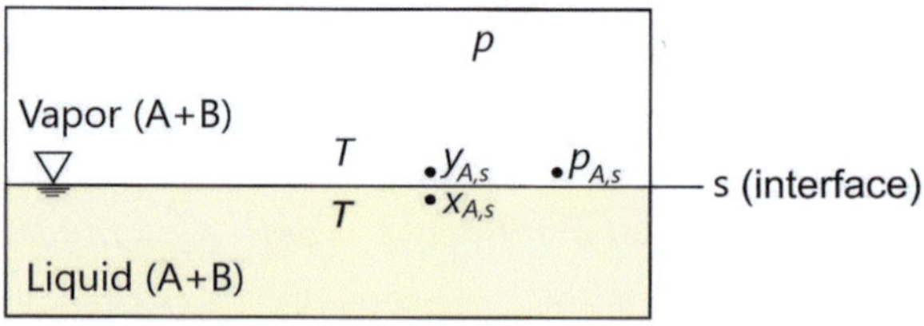

Fig. 12.6 Gas–liquid interface

where $p_{A,s}$ is the partial pressure of A at the gas-side interface, p_A^{vap} is the vapor partial pressure of pure A evaluated at temperature T. Thus, combining the terms, we have:

$$y_{A,s} = \frac{p_A^{vap}}{p} x_{A,s} \tag{12.46}$$

This ideal approximation usually occurs in diluted solutions, considering that, in situations of greater accuracy, one should find the equilibrium relationship experimentally or by applying more realistic thermodynamic equilibrium models. In general, the mass fractions of A at the interface are different. Figure 12.7 shows a scheme of molar fraction behavior in liquid and vapor and at the interface (boundary).

In the case of a pure solid, such as naphthalene, in *sublimation* at low pressures, it can be assumed that (Smith et al. 2005):

$$y_{A,s} = \frac{p_A^{\text{vap}}}{p} \tag{12.47}$$

Another boundary condition can be the knowledge of the mass flow of A at the boundary, which includes walls impermeable to species A ($n_{A,s} = 0$). Chemical reactions (r''_A) can also be specified; they can occur at the border and, in this case, it is treated as known mass (or molar) flow $n_A = r''_A$.

Example 12.6 (*One-dimensional diffusion in cartesian coordinates*)

A test tube with a diameter of 2 cm and a height of 10 cm contains 2 cm of liquid water, as shown in the figure below. Assuming a steady state and that the air inside the tube is initially dry and stagnant at 1 atm, evaluate: (a) the symbolic variation (or distribution) of the molar fraction of water vapor in the dry air; and (b) the molar flow rate and mass flow rate of water at 30 °C. The mass fraction of water at the mouth of the tube is 0.001.

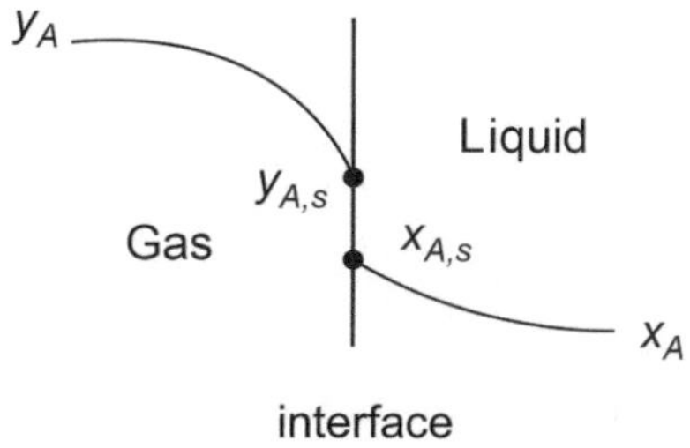

Fig. 12.7 Molar fraction profile in absorption and gas–liquid equilibrium at the interface (boundary)

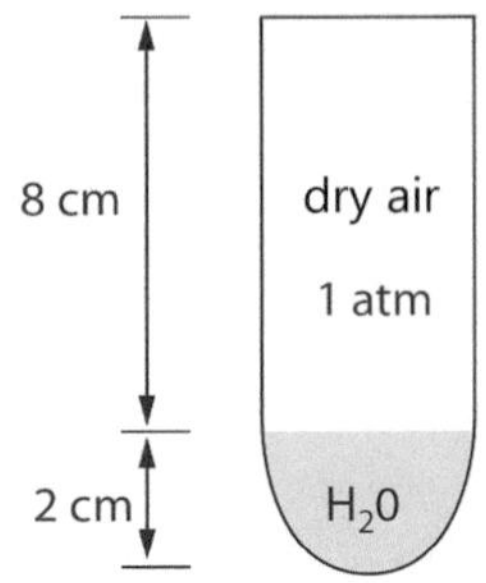

Solution

A diagram of the molar flow is indicated in the figure below.

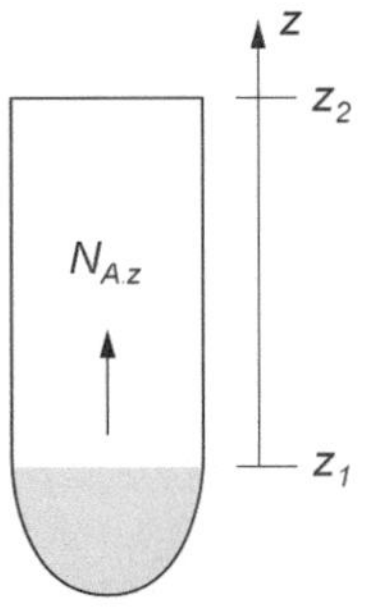

(a) Although the regime is not truly steady since the liquid level is decreasing, for simplicity, we will assume a steady-state condition, where $A =$ water and $B =$ dry air, without any chemical reaction. Additionally, in cylindrical coordinates, it is assumed that mass transfer is one-dimensional in the z-direction, and thus, there will be no change in fraction in the radial or angular directions. The following equations apply only to the gaseous phase.

From Eq. (12.37) for the z-axis: $N_{A,z} = -CD_{AB}\frac{dy_A}{dz} + y_A\left(N_{A,z} + N_{B,z}\right)$, as $N_{B,z} = 0$, so:

$N_{A,z} = -\frac{CD_{AB}}{1-y_A}\frac{dy_A}{dz}$, C and D_{AB} are constants because T and p are constants. From the species conservation balance equation, Eqs. (12.41a) and (12.41b) in steady state, cylindrical coordinates and in molar terms:

$\frac{\partial C_A}{\partial t} + \left[\frac{1}{r}\frac{\partial}{\partial r}\left(rN_{A,r}\right) + \frac{1}{r}\frac{\partial N_{A,\theta}}{\partial \theta} + \frac{\partial N_{A,z}}{\partial z}\right] = R_A''' \rightarrow \frac{dN_{A,z}}{dz} = 0$, thus:

$\frac{dN_{A,z}}{dz} = -CD_{AB}\frac{d}{dz}\left(\frac{1}{1-y_A}\frac{dy_A}{dz}\right) = 0$, integrating indefinitely twice:

$$\int d\left(\frac{1}{1-y_A}\frac{dy_A}{dz}\right) = 0\int dz \rightarrow \frac{1}{1-y_A}\frac{dy_A}{dz} = k_1 \rightarrow \int \frac{dy_A}{1-y_A} = k_1\int dz$$

$$\ln(1-y_A) = -k_1 z + k_2$$

The problem has two boundary conditions (BC):
BC_1: at $z = z_1$, $y_A = y_{A1}$.
BC_2: at $z = z_2$, $y_A = y_{A2}$.
Substituting these BCs into the equation: $\ln(1 - y_A) = -k_1 z + k_2$;

$$\left.\begin{array}{l}\ln(1 - y_{A1}) = -k_1 z_1 + k_2\\ \ln(1 - y_{A2}) = -k_1 z_2 + k_2\end{array}\right\} \rightarrow \begin{array}{r} k_1 = \dfrac{\ln\left(\frac{1-y_{A1}}{1-y_{A2}}\right)}{z_2 - z_1}\\ k_2 = \ln(1 - y_{A1})\left(\frac{1-y_{A1}}{1-y_{A2}}\right)^{\frac{z_1}{z_2-z_1}}\end{array}$$

Substituting the constants k_1 and k_2 into: $\ln(1 - y_A) = -k_1 z + k_2$, we obtain:

$$y_A = 1 - (1 - y_{A1})\left(\frac{1 - y_{A2}}{1 - y_{A1}}\right)^{\frac{z-z_1}{z_2-z_1}}$$

(b) The molar flow is obtained from: $N_{A,z} = -\frac{CD_{AB}}{1-y_A}\frac{d}{dz}\left[1 - (1 - y_{A1})\left(\frac{1-y_{A2}}{1-y_{A1}}\right)^{\frac{z-z_1}{z_2-z_1}}\right]$.

Deriving and evaluating at z_1 or at z_2, we obtain: $N_{A,z} = \frac{CD_{AB}}{z_2-z_1}\ln\left(\frac{1-y_{A2}}{1-y_{A1}}\right)$. It can be noted in this equation that the molar flow is constant at z.

At 30 °C (303.2 K), the concentration of the mixture is: $C = \frac{P}{RT} = \frac{1}{82.05 \times 303.2} = 4.02 \times 10^{-5}$ mol/cm^3 ($R = 82.05$ cm^{3}atm/molK universal ideal gas constant).

$D_{AB} = 0.292$ cm^2/s a 313 K (Table 12.4), correcting for another temperature (Eq. 12.20)

$$D_{AB} = 0.292\left(\frac{303.2}{313}\right)^{1.75}\left(\frac{1}{1}\right) = 0.276\ \text{cm}^2/\text{s}$$

In the interface, $x_{A1} = 1$, $p_A^{vap} = 0.042$ atm (Table A.7 at 30 °C). Thus,

$$y_{A1} = \frac{p_A^{vap}}{p} x_{A1} = \frac{0.042}{1}(1) = 0.042$$

$$N_{A,z} = \frac{CD_{AB}}{z_2 - z_1}\ln\left(\frac{1 - y_{A2}}{1 - y_{A1}}\right) = \frac{4.02 \times 10^{-5}(0.276)}{8 - 0}\ln\left(\frac{1 - 0.001}{1 - 0.042}\right)$$

$$N_{A,z} = 5.81 \times 10^{-8}\ \text{mol/cm}^2\text{s}$$

$$n_{A,z} = N_{A,z} M_A = 5.81 \times 10^{-8}(18) = 1.05 \times 10^{-6}\ \text{g/cm}^2\text{s}$$

$$\dot{m}_{A,z} = n_{A,z} A = 1.05 \times 10^{-6}\left(\frac{\pi 2^2}{4}\right)$$

$$\dot{m}_{A,z} = 3.3 \times 10^{-6} \quad \text{g/s}$$

Example 12.7 (*Pseudo-steady diffusion in a sphere*)

A 1 cm diameter naphthalene sphere is located in stagnant air at 30 °C and 0.82 atm, as illustrated in the figure below. After 14 days, its measured diameter was 0.72 cm. Knowing that its molar mass is 128.2 g/mol, its specific mass is 1.14 g/cm^3 (solid), the molar fraction of naphthalene in the distance is negligible, and its sublimation pressure can be evaluated by $\log p = 10.56 - 3472/T$ with T (K) and p (mmHg), evaluate the diffusivity of naphthalene in the air assuming pseudo-steady diffusion.

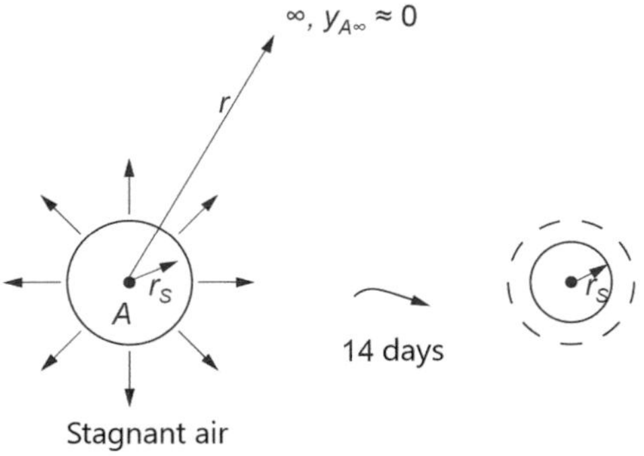

Solution

Pseudo-steady diffusion considers a moving interface while keeping other variables in a steady state. The volume of the naphthalene is $\forall = 4\pi r_s^3/3$, where r_s is the outer radius of the sphere, which decreases over time. The molar flow of naphthalene being sublimated in the sphere is $W_{A,r} = -\frac{\rho_A}{M_A}\frac{d\forall}{dt}$. The negative signal is used to obtain a positive molar flow, given that the volume decreases over time: $W_{A,r} = -4\pi r_s^2 \frac{\rho_A}{M_A}\frac{dr_s}{dt}$.

On the other hand, on the gaseous side, the air (B) remains stagnant, and thus the molar flow at radius r will be: $N_{A,r} = -\frac{CD_{AB}}{1-y_A}\frac{dy_A}{dr}$. From the conservation of species in spherical coordinates and assuming that there are no flows in the angular directions, there is only flow of A in r-direction, without chemical reaction and in steady state: $\frac{\partial C_A}{\partial t} + \left[\frac{1}{r^2}\frac{\partial}{\partial r}\left(r^2 N_{A,r}\right) + \frac{1}{r\sin\theta}\frac{\partial}{\partial\theta}\left(\sin\theta N_{A,\theta}\right) + \frac{1}{r\sin\theta}\frac{\partial N_{A,\phi}}{\partial\phi}\right] = R_A'''$, is obtained $\frac{d}{dr}\left(r^2 N_{A,r}\right) = 0$, then, the molar flow rate will be:

$W_{A,r} = N_{A,r}A = N_{A,r}\left(4\pi r^2\right) \equiv$ cte, therefore, replacing the molar flow,

$W_{A,r} = -\left(\frac{4\pi r^2 CD_{AB}}{1-y_A}\right)\frac{dy_A}{dr} \rightarrow \frac{W_{A,r}}{4\pi CD_{AB}}\int\limits_{r_s}^{\infty}\frac{dr}{r^2} = \int\limits_{y_{As}}^{y_{A\infty}}\frac{dy_A}{y_A-1}$. Integrating twice, recalling that C and D_{AB} are constants because the temperature and pressure are constant, we have: $W_{A,r} = 4\pi r_s CD_{AB}\ln\left(\frac{1-y_{A\infty}}{1-y_{As}}\right)$. Equating the molar flow terms at radius r_s:

$$-4\pi r_s^2 \frac{\rho_A}{M_A}\frac{dr_s}{dt} = 4\pi r_s C D_{AB} \ln\left(\frac{1-y_{A\infty}}{1-y_{As}}\right)$$

$$-\int_{r_{s0}}^{r_s} r_s dr_s = \frac{C D_{AB} M_A}{\rho_A} \ln\left(\frac{1-y_{A\infty}}{1-y_{As}}\right)$$

$$\frac{r_{s0}^2 - r_s^2}{2} = \frac{C D_{AB} M_A}{\rho_A} \ln\left(\frac{1-y_{A\infty}}{1-y_{As}}\right)(t - t_0)$$

$$C = \frac{p}{RT} = \frac{0.72}{82.05(303.2)} = 2.90 \times 10^{-5}$$

$$\log p_A^{vap} = 10.56 - 3472/T = 10.56 - 3472/303.2 \rightarrow p_A^{vap} = 0.1285 \text{ mmHg}.$$

$$y_{As} = \frac{0.1285}{0.82(760)} = 2.1 \times 10^{-4}, \quad y_{A\infty} = 0$$

$$\frac{0.5^2 - 0.36^2}{2} = \frac{2.90 \times 10^{-5} D_{AB} 128.2}{1.14} \ln\left(\frac{1-0}{1-2.1 \times 10^{-4}}\right)(14(3600) - 0)$$

$$D_{AB} = 0.073 \text{ cm}^2/\text{s}$$

12.3 Transient Diffusion

In some situations, diffusion occurs in a transient regime, such as in the case of clothing or material drying, where water mass diffuses from the interior of the solid to its surface and is then transferred to the environment air. Figure 12.8 illustrates two scenarios: (a) where the process is assumed to occur in a steady state, and (b) in a transient state (note that the variation of concentrations over time may exhibit different behavior than what is indicated in this diagram). In the first case, there will be a variation of concentration in space, and in the second, a variation of concentration in both, space and time.

Assuming that the medium (solvent) is stagnant, $N_{B,z} = 0$, the molar flux of component A in the z-direction wil be $N_{A,z} = -\frac{D_{AB}}{1-y_A}\frac{\partial C_A}{\partial z}$. Furthermore, assuming solute A is diluted, this relationship can be expressed in terms of the effective diffusivity (D_{ef}), as:

$$N_{A,z} = -D_{ef}\frac{\partial C_A}{\partial z} \tag{12.48}$$

According to the conservation equations of species A, in situations where the temperature and pressure are constant, the mixture velocity negligible (stagnant

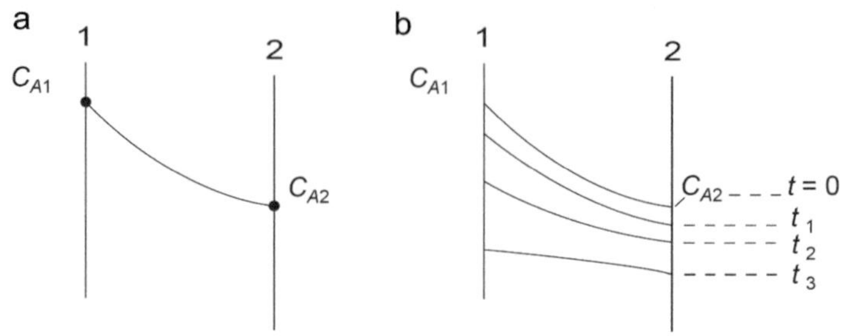

Fig. 12.8 Mass transfer in **a** steady state and **b** transient regime

Fig. 12.9 Molar concentration profile in transient drying regime

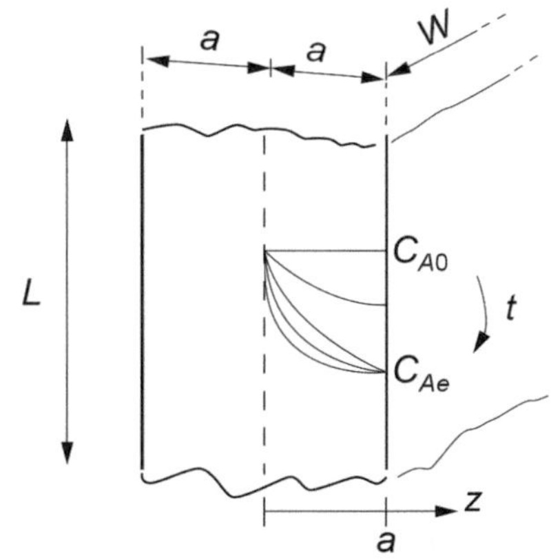

medium and diluted solute) and without chemical reaction, Fick's Second Law is obtained (Eqs. 12.43a and 12.43b):

$$\frac{\partial C_A}{\partial t} = D_{ef} \nabla^2 C_A \tag{12.49}$$

This equation can be analytically solved for specific situations involving elementary geometries, such as the case of a flat plate with one-dimensional diffusion, which will be the subject next.

Figure 12.9 shows a flat plate of thickness $2a$, where the initial molar concentration of A is C_{A0}. Over time, this concentration decreases. On the external surface wall, the concentration decreases until it reaches the equilibrium condition (C_{Ae}) with the adjacent external medium, such as air. It is assumed that the time required for the external part to reach equilibrium is negligible. Note that the behavior is symmetrical with respect to the central plane.

Assuming one-dimensional mass transfer in the z-direction with $L \gg 16a$ and $W \gg 16a$. The presented problem has an initial condition and two boundary conditions,

IC: $t = 0$, $C_A = C_{A0}$ for any z.

BC_1: $t > 0, z = 0 \rightarrow \left.\frac{\partial C_A}{\partial z}\right|_{z=0} = 0$ (symmetry, there is no net mass transfer at $z = 0$).

BC_2: $t > 0, z = a \rightarrow C_A = C_{Ae}$ (the equilibrium condition in $z = a$ occurs quickly, due to the drying process).

To solve Fick's Second Law with the initial and boundary conditions shown, initially the dimensionless concentration $\theta = \theta(z, t)$ is defined as:

$$\theta = \frac{C_A - C_{Ae}}{C_{A0} - C_{Ae}} \tag{12.50}$$

Thus, Fick's Second Law can be expressed in terms of this dimensionless concentration:

$$\frac{\partial \theta}{\partial t} = D_{ef} \frac{\partial^2 \theta}{\partial z^2} \tag{12.51}$$

The new initial and boundary conditions will be:

IC: $t = 0, \theta = 1$ for any z.

BC_1: $t > 0, z = 0 \rightarrow \left.\frac{\partial \theta}{\partial z}\right|_{z=0} = 0.$

BC_2: $t > 0, z = a \rightarrow \theta = 0.$

The solution by separation of variables allows us to obtain the concentration distribution,

$$\theta = \theta(\eta, Fo_m) = 2\sum_{n=0}^{\infty}\left[\frac{(-1)^n}{\gamma_n}\right]\cos(\eta\gamma_n)e^{-\gamma_n^2 Fo_m} \tag{12.52}$$

where $\gamma_n = (2n+1)\pi/2$, $\eta = z/a$, $Fo_m = D_{ef}t/a^2$, which is the mass Fourier number. If the considered time is high, $Fo_m > 0.2$, the series can be truncated in the first expansion term, that is, at $n = 0$; if the time is short, $Fo_m < 0.2$, the series should be truncated at least to the third term (consider in the summation $n = 0$, 1, and 2, at least).

As seen, the concentration (of water, for example) varies with position and time. However, it is possible to define an average concentration of A, representative of the solid, which depends on time only.

$$\bar{C_A} = \frac{\int_0^a C_A dz}{\int_0^a dz} = \frac{\int_0^a \left\{\frac{4(C_{A0}-C_{Ae})}{\pi}\sum_{n=0}^{\infty}\left[\frac{(-1)^n}{\gamma_n}\right]\cos(\eta\gamma_n)e^{-\gamma_n^2 Fo_m} + C_{Ae}\right\}}{\int_0^a dz}$$

By carrying out this integration, we obtain:

$$\bar{\theta} = \frac{\bar{C_A} - C_{Ae}}{C_{A0} - C_{Ae}} = 2\sum_{n=0}^{\infty}\left(\frac{1}{\gamma_n^2}\right)e^{-\gamma_n^2 Fo_m} \tag{12.53}$$

The molar flow, $N_A = N_A(z, t)$, for any position (cross-sectional area to the solute movement) and time, can be obtained from Fick's Law,

$$N_A = -D_{ef}\frac{\partial C_A}{\partial z} = -D_{ef}\frac{\partial}{\partial z}\left\{2(C_{A0} - C_{Ae})\sum_{n=0}^{\infty}\left[\frac{(-1)^n}{\gamma_n}\right]\cos(z\gamma_n/a)e^{-\gamma_n^2 Fo_m} + C_{Ae}\right\}$$

From which we obtain:

$$N_A = \frac{2D_{ef}(C_{A0} - C_{Ae})}{a}\sum_{n=0}^{\infty}(-1)^n \sin(\gamma_n\eta)e^{-\gamma_n^2 Fo_m} \tag{12.54}$$

Example 12.8 (*Transient diffusion—wood drying*)

A square wooden board of 50 cm side and 1 cm thick is at 45 °C and 17% humidity ($w_A = \frac{m_A}{m}$). This board is subjected to drying. Under the conditions of the air around the wood, the equilibrium moisture on the wood surface is 9.5%. How long will it take for its average moisture to be 12%, assuming that the water diffusivity coefficient in wood is 5.1 × 10^{-6} cm^2/s?

Solution

First, it is be necessary to make a change of variables in Eq. (12.53) to humidity w (mass-based) on a dry mass basis ($X_A = \frac{m_A}{m_s}$). It is preferable to work with this variable as the dry mass does not change over time.

$w_A = \frac{m_A}{m_A + m_s} = \frac{X_A m_s}{X_A m_s + m_s} = \frac{X_A}{X_A + 1}$, or $X_A = \frac{w_A}{1 - w_A}$. On the other hand,

$$\bar{\theta} = \frac{\overline{C}_A - C_{Ae}}{C_{A0} - C_{Ae}} = \frac{\frac{\overline{C}_A M_A}{\rho_s} - C_{Ae}\frac{M_A}{\rho_s}}{C_{A0}\frac{M_A}{\rho_s} - C_{Ae}\frac{M_A}{\rho_s}} = \frac{\frac{\overline{\rho}_A}{\rho_s} - \frac{\rho_{Ae}}{\rho_s}}{\frac{\rho_{A0}}{\rho_s} - \frac{\rho_{Ae}}{\rho_s}} \approx \frac{\overline{X}_A - X_{Ae}}{X_{A0} - X_{Ae}}$$

The last term is an approximation, since, although the volumes of the mixture are not equal for each case, they are close. Thus,

$$\overline{X}_A = \frac{\overline{w}_A}{1 - \overline{w}_A} = \frac{0.12}{1 - 0.12} = 0.1364$$

$$X_{Ae} = \frac{w_{Ae}}{1 - w_{Ae}} = \frac{0.095}{1 - 0.095} = 0.1050$$

$$X_{A0} = \frac{w_{A0}}{1 - w_{A0}} = \frac{0.17}{1 - 0.17} = 0.2048$$

Assuming that $Fo_m > 0.2$, $\gamma_1 = [2(0) + 1]\pi/2 = 1.57$.

$\frac{0.1364 - 0.1050}{0.2048 - 0.1050} = 2\left(\frac{1}{1.57^2}\right)e^{-1.57^2 Fo_m}$, solving this equation, we find that $Fo_m =$ 0.3843, so the time will be $0.3843 = \frac{5.1 \times 10^{-6} t}{0{,}5^2}$.

$t = 18838$ s (5 h 14 min).

The hypothesis of $Fo_m > 0.2$ was correct. If the considered time was shorter, $Fo_m < 0.2$, at least the first three terms of the summation should be considered.

12.4 Mass Transfer Convection

Mass transfer by convection differs from mass transfer by diffusion (or molecular diffusion): in the former, the medium is in natural or forced motion, while in the latter, diffusion primarily occurs in quiescent media. Therefore, in convection, mass transfer occurs due to both the concentration gradient and the macroscopic movement of the medium, whereas in diffusion, it mainly occurs due to the concentration gradient alone.

Similar to heat convection, mass convection can be either natural or forced. Natural convection occurs when the solute causes a sufficient density variation to induce convective movements within the stagnant medium. If the medium's movement is driven by external forces (such as fans or pumps), the convection is said to be forced. Generally, whether natural or forced, the mass transfer of species A by convection is defined as:

$$n_A = k_m(\rho_{Aw} - \rho_{A\infty}) \tag{12.55}$$

where k_m [m/s] is the mass transfer coefficient by convection of species A in a fluid medium (gas or liquid), ρ_{Aw} is the mass concentration of A at the wall (liquid, stagnant or moving, or solid) and $\rho_{A\infty}$ is the mass concentration of A far from this wall. The difference in concentrations drives the mass transfer process, this being transferred from the region of higher concentration to the one of lower concentration. Equation (12.55) can also be expressed on a molar basis.

Mass transfer phenomena by convection and diffusion often act in a combined way in different media. For example, in grains drying process, water diffusion acts inside the grain (solid), while water vapor is transferred by convection to the air at the grain surface.

Mass convection occurs analogously to heat transfer by convection and can be explained by the concept of concentration boundary layer, as illustrated in Fig. 12.10.

Referring to Fig. 12.10, the fluid upstream of the plate has a velocity u_∞ and a mass concentration of A, $\rho_{A\infty}$ (which may or not be zero). Upon reaching the plate, the hydrodynamic and mass boundary layers form—the former due to the viscosity or turbulent action on the fluid and the latter due to concentration difference of A between the fluid far from and at the surface of the plate. The mass boundary layer is the region of the fluid perpendicular to the plate where the mass concentration of A is $\rho_{A\delta}$, such that $\frac{(\rho_{A\delta}-\rho_{Aw})}{(\rho_{A\infty}-\rho_{Aw})} = 0.99$. In this figure, since $\rho_{Aw} > \rho_{A\infty}$, there will be mass transfer of A from the plate to the fluid. The opposite happens if $\rho_{Aw} < \rho_{A\infty}$. The plate represents a solid surface or that of a liquid. The boundary layer starts as laminar, and if the plate is long enough, where $Re_{x,crit} = u_\infty x_{crit}/\upsilon > 5 \times 10^5$ (sometimes, it is also used 3×10^5), the transition occurs and then it becomes a turbulent boundary layer. Obviously, if there is a temperature difference, there will also be the formation of the thermal boundary layer, as studied in Chaps. 5 and 6.

One important difference in the convective analysis of mass transfer compared to heat transfer is related to the average velocity of the fluid next to the plate in the

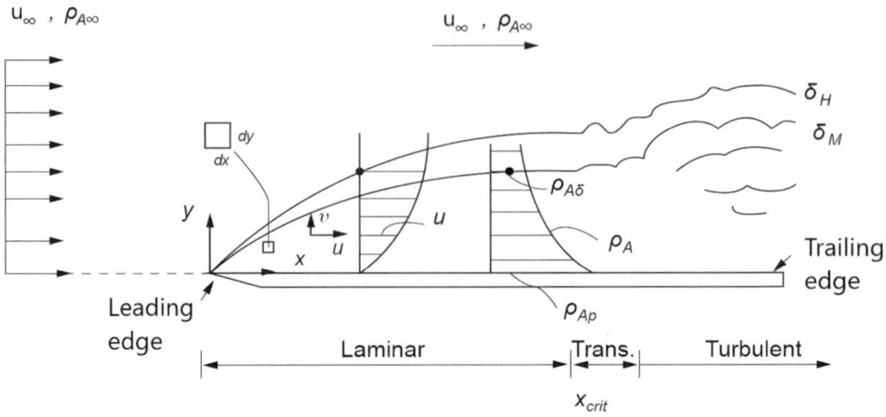

Fig. 12.10 Mass and hydrodynamic boundary layer

y-direction, denoted as v, which is not zero, due to the mass transfer of species A. In mass convection, perhaps the most important parameter is the convective mass transfer coefficient (k_m), which can be obtained through the solution of the mass boundary layer (in the laminar case). The solution for the mass boundary layer in the laminar case assumes steady state, Newtonian and incompressible fluid, negligible gravity effects, no resistance to mass transfer at the interface, no chemical reaction in the medium, two-dimensional flow, constant transport properties and that $\frac{\partial^2 \rho_A}{\partial y^2}$ is much greater than $\frac{\partial^2 \rho_A}{\partial x^2}$. The total mass balance in the differential element (see Fig. 12.10) within the boundary layer under these assumptions provides the following relation:

$$\frac{\partial u}{\partial x} + \frac{\partial v}{\partial y} = 0 \tag{12.56}$$

The mass balance of species A in the differential control volume using Eq. (12.41) allows us to find that:

$$u\frac{\partial \rho_A}{\partial x} + v\frac{\partial \rho_A}{\partial y} = D_{AB}\frac{\partial^2 \rho_A}{\partial y^2} \tag{12.57}$$

As mentioned, the mass transport of A in the y-direction causes the velocity v to be non-zero. However, in situations of low mass transfer, this velocity can be considered very low or negligible ($v \approx 0$). As a result, the boundary conditions to solve Eqs. (12.56) and (12.57) are:

BC $_1$: $y = 0 \rightarrow u = 0$.
BC $_2$: $y = 0 \rightarrow v \approx 0$.
BC $_3$: $y \rightarrow \infty \rightarrow u = u_\infty$.
BC $_4$: $y = 0 \rightarrow \rho_A = \rho_{Aw}$.

BC $_5$: $y \to \infty \to \rho_A = \rho_{A\infty}$.

which have a solution analogous to that of Blasius (see Appendix B.1), in which the concentration gradient at $y = 0$ is:

$$\left.\frac{\partial \rho_A}{\partial y}\right|_{y=0} = \frac{0.332(\rho_{A\infty} - \rho_{Aw})Re_x^{1/2}Sc^{1/3}}{x} \tag{12.58}$$

where Sc is the Schmidt number (analogous to the Prandtl number in heat transfer), defined as:

$$Sc = \frac{\nu}{D_{AB}} \tag{12.59}$$

Where v is the kinematic viscosity. From Eq. (12.34), where $v \approx 0$, the mass flux of A at the wall is:

$$n_{Aw} = -D_{AB}\left.\frac{\partial \rho_A}{\partial y}\right|_{y=0} \tag{12.60}$$

Thus,

$$n_{Aw} = \frac{0.332D_{AB}(\rho_{Aw} - \rho_{A\infty})Re_x^{1/2}Sc^{1/3}}{x} \tag{12.61}$$

Note that the mass flux of A has the form $n_{Aw} = \frac{\text{const}}{x^{2/3}}$, being infinitely large at the beginning and then decreasing, as shown in Fig. 12.11.

By equating the definition of mass transfer by convection, Eq. (12.55), with Eq. (12.61), we obtain the local coefficient of mass transfer by convection from A,

$$k_{mx}(\rho_{Aw} - \rho_{A\infty}) = \frac{0.332D_{AB}(\rho_{Aw} - \rho_{A\infty})Re_x^{1/2}Sc^{1/3}}{x}$$

Fig. 12.11 Local mass flux

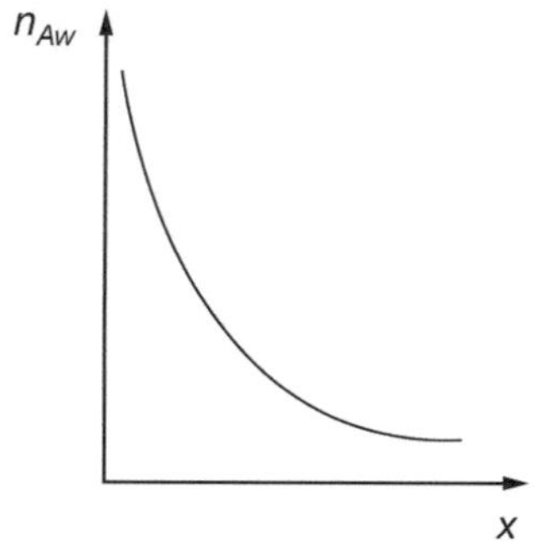

$$k_{mx} = 0.332 D_{AB} \frac{Re_x^{1/2}}{x} Sc^{1/3} \tag{12.62}$$

The local Sherwood number is defined (analogous to the Nusselt number in heat transfer) as:

$$Sh_x = \frac{k_{mx} x}{D_{AB}} \tag{12.63}$$

Thus, in this case,

$$Sh_x = \frac{x}{D_{AB}} 0.332 D_{AB} \frac{Re_x^{1/2}}{x} Sc^{1/3}$$

$$Sh_x = 0.332 Re_x^{1/2} Sc^{1/3} \tag{12.64}$$

One is usually interested in average values for the entire plate length L, so the average convection mass transfer coefficient is $k_m = \frac{1}{L} \int_0^L k_{mx} dx$. Substituting Eq. (12.62) into the previous equation and carrying out the integral, we obtain:

$$\bar{k}_m = 0.664 D_{AB} \frac{Re_L^{1/2}}{L} Sc^{1/3} \tag{12.65}$$

The average Sherwood number for the entire plate of length L is defined as:

$$\overline{Sh_L} = \frac{\bar{k}_m L}{D_{AB}} \tag{12.66}$$

Substituting the mass transfer coefficient from Eq. (12.65), we get:

$$\overline{Sh_L} = 0.664 Re_L^{1/2} Sc^{1/3} \tag{12.67}$$

The total mass flow rate of species A transferred will be:

$$\dot{m}_A = \bar{k}_m A (\rho_{Aw} - \rho_{A\infty}) \tag{12.68}$$

where A is the mass exchange area.

The laminar mass boundary layer thickness is also analogous to heat transfer, being defined as:

$$\delta_M = \frac{4.92x}{Re_x^{1/2} Sc^{1/3}} \tag{12.69}$$

12.5 Chilton-Colburn Analogy (1934)

There are some similarities between the processes of mass and heat transfer. These similarities serve to anticipate mass transfer coefficients by knowing from heat transfer or momentum, in which it should be noted that: there is no mass or thermal energy produced or consumed within the medium; radiant energy exchange is not considered; also, the mass transfer rate is low enough such that the velocity profile is not affected; finally, transport properties are considered constant. It is important to note that, to apply the similarity, the thermal and mass boundary conditions must be analogous. Under these restrictions, the *Chilton-Colburn* analogy is valid, which can be used for gases and liquids with $0.6 < Sc < 2500$, holding valid for both local and average values.

$$j_M = \frac{\overline{k}_m}{u_\infty} Sc^{2/3} = \frac{\overline{C}_{f,L}}{2} \tag{12.70}$$

where j_M is the j factor for mass transfer, which, according to Chilton-Colburn (1934), is equal to the analogy in heat transfer (Eq. 5.54) and can be used for project calculations. In addition, it is valid for various geometries such as: flow parallel to a flat plate (laminar and turbulent), fully developed flow inside tubes, external flow in tubes, and in tube banks.

For example, the local and average coefficients of mass transfer by convection for turbulent flow, in analogy with those of heat transfer, are:

$$Sh_x = 0.0296 Re_x^{0.8} Sc^{1/3} \tag{12.71}$$

and

$$\overline{Sh_L} = 0.0365 Re_L^{0.8} Sc^{1/3} \tag{12.72}$$

12.6 Mass Transfer Correlations

Experimental studies have been developed for specific geometries, such as the case of internal flow. This correlation is similar to that of Dittus-Boelter and was developed by Linton and Sherwood (1950) for turbulent internal flow in a circular tube of internal diameter, d_i, which is presented in Eq. (12.73) and valid for dilute liquid solutions with $2000 < Re_D < 35,000$ and $1000 < Sc < 2260$.

$$\overline{Sh_D} = \frac{\overline{k}_m d_i}{D_{AB}} = 0.023 Re_D^{0.83} Sc^{1/3} \tag{12.73}$$

In the case of an external surface to a cylinder (with an external diameter d_o) sublimating, Bedingfield and Drew (1950) proposed the following correlation valid for $400 < Re_D < 25,000$ and $0.6 < Sc < 2.6$.

$$\overline{Sh_D} = \frac{\overline{k}_m d_o}{D_{AB}} = 0.281 Re_D^{0.6} Sc^{0.44} \tag{12.74}$$

For isolated spheres (not grouped) of external diameter d_o, the Sherwood number for gaseous streams in the range $2 < Re_D < 800$ and $0.6 < Sc < 2.7$ was proposed by Fröessling (1938):

$$\overline{Sh_D} = \frac{\overline{k_m} d_o}{D_{AB}} = 2 + 0.552 Re_D^{1/2} Sc^{1/3} \tag{12.75}$$

For liquid streams, the Sherwood number given by Garner and Suckling (1958) is recommended:

$$\overline{Sh_D} = \frac{\bar{k}_m\, d_o}{D_{AB}} = 2 + 0.95 Re_D^{0.5} Sc^{0.33} \tag{12.76}$$

valid for $100 < Re_D < 700$ and $100 < Re_D^{0.5} Sc^{0.33} < 300$.

Example 12.9 (*Mass convection on a flat plate*)

An air stream at 30 °C, 95 kPa, velocity of 1.5 m/s and relative humidity $\phi = 60\%$ ($\phi = \frac{p_{A\infty}}{p_A^{vap}}$) flows over a flat plate 50 cm long and 20 cm wide covered by a thin layer of water. Evaluate the thickness of the mass boundary layer at the end of the plate, the average Sherwood number, the average mass transfer coefficient, and the total mass flow rate of water vapor transferred to the air. Assume that the molar mass of dry air is 28.85 g/mol and the average kinematic viscosity is 0.167 cm^2/s.

Solution

The vapor or saturation pressure at 30 °C is 4.2 kPa (Table A.7).

$y_A^{vap} = \frac{p_A^{vap}}{p} = \frac{4.2}{95} = 0.0442$, this will also be the molar fraction of water vapor in the gas phase at the liquid–gas interface.

The relative humidity (ϕ) can be expressed in terms of molar fractions:

$$\phi = \frac{p_{A\infty}}{p_A^{vap}} = \frac{p_{A\infty}/p}{p_A^{vap}/p} = \frac{y_{A\infty}}{y_A^{vap}}$$

$$y_{A\infty} = \phi y_A^{vap} = 0.6 \times 0.0442 = 0.0265$$

The viscosity depends on the concentration of the components and, as the concentration of water varies from the wall to the free flow, its average value is usually considered. In this example, the kinematic viscosity is given; thus,

$$Re_L = \frac{u_\infty L}{\nu} = \frac{150 \times 50}{0.167} = 44{,}910 \quad \text{(laminar)}$$

The diffusivity of water vapor in dry air is obtained from Table 12.3:

$$D_{AB} = 0.292\left(\frac{303.2}{313}\right)^{1.75}\left(\frac{101.3}{95}\right) = 0.295\ \text{cm}^2/\text{s}$$

$$Sc = \frac{\nu}{D_{AB}} = \frac{0.167}{0.295} = 0.5661$$

Thickness of the MBL at the trailing edge:

$$\delta_M = \frac{4.92x}{Re_x^{1/2} Sc^{1/3}} = \frac{4.92 \times 50}{44{,}910^{1/2} 0.5661^{1/3}} = 1.4\ \text{cm}$$

Average Sherwood number:

$$Sh_L = 0.664 Re_L^{1/2} Sc^{1/3} = 0.664 \times 44{,}910^{1/2} \times 0.5661^{1/3} = 116.4$$

Average mass transfer coefficient:

$$Sh_L = \frac{k_m L}{D_{AB}} \rightarrow k_m = Sh_L \frac{D_{AB}}{L} = 116.4 \frac{0.295}{50} = 0.687\ \text{cm/s}$$

Mass flow rate of water vapor transferred to the air:
It is necessary to know the concentration of water at the interface and far away.

$$w_A = y_A \frac{M_A}{M},\ \rho = \frac{pM}{RT} \rightarrow \rho_A = w_A \rho = y_A \frac{M_A}{M}\frac{pM}{RT} \rightarrow \rho_A = y_A \frac{pM_A}{RT}$$

$$\rho_{Aw} = 0.0442 \frac{95(18)}{8.314(303.2)} = 0.030\ \text{kg/m}^3$$

$$\rho_{A\infty} = 0.0265 \frac{95(18)}{8.314(303.2)} = 0.018\ \text{kg/m}^3$$

$$\dot{m}_A = k_m A(\rho_{Aw} - \rho_{A\infty}) = 0.00687(0.5 \times 0.2)(0.03 - 0.018)$$

$$\dot{m}_A = 0.0082\ \text{g/s}$$

12.7 Mass Transfer Inside Tubes

Species A in a tube has a constant concentration at the wall, ρ_{Aw}, and is subjected to a mass transfer process. At the tube entrance, the concentration of A is ρ_{Ai} and at the exit is ρ_{Ao}, as shown in Fig. 12.12. In case that $\rho_{Ai} < \rho_{Ao} < \rho_{Aw}$, there will be mass transfer from A from the tube wall to the fluid. The analysis for mass transfer is analogous to that of heat transfer.

By conservation of mass of A, the rate of mass transferred to the fluid is:

$$\dot{m}_{Aw} = \dot{m}_{Ao} - \dot{m}_{Ai} \tag{12.77}$$

The transferred mass rate is also equal to:

$$\dot{m}_{Aw} = \overline{k}_m A_w LMCD \tag{12.78}$$

where A_w is the mass transfer area. Let us call the potential that promotes mass transfer logarithmic mean concentration difference ($LMCD$) in analogy to $LMTD$ (see Sect. 6.3). Based on Fig. 12.12, we can define the $LMCD$ as:

$$LMCD = \frac{\Delta\rho_{Ai} - \Delta\rho_{Ao}}{\ln\frac{\Delta\rho_{Ai}}{\Delta\rho_{Ao}}} \tag{12.79}$$

Example 12.10 (*Mass convection inside a tube*)

Air is humidified inside a 6 cm diameter tube. A thin layer of water at 60 °C flows at the wall. The air enters the tube with a relative humidity of $\phi = 25$ %, velocity of 7 m/s, temperature and pressure of 60 °C and 1 atm, respectively, and leaves with 75% relative humidity. Assuming a steady state, evaluate the rate of water evaporation and the necessary tube length to achieve $\phi = 75$ %.

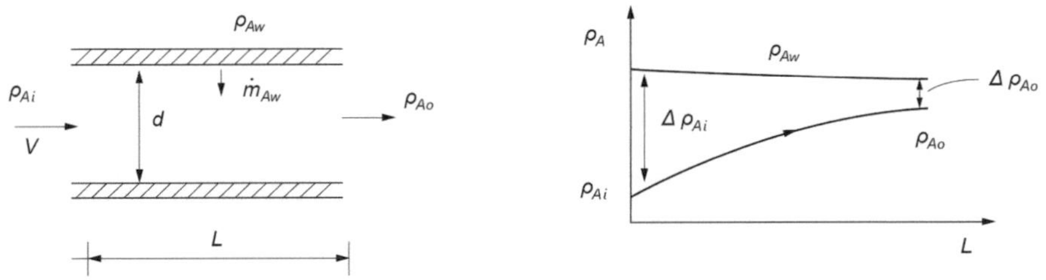

Fig. 12.12 Mass transfer inside a tube with a constant wall concentration

Solution

At 60 °C, the vapor pressure of water is $p_A^{vap} = 19.94$ kPa (Table A.7), therefore the partial vapor pressures at the inlet and outlet will be:

$$p_{Ai} = \phi_i \times p_A^{\text{vap}} = 0.25 \times 19.94 = 5.0 \ \text{kPa}$$

$$p_{Ao} = \phi_o \times p_A^{\text{vap}} = 0.75 \times 19.94 = 15 \ \text{kPa}$$

The concentrations of water vapor at the inlet, outlet, and on the wet wall will be:

$$\rho_{Ai} = \frac{p_{Ai} M_A}{RT} = \frac{25 \times 18}{8.314 \times 333.2} = 0.16245 \ \text{kg/m}^3$$

$$\rho_{Ao} = \frac{p_{Ao} M_A}{RT} = \frac{15 \times 18}{8.314 \times 333.2} = 0.09747 \ \text{kg/m}^3$$

$$\rho_{Aw} = \frac{p_{Aw} M_A}{RT} = \frac{19.94 \times 18}{8.314 \times 333.2} = 0.12956 \ \text{kg/m}^3$$

$$\Delta\rho_{Ai} = \rho_{Aw} - \rho_{Ai} = 0.12956 - 0.03249 = 0.09707 \ \text{kg/m}^3$$

$$\Delta\rho_{Ao} = \rho_{Aw} - \rho_{Ao} = 0.12956 - 0.09738 = 0.03209 \ \text{kg/m}^3$$

The evaporated mass flow rate will be: $\dot{m}_{Aw} = \dot{m}_{Ao} - \dot{m}_{Ai} = VA(\rho_{Ao} - \rho_{Ai})$

$$\dot{m}_{Aw} = 7 \times \pi \frac{0.06^2}{4}(0.09747 - 0.03249) = 1.285 \times 10^{-3} \ \text{kg/s or 4.6 kg/h}$$

$$LMCD = \frac{0.09707 - 0.03209}{\ln\left(\frac{0.09707}{0.03209}\right)} = 0.05871 \ \text{kg/m}^3$$

Assuming the influence of water on air viscosity is negligible, from Table A.7, $\nu = 1.896 \times 10^{-5} \ \text{m}^2/\text{s}$

$$Re_D = \frac{Vd}{\nu} = \frac{7 \times 0.06}{1.896 \times 10^{-5}} = 22,152 \quad \text{(turbulent)}$$

From Table 12.4, at 313 K, $D_{AB} = 2.92 \times 10^{-5} \ \text{m}^2/\text{s}$, correcting by Eq. (12.20): $D_{AB} = 2.92 \times 10^{-5}\left(\frac{333.2}{313}\right)^{1.75}\left(\frac{1}{1}\right) = 3.26 \times 10^{-5} \ \text{m}^2/\text{s}$

$$Sc = \frac{\nu}{D_{AB}} = \frac{1.896 \times 10^{-5}}{3.26 \times 10^{-5}} = 0.5816$$

The Sherwood number is evaluated by Eq. (12.72):

$$\overline{Sh}_D = 0.023 \times 22,152^{0.83} \times 0.5816^{1/3} = 77.6$$

The average mass transfer coefficient by convection:

$$\overline{k}_m = \frac{\overline{Sh}_D D_{AB}}{d_i} = \frac{77.6 \times 3.26 \times 10^{-5}}{0.06} = 0.042 \ \ \mathrm{m/s}$$

The necessary length will be: $\dot{m}_{Aw} = \overline{k}_m \times A_w \times LMCD$

$$1.285 \times 10^{-3} = 0.042 \times \pi \times 0.06L \times 0.05871$$

$$L = 2.8 \ \mathrm{m}$$

Many industrial equipment, such as cooling towers and evaporative condensers, involve the process of mass transfer through the evaporation of water into the atmospheric air. The evaporative process of water into an air stream is very effective in lowering the water temperature and the study of this process is one of the objects of Psychrometry. Specific work on the subject with the theory and information on processes involving the mixture of dry air and water vapor is the reference Simões-Moreira and Hernandez Neto (2019).

12.8 Proposed Problems

12.1 The pressure of a gas is 200 kPa and it has the following molar composition: 60% of N_2, 21% of O_2, 15% of CO_2, 3% of CO, and 1% of H_2O. Evaluate (a) the mass fractions; and (b) the partial pressures of each gas.

12.2 Evaluate the mass diffusion coefficient in gases for the systems at 1 atm of total pressure (a) hydrogen (H_2) in oxygen (O_2) at 55 °C; and (b) benzene in N_2 at 40 °C.

12.3 From experimental values of diffusion coefficients in gases (see Table 12.4) find the diffusion coefficients for (a) SF_6 in N_2 at 80 °C and 2 atm; (b) CO_2 in N_2 at 45 °C and 95 kPa.

12.4 For a gas system with a mass composition of 65% of N_2, 25% of CO_2, and 10% of O_2, evaluate the diffusion coefficient of O_2 in the mixture, if the system is at 1 atm and 30 °C.

12.5 Evaluate the diffusion coefficient of a liquid mixture of CCl_4 diluted in hexane at 1.1 atm and at (a) 15 °C and (b) 50 °C, knowing that the dynamic viscosity of the mixture is 0.3348×10^{-3} kg/ms at 15 °C and 0.2350×10^{-3} kg/ms at 50 °C. Also, consider that the normal liquid volume of CCl_4 is 102 cm^3/mol and that the association factor of hexane is 1.

12.6 Determine the liquid diffusion coefficient of ethanol in water, considering the molar fraction of ethanol of 0.35, and that the system is at 40 °C and 1.2 atm. The thermodynamic factor can be evaluated by $\alpha = 1 - \frac{3.9358 x_A x_B}{0.9609 x_A + 1.4599 x_B}$. The viscosities of pure liquid substances can be obtained from the Vogel adjustment equation: $\ln \mu = A + \frac{B}{C+T}$, with T in K and μ in 10^{-3} Pa s (mPa s); the constants are found in the following table:

Substance	A	B	C
Ethanol	–3.7188	578.919	–137.546
Water	–7.37146	2770.25	74.6787

The dynamic viscosity of the ethanol and water mixture under these conditions is 1.416×10^{-3} Pa s.

12.7 Evaluate the diffusion coefficient of solids at temperatures of 25 °C and 700 °C for: (a) carbon in α-Fe, (b) carbon in γ-Fe, (c) iron in α-Fe and (d) nickel in γ-Fe.

12.8 A nickel cylinder with an external diameter of 10 cm, height of 50 cm, and thickness of 2.1 mm is used to store hydrogen at 85 °C and 280 kPa. Evaluate (a) the rate of hydrogen loss, knowing that its mass diffusivity is 1.2×10^{-8} cm/s and the concentration of hydrogen on the inner wall is $C_A = S p_A$ (see Table A.13 in Appendix A), where S is its solubility; the concentration on the outer wall can be disregarded. (b) What percentage of the mass of hydrogen will be lost in a week?

12.9 Methanol is placed in a 5 cm diameter and 10 cm high glass flask, filling it to a height of 2 cm. The system is in an air environment at 30 °C and 1 atm. Knowing that the mass diffusivity of methanol vapor in air at 25 °C and 1 atm is 0.162 cm²/s, $\rho_A = 0.782$ g/cm³ (liquid) and that the vapor pressure of methanol at 30 °C is 38 kPa, evaluate: (a) the rate of methanol evaporated, assuming steady state and zero concentration at the top of the tube and (b) the time required for all the methanol to be transferred to the environment and, (c) assuming pseudo-stationary regime, evaluate item (b).

12.10 To repel cockroaches from a newly purchased room, solid naphthalene spheres of 15 mm in diameter were placed in various rooms of the house where the air is stagnant at 25 °C and 1 atm. The density of solid naphthalene is 1.16 g/cm³, its molar mass of 128.2 kg/kmol, its sublimation pressure is $\log p = 10.56 - 3472/T$, T in K and p in torr, the mass diffusivity is 0.070 cm²/s and, far away its molar fraction is 3% of the solid-air interface value. Estimate: (a) the time required for its diameter to be reduced to 5 mm; and (b) the mass flow rate of the naphthalene sublimating at the beginning and when its diameter is 5 mm. Assume that the gases behave as ideal.

12.11 In a drying process, a wooden board of 1 cm × 50 cm × 100 cm at 50 °C and 1 atm has an initial moisture content of 20%, its effective mass diffusivity is 5.2×10^{-6} cm²/s and its equilibrium moisture content at the surface is 9.5%. When its average humidity is 15%, evaluate the humidity (a) in the center of the plate; and (b) 0.2 cm from the edge of the plate.

12.12 A circular gel of 5 cm in diameter and 0.4 cm thick contains a solute in its matrix. The gel is immersed in a liquid that also contains this solute at a higher concentration than in the gel. Thus, the solute will diffuse mainly through the flat parts of the gel. The initial concentration of the solute in the gel is 20 mol/(L of gel), the mass diffusivity of the solute in the gel is 3.6×10^{-7} cm^2/s and the equilibrium concentration of the solute on the surface of the gel is 75 mol/(L of gel). Evaluate: (a) the time required for the concentration of the solute on the central surface of the gel (parallel to the faces) to be 50 mol/(L of gel); and (b) at this moment from item (a) determine the average concentration of the solute in the gel.

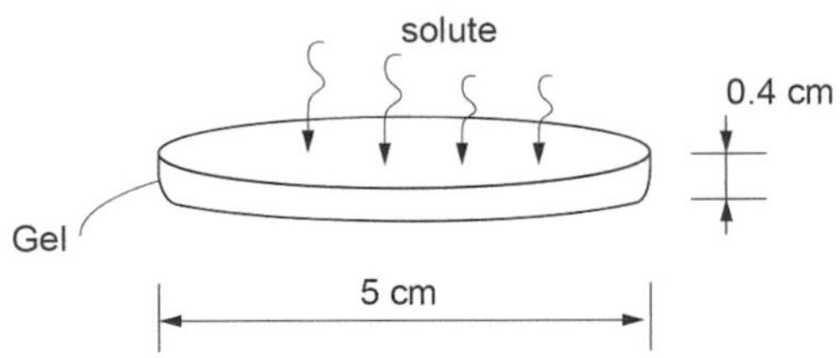

12.13 In a drying experiment, 2 mm diameter rice grains and dry base moisture of 0.30 is put in contact with dry air at 55 °C and 1 atm. Under these conditions, its equilibrium moisture at its surface on a dry basis is 0.038. If after 1 h of drying, its average moisture on a dry basis is 0.15, evaluate the diffusion coefficient of water in the rice. Consider the rice an infinite length cylinder and, under these conditions, the average dimensionless moisture can be evaluated by: $\overline{\theta} = 4\sum_{n=1}^{\infty} \frac{1}{\gamma_n^2} e^{-\gamma_n^2 Fo_m}$, where the first six coefficients γ are: $\gamma_1 = 2.4048$, $\gamma_2 = 5.5201$, $\gamma_3 = 8.6537$, $\gamma_4 = 11.7915$, $\gamma_5 = 14.9309$ and $\gamma_6 = 18.0711$ and $Fo_m = \frac{D_{ef} t}{R^2}$.

12.14 A 1 m long wet plate is at 30 °C. Air at the same temperature, 1 atm and 4 m/s flows over the plate. Assuming the kinematic viscosity ν of pure air (Table A.5), find the Sherwood number and the local mass transfer coefficient at the middle and end of the plate, as well as their average values for the entire plate.

12.15 A sports court of 28 m × 15 m was wetted by rainwater at 35 °C to a thickness of 0.5 mm. A strong and constant wind of 10 m/s at 35 °C and 1 atm blows over the surface in the direction of the 15 m. Assuming that this wind is dry and that there is a turbulence generator at the leading edge, determine the time required to dry the court. The analogy with the heat transfer correlation can be used.

12.16 A naphthalene sphere 19 mm in diameter is exposed to air at 25 °C, 1 atm, and 0.6 m/s. Evaluate the rate of sublimation of the naphthalene, considering that, at a distance, its concentration is negligible. Consider the molar mass of naphthalene to be 128.2 kg/kmol, its pressure of $\log p = 10.56 - 3472/T$, T in K and p in Torr, consider its mass diffusivity to be 0.070 cm^2/s and the kinematic viscosity as that of dry air.

References

Bedingfield, C.H., and T.B. Drew. 1950. Analogy between heat transfer and mass transfer. *Industrial and Engineering Chemistry* 42: 1164–1173.

Brakel, J.V. 1975. Pore space models for transport phenomena in porous media review and evaluation with special emphasis on capillary liquid transport. *Powder Technology* 11: 205–236.

Chilton, T.H., and A.P. Colburn. 1934. Mass transfer (absorption) coefficients prediction from data on heat transfer and fluid friction. *Industrial and Engineering Chemistry* 26 (11): 1183–1187.

Colburn, A.P. 1933. A method of correlating forced convection heat transfer data and a comparison with fluid friction. *Transactions of the American Institute of Chemical Engineers* 29: 174–210.

Epstein, N. 1989. On tortuosity and the tortuosity factor in flow and diffusion through porous media. *Chemical Engineering Science* 44: 777–779.

Fairbanks, D.F., and C.R. Wilke. 1950. Diffusion coefficients in multicomponent gas mixtures. *Industrial and Engineering Chemistry* 42: 471–475.

Foust A.S., L.A. Wenzel, and C.W. Clump et al. 2006. *Operações unitárias*, 2nd ed. México: Compañia Editorial Continental.

Fröessling, N. 1938. The evaporation of falling drops. *Gerlands Beitrage Zur Geophysik* 52: 120.

Fuller, E.N., P.D. Schettler, and J.C. Giddings. 1966. A new method for prediction of binary gas-phase diffusion coefficients. *Industrial and Engineering Chemistry* 58 (5): 18–27.

Garner, F.H., and R.D. Suckling. 1958. Mass transfer from a soluble solid sphere. *AIChE Journal* 4: 114–124.

Leffler, J., and H.T. Cullinan. 1970. Variation of liquid diffusion coefficients with composition. *Industrial and Engineering Chemistry Fundamentals* 9: 84–88.

Linton, W.H., and T.K. Sherwood. 1950. Mass transfer from solid shapes to water in streamline and turbulent flow. *Chemical Engineering Progress* 46: 258.

NR 15. Unhealthy activities and operations, Annex 13A (*Atividades e operações insalubres, anexo 13A, in Portuguese*). Available at: https://www.gov.br/trabalho/pt-br/inspecao/seguranca-e-saude-no-trabalho/normas-regulamentadoras/NR15_Anexo_13A.pdf. Accessed 29th June 2021.

Reid, R.C., J.M. Prausnitz, and B.E. Poling. 1987. *The properties of gases and liquids*, 4th ed. New York: McGraw-Hill.

Simões-Moreira, J.R., and A. Hernandez Neto. 2019. *Fundaments and applications of psychrometry* (*Fundamentos e Aplicações da Psicrometria, in Portuguese*), 2nd ed. São Paulo: Blucher.

Smith, J.M., H.C. van Ness, and M.M. Abbott. 2005. *Introduction to chemical engineering thermodynamics*, 7th ed. Boston: McGraw-Hill Education.

Vlack V. 1964. *Elements of materials science*, 2nd ed. Addison-Wesley Educational Publishers Inc.

Wang, Z., J. Liu, F. Huang, Y. Bi, and S. Zhang. 2020. Hydrogen diffusion and its effect on hydrogen embrittlement in DP steels with different martensite content. *Frontiers in Materials* 7: 620000.

Wilke, C.R., and P. Chang. 1955. Correlation of diffusion coefficients in dilute solutions. *AIChE Journal* 1 (2): 264–270.

Appendix A

Table A.1 Unit conversion

Magnitude	From	To	Multiply by
Length	ft	m	0.3048
	in	cm	2.54
	ft	in	12
	mile	km	1.609
Area	ft^2	m^2	0.929
	in^2	cm^2	6.452
Volume	ft^3	m^3	2.832×10^{-2}
	in^3	cm^3	16.3871
Mass	lbm	kg	0.4536
Speed*	mile/h	km/h	1.609
	ft/s	cm/s	30.48
Density	lbm/ft^3	kg/m^3	16.0186
	lbm/in^3	g/cm^3	27.6802
Force	kgf	N	9.807
	lbf	N	4.448
Energy	cal	J	4.1868
	BTU	kJ	1.055
	ft lbf	J	1.356
	kWh	kJ	3600
Specific energy	cal/g	J/g	4.1868
	kcal/kg	kJ/kg	4.186
	Btu/lbm	kJ/kg	2.3261
	ft lbf/lbm	kJ/kg	2.9894
Power	W	J/s	1
	HP	kW	0.7457
	ft lbf/s	kW	1.356×10^{-3}

(continued)

J. R. Simões-Moreira and E. W. Zavaleta-Aguilar, *Heat Transfer*, Heat and Mass Transfer, https://doi.org/10.1007/978-3-031-80318-5

Table A.1 (continued)

Magnitude	From	To	Multiply by
	Btu/h	kW	2.9307×10^{-4}
Pressure	N/m^2	Pa	1
	atm	kPa	101.325
	bar	kPa	100
	bar	atm	0.98692
	kgf/cm^2	atm	0.96784
	mm Hg	kPa	0.13332
	in Hg	kPa	3.3864
	lbf/in^2	kPa	6.8944
	kgf/cm^2	kPa	98.0665
	mH_2O	kPa	9.8062
Convective heat transf. coefficient	$kcal/hm^2°C$	$kW/m^2°C$	1.163×10^{-3}
	$kcal/hft^2°C$	$kW/m^2°C$	1.252×10^{-2}
	$BTU/hft^2°F$	$kW/m^2°C$	5.6486×10^{-3}
Specific Heat	cal/g°C	J/kg°C	4186.8
	cal/g°F	J/kg°C	7536.2
Thermal conductivity	W/mK	W/m°C	1
	kW/mK	BTU/hfoot°F	577.82
	cal/scm°C	kW/mK	0.4187
Kinematic viscosity**	m^2/s	cSt	1.0×10^6
	cm^2/s	St	1
	m^2/s	$foot^2/s$	10.764
Dynamic viscosity	kg/ms	P (poise)	10
	Pa s	lbm/ft s	2419.1
	$lbf\ s/ft^2$	kg/ms	47.879

*The mass transfer coefficient by convection has the same units
**Thermal and mass diffusivities have the same units

Example 1 To convert the power unit from HP to kW, look for the conversion factor 0.7457 in the table. Then, multiply by this magnitude. Example: 10 HP = 10 × 0.7457 = 7.457 kW.

Example 2 To convert the power unit from kW to HP, look for the conversion factor 0.7457 in the table. Note that, in this case, you should DIVIDE by this magnitude. Example: 10 kW = 10/0.7457 = 13.41 HP.

Table A.2 Properties of solid metals

Material	Melting point temp. (°C)	ρ (kg/m^3)	C_p (J/kgK)	k (W/mK)	α (m^2/s)	C_p (J/kgK) / k (W/mK)				
		25 °C				−100 °C	0 °C	100 °C	200 °C	400 °C
Aluminum	660	2699	901	234.9	9.658×10^{-5}	739.7/238.6	883.2/233.4	928.6/240	979.8/237.8	1076/227.6
Aluminum 1100	660	2659	901	211.8	8.844×10^{-5}	739.7/217.8	883.2/212.4	928.6/210	979.8/206.6	1076/195.8
Aluminum 6063	616	2900	900	200.8	7.695×10^{-5}					
Carbon steel	1410	7849	462	65.58	1.809×10^{-5}	/69.87	/69.22	513.5/51.71	568.1/39.01	663.5/24.69
Carbon steel AISI 1010	1410	7849	434	47.47	1.394×10^{-5}	/45.9	/47.2	467/47.8	506.6/46.4	583.4/40.8
Stainless steel AISI 304	1399	7997	476	14.86	3.906×10^{-6}	367.1/11.69	456.9/14.28	504.8/16.14	530.4/17.77	566.1/20.82
Stainless steel AISI 316	1385	8025	489	13.41	3.417×10^{-6}	384.1/10.7	474/12.94	504.8/14.71	524.7/16.33	559.5/19.47
Beryllium	1287	1850	1814	201.3	5.998×10^{-5}	869.4/458.4	1645/224.4	2096/170.7	2342/147.6	2684/118.7
Bismuth	272	9778	122	7.926	6.646×10^{-6}	117/12.2	121.5/8.351	122.7/7.242		
Cadmium	321	8646	231	96.84	4.852×10^{-5}	214.8/100.3	228.9/97.41	239/95.26		
Lead	328	11,335	129	35.33	2.417×10^{-5}	123.1/37.51	128.1/35.68	131.2/34.35	135.7/33.05	
Cobalt	1495	8898	425	100.4	2.653×10^{-5}	337.9/132.1	415.3/105.4	444.3/89.03	472.2/78.42	521.4/63.71

(continued)

Table A.2 (continued)

Material	Melting point temp. (°C)	ρ (kg/m^3)	C_p (J/kgK)	k (W/mK)	α (m^2/s)	C_p (J/kgK) / k (W/mK)				
		25 °C				−100 °C	0 °C	100 °C	200 °C	400 °C
Copper	1085	8958	389	396.5	1.137×10^{-4}	340.3/413	385.4/398.8	368/392.8	374.4/389.1	400.2 / 378.6
Cu70%-Ni30%	1200	8947	377	29.75	8.822×10^{-6}	/26	/29	/32	/35.5	/49.02
Cu90%-Ni10%	1117	8897	377	51.76	1.543×10^{-5}	/39.02	/49.62	/57.64	/69.27	/88
Chromium	1907	7189	449	93.85	2.907×10^{-5}	330.4/123.6	436/96.66	487.3/91.08	524.4/87.17	581.8/77.26
Tin	232	7362	227	68.29	4.090×10^{-5}	208/76.52	223.8/69.57	238.7/63.81	/60.3	
Iron	1538	7873	441	80.52	2.317×10^{-5}	352.4/99.37	435.4/83.51	476.6/72.09	517.7/63.47	602.9/50.31
Germanium	938	5350	322	60.47	3.507×10^{-5}	260.8/132.7	313.3/67.79	337.6/46.65	350.7/37.15	368.8/24.44
Inconel 740H		8060	450	10.24	2.825×10^{-6}			476/11.7	489/13	503/15.7
Iridium	2466	22,558	130	147.1	5.015×10^{-5}	115.6/156.9	129.2/148	132.5/144.7	135.2/141.8	140.2/135.8
Magnesium	650	1737	1022	156.1	8.787×10^{-5}	857.5/161.7	1000/156.8	1061/153.8	1073/151.5	1142/147.9
Molybdenum	2623	10,279	251	138.1	5.363×10^{-5}	201.7/152.7	243.8/139.3	258.3/135.1	266.1/131.1	278.7/123.1

(continued)

Table A.2 (continued)

Material	Melting point temp. (°C)	ρ (kg/m^3)	C_p (J/kgK)	k (W/mK)	α (m^2/s)	C_p (J/kgK) / k (W/mK)				
		25 °C				−100 °C	0 °C	100 °C	200 °C	400 °C
Nichrome	1400	8400	420	12.00	3.401×10^{-6}			463.9/13.46	496.5/14.73	532.3/17.83
Niobium	2477	8569	265	53.68	2.367×10^{-5}	232.6/53.3	260.7/53.4	271.6/54.8	277.3/56.3	286.3/59.33
Nickel	1455	8901	443	91.00	2.309×10^{-5}	342.5/120.7	427.6/95.08	474/83.02	496.3/74.86	521.1/66.33
Gold	1064	19,288	129	315.2	1.264×10^{-4}	121.4/331.3	127.7/318	132.4/312.5	135.6/309.8	147.2/299.6
Palladium	1555	12,021	245	72.0	2.449×10^{-5}	233.2/73.34	243.3/72	255.6/73.46	264.7/76.19	274.7/82.56
Platinum	1768	21,447	133	73.03	2.563×10^{-5}	118.3/75.93	130.9/73.48	135.2/72.41	137.8/72.13	142.8/72.33
Silver	962	10,487	235	429.0	1.742×10^{-4}	214.8/435.4	232.3/429.3	237.9/426.1	243/420.2	254.4/406.1
Rhenium	3186	21,018	136	47.96	1.680×10^{-5}	118.9/53.12	133.6/48.73	138.2/46.58	141.2/45.41	147.2/44.16
Rhodium	1964	12,450	243	150.1	4.969×10^{-5}	200.4/162.6	236.8/151.1	250.3/147.1	260.7/142.3	280.9/132.7
Silicon	1414	2329	709	149.6	9.058×10^{-5}	476/341.9	670.1/171.1	769.1/112.1	818.5/82.29	883.8/53.78
Tantalum	3017	16,688	140	57.50	2.463×10^{-5}	126.3/57.96	138.1/57.5	142.9/57.72	144.7/58.09	147.1/58.89

(continued)

Table A.2 (continued)

Material	Melting point temp. (°C)	ρ (kg/m^3)	C_p (J/kgK)	k (W/mK)	α (m^2/s)	C_p (J/kgK) / k (W/mK)				
		25 °C				−100 °C	0 °C	100 °C	200 °C	400 °C
Titanium	1668	4505	521	21.95	9.351×10^{-6}	420.7/26.11	506.7/22.6	543.2/20.8	565.6/20.03	606.4/19.51
Thorium	1842	11,697	118	54.00	3.916×10^{-5}	108.5/55.54	116.4/54	122.4/54.73	127.7/55	138/56.37
Tungsten	3422	19,249	132	174.4	6.875×10^{-5}	112.6/207.2	129.3/180.2	135.7/163	138.8/151	143.1/132.6
Uranium	1132	19,096	116	27.93	1.262×10^{-5}	104.2/24.46	113.9/26.93	122.6/29.46	132.7/32.19	157/35.83
Vanadium	1890	5999	482	31.0	1.072×10^{-5}	404.5/31.54	471.7/31	504.2/31	523.7/31.73	548.4/34.1
Zinc	420	7137	389	121.1	4.366×10^{-5}	354.7/127.6	384.2/122.3	400/117.3	415/112	
Zirconium	1855	6519	278	22.75	1.256×10^{-5}	248.2/27.35	274.2/23.37	294.1/21.9	308/21.27	329.3/21.03

Properties generated by the Engineering Equation Solver—EES, version V10.833-3D, with permission

Table A.3 Properties of non-metallic solids

Material	Melting point temp. (°C)	ρ (kg/m^3)	C_p (J/kgK)	k (W/mK)	α (m^2/s)	C_p (J/kgK) / k (W/mK)				
		25 °C				−100 °C	0 °C	100 °C	200 °C	400 °C
Boron	2075	2460	1098	27.1	1.00×10^{-5}	473.3 / 91.61	973.9 / 32.81	1369/ 19.34	1620/ 14.53	1990/10.23
Titanium dioxide		4156	710	8.4	2.847×10^{-6}			779.5/ 7.383	832.4/ 6.282	891/4.625
Thorium dioxide		9109	235	13.0	6.073×10^{-6}			249.6/ 10.95	261.9/ 8.883	278/5.905
Polycrystalline aluminum oxide	2052	3970	765	38.67	1.273×10^{-5}	/50.27	/41.63	893/31.22	1002/ 23.45	1136/12.27
Beryllium oxide	2507	3000	1023	273.7	8.919×10^{-5}			1266/216	1474/ 164.9	1754/96
Teflon	321	2192	1028	0.2727	1.211×10^{-7}	645/0.2626	965.7/ 0.272	1123/ 0.3074	/0.3752	

Properties generated by the Engineering Equation Solver—EES, version V10.833-3D, with permission

Table A.4 Properties of common materials at ~25 °C

Material	ρ (kg/m^3)	C_p (J/kgK)	k (W/mK)	α (m^2/s)
High-density particleboard	1000	1300	0.170	1.308×10^{-7}
Low-density particleboard	590	1300	0.078	1.017×10^{-7}
Clay	1515	800	0.027	2.228×10^{-8}
Cement mortar	1860	780	0.720	4.963×10^{-7}
Clay	1460	880	1.300	1.012×10^{-6}
Asphalt	2115	920	0.062	3.186×10^{-8}
High-density plywood	1010	1380	0.150	1.076×10^{-7}
Wood plywood	545	1215	0.120	1.812×10^{-7}
External hard plywood	640	1170	0.094	1.255×10^{-7}
Conventional concrete	2300	880	1.400	6.917×10^{-7}
Acoustic lining	290	1340	0.058	1.493×10^{-7}
Granite	2630	775	2.790	1.369×10^{-6}
Red acer wood	626.5	1652	0.1456	1.407×10^{-7}
Poplar wood	447.9	1652	0.1194	1.614×10^{-7}
White ash wood	699.9	1652	0.1692	1.463×10^{-7}
Oakwood	738.9	1652	0.1792	1.468×10^{-7}
Cypress wood	514.9	1652	0.1294	1.521×10^{-7}
Spruce wood	436.9	1652	0.1145	1.586×10^{-7}
Pinewood	447.9	1652	0.1194	1.614×10^{-7}
Walnut wood	795.9	1652	0.2090	1.590×10^{-7}
Picea wood	469.9	1652	0.1194	1.538×10^{-7}
Marble	2680	830	2.800	1.259×10^{-6}
Paper	930	1400	0.0500	3.840×10^{-8}
Cardboard (12 juxtaposed)*			0.0530	
Plaster and sand plaster	1680	1085	0.2200	1.207×10^{-7}
Regular density coating	290	1300	0.0550	1.459×10^{-7}
Common brick	1920	835	0.72	4.491×10^{-7}

Properties generated by the Engineering Equation Solver—EES, version V10.833-3D, with permission

*Asdrubali et al. Innovative cardboard based panels with recycled materials from the packaging industry: thermal and acoustic performance analysis. *Energy Procedia.* 2015; 78: 321–26

Insulating material	ρ^* (kg/m^3)	Thermal conductivity k (W/mK) at various temperatures (°C)											
		–70 °C	–50 °C	–25 °C	0 °C	25 °C	50 °C	75 °C	100 °C	200 °C	300 °C	400 °C	470 °C
Asbestos duct	190					0.07963	0.08583	0.0931					
Asbestos duct	255					0.07222	0.07663	0.08163					
Asbestos duct	300					0.06922	0.07363	0.07863					
Polystyrene bead duct	16	0.03975	0.02954	0.03409	0.03642	0.03975							
Diatomaceous duct	345										0.09425	0.09961	0.1036
Magnesia duct	185						0.05196	0.05377	0.05589				
Extruded polystyrene duct	35	0.02875	0.023	0.025	0.02621	0.02875							
Extruded polystyrene duct	56	0.02688	0.02246	0.023	0.02521	0.02688							
Silica duct	385										0.1006	0.104	0.114
Calcium silicate duct	190					0.055	0.05596	0.05777	0.05959	0.0688	0.08025	0.09302	0.103
Cellular glass duct	145			0.04963	0.05263	0.05763	0.06367	0.06686	0.07048				
Fiberglass duct lining	32					0.038							
Ceramic fiber blanket (silica-alumina)	128										0.05613	0.07417	0.0895

(continued)

(continued)

Insulating material	ρ^* (kg/m^3)	Thermal conductivity k (W/mK) at various temperatures (°C)											
		–70 °C	–50 °C	–25 °C	0 °C	25 °C	50 °C	75 °C	100 °C	200 °C	300 °C	400 °C	470 °C
Ceramic fiber blanket (silica-alumina)	48										0.08376	0.1171	0.1471
Ceramic fiber blanket (silica-alumina)	64										0.06951	0.09719	0.1225
Ceramic fiber blanket (silica-alumina)	96										0.06101	0.08243	0.09843
Fiberglass blanket	12			0.03554	0.03963	0.04551	0.05378	0.06287					

(continued)

(continued)

Insulating material	ρ^* (kg/m^3)	Thermal conductivity k (W/mK) at various temperatures (°C)											
		–70 °C	–50 °C	–25 °C	0 °C	25 °C	50 °C	75 °C	100 °C	200 °C	300 °C	400 °C	470 °C
Fiberglass blanket	16			0.03409	0.03663	0.04163	0.04983	0.0571					
Fiberglass blanket	24			0.03109	0.03363	0.03863	0.04311	0.04902					
Fiberglass blanket	32			0.02954	0.03221	0.03563	0.04039	0.04494					
Fiberglass blanket	48			0.02809	0.03042	0.03288	0.03739	0.04194					
Loose wood pieces	45					0.03888							
Loose perlite	105	0.05275	0.04063	0.04463	0.04942	0.05275							
Fiberglass board	16					0.046							
Fiberglass board	28					0.038							
Fiberglass board	40					0.035							
Polystyrene	51	0.02488	0.01692	0.01936	0.02202	0.02488							
Molded polystyrene beads	16	0.03975	0.02954	0.03409	0.03642	0.03975							
Extruded polystyrene R12	56	0.02688	0.02246	0.023	0.02521	0.02688							
Polyurethane	50	0.03538	0.0317	0.03349	0.03475	0.03538							
Loose vermiculite	122			0.05963	0.06342	0.06763							

(continued)

(continued)

Insulating material	ρ^* (kg/m^3)	Thermal conductivity k (W/mK) at various temperatures (°C)											
		–70 °C	–50 °C	–25 °C	0 °C	25 °C	50 °C	75 °C	100 °C	200 °C	300 °C	400 °C	470 °C
Loose vermiculite	80			0.05317	0.05863	0.06275							
Refractory brick**									170 °C	300 °C	500 °C	900 °C	1000 °C
Clay	2418										1.6680	1.7874	1.8399
Chromed	3107								2.1641	2.2019	2.2336	2.1335	2.0574
Magnesia	1787								10.007	8.2055	6.2571	4.32	
Silica	1874								1.2835	1.2913	1.3660	1.6192	1.7099
Alumina (99%)	2867								5.0617	4.1684	3.4526	2.5897	2.5276

*Typical density in the temperature range filled to the right. Properties generated by the Engineering Equation Solver—EES, version V10.833-3D, with permission. **Ruth, E. Refractory materials, metallurgical. In: Brook, R. J. Concise Encyclopedia of Advanced Ceramic Materials. Pergamon. 1991; p. 394–402 (evaluated by interpolation)

Table A.5 Properties of dry air at 1 atm pressure

T (°C)	ρ (kg/m^3)	C_p (J/kgK)	k (W/mK)	μ (kg/ms)	ν (m^2/s)	α (m^2/s)	Pr (–)
–150	2.8991	1021	0.01168	8.665×10^{-6}	2.989×10^{-6}	3.948×10^{-6}	0.7571
–100	2.0467	1009	0.01621	1.178×10^{-5}	5.757×10^{-6}	7.850×10^{-6}	0.7334
–50	1.5843	1006	0.02042	1.462×10^{-5}	9.225×10^{-6}	1.281×10^{-5}	0.7200
–40	1.5160	1006	0.02123	1.515×10^{-5}	9.996×10^{-6}	1.392×10^{-5}	0.7179
–30	1.4533	1006	0.02203	1.568×10^{-5}	1.079×10^{-5}	1.507×10^{-5}	0.7160
–20	1.3957	1006	0.02281	1.620×10^{-5}	1.161×10^{-5}	1.626×10^{-5}	0.7142
–10	1.3424	1006	0.02359	1.672×10^{-5}	1.245×10^{-5}	1.748×10^{-5}	0.7124
0	1.2931	1006	0.02436	1.722×10^{-5}	1.332×10^{-5}	1.873×10^{-5}	0.7108
5	1.2697	1006	0.02474	1.747×10^{-5}	1.376×10^{-5}	1.938×10^{-5}	0.7101
10	1.2473	1006	0.02512	1.772×10^{-5}	1.421×10^{-5}	2.003×10^{-5}	0.7093
15	1.2255	1006	0.02550	1.796×10^{-5}	1.466×10^{-5}	2.068×10^{-5}	0.7086
20	1.2046	1006	0.02588	1.821×10^{-5}	1.512×10^{-5}	2.135×10^{-5}	0.7080
25	1.1843	1006	0.02625	1.845×10^{-5}	1.558×10^{-5}	2.203×10^{-5}	0.7073
30	1.1647	1006	0.02662	1.869×10^{-5}	1.605×10^{-5}	2.271×10^{-5}	0.7067
35	1.1458	1007	0.02699	1.893×10^{-5}	1.652×10^{-5}	2.340×10^{-5}	0.7061
40	1.1275	1007	0.02736	1.917×10^{-5}	1.700×10^{-5}	2.410×10^{-5}	0.7055
45	1.1097	1007	0.02772	1.940×10^{-5}	1.749×10^{-5}	2.480×10^{-5}	0.7049
50	1.0925	1007	0.02809	1.964×10^{-5}	1.798×10^{-5}	2.552×10^{-5}	0.7044
60	1.0596	1008	0.02881	2.010×10^{-5}	1.897×10^{-5}	2.697×10^{-5}	0.7034
70	1.0287	1009	0.02952	2.056×10^{-5}	1.999×10^{-5}	2.845×10^{-5}	0.7025
80	0.9995	1009	0.03023	2.101×10^{-5}	2.102×10^{-5}	2.996×10^{-5}	0.7017

(continued)

Table A.5 (continued)

T (°C)	ρ (kg/m^3)	C_p (J/kgK)	k (W/mK)	μ (kg/ms)	ν (m^2/s)	α (m^2/s)	Pr (–)
90	0.9720	1010	0.03093	2.146×10^{-5}	2.208×10^{-5}	3.150×10^{-5}	0.7009
100	0.9459	1011	0.03162	2.190×10^{-5}	2.315×10^{-5}	3.306×10^{-5}	0.7003
120	0.8977	1013	0.03299	2.277×10^{-5}	2.536×10^{-5}	3.627×10^{-5}	0.6992
140	0.8542	1016	0.03434	2.361×10^{-5}	2.764×10^{-5}	3.958×10^{-5}	0.6985
160	0.8147	1019	0.03566	2.444×10^{-5}	3.000×10^{-5}	4.298×10^{-5}	0.6980
180	0.7787	1022	0.03697	2.525×10^{-5}	3.243×10^{-5}	4.647×10^{-5}	0.6979
200	0.7458	1025	0.03825	2.605×10^{-5}	3.493×10^{-5}	5.004×10^{-5}	0.6980
250	0.6745	1034	0.04139	2.797×10^{-5}	4.147×10^{-5}	5.932×10^{-5}	0.6992
300	0.6157	1045	0.04442	2.981×10^{-5}	4.843×10^{-5}	6.904×10^{-5}	0.7014
350	0.5663	1057	0.04737	3.158×10^{-5}	5.578×10^{-5}	7.918×10^{-5}	0.7044
400	0.5242	1069	0.05025	3.329×10^{-5}	6.350×10^{-5}	8.971×10^{-5}	0.7079
450	0.4879	1081	0.05305	3.494×10^{-5}	7.160×10^{-5}	1.006×10^{-4}	0.7116
500	0.4564	1092	0.05580	3.653×10^{-5}	8.005×10^{-5}	1.119×10^{-4}	0.7152
600	0.4041	1115	0.06114	3.960×10^{-5}	9.799×10^{-5}	1.357×10^{-4}	0.7222
700	0.3626	1136	0.06632	4.252×10^{-5}	1.173×10^{-4}	1.610×10^{-4}	0.7283
800	0.3288	1154	0.07136	4.532×10^{-5}	1.378×10^{-4}	1.880×10^{-4}	0.7331
900	0.3008	1170	0.07628	4.802×10^{-5}	1.596×10^{-4}	2.166×10^{-4}	0.7369
1000	0.2772	1185	0.08111	5.064×10^{-5}	1.827×10^{-4}	2.470×10^{-4}	0.7397
1500	0.1990	1235	0.10430	6.285×10^{-5}	3.158×10^{-4}	4.244×10^{-4}	0.7441
2000	0.1553	1265	0.12663	7.421×10^{-5}	4.780×10^{-4}	6.446×10^{-4}	0.7415

Properties generated by the Engineering Equation Solver—EES, version V10.833-3D, with permission

Table A.6 Properties of gases at 1 atm pressure

T (°C)	ρ (kg/m^3)	C_p (J/kgK)	k (W/mK)	μ (kg/ms)	ν (m^2/s)	α (m^2/s)	Pr (–)
Ammonia (NH_3)							
0	0.7715	2179	0.02292	9.193×10^{-6}	1.192×10^{-5}	1.363×10^{-5}	0.8743
20	0.7161	2165	0.02449	9.910×10^{-6}	1.384×10^{-5}	1.580×10^{-5}	0.8758
40	0.6686	2169	0.02636	1.065×10^{-5}	1.592×10^{-5}	1.817×10^{-5}	0.8762
60	0.6273	2186	0.02849	1.140×10^{-5}	1.817×10^{-5}	2.078×10^{-5}	0.8745
80	0.5909	2210	0.03085	1.216×10^{-5}	2.058×10^{-5}	2.363×10^{-5}	0.8707
100	0.5587	2239	0.03344	1.293×10^{-5}	2.314×10^{-5}	2.674×10^{-5}	0.8655
120	0.5299	2271	0.03622	1.370×10^{-5}	2.587×10^{-5}	3.010×10^{-5}	0.8594
140	0.5039	2307	0.03917	1.448×10^{-5}	2.874×10^{-5}	3.369×10^{-5}	0.8531
160	0.4804	2345	0.04225	1.526×10^{-5}	3.177×10^{-5}	3.751×10^{-5}	0.8472
180	0.4590	2385	0.04544	1.604×10^{-5}	3.495×10^{-5}	4.151×10^{-5}	0.8420
200	0.4395	2426	0.04870	1.682×10^{-5}	3.828×10^{-5}	4.568×10^{-5}	0.8380
220	0.4215	2468	0.05199	1.760×10^{-5}	4.175×10^{-5}	4.997×10^{-5}	0.8354
240	0.4050	2511	0.05528	1.837×10^{-5}	4.536×10^{-5}	5.436×10^{-5}	0.8345
260	0.3898	2555	0.05854	1.914×10^{-5}	4.911×10^{-5}	5.878×10^{-5}	0.8354
280	0.3756	2600	0.06172	1.991×10^{-5}	5.300×10^{-5}	6.321×10^{-5}	0.8385
300	0.3624	2645	0.06477	2.067×10^{-5}	5.702×10^{-5}	6.757×10^{-5}	0.8439
Carbon Monoxide (CO)							
–50	1.5317	1043	0.02085	1.392×10^{-5}	9.087×10^{-6}	1.306×10^{-5}	0.6960
0	1.2500	1041	0.02459	1.639×10^{-5}	1.311×10^{-5}	1.890×10^{-5}	0.6937
50	1.0561	1040	0.02807	1.869×10^{-5}	1.769×10^{-5}	2.556×10^{-5}	0.6921

(continued)

Table A.6 (continued)

Carbon Monoxide (CO)							
100	0.9143	1039	0.03131	2.082×10^{-5}	2.277×10^{-5}	3.295×10^{-5}	0.6911
150	0.8062	1039	0.03434	2.281×10^{-5}	2.829×10^{-5}	4.099×10^{-5}	0.6903
200	0.7209	1039	0.03717	2.467×10^{-5}	3.422×10^{-5}	4.961×10^{-5}	0.6897
250	0.6520	1040	0.03985	2.641×10^{-5}	4.051×10^{-5}	5.878×10^{-5}	0.6892
300	0.5951	1040	0.04238	2.806×10^{-5}	4.716×10^{-5}	6.845×10^{-5}	0.6889
350	0.5473	1041	0.04480	2.963×10^{-5}	5.414×10^{-5}	7.861×10^{-5}	0.6887
400	0.5067	1042	0.04713	3.114×10^{-5}	6.146×10^{-5}	8.925×10^{-5}	0.6886
450	0.4716	1043	0.04940	3.260×10^{-5}	6.911×10^{-5}	1.004×10^{-4}	0.6885
500	0.4411	1045	0.05163	3.402×10^{-5}	7.712×10^{-5}	1.120×10^{-4}	0.6885
550	0.4143	1046	0.05385	3.543×10^{-5}	8.551×10^{-5}	1.242×10^{-4}	0.6885
600	0.3906	1048	0.05608	3.684×10^{-5}	9.432×10^{-5}	1.370×10^{-4}	0.6884
650	0.3695	1050	0.05835	3.827×10^{-5}	1.036×10^{-4}	1.505×10^{-4}	0.6883
700	0.3505	1051	0.06069	3.973×10^{-5}	1.134×10^{-4}	1.647×10^{-4}	0.6881
Carbon Dioxide (CO_2)							
–50	2.4357	783	0.01110	1.122×10^{-5}	4.607×10^{-6}	5.821×10^{-6}	0.7914
0	1.9768	827	0.01466	1.371×10^{-5}	6.936×10^{-6}	8.968×10^{-6}	0.7734
50	1.6661	875	0.01867	1.613×10^{-5}	9.683×10^{-6}	1.281×10^{-5}	0.7558
100	1.4407	919	0.02288	1.848×10^{-5}	1.282×10^{-5}	1.727×10^{-5}	0.7425
150	1.2694	960	0.02712	2.073×10^{-5}	1.633×10^{-5}	2.225×10^{-5}	0.7339
200	1.1346	997	0.03131	2.289×10^{-5}	2.018×10^{-5}	2.768×10^{-5}	0.7289
250	1.0258	1031	0.03544	2.497×10^{-5}	2.434×10^{-5}	3.352×10^{-5}	0.7261

(continued)

Table A.6 (continued)

Carbon Dioxide (CO_2)							
300	0.9361	1061	0.03947	2.696×10^{-5}	2.880×10^{-5}	3.974×10^{-5}	0.7248
350	0.8609	1089	0.04341	2.888×10^{-5}	3.354×10^{-5}	4.631×10^{-5}	0.7243
400	0.7968	1114	0.04726	3.072×10^{-5}	3.855×10^{-5}	5.323×10^{-5}	0.7242
450	0.7417	1137	0.05102	3.249×10^{-5}	4.380×10^{-5}	6.049×10^{-5}	0.7242
500	0.6937	1159	0.05470	3.420×10^{-5}	4.930×10^{-5}	6.807×10^{-5}	0.7242
550	0.6515	1178	0.05831	3.584×10^{-5}	5.502×10^{-5}	7.598×10^{-5}	0.7242
600	0.6142	1196	0.06184	3.744×10^{-5}	6.096×10^{-5}	8.420×10^{-5}	0.7239
650	0.5809	1212	0.06530	3.898×10^{-5}	6.711×10^{-5}	9.275×10^{-5}	0.7235
700	0.551	1227	0.06869	4.048×10^{-5}	7.346×10^{-5}	1.016×10^{-4}	0.7229
Helium (He)							
–200	0.6656	5197	0.05986	8.044×10^{-6}	1.208×10^{-5}	1.731×10^{-5}	0.6983
–150	0.3956	5194	0.08480	1.098×10^{-5}	2.775×10^{-5}	4.127×10^{-5}	0.6724
–100	0.2815	5194	0.10690	1.375×10^{-5}	4.885×10^{-5}	7.313×10^{-5}	0.6681
–50	0.2184	5193	0.12719	1.630×10^{-5}	7.462×10^{-5}	1.121×10^{-4}	0.6655
0	0.1785	5193	0.14620	1.869×10^{-5}	1.047×10^{-4}	1.577×10^{-4}	0.6641
50	0.1509	5193	0.16420	2.097×10^{-5}	1.390×10^{-4}	2.096×10^{-4}	0.6632
100	0.1307	5193	0.18141	2.315×10^{-5}	1.772×10^{-4}	2.673×10^{-4}	0.6628
200	0.1031	5193	0.21393	2.729×10^{-5}	2.648×10^{-4}	3.997×10^{-4}	0.6626
300	0.08510	5193	0.24447	3.120×10^{-5}	3.667×10^{-4}	5.533×10^{-4}	0.6627
400	0.07240	5193	0.27348	3.492×10^{-5}	4.820×10^{-4}	7.269×10^{-4}	0.6631
500	0.06310	5193	0.30125	3.849×10^{-5}	6.102×10^{-4}	9.196×10^{-4}	0.6636

(continued)

Table A.6 (continued)

Helium (He)							
600	0.05590	5193	0.32797	4.194×10^{-5}	7.509×10^{-4}	1.131×10^{-3}	0.6641
700	0.05010	5193	0.35380	4.528×10^{-5}	9.035×10^{-4}	1.359×10^{-3}	0.6646
800	0.04540	5193	0.37886	4.853×10^{-5}	1.068×10^{-3}	1.605×10^{-3}	0.6651
900	0.04160	5193	0.40324	5.169×10^{-5}	1.243×10^{-3}	1.868×10^{-3}	0.6657
1000	0.03830	5193	0.42701	5.478×10^{-5}	1.430×10^{-3}	2.146×10^{-3}	0.6662
1100	0.03550	5193	0.45024	5.780×10^{-5}	1.627×10^{-3}	2.441×10^{-3}	0.6667
1200	0.03310	5193	0.47297	6.076×10^{-5}	1.835×10^{-3}	2.751×10^{-3}	0.6671
Hydrogen (H_2)							
–250	1.1372	11,365	0.02047	1.163×10^{-6}	1.023×10^{-6}	1.584×10^{-6}	0.6457
–200	0.3365	10,617	0.05220	3.259×10^{-6}	9.682×10^{-6}	1.461×10^{-5}	0.6628
–150	0.1994	11,896	0.08343	4.793×10^{-6}	2.403×10^{-5}	3.517×10^{-5}	0.6833
–100	0.1418	13,099	0.11634	6.102×10^{-6}	4.303×10^{-5}	6.264×10^{-5}	0.6870
–50	0.1100	13,818	0.14666	7.282×10^{-6}	6.619×10^{-5}	9.647×10^{-5}	0.6861
0	0.08990	14,197	0.17351	8.377×10^{-6}	9.320×10^{-5}	1.360×10^{-4}	0.6855
50	0.07600	14,379	0.19762	9.410×10^{-6}	1.238×10^{-4}	1.809×10^{-4}	0.6847
100	0.06580	14,458	0.21979	1.040×10^{-5}	1.580×10^{-4}	2.310×10^{-4}	0.6839
200	0.05190	14,506	0.26060	1.226×10^{-5}	2.362×10^{-4}	3.461×10^{-4}	0.6823
300	0.04280	14,537	0.29896	1.401×10^{-5}	3.270×10^{-4}	4.800×10^{-4}	0.6813
400	0.03650	14,593	0.33641	1.568×10^{-5}	4.298×10^{-4}	6.318×10^{-4}	0.6802
500	0.03180	14,681	0.37380	1.728×10^{-5}	5.441×10^{-4}	8.015×10^{-4}	0.6788
600	0.02810	14,799	0.41157	1.883×10^{-5}	6.695×10^{-4}	9.886×10^{-4}	0.6771
700	0.02520	14,947	0.44997	2.033×10^{-5}	8.055×10^{-4}	1.193×10^{-3}	0.6753

(continued)

Table A.6 (continued)

Nitrogen (N_2)							
–150	2.8021	1056	0.01157	8.443×10^{-6}	3.013×10^{-6}	3.909×10^{-6}	0.7707
–100	1.9789	1045	0.01603	1.142×10^{-5}	5.773×10^{-6}	7.746×10^{-6}	0.7453
–50	1.5320	1042	0.02015	1.414×10^{-5}	9.228×10^{-6}	1.262×10^{-5}	0.7313
0	1.2504	1041	0.02400	1.663×10^{-5}	1.330×10^{-5}	1.843×10^{-5}	0.7215
50	1.0564	1042	0.02762	1.894×10^{-5}	1.793×10^{-5}	2.510×10^{-5}	0.7144
100	0.91470	1043	0.03104	2.110×10^{-5}	2.307×10^{-5}	3.252×10^{-5}	0.7093
200	0.72120	1053	0.03742	2.507×10^{-5}	3.475×10^{-5}	4.929×10^{-5}	0.7051
300	0.59540	1070	0.04332	2.866×10^{-5}	4.814×10^{-5}	6.803×10^{-5}	0.7076
400	0.50690	1092	0.04887	3.198×10^{-5}	6.309×10^{-5}	8.830×10^{-5}	0.7145
500	0.44140	1116	0.05414	3.508×10^{-5}	7.949×10^{-5}	1.099×10^{-4}	0.7231
600	0.39080	1140	0.05919	3.802×10^{-5}	9.727×10^{-5}	1.329×10^{-4}	0.7319
700	0.35070	1162	0.06408	4.081×10^{-5}	1.164×10^{-4}	1.573×10^{-4}	0.7400
1000	0.26810	1215	0.07799	4.860×10^{-5}	1.813×10^{-4}	2.394×10^{-4}	0.7573
1300	0.21700	1251	0.09119	5.578×10^{-5}	2.571×10^{-4}	3.358×10^{-4}	0.7655
Oxygen (O_2)							
–150	3.2085	927	0.01129	9.452×10^{-6}	2.946×10^{-6}	3.795×10^{-6}	0.7761
–100	2.2629	916	0.01588	1.296×10^{-5}	5.726×10^{-6}	7.657×10^{-6}	0.7479
–50	1.7512	914	0.02022	1.617×10^{-5}	9.235×10^{-6}	1.264×10^{-5}	0.7309
0	1.4290	916	0.02435	1.914×10^{-5}	1.340×10^{-5}	1.860×10^{-5}	0.7203
50	1.2073	923	0.02829	2.191×10^{-5}	1.815×10^{-5}	2.539×10^{-5}	0.7149
100	1.0452	934	0.03206	2.451×10^{-5}	2.345×10^{-5}	3.286×10^{-5}	0.7136
200	0.8241	962	0.03921	2.928×10^{-5}	3.554×10^{-5}	4.945×10^{-5}	0.7187

(continued)

Table A.6 (continued)

Oxygen (O_2)							
300	0.6802	993	0.04592	3.362×10^{-5}	4.943×10^{-5}	6.798×10^{-5}	0.7271
400	0.5792	1021	0.05230	3.763×10^{-5}	6.497×10^{-5}	8.845×10^{-5}	0.7345
500	0.5042	1045	0.05842	4.136×10^{-5}	8.203×10^{-5}	1.109×10^{-4}	0.7398
600	0.4465	1064	0.06431	4.489×10^{-5}	1.005×10^{-4}	1.354×10^{-4}	0.7427
700	0.4006	1080	0.07004	4.824×10^{-5}	1.204×10^{-4}	1.619×10^{-4}	0.7437
1000	0.3062	1111	0.08640	5.752×10^{-5}	1.878×10^{-4}	2.541×10^{-4}	0.7393
1300	0.2478	1127	0.10191	6.598×10^{-5}	2.662×10^{-4}	3.649×10^{-4}	0.7296
Water vapor (H_2O)							
100	0.5976	2080	0.02460	1.227×10^{-5}	2.053×10^{-5}	1.979×10^{-5}	1.0372
150	0.5233	1986	0.02885	1.418×10^{-5}	2.710×10^{-5}	2.777×10^{-5}	0.9762
200	0.4664	1976	0.03344	1.618×10^{-5}	3.468×10^{-5}	3.628×10^{-5}	0.9558
250	0.4211	1990	0.03834	1.822×10^{-5}	4.326×10^{-5}	4.576×10^{-5}	0.9454
300	0.384	2013	0.04353	2.029×10^{-5}	5.284×10^{-5}	5.633×10^{-5}	0.9381
350	0.3529	2040	0.04898	2.237×10^{-5}	6.338×10^{-5}	6.802×10^{-5}	0.9318
400	0.3266	2070	0.05465	2.445×10^{-5}	7.487×10^{-5}	8.085×10^{-5}	0.9261
450	0.3039	2102	0.06053	2.652×10^{-5}	8.727×10^{-5}	9.477×10^{-5}	0.9208
500	0.2842	2134	0.06659	2.857×10^{-5}	1.006×10^{-4}	1.098×10^{-4}	0.9160
550	0.2669	2168	0.07281	3.061×10^{-5}	1.147×10^{-4}	1.258×10^{-4}	0.9116
600	0.2516	2203	0.07917	3.262×10^{-5}	1.297×10^{-4}	1.429×10^{-4}	0.9075
650	0.2379	2238	0.08567	3.460×10^{-5}	1.454×10^{-4}	1.609×10^{-4}	0.9038

(continued)

Table A.6 (continued)

Water vapor (H_2O)							
700	0.2257	2273	0.09228	3.655×10^{-5}	1.620×10^{-4}	1.799×10^{-4}	0.9004
800	0.2046	2343	0.10581	4.038×10^{-5}	1.973×10^{-4}	2.207×10^{-4}	0.8942
900	0.1872	2412	0.11967	4.408×10^{-5}	2.355×10^{-4}	2.651×10^{-4}	0.8885
1000	0.1725	2478	0.13380	4.767×10^{-5}	2.764×10^{-4}	3.131×10^{-4}	0.8829

Properties generated by the Engineering Equation Solver—EES, version V10.833-3D, with permission

Table A.7 Properties of saturated water in the liquid (L) and vapor (V) phases

Temperature (°C)	Pressure (kPa)	Density (kg/m^3)		Specific enthalpy of vaporization (kJ/kg)	Specific heat (J/kgK)		Thermal conductivity (W/mK)		Dynamic viscosity (kg/ms)		Prandtl number (-)		Volumetric expansion coefficient (1/K)		Surface tension (N/m)
T	P_{sat}	ρ_L	ρ_V	h_{LV}	C_{pL}	C_{pV}	k_L	k_V	μ_L	μ_V	Pr_L	Pr_V	β_L	β_V	σ_{LV}
0.01	0.611	999.8	0.004854	2501	4220	1884	0.5564	0.01676	1.791×10^{-3}	9.216×10^{-6}	13.6	1.036	-6.799×10^{-5}	3.681×10^{-3}	0.0756
5	0.873	999.9	0.006801	2489	4205	1889	0.5686	0.01708	1.518×10^{-3}	9.336×10^{-6}	11.2	1.033	1.572×10^{-5}	3.619×10^{-3}	0.0749
10	1.228	999.7	0.009406	2477	4196	1895	0.5798	0.01741	1.306×10^{-3}	9.461×10^{-6}	9.45	1.029	8.768×10^{-5}	3.559×10^{-3}	0.0742
15	1.706	999.1	0.01284	2465	4189	1900	0.5900	0.01775	1.138×10^{-3}	9.592×10^{-6}	8.08	1.027	1.506×10^{-4}	3.502×10^{-3}	0.0735
20	2.339	998.2	0.01731	2454	4184	1906	0.5994	0.01809	1.002×10^{-3}	9.727×10^{-6}	6.99	1.025	2.066×10^{-4}	3.447×10^{-3}	0.0727
25	3.170	997.0	0.02307	2442	4182	1912	0.6081	0.01843	8.901×10^{-4}	9.867×10^{-6}	6.12	1.023	2.572×10^{-4}	3.395×10^{-3}	0.0720
30	4.247	995.6	0.03041	2430	4180	1918	0.6162	0.01879	7.974×10^{-4}	1.001×10^{-5}	5.41	1.022	3.033×10^{-4}	3.345×10^{-3}	0.0712
35	5.629	994.0	0.03967	2418	4180	1925	0.6237	0.01915	7.193×10^{-4}	1.016×10^{-5}	4.82	1.021	3.458×10^{-4}	3.298×10^{-3}	0.0704
40	7.385	992.2	0.05124	2406	4180	1931	0.6307	0.01951	6.530×10^{-4}	1.031×10^{-5}	4.33	1.020	3.854×10^{-4}	3.253×10^{-3}	0.0696
45	9.595	990.2	0.06557	2394	4180	1939	0.6372	0.01988	5.961×10^{-4}	1.046×10^{-5}	3.91	1.020	4.226×10^{-4}	3.210×10^{-3}	0.0688
50	12.35	988.0	0.08315	2382	4182	1947	0.6433	0.02027	5.468×10^{-4}	1.062×10^{-5}	3.55	1.020	4.578×10^{-4}	3.170×10^{-3}	0.0679

(continued)

Table A.7 (continued)

Temperature (°C)	Pressure (kPa)	Density (kg/m³)		Specific enthalpy of vaporization (kJ/kg)	Specific heat (J/kgK)		Thermal conductivity (W/mK)		Dynamic viscosity (kg/ms)		Prandtl number (-)		Volumetric expansion coefficient (1/K)		Surface tension (N/m)
T	P_{sat}	ρ_L	ρ_V	h_{LV}	C_{pL}	C_{pV}	k_L	k_V	μ_L	μ_V	Pr_L	Pr_V	β_L	β_V	σ_{LV}
55	15.76	985.7	0.1046	2370	4183	1955	0.6490	0.02065	5.040×10^{-4}	1.077×10^{-5}	3.25	1.020	4.913×10^{-4}	3.132×10^{-3}	0.0671
60	19.95	983.2	0.1304	2358	4185	1965	0.6542	0.02105	4.664×10^{-4}	1.093×10^{-5}	2.98	1.021	5.233×10^{-4}	3.096×10^{-3}	0.0662
65	25.04	980.5	0.1615	2345	4188	1975	0.6591	0.02146	4.332×10^{-4}	1.110×10^{-5}	2.75	1.021	5.542×10^{-4}	3.063×10^{-3}	0.0654
70	31.20	977.7	0.1984	2333	4190	1986	0.6636	0.02187	4.039×10^{-4}	1.126×10^{-5}	2.55	1.023	5.840×10^{-4}	3.033×10^{-3}	0.0645
75	38.60	974.8	0.2422	2321	4193	1999	0.6676	0.02230	3.777×10^{-4}	1.143×10^{-5}	2.37	1.024	6.131×10^{-4}	3.004×10^{-3}	0.0636
80	47.42	971.8	0.2937	2308	4197	2012	0.6714	0.02273	3.543×10^{-4}	1.159×10^{-5}	2.22	1.026	6.414×10^{-4}	2.979×10^{-3}	0.0627
85	57.87	968.6	0.3539	2295	4201	2027	0.6748	0.02318	3.333×10^{-4}	1.176×10^{-5}	2.08	1.028	6.693×10^{-4}	2.955×10^{-3}	0.0618
90	70.18	965.3	0.4239	2283	4205	2043	0.6778	0.02364	3.144×10^{-4}	1.193×10^{-5}	1.95	1.031	6.967×10^{-4}	2.935×10^{-3}	0.0608
95	84.61	961.9	0.5049	2270	4210	2061	0.6805	0.02412	2.973×10^{-4}	1.210×10^{-5}	1.84	1.034	7.238×10^{-4}	2.917×10^{-3}	0.0599
100	101.325	958.4	0.5981	2256	4216	2080	0.6828	0.02460	2.817×10^{-4}	1.227×10^{-5}	1.74	1.037	7.506×10^{-4}	2.902×10^{-3}	0.0589

(continued)

Table A.7 (continued)

Temperature (°C)	Pressure (kPa)	Density (kg/m^3)		Specific enthalpy of vaporization (kJ/kg)	Specific heat (J/kgK)		Thermal conductivity (W/mK)		Dynamic viscosity (kg/ms)		Prandtl number (-)		Volumetric expansion coefficient (1/K)		Surface tension (N/m)
T	P_{sat}	ρ_L	ρ_V	h_{LV}	C_{pL}	C_{pV}	k_L	k_V	μ_L	μ_V	Pr_L	Pr_V	β_L	β_V	σ_{LV}
110	143.38	951.0	0.8268	2230	4228	2124	0.6866	0.02562	2.547×10^{-4}	1.261×10^{-5}	1.57	1.046	8.041×10^{-4}	2.881×10^{-3}	0.0570
120	198.67	943.1	1.122	2202	4244	2177	0.6891	0.02671	2.321×10^{-4}	1.296×10^{-5}	1.43	1.056	8.578×10^{-4}	2.872×10^{-3}	0.0550
130	270.28	934.8	1.497	2174	4261	2239	0.6905	0.02788	2.129×10^{-4}	1.330×10^{-5}	1.31	1.068	9.123×10^{-4}	2.876×10^{-3}	0.0529
140	361.53	926.1	1.967	2144	4283	2311	0.6907	0.02913	1.965×10^{-4}	1.365×10^{-5}	1.22	1.083	9.683×10^{-4}	2.894×10^{-3}	0.0509
150	476.16	917.0	2.548	2114	4307	2394	0.6898	0.03047	1.825×10^{-4}	1.399×10^{-5}	1.14	1.099	1.027×10^{-3}	2.926×10^{-3}	0.0487
160	618.23	907.5	3.259	2082	4335	2488	0.6879	0.03192	1.702×10^{-4}	1.434×10^{-5}	1.07	1.118	1.088×10^{-3}	2.973×10^{-3}	0.0466
170	792.18	897.5	4.122	2049	4368	2594	0.6850	0.03347	1.596×10^{-4}	1.468×10^{-5}	1.02	1.138	1.153×10^{-3}	3.036×10^{-3}	0.0444
180	1002.8	887.0	5.159	2014	4405	2713	0.6811	0.03515	1.501×10^{-4}	1.503×10^{-5}	0.971	1.160	1.222×10^{-3}	3.116×10^{-3}	0.0422
190	1255.2	876.1	6.395	1978	4447	2844	0.6763	0.03697	1.418×10^{-4}	1.537×10^{-5}	0.932	1.183	1.297×10^{-3}	3.215×10^{-3}	0.0399
200	1554.9	864.7	7.861	1940	4496	2990	0.6705	0.03893	1.343×10^{-4}	1.571×10^{-5}	0.901	1.207	1.379×10^{-3}	3.334×10^{-3}	0.0377

(continued)

Table A.7 (continued)

Temperature (°C)	Pressure (kPa)	Density (kg/m^3)		Specific enthalpy of vaporization (kJ/kg)	Specific heat (J/kgK)		Thermal conductivity (W/mK)		Dynamic viscosity (kg/ms)		Prandtl number (-)		Volumetric expansion coefficient (1/K)		Surface tension (N/m)
T	P_{sat}	ρ_L	ρ_V	h_{LV}	C_{pL}	C_{pV}	k_L	k_V	μ_L	μ_V	Pr_L	Pr_V	β_L	β_V	σ_{LV}
220	2319.6	840.3	11.62	1857	4614	3329	0.6563	0.04337	1.215×10^{-4}	1.641×10^{-5}	0.854	1.260	1.569×10^{-3}	3.644×10^{-3}	0.0331
240	3347.0	813.4	16.75	1765	4771	3754	0.6386	0.04861	1.109×10^{-4}	1.712×10^{-5}	0.828	1.322	1.809×10^{-3}	4.084×10^{-3}	0.0284
260	4692.3	783.7	23.71	1662	4984	4307	0.6174	0.05491	1.017×10^{-4}	1.788×10^{-5}	0.821	1.402	2.128×10^{-3}	4.714×10^{-3}	0.0237
280	6416.6	750.4	33.16	1543	5286	5073	0.5925	0.06266	9.354×10^{-5}	1.870×10^{-5}	0.835	1.514	2.577×10^{-3}	5.657×10^{-3}	0.0190
300	8587.9	712.3	46.17	1405	5746	6220	0.5635	0.07259	8.593×10^{-5}	1.965×10^{-5}	0.876	1.684	3.267×10^{-3}	7.167×10^{-3}	0.0144
320	11,284	667.3	64.64	1239	6532	8160	0.5294	0.08614	7.843×10^{-5}	2.085×10^{-5}	0.968	1.975	4.477×10^{-3}	9.873×10^{-3}	0.0099
340	14,601	610.6	92.74	1027	8211	12,229	0.4878	0.1069	7.042×10^{-5}	2.255×10^{-5}	1.19	2.581	7.181×10^{-3}	1.587×10^{-2}	0.0056
360	18,666	527.7	143.9	720	14,976	27,336	0.4308	0.1488	6.034×10^{-5}	2.572×10^{-5}	2.10	4.726	1.907×10^{-2}	3.941×10^{-2}	0.0019
373.95*	22,064	322.0	322.0	0					8.031×10^{-5}	8.031×10^{-5}	354.4				0.0000

Properties generated by the Engineering Equation Solver—EES, version V10.833-3D, with permission. *Critical values

Table A.8 Properties of various saturated fluids of the liquid (L) and vapor (V) phases

R134a ($C_2H_2F_4$)

Temperature (°C)	Pressure (kPa)	Density (kg/m³)		Specific enthalpy of vaporization (kJ/kg)	Specific heat (J/kgK)		Thermal conductivity (W/mK)		Dynamic viscosity (kg/ms)		Prandtl number (−)		Volumetric expansion coefficient (1/K)		Surface tension (N/m)
T	p_{sat}	ρ_L	ρ_V	h_{LV}	C_{pL}	C_{pV}	k_L	k_V	μ_L	μ_V	Pr_L	Pr_V	β_L	β_V	σ_{LV}
−40	51.25	1418	2.772	225.9	1255	749	0.1101	0.00811	4.653×10^{-4}	9.335×10^{-6}	5.3	0.862	2.045×10^{-3}	4.797×10^{-3}	0.0173
−35	66.19	1403	3.523	222.7	1264	765	0.1083	0.00863	4.305×10^{-4}	9.531×10^{-6}	5.0	0.845	2.094×10^{-3}	4.794×10^{-3}	0.0165
−30	84.43	1388	4.429	219.5	1273	781	0.1065	0.00913	3.994×10^{-4}	9.727×10^{-6}	4.77	0.832	2.146×10^{-3}	4.806×10^{-3}	0.0158
−25	106.5	1373	5.510	216.3	1283	798	0.1047	0.00963	3.714×10^{-4}	9.923×10^{-6}	4.55	0.822	2.203×10^{-3}	4.833×10^{-3}	0.0150
−20	132.8	1358	6.789	212.9	1293	816	0.1028	0.01013	3.462×10^{-4}	1.012×10^{-5}	4.36	0.815	2.266×10^{-3}	4.875×10^{-3}	0.0143
−15	164.0	1343	8.293	209.5	1304	835	0.1008	0.01063	3.232×10^{-4}	1.032×10^{-5}	4.18	0.810	2.334×10^{-3}	4.934×10^{-3}	0.0136
−10	200.7	1327	10.05	206.0	1316	854	0.0988	0.01112	3.023×10^{-4}	1.052×10^{-5}	4.03	0.808	2.409×10^{-3}	5.011×10^{-3}	0.0128
−5	243.5	1311	12.09	202.3	1328	875	0.0967	0.01161	2.832×10^{-4}	1.071×10^{-5}	3.89	0.808	2.492×10^{-3}	5.107×10^{-3}	0.0121
0	293.0	1295	14.44	198.6	1341	897	0.0946	0.01210	2.655×10^{-4}	1.092×10^{-5}	3.76	0.810	2.584×10^{-3}	5.224×10^{-3}	0.0114
5	349.9	1278	17.14	194.7	1355	921	0.0925	0.01259	2.492×10^{-4}	1.112×10^{-5}	3.65	0.813	2.686×10^{-3}	5.365×10^{-3}	0.0107

(continued)

Table A.8 (continued)

R134a ($C_2H_2F_4$)															
Temperature (°C)	Pressure (kPa)	Density (kg/m^3)		Specific enthalpy of vaporization (kJ/kg)	Specific heat (J/kgK)		Thermal conductivity (W/mK)		Dynamic viscosity (kg/ms)		Prandtl number (−)		Volumetric expansion coefficient (1/K)		Surface tension (N/m)
T	p_{sat}	ρ_L	ρ_V	h_{LV}	C_{pL}	C_{pV}	k_L	k_V	μ_L	μ_V	Pr_L	Pr_V	β_L	β_V	σ_{LV}
10	414.9	1261	20.24	190.7	1370	946	0.0902	0.01308	2.341×10^{-4}	1.132×10^{-5}	3.56	0.819	2.799×10^{-3}	5.531×10^{-3}	0.0100
15	488.7	1243	23.78	186.6	1387	972	0.0880	0.01357	2.200×10^{-4}	1.153×10^{-5}	3.47	0.826	2.927×10^{-3}	5.727×10^{-3}	0.0094
20	572.1	1225	27.80	182.3	1405	1001	0.0856	0.01406	2.068×10^{-4}	1.175×10^{-5}	3.39	0.836	3.072×10^{-3}	5.957×10^{-3}	0.0087
25	665.8	1207	32.37	177.8	1425	1032	0.0832	0.01456	1.944×10^{-4}	1.197×10^{-5}	3.33	0.848	3.236×10^{-3}	6.227×10^{-3}	0.0080
30	770.6	1187	37.56	173.1	1446	1066	0.0808	0.01507	1.827×10^{-4}	1.219×10^{-5}	3.27	0.863	3.425×10^{-3}	6.545×10^{-3}	0.0074
35	887.5	1168	43.45	168.2	1471	1103	0.0783	0.01558	1.717×10^{-4}	1.243×10^{-5}	3.23	0.880	3.643×10^{-3}	6.921×10^{-3}	0.0067
40	1017.1	1147	50.12	163.0	1498	1145	0.0757	0.01610	1.612×10^{-4}	1.268×10^{-5}	3.19	0.902	3.898×10^{-3}	7.368×10^{-3}	0.0061
45	1160.5	1125	57.70	157.6	1530	1192	0.0731	0.01664	1.512×10^{-4}	1.295×10^{-5}	3.17	0.928	4.200×10^{-3}	7.906×10^{-3}	0.0055
50	1318.6	1102	66.32	151.8	1566	1247	0.0704	0.01720	1.416×10^{-4}	1.324×10^{-5}	3.15	0.960	4.561×10^{-3}	8.560×10^{-3}	0.0049

(continued)

Table A.8 (continued)

R134a ($C_2H_2F_4$)															
Temperature (°C)	Pressure (kPa)	Density (kg/m^3)		Specific enthalpy of vaporization (kJ/kg)	Specific heat (J/kgK)		Thermal conductivity (W/mK)		Dynamic viscosity (kg/ms)		Prandtl number (−)		Volumetric expansion coefficient (1/K)		Surface tension (N/m)
T	p_{sat}	ρ_L	ρ_V	h_{LV}	C_{pL}	C_{pV}	k_L	k_V	μ_L	μ_V	Pr_L	Pr_V	β_L	β_V	σ_{LV}
55	1492.3	1078	76.17	145.7	1609	1311	0.0676	0.01778	1.325×10^{-4}	1.355×10^{-5}	3.15	0.999	5.002×10^{-3}	9.369×10^{-3}	0.0043
60	1682.8	1053	87.46	139.1	1660	1388	0.0647	0.01838	1.236×10^{-4}	1.390×10^{-5}	3.17	1.050	5.551×10^{-3}	1.039×10^{-2}	0.0037
65	1891.0	1026	100.5	132.0	1723	1484	0.0618	0.01903	1.150×10^{-4}	1.430×10^{-5}	3.21	1.115	6.254×10^{-3}	1.171×10^{-2}	0.0032
70	2118.2	996.3	115.7	124.3	1804	1607	0.0587	0.01972	1.065×10^{-4}	1.477×10^{-5}	3.27	1.204	7.184×10^{-3}	1.347×10^{-2}	0.0026
75	2365.8	964.1	133.7	115.8	1911	1775	0.0555	0.02048	9.823×10^{-5}	1.532×10^{-5}	3.38	1.328	8.479×10^{-3}	1.593×10^{-2}	0.0021
80	2635.2	928.3	155.4	106.3	2064	2018	0.0521	0.02133	8.992×10^{-5}	1.601×10^{-5}	3.56	1.515	1.041×10^{-2}	1.959×10^{-2}	0.0016
85	2928.2	887.3	182.3	95.44	2305	2409	0.0485	0.02233	8.145×10^{-5}	1.692×10^{-5}	3.88	1.825	1.359×10^{-2}	2.558×10^{-2}	0.0012
90	3246.9	838.1	217.4	82.35	2751	3148	0.0444	0.02356	7.252×10^{-5}	1.818×10^{-5}	4.49	2.430	1.982×10^{-2}	3.709×10^{-2}	0.0007
95	3594.1	773.2	268.4	65.21	3916	5126	0.0396	0.02540	6.244×10^{-5}	2.024×10^{-5}	6.17	4.084	3.703×10^{-2}	6.821×10^{-2}	0.0004

(continued)

Table A.8 (continued)

R1234yf (HFO-1234yf, $C_3H_2F_4$)															
–40	62.36	1292	3.785	185.5	1158	778	0.0852	0.00843	3.267×10^{-4}	8.230×10^{-6}	4.44	0.759	2.093×10^{-3}	4.844×10^{-3}	0.0155
–35	79.03	1278	4.721	183.1	1173	794	0.0834	0.00882	3.078×10^{-4}	8.495×10^{-6}	4.33	0.765	2.146×10^{-3}	4.847×10^{-3}	0.0147
–30	99.05	1265	5.837	180.6	1188	811	0.0817	0.00922	2.902×10^{-4}	8.753×10^{-6}	4.22	0.770	2.205×10^{-3}	4.864×10^{-3}	0.0139
–25	122.9	1251	7.154	178.0	1204	828	0.0799	0.00961	2.737×10^{-4}	9.006×10^{-6}	4.12	0.777	2.269×10^{-3}	4.897×10^{-3}	0.0131
–20	150.9	1237	8.699	175.2	1220	847	0.0782	0.01000	2.582×10^{-4}	9.254×10^{-6}	4.03	0.783	2.340×10^{-3}	4.947×10^{-3}	0.0124
–15	183.7	1222	10.50	172.4	1236	865	0.0764	0.01039	2.437×10^{-4}	9.497×10^{-6}	3.94	0.791	2.418×10^{-3}	5.013×10^{-3}	0.0116
–10	221.8	1207	12.57	169.5	1253	885	0.0747	0.01079	2.300×10^{-4}	9.738×10^{-6}	3.86	0.799	2.505×10^{-3}	5.098×10^{-3}	0.0109
–5	265.6	1192	14.96	166.4	1271	905	0.0730	0.01118	2.172×10^{-4}	9.976×10^{-6}	3.78	0.807	2.600×10^{-3}	5.203×10^{-3}	0.0102
0	315.8	1177	17.69	163.3	1289	926	0.0713	0.01158	2.050×10^{-4}	1.021×10^{-5}	3.70	0.817	2.707×10^{-3}	5.331×10^{-3}	0.0095
5	372.9	1161	20.80	160.0	1307	949	0.0697	0.01198	1.935×10^{-4}	1.045×10^{-5}	3.63	0.827	2.826×10^{-3}	5.484×10^{-3}	0.0088
10	437.5	1144	24.33	156.6	1327	972	0.0680	0.01239	1.826×10^{-4}	1.069×10^{-5}	3.56	0.838	2.959×10^{-3}	5.667×10^{-3}	0.0081
15	510.2	1127	28.32	153.0	1347	997	0.0664	0.01281	1.722×10^{-4}	1.093×10^{-5}	3.50	0.851	3.109×10^{-3}	5.884×10^{-3}	0.0074

(continued)

Table A.8 (continued)

R1234yf (HFO-1234yf, $C_3H_2F_4$)															
20	591.7	1110	32.84	149.3	1369	1024	0.0648	0.01323	1.624×10^{-4}	1.117×10^{-5}	3.43	0.865	3.280×10^{-3}	6.144×10^{-3}	0.0068
25	682.5	1092	37.94	145.4	1392	1053	0.0632	0.01367	1.530×10^{-4}	1.143×10^{-5}	3.37	0.881	3.473×10^{-3}	6.454×10^{-3}	0.0062
30	783.5	1073	43.70	141.3	1416	1086	0.0616	0.01412	1.440×10^{-4}	1.169×10^{-5}	3.31	0.899	3.696×10^{-3}	6.828×10^{-3}	0.0056
35	895.2	1054	50.22	136.9	1443	1123	0.0600	0.01458	1.354×10^{-4}	1.196×10^{-5}	3.26	0.921	3.954×10^{-3}	7.280×10^{-3}	0.0050
40	1018.4	1034	57.63	132.3	1473	1166	0.0585	0.01507	1.272×10^{-4}	1.225×10^{-5}	3.21	0.948	4.258×10^{-3}	7.832×10^{-3}	0.0044
45	1153.9	1012	66.06	127.4	1508	1216	0.0569	0.01558	1.192×10^{-4}	1.255×10^{-5}	3.16	0.980	4.626×10^{-3}	8.514×10^{-3}	0.0039
50	1302.4	990.0	75.70	122.2	1549	1276	0.0554	0.01612	1.116×10^{-4}	1.289×10^{-5}	3.12	1.020	5.083×10^{-3}	9.370×10^{-3}	0.0033
55	1464.8	966.1	86.80	116.5	1599	1349	0.0538	0.01670	1.041×10^{-4}	1.326×10^{-5}	3.09	1.071	5.666×10^{-3}	1.047×10^{-2}	0.0028
60	1642.0	940.6	99.67	110.4	1660	1441	0.0522	0.01733	9.684×10^{-5}	1.367×10^{-5}	3.08	1.137	6.434×10^{-3}	1.191×10^{-2}	0.0023
65	1834.9	913.0	114.7	103.6	1738	1561	0.0505	0.01802	8.970×10^{-5}	1.415×10^{-5}	3.08	1.226	7.475×10^{-3}	1.387×10^{-2}	0.0019
70	2044.5	882.7	132.6	96.2	1841	1728	0.0488	0.01878	8.260×10^{-5}	1.471×10^{-5}	3.11	1.353	8.941×10^{-3}	1.667×10^{-2}	0.0015
75	2272.2	848.8	154.1	87.9	1987	1974	0.0470	0.01966	7.547×10^{-5}	1.539×10^{-5}	3.19	1.546	1.114×10^{-2}	2.092×10^{-2}	0.0011

(continued)

Table A.8 (continued)

R1234yf (HFO-1234yf, $C_3H_2F_4$)															
80	2519.4	809.4	180.8	78.3	2221	2377	0.0450	0.02068	6.813×10^{-5}	1.627×10^{-5}	3.36	1.870	1.482×10^{-2}	2.801×10^{-2}	0.0007
85	2787.8	761.4	215.6	66.69	2682	3165	0.0427	0.02194	6.029×10^{-5}	1.747×10^{-5}	3.79	2.520	2.240×10^{-2}	4.204×10^{-2}	0.0004
R32 (HFO-32, CH_2F_2)															
–40	177.40	1180	5.065	368.8	1608	940.1	0.1799	0.01046	2.451×10^{-4}	9.697×10^{-6}	2.19	0.872	2.444×10^{-3}	5.609×10^{-3}	0.0184
–35	221.36	1166	6.247	362.9	1619	970.9	0.1755	0.01083	2.307×10^{-4}	9.908×10^{-6}	2.13	0.889	2.523×10^{-3}	5.692×10^{-3}	0.0174
–30	273.42	1151	7.638	356.8	1631	1003	0.1712	0.01121	2.175×10^{-4}	1.012×10^{-5}	2.07	0.906	2.609×10^{-3}	5.795×10^{-3}	0.0165
–25	334.54	1136	9.265	350.5	1645	1038	0.1668	0.01162	2.054×10^{-4}	1.034×10^{-5}	2.03	0.924	2.704×10^{-3}	5.919×10^{-3}	0.0155
–20	405.73	1121	11.16	344.0	1661	1075	0.1625	0.01205	1.941×10^{-4}	1.056×10^{-5}	1.98	0.942	2.809×10^{-3}	6.067×10^{-3}	0.0146
–15	488.07	1105	13.35	337.3	1678	1114	0.1581	0.01250	1.836×10^{-4}	1.078×10^{-5}	1.95	0.961	2.924×10^{-3}	6.242×10^{-3}	0.0137
–10	582.64	1089	15.87	330.3	1698	1156	0.1538	0.01298	1.737×10^{-4}	1.100×10^{-5}	1.92	0.980	3.054×10^{-3}	6.448×10^{-3}	0.0128
–5	690.60	1072	18.77	322.9	1720	1202	0.1495	0.01349	1.645×10^{-4}	1.123×10^{-5}	1.89	1.000	3.198×10^{-3}	6.689×10^{-3}	0.0119
0	813.15	1055	22.09	315.3	1745	1251	0.1451	0.01404	1.557×10^{-4}	1.147×10^{-5}	1.87	1.022	3.362×10^{-3}	6.972×10^{-3}	0.0110

(continued)

Table A.8 (continued)

R32 (HFO-32, CH_2F_2)														
5	951.52	1038	25.89	307.3	1773	1306	0.1408	0.01462	1.474×10^{-4}	1.171×10^{-5}	1.86	1.046	3.548×10^{-3}	7.305×10^{-3}	0.0101
10	1107.0	1020	30.24	298.9	1806	1367	0.1364	0.01525	1.395×10^{-4}	1.196×10^{-5}	1.85	1.072	3.762×10^{-3}	7.697×10^{-3}	0.0093
15	1280.9	1001	35.19	290.1	1843	1435	0.1320	0.01593	1.319×10^{-4}	1.222×10^{-5}	1.84	1.101	4.010×10^{-3}	8.162×10^{-3}	0.0084
20	1474.7	981.4	40.86	280.8	1886	1514	0.1275	0.01666	1.246×10^{-4}	1.249×10^{-5}	1.84	1.134	4.302×10^{-3}	8.719×10^{-3}	0.0076
25	1689.7	961.0	47.34	270.9	1937	1605	0.1230	0.01746	1.176×10^{-4}	1.276×10^{-5}	1.85	1.172	4.650×10^{-3}	9.391×10^{-3}	0.0068
30	1927.5	939.6	54.78	260.4	1997	1712	0.1185	0.01834	1.108×10^{-4}	1.305×10^{-5}	1.87	1.218	5.072×10^{-3}	1.021×10^{-2}	0.0060
35	2189.8	917.1	63.34	249.2	2071	1841	0.1138	0.01931	1.042×10^{-4}	1.336×10^{-5}	1.90	1.274	5.596×10^{-3}	1.124×10^{-2}	0.0052
40	2478.2	893.0	73.26	237.1	2163	2001	0.1091	0.02039	9.769×10^{-5}	1.371×10^{-5}	1.94	1.345	6.263×10^{-3}	1.255×10^{-2}	0.0045
45	2794.6	867.3	84.85	224.0	2281	2205	0.1041	0.02161	9.129×10^{-5}	1.409×10^{-5}	2.00	1.438	7.140×10^{-3}	1.426×10^{-2}	0.0038
50	3141.0	839.3	98.54	209.6	2439	2477	0.09904	0.02301	8.494×10^{-5}	1.453×10^{-5}	2.09	1.564	8.344×10^{-3}	1.659×10^{-2}	0.0031
55	3519.6	808.3	115.0	193.7	2661	2859	0.09368	0.02464	7.856×10^{-5}	1.504×10^{-5}	2.23	1.745	1.010×10^{-2}	1.992×10^{-2}	0.0024
60	3933.0	773.3	135.2	175.5	3001	3440	0.08796	0.02661	7.208×10^{-5}	1.567×10^{-5}	2.46	2.026	1.286×10^{-2}	2.510×10^{-2}	0.0018

(continued)

Table A.8 (continued)

R32 (HFO-32, CH_2F_2)															
65	4384.3	732.3	161.1	154.2	3588	4447	0.08170	0.02910	6.531×10^{-5}	1.649×10^{-5}	2.87	2.519	1.784×10^{-2}	3.420×10^{-2}	0.0012
70	4877.2	681.0	196.8	127.8	4864	6648	0.07455	0.03251	5.797×10^{-5}	1.765×10^{-5}	3.78	3.609	2.919×10^{-2}	5.435×10^{-2}	0.0007
Propane (R290, C_3H_8)															
–80	13.01	623.3	0.360	462.6	2097	1265	0.1567	0.00844	3.289×10^{-4}	5.409×10^{-6}	4.40	0.811	1.700×10^{-3}	5.339×10^{-3}	0.0206
–70	24.34	612.6	0.644	453.4	2130	1309	0.1503	0.00923	2.868×10^{-4}	5.680×10^{-6}	4.06	0.806	1.764×10^{-3}	5.167×10^{-3}	0.0192
–60	42.61	601.7	1.08	443.9	2168	1359	0.1440	0.01004	2.522×10^{-4}	5.956×10^{-6}	3.80	0.806	1.838×10^{-3}	5.045×10^{-3}	0.0179
–50	70.47	590.5	1.73	434.0	2211	1414	0.1377	0.01089	2.232×10^{-4}	6.239×10^{-6}	3.59	0.810	1.924×10^{-3}	4.972×10^{-3}	0.0165
–40	111.0	579.0	2.63	423.6	2259	1474	0.1315	0.01177	1.986×10^{-4}	6.529×10^{-6}	3.41	0.818	2.025×10^{-3}	4.949×10^{-3}	0.0152
–35	137.1	573.1	3.20	418.2	2285	1507	0.1284	0.01222	1.877×10^{-4}	6.677×10^{-6}	3.34	0.823	2.082×10^{-3}	4.958×10^{-3}	0.0145
–30	167.8	567.2	3.87	412.6	2312	1541	0.1254	0.01269	1.775×10^{-4}	6.828×10^{-6}	3.27	0.829	2.143×10^{-3}	4.980×10^{-3}	0.0138
–25	203.4	561.1	4.63	406.8	2340	1577	0.1224	0.01316	1.680×10^{-4}	6.981×10^{-6}	3.21	0.836	2.209×10^{-3}	5.016×10^{-3}	0.0132
–20	244.5	554.9	5.51	400.8	2371	1615	0.1194	0.01365	1.591×10^{-4}	7.136×10^{-6}	3.16	0.844	2.282×10^{-3}	5.067×10^{-3}	0.0125

(continued)

Table A.8 (continued)

Propane (R290, C_3H_8)															
–15	291.6	548.5	6.51	394.6	2402	1655	0.1165	0.01415	1.507×10^{-4}	7.295×10^{-6}	3.11	0.853	2.360×10^{-3}	5.134×10^{-3}	0.0119
–10	345.3	542.1	7.64	388.2	2436	1696	0.1137	0.01467	1.429×10^{-4}	7.457×10^{-6}	3.06	0.862	2.446×10^{-3}	5.218×10^{-3}	0.0112
–5	406.1	535.5	8.92	381.6	2471	1740	0.1108	0.01521	1.355×10^{-4}	7.624×10^{-6}	3.02	0.872	2.540×10^{-3}	5.322×10^{-3}	0.0106
0	474.6	528.7	10.36	374.7	2509	1787	0.1081	0.01577	1.285×10^{-4}	7.794×10^{-6}	2.98	0.883	2.644×10^{-3}	5.446×10^{-3}	0.0100
5	551.2	521.8	11.98	367.5	2549	1836	0.1053	0.01635	1.218×10^{-4}	7.970×10^{-6}	2.95	0.895	2.758×10^{-3}	5.594×10^{-3}	0.0093
10	636.7	514.7	13.80	359.9	2591	1889	0.1027	0.01696	1.156×10^{-4}	8.151×10^{-6}	2.92	0.908	2.884×10^{-3}	5.769×10^{-3}	0.0087
15	731.7	507.4	15.83	352.1	2636	1945	0.1000	0.01759	1.096×10^{-4}	8.339×10^{-6}	2.89	0.922	3.025×10^{-3}	5.976×10^{-3}	0.0081
20	836.6	499.8	18.10	343.9	2684	2006	0.0975	0.01826	1.039×10^{-4}	8.534×10^{-6}	2.86	0.937	3.184×10^{-3}	6.219×10^{-3}	0.0075
25	952.2	492.1	20.64	335.3	2737	2072	0.0950	0.01897	9.846×10^{-5}	8.737×10^{-6}	2.84	0.954	3.362×10^{-3}	6.506×10^{-3}	0.0069
30	1079.1	484.1	23.48	326.2	2794	2145	0.0925	0.01972	9.326×10^{-5}	8.951×10^{-6}	2.82	0.974	3.567×10^{-3}	6.847×10^{-3}	0.0063
35	1218.0	475.7	26.65	316.6	2857	2226	0.0901	0.02052	8.827×10^{-5}	9.176×10^{-6}	2.80	0.995	3.803×10^{-3}	7.253×10^{-3}	0.0058
40	1369.5	467.1	30.20	306.5	2927	2317	0.0877	0.02137	8.347×10^{-5}	9.415×10^{-6}	2.78	1.021	4.080×10^{-3}	7.741×10^{-3}	0.0052

(continued)

Table A.8 (continued)

Propane (R290, C_3H_8)														
50	1713.4	448.5	38.68	284.3	3099	2543	0.0831	0.02327	7.433×10^{-5}	9.947×10^{-6}	2.77	1.087	4.814×10^{-3}	9.069×10^{-3}	0.0041
60	2116.9	427.8	49.56	258.8	3342	2868	0.0787	0.02550	6.566×10^{-5}	1.058×10^{-5}	2.79	1.190	5.967×10^{-3}	1.118×10^{-2}	0.0030
70	2587.2	403.7	64.01	228.3	3736	3405	0.0742	0.02822	5.717×10^{-5}	1.138×10^{-5}	2.88	1.374	8.049×10^{-3}	1.493×10^{-2}	0.0020
80	3132.7	373.6	84.51	189.8	4544	4534	0.0695	0.03171	4.842×10^{-5}	1.251×10^{-5}	3.17	1.789	1.284×10^{-2}	2.326×10^{-2}	0.0011
90	3765.4	329.2	119.0	133.2	7625	8923	0.0669	0.03691	3.831×10^{-5}	1.451×10^{-5}	4.37	3.508	3.356×10^{-2}	5.659×10^{-2}	0.0004
Isobutane (R600a, C_4H_{10})															
–80	2.234	665.0	0.081	421.0	1944	1213	0.1323	0.00751	6.226×10^{-4}	4.943×10^{-6}	9.15	0.799	1.467×10^{-3}	5.226×10^{-3}	0.0230
–70	4.710	655.1	0.163	413.5	1980	1256	0.1281	0.00828	5.214×10^{-4}	5.176×10^{-6}	8.06	0.785	1.511×10^{-3}	5.004×10^{-3}	0.0216
–60	9.164	645.2	0.303	405.9	2017	1300	0.1238	0.00907	4.428×10^{-4}	5.412×10^{-6}	7.21	0.776	1.561×10^{-3}	4.818×10^{-3}	0.0203
–50	16.64	635.0	0.527	398.1	2057	1347	0.1195	0.00989	3.803×10^{-4}	5.654×10^{-6}	6.55	0.770	1.617×10^{-3}	4.669×10^{-3}	0.0190
–40	28.50	624.6	0.869	390.1	2100	1397	0.1152	0.01073	3.298×10^{-4}	5.901×10^{-6}	6.01	0.768	1.680×10^{-3}	4.556×10^{-3}	0.0177
–35	36.56	619.4	1.10	386.0	2122	1424	0.1131	0.01116	3.080×10^{-4}	6.027×10^{-6}	5.78	0.769	1.715×10^{-3}	4.514×10^{-3}	0.0171

(continued)

Table A.8 (continued)

Isobutane (R600a, C_4H_{10})														
−30	46.37	614.0	1.37	381.9	2145	1451	0.1109	0.01160	2.882×10^{-4}	6.155×10^{-6}	5.57	0.770	1.752×10^{-3}	4.481×10^{-3}	0.0165
−25	58.15	608.6	1.69	377.7	2169	1479	0.1088	0.01204	2.701×10^{-4}	6.285×10^{-6}	5.39	0.772	1.791×10^{-3}	4.457×10^{-3}	0.0158
−20	72.19	603.2	2.06	373.4	2194	1508	0.1067	0.01249	2.535×10^{-4}	6.417×10^{-6}	5.21	0.775	1.833×10^{-3}	4.442×10^{-3}	0.0152
−15	88.75	597.6	2.50	368.9	2219	1538	0.1046	0.01295	2.382×10^{-4}	6.551×10^{-6}	5.05	0.778	1.878×10^{-3}	4.437×10^{-3}	0.0146
−10	108.1	592.0	3.00	364.4	2245	1569	0.1025	0.01341	2.242×10^{-4}	6.688×10^{-6}	4.91	0.782	1.926×10^{-3}	4.443×10^{-3}	0.0140
−5	130.7	586.3	3.59	359.8	2272	1601	0.1005	0.01388	2.111×10^{-4}	6.827×10^{-6}	4.77	0.787	1.977×10^{-3}	4.458×10^{-3}	0.0134
0	156.7	580.5	4.25	355.0	2299	1634	0.0985	0.01436	1.991×10^{-4}	6.970×10^{-6}	4.65	0.793	2.032×10^{-3}	4.484×10^{-3}	0.0128
5	186.4	574.6	5.01	350.1	2327	1668	0.0965	0.01486	1.879×10^{-4}	7.116×10^{-6}	4.53	0.799	2.091×10^{-3}	4.522×10^{-3}	0.0122
10	220.3	568.6	5.86	345.1	2357	1704	0.0946	0.01536	1.774×10^{-4}	7.265×10^{-6}	4.42	0.806	2.154×10^{-3}	4.571×10^{-3}	0.0116
15	258.7	562.4	6.83	339.9	2387	1741	0.0927	0.01587	1.677×10^{-4}	7.419×10^{-6}	4.32	0.814	2.222×10^{-3}	4.633×10^{-3}	0.0110
20	302.0	556.2	7.91	334.6	2418	1779	0.0908	0.01640	1.586×10^{-4}	7.577×10^{-6}	4.22	0.822	2.295×10^{-3}	4.709×10^{-3}	0.0104
25	350.4	549.9	9.12	329.1	2450	1819	0.0889	0.01694	1.500×10^{-4}	7.739×10^{-6}	4.13	0.831	2.375×10^{-3}	4.799×10^{-3}	0.0099

(continued)

Table A.8 (continued)

Isobutane (R600a, C_4H_{10})															
30	404.5	543.4	10.48	323.4	2483	1860	0.0872	0.01749	1.420×10^{-4}	7.907×10^{-6}	4.05	0.841	2.461×10^{-3}	4.907×10^{-3}	0.0093
35	464.5	536.8	11.99	317.5	2518	1904	0.0854	0.01807	1.345×10^{-4}	8.080×10^{-6}	3.97	0.851	2.554×10^{-3}	5.033×10^{-3}	0.0087
40	530.9	530.0	13.66	311.4	2553	1949	0.0837	0.01866	1.274×10^{-4}	8.260×10^{-6}	3.89	0.863	2.656×10^{-3}	5.180×10^{-3}	0.0082
50	684.4	516.0	17.59	298.5	2630	2048	0.0804	0.01991	1.144×10^{-4}	8.642×10^{-6}	3.74	0.889	2.890×10^{-3}	5.551×10^{-3}	0.0071
60	868.4	501.2	22.42	284.5	2714	2160	0.0773	0.02127	1.027×10^{-4}	9.059×10^{-6}	3.61	0.920	3.180×10^{-3}	6.055×10^{-3}	0.0061
70	1086.4	485.6	28.36	269.2	2812	2292	0.0743	0.02276	9.210×10^{-5}	9.521×10^{-6}	3.49	0.959	3.554×10^{-3}	6.751×10^{-3}	0.0051
80	1342.2	468.8	35.71	252.2	2932	2456	0.0715	0.02442	8.244×10^{-5}	1.005×10^{-5}	3.38	1.010	4.068×10^{-3}	7.743×10^{-3}	0.0041
90	1640.0	450.5	44.9	233.1	3091	2672	0.0689	0.02628	7.345×10^{-5}	1.065×10^{-5}	3.30	1.083	4.837×10^{-3}	9.224×10^{-3}	0.0032
Ethanol (C_2H_6O)															
0	1.733	806.4	0.033	949.2	2257	1341	0.1763	0.01270	1.758×10^{-3}	8.215×10^{-6}	22.51	0.867	1.050×10^{-3}	3.694×10^{-3}	0.0243
10	3.308	797.9	0.062	938.7	2324	1377	0.1736	0.01361	1.418×10^{-3}	8.498×10^{-6}	18.98	0.859	1.065×10^{-3}	3.583×10^{-3}	0.0235
20	6.049	789.4	0.111	927.8	2396	1415	0.1708	0.01455	1.162×10^{-3}	8.780×10^{-6}	16.29	0.854	1.084×10^{-3}	3.489×10^{-3}	0.0227

(continued)

Table A.8 (continued)

Ethanol (C_2H_6O)														
30	10.63	780.7	0.191	916.4	2475	1457	0.1680	0.01550	9.645×10^{-4}	9.062×10^{-6}	14.20	0.852	1.110×10^{-3}	3.412×10^{-3}	0.0219
40	18.01	772.0	0.317	904.5	2559	1502	0.1652	0.01647	8.100×10^{-4}	9.344×10^{-6}	12.55	0.852	1.141×10^{-3}	3.354×10^{-3}	0.0211
50	29.49	763.1	0.51	891.7	2649	1551	0.1623	0.01747	6.868×10^{-4}	9.625×10^{-6}	11.21	0.855	1.181×10^{-3}	3.316×10^{-3}	0.0202
60	46.77	754.0	0.79	878.0	2744	1604	0.1594	0.01849	5.872×10^{-4}	9.907×10^{-6}	10.11	0.860	1.230×10^{-3}	3.300×10^{-3}	0.0193
70	72.03	744.6	1.19	863.2	2844	1664	0.1564	0.01954	5.054×10^{-4}	1.019×10^{-5}	9.19	0.867	1.289×10^{-3}	3.308×10^{-3}	0.0184
80	107.9	734.9	1.75	847.2	2948	1729	0.1533	0.02063	4.374×10^{-4}	1.047×10^{-5}	8.41	0.878	1.360×10^{-3}	3.342×10^{-3}	0.0175
90	157.6	724.7	2.51	829.9	3058	1802	0.1501	0.02175	3.804×10^{-4}	1.076×10^{-5}	7.75	0.891	1.446×10^{-3}	3.404×10^{-3}	0.0165
100	224.7	714.0	3.52	811.0	3174	1882	0.1469	0.02292	3.319×10^{-4}	1.105×10^{-5}	7.17	0.908	1.550×10^{-3}	3.499×10^{-3}	0.0156
110	313.5	702.8	4.86	790.5	3296	1973	0.1436	0.02414	2.905×10^{-4}	1.135×10^{-5}	6.67	0.927	1.675×10^{-3}	3.629×10^{-3}	0.0145
120	428.4	690.8	6.58	768.1	3425	2074	0.1402	0.02541	2.548×10^{-4}	1.165×10^{-5}	6.22	0.951	1.825×10^{-3}	3.801×10^{-3}	0.0135
130	574.5	678.0	8.77	743.8	3559	2188	0.1366	0.02675	2.237×10^{-4}	1.196×10^{-5}	5.83	0.978	2.006×10^{-3}	4.022×10^{-3}	0.0124
140	757.0	664.2	11.53	717.3	3696	2318	0.1330	0.02817	1.966×10^{-4}	1.228×10^{-5}	5.46	1.010	2.221×10^{-3}	4.303×10^{-3}	0.0113

(continued)

Table A.8 (continued)

Ethanol (C_2H_6O)															
150	981.6	649.4	14.99	688.5	3833	2469	0.1292	0.02968	1.728×10^{-4}	1.261×10^{-5}	5.13	1.049	2.475×10^{-3}	4.661×10^{-3}	0.0102
160	1254.2	633.4	19.29	657.4	3968	2646	0.1251	0.03130	1.518×10^{-4}	1.296×10^{-5}	4.81	1.095	2.773×10^{-3}	5.121×10^{-3}	0.0091
170	1581.1	616.2	24.62	623.7	4103	2859	0.1209	0.03305	1.333×10^{-4}	1.333×10^{-5}	4.52	1.153	3.129×10^{-3}	5.723×10^{-3}	0.0079
180	1969.1	597.6	31.23	587.3	4249	3124	0.1164	0.03497	1.169×10^{-4}	1.372×10^{-5}	4.27	1.226	3.572×10^{-3}	6.535×10^{-3}	0.0067
190	2425.7	577.3	39.46	547.6	4429	3472	0.1116	0.03709	1.023×10^{-4}	1.416×10^{-5}	4.06	1.325	4.173×10^{-3}	7.677×10^{-3}	0.0055
200	2959.3	554.8	49.80	503.9	4690	3959	0.1063	0.03948	8.941×10^{-5}	1.466×10^{-5}	3.95	1.470	5.091×10^{-3}	9.389×10^{-3}	0.0043
Ammonia (NH_3)															
–50	40.82	701.9	0.380	1416	4361	2178	0.7218	0.02024	3.283×10^{-4}	7.573×10^{-6}	1.98	0.815	1.676×10^{-3}	4.816×10^{-3}	0.0380
–40	71.66	690.0	0.644	1388	4415	2244	0.6876	0.02064	2.807×10^{-4}	7.859×10^{-6}	1.80	0.854	1.763×10^{-3}	4.746×10^{-3}	0.0357
–35	93.07	683.9	0.822	1374	4441	2283	0.6708	0.02088	2.611×10^{-4}	8.005×10^{-6}	1.73	0.875	1.807×10^{-3}	4.728×10^{-3}	0.0345
–30	119.40	677.7	1.04	1360	4466	2326	0.6543	0.02115	2.438×10^{-4}	8.152×10^{-6}	1.66	0.896	1.852×10^{-3}	4.720×10^{-3}	0.0334
–25	151.46	671.4	1.30	1344	4490	2373	0.6380	0.02144	2.282×10^{-4}	8.300×10^{-6}	1.61	0.918	1.898×10^{-3}	4.724×10^{-3}	0.0322

(continued)

Table A.8 (continued)

Ammonia (NH_3)															
–20	190.09	665.1	1.60	1329	4514	2425	0.6218	0.02177	2.143×10^{-4}	8.450×10^{-6}	1.56	0.941	1.945×10^{-3}	4.740×10^{-3}	0.0310
–15	236.21	658.7	1.97	1313	4538	2481	0.6059	0.02212	2.017×10^{-4}	8.600×10^{-6}	1.51	0.964	1.995×10^{-3}	4.768×10^{-3}	0.0298
–10	290.79	652.1	2.39	1297	4563	2542	0.5902	0.02250	1.903×10^{-4}	8.751×10^{-6}	1.47	0.989	2.047×10^{-3}	4.808×10^{-3}	0.0287
–5	354.9	645.5	2.89	1280	4589	2608	0.5747	0.02292	1.798×10^{-4}	8.903×10^{-6}	1.44	1.013	2.102×10^{-3}	4.861×10^{-3}	0.0275
0	429.6	638.7	3.46	1262	4615	2680	0.5594	0.02337	1.702×10^{-4}	9.056×10^{-6}	1.40	1.039	2.161×10^{-3}	4.928×10^{-3}	0.0263
5	516.0	631.8	4.12	1244	4644	2758	0.5443	0.02385	1.614×10^{-4}	9.209×10^{-6}	1.38	1.065	2.224×10^{-3}	5.009×10^{-3}	0.0251
10	615.3	624.8	4.87	1226	4674	2842	0.5294	0.02437	1.532×10^{-4}	9.364×10^{-6}	1.35	1.092	2.293×10^{-3}	5.105×10^{-3}	0.0240
15	728.8	617.6	5.73	1206	4707	2932	0.5147	0.02492	1.456×10^{-4}	9.519×10^{-6}	1.33	1.120	2.367×10^{-3}	5.218×10^{-3}	0.0228
20	857.8	610.3	6.71	1187	4743	3030	0.5001	0.02552	1.384×10^{-4}	9.676×10^{-6}	1.31	1.149	2.449×10^{-3}	5.349×10^{-3}	0.0216
25	1003.5	602.9	7.81	1166	4783	3136	0.4857	0.02616	1.318×10^{-4}	9.835×10^{-6}	1.30	1.179	2.538×10^{-3}	5.500×10^{-3}	0.0205
30	1167.4	595.3	9.06	1145	4827	3250	0.4714	0.02684	1.255×10^{-4}	9.995×10^{-6}	1.29	1.210	2.637×10^{-3}	5.673×10^{-3}	0.0194

(continued)

Table A.8 (continued)

Ammonia (NH_3)														
35	1350.8	587.5	10.46	1122	4876	3375	0.4574	0.02758	1.196×10^{-4}	1.016×10^{-5}	1.28	1.243	2.747×10^{-3}	5.870×10^{-3}	0.0182
40	1555.3	579.5	12.03	1099	4931	3510	0.4434	0.02837	1.141×10^{-4}	1.033×10^{-5}	1.27	1.278	2.870×10^{-3}	6.097×10^{-3}	0.0171
45	1782.3	571.3	13.80	1075	4993	3659	0.4296	0.02922	1.088×10^{-4}	1.050×10^{-5}	1.26	1.314	3.007×10^{-3}	6.356×10^{-3}	0.0160
50	2033.5	562.8	15.78	1050	5064	3822	0.4159	0.03013	1.038×10^{-4}	1.067×10^{-5}	1.26	1.354	3.163×10^{-3}	6.653×10^{-3}	0.0149
55	2310.3	554.1	18.00	1024	5145	4004	0.4023	0.03112	9.898×10^{-5}	1.086×10^{-5}	1.27	1.397	3.340×10^{-3}	6.995×10^{-3}	0.0138
60	2614.5	545.1	20.48	997.2	5238	4206	0.3888	0.03219	9.442×10^{-5}	1.105×10^{-5}	1.27	1.444	3.543×10^{-3}	7.393×10^{-3}	0.0127
70	3312.0	526.1	26.39	938.8	5469	4696	0.3620	0.03462	8.586×10^{-5}	1.147×10^{-5}	1.30	1.556	4.052×10^{-3}	8.402×10^{-3}	0.0106
80	4140.6	505.5	33.87	873.9	5789	5352	0.3353	0.03754	7.792×10^{-5}	1.195×10^{-5}	1.35	1.704	4.761×10^{-3}	9.836×10^{-3}	0.0086
90	5116.3	482.7	43.48	800.5	6252	6290	0.3083	0.04117	7.045×10^{-5}	1.255×10^{-5}	1.43	1.917	5.812×10^{-3}	1.200×10^{-2}	0.0067
100	6257.1	456.8	56.15	715.6	6984	7771	0.2806	0.04587	6.326×10^{-5}	1.332×10^{-5}	1.58	2.257	7.516×10^{-3}	1.555×10^{-2}	0.0048

Properties generated by the Engineering Equation Solver—EES, version V10.833-3D, with permission

T (°C)	ρ (kg/m³)	C_p (J/kgK)	k (W/mK)	μ (kg/ms)	ν (m²/s)	α (m²/s)	Pr (–)	β (1/K)
Unused motor oil								
0	899.0	1797	0.1466	3.814	4.242×10^{-3}	9.076×10^{-8}	46,743	7.0×10^{-4}
10	893.6	1840	0.1456	1.801	2.016×10^{-3}	8.857×10^{-8}	22,756	7.0×10^{-4}
20	888.1	1881	0.1447	8.374×10^{-1}	9.429×10^{-4}	8.659×10^{-8}	10,889	7.0×10^{-4}
30	882.2	1922	0.1437	4.126×10^{-1}	4.677×10^{-4}	8.474×10^{-8}	5520	7.0×10^{-4}
40	876.0	1964	0.1427	2.177×10^{-1}	2.485×10^{-4}	8.294×10^{-8}	2997	7.0×10^{-4}
50	869.9	2006	0.1417	1.229×10^{-1}	1.413×10^{-4}	8.122×10^{-8}	1740	7.0×10^{-4}
60	863.9	2048	0.1408	7.399×10^{-2}	8.565×10^{-5}	7.956×10^{-8}	1076	7.0×10^{-4}
70	858.0	2089	0.1398	4.759×10^{-2}	5.546×10^{-5}	7.799×10^{-8}	711.2	7.0×10^{-4}
80	852.0	2132	0.1388	3.232×10^{-2}	3.794×10^{-5}	7.645×10^{-8}	496.3	7.0×10^{-4}
90	845.9	2175	0.1379	2.312×10^{-2}	2.733×10^{-5}	7.492×10^{-8}	364.8	7.0×10^{-4}
100	840.0	2220	0.1369	1.718×10^{-2}	2.046×10^{-5}	7.341×10^{-8}	278.6	7.0×10^{-4}
110	834.3	2264	0.1359	1.312×10^{-2}	1.573×10^{-5}	7.196×10^{-8}	218.6	7.0×10^{-4}
120	828.9	2308	0.1349	1.029×10^{-2}	1.241×10^{-5}	7.055×10^{-8}	175.9	7.0×10^{-4}
130	823.2	2351	0.1340	8.186×10^{-3}	9.944×10^{-6}	6.923×10^{-8}	143.6	7.0×10^{-4}
140	816.8	2395	0.1330	6.558×10^{-3}	8.029×10^{-6}	6.798×10^{-8}	118.1	7.0×10^{-4}
150	810.3	2441	0.1320	5.344×10^{-3}	6.595×10^{-6}	6.675×10^{-8}	98.79	7.0×10^{-4}
Therminol 59								
–49	1024.4	1458	0.1256	2.326	2.271×10^{-3}	8.408×10^{-8}	27,005	–
–40	1018.0	1490	0.1251	4.540×10^{-1}	4.460×10^{-4}	8.248×10^{-8}	5407	–

(continued)

(continued)

Therminol 59								
–20	1003.5	1553	0.1239	5.819×10^{-2}	5.799×10^{-5}	7.954×10^{-8}	729	–
0	988.91	1619	0.1227	1.588×10^{-2}	1.606×10^{-5}	7.660×10^{-8}	210	–
10	982.00	1650	0.1220	1.025×10^{-2}	1.044×10^{-5}	7.533×10^{-8}	139	–
20	975.09	1681	0.1214	7.113×10^{-3}	7.294×10^{-6}	7.407×10^{-8}	98.5	–
30	967.82	1711	0.1207	5.162×10^{-3}	5.333×10^{-6}	7.287×10^{-8}	73.2	–
40	960.36	1745	0.1199	3.927×10^{-3}	4.089×10^{-6}	7.150×10^{-8}	57.2	–
50	952.27	1774	0.1191	3.092×10^{-3}	3.247×10^{-6}	7.053×10^{-8}	46.0	–
60	945.00	1810	0.1183	2.510×10^{-3}	2.656×10^{-6}	6.916×10^{-8}	38.4	–
70	937.73	1846	0.1175	2.092×10^{-3}	2.231×10^{-6}	6.785×10^{-8}	32.9	–
80	930.45	1875	0.1166	1.772×10^{-3}	1.904×10^{-6}	6.684×10^{-8}	28.5	–
90	923.18	1909	0.1157	1.521×10^{-3}	1.647×10^{-6}	6.563×10^{-8}	25.1	–
100	915.91	1945	0.1148	1.316×10^{-3}	1.437×10^{-6}	6.441×10^{-8}	22.3	–
120	901.09	2005	0.1130	1.032×10^{-3}	1.145×10^{-6}	6.256×10^{-8}	18.3	–
140	886.36	2075	0.1110	8.216×10^{-4}	9.270×10^{-7}	6.033×10^{-8}	15.4	–
160	871.00	2140	0.1088	6.710×10^{-4}	7.704×10^{-7}	5.837×10^{-8}	13.2	–
180	854.64	2205	0.1065	5.610×10^{-4}	6.564×10^{-7}	5.655×10^{-8}	11.6	–
200	839.27	2275	0.1042	4.763×10^{-4}	5.675×10^{-7}	5.457×10^{-8}	10.4	–
220	823.73	2341	0.1019	4.124×10^{-4}	5.006×10^{-7}	5.284×10^{-8}	9.47	–
240	807.36	2407	0.0993	3.585×10^{-4}	4.440×10^{-7}	5.111×10^{-8}	8.69	–
260	790.00	2480	0.0966	3.150×10^{-4}	3.987×10^{-7}	4.931×10^{-8}	8.09	–

(continued)

(continued)

Therminol 59								
280	772.82	2553	0.0939	2.795×10^{-4}	3.616×10^{-7}	4.758×10^{-8}	7.60	–
300	754.64	2625	0.0909	2.505×10^{-4}	3.319×10^{-7}	4.589×10^{-8}	7.23	–
315	740.92	2677	0.0888	2.322×10^{-4}	3.134×10^{-7}	4.479×10^{-8}	7.00	–
Therminol VP1								
12	1071.0	1520	0.1370	5.480×10^{-3}	5.117×10^{-6}	8.416×10^{-8}	60.8	–
20	1064.6	1544	0.1363	4.336×10^{-3}	4.073×10^{-6}	8.291×10^{-8}	49.1	–
40	1047.6	1604	0.1344	2.616×10^{-3}	2.497×10^{-6}	7.998×10^{-8}	31.2	–
60	1031.6	1664	0.1322	1.768×10^{-3}	1.714×10^{-6}	7.703×10^{-8}	22.3	–
80	1015.6	1716	0.1300	1.290×10^{-3}	1.270×10^{-6}	7.461×10^{-8}	17.0	–
100	998.60	1774	0.1276	9.876×10^{-4}	9.890×10^{-7}	7.205×10^{-8}	13.7	–
120	982.60	1834	0.1252	7.856×10^{-4}	7.995×10^{-7}	6.945×10^{-8}	11.5	–
140	965.60	1884	0.1226	6.426×10^{-4}	6.655×10^{-7}	6.737×10^{-8}	9.88	–
160	948.60	1944	0.1198	5.374×10^{-4}	5.665×10^{-7}	6.495×10^{-8}	8.72	–
180	930.80	1994	0.1168	4.572×10^{-4}	4.912×10^{-7}	6.293×10^{-8}	7.81	–
200	912.80	2046	0.1138	3.948×10^{-4}	4.325×10^{-7}	6.093×10^{-8}	7.10	–
220	894.80	2104	0.1106	3.456×10^{-4}	3.862×10^{-7}	5.876×10^{-8}	6.57	–
240	876.80	2154	0.1072	3.048×10^{-4}	3.476×10^{-7}	5.678×10^{-8}	6.12	–
260	857.80	2206	0.1038	2.720×10^{-4}	3.171×10^{-7}	5.483×10^{-8}	5.78	–
280	838.00	2264	0.1002	2.446×10^{-4}	2.919×10^{-7}	5.279×10^{-8}	5.53	–
300	817.00	2314	0.0964	2.212×10^{-4}	2.707×10^{-7}	5.098×10^{-8}	5.31	–
320	795.20	2366	0.0925	2.018×10^{-4}	2.538×10^{-7}	4.916×10^{-8}	5.16	–

(continued)

(continued)

Therminol VP1								
340	772.40	2424	0.0885	1.846×10^{-4}	2.390×10^{-7}	4.727×10^{-8}	5.06	–
360	748.60	2484	0.0843	1.704×10^{-4}	2.276×10^{-7}	4.534×10^{-8}	5.02	–
380	722.60	2552	0.0800	1.572×10^{-4}	2.175×10^{-7}	4.340×10^{-8}	5.01	–
400	694.00	2630	0.0756	1.460×10^{-4}	2.104×10^{-7}	4.142×10^{-8}	5.08	–
Glycerin (glycerol, $C_3H_5(OH)_3$)								
0	1272.7	2288	0.2820	12.03	9.452×10^{-3}	9.683×10^{-8}	97,611	–
10	1267.0	2340	0.2829	3.888	3.068×10^{-3}	9.540×10^{-8}	32,165	–
20	1261.3	2392	0.2837	1.410	1.118×10^{-3}	9.403×10^{-8}	11,888	–
30	1255.1	2444	0.2861	8.456×10^{-1}	6.737×10^{-4}	9.327×10^{-8}	7223	–
40	1248.9	2495	0.2884	2.835×10^{-1}	2.270×10^{-4}	9.255×10^{-8}	2453	–
50	1241.9	2547	0.2908	1.824×10^{-1}	1.469×10^{-4}	9.193×10^{-8}	1598	–
60	1234.8	2598	0.2931	8.130×10^{-2}	6.584×10^{-5}	9.135×10^{-8}	721	–
70	1227.8	2649	0.2946	5.654×10^{-2}	4.605×10^{-5}	9.056×10^{-8}	509	–
80	1220.7	2700	0.2960	3.186×10^{-2}	2.610×10^{-5}	8.980×10^{-8}	291	–
90	1213.9	2752	0.2975	2.333×10^{-2}	1.922×10^{-5}	8.905×10^{-8}	216	–
100	1207.0	2804	0.2990	1.480×10^{-2}	1.226×10^{-5}	8.833×10^{-8}	139	–
Glycerin 50% – water 50% (by mass)								
–40	1150.1	2889	0.3778	2.849×10^{-1}	2.477×10^{-4}	1.137×10^{-7}	2179	–
–30	1147.1	2943	0.3838	1.063×10^{-1}	9.264×10^{-5}	1.137×10^{-7}	815	–
–20	1143.7	3001	0.3900	4.686×10^{-2}	4.097×10^{-5}	1.137×10^{-7}	360	–
–10	1139.8	3061	0.3965	2.387×10^{-2}	2.094×10^{-5}	1.137×10^{-7}	184	–
0	1135.5	3122	0.4033	1.372×10^{-2}	1.209×10^{-5}	1.138×10^{-7}	106	–

(continued)

(continued)

Glycerin 50% – water 50% (by mass)								
10	1130.9	3184	0.4102	8.708×10^{-3}	7.700×10^{-6}	1.139×10^{-7}	68	–
20	1126.1	3247	0.4173	5.958×10^{-3}	5.291×10^{-6}	1.141×10^{-7}	46	–
30	1121.0	3309	0.4246	4.295×10^{-3}	3.832×10^{-6}	1.145×10^{-7}	33	–
40	1115.8	3370	0.4321	3.189×10^{-3}	2.858×10^{-6}	1.149×10^{-7}	25	–
Ethylene glycol ($C_2H_4(OH)_2$)								
–10	1133.4	2260	0.2497	9.541×10^{-2}	8.418×10^{-5}	9.749×10^{-8}	863.5	–
–5	1130.0	2278	0.2507	8.300×10^{-2}	7.345×10^{-5}	9.740×10^{-8}	754.1	–
0	1126.7	2297	0.2518	7.059×10^{-2}	6.265×10^{-5}	9.731×10^{-8}	643.8	–
5	1123.4	2315	0.2528	5.817×10^{-2}	5.179×10^{-5}	9.723×10^{-8}	532.6	–
10	1120.0	2333	0.2539	4.576×10^{-2}	4.086×10^{-5}	9.715×10^{-8}	420.6	–
15	1116.7	2352	0.2550	3.335×10^{-2}	2.987×10^{-5}	9.708×10^{-8}	307.6	–
20	1113.4	2370	0.2560	2.094×10^{-2}	1.881×10^{-5}	9.702×10^{-8}	193.9	–
25	1109.9	2392	0.2567	1.863×10^{-2}	1.678×10^{-5}	9.670×10^{-8}	173.6	–
30	1106.5	2413	0.2574	1.632×10^{-2}	1.475×10^{-5}	9.638×10^{-8}	153.0	–
35	1103.0	2435	0.2581	1.400×10^{-2}	1.270×10^{-5}	9.608×10^{-8}	132.2	–
40	1099.6	2457	0.2587	1.169×10^{-2}	1.063×10^{-5}	9.578×10^{-8}	111.0	–
45	1096.1	2478	0.2594	9.381×10^{-3}	8.559×10^{-6}	9.550×10^{-8}	89.62	–
50	1092.7	2500	0.2601	7.069×10^{-3}	6.470×10^{-6}	9.522×10^{-8}	67.95	–
55	1089.1	2522	0.2603	6.413×10^{-3}	5.888×10^{-6}	9.480×10^{-8}	62.12	–
60	1085.5	2543	0.2606	5.756×10^{-3}	5.303×10^{-6}	9.438×10^{-8}	56.19	–
65	1081.9	2565	0.2608	5.100×10^{-3}	4.714×10^{-6}	9.398×10^{-8}	50.16	–

(continued)

(continued)

Ethylene glycol ($C_2H_4(OH)_2$)								
70	1078.3	2587	0.2610	4.443×10^{-3}	4.120×10^{-6}	9.359×10^{-8}	44.03	–
75	1074.7	2608	0.2613	3.786×10^{-3}	3.523×10^{-6}	9.320×10^{-8}	37.80	–
80	1071.1	2630	0.2615	3.130×10^{-3}	2.922×10^{-6}	9.283×10^{-8}	31.48	–
85	1067.3	2653	0.2613	2.885×10^{-3}	2.703×10^{-6}	9.226×10^{-8}	29.29	–
90	1063.6	2677	0.2611	2.640×10^{-3}	2.482×10^{-6}	9.170×10^{-8}	27.06	–
95	1059.8	2700	0.2608	2.394×10^{-3}	2.259×10^{-6}	9.116×10^{-8}	24.79	–
100	1056.1	2723	0.2606	2.149×10^{-3}	2.035×10^{-6}	9.062×10^{-8}	22.46	–
Ethylene glycol 50%—water 50% (by mass)								
−50	1088.9	2904	0.3475	1.956×10^{-1}	1.796×10^{-4}	1.099×10^{-7}	1634.3	–
−45	1088.2	2935	0.3503	1.281×10^{-1}	1.177×10^{-4}	1.097×10^{-7}	1073.4	–
−40	1087.3	2966	0.3531	8.607×10^{-2}	7.916×10^{-5}	1.095×10^{-7}	723.0	–
−35	1086.3	2997	0.3560	5.927×10^{-2}	5.456×10^{-5}	1.094×10^{-7}	498.9	–
−30	1085.1	3027	0.3589	4.177×10^{-2}	3.850×10^{-5}	1.093×10^{-7}	352.4	–
−25	1083.7	3057	0.3618	3.011×10^{-2}	2.779×10^{-5}	1.092×10^{-7}	254.5	–
−20	1082.2	3087	0.3647	2.218×10^{-2}	2.049×10^{-5}	1.092×10^{-7}	187.7	–
−15	1080.5	3116	0.3677	1.667×10^{-2}	1.543×10^{-5}	1.092×10^{-7}	141.3	–
−5	1076.7	3174	0.3738	9.979×10^{-3}	9.268×10^{-6}	1.094×10^{-7}	84.8	–
0	1074.6	3203	0.3768	7.930×10^{-3}	7.379×10^{-6}	1.095×10^{-7}	67.4	–
10	1070.0	3258	0.3830	5.257×10^{-3}	4.913×10^{-6}	1.098×10^{-7}	44.7	–
20	1064.9	3312	0.3891	3.693×10^{-3}	3.468×10^{-6}	1.103×10^{-7}	31.4	–
30	1059.4	3364	0.3953	2.729×10^{-3}	2.576×10^{-6}	1.109×10^{-7}	23.2	–

(continued)

(continued)

Ethylene glycol 50%—water 50% (by mass)									
40	1053.4	3413	0.4015	2.103×10^{-3}	1.997×10^{-6}	1.117×10^{-7}	17.9	–
50	1047.1	3459	0.4077	1.678×10^{-3}	1.603×10^{-6}	1.125×10^{-7}	14.2	–
60	1040.5	3503	0.4138	1.375×10^{-3}	1.321×10^{-6}	1.135×10^{-7}	11.6	–
70	1033.6	3544	0.4198	1.148×10^{-3}	1.110×10^{-6}	1.146×10^{-7}	9.69	–
80	1026.4	3582	0.4257	9.685×10^{-4}	9.435×10^{-7}	1.158×10^{-7}	8.15	–
90	1019.0	3616	0.4315	8.195×10^{-4}	8.042×10^{-7}	1.171×10^{-7}	6.87	–
100	1011.5	3646	0.4371	6.900×10^{-4}	6.821×10^{-7}	1.185×10^{-7}	5.76	–
110	1003.9	3673	0.4425	5.735×10^{-4}	5.712×10^{-7}	1.200×10^{-7}	4.76	–
120	996.2	3695	0.4477	4.668×10^{-4}	4.686×10^{-7}	1.216×10^{-7}	3.85	–

Properties generated by the Engineering Equation Solver—EES, version V10.833-3D, with permission

Table A.9 Properties of liquid metals

T (°C)	ρ (kg/m^3)	C_p (J/kgK)	k (W/mK)	μ (kg/ms)	ν (m^2/s)	α (m^2/s)	Pr (–)	β (1/K)
Bismuth (Bi), melting temperature: 271.4 °C								
300	10,027	143.6	12.28	1.735×10^{-3}	1.730×10^{-7}	8.530×10^{-6}	0.020	–
400	9904	138.1	13.28	1.415×10^{-3}	1.429×10^{-7}	9.709×10^{-6}	0.015	–
500	9782	134.8	14.28	1.218×10^{-3}	1.245×10^{-7}	1.083×10^{-5}	0.011	–
600	9660	132.8	15.28	1.086×10^{-3}	1.124×10^{-7}	1.191×10^{-5}	0.009	–
700	9538	131.6	16.28	9.902×10^{-4}	1.038×10^{-7}	1.297×10^{-5}	0.008	–
Lead (Pb), melting temperature: 327.5 °C								
350	10,623	147.5	16.05	2.542×10^{-3}	2.393×10^{-7}	1.025×10^{-5}	0.023	–
400	10,563	146.7	16.60	2.235×10^{-3}	2.116×10^{-7}	1.071×10^{-5}	0.020	–
600	10,324	143.5	18.80	1.550×10^{-3}	1.501×10^{-7}	1.269×10^{-5}	0.012	–
800	10,085	140.8	21.00	1.233×10^{-3}	1.222×10^{-7}	1.479×10^{-5}	0.008	–
1000	9846	138.9	23.20	1.054×10^{-3}	1.070×10^{-7}	1.697×10^{-5}	0.006	–
Mercury (Hg), melting temperature: –38.9 °C								
0	13,595	140.4	8.18	1.687×10^{-3}	1.241×10^{-7}	4.287×10^{-6}	0.029	–
200	13,112	135.5	10.64	1.040×10^{-3}	7.930×10^{-8}	5.990×10^{-6}	0.013	–
400	12,633	136.4	12.60	8.629×10^{-4}	6.831×10^{-8}	7.313×10^{-6}	0.009	–
600	12,130	142.8	14.04	7.844×10^{-4}	6.467×10^{-8}	8.106×10^{-6}	0.008	–
800	11,584	154.3	14.98	7.414×10^{-4}	6.400×10^{-8}	8.381×10^{-6}	0.008	–

(continued)

Table A.9 (continued)

Potassium (K), melting temperature: 63.6 °C								
150	808	801.1	49.63	3.894×10^{-4}	4.818×10^{-7}	7.665×10^{-5}	0.0063	–
300	774	775.0	44.15	2.536×10^{-4}	3.277×10^{-7}	7.360×10^{-5}	0.0045	–
400	751	765.5	40.81	2.080×10^{-4}	2.769×10^{-7}	7.096×10^{-5}	0.0039	–
500	727	761.7	37.70	1.777×10^{-4}	2.445×10^{-7}	6.811×10^{-5}	0.0036	–
700	678	773.6	31.97	1.399×10^{-4}	2.065×10^{-7}	6.097×10^{-5}	0.0034	–
Sodium (Na), melting temperature: 97.8 °C								
100	927	1378	88.15	6.788×10^{-4}	7.322×10^{-7}	6.899×10^{-5}	0.0106	–
200	903	1344	82.00	4.644×10^{-4}	5.143×10^{-7}	6.757×10^{-5}	0.0076	–
400	858	1283	69.53	2.793×10^{-4}	3.256×10^{-7}	6.315×10^{-5}	0.0052	–
500	834	1265	64.27	2.369×10^{-4}	2.839×10^{-7}	6.091×10^{-5}	0.0047	–
700	787	1252	55.34	1.864×10^{-4}	2.367×10^{-7}	5.613×10^{-5}	0.0042	–
Sodium 22%—Potassium 78% (by mass), melting temperature: –12.6 °C								
100	841	937	23.40	4.481×10^{-4}	5.328×10^{-7}	2.970×10^{-5}	0.0179	–
200	818	911	24.70	3.189×10^{-4}	3.898×10^{-7}	3.315×10^{-5}	0.0118	–
400	771	877	26.00	2.005×10^{-4}	2.600×10^{-7}	3.845×10^{-5}	0.0068	–
600	724	873	25.60	1.578×10^{-4}	2.179×10^{-7}	4.050×10^{-5}	0.0054	–
800	676	901	23.50	1.278×10^{-4}	1.890×10^{-7}	3.858×10^{-5}	0.0049	–
1000	626	962	19.80	1.108×10^{-4}	1.770×10^{-7}	3.287×10^{-5}	0.0054	–

(continued)

Table A.9 (continued)

$NaNO_3$ 40%—KNO_3 60% (by mass)—solar salt, melting temperature: 220.9 °C								
260	1924	1492	0.4922	4.343×10^{-3}	2.257×10^{-6}	1.714×10^{-7}	13.17	–
300	1898	1499	0.4999	3.288×10^{-3}	1.732×10^{-6}	1.756×10^{-7}	9.862	–
350	1866	1508	0.5098	2.343×10^{-3}	1.256×10^{-6}	1.812×10^{-7}	6.932	–
400	1834	1516	0.5195	1.779×10^{-3}	9.699×10^{-7}	1.868×10^{-7}	5.193	–
500	1770	1534	0.5390	1.315×10^{-3}	7.431×10^{-7}	1.986×10^{-7}	3.743	–
593	1710	1550	0.5572	1.026×10^{-3}	5.999×10^{-7}	2.102×10^{-7}	2.854	–

Properties generated by the Engineering Equation Solver—EES, version V10.833-3D, with permission

Table A.10 Total hemispherical emissivity of some surfaces for some temperatures (in order according to temperature)

Metals		
Material	Temperature (°C)	Emissivity, ε (–)
Polished AISI 304 steel	64–447–830	0.1043–0.1594–0.1963
Polished aluminum	–100–165–429	0.01–0.0243–0.037
Smooth cobalt	474–727–979	0.125–0.175–0.23
Polished copper	–226–314–854	0.022–0.0359–0.061
Polished chrome	37–92	0.08–0.11
Polished tin	21–102	0.04–0.06
Polished iron	24–427–829	0.07–0.2051–0.34
Polished germanium	724–929	0.56–0.53
Smooth iridium	724–1477–2230	0.11–0.1575–0.21
Steam-blasted magnesium	79–208	0.31–0.26
Clean mercury	37–102	0.09–0.12
Polished molybdenum	824–1627–2430	0.1–0.2–0.276
Smooth niobium	724–1477–2230	0.07–0.19–0.252
Smooth oxidized nickel	–226–202–629	0.385–0.4743–0.6
Polished nickel	–26–602–1230	0.053–0.13–0.195
Polished gold	–226–314–854	0.013–0.0385–0.062
Smooth platinum	24–777–1530	0.04–0.1316–0.196
Smooth silver	–226–302–829	0.013–0.028–0.048
Smooth rhodium	974–1855	0.12–0.18
Smooth tantalum	724–1877–3030	0.132–0.2475–0.316
Smooth titanium	599–1040–1480	0.505–0.5546–0.613
Smooth tungsten	37–1584–3130	0.04–0.2327–0.348
Smooth vanadium	849–1202–1555	0.168–0.2102–0.246
Non-metals		
Clean water	0–99	0.95–0.96
Smooth clay	17–72	0.86
Smooth soft grey rubber	17–22	0.89
Smooth hard dark rubber	20–39	0.94
Melted silicon dioxide	–201–41	0.753–0.76
Smooth plaster	20–22	0.9
Smooth graphite	924–1927–2930	0.77–0.8075–0.83
Smooth wood	17–42	0.90–0.94
Polished silicon	20–42	0.72
Smooth common glass	24–627–1230	0.93–0.8599–0.56

Properties generated by the Engineering Equation Solver—EES, version V10.833-3D, with permission

Table A.11 Solar radiation properties of some materials: solar absorptivity (α_s), total hemispherical emissivity (ε) (25 °C), ratio between these quantities and solar transmissivity (τ_s)

Material	Treatment or surface condition	α_s	ε	α_s/ε	τ_s
AISI 321 steel, Light blue	Mirrored and chemically oxidized for 12 min, 0.6 M of chromic and sulfurous acid at 90 °C	0.85	0.18	4.72	–
AISI 321 steel, Light blue	Thermally oxidized for 10 min at 1043 K under normal atmospheric conditions	0.85	0.14	6.07	–
AISI 321 steel, Silver gray	Mirrored	0.38	0.15	2.53	–
AISI 321 steel, Matte silver	Without polishing	0.42	0.23	1.83	–
Steel, Bright gray	Mirrored	0.41	0.05	8.20	–
Steel, Dark brown	Worn and heavily rusted over time	0.89	0.92	0.97	–
Anodized aluminum, Light green	Anodized in 2 to 4% of oxalic acid, 20 min, 2.2 A/dm^2, 5 to 12 V	0.55	0.29	1.90	–
Aluminum, Bright silver	Mirrored	0.24	0.04	6.00	–
Aluminum, Matte silver	As received	0.28	0.07	4.00	–
Asbestos, Gray	Dry	0.73	0.89	0.82	–
Asbestos, Gray	Wet	0.92	0.92	1.00	–
Sand, Whitish	Dry	0.52	0.82	0.63	–
Sand, Matte red	Dry	0.73	0.86	0.85	–
Clay, Dark gray	A thin dry layer on a mirrored aluminum plate with $\varepsilon = 0.04$	0.76	0.92	0.83	–
Cement, Light gray	A thin dry layer on a mirrored aluminum plate with $\varepsilon = 0.04$	0.67	0.88	0.76	–
Copper, Light red	Mirrored	0.27	0.03	9.00	–
Concrete, Light pink	Smooth, Non-reflective surface	0.65	0.87	0.75	–
Enamel, Yellow	Hand-coated on a mirrored aluminum plate with $\varepsilon = 0.04$	0.46	0.88	0.52	–
Enamel, Blue		0.68	0.87	0.78	–
Enamel, White		0.28	0.90	0.31	–
Enamel, Black		0.93	0.90	1.03	–
Enamel, Green		0.78	0.90	0.87	–
Enamel, Red		0.65	0.87	0.75	–
Galvanized iron, Silver gray	Bright	0.39	0.05	7.80	–
Galvanized iron, Dark brown	Highly rusted and worn	0.90	0.90	1.00	–
Fiberglass "Sun-lite", transparent	As received		0.87		0.88

(continued)

Table A.11 (continued)

Material	Treatment or surface condition	α_s	ε	α_s/ε	τ_s
Lacquer, Transparent	Hand-coated on a mirrored aluminum plate with ε = 0.04		0.88		–
Tin, Bright silver	Mirrored	0.30	0.04	7.50	–
Plywood, Dark brown	As received	0.67	0.80	0.84	–
Wood, Light brown	Planed and tuned	0.59	0.90	0.66	–
Marble, Slightly whitish	Non-reflective	0.40	0.88	0.45	–
Paper, White		0.27	0.83	0.33	–
Stone, Light pink	Smooth and non-reflective	0.65	0.87	0.75	–
Aluminum paint, Bright silver	Hand-painted	0.35	0.56	0.63	–
Transparent Makrolon® polycarbonate	As received		0.88		0.88
Roof tile, Bright red	Dry	0.65	0.85	0.76	–
Roof tile, Bright red	Wet	0.88	0.91	0.97	–
Porcelain tile, White	Glazed reflective	0.26	0.85	0.31	–
Mosaic tiles, Chocolate	Non-reflective	0.82	0.82	1.00	–
Brick, Bright red	Tuned and smoothed, Dry	0.65	0.85	0.76	–
Brick, Bright red	Tuned and smoothed, Wet	0.88	0.91	0.97	–
Varnish, Transparent	Hand-coated on a mirrored aluminum plate with ε = 0.04		0.90		–
Window glass, Transparent			0.86		0.88

Adapted from Sharma V. C. and Sharma A. Solar properties of some building elements. Energy. 1989;14(12): 805–810

Table A.12 Henry's constant, *H*, for gases in water at moderate pressure at various temperatures

H (kPa)

Gas	0 °C	10 °C	20 °C	30 °C	40 °C	50 °C
Air	4.440×10^6	5.475×10^6	6.656×10^6	7.988×10^6	9.476×10^6	1.112×10^7
CO	3.827×10^6	4.528×10^6	5.295×10^6	6.129×10^6	7.029×10^6	7.992×10^6
CO_2	80,255	109,455	146,152	191,466	246,541	312,528
CH_4	2.206×10^6	2.820×10^6	3.545×10^6	4.390×10^6	5.363×10^6	6.470×10^6
H_2	6.028×10^6	6.456×10^6	6.882×10^6	7.304×10^6	7.724×10^6	8.139×10^6
H_2S	29,039	38,098	49,065	62,144	77,530	95,410
N_2	5.290×10^6	6.506×10^6	7.889×10^6	9.445×10^6	1.118×10^7	1.309×10^7
NH_3	25.83	44.46	73.75	118.3	184.1	278.9
O_2	2.736×10^6	3.409×10^6	4.183×10^6	5.065×10^6	6.059×10^6	7.167×10^6
SO_2	1747	2542	3605	4997	6782	9032

Properties generated by the Engineering Equation Solve—EES, version V10.833-3D, with permission

Table A.13 Solubility, *S*, of gases in solids

Gas	Solid	*T* (°C)	*S* (kmol/m^3 bar)
H_2	Vulcanized rubber	25	0.00176
CO_2	Vulcanized rubber	25	0.03963
N_2	Vulcanized rubber	25	0.00154
O_2	Vulcanized rubber	25	0.00308
H_2	Ni	85	0.00889
H_2	Ni	165	0.00845
H_2	SiO_2	300	0.00024
He	SiO_2	20	0.00044
He	Pire ×	20	0.00037

Adapted from Barrer R. M. Diffusion in and through solids. New York: Macmillan, 1941

Appendix B

Exact Solution of the Laminar Boundary Layer on a Flat Plate

The conservation equations of mass (Eq. 5.8), momentum (Eq. 5.13) and energy (Eq. 5.14) in the laminar boundary layer on a flat surface (also see Table 5.2) are:

$$\frac{\partial u}{\partial x} + \frac{\partial v}{\partial y} = 0 \tag{5.8}$$

$$u\frac{\partial u}{\partial x} + v\frac{\partial u}{\partial y} = \nu\frac{\partial^2 u}{\partial y^2} \tag{5.13}$$

$$u\frac{\partial T}{\partial x} + v\frac{\partial T}{\partial y} = \alpha\frac{\partial^2 T}{\partial y^2} \tag{5.14}$$

Initially, the conservation equations of mass and momentum are solved. Blasius proposed, in 1908, to reduce these two equations to one by the following three considerations:

1. He defined the similarity variable η that reduces the two variables (x, y) to a single variable:

$$\eta = y\sqrt{\frac{u_\infty}{\nu x}} \tag{B.1}$$

2. He introduced the stream function $\psi(x, y)$, where it is fulfilled that:

$$u = \frac{\partial \psi}{\partial y} \textit{and } v = -\frac{\partial \psi}{\partial x} \tag{B.2}$$

J. R. Simões-Moreira and E. W. Zavaleta-Aguilar, *Heat Transfer*, Heat and Mass Transfer, https://doi.org/10.1007/978-3-031-80318-5

Due to the definitions of Eq. (B.2), the conservation of mass (Eq. 5.8) is naturally satisfied.

3. Defined the similarity function $f = f(\eta)$:

$$f = \frac{\psi}{u_\infty\sqrt{\frac{\nu x}{u_\infty}}} \; or \; \psi = f u_\infty \sqrt{\frac{\nu x}{u_\infty}} \tag{B.3}$$

The velocities and their derivatives in terms of the variable and the similarity function are as follows:

$$\begin{aligned} u &= \frac{\partial \psi}{\partial y} = \frac{\partial \psi}{\partial \eta}\frac{\partial \eta}{\partial y} = \left(f' u_\infty \sqrt{\frac{\nu x}{u_\infty}}\right)\sqrt{\frac{u_\infty}{\nu x}} \\ u &= u_\infty f' \end{aligned} \tag{B.4}$$

$$\begin{aligned} v &= -\frac{\partial \psi}{\partial x} = \frac{\partial \psi}{\partial \eta}\frac{\partial \eta}{\partial x} \\ v &= \frac{1}{2}\sqrt{\frac{u_\infty \nu}{x}}(\eta f' - f) \end{aligned} \tag{B.5}$$

Similarly, the other derivatives are obtained:

$$\frac{\partial u}{\partial x} = -\frac{u_\infty}{2x}\eta f'' \tag{B.6}$$

$$\frac{\partial u}{\partial y} = u_\infty \sqrt{\frac{u_\infty}{\nu x}} f'' \tag{B.7}$$

$$\frac{\partial^2 u}{\partial y^2} = \frac{u_\infty^2}{\nu x} f''' \tag{B.8}$$

Substituting into the momentum equation (Eq. 5.13):

$$2f''' + f'' f = 0 \tag{B.9}$$

The boundary conditions, according to the new variables, are:
$y = 0$, $u = 0$, $\eta = 0$, from Eq. (B.4) $\rightarrow f'(0) = 0$.
$y = 0$, $v = 0$, $\eta = 0$, from Eq. (B.5) $\rightarrow f(0) = 0$.
$y \rightarrow \infty$, $u = u_\infty$, $\eta \rightarrow \infty$, from Eq. (B.4) $\rightarrow f'(\infty) = 1$.

Equation (B.9), along with the indicated boundary conditions, was solved by Blasius in 1908, with no explicit equation for velocity. Table B.1 presents the values of some variables of interest.

Table B.1 Variables of interest from the boundary layer solution

η	f	f'	f''
0.0	0.00000	0.00000	0.33206
0.4	0.02656	0.13277	0.33147
0.8	0.10611	0.26471	0.32739
1.2	0.23795	0.39378	0.31659
1.6	0.42032	0.51676	0.29667
2.0	0.65003	0.62977	0.26675
2.4	0.92230	0.72899	0.22809
2.8	1.23099	0.81162	0.18401
3.2	1.56911	0.87609	0.13913
3.6	1.92954	0.92333	0.09809
4.0	2.30576	0.95552	0.06424
4.4	2.69238	0.97587	0.03897
4.8	3.08534	0.98779	0.02187
5.0	3.28329	0.99155	0.01591
5.4	3.68094	0.99616	0.00793
5.8	4.07990	0.99838	0.00365
6.2	4.47948	0.99937	0.00155
6.6	4.87931	0.99977	0.00061
7.0	5.27926	0.99992	0.00022
7.4	5.67924	0.99998	0.00007
7.8	6.07923	1.00000	0.00002
8.2	6.47923	1.00000	0.00001
8.6	6.87923	1.00000	0.00000

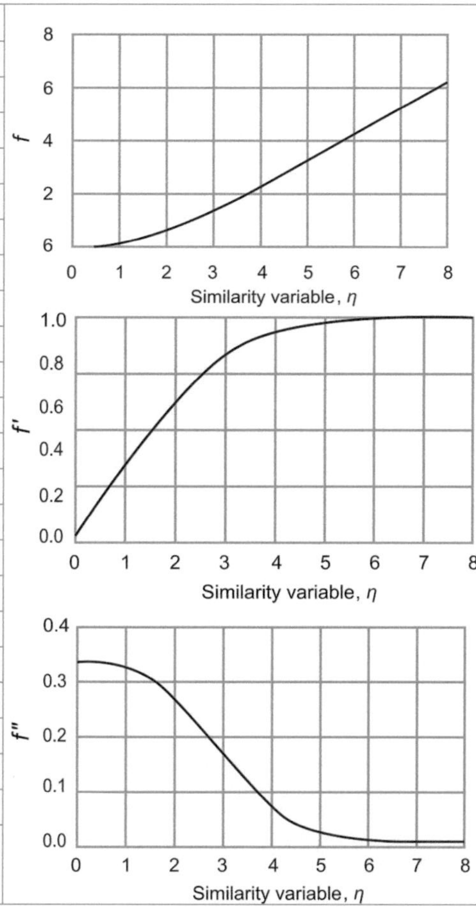

Aadapted from Howarth (1938)

According to Eq. (5.6), the thickness of the hydrodynamic boundary layer is defined as that in which the horizontal speed is: $u = 0.99u_\infty$. This corresponds, according to Eq. (B.4), to $f' = 0.99$. By interpolating in Table B.1, the value of the similarity variable is equal to $\eta = 4.9176$ or, approximately, $\eta = 4.92$. Thus, the thickness of the hydrodynamic boundary layer is discovered:

$$\eta = y\sqrt{\frac{u_\infty}{\nu x}} \rightarrow 4.92 = \delta_H\sqrt{\frac{u_\infty}{\nu x}} \rightarrow \delta_H = \frac{4.92x}{\sqrt{\frac{u_\infty x}{\nu}}}, ou \qquad \text{(B.10)}$$

$$\delta_H = \frac{4.92x}{\sqrt{Re_x}}$$

where Re_x is the local Reynolds number measured from the leading edge:

$$Re_x = \frac{u_\infty x}{\nu} \tag{B.11}$$

The local shear stress on the wall of the flat plate is defined by:

$$\tau_{p,x} = \mu \left.\frac{\partial u}{\partial y}\right|_{y=0} \tag{B.12}$$

From Eq. (B.7) and Table B.1:

$$\begin{aligned}\tau_{p,x} &= \mu \left.\frac{\partial u}{\partial y}\right|_{y=0} = \mu u_\infty \sqrt{\frac{u_\infty}{\nu x}} f''(0) = \mu u_\infty \sqrt{\frac{u_\infty}{\nu x}} 0.332, ou \\ \tau_{p,x} &= \frac{0.332 \rho u_\infty^2}{\sqrt{Re_x}}\end{aligned} \tag{B.13}$$

The local friction coefficient is defined by:

$$C_{f,x} = \frac{\tau_{p,x}}{\rho u_\infty^2/2} \tag{B.14}$$

Substituting from Eq. (B.13):

$$C_{f,x} = \frac{0.664}{\sqrt{Re_x}} \tag{B.15}$$

Now, energy conservation will be analyzed. For this, the dimensionless temperature $\theta = \theta(x, y)$ is defined:

$$\theta = \frac{T - T_w}{T_\infty - T_w} \tag{B.16}$$

where $T = T(x, y)$ is the flow temperature at position (x, y) and T_w is the surface temperature of the plate considered constant (isothermal plate). Substituting the dimensionless temperature into the energy conservation equation (Eq. 5.14) results in:

$$u \frac{\partial \theta}{\partial x} + v \frac{\partial \theta}{\partial y} = \alpha \frac{\partial^2 \theta}{\partial y^2} \tag{B.17}$$

The dimensionless temperature can be defined in terms of the similarity variable, $\theta = \theta(\eta)$. Substituting the velocities u and v from Eqs. (B.4) and (B.5) into (B.17) results in:

$$u_\infty f' \frac{\partial \theta}{\partial \eta}\frac{\partial \eta}{\partial x} + \frac{1}{2}\sqrt{\frac{u_\infty \upsilon}{x}}\left(\eta f' - f\right)\frac{\partial \theta}{\partial \eta}\frac{\partial \eta}{\partial y} = \alpha \frac{\partial^2 \theta}{\partial \eta^2}\left(\frac{\partial \eta}{\partial y}\right)^2$$

and thus, we obtain:

$$2\theta'' + \Pr \theta' f = 0 \tag{B.18}$$

The boundary conditions for the new variables are:
$y = 0$, $T = T_w$, $\eta = 0$, from Eq. (B.16) $\rightarrow \theta(0) = 0$.
$y \rightarrow \infty$, $T \rightarrow T_\infty$, $\eta \rightarrow \infty$, from Eq. (B.16) $\rightarrow \theta(\infty) = 1$.
Equation (B.18), along with these boundary conditions and the condition of Eq. (5. 7) was solved numerically and, for $Pr > 0.6$, it was obtained:

$$\left.\frac{\partial \theta}{\partial \eta}\right|_{\eta=0} = 0.332 Pr^{1/3} \text{ and } \delta_{T,x} = \frac{\delta_{H,x}}{Pr^{1/3}} \tag{B.19}$$

The convective heat transfer flow from the plate to the flow must be equal to the conduction in the first fluid layer, which is stationary, and Fourier's Law is valid:

$$q = h_x(T_w - T_\infty) = -k \left.\frac{\partial T}{\partial y}\right|_{y=0}$$

From Eqs. (B.16) and (B.19):

$$\left.\frac{\partial T}{\partial y}\right|_{y=0} = (T_\infty - T_w)\left.\frac{\partial \theta}{\partial y}\right|_{y=0} = (T_\infty - T_w)\left.\frac{\partial \theta}{\partial \eta}\right|_{\eta=0}\left.\frac{\partial \eta}{\partial y}\right|_{y=0}$$
$$= (T_\infty - T_w) 0.332 Pr^{1/3}\sqrt{\frac{u_\infty}{\nu x}}$$

Substituting in the heat flux equation:

$$h_x(T_w - T_\infty) = -k(T_\infty - T_w) 0.332 Pr^{1/3}\sqrt{\frac{u_\infty}{\nu x}}$$

From which the local convective heat transfer coefficient is obtained for flat plate in laminar flow:

$$h_x = 0.332 \frac{k}{x} Re_x^{1/2} Pr^{1/3} \tag{B.20}$$

and the local Nusselt number: $Nu_x = \frac{h_x x}{k}$,

$$Nu_x = 0.332 Re_x^{1/2} Pr^{1/3} \tag{B.21}$$

Boundary Layer Equations in Natural Convection

Natural convection can occur in a stationary fluid medium in contact with a surface that is at a different temperature. Here is analyzed the boundary layer on a vertical flat wall in a steady state, two-dimensional, Newtonian fluid, constant properties, constant surface temperature (T_w) and constant medium temperature (T_∞). Figure B.1 indicates the case where $T_w > T_\infty$.

Analogously to the analysis of the boundary layer in forced flow (Sect. 5.2.1), three conservation equations can be defined in the boundary layer in natural convection: mass (Eq. 5.8), momentum (Eq. 7.7) and energy conservation (Eq. 5.14), which are indicated below, respectively:

$$\frac{\partial u}{\partial x} + \frac{\partial v}{\partial y} = 0 \tag{5.8}$$

$$u\frac{\partial u}{\partial x} + v\frac{\partial u}{\partial y} = g\beta(T - T_\infty) + \nu\frac{\partial^2 u}{\partial y^2} \tag{7.7}$$

$$u\frac{\partial T}{\partial x} + v\frac{\partial T}{\partial y} = \alpha\frac{\partial^2 T}{\partial y^2} \tag{5.11}$$

Before analyzing the solution of the equations, Eq. (7.7) will be made dimensionless to obtain an important dimensionless number in natural convection. Substituting $x^* = \frac{x}{L}$, $y^* = \frac{y}{L}$, $u^* = \frac{u}{V}$, $v^* = \frac{v}{V}$, where L is a characteristic length and V an arbitrary reference speed in terms of the Reynolds number $\left(Re_L = \frac{VL}{\nu}\right)$. Note that,

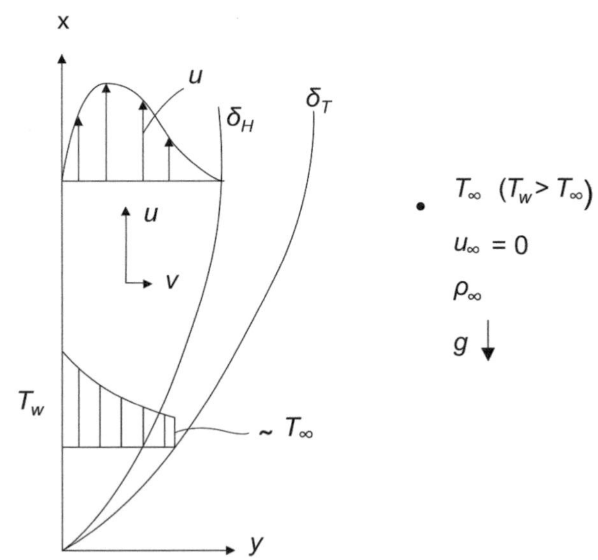

Fig. B.1 Scheme of the hydrodynamic and thermal boundary layers in natural convection

in natural convection (pure, not mixed, see Sect. 7.3), there is no constant non-zero reference speed, since the speed varies from 0 at the wall, increasing and decreasing in the boundary layer, to then become 0 again, in the medium that is at rest (see Fig. B.1). In addition, the dimensionless temperature is defined as:

$$\theta = \frac{T - T_\infty}{T_w - T_\infty} \tag{B.22}$$

Substituting these dimensionless variables in Eq. (7.7), we get:

$$u^* \frac{\partial u^*}{\partial x^*} + v^* \frac{\partial u^*}{\partial y^*} = \left[\frac{g\beta(T_w - T_\infty)L^3}{\nu^2}\right]\frac{\theta}{Re^2} + \frac{1}{Re}\frac{\partial^2 u^*}{\partial y^{*2}} \tag{B.23}$$

The term in brackets is known as the Grashof number, which represents the relationship between buoyancy forces and viscous forces. Thus, this number plays a role similar to the Reynolds number in forced convection, for example, for vertical plates, the flow is laminar for $Gr \lesssim 10^9$.

Returning to the solution of the differential Eqs. (5.8), (7.7), and (5.14), Ostrach (1952) proposed a solution similar to that of Blasius (forced convection), when the following three considerations were made:

1. Defined the similarity variable η that reduces two variables (x, y) to one:

$$\eta = \frac{y}{x}\left(\frac{Gr_x}{4}\right)^{1/4} \tag{B.24}$$

2. Introduced the stream function $\psi = \psi(x, y)$ in which it is fulfilled that:

$$u = \frac{\partial \psi}{\partial y} \textit{ and } v = -\frac{\partial \psi}{\partial x} \tag{B.25}$$

3. Defined the similarity function $f = f(\eta)$:

$$f = \frac{\psi}{4\nu\left(\frac{Gr_x}{4}\right)^{1/4}} \Rightarrow \psi = 4f\nu\left(\frac{Gr_x}{4}\right)^{1/4} \tag{B.26}$$

Then, the velocities and their derivatives must be represented in terms of the new variables (η, f). The velocity u is as follows:

$$u = \frac{\partial \psi}{\partial y} = \frac{\partial \psi}{\partial \eta}\frac{\partial \eta}{\partial y} = \left[4\nu\left(\frac{Gr_x}{4}\right)^{1/4} f'\right]\left[\frac{1}{x}\left(\frac{Gr_x}{4}\right)^{1/4}\right]$$
$$u = \frac{2\nu}{x} Gr_x^{1/2} f' \tag{B.27}$$

Table B.2 Boundary conditions for Eqs. (B.28) and (B.29)

In terms of x and y		In terms of η and f	
$y = 0$	$u = 0$	$\eta = 0$	$f'(0) = 0$
$y = 0$	$v = 0$	$\eta = 0$	$f(0) = 0$
$y \to \infty$	$u = 0$	$\eta \to \infty$	$f'(\infty) \to 0$
$y = 0$	$T = T_w$	$\eta = 0$	$\theta = 1$
$y \to \infty$	$T \to T_\infty$	$\eta \to \infty$	$\theta = 0$

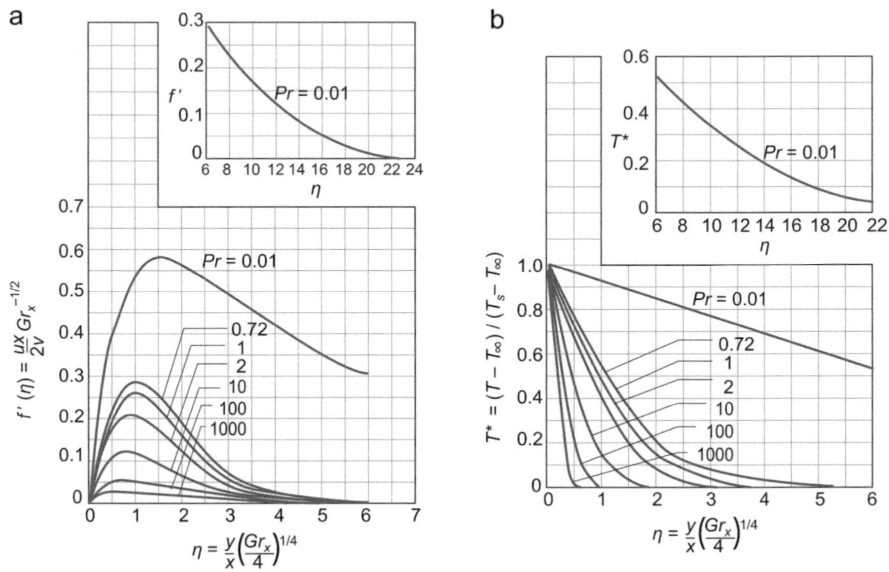

Fig. B.2 **a** Velocity profiles and **b** temperature profiles for natural convection on a vertical plate. Adapted from Ostrach (1952)

In a similar way, the variables: v, $\frac{\partial u}{\partial x}$, $\frac{\partial u}{\partial y}$, $\frac{\partial^2 u}{\partial y^2}$ are obtained, which, along with Eq. (B.22), must be substituted into Eqs. (5.8), (7.7) and (5.14). Note that mass conservation (Eq. 5.8) is satisfied using Eq. (B.25). Equations (7.7) and (5.14) can be written, respectively:

$$f''' + 3f''f - 2f'^2 + \theta = 0 \tag{B.28}$$

$$\theta'' + 3\,Prf\theta' = 0 \tag{B.29}$$

The boundary conditions to solve these equations are indicated in Table B.2.

Ostrach (1952) numerically solved Eqs. (B.27) and (B.28), together with the boundary conditions from Table B.2. Some of the results are shown in the graphs of Fig. B.2.

By equating Newton's Cooling Law with Fourier's Law at $y = 0$, the local heat transfer coefficient by convection (h_x) can be found.

$$q_w = h_x(T_w - T_\infty) = -k \left.\frac{\partial T}{\partial y}\right|_{y=0}$$

In which $\left.\frac{\partial T}{\partial y}\right|_{y=0} = (T_w - T_\infty)\left.\frac{\partial \theta}{\partial y}\right|_{y=0} = (T_w - T_\infty)\left.\frac{\partial \theta}{\partial \eta}\right|_{y=0}\left.\frac{\partial \eta}{\partial y}\right|_{y=0} = (T_w - T_\infty)\theta'(0)\frac{1}{x}\left(\frac{Gr_x}{4}\right)^{1/4}$.

thus:

$$h_x(T_w - T_\infty) = -k \left.\frac{\partial T}{\partial y}\right|_{y=0} = -k(T_w - T_\infty)\frac{1}{x}\left(\frac{Gr_x}{4}\right)^{1/4}\theta'(0).$$

It can be noted, in Fig. B.2b, that $\theta'(0)$ is negative and depends on the Prandtl number, thus, it is possible to say that $F(Pr) = -\theta'(0)$, therefore,

$$h_x = \frac{k}{x}\left(\frac{Gr_x}{4}\right)^{1/4} F(Pr) \tag{B.30}$$

in which $F(Pr)$ was correlated by curve fitting by LeFevre (1956) as:

$$F(Pr) = \frac{0.75Pr^{1/2}}{\left(0.609 + 1.221Pr^{1/2} + 1.238Pr\right)^{1/4}} \tag{B.31}$$

The local Nusselt number will be:

$$\begin{aligned} Nu_x &= \frac{h_x x}{k} \\ Nu_x &= \left(\frac{Gr_x}{4}\right)^{1/4} F(Pr) \end{aligned} \tag{B.32}$$

The average heat transfer coefficient by convection can be found by $\overline{h}_L = \frac{1}{L}\int_0^L h_x dx$. Performing this integral, we get:

$$\overline{h}_L = \frac{4}{3}\frac{k}{L}\left(\frac{Gr_L}{4}\right)^{1/4} F(Pr) \tag{B.33}$$

and the average Nusselt number is defined as $\overline{Nu_L} = \frac{\overline{h}_L L}{k}$,

$$\overline{N}u_L = \frac{4}{3}\left(\frac{Gr_L}{4}\right)^{1/4} F(Pr) \tag{B.34}$$

or

$$\overline{Nu_L} = \left(\frac{Ra_L Pr}{2.436 + 4.884 Pr^{1/2} + 4.952 Pr} \right)^{1/4} \tag{B.35}$$

in which, Ra_L is the Rayleigh number (see Eq. 7.12).

References

1. Howarth, L. 1938. On the solution of the laminar boundary layer equations. *Proceedings of the Royal Society of London, Series a: Mathematical and Physical Sciences* 164: 547–579.
2. LeFevre E.J. 1956. Laminar free convection from a vertical plane surface. In *Proceedings of 9th international congress of applied mechanics*, 168–174. Belgium: Brussels.
3. Ostrach S. 1952. *An analysis of laminar free convection flow and heat transfer about a flat plate parallel to the direction of the generating body force.* National Advisory Committee for Aeronautics, Technical Note 2635.

Appendix C

Bessel Functions

Given the differential equation $x^2\frac{d^2y}{dx^2} + x\frac{dy}{dx} + \left(x^2 - m^2\right)y = 0$, known as the Bessel equation, the Bessel function (J) is the solution ($y(x) = J_m(x)$) of this equation, which is initially expressed in a sum of an infinite series of powers:

$y = \sum_{n=0}^{\infty} a_n x^{n+r}$, with $a_0 \neq 0$. Deriving this function:

$$\frac{dy}{dx} = \sum_{n=0}^{\infty}(n+r)a_n x^{n+r-1} \text{ and } \frac{d^2y}{dx^2} = \sum_{n=0}^{\infty}(n+r)(n+r-1)a_n x^{n+r-2}$$

Substituting in the differential equation, we get:

$$x^2\sum_{n=0}^{\infty}(n+r)(n+r-1)a_n x^{n+r-2} + x\sum_{n=0}^{\infty}(n+r)a_n x^{n+r-1} + \left(x^2 - m^2\right)\sum_{n=0}^{\infty} a_n x^{n+r} = 0$$

Eliminating x^r, simplifying and grouping:

$$\sum_{n=0}^{\infty} a_n\left[(n+r)^2 - m^2\right]x^n + \sum_{n=0}^{\infty} a_n x^{n+2} = 0$$

Redefining the second summation:

$$\sum_{n=0}^{\infty} a_n\left[(n+r)^2 - m^2\right]x^n + \sum_{n=2}^{\infty} a_{n-2} x^n = 0$$

J. R. Simões-Moreira and E. W. Zavaleta-Aguilar, *Heat Transfer*, Heat and Mass Transfer, https://doi.org/10.1007/978-3-031-80318-5

Performing the partial summation:

$$a_0\left(r^2 - m^2\right) + a_1\left[(1+r)^2 - m^2\right]x + \sum_{n=2}^{\infty}\left\{a_n\left[(n+r)^2 - m^2\right] + a_{n-2}\right\}x^n = 0$$

$n = 0 \rightarrow$ since $a_0 \neq 0$, therefore, $r = m$ or $r = -m$. Assuming $r = m$.
$n = 1 \rightarrow$ since $(1+r)^2 - m^2 \neq 0$, therefore, $a_1 = 0$

$$n \geq 2 \rightarrow a_n = -\frac{a_{n-2}}{n(n+2m)}$$

From this last expression, since $a_1 = 0$, therefore: $a_3 = 0$, $a_5 = 0$ etc.

$$n = 2 \rightarrow a_2 = -\frac{a_0}{2(2+2m)} = -\frac{(-1)^1 a_0}{2^2 1!(1+m)}$$
$$n = 4 \rightarrow a_4 = -\frac{a_2}{4(4+2m)} = \frac{(-1)^2 a_0}{2^4 2!(2+m)(1+m)}$$
$$n = 6 \rightarrow a_6 = -\frac{a_4}{6(6+2m)} = \frac{(-1)^3 a_0}{2^6 3!(3+m)(2+m)(1+m)}$$

Generalizing: $a_{2n} = \frac{(-1)^n a_0 m!}{2^{2n} n!(n+m)!}$, thus, the power series will be:

$$y = a_0 x^m + a_1 x^{m+1} + a_2 x^{m+2} + a_3 x^{m+3} + a_4 x^{m+4} + \cdots$$
$$y = a_0 x^m + a_2 x^{m+2} + a_4 x^{m+4} + \cdots$$

$y = \sum_{n=0}^{\infty} \frac{(-1)^n a_0 m! 2^m}{n!(n+m)!}\left(\frac{x}{2}\right)^{2n+m}$, since a_0 is arbitrary, it can be: $a_0 = \frac{1}{2^m m!}$. The final solution is:

$$J_m(x) = \sum_{n=0}^{\infty} \frac{(-1)^n}{n!(n+m)!}\left(\frac{x}{2}\right)^{2n+m}$$

which is known as the *Bessel function of the first kind of order m*.

There is also a similar differential equation: $x^2 \frac{d^2y}{dx^2} + x\frac{dy}{dx} - \left(x^2 + m^2\right)y = 0$, whose solution is: $I_m(x) = \sum_{n=0}^{\infty} \frac{1}{n!(n+m)!}\left(\frac{x}{2}\right)^{2n+m}$, known as the *modified Bessel function of the first kind of order m*. The second solution to the equation is: $K_m(x) = \frac{\pi}{2}\frac{I_{-m}(x) - I_m(x)}{\sin(m\pi)}$, known as the *modified Bessel function of the second kind of order m*. Table C.1 shows some values of Bessel functions, of the first kind of order 0 (J_0), modified of the first kind of order 0 and 1, I_0 and I_1, respectively, and, modified of the second kind of order 0 and 1, K_0 and K_1, respectively.

Table C.1 Modified Bessel functions

x	J_0	J_1	I_0	I_1	K_0	K_1
0.0	1.0000	0.0000	1.0000	0.0000	–	–
0.2	0.9900	0.0995	1.0100	0.1005	1.75	4.78
0.4	0.9604	0.1960	1.0404	0.2040	1.11	2.18
0.6	0.9120	0.2867	1.0920	0.3137	7.78×10^{-1}	1.30
0.8	0.8463	0.3688	1.1665	0.4329	5.65×10^{-1}	8.62×10^{-1}
1.0	0.7652	0.4401	1.2661	0.5652	4.21×10^{-1}	6.02×10^{-1}
1.2	0.6711	0.4983	1.3937	0.7147	3.19×10^{-1}	4.35×10^{-1}
1.4	0.5669	0.5419	1.5534	0.8861	2.44×10^{-1}	3.21×10^{-1}
1.6	0.4554	0.5699	1.7500	1.0848	1.88×10^{-1}	2.41×10^{-1}
1.8	0.3400	0.5815	1.9896	1.3172	1.46×10^{-1}	1.83×10^{-1}
2.0	0.2239	0.5767	2.2796	1.5906	1.14×10^{-1}	1.40×10^{-1}
2.2	0.1104	0.5560	2.6291	1.9141	8.93×10^{-2}	1.08×10^{-1}
2.4	0.0025	0.5202	3.0493	2.2981	7.02×10^{-2}	8.37×10^{-2}
2.6	–	–	3.5533	2.7554	5.54×10^{-2}	6.53×10^{-2}
2.8	–	–	4.1573	3.3011	4.38×10^{-2}	5.11×10^{-2}
3.0	–	–	4.8808	3.9534	3.47×10^{-2}	4.02×10^{-2}
3.2	–	–	5.7472	4.7343	2.76×10^{-2}	3.16×10^{-2}
3.4	–	–	6.7848	5.6701	2.20×10^{-2}	2.50×10^{-2}
3.6	–	–	8.0277	6.7927	1.75×10^{-2}	1.98×10^{-2}
3.8	–	–	9.5169	8.1404	1.40×10^{-2}	1.57×10^{-2}
4.0	–	–	11.3019	9.7595	1.12×10^{-2}	1.25×10^{-2}
4.2	–	–	13.4425	11.7056	8.93×10^{-3}	9.94×10^{-3}
4.4	–	–	16.0104	14.0462	7.15×10^{-3}	7.92×10^{-3}
4.6	–	–	19.0926	16.8626	5.73×10^{-3}	6.33×10^{-3}
4.8	–	–	22.7937	20.2528	4.60×10^{-3}	5.06×10^{-3}
5.0	–	–	27.2399	24.3356	3.69×10^{-3}	4.04×10^{-3}
5.2	–	–	32.5836	29.2543	2.97×10^{-3}	3.24×10^{-3}
5.4	–	–	39.0088	35.1821	2.38×10^{-3}	2.60×10^{-3}
5.6	–	–	46.7376	42.3283	1.92×10^{-3}	2.08×10^{-3}
5.8	–	–	56.0381	50.9462	1.54×10^{-3}	1.67×10^{-3}
6.0	–	–	67.2344	61.3419	1.24×10^{-3}	1.34×10^{-3}
6.2	–	–	80.7179	73.8859	1.00×10^{-3}	1.08×10^{-3}
6.4	–	–	96.9616	89.0261	8.08×10^{-4}	8.69×10^{-4}
6.6	–	–	116.5373	107.3047	6.52×10^{-4}	7.00×10^{-4}
6.8	–	–	140.1362	129.3776	5.26×10^{-4}	5.64×10^{-4}
7.0	–	–	168.5939	156.0391	4.25×10^{-4}	4.54×10^{-4}
7.2	–	–	202.9213	188.2503	3.43×10^{-4}	3.66×10^{-4}

(continued)

Table C.1 (continued)

x	J_0	J_1	I_0	I_1	K_0	K_1
7.4	–	–	244.3410	227.1750	2.77×10^{-4}	2.95×10^{-4}
7.6	–	–	294.3322	274.2225	2.24×10^{-4}	2.38×10^{-4}
7.8	–	–	354.6845	331.0995	1.81×10^{-4}	1.92×10^{-4}
8.0	–	–	427.5641	399.8731	1.46×10^{-4}	1.55×10^{-4}
8.2	–	–	515.5927	483.0477	1.18×10^{-4}	1.26×10^{-4}
8.4	–	–	621.9441	583.6570	9.59×10^{-5}	1.01×10^{-4}
8.6	–	–	750.4612	705.3773	7.76×10^{-5}	8.20×10^{-5}
8.8	–	–	905.7973	852.6635	6.28×10^{-5}	6.63×10^{-5}
9.0	–	–	1093.5884	1030.9147	5.09×10^{-5}	5.36×10^{-5}
9.2	–	–	1320.6608	1246.6755	4.12×10^{-5}	4.34×10^{-5}
9.4	–	–	1595.2844	1507.8794	3.34×10^{-5}	3.51×10^{-5}
9.6	–	–	1927.4788	1824.1447	2.71×10^{-5}	2.84×10^{-5}
9.8	–	–	2329.3851	2207.1337	2.19×10^{-5}	2.30×10^{-5}
10.0	–	–	2815.7167	2670.9883	1.78×10^{-5}	1.86×10^{-5}

Properties of the Bessel function:
If $y = f(x)$, then $\frac{dJ_0(y)}{dx} = -J_1(y)\frac{dy}{dx}$.

Index

J. R. Simões-Moreira and E. W. Zavaleta-Aguilar, *Heat Transfer*, Heat and Mass Transfer, https://doi.org/10.1007/978-3-031-80318-5

MIX
Papier aus verantwortungsvollen Quellen
Paper from responsible sources
FSC® C105338

If you have any concerns about our products,
you can contact us on
ProductSafety@springernature.com

In case Publisher is established outside the EU,
the EU authorized representative is:
Springer Nature Customer Service Center GmbH
Europaplatz 3, 69115 Heidelberg, Germany

Printed by Libri Plureos GmbH
in Hamburg, Germany